AF360725

Edible Insects as Innovative Foods: Nutritional, Functional and Acceptability Assessments II

Edible Insects as Innovative Foods: Nutritional, Functional and Acceptability Assessments II

Editors

Victor Benno Meyer-Rochow
Chuleui Jung

Basel • Beijing • Wuhan • Barcelona • Belgrade • Novi Sad • Cluj • Manchester

Editors
Victor Benno Meyer-Rochow
Oulu University
Oulu
Finland

Chuleui Jung
Andong National University
Andong
Republic of Korea

Editorial Office
MDPI
St. Alban-Anlage 66
4052 Basel, Switzerland

This is a reprint of articles from the Special Issue published online in the open access journal *Foods* (ISSN 2304-8158) (available at: https://www.mdpi.com/journal/foods/special_issues/edible_insects_foods).

For citation purposes, cite each article independently as indicated on the article page online and as indicated below:

Lastname, A.A.; Lastname, B.B. Article Title. *Journal Name* **Year**, *Volume Number*, Page Range.

ISBN 978-3-0365-9412-5 (Hbk)
ISBN 978-3-0365-9413-2 (PDF)
doi.org/10.3390/books978-3-0365-9413-2

Cover image courtesy of Victor Benno Meyer-Rochow

Contents

About the Editors

Victor Benno Meyer-Rochow

Victor Benno Meyer-Rochow is a New Zealander with a PhD in Neurobiology and a DSc in Ethnobiological Studies (both from the Australian National University in Canberra). Dr. Meyer-Rochow's neurobiological research took him several times to Antarctica and the Arctic, but his ethnobiological work focused on Central and Western Australia (1973), Papua New Guinea (1972, 1975, 2002, 2005), North-East India (1990, 2012, 2014, 2015, 2016), and the DPRK (North Korea: 2012). He was Managing Director of the Research Institute of Luminous Organisms on the Japanese Pacific island of Hachijojima from 2014 to 2019, and thereafter was Visiting Professor at Andong National University's Department of Plant Medicals in the Republic of Korea, before returning to the Department of Genetics and Ecology of Oulu University, Oulu, Finland, in 2022.

Chuleui Jung

Chuleui Jung studied Biology at Seoul National University in Korea and finished his PhD from Oregon State University with "Dispersal behaviors of predatory mites", a group of acari which are important natural enemies of spider mites in agriculture. Since being accepted at Andong National University, he has taught entomology, ecology, acarology, honey bee science and pollination, and ecosystem modeling. His research activities are oriented toward the ecological processes and interactions of beneficial insects and arthropods with mutualistic ecosystem services, such as biocontrol and pollination. At the same time, he believes that developing insects as generic food or functional sources could alleviate global food-related problems. Prof. Jung has published more than 200 scientific papers and several book chapters. He has also served in many scientific journals as an Editor or Editorial Board Member.

Preface

There is no doubt that, throughout the history of mankind, humans have consumed insects, either by ingesting them more or less accidentally within fruit or by seeking to eat them deliberately (Bequaert, 1921; Bergier, 1941; Bodenheimer, 1951). Even monkeys, our closest animal relatives, have been observed to actively collect insects and other arthropods in order to eat them (e.g., Marshall, 1902; Carpenter, 1921; Nickle and Heymann, 1996; Sanz et al., 2009). Over the last 50 years, there has been a renewed interest in insects as a source of human food. Conferences have begun to focus on edible insects since the XVI International Pacific Science Congress in Seoul in 1987 and the International Conference on Minilivestock in Beijing in 1995 brought this topic to a wider audience. Numerous scientific publications over the last 20–30 years have praised the advantages of insect-based diets over conventional diets like poultry and especially ruminants, and have highlighted the environmentally advantageous aspects of reared insects over traditionally farmed animals. Various edible insect species have been examined for their acceptability as a novelty food (or feed in animal husbandry and fish culture), as well as undergoing scrutiny with regard to their potential risk of carrying diseases or undesired microbes. Although insects should not merely be seen as a food item for humans in order to survive periods of starvation, there is no way to deny that the global food security situation for the human population is precarious.

Despite earlier reports of people living in different parts of the world who traditionally engage in entomophagy (the consumption of insects), it was not until Meyer-Rochow (1975) linked global food security to the universal and extensive use of insects around the world as a possible and potent way to ease global food shortages. We now possess a myriad of information on the kinds of insects that serve as food sources to various people worldwide; we know that most insects are nutritious, consist of valuable protein and easily digestible fatty acids, and contain important minerals and vitamins, and recommendations exist regarding how to breed the most lucrative species optimally. However, there are still gaps to be filled with regard to the processing of cultured insects, the preparation and conservation of insect-based foods, economics and marketing, and the potential of insects as suppliers of health-promoting drugs and medicines. It is with these thoughts in mind that we accepted to serve as Guest Editors for this Special Issue of Foods on edible insects, focusing on their role as food sources and as raw materials for a variety of products.

1. Bequaert, J. Insects as food: How they have augmented the food supply of mankind in early and recent years. *Nat. Hist. J.* **1921**, *21*, 191–200.
2. Bergier, E. *Peuples Entomophages et Insectes Comestibles: Etude sur les Moeurs de L'homme et de L'insecte*; Imprimerie Rulliere Freres: Avignon, France, 1941.
3. Bodenheimer, F.S. *Insects as Human Food*; W. Junk Publishers: The Hague, The Netherlands, 1951.
4. Carpenter, G.D.H. Experiment on the relative edibility of insects with special reference to their coloration. *Trans. R. Entomol. Soc. Lond.* **1921**, *45*, 1–105.
5. Marshall, G.A.K. Five years' observations and experiments (1896–1901) on the bionomics of South African insects, chiefly mimicry and warning colours. *Trans. R. Entomol. Soc. Lond.* **1902**, *50*, 287–584.
6. Nickle, D.A.; Heymann, E.W. Predation on orthoptera and other orders of insects by tamarind monkeys, Saguinus mystax mystax and S. fuscicollis nigrifrons (Primates: Callitrichidae) in north-eastern Peru. *J. Zool. (Lond.)* **1996**, *239*, 799–819.

7. Sanz, C.; Schoning, C.; Morgan, D. Chimpanzees prey on army ants with specialized tool set. *Am. J. Primatol.* **2009**, *71*, 1e8.

8. Meyer-Rochow, V.B. Can insects help to ease the problem of world food shortage? *Search* **1975**, *6*, 261–262.

Victor Benno Meyer-Rochow and Chuleui Jung
Editors

Editorial

Interest in Insects as Food and Feed: It Does Not Wane in the Public Domain

Victor Benno Meyer-Rochow [1,2,*] and Chuleui Jung [1,3,*]

1   Agricultural Science and Technology Research Institute, Andong National University, Andong 36729, Korea
2   Department of Ecology and Genetics, Oulu University, SF-90140 Oulu, Finland
3   Department of Plant Medicals, Andong National University, Andong 36729, Korea
*   Correspondence: meyrow@gmail.com (V.B.M.-R.); cjung@andong.ac.kr (C.J.)

Citation: Meyer-Rochow, V.B.; Jung, C. Interest in Insects as Food and Feed: It Does Not Wane in the Public Domain. *Foods* **2022**, *11*, 3184. https://doi.org/10.3390/foods11203184

Received: 13 September 2022
Accepted: 26 September 2022
Published: 12 October 2022

Publisher's Note: MDPI stays neutral with regard to jurisdictional claims in published maps and institutional affiliations.

This Special Issue of *Foods* represents Volume 2 of the topic "Edible Insects as Innovative Foods: Nutritional, Functional and Acceptability Assessments". Although some of the 20 contributions in this volume deal with hitherto unreported food insects and some explore the effects that a diet containing insects or insect products has on the gut microbiota of the consumer, whether human or non-human, most of the articles deal with improvements related to processing and rearing food insects. Food safety questions are not ignored and questions of the acceptability of insect containing food stuffs are not either. What affects the palatability and the chemical composition of farmed insects most, is explored and how protein and mineral levels in edible insects can be increased is dealt with in several articles. Finally, the plea is made not to focus only on the reasons why some people reject insects as food, but to provide convincing reasons for the advantages to health and the environment that a greater use of insects as food and feed would present.

It is exactly 47 years ago that for the first time it had been suggested by Meyer-Rochow that the use of insects as food and feed could ease the problem of global nutritional shortfalls and that WHO and FAO should be encouraged to support the use of insects as food [1]. Ridiculed at first, and not taken seriously even by some science magazines and often accompanied by funny cartoons (Figure 1), the idea gradually gained momentum and the suggestion began to be dealt with at conferences when the topic of Food Security was discussed. Recognition by the FAO finally came when a renewed call for support was published by Van Huis et al. [2] and the need to increase global food production to feed the increasing world population could no longer be overlooked.

The extraordinary increase in publications dealing with edible insects and entomophagy from 1900 to 2015 was documented in 2015 by Evans et al. [3] and Müller et al. [4]. Since then, hundreds of additional papers, too many to list individually, covering a huge range of issues all related to the use of insects as food and feed, have appeared. Yet, there seems to be no sign of a waning interest in insects that are being promoted as a 'novel food for humans' (which is actually wrong as insects represent a very ancient kind of food for humans) [5]. In fact, products that contain insect material are now becoming increasingly available and insects reared to be fed to livestock and poultry and to be used in fish culture have become popular alternatives to conventional feed [6–10].

Volume one of our Special Issue on "Edible Insects as Innovative Foods: Nutritional, Functional and Acceptability Assessments" appeared in 2020 and contained 20 articles [11]. Volume two also contains 20 articles, but the emphasis now seems to have shifted somewhat and this time the Special Issue includes more papers related to the effects of rearing conditions on the composition and the nutritive status of commercially reared insets and how to best utilize these insects. Authors from 13 different countries were involved in the articles that make up volume 2 and while it is obvious that food insects can no longer be ignored as a food or feed item, a number of issues remain to be explored.

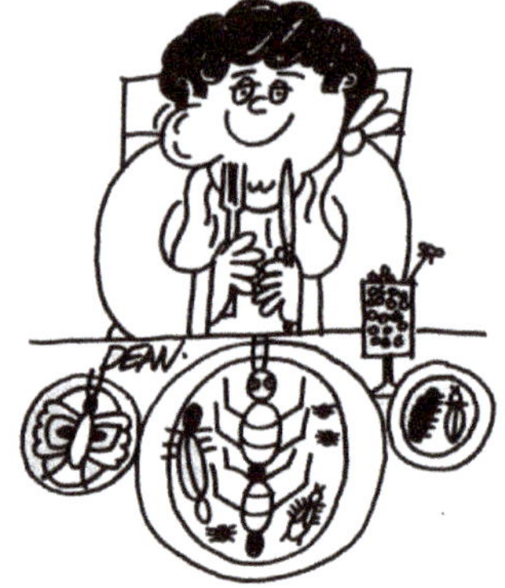

Insects should be seen as a high protein food source, and not just as a threat to food crops for man and his animals, according to V. B. Meyer-Rochow of the Department of Zoology, University of Western Australia.

Writing in the current issue of *Search* (vol 6, p 261), the Journal of the Australian and New Zealand Association for the Advancement of Science, Meyer-Rochow says that "Insects are extremely nutritious. They consist of easily-digestible proteins and fats, and small but significant amounts of carbohydrates, minerals and vitamins. One hundred grams of fried termites have a caloric value of 561, which places them among the richest foods." He goes on to suggest that breeding programmes could lead to even more nutritious insects, in the same way as milk and egg production have been increased by selective breeding of cows and poultry.

He does not advocate the conversion of Europeans away from their present diets—repugnant though these are to some vociferous journalists—towards entomophagy, but suggests that its revival "could ease the hazards of malnutrition in countries where the consumption of insects has only recently been given up", not only through the use of insects as a direct food source, but by feeding them to poultry or other domestic animals as well.

Figure 1. One of many cartoons that appeared in newspapers and magazines [12] after it had been suggested that insects could ease the problem of food shortages.

Grinding pupae of honeybee drones and using that powder as an ingredient in puffed-rice snacks has been carried out by Woo-Hee Cho in Korea [13]. To what extent different heating and drying conditions affected the puffed-rice snack product and how the chemical composition of the latter differed from the control were subjects of the study. The product enriched by pupal drone bee powder showed a higher content of proteins, fats, amino acids and fatty acids and it was concluded that the developed product could be consumed as a nutritional snack, although the quality characteristics could still be improved through optimal processing.

Effects of various de-fattening methods on the physicochemical properties of proteins extracted from *Hermetia illucens* larvae were studies by Tae-Kyung Kim who found that the total essential amino acid contents were higher with cold pressure protein extraction than with other treatments [14]. Cold pressure-defatted protein showed the highest emulsifying capacity while water extracted protein showed the lowest emulsifying capacity. Although organic solvents may be efficient for defatting proteins extracted from insects, they have detrimental effects on the human body. Moreover, the organic solvent extraction method requires a considerable amount of time for lipid extraction. Using cold pressure protein extraction on edible insect proteins is eco-friendly and economical due to the reduced degreasing time and its potential industrial applications.

Maiyo et al. [15] investigated whether the nutritional quality of four novel porridge products blended with edible cricket (*Scapsipedus icipe*) meal is improved to an extent that it can be recommended to reduce incidences of malnutrition in low and middle-income countries. Porridge enriched with the meal of the newly discovered cricket *Scapsipedus icipe* had significantly higher protein (2-fold), crude fat (3.4–4-fold), and energy (1.1–1.2-fold) levels than the commercial porridge flour. Fermented cereal porridge fortified with cricket meal had all three types of omega-3 fatty acids and germinated cereal porridge with cricket meal had a significantly higher iron content (19.5 mg/100 g). In conclusion, to fortify porridge products with cricket meal is to be recommended.

In a study by Vanqa et al. [16] edible insect flours from *Gonimbrasia belina* (Mashonzha), *Hermetia illucens* (black soldier fly larvae) and *Macrotermes subhylanus* (Madzhulu) were assessed in terms of proximal, physicochemical, techno-functional and antioxidant properties. Crude protein of the edible insect flours varied between 34.90–52.74% but there were no significant differences ($p > 0.05$) in foam capacity and foam stability of all three edible insect flours. The findings revealed that the flours of the three edible insects are a good source of antioxidants and can be used as an alternative protein source and a potential novel food additive due to their techno-functional qualities.

Investigating the nutritional, techno-functional and structural properties of Black Soldier Fly (*Hermetia illucens*) larval flours (BSFL) and protein concentrates has been the subject of the paper by Mshayisa et al. [17]. The highest protein content (73.35%) was obtained under alkaline and acid precipitation extraction (BSFL-PC1). The sum of essential amino acids significantly increased ($p < 0.05$) from 24.98% to 38.20% due to the defatting process during extraction. The protein extraction method influenced the structural properties of the protein concentrates and in conclusion it can be stated that BSFL flour fractions and protein concentrates show promise as novel functional ingredients for use in food applications.

Gan et al. [18] explored effects of different processing and packaging methods on lipid oxidation of deep-fried crickets (*Gryllus bimaculatus*) during storage. The composition of fatty acids changed and the content of FFA, PV, and TBAR values also increased with the extension of storage time, indicating that the lipid oxidation dominated by oxidation of unsaturated fatty acids could occur in deep fried *Gryllus bimaculatus* during storage. The study revealed that dibutyl hydroxyl toluene (BHT) was the most effective antioxidant.

An experiment by Selaledi et al. [19] was conducted to examine the effects of larval yellow mealworm meal (*Tenebrio molitor*) added to the diets of indigenous chicken. Boschveld chickens were randomly divided into four categories. Controls were given only soybean meal (SBM) and yellow mealworm with percentage levels of 5, 10 and 15 (TM5, TM10 and TM15, respectively) were used in the experiment. The authors demonstrated that dietary *Tenebrio molitor* in growing Boschveld chickens could be regarded as a promising added protein source, improving quality rearing of Boschveld chickens.

A more traditional investigation was that by Dürr and Ratompoarison [20], who provided evidence for the importance of food insects in the local diet of rural communities in the central highlands of Madagascar. The investigation showed that the insects contributed significantly to the animal protein uptake of the local population, especially in the humid season, when other protein sources were scarce. The authors then discussed how traditionally appreciated insects could be promoted in food-insecure rural areas and how entomophagy could support food and nutrition security in a growing population.

Ying Su [21] report that the larvae of the sphingid moth *Clanis bilineata tsingtauica* Mell 1922 are commonly used as food for humans in the eastern part of China. This insect was found to be high in nutrients, particularly in the epidermis where protein total was 71.82%. In this case, 16 different amino acids were quantified and the ratio of essential to nonessential amino acids in the epidermis and meat was 68.14% and 59.27%, respectively. Thestudy confirmed that *C. bilineata tsingtauica* was a highly nutritious food source for human consumption, and the results provide a basis for further uses and industrialization of this edible insect.

Storing edible insects and increasing their shelf lives are important issues. However, so are storage conditions for mass reared edible species, which is why Zhu et al. [22] set out to determine the optimal temperature under which eggs of the sphingid *Clanis bilineata tsingtauica* Mell, 1922 should be kept. Considering various combinations, the authors found that optimal egg hatching occurred if eggs were stored at 15 °C for 11 days, and then held at 15–20 °C under dark conditions. The conditions described allow for easier mass rearing of *C. bilineata tsingtauica* by providing a stable supply of eggs throughout the year.

Researchers from Iceland headed by Rune Thrastardottir et al. [23] presented an overview of the most popular insects farmed in Europe, namely the yellow mealworm, *Tenebrio molitor* and the black soldier fly (BSF) *Hermetia illucens*, focusing on the main obstacles and risks associated with maintenance. The results showed that the insect farming industry was increasing in Europe, and that the success of the frontrunners of the farmed insect industry was based on large investments in technology, automation and economy.

However, more information still had to be forthcoming regarding risks posed by edible insects in terms of food safety.

Food safety and risk assessment was also the topic of the study by Anja et al. [24]. These researchers investigated samples of edible insect species for the presence of antimicrobial-resistant and Shiga toxin-producing *Escherichia coli* (STEC). The presence of genes associated with antimicrobial resistance or virulence, including *stx1*, *stx2*, and *eae*, was investigated by PCR. The study showed that STEC can be present in edible insects, representing a potential health hazard. By contrast, the low resistance rate among the isolates indicated a low risk for the transmission of antimicrobial-resistant *E. coli* to consumers.

In the study by Kipkoech et al. [25] cricket chitin was deacetylated to chitosan and the latter was added in various concentrations to *Salmonella/Shigella* growth media. Growth of the probiotic bacteria was monitored on chitosan-supplemented media after 6, 12, 24, and 48 h upon incubation at 37 °C. The good news is that all chitosan concentrations significantly increased the populations of probiotic bacteria and decreased the populations of pathogenic bacteria. This study suggests that cricket-derived chitosan can function as a prebiotic, with an ability to eliminate pathogenic bacteria in the presence of probiotic bacteria.

That a partial substitution of meat with insects (*Alphitobius diaperinus*) in a diet for rats can change the gut microbiome was shown by Lanng et al. [26]. These researchers following a four-week dietary intervention in a healthy rat model could show that metabolomics analyses revealed a larger escape of protein residues into the colon and a different microbial metabolization pattern of aromatic amino acids when pork was partly substituted by insects. It could be shown that the introduction of insects in a common Western omnivorous diet altered the gut microbiome diversity with consequences for the endogenous metabolism.

The food an insect consumes can affect the insect's chemical constitution and when leftovers of food originally for humans are the given to insects, this can affect the insects in many ways. The Black Soldier Fly (BSF) is able to thrive on a wide range of leftovers and especially vegetable processing industries generate huge amounts of by-products that can be used to rear BSF. Significant lower protein contents were detected by Fuso et al. [27] in BSF grown on fruit by-products, while higher contents were observed when autumnal leftovers were employed. Lysine, valine and leucine were amino acids most affected by the diet, but essential amino acids generally satisfied the Food and Agricultural Organization (FAO) requirements for human nutrition with the exception of lysine.

Placentino et al. [28] focused in their study on the diet of professional athletes and tried to discern what motivated professional athletes to accept an energy protein bar fortified with cricket flour. A second aim was how information on the benefits of edible insects impacted their acceptance as food by the athletes. The researchers' results showed that the protein content and the curiosity about texture were the main drivers to taste the cricket energy bar; but the feeling of disgust justified the rejection of tasting insects. Although male athletes were more likely than females to endorse the product that contained cricket flour, the authors point out that a relatively small sample was involved, and therefore advise caution not to draw hasty conclusions.

The aim of the study by Son et al. [29] was to investigate the carbohydrate content and composition of mealworms and to determine the amount of chitin. The crude carbohydrate content of mealworms was 11.5%, but the total soluble sugar content was only 30% of the total carbohydrate content and fructose was identified as the most abundant free sugar in mealworms. With a yield of 4.7%, chitin derivatives were the key components of mealworm carbohydrate. Although similar to crustacean chitin, mealworm chitin exhibited a significantly softer texture than crustacean chitin and showed superior anti-inflammatory effects.

Hornets have become increasingly popular as a food item because of their nutritional value and abundance. This is why Ghosh et al. [30] analyzed the nutrient compositions of the edible broods of *Vespa velutina*, *V. mandarinia*, and *V. basalis*. Farmed *V. velutina* and *V. mandarinia* were found to have similar protein contents, i.e., total amino acids, but *V. basalis* contained less. In all three species: leucine followed by tyrosine and lysine were predominant among the essential and glutamic acid among the non-essential amino acids.

Polyunsaturated fatty acids were dominant in *V. mandarinia* and *V. basalis*, but saturated fatty acids were most abundant in *V. velutina.* It is concluded that the high content, especially of micro minerals such as iron, zinc, and the high K/Na ratio in hornets could help mitigate mineral deficiencies among those with inadequate nutrition.

Aquatic insects, although consumed by many, feature less often in research on human entomophagy. This is why the paper by Min Zhao et al. [31] is important. The authors reviewed what is known about edible aquatic insects and pointed out that the vast majority of the latter, in contrast to the phytophagous terrestrial edible species, are carnivorous. There are differences in, for example, fat, fatty acids and mineral content between terrestrial and aquatic insects and regarding food safety, it is advisable not to consume large quantities of wild aquatic species as they could contain higher amounts of heavy metals, pest residues and uric acid than phytophagous species.

A review covering differences in the chemical composition and nutrient quality (and thereby acceptability of edible insects) by Meyer-Rochow et al. [32], emphasized that multiple reasons can lead to such differences. Choice of the insect species, collection site, processing method, insect life stage, rearing technology and insect feed can affect an individual's palatability and acceptance. According to the authors, the review can assist the food insect industry to select the most suitable species as well as processing methods for insect-based food products.

In the past a great deal of effort has been spent on recording which kinds of insects were consumed where and by whom [5,33–35], but as of late the emphasis has been shifting to analyzing the chemical composition of the edible species and, as this Special Issue has demonstrated, on determining optimal methods to rear and process the insects. Obviously, there are still communities and areas which have not been visited and interviewed, so that a complete picture where and how people eat which kinds of insect is still incomplete. Furthermore, as some articles in this Special Issue have shown, we still do not yet have information on the chemical contents of all known edible insects and do not yet know what affects their composition. However, the majority of the papers in this Special Issue deal with assessing quality questions of the insect-containing product and that presently seems to be one of the main areas of interest.

Food safety issues, of course, came up as well and, no doubt, will continue to be one of the foci of research in the future, similar to how insect chitin will be with its many possible applications in the health sector [36]. Insects have been used therapeutically since time immemorial and microbiological tests have been used to demonstrate the effectiveness of certain insect-derived substances to fight infections and other illnesses [37–40]. What to some extent seems surprising, is that only a handful of insect species, i.e., primarily crickets, black soldier fly and mealworms, receive the bulk of attention. Perhaps they do represent species that are easiest to culture as they accept a variety of food stuffs, feature short generation times, are unproblematic in their requirements, can be used in multiple ways and provide acceptable returns of the investment put into their maintenance. However, there may well be other species whose potential has not yet been fully realized, e.g., drone bees, hornets and wasps come to mind.

As we have already pointed out in the first volume of the Special Issue on "Edible Insects as Innovative Foods: Nutritional, Functional and Acceptability Assessments", insects have many advantages over conventional meat-supplying species. The former requires less space than the latter, their rate of reproduction is considerably higher, a much greater percentage of an insect's body mass can be used as food and the so-called "carbon footprint" of cultured insects is considered to be much lower than that of conventional animals used as food [41–43], which is due largely to the insects' much lower water and food needs and their significantly higher food conversion rate. Thus, there is less "waste" generated by farmed insects. What needs to be entered into the equation, however, is the need to keep farmed insects warm and pest free, but this is not likely to change the conclusion that overall farming insects can be achieved with fewer negative side effects than rearing big mammals and poultry.

As people in countries originally famous for entomophagy increasingly abandon insects as a traditional food and begin to reject insect-containing food [44], there is a need to promote edible insects and food items that contain insect products. It is not only important to learn why people reject insects as food [45–48], it is more important to convince people to accept farmed edible insects as the latter are wholesome and nutritionally valuable. Education, clever marketing, turning to traditional recipes, tapping into the trend of buying healthy food and highlighting nutritional and medical benefits, can all help consumers to consider buying insect-containing food. Curiosity should be encouraged. In addition, as with the first volume of this Special Issue topic [49], we end this paper with the suggestion "Mealworms and spaghetti is food that makes you happy" and advise the readers to "Forget about the pork and put a cricket on your fork!".

Author Contributions: Conceptualization, V.B.M.-R. and C.J.; methodology, V.B.M.-R. and C.J.; validation, V.B.M.-R. and C.J.; formal analysis, V.B.M.-R. and C.J.; investigations, V.B.M.-R. and C.J.; resources, V.B.M.-R. and C.J.; writing—original draft preparation, V.B.M.-R.; writing—review and editing, V.B.M.-R. and C.J.; visualization, V.B.M.-R. and C.J.; project administration, C.J. and V.B.M.-R. All authors have read and agreed to the published version of the manuscript.

Funding: This research was supported by a grant to C.J., Rural Development Administration Agenda project PJ01480803.

Conflicts of Interest: The authors declare no conflict of interest.

References

1. Meyer-Rochow, V.B. Can insects help to ease the problem of world food shortage? *Search* **1975**, *6*, 261–262.
2. Van Huis, A.; Van Itterbeeck, J.; Klunder, H.; Mertens, E.; Halloran, A.; Muir, G.; Vantomme, P. *Edible Insects: Future Prospects for Food and Feed*; Food and Agriculture Organisation of the United Nations: Rome, Italy, 2013; pp. 1–190.
3. Evans, J.; Alemu, M.H.; Flore, R.; Frost, M.B.; Halloran, A.; Jensen, A.B.; Maciel-Vergara, G.; Meyer-Rochow, V.B.; Münke-Svendsen, C.; Olsen, S.B.; et al. 'Entomophagy': An evolving terminology in need of review. *J. Insects Food Feed* **2015**, *1*, 293–305. [CrossRef]
4. Müller, A.; Evans, J.; Payne, C.L.R.; Roberts, R. Entomophagy and power. *J. Insects Food Feed* **2016**, *2*, 121–136. [CrossRef]
5. Bodenheimer, F.S. *Insects as Human Food*; W. Junk Publishers: The Hague, The Netherlands, 1951.
6. Khan, S.H. Recent advances in the role of insects as an alternative protein source in poultry nutrition. *J. Appl. Anim. Res.* **2018**, *46*, 1144–1157. [CrossRef]
7. Veldkamp, T.; Bosch, G. Insects: A protein-rich feed ingredient in pig and poultry diets. *Anim. Front.* **2015**, *5*, 45–50.
8. Arru, B.; Furesi, R.; Gasco, L.; Madau, F.A.; Pulina, P. The first economic analysis on European Sea Bass farming. *Sustainability* **2019**, *11*, 1697. [CrossRef]
9. Szendrö, K.; Nagy, M.Z.; Toth, K. Consumer acceptance of meat from animals reared on insect meal as feed. *Animals* **2020**, *10*, 1312. [CrossRef]
10. Penazzi, L.; Schiavone, A.; Russo, N.; Nery, J.; Valle, E.; Madrid, J.; Martinez, S.; Hernandez, F. In vivo and in vitro digestibility of an extruded complete dog food containing black soldier fly (*Hermetia illucens*) larvae meal as protein source. *Front. Vet. Sci.* **2021**, *8*, 542. [CrossRef]
11. Jung, C.; Meyer-Rochow, V.B. Edible Insects as Innovative Foods—Nutritional, Functional and Acceptability Assessments. 2020 Foods (ISSN 2304-8158). Available online: https://www.mdpi.com/journal/foods/special_issues/edible_insects_innovative_food (accessed on 2 August 2022).
12. New Scientist. Available online: https://books.google.fi/books?id=txDVQ-vzXMQC&pg=PA307&source=gbs_toc&cad=2#v=onepage&q&f=false (accessed on 12 September 2022).
13. Cho, W.-H.; Park, J.-M.; Kim, E.-J.; Mohibbullah, M.; Choi, J.-S. Evaluation of the Quality Characteristics and Development of a Puffed-Rice Snack Enriched with Honeybee (*Apis mellifera* L.) Drone Pupae Powder. *Foods* **2022**, *11*, 1599. [CrossRef]
14. Kim, T.-K.; Lee, J.-H.; Yong, H.I.; Kang, M.-C.; Cha, J.Y.; Chun, J.Y.; Choi, Y.-S. Effects of Defatting Methods on the Physicochemical Properties of Proteins Extracted from *Hermetia illucens* Larvae. *Foods* **2022**, *11*, 1400. [CrossRef]
15. Maiyo, N.C.; Khamis, F.M.; Okoth, M.W.; Abong, G.O.; Subramanian, S.; Egonyu, J.P.; Xavier, C.; Ekesi, S.; Omuse, E.R.; Nakimbugwe, D.; et al. Nutritional Quality of Four Novel Porridge Products Blended with Edible Cricket (*Scapsipedus icipe*) Meal for Food. *Foods* **2022**, *11*, 1047. [CrossRef]
16. Vanqa, N.; Mshayisa, V.V.; Basitere, M. Proximate, Physicochemical, Techno-Functional and Antioxidant Properties of Three Edible Insect (*Gonimbrasia belina*, *Hermetia illucens* and *Macrotermes subhylanus*) Flours. *Foods* **2022**, *11*, 976. [CrossRef]
17. Mshayisa, V.V.; Van Wyk, J.; Zozo, B. Nutritional, Techno-Functional and Structural Properties of Black Soldier Fly (*Hermetia illucens*) Larvae Flours and Protein Concentrates. *Foods* **2022**, *11*, 724. [CrossRef]

18. Gan, J.; Zhao, M.; He, Z.; Sun, L.; Li, X.; Feng, Y. The Effects of Antioxidants and Packaging Methods on Inhibiting Lipid Oxidation in Deep Fried Crickets (*Gryllus bimaculatus*) during Storage. *Foods* **2022**, *11*, 326. [CrossRef]
19. Selaledi, L.; Baloyi, J.; Mbajiorgu, C.; Sebola, A.N.; Kock, H.D.; Mabelebele, M. Meat Quality Parameters of Boschveld Indigenous Chickens as Influenced by Dietary Yellow Mealworm Meal. *Foods* **2021**, *10*, 3094. [CrossRef]
20. Dürr, J.; Ratompoarison, C. Nature's "Free Lunch": The Contribution of Edible Insects to Food and Nutrition Security in the Central Highlands of Madagascar. *Foods* **2021**, *10*, 2978. [CrossRef]
21. Su, Y.; Lu, M.-X.; Jing, L.-Q.; Qian, L.; Zhao, M.; Du, Y.-Z.; Liao, H.-J. Nutritional Properties of Larval Epidermis and Meat of the Edible Insect *Clanis bilineata tsingtauica* (Lepidoptera: Sphingidae). *Foods* **2021**, *10*, 2895. [CrossRef]
22. Zhu, C.; Zhao, M.; Zhang, H.; Zhang, F.; Du, Y.; Lu, M. Extending the Storage Time of *Clanis bilineata tsingtauica* (Lepidoptera; Sphingidae) Eggs through Variable-Temperature Cold Storage. *Foods* **2021**, *10*, 2820. [CrossRef]
23. Thrastardottir, R.; Olafsdottir, H.T.; Thorarinsdottir, R.I. Yellow Mealworm and Black Soldier Fly Larvae for Feed and Food Production in Europe, with Emphasis on Iceland. *Foods* **2021**, *10*, 2744. [CrossRef]
24. Müller, A.; Seinige, D.; Grabowski, N.T.; Ahlfeld, B.; Yue, M.; Kehrenberg, C. Characterization of *Escherichia coli* from Edible Insect Species: Detection of Shiga Toxin-Producing Isolate. *Foods* **2021**, *10*, 2552. [CrossRef]
25. Kipkoech, C.; Kinyuru, J.N.; Imathiu, S.; Meyer-Rochow, V.B.; Roos, N. In Vitro Study of Cricket Chitosan's Potential as a Prebiotic and a Promoter of Probiotic Microorganisms to Control Pathogenic Bacteria in the Human Gut. *Foods* **2021**, *10*, 2310. [CrossRef]
26. Lanng, S.K.; Zhang, Y.; Christensen, K.R.; Hansen, A.K.; Nielsen, D.S.; Kot, W.; Bertram, H.C. Partial Substitution of Meat with Insect (*Alphitobius diaperinus*) in a Carnivore Diet Changes the Gut Microbiome and Metabolome of Healthy Rats. *Foods* **2021**, *10*, 1814. [CrossRef]
27. Fuso, A.; Barbi, S.; Macavei, L.I.; Luparelli, A.V.; Maistrello, L.; Montorsi, M.; Sforza, S.; Caligiani, A. Effect of the Rearing Substrate on Total Protein and Amino Acid Composition in Black Soldier Fly. *Foods* **2021**, *10*, 1773. [CrossRef]
28. Placentino, U.; Sogari, G.; Viscecchia, R.; De Devitiis, B.; Monacis, L. The New Challenge of Sports Nutrition: Accepting Insect Food as Dietary Supplements in Professional Athletes. *Foods* **2021**, *10*, 1117. [CrossRef]
29. Son, Y.-J.; Hwang, I.-K.; Nho, C.W.; Kim, S.M.; Kim, S.H. Determination of Carbohydrate Composition in Mealworm (*Tenebrio molitor* L.) Larvae and Characterization of Mealworm Chitin and Chitosan. *Foods* **2021**, *10*, 640. [CrossRef]
30. Ghosh, S.; Namin, S.M.; Meyer-Rochow, V.B.; Jung, C. Chemical Composition and Nutritional Value of Different Species of *Vespa* Hornets. *Foods* **2021**, *10*, 418. [CrossRef]
31. Zhao, M.; Wang, C.-Y.; Sun, L.; He, Z.; Yang, P.-L.; Liao, H.-J.; Feng, Y. Edible Aquatic Insects: Diversities, Nutrition, and Safety. *Foods* **2021**, *10*, 3033. [CrossRef]
32. Meyer-Rochow, V.B.; Gahukar, R.T.; Ghosh, S.; Jung, C. Chemical Composition, Nutrient Quality and Acceptability of Edible Insects Are Affected by Species, Developmental Stage, Gender, Diet, and Processing Method. *Foods* **2021**, *10*, 1036. [CrossRef]
33. Bequaert, J. Insects as food: How they have augmented the food supply of mankind in early and recentyears. *Nat. Hist. J.* **1921**, *21*, 191–200.
34. Bergier, E. *Peuples Entomophages et Insectes Comestibles: Étude sur les Moeurs de L'homme et de L'insecte*; Imprimérie Rullière Frères: Avignon, France, 1941.
35. McGrew, W.C. The 'other faunivory' revisited: Insectivory in human and non-human primates and the evolution of human diet. *J. Human Evol.* **2014**, *71*, 4–11. [CrossRef]
36. Kipkoech, C.; Kinyuru, J.N.; Imathiu, S.; Roos, N. Use of house cricket to address food security in Kenya; nutritional and chitin composition of farmed crickets as influenced by age. *Afr. J. Agric. Res.* **2017**, *12*, 3189–3197. [CrossRef]
37. Seabrooks, L.; Hu, L. Insects: An underrepresented resource for tye discovery of biologically active natural products. *Acta Pharm. Sin. B* **2017**, *7*, 409–427. [CrossRef]
38. Meyer-Rochow, V.B. Therapeutic arthropods and other, largely terrestrial, folk medicinally important invertebrates: Acomparative survey and review. *J. Ethnobiol. Ethnomed.* **2017**, *13*, 9. [CrossRef]
39. Groib, F. Ameise und Volkskultur. *Denisia Neue Ser.* **2009**, *85*, 165–188.
40. Zengin, E.; Karaca, I. Böceklerin Ilaçolarak kullanılması. *Adü Zirrat Derg.* **2017**, *14*, 71–78.
41. Oonincx, D.G.A.B.; De Boer, I.J.M. Environmental impact of the production of mealworms as a protein source for humans—A life cycle assessment. *PLoS ONE* **2012**, *7*, e51145. [CrossRef]
42. Abbasi, T.; Abbasi, T.; Abbasi, S.A. Reducing the global environmental impact of livestock production: The minilivestock option. *J. Clean. Prod.* **2015**, *112*, 1754–1766. [CrossRef]
43. Halloran, A.; CaparrosMegido, R.; Oloo, J.; Weigel, T.; Nsevolo, P.; Francis, F. Comparative aspect of cricket framing in Thailand, Cambodia, Lao People's Democratic Republic, Democratic Republic of the Congo, and Kenya. *J. Insects Food Feed* **2018**, *4*, 101–114. [CrossRef]
44. Müller, A. Insects as food in Laos and Thailand—A case of "Westernisation"? *Asian J. Soc. Sci.* **2019**, *47*, 204–223. [CrossRef]
45. Ghosh, S.; Jung, C.; Meyer-Rochow, V.B. What governs selection and acceptance of edible insect species? In *Edible Insects in Sustainable Food Systems*; Halloran, A., Flore, R., Vantomme, P., Roos, N., Eds.; Springer: Cham, Germany, 2018; pp. 331–351. [CrossRef]
46. Tso, R.; Lim, A.J.Y.; Forde, G.C. A Critical Appraisal of the Evidence Supporting Consumer Motivations for Alternative Proteins. *Foods* **2021**, *10*, 24. [CrossRef]

47. Castro, M.; Chambers, E.I.V. Consumer avoidance of insect containing food: Primary emotions, perceptions and sensory characteristics driving consumers considerations. *Foods* **2019**, *8*, 351. [CrossRef] [PubMed]
48. Mancini, S.; Sogari, G.; Menozzi, D.; Nuvoloni, R.; Torracca, B.; Moruzzo, R.; Paci, G. Factors predicting the intention of eating an insect-based product. *Foods* **2019**, *8*, 270. [CrossRef] [PubMed]
49. Meyer-Rochow, V.B.; Jung, C. Insects used as food and feed: Isn't that what we all need? *Foods* **2020**, *9*, 1003. [CrossRef] [PubMed]

 *foods*

Article

The New Challenge of Sports Nutrition: Accepting Insect Food as Dietary Supplements in Professional Athletes

Umberto Placentino [1], Giovanni Sogari [2], Rosaria Viscecchia [3], Biagia De Devitiis [3,*] and Lucia Monacis [4]

[1] ITAF Sports Centre, Office for the Coordination and Management of Professional Athletes, Vigna di Valle, 00062 Roma, Italy; uplacentino@gmail.com
[2] Department of Food and Drug, University of Parma, 43124 Parma, Italy; giovanni.sogari@unipr.it
[3] Department of Agriculture, Food, Natural Resources and Engineering, University of Foggia, 71122 Foggia, Italy; rosaria.viscecchia@unifg.it
[4] Department of Humanities, Literature, Cultural Heritage, Education Sciences, University of Foggia, 71122 Foggia, Italy; lucia.monacis@unifg.it
* Correspondence: biagia.dedevitiis@unifg.it

Abstract: Background: The dietary supplements market is growing, and their use is increasing among professional athletes. Recently, several new protein supplements have been placed in the marketplace, including energy bars enriched with insect flour. Edible insects, which are rich in protein content, have been promoted as the food of the future and athletes could be a reference sample for their continued emphasis on higher protein demand. The present study investigated the potential motivations to accept an energy protein bar with cricket flour, among a group of selected Italian professional athletes. A second aim was also to measure how an information treatment about the benefits of edible insects would have impact on acceptance. Methods: 61 Italian professional athletes (27 females) completed a structured questionnaire regarding supplements and eating habits, food neophobia, nutrition knowledge, willingness to taste edible insects and the associated factors. A question about sports endorsement was also posed at the end of the survey. Results: all subjects consumed supplements, generally recommended by medical personnel, even though their general knowledge of nutrition was poor (47.8%). Our main results shown that on a seven-point Likert scale, the protein content (5.74 ± 1.01) and the curiosity about texture (5.24 ± 0.98) were the main drivers to taste the cricket energy bar; whereas the feeling of disgust (5.58 ± 1.08) justified the rejection of tasting insects. In addition, the level of food neophobia increases with age ($p < 0.05$) and reduces willingness to endorse the cricket bar ($p < 0.05$). Male athletes (4.47 ± 1.69) were more likely to endorse the product than females (3.3 ± 1.49). An increase in willingness to taste was observed after the information treatment ($z = 4.16$, $p < 0.001$). Even though the population under investigation is unique, it is important to mention that this study involves a relatively small and convenience sample, and therefore generalizability of the results should be done with caution.

Keywords: food neophobia; disgust; protein source; sport endorsement

Citation: Placentino, U.; Sogari, G.; Viscecchia, R.; De Devitiis, B.; Monacis, L. The New Challenge of Sports Nutrition: Accepting Insect Food as Dietary Supplements in Professional Athletes. *Foods* **2021**, *10*, 1117. https://doi.org/10.3390/foods10051117

Academic Editors: V. B. Meyer-Rochow, Chuleui Jung and Massimo Mozzon

Received: 7 April 2021
Accepted: 16 May 2021
Published: 18 May 2021

1. Introduction

Food plays a key role in acquiring the best physical condition and in ensuring optimal athletic performance. Physical activity, sports performance, and recovery after exercise are favored by optimal nutrition and professional athletes should be able to fully cover their nutritional needs through the consumption of foods, in adequate quantity and quality [1]. However, it is common among professional athletes to look for dietary supplements to ensure an optimal performance. Frequent competitive commitments and high intensity daily workouts could justify the use of supplements in professional athletes, especially in the case of inadequate diets or situations of temporary inability to maintain correct eating habits [2].

The global sports nutrition market accounted for several billion dollars and is expected to grow significantly because of its increasing demand from athletes and sportspersons in the near future [3]. The size of sport nutrition market in Europe is expected to achieve USD 15.12 billion by 2025. It refers to the consumption of sports drinks, bars, powders and other food supplements to improve physical performance [4].

However, a high proportion of these marketed supplements lack available evidence of their efficacy [5] and these products are often used without a full understanding of the potential benefits, negative side effects, and risks associated. As reported by the World Anti-Doping Agency (WADA), between 10 and 25% of currently marketed supplements reportedly contain prohibited substances, contributing to 6–9% of the total doping offenses [6,7]. Previous literature has shown gaps in knowledge about effective nutrition and supplementation among coaches and athletes [1]. It goes without saying that greater nutritional knowledge may improve dietary practices and food choices [8], reducing the risk of using prohibited substances.

The prevalence rate of the intake of sports supplements is generally high and similar among different countries. However, several demographic variables affect the proportion of athletes that consumed dietary supplements. Gender, age, level of competition, type of sport and professionalism influenced this proportion. It exceeds 60% and increases with professionalism [2,9,10].

The main reasons for consuming supplements are to enhance performance, speed, strength, and power, or to simply improve health. In a recent study, most of the athletes reported that supplements are safe and can be consumed without any risk. Male athletes were more likely to obtain information about the use of supplements mainly from a coach or physician [11].

Professional athletes are often used as brand ambassadors in marketing campaigns due to their high popularity and power to promote consumer engagement. They can use their highly visibility to endorse healthy and environmentally friendly foods, including supplements. Recently on the market of food supplements, there has been focus drawn to products derived from "novel foods", including insects. In particular there has been a growing prevalence of energy bars enriched with insect protein, in particular cricket flour [12,13].

Factors Influencing Edible Insects' Acceptance

In the last decades, there has been a growing interest among the scientific community, private industry, the general public, and the media about edible insects as human food [14,15]. Especially, after the publication of the report on edible insects as food and feed in 2013 by the Food and Agriculture Organization of the United Nations [16], and the new Regulation of the European Union 2015/2283 on novel foods [17], there has been an increasing number of scientific publications investigating the European consumer acceptance towards entomophagy [18,19], especially in Italy [20]. In addition, more recently the European Food Safety Authority (EFSA) published a scientific opinion on the safety of the first edible insect, i.e., *Tenebrio molitor larva* [21]. This will likely favor the development of the edible insect market also in the EU. Globally, a recent study forecasts that this market will reach almost USD 8 billion, with a volume of 730,000 tons by 2030 [22]. Thus, it seems plausible to predict a potential increase in the consumption, especially with regard to products containing invisible insects as ingredients such as in energy protein bars [23]. In particular, a potential target could be individuals who engage in sports regularly or follow balanced diets, who search for varied protein sources based on their origin [22].

The main components of insects are protein and fat, followed by fiber and carbohydrates. In particular, species from the order Orthoptera (grasshoppers, crickets, locusts) are rich in proteins and represent a valuable alternative protein source [24–26]. The nutrient quality of insect protein is promising in comparison to casein and soy, but varies and can be improved by the removal of the chitin [27]. Furthermore, most edible insects sufficiently provide the required essential amino acids [26,28]. Fat represents the second largest portion of the nutrient composition of edible insects, ranging from 13.41% for Orthoptera (grasshoppers, crickets, locusts) to 33.40% for Coleoptera (beetles, grubs). The fatty acids of insects are generally comparable to those of poultry and fish in their degree of saturation, but contain more polyunsaturated fatty acids (PUFA) [25,26]. The ratio of ω-6 and ω-3 fatty acids of edible insects is mostly 5:1 to 5.7:1 [24,29], reducing proinflammatory profile, which contributes to the prevalence of atherosclerosis, obesity, and diabetes [30]. Carbohydrates are mostly formed from chitin. It is still unclear if chitin can be digested by humans [27] and its content depend on the species and the developmental stage [25]. Edible insects have the potential to provide with specific micronutrients such as copper, iron, magnesium, manganese, phosphorous, selenium, and zinc; 100 g of edible insects generally lack enough calcium and potassium and they can be utilized in low-sodium diets. In addition to minerals, edible insects can be rich in vitamins, but species have to be specifically selected for the provision of desired vitamins [25,26].

Notably, different kind of exercise produce acute changes on metabolism of several nutrients. Dietary requirements for athletes are widely studied [1], as well as nutritional strategies that combine different forms of supplements to maximize the bioavailability of nutrients within the periodized training program [31]. The nutritional profile of some species of insects could cover the nutritional needs of athletes, representing an alternative daily diet to traditional animal-based food. Energy-protein insect bars, properly balanced and enriched with carbohydrates, could represent an alternative to those already widely used on the market as supplements for sport.

As a result, the opportunity promoted in this study could be a new challenge in nutritional supplements for sportsmen, including professional athletes [22,32,33].

However, a consumer's acceptance of insects as food is crucial in order to include this novel food in the diet. The most common barriers in explaining the aversion of Italian consumers towards eating insects are neophobia and disgust [20,23,34,35]. Although disgust is a universal emotion, it is important to note that the factors eliciting disgust can be different across individuals and cultures [36–38]. There is reason to believe that insects presented in different meal formats such as processed insect-based foods (e.g., snacks) might elicit more positive associations and taste expectations. Considering that negative taste expectation is a strong barrier to include a novel food [23], invisible insects as ingredients might reduce neophobic reactions and thus increase consumers' willingness to try [23,34,39]. Moreover, several studies with Italian consumers [20,40] also showed that also curiosity about the sensory attributes and a focus on environmental benefits might be motivating factors to promote entomophagy.

To the best of the authors' knowledge, no studies have focused on psychological and demographic predictors of sportsmen influencing the willingness to accept insect-based food.

In line with the growing interest in insects, consumption as food, and athletes' attitude for consuming sports dietary supplements, the present study introduces a protein insect-bar to investigate dispositional traits and emotional factors in accepting insect-edible food as dietary supplements among professional athletes.

In line with previous studies [9,41–44], several factors have been investigated: dietary supplements consumption, nutritional knowledge, food neophobia, and individual factors influencing the willingness to consume insect-based protein bars as dietary supplements. Finally, we evaluated how a brief informative text on the environmental and nutritional benefits of using insects as food [34,45] impacts the acceptance of eating insect food, i.e., protein bars enriched with cricket-flour.

2. Materials and Methods

2.1. Partecipants

Sixty-one professional athletes (27 females) aged 19 to 39 years (M = 27.8; SD = ±5.03) were recruited on a voluntary basis from the Italian Air Force Sports Centre in Rome. The average educational level was mainly high school (65.6%). The target population is made by professional athletes selected by the Italian National Sports Federations and most of them were currently competing at the international level (91.8%). We recruited mainly athletes from track and field (31.1%), fencing (23.0%), and beach-volleyball (18%); athletes from archery, sailing, skeet shooting, tennis table and artistic gymnastic were also recruited but in smaller numbers. The only inclusion criterion was that they had to be consumers of nutritional supplements. There were no age or gender restrictions. The exclusion criteria were (1) being vegetarians, (2) being in any concomitant nutritional counselling programs, (3) having any disease or health condition that required specialized dietary planning. Participants who had retired from a sport or had not participated in a competitive game or competition in the last year were excluded. Each participant was identified with an ID number to guarantee his/her anonymity.

2.2. Measures

A self-administered questionnaire was used with different sections:

(a) Information on socio-demographic characteristics, sports characteristics, knowledge and motivational aspects on nutritional supplement consumption, and dietary habits mainly related to animal protein food.

(b) The personal knowledge about general and sport nutrition (general nutrition knowledge—GNK and sport nutrition knowledge-SNK) by an adapted version of the Nutrition for Sport Knowledge Questionnaire (A-NSKQ) [46,47]. Total scores were assessed using one point for each correct answer, no negative points, and coding "unsure" answers as incorrect. The total score was out of 37. All domains were weighted equally during scoring, and percentages were determined. The following cut-off points were used poor knowledge (0–49%), average knowledge (50–65%), good knowledge (66–75%), and excellent knowledge (over 75%).

(c) Food neophobia was evaluated by the Food Neophobia Scale (FNS) [48]. It consists of 10 statements, five neophilic and five neophobic statements about food or situations related to food consumption, rated on a 7-point scale ranging from 1 = strongly disagree to 7 = strongly agree. After reverse coding the responses for the neophilic statements, a total FNS score ranging from 10 to 70 was then calculated by summing the ratings for each item; the higher the FNS score, the higher the food-neophobia level. According to recent studies, the consumers were categorized as having a low, medium or high level of food neophobia, and sustainable behavior. The frequency distribution of the FNS scores was calculated and the subjects were divided into the following three groups: "low neophobia" (subjects in the lowest quartile, FNS scores $\leq$ 23), "medium neophobia" (subjects in the second and third quartile, FNS scores $\geq$ 24 and $\leq$41) and "high neophobia" (subjects in the highest quartile, FNS scores $\geq$ 42) [49].

(d) Athletes' willingness to taste an insect-based protein bar was evaluated using a 7-point Likert scale (1 = strongly disagree, 7 = strongly agree). The population was split between groups: willing (from 5 to 7 points), uncertain (point 4) and unwilling (from 1 to 3 points).

(e) A brief informative text on the environmental and nutritional benefits of edible insects was provided to the participants (Table 1). After reading the text, participants were asked to assess their degree of agreement/disagreement in tasting the product.

(f) After the athletes expressed their willingness to taste the insects, two separate groups were identified (taster/no taster) and the factors which influenced their choice, i.e., curiosity about the texture, palatability, and alternative protein source for the tasters, and disgust, suitability for society, personal diet, poor hygiene and fear of unpleasant taste for the no tasters, were investigated using a 7-point Likert scale.

(g) Finally, athlete-endorsements in the food market were also explored using the following question: "from what it has been described above, how much would you be willing to promote this novel food product?" A 7-point Likert scale was used (1 = strongly disagree, 7 = strongly agree). The population was split between groups:

willing to endorse (from 5 to 7 points), uncertain (point 4) and unwilling to endorse (from 1 to 3 points).

Table 1. The information text provided to the sample.

In recent years, several European countries have begun to sell edible insects in supermarkets. Energy or protein bars are produced in a certified way, using insect's flour as supplements and their use is already widespread among athletes all over the world. From a nutritional point of view, insects are rich in proteins, minerals and vitamins, have a low-fat content and a reduced caloric intake, all elements that identify them as complete and healthy foods. Furthermore, insects farming has a lower environmental impact (e.g., few resources needed to raise them and reduced emission of carbon dioxide) compared to domestic animals.

Informed consent was obtained from all human research participants. The study was conducted in accordance with the Helsinki declaration and the ethical rules of the Italian Psychological Association.

2.3. Statistical Analysis

The Statistical Package for Social Science (SPSS 25) was used to analyze the data for all variables. Descriptive statistics were run to summarize the data collected and the results were displayed in frequencies and percentages. Friedman's and Cochran's Q test were run to determine if response for each category differed significantly. The internal consistency of the multi-scales was measured with Cronbach's α coefficient. Association between groups were calculated by Pearson's correlation or Spearman' correlation (if normal data distribution was not obtained or for ordinal variables). Differences between groups were calculated by independent-sample *t*-test, one-way-ANOVA, and a Mann–Whitney U test (if normal data distribution was not obtained) and one-way Welch ANOVA if there was heterogeneity of variances, assessed by Levene's test. The influence of various socio-demographics variables on food neophobia and A-NSKQ was examined using multiple linear regression analysis. A two-way repeated measures ANOVA was run to determine the effect of information treatment on the willingness to taste a cricket bar over time. The Shapiro-Wilk test ($p < 0.05$) was conducted on the difference scores to ensure normality for all variables with significant main effects. For normally distributed variables with significant main effects, post hoc dependent t-tests were conducted and effect sizes (Cohen's d) were calculated. Effect sizes were interpreted as small (0.20), medium (0.50), and large (0.80). For any variables that was not be normally distributed, the Wilcoxon signed-rank test and Glass's delta (effect size) were utilized for post hoc contrasts. In all tests, $p < 0.05$ was considered statistically significant.

3. Results

3.1. Dietary Habits and Nutrition Supplements

All 61 athletes consumed nutritional supplements, such as protein-amino acid supplements (68.6%), multivitamins (64%), minerals (44.3%), sport bars (42.6%) and fish oils (39.3%), χ^2 (3) = 12.7, $p < 0.001$. The prevalence of supplement consumption among professional athletes in the current study is similar compared to previous reports [2,9,10]. The qualified personnel (nutritionist/dietician/physician/pharmacy) were the most common source of information regarding nutritional supplements (88.5%) among athletes, χ^2 (4) = 52.61, $p < 0.001$. Most of the athletes used supplements to improve their recovery (72.1%), improve performance (49.2%), and health (45.9%) and to prevent deficiencies (36.1%) (Supplementary Table S1).

3.2. Neophobic Scale

Participants showed a mean value in FNS of 33.6 (SD 11.4). Subjects with "low neophobia" comprised 19.7% of the sample, "medium neophobia" comprised 55.7%, and "high neophobia" comprised 24.6%. Internal consistency was acceptable for food neophobia multi-scales (Cronbach's α = 0.85). Moderate correlation was found between age and FNS, r (59) = 0.3, $p < 0.05$. A linear regression established that age could statistically significantly predict FNS, F (1, 25) = 4.83, $p < 0.05$ and age accounted for 16% of the explained variability in FNS. The regression equation was: FNS = 8.8 + 0.99× (age).

3.3. Abridged–Nutritional Sport Knowledge Questionnaire (A-NSKQ)

Professional athletes in this study showed a poor nutritional sport knowledge (47.8%). Internal consistency was acceptable for A-NSKQ multi-scales (Cronbach's α = 0.82). There was a large variability amongst participants and between SNK and GNK and several misconceptions were evident in A-NSKQ, especially with regards to hydration, micronutrients and proteins (Supplementary Table S2).

Weak correlation was found between age and A-NSKQ, r (59) = 0.29, $p < 0.05$. An ANCOVA was run to determine the effect of level of education and ANSKQ after controlling for age. After adjustment for age, there is not a statistically significant difference in ANSKQ score, F(1, 39) = 2.34, p = 0.134, partial η^2 = 0.057.

3.4. Willingness to Taste the Cricket Flour Enriched Bar

Participants were classified into three groups based on their willingness to taste the cricket flour enriched bar: willing (n = 26), uncertain (n = 7), unwilling (n = 28), with a total mean value of 3.9 (SD 2.1). There were statistically significant differences in food neophobia between the different groups, F(2, 58) = 7.045, p = 0.002. Regression analysis indicated that 29% of the variability on willingness of tasting before the information was significantly accounted for by FNS, F(1, 59) = 24.3, $p < 0.001$, adj. R^2 = 0.28.

Following the information treatment, there was a statistically significant median increase in acceptance of the insect-bar (z = 4.16, $p < 0.001$) in number of respondents, in particular among the unwilling athletes, z = 3.88, $p < 0.001$ (Figure 1). A Wilcoxon signed-rank test determined that there was a statistically significant median increase in willingness to taste among the participants, rated on a 7-point scale (0.71), z = 4.16, $p < 0.001$ (Figure 2).

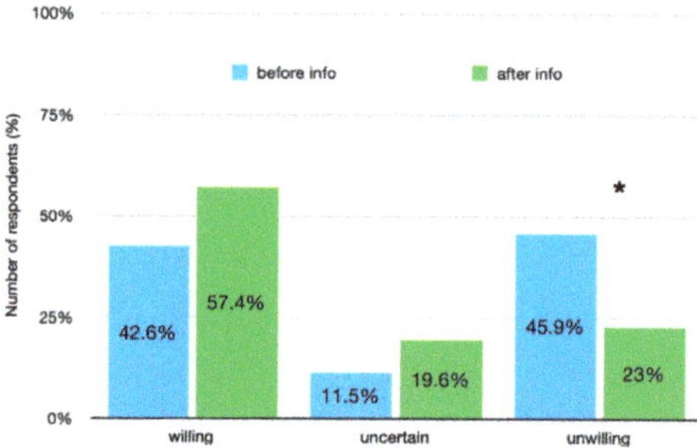

Figure 1. Willingness to taste an energy bar enriched with cricket flour. Difference in number of respondents (%) on the willingness to taste the cricket bar, before and after the information, among three different groups. Groups were defined using a 7-point Likert scale (1 = strongly disagree, 7 = strongly agree): willing (from 5 to 7), uncertain (point 4) and unwilling (from 1 to 3); number of respondents of the unwilling group decreased significantly * $p < 0.001$.

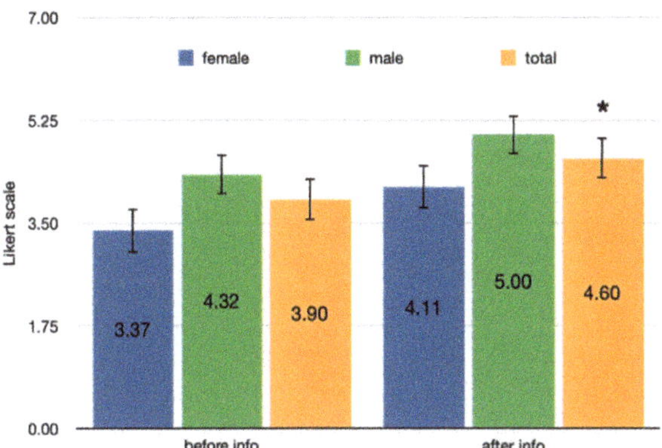

Figure 2. Willingness to taste an energy bar enriched with cricket flour, pre and post treatment. Difference in mean score in Likert scale, ranging from 1 = strongly disagree to 7 = strongly agree on the willingness to taste the cricket bar, before and after the information, among gender and the whole sample; willing to taste significantly increases in the reference sample * $p < 0.001$.

Finally, two groups were identified (taster/refusing): 43 athletes (70.5%) were willing to taste the cricket-bar, 15 females and 28 males (.3, $p = 0.011$) and the remaining 18 were not willing. The reasons to taste were investigated and the response for each category differed significantly, χ^2 (3) = 18.1, $p < 0.001$. A regression analysis indicated that 31% of the variability in the willingness to taste the product was significantly accounted for by "alternative research of a protein source", F(1, 11) = 6.35, $p < 0.05$, adj. $R^2 = 0.3$. Afterwards, factors influencing "no taster" group were investigated, the response for each category differed significantly, χ^2 (4) = 41, $p < 0.001$. Regression analysis indicated that 38% of the variability in the rejection to taste the product was significantly accounted for by "disgust", F(1, 8) = 6.5, $p < 0.05$, adj. $R^2 = 0.38$ (Figure 3).

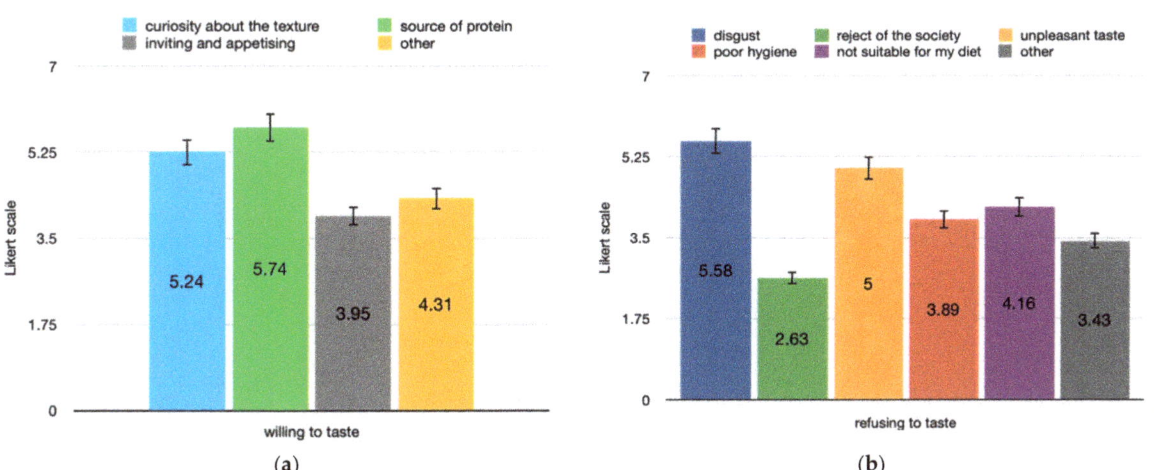

Figure 3. Reasons to taste or refuse cricket bar among the athletes. Difference in mean score in Likert scale, ranging from 1 = strongly disagree to 7 = strongly agree on the motivations (**a**) to taste or (**b**) refuse the cricket bar.

3.5. Athletes Endorsement in Cricket-Bar Marketing

In this study, 37.7% were willing to endorse the cricket bar. The taster group was more willing to endorse (4.67 ± 1.34) the cricket bar than no-taster group (2.22 ± 1.11), a statistically significant difference of 2.45 (95% CI, -3.17 to -1.73), t(59) $= -6.828$, $p < 0.001$. Moderate negative correlation was found between athletes-endorsement and FNS score, r (59) $= 0.3$, $p < 0.05$. Regression analysis indicated that 10% of the variability in the willingness to sponsor the product was significantly accounted for by NS, F(1, 59) $= 6.78$, $p < 0.05$, adj. $R^2 = 0.9$. Males were more willing to endorse (4.47 ± 1.69) the cricket bar than females (3.3 ± 1.49), a statistically significant difference of 1.17 (95% CI, -2.01 to -0.35), t(59) $= -2.365$, $p = 0.006$ (Figure 4).

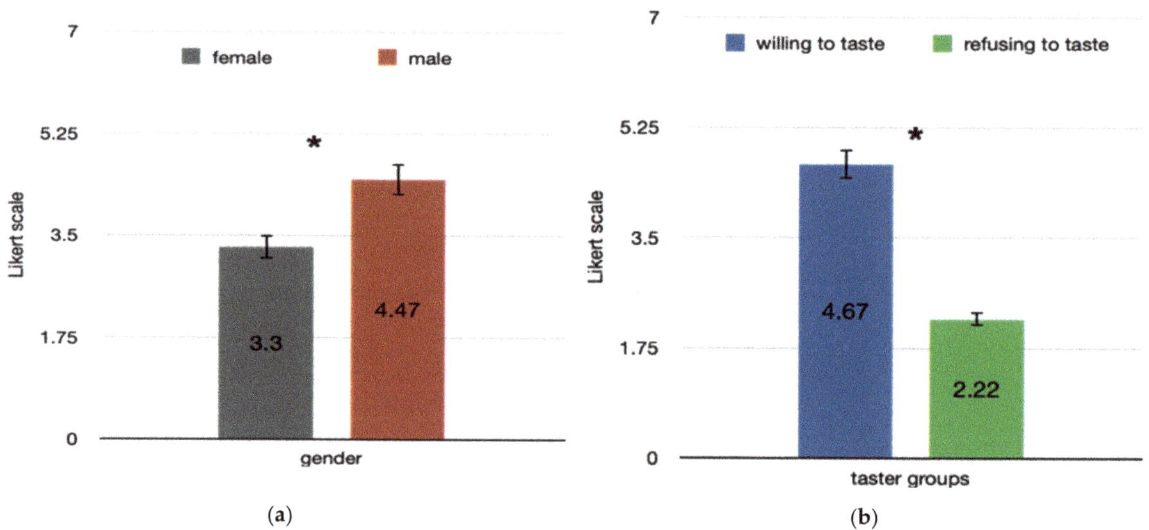

Figure 4. Willingness to endorse an energy bar enriched with cricket flour. Difference in mean score in Likert scale, ranging from 1 = strongly disagree to 7 = strongly agree on the willingness to endorse the cricket bar among (**a**) gender and (**b**) taster groups; males and willing to taste group were significantly prone to endorse the product * $p < 0.01$.

4. Discussion

This study investigated the potential motivations to accept eating an energy protein bar enriched with cricket flour among athletes, considering that motivation is a strong determinant in the type of supplement used [50,51]. In line with previous studies, the main motivations among athletes for the use of supplements in general included improving performance and recovery from exercise and preventing deficiencies. Protein, vitamin, minerals, and fish oils supplements were the most frequently used by athletes and recommended by acquired from qualified personnel (nutritionists/dieticians and physicians). This is important if we consider the vulnerability of the participants and their limited knowledge in sports nutrition [2]. Moreover, the high number of products available, often without a guarantee of accurate information, may increase the risk of using illegal substances, i.e., doping [6,7,41,52]. Our results support what has already been underlined in previous studies, i.e., the promotion of additional nutrition training as part of continued professional development of coaches and trainers would ensure better preparation to address nutritional concerns for athletes [53,54].

When asked about food neophobia, only 24.6% of athletes showed high neophobia and it could reflect exposure to different dishes during numerous contacts with foreign cultures during international events they attend [55]. In line with past studies [20,23,38], age was confirmed to be positively associated with food neophobia. No significant difference was found in food neophobia between gender; however, women seemed to be less likely to accept the energy bar with insect flour, probably due to their higher insect aversions than men [18,56,57]. It is likely that the high level of competition tends to make the difference between genders in supplement consumption less significant. Nevertheless, female athletes in this study report the same adversity towards insects as the common population.

Our results shown that food neophobia negatively contributes to the intention to taste and endorse the insect bar, and disgust was the main factor that determined the rejection of tasting. According to our data, food neophobia, even if it corresponds to the refusal to introduce insects into the diet, does not represent a characteristic of this type of population. Rather, it would be more relevant to investigate the disgust factor in athletes. After all, La Barbera et al. [35] demonstrated that the explanatory power of disgust towards the intention to consume insects was even higher than the explanatory power of food neophobic tendencies. Professional athletes have an important opportunity to promote the public's health, particularly for youth, by refusing endorsement contracts that involve promotion of energy-dense, nutrient-poor foods and beverages. Environmental sustainability and health promotion represents the main benefits of insects as food. However, only 37.7% of athletes willing to endorse the product, mainly males. Nevertheless, athletes who expressed the willingness to taste the product were more likely to endorse it. Reducing food neophobia could increase their willingness to endorse cricket bars as a new form of supplement. We believe that this sector should invest in sensory and gastronomic features, as well as advertising messages to reduce food neophobia and disgust about insects as food [23,39,45].

In line with previous consumer studies [37,40], our data indicated that the main factors for trying the insect-based product are the high protein content and the curiosity about the texture. Reinforcing the association between the relevant protein intake, i.e., nutrition profile, in a familiar product (energy bar) could increase the acceptability of insects as supplements in this target population.

Athletes have always shown a greater focus on high protein foods and strategies aimed at the consumption of protein-enriched foods or supplements. From a managerial point of view, planning proper marketing campaigns is supposed to reinforce the association between high protein content or alternative protein source and insect-based foods in the minds of this type of population.

As for the general population, taste has a strong influence on food choice, however, it may become less critical prior to an important game or event when foods that benefit performance are preferred, particularly in a highly competitive sport. The limited number of studies with athletes has reported performance or competition are one of the most important influences on food choice [8].

Our results shown a significant variation in the willingness to taste before and after the information treatment, in line with previous studies [45,49,58,59]. The athlete acceptability of insect-based foods was systematically increased after receiving information about the potential environmental and nutritional benefits of including insects in the diet.

Consequently, although recent technological developments in assuring food security for edible insects, e.g., automation and reduction in microbial contamination by personnel [16,19], the future marketing of these products in professional sport must also ensure that the products do not lead to any anti-doping rule violation.

This study has several limitations. First, it is a small sample for a quantitative study, and therefore, we are aware that generalization of the findings is difficult and should be done with caution. However, this is due to the specific target under investigation (professional athletes) which is not allowing to recruit a higher number of respondents. Thus, an extension of the dataset, including the widest variation possible for the sample, may increase the quality of future studies. Moreover, the comparison in nutritional knowledge with other studies was difficult, due to the heterogeneity of the measures used and reduced number of articles. The questionnaire only measured self-reported willingness to try insects and did not observe actual behavior to eat insects. In fact, in our study, it was not possible to carry out a "bug experience" that they might be categorized as "hidden substances" considered as high risk of doping in this specific category of participants. It is important to identify the species of insect best suited to cover the nutritional needs of athletes, reducing the risk of doping and optimizing training sessions as well as recovery and metabolic adaptation.

Future studies should consider using real products (e.g., bars with cricket flour) to measure actual consumption of insect food [18] and also include the Entomophagy Attitude Questionnaire Scale [60] instead of the FNS scale, to better investigate the individual aversion to eating insects.

5. Conclusions

This study represents a novelty in the evaluation of the acceptance of edible insects, for the choice of professional athletes as a sample and for the use of informative treatments in this type of subject. Investigating endorsements for edible insect among athletes could represent a new strategy to promote environmental and nutritional education in ordinary people. Insect protein could represent a new option to deliver nutrients to individuals who practice exercise and sport activity for work. Moreover, usually society holds a positive image towards professional athletes, and thus, these individuals could represent a target to promote the integration of insect-based products into sports nutrition.

Providing information about the environmental and nutritious benefits of edible insects will reduce food neophobia and disgust and favor acceptance of insects as food among individuals.

Supplementary Materials: The following are available online at https://www.mdpi.com/article/10.3390/foods10051117/s1, Table S1: responses (percent correct) of individual items in the use of nutritional supplements; Table S2: Responses (percent correct) of individual items in the A-NSKQ.

Author Contributions: Conceptualization, U.P., R.V., B.D.D. and L.M.; methodology, U.P., L.M., R.V. and B.D.D.; formal analysis, U.P.; data curation, U.P.; writing—original draft preparation, U.P. and G.S.; writing—review and editing, B.D.D., G.S., L.M., R.V. and U.P.; supervision, G.S. and L.M. All authors have read and agreed to the published version of the manuscript.

Funding: This research received no external funding.

Institutional Review Board Statement: The study was conducted according to the guidelines of the Declaration of Helsinki and the ethical rules of the Italian Psychological Association. Since neither biological substances nor "tasting experience" were carried out, no ethical approval was requested. The research did not involve any minimal risk to subjects, since participants voluntary took part in the online survey and gave an informed written consent form.

Informed Consent Statement: Informed consent was obtained from all subjects involved in the study.

Data Availability Statement: The data presented in this study are available upon reasonable request from the corresponding author.

Acknowledgments: We want to acknowledge Italian Air Force Sports Centre (ITAISC) personnel of Vigna di Valle–Bracciano (RM), which has undertaken to recruit the athletes interested in the study.

Conflicts of Interest: The authors declare no conflict of interest.

References

1. Thomas, D.T.; Erdman, K.A.; Burke, L.M. American College of Sports Medicine Joint Position Statement. Nutrition and Athletic Performance. *Med. Sci. Sports Exerc.* **2016**, *48*, 543–568. [CrossRef]
2. Knapik, J.J.; Steelman, R.A.; Hoedebecke, S.S.; Austin, K.G.; Farina, E.K.; Lieberman, H.R. Prevalence of Dietary Supplement Use by Athletes: Systematic Review and Meta-Analysis. *Sports Med.* **2016**, *46*, 103–123. [CrossRef]
3. Sarasota, F.L. *Sports Nutrition Market (Sports Food, Sports Drink & Sports Supplements): Global Industry Perspective, Comprehensive Analysis, and Forecast, 2016–2022)*; Zion Market Research: Pune, India, 2017.
4. Market Data Forecast Europe Sports Nutrition Market by Product Type, End-User, Distribution Channel and Industry Forecast to 2025. Available online: https://www.researchandmarkets.com/reports/3976185/europe-sports-nutrition-market-by-type (accessed on 1 February 2021).

5. Burke, L.M.; Peeling, P. Methodologies for Investigating Performance Changes with Supplement Use. *Int. J. Sport Nutr. Exerc. Metab.* **2018**, *28*, 159–169. [CrossRef] [PubMed]
6. Outram, S.; Stewart, B. Doping through supplement use: A review of the available empirical data. *Int. J. Sport Nutr. Exerc. Metab.* **2015**, *25*, 54–59. [CrossRef] [PubMed]
7. Pipe, A.; Ayotte, C. Nutritional supplements and doping. *Clin. J. Sport Med. Off. J. Can. Acad. Sport Med.* **2002**, *12*, 245–249. [CrossRef] [PubMed]
8. Birkenhead, K.L.; Slater, G. A Review of Factors Influencing Athletes' Food Choices. *Sports Med.* **2015**, *45*, 1511–1522. [CrossRef]
9. Aljaloud, S.O. Understanding the Behaviors and Attitudes of Athletes Participating in the 2016 Rio Olympics Regarding Nutritional Supplements, Energy Drinks, and Doping. *Int. J. Sport. Exerc. Med.* **2018**, *4*, 99. [CrossRef]
10. Maughan, R.J. IOC Medical and Scientific Commission reviews its position on the use of dietary supplements by elite athletes. *Br. J. Sports Med.* **2018**, *52*, 418–419. [CrossRef]
11. Aguilar-Navarro, M.; Baltazar-Martins, G.; Brito de Souza, D.; Muñoz-Guerra, J.; Del Mar Plata, M.; Del Coso, J. Gender Differences in Prevalence and Patterns of Dietary Supplement Use in Elite Athletes. *Res. Q. Exerc. Sport* **2020**, 1–10. [CrossRef] [PubMed]
12. Reverberi, M. Edible insects: Cricket farming and processing as an emerging market. *J. Insects Food Feed* **2020**, *6*, 211–220. [CrossRef]
13. Reverberi, M. The new packaged food products containing insects as an ingredient. *J. Insects Food Feed* **2021**, 1–8. [CrossRef]
14. Payne, C.; Caparros Megido, R.; Dobermann, D.; Frédéric, F.; Shockley, M.; Sogari, G. Insects as Food in the Global North—The Evolution of the Entomophagy Movement. In *Edible Insects in the Food Sector*; Springer International Publishing: Cham, Switzerland, 2019; pp. 11–26.
15. Meyer-Rochow, V.B.; Jung, C. Insects Used as Food and Feed: Isn't That What We All Need? *Foods* **2020**, *9*, 1003. [CrossRef]
16. van Huis, A.; Van Itterbeeck, J.; Klunder, H.; Mertens, E.; Halloran, A.; Muir, G.; Vantomme, P. *Edible Insects: Future Prospects for Food and Feed Security*; Food and Agriculture Organization of the United Nations: Rome, Italy, 2013.
17. EFSA. Scientific Committee Regulation (EU) 2015/2283 of the European Parliament and of the Council of 25 November 2015 on Novel Foods. Available online: http://data.europa.eu/eli/reg/2015/2283/oj (accessed on 19 March 2021).
18. Sogari, G.; Menozzi, D.; Hartmann, C.; Mora, C. How to Measure Consumers Acceptance Towards Edible Insects?—A Scoping Review About Methodological Approaches. In *Edible Insects in the Food Sector*; Springer International Publishing: Cham, Switzerland, 2019; pp. 27–44.
19. Raheem, D.; Carrascosa, C.; Oluwole, O.B.; Nieuwland, M.; Saraiva, A.; Millán, R.; Raposo, A. Traditional consumption of and rearing edible insects in Africa, Asia and Europe. *Crit. Rev. Food Sci. Nutr.* **2019**, *59*, 2169–2188. [CrossRef] [PubMed]
20. Toti, E.; Massaro, L.; Kais, A.; Aiello, P.; Palmery, M.; Peluso, I. Entomophagy: A Narrative Review on Nutritional Value, Safety, Cultural Acceptance and A Focus on the Role of Food Neophobia in Italy. *Eur. J. Investig. Health Psychol. Educ.* **2020**, *10*, 628–643. [CrossRef]
21. Turck, D.; Castenmiller, J.; De Henauw, S.; Hirsch-Ernst, K.I.; Kearney, J.; Maciuk, A.; Mangelsdorf, I.; McArdle, H.J.; Naska, A.; Pelaez, C.; et al. Safety of dried yellow mealworm (*Tenebrio molitor* larva) as a novel food pursuant to Regulation (EU) 2015/2283. *EFSA J.* **2021**, *19*, e06343. [CrossRef]
22. Pippinato, L.; Gasco, L.; Di Vita, G.; Mancuso, T. Current scenario in the European edible-insect industry: A preliminary study. *J. Insects Food Feed* **2020**, *6*, 371–381. [CrossRef]
23. Sogari, G.; Menozzi, D.; Mora, C. The food neophobia scale and young adults' intention to eat insect products. *Int. J. Consum. Stud.* **2019**, *43*, 68–76. [CrossRef]
24. Kinyuru, J.N.; Mogendi, J.B.; Riwa, C.A.; Ndung'u, N.W. Edible insects—A novel source of essential nutrients for human diet: Learning from traditional knowledge. *Anim. Front.* **2015**, *5*, 14–19. [CrossRef]
25. Dobermann, D.; Swift, J.A.; Field, L.M. Opportunities and hurdles of edible insects for food and feed. *Nutr. Bull.* **2017**, *42*, 293–308. [CrossRef]
26. Raheem, D.; Raposo, A.; Oluwole, O.B.; Nieuwland, M.; Saraiva, A.; Carrascosa, C. Entomophagy: Nutritional, ecological, safety and legislation aspects. *Food Res. Int.* **2019**, *126*, 108672. [CrossRef]
27. Payne, C.L.R.; Scarborough, P.; Rayner, M.; Nonaka, K. Are edible insects more or less 'healthy' than commonly consumed meats? A comparison using two nutrient profiling models developed to combat over- and undernutrition. *Eur. J. Clin. Nutr.* **2016**, *70*, 285–291. [CrossRef] [PubMed]
28. Kim, T.-K.; Yong, H.I.; Kim, Y.-B.; Kim, H.-W.; Choi, Y.-S. Edible Insects as a Protein Source: A Review of Public Perception, Processing Technology, and Research Trends. *Food Sci. Anim. Resour.* **2019**, *39*, 521–540. [CrossRef] [PubMed]
29. Adámková, A.; Mlček, J.; Kouřimská, L.; Borkovcová, M.; Bušina, T.; Adámek, M.; Bednářová, M.; Krajsa, J. Nutritional Potential of Selected Insect Species Reared on the Island of Sumatra. *Int. J. Environ. Res. Public Health* **2017**, *14*, 521. [CrossRef] [PubMed]
30. Simopoulos, A.P. An Increase in the Omega-6/Omega-3 Fatty Acid Ratio Increases the Risk for Obesity. *Nutrients* **2016**, *8*, 128. [CrossRef]
31. Kerksick, C.M.; Wilborn, C.D.; Roberts, M.D.; Smith-Ryan, A.; Kleiner, S.M.; Jäger, R.; Collins, R.; Cooke, M.; Davis, J.N.; Galvan, E.; et al. ISSN exercise & sports nutrition review update: Research & recommendations. *J. Int. Soc. Sports Nutr.* **2018**, *15*, 38. [CrossRef] [PubMed]

32. Meyer, N.; Reguant-Closa, A. "Eat as If You Could Save the Planet and Win!" Sustainability Integration into Nutrition for Exercise and Sport. *Nutrients* **2017**, *9*, 412. [CrossRef]
33. Nowakowski, A.C.; Miller, A.C.; Miller, M.E.; Xiao, H.; Wu, X. Potential health benefits of edible insects. *Crit. Rev. Food Sci. Nutr.* **2021**, 1–10. [CrossRef]
34. Mancini, S.; Sogari, G.; Menozzi, D.; Nuvoloni, R.; Torracca, B.; Moruzzo, R.; Paci, G. Factors Predicting the Intention of Eating an Insect-Based Product. *Foods* **2019**, *8*, 270. [CrossRef] [PubMed]
35. La Barbera, F.; Verneau, F.; Amato, M.; Grunert, K. Understanding Westerners' disgust for the eating of insects: The role of food neophobia and implicit associations. *Food Qual. Prefer.* **2018**, *64*, 120–125. [CrossRef]
36. Meyer-Rochow, V.B. Food taboos: Their origins and purposes. *J. Ethnobiol. Ethnomed.* **2009**, *5*, 18. [CrossRef]
37. Ghosh, S.; Jung, C.; Meyer-Rochow, V.B. What Governs Selection and Acceptance of Edible Insect Species? In *Edible Insects in Sustainable Food Systems*; Springer International Publishing: Cham, Switzerland, 2018; pp. 331–351.
38. Mascarello, G.; Pinto, A.; Rizzoli, V.; Tiozzo, B.; Crovato, S.; Ravarotto, L. Ethnic Food Consumption in Italy: The Role of Food Neophobia and Openness to Different Cultures. *Foods* **2020**, *9*, 112. [CrossRef] [PubMed]
39. Tan, H.S.G.; Fischer, A.R.H.; Tinchan, P.; Stieger, M.; Steenbekkers, L.P.A.; van Trijp, H.C.M. Insects as food: Exploring cultural exposure and individual experience as determinants of acceptance. *Food Qual. Prefer.* **2015**, *42*, 78–89. [CrossRef]
40. Sogari, G.; Menozzi, D.; Mora, C. Exploring young foodies' knowledge and attitude regarding entomophagy: A qualitative study in Italy. *Int. J. Gastron. Food Sci.* **2017**, *7*, 16–19. [CrossRef]
41. Nieper, A. Nutritional supplement practices in UK junior national track and field athletes. *Br. J. Sports Med.* **2005**, *39*, 645–649. [CrossRef]
42. Aljaloud, S.O.; Ibrahim, S.A. Use of Dietary Supplements among Professional Athletes in Saudi Arabia. *J. Nutr. Metab.* **2013**, *2013*, 245349. [CrossRef]
43. Bianco, A.; Mammina, C.; Paoli, A.; Bellafiore, M.; Battaglia, G.; Caramazza, G.; Palma, A.; Jemni, M. Protein supplementation in strength and conditioning adepts: Knowledge, dietary behavior and practice in Palermo, Italy. *J. Int. Soc. Sports Nutr.* **2011**, *8*, 25. [CrossRef]
44. El Khoury, D.; Antoine-Jonville, S. Intake of Nutritional Supplements among People Exercising in Gyms in Beirut City. *J. Nutr. Metab.* **2012**, *2012*, 703490. [CrossRef]
45. Verneau, F.; La Barbera, F.; Kolle, S.; Amato, M.; Del Giudice, T.; Grunert, K. The effect of communication and implicit associations on consuming insects: An experiment in Denmark and Italy. *Appetite* **2016**, *106*, 30–36. [CrossRef]
46. Trakman, G.L.; Forsyth, A.; Hoye, R.; Belski, R. The nutrition for sport knowledge questionnaire (NSKQ): Development and validation using classical test theory and Rasch analysis. *J. Int. Soc. Sports Nutr.* **2017**, *14*, 26. [CrossRef]
47. Trakman, G.L.; Forsyth, A.; Hoye, R.; Belski, R. Development and validation of a brief general and sports nutrition knowledge questionnaire and assessment of athletes' nutrition knowledge. *J. Int. Soc. Sports Nutr.* **2018**, *15*, 17. [CrossRef]
48. Pliner, P.; Hobden, K. Development of a scale to measure the trait of food neophobia in humans. *Appetite* **1992**, *19*, 105–120. [CrossRef]
49. Laureati, M.; Proserpio, C.; Jucker, C.; Savoldelli, S. New sustainable protein sources: Consumers' willingness to adopt insects as feed and food. *Ital. J. Food Sci.* **2016**, *28*, 652–668. [CrossRef]
50. Giannopoulou, I.; Noutsos, K.; Apostolidis, N.; Bayios, I.; Nassis, G.P. Performance level affects the dietary supplement intake of both individual and team sports athletes. *J. Sports Sci. Med.* **2013**, *12*, 190–196. [PubMed]
51. Heikkinen, A.; Alaranta, A.; Helenius, I.; Vasankari, T. Dietary supplementation habits and perceptions of supplement use among elite Finnish athletes. *Int. J. Sport Nutr. Exerc. Metab.* **2011**, *21*, 271–279. [CrossRef]
52. Tian, H.H.; Ong, W.S.; Tan, C.L. Nutritional supplement use among university athletes in Singapore. *Singap. Med. J.* **2009**, *50*, 165–172.
53. Torres-McGehee, T.M.; Pritchett, K.L.; Zippel, D.; Minton, D.M.; Cellamare, A.; Sibilia, M. Sports nutrition knowledge among collegiate athletes, coaches, athletic trainers, and strength and conditioning specialists. *J. Athl. Train.* **2012**, *47*, 205–211. [CrossRef] [PubMed]
54. Cockburn, E.; Fortune, A.; Briggs, M.; Rumbold, P. Nutritional knowledge of UK coaches. *Nutrients* **2014**, *6*, 1442–1453. [CrossRef]
55. Olabi, A.; Najm, N.E.O.; Baghdadi, O.K.; Morton, J.M. Food neophobia levels of Lebanese and American college students. *Food Qual. Prefer.* **2009**, *20*, 353–362. [CrossRef]
56. Ghosh, S.; Jung, C.; Meyer-Rochow, V.B.; Dekebo, A. Perception of entomophagy by residents of Korea and Ethiopia revealed through structured questionnaire. *J. Insects Food Feed* **2020**, *6*, 59–64. [CrossRef]
57. Menozzi, D.; Sogari, G.; Veneziani, M.; Simoni, E.; Mora, C. Eating novel foods: An application of the Theory of Planned Behaviour to predict the consumption of an insect-based product. *Food Qual. Prefer.* **2017**, *59*, 27–34. [CrossRef]
58. Tuorila, H.; Meiselman, H.L.; Bell, R.; Cardello, A.V.; Johnson, W. Role of sensory and cognitive information in the enhancement of certainty and liking for novel and familiar foods. *Appetite* **1994**, *23*, 231–246. [CrossRef]
59. McFarlane, T.; Pliner, P. Increasing willingness to taste novel foods: Effects of nutrition and taste information. *Appetite* **1997**, *28*, 227–238. [CrossRef]
60. La Barbera, F.; Verneau, F.; Videbæk, P.N.; Amato, M.; Grunert, K.G. A self-report measure of attitudes toward the eating of insects: Construction and validation of the Entomophagy Attitude Questionnaire. *Food Qual. Prefer.* **2020**, *79*, 103757. [CrossRef]

 *foods*

Article

Nature's "Free Lunch": The Contribution of Edible Insects to Food and Nutrition Security in the Central Highlands of Madagascar

Jochen Dürr [1],* and Christian Ratompoarison [2]

[1] Center for Development Research (ZEF), University of Bonn, 53113 Bonn, Germany

[2] Department of Food Sciences and Technology, Campus of Ambohitsaina, College of Agricultural Sciences, University of Antananarivo, P.O. Box 175, Antananarivo 101, Madagascar; christianratompoarison@gmail.com

* Correspondence: jochenduerr@gmx.de

Abstract: Edible insects are a healthy, sustainable, and environmentally friendly protein alternative. Thanks to their quantitative and qualitative protein composition, they can contribute to food security, especially in Africa, where insects have been consumed for centuries. Most insects are still harvested in the wild and used for household consumption. So far, however, little attention has been paid to insects' real contribution to food security in low-income countries. Entomophagy, the human consumption of insects, is widespread in many rural areas of Madagascar, a country, at the same time, severely affected by chronic malnutrition. This case study was carried out in a region where entomophagy based on wild harvesting is a common practice and malnutrition is pervasive. The data were obtained in 2020 from a survey among 216 households in the rural commune of Sandrandahy in the central highlands of Madagascar. Descriptive statistics, correlation, and regression analysis were used to show the relative importance of insects for the local diet and to test various hypotheses related to food security. Results show that insects contribute significantly to animal protein consumption, especially in the humid season, when other protein sources are scarce. They are a cheap protein source, as much esteemed as meat by the rural population. There are no significant differences in the quantities of insects consumed by poorer versus richer households, nor between rural and urban households. Insect consumption amounts are strongly related to the time spent on wild harvesting. The importance of edible insects for poor, food-insecure rural areas and how entomophagy can be promoted for better food and nutrition security are discussed.

Keywords: entomophagy; insect consumption; protein intake; rural areas; Sandrandahy

Citation: Dürr, J.; Ratompoarison, C. Nature's "Free Lunch": The Contribution of Edible Insects to Food and Nutrition Security in the Central Highlands of Madagascar. *Foods* **2021**, *10*, 2978. https://doi.org/10.3390/foods10122978

Academic Editors: Victor Benno Meyer-Rochow and Chuleui Jung

Received: 2 November 2021
Accepted: 26 November 2021
Published: 3 December 2021

Publisher's Note: MDPI stays neutral with regard to jurisdictional claims in published maps and institutional affiliations.

1. Introduction

Already, in the 1970s, insects were proposed by Meyer-Rochow as a solution to the world protein shortage [1]. At the same time, Chavunduka [2] claimed that insects are a cheap source of protein in Africa. In the early 21st century, there has been a renewed interest in edible insects' possible contribution to food and nutrition security [3], especially as an important protein source [4]. Worldwide, insects are mostly harvested in the wild and are predominantly consumed within the household. Depending on the local context and traditions, insects play a significant role for local diets, especially in times of food shortages [3,5,6]. Edible insects are healthy, sustainable, and environmentally friendly protein alternatives [7]. Thanks to their quantitative and qualitative protein composition, they can contribute significantly to food security, especially in Africa, where insects have been consumed for centuries [8,9], with exceptions such as Ethiopia, where insects appear much less appreciated than elsewhere in Africa [10]. Proteins are the main components of an insect's body, representing 35% to 77% of its dry matter [11]. In addition, the digestibility of insect proteins is very high for many preparations. As an indication, for the house cricket, which is popular for rearing (but not very popular as a food item in Madagascar [12]), the digestibility is between 78% and 83% [13]. Some insect species contain considerable amounts of essential minerals (K, Na, Ca, Cu, Fe, Zn, Mn, and P), as well as vitamins from group B and vitamins A, C, D, E, and K [14–16]. Moreover, insects contain important mono- or polyunsaturated fatty acids recommended by nutritionists. However, when considering entomophagy for nutrition security, one also has to bear in mind the possible contamination of edible insects by pathogens or pesticides, and the presence of allergic proteins and anti-nutrients [17].

Nevertheless, the importance of insects for food and nutrition security cannot be estimated yet on a global scale, and there are only a few specific studies available for the local level [3]. Moreover, insects are not only consumed in the case of food scarcity, but also because, and perhaps more importantly, they are considered a traditional and delicious food item [18].

Madagascar still faces severe problems of chronic malnutrition and stunting. Average protein supply is only 43 g/capita/day [19], one of the lowest per capita supplies worldwide, and not enough to ensure the recommended daily intake (of 0.75 g proteins per kg body weight of adults, see [20]) for all Malagasy. Insufficient intake of animal proteins is a major nutritional problem in the country [21]. Moreover, micronutrients, such as iron, are not sufficiently consumed; around one third of women and men in Madagascar are anaemic [22]. At the same time, entomophagy is still a widespread phenomenon in many parts of Madagascar. Insects are a commonly used food of the rural population. Nevertheless, there are no studies showing the actual amounts of consumed insects [12]. As there are no public statistics of insect consumption in terms of quantities or nutritional values, or of the socioeconomic background of insect consumers, it is hardly possible to know the real contribution of insects to food security in Madagascar. Besides, there are concerns that the potential of entomophagy is in danger because of decreasing insect supply, possibly because of habitat loss due to slash-and-burn practices and forest overuse [23], but there are also no data available to confirm this declining trend.

Some studies have focused on identifying the insect species consumed by local communities, using survey methods with questionnaires or semi-structured interviews and taxonomy. For example, Agbydye et al. [24] found that termites, the large African cricket, and the pallid emperor moth were the most frequently consumed insect species in Benue state of Nigeria. Bomolo et al. [25] found 11 species consumed in Haut-Katanga Province of Congo, and different proportions of groups eating these species. Caterpillars and termites were among the most consumed insects, but the results depended significantly on the age, ethnicity, family status, and education level of the respondents. These studies showed the wide variety of insect species consumed, but did not quantify the respective amounts.

So far, edible insects have not been included in national consumption surveys [26], and only one nationally representative survey was carried out accounting for insect consumption. This survey [27] showed that, in Laos, the most frequently consumed insects are weaver ant eggs, crickets, grasshoppers, cicadas, and bamboo worms. Almost 97% of people consume insects, and 44% of the population does this (very) frequently. However, the survey did not quantify the consumption amounts and, hence, cannot specify the nutritional intake of insects and their importance to food and nutrition security.

There are some studies on the regional or local level that state the amounts of insects consumed. Using 24-h recall, Yhoung-Aree et al. [28] found that between 2 and 32% of schoolchildren in Pana District, Northeastern Thailand, consumed, depending on the season, between 2 g and 26 g of insects per person per day, mainly silkworm pupae, crickets, and beetles. Acuña et al. [29] reported approximate monthly amounts of edible insects consumed as part of the traditional food system of a Popoloca village in Puebla State, Mexico. The quantities vary according to weather conditions, individual preferences, and the prevalence of specific species and are, therefore, difficult to estimate. Between February and September, a whole family gathers around one to two litres of different species from different orders, once or twice per season, and individual persons collect around 12–15 larvae, also once or twice per season. However, the nutrient intake, especially the protein intake, resulting from the amounts consumed were not calculated by these studies. One exception is the study of N'Gasse et al. [30], who report that, in the district of Bangui in the Central African Republic, 29% of the annual animal protein intake comes from the consumption of caterpillars.

Given the scarce evidence on the importance of insects as a protein source, our central objective was to use a case study in the central highlands of Madagascar to analyse how much entomophagy actually contributes to local food and nutrition security. The originality of this case study lies, on the one hand, in its comprehensive empirical assessment of the above-mentioned food security topics: firstly, by showing how much insects are contributing to local diets as a protein source, and how this varies seasonally; secondly, by examining the relative importance of the insect protein source in relation to other animal and plant protein sources; and, thirdly, by evaluating if insects are a cheaper protein source compared to other protein sources. On the other hand, we tested three hypotheses related to important food security issues: 1. Poorer households consume more insect protein compared to richer households. 2. Significant differences in protein consumption patterns exist between villages, and between rural and urban sites. 3. Insect protein consumption is positively related to farm size and harvesting time.

Better knowledge about the contribution of insects to food and nutrition security could foster sustainable ways of promoting entomophagy in different parts of Madagascar. Even if between regions, and even within regions, different sorts of edible insects are consumed in different numbers [23], and a generalisation of specific case study results is therefore difficult, there are some general conclusions this paper will draw, which can be tested in further studies: firstly, the importance of edible insects as a protein source in poor, rural areas where entomophagy is common; and, secondly, the fact that, as long as entomophagy is based on wild harvesting, its contribution to food and nutrition security remains restricted by seasonality and natural conditions, and by the time that can be spent on the collection of insects.

The rest of the article is organised as follows. In chapter two, we describe the study site and explain our data sampling process. Chapter three presents the descriptive results concerning consumption amounts of insects, protein intake, and the cost of insect proteins compared to other animal and plant proteins. In chapter four, the hypotheses are tested. Then, we interpret the results in the discussion section (chapter five). Conclusions for the promotion of insect consumption for better nutrition are finally drawn (chapter six).

2. Materials and Methods

2.1. Research Site

The research was carried out in the southern zone of the central highlands of Madagascar, in the Amoron'i Mania region, district of Fandriana, rural commune of Sandrandahy (20°21′00″ S, 47°17′42″ E, see the map in the Appendix A), where the ProciNut project "Production and Processing of Edible Insects for Improved Nutrition" is being undertaken. Sandrandahy consists of 38 fokontany (villages), with a total of 5055 households. The altitudes vary between 1200 and 1500 m.a.s.l., and the tropical climate has two well-marked seasons—a hot and humid one, from October to April, and a cool, dry one, from May to September [31]. Agriculture is the chief economic activity there, rice being the main staple crop. Poverty is widespread and per capita income in 2010 was only Madagascar Ariary (MGA) 672 thousand (at the time, around Euro (EUR) 272) per annum (or EUR 0.65 per day; [32]). Stunting and chronic malnutrition is pervasive, reaching 64% and 69%, respectively, of the population [33]. The widespread and chronic food and nutritional insecurity is aggravated after October, and is especially strong in the months of January and February, when rice reserves are dwindling. The main dish consists of rice, complemented by beans, vegetables, and sometimes meat or fish. According to FAO [33], the underlying problems are the small landholdings coupled with demographic pressure on land, low number of livestock, low prices of agricultural produce, and an increase in food prices during the lean season, when the availability of rice is low. Given the characteristics of the commune, it represents an interesting case for studying the contribution of entomophagy to food security in a context of malnutrition and protein scarcity, a situation which is common in most parts of Madagascar.

2.2. Sampling Design

We used cluster sampling as a common probability sampling procedure. Each fokontany was considered as a cluster. As we knew the number of households of each of the 38 clusters, systematic sampling using probability proportional to size was possible, i.e., larger clusters have a proportionally higher likelihood of being selected than smaller ones. In order to keep the design effect at low levels, but also taking into account the difficulties of reaching the villages, we decided to take interviews in 12 of the 38 villages, so that the relation of number of villages (clusters) to number of households was in the range considered as "relatively safe" (maximum 40–50 households per cluster; see Magnani 1999, p. 18). We used the procedure described in Magnani [34] (p. 24) to select the 12 fokontany.

In January 2020, we conducted 18 interviews in each of the 12 villages—216 interviews in total. The measurement entity was the household, which was our unit of interest. For each of the 12 randomly chosen clusters, we were able to access a list of all households of the village. First, we randomly and systematically drew 18 households from the list. This was achieved by dividing the total number of households by 18 (planned number of interviews considered manageable given the available time and number of enumerators). The result was the sampling interval (SI). We started randomly at the top of the list at one of the households, number 1 to number (SI), and then chose every (SI)th household. We applied the same procedure in a second step to draw 12 additional households in the event that some of the 18 original households were absent or would decline to co-operate (which has happened in only two cases).

Data collection was carried out by face-to-face interviews at households with a standardised questionnaire. Preferentially, the head of each household and/or his/her spouse was interviewed to ensure that the respondents were aware of the consumption of the whole household. The sample comprised 66% of male and 34% of female respondents, 72% of whom were heads of household. Six local enumerators were recruited and trained on the questions before conducting the survey. The questionnaire was pretested to ensure a general understanding and a uniform questioning technique. The survey was announced beforehand by the village chiefs to instil confidence. Participation was voluntary and took place without remuneration. The survey software KoBoToolbox from the Harvard Humanitarian Initiative was used to collect the data. The answers were entered into a data acquisition app via smartphones, stored in a cloud, and then transmitted to computers. The questionnaire was initially developed in English, and then translated into Malagasy by the local project partners. The analyses were carried out with SPSS (version 27.0, IBM, Böblingen, Germany) and Microsoft Excel.

The main focus of the questionnaire was on quantitative data on the agro-socioeconomic context, insect harvest and consumption, and nutrition. In addition, qualitative data were collected to better understand local entomophagy practices. Qualitative questions included dichotomous questions ("Would you like to harvest more insects?"), as well as open-ended questions ("Why would you like to harvest more insects?"). The questionnaire was divided into five main parts: 1. General information; 2. Socio-economic status and agriculture; 3. Insect harvesting; 4. Insect consumption; 5. Food frequencies.

The survey was conducted with the approval of the Head of Amoron'i Mania Region, the prefecture of Ambositra District, the Mayor of Sandrandahy Commune, and the Chiefs of each fokontany. We informed the respondents that the data collected remain confidential, that they have the right to stop the interview at any point in time, and that all gathered information will be used only for research purposes. Respondents gave their oral consent to the data collection.

2.3. Seasonal Consumption and Harvesting Amounts

For the analysis of the contribution of insects to food and nutrition security, we used food frequency tables to gather consumption data of insects, as well as of other food items, for a whole year. First, it was examined in which months of the year (from January to December) every item is consumed. Then, it was asked how often and how much of the food is consumed each time, i.e., the quantities per day, week, or month, so that average monthly amounts could be calculated. Multiplying the monthly amounts by the number of months the food item is consumed gives the total annual amount. This procedure also allowed us to assess seasonal patterns of consumption. (Note that the monthly consumption amounts per individual household are the same for all months when the item is consumed by this household (zero otherwise); collecting monthly consumption data was considered too difficult. The seasonal consumption pattern is the result of adding the amounts of all households.) However, in the case of insects, the seasonality pattern relies on harvesting amounts, not on consumption amounts. (Again, monthly data on consumption were considered difficult to collect but, as there is hardly any buying or selling of insects, monthly harvesting amounts nearly equal consumption amounts.)

The amounts consider the consumption of the whole household; despite possible individual differences, when calculating per capita consumption, we simply divided the household amount by the number of household members. Similar to consumption amounts, harvesting amounts were calculated by the number of months during the year when insects are collected, multiplied by the average monthly harvesting amounts. For transformation into kg, we used a standardised transformation factor for each local unit (e.g., one cup of insects = 200 g).

In different parts of the questionnaire, we independently collected data on the harvesting, buying, and selling of insects. Considering that total consumption amount = amount harvested + purchased − sold, it was possible to cross-check the consumption with the harvesting data (excluding the possibility of transferring insects between households as gifts). Only in 5% of all cases is the difference between the two higher than the average consumption amount. For all insect groups except one, average consumption amounts differ less than one kilogram from harvesting amounts, and the total average harvesting amount is 10% (0.9 kg per household) higher than the consumption amount, which results in rather conservative estimations about the consumption of edible insects.

2.4. Nutritional Composition and Costs of Insects and Other Foods

In the survey, people used local terms for insects, which normally do not denominate specific species, but often a certain stage in the life cycle of a certain insect family or order. For example, *Sakivy* denominates larvae, and *Abado* pupae of Coleoptera. The insect species, families, and orders belonging to each local name were identified by an entomologist of the ProciNut project. To determine the nutritional composition of the groups of species identified, bibliographic research was carried out. The keywords used were the scientific names of the species. The nutritional composition used for the locally denominated insects is given in Table A1 in the Appendix A. In some cases, an average protein content was calculated when data were available from different sources. In two cases (snout beetles and wild silkworm), where no data could be found, nutritional composition was used from different families belonging to the same order. In cases where nutritional information was only available for dry matter (beetle larvae and cicadas), an average moisture content of 70% was applied to calculate the protein content for fresh matter. In order to compare the protein content of commonly consumed animal- and plant-based food items, we used the tables given by the National Bureau of Nutrition (ONN) of Madagascar (see Table A2 in the Appendix A).

The protein costs were calculated by the average prices paid by the respondents per kg of purchased food item, and its protein content, described above. As most insects are not bought, we only obtained sufficient price data for two types of insects (adult scarab beetles and silkworms).

3. Descriptive Results

3.1. Socio-Demographic Characteristics

The average size of the interviewed households is 5.5 members. Nearly half of household members (49%) belong to the working age population, defined as 16–64 years old, meaning that the dependency ratio is around one. Ethnically, the population is quite homogenous: 95% of the interviewees are members of the Betsileo group; the rest belong to the Merina group, except for one respondent from Sakalava group. Most of the communities' members (75%) have lived already for a long time in their villages; 25% of households have moved to their community only after the year 2000. With regard to the level of education of the interviewees, it can be stated that 4% were not enrolled in school, 56% have completed primary education, 25% have passed college, 12% achieved high school level, and 3% went beyond this level. Regarding religious beliefs, 61% stated to be Catholic, 26% Protestant, 12% Lutheran, and 2% attend the free churches. In total, 10% of the sample are female-headed households. However, all of these socio-demographic variables are not significantly correlated with insect protein consumption (household size, dependency ratio) or the mean values of insect protein consumption per capita of the different groups (long vs. recently established, female vs. male-headed, Betsileo vs. non-Betsileo, education levels, and religious beliefs) are not statistically different.

3.2. Agriculture and Income

Subsistence agriculture is predominant in Sandrandahy. According to our survey, farm households own, on average, 1.8 ha of land, of which 0.5 ha is irrigated for rice production. Paddy is the main crop in the region; on average, farmers produce 836 kg per year. One quarter of households sell rice, while the large majority (76%) only use it for home consumption. Rice, in terms of production value, accounts for 50%, followed by cassava (19%), sweet potato (9%), and corn (5%). In other words, more than 80% of production value comes from staple crops, whereas legumes, such as groundnuts, Bambara nuts (*Vigna subterranea*), and beans, together cover around 10%, and fruit and vegetables make up 4%. Most households own livestock: on average, they have two zebus and one pig, around 15 chicken, and three ducks. Moreover, 58% of households catch fish in their rice fields.

Households complement their agricultural activities with non-farm activities to make ends meet. Actually, 96% of households interviewed earn some sort of non-farm income, mainly from low remunerated casual work. Non-farm activities account for 53% of cash income, whereas livestock selling contributes 39% and crops around 7% of total cash income. On average, cash income is around EUR 120 per capita per year (EUR 0.32 per day), of which only around EUR 20 is spent on food.

3.3. Harvesting and Consumption of Edible Insects

In the rural commune of Sandrandahy, wild harvesting for home consumption dominates. Insects are rarely bought (2% of total consumption amount) or sold (0.2%), so that consumption nearly exclusively depends on the collection of insects. The preferred insects are adult beetles from the family of Scarabaeidae (*Voangory*), harvested and eaten by 87% of all households, followed by cicada (30%), pupae of Coleoptera (*Abado*, 29%), and locusts (21%) (see Table 1). The *Voangory* are adult scarab beetles that stay in the ground during the day, from which they crawl out and fly away after sunset. They are then found in larger clusters near rice fields or in the grass or shrubs where they can be easily caught. Larvae and pupae of silkworm are not much consumed (3%), as no tapia trees, the natural host plant of the wild silkworm (*Borocera cajani*), are found nearby, and domesticated silkworm (*Bombyx mori*) production is not important in the region. Locusts and crickets are collected on a rather irregular basis, for example, when there is an infestation. In quantitative terms, adult scarab beetles dominate consumption, as 66% of the total amount of insects consumed (as measured in kg) come from the *Voangory* type. The average and total amounts of other species are comparatively low, because only a few households consume them. However, those households which consume these species can consume relatively large amounts. Almost all (95%) of the households consume insects. The average consumption is 9.0 kg per household per year, which corresponds to 1.7 kg per capita, but varies greatly (Standard Deviation (SD): 10.6/2.0 kg).

Table 1. Annual consumption of insects by households ($n = 216$).

	Beetle (Larva) *Sakivy*	Beetle (Pupae) *Abado*	Beetle (Adult) *Voangory*	Beetle (Adult) *Voanosy*	Silkworm *Landibe* *Zana-dandy*	Locust *Valala*	Cricket *Akitra*	Cicada *Jorery*	Diving beetle *Tsikovoka*	Total
No. of consuming hh	8	62	187	27	6	45	2	65	4	205
% of all hh	4%	29%	87%	13%	3%	21%	1%	30%	2%	95%
Total amount in kg	34.2	114.2	1279.1	107.6	7.8	195.6	0.6	200.5	1.5	1941.0
% of total amount	2%	6%	66%	6%	0%	10%	0%	10%	0%	100%
Amount per hh in kg	0.2	0.5	5.9	0.5	0.0	0.9	0.0	0.9	0.0	9.0
Per consuming hh in kg	4.3	1.8	6.8	4.0	1.3	4.3	0.3	3.1	0.4	9.5

Source: own calculations based on survey data. hh = household(s).

In general, insects are consumed for two main reasons: people find them tasty (88% stated that they have a good taste) and it is a traditional food (65%, multiple answers possible). Moreover, insects are considered by the local population as a full meat substitute: they are predominantly consumed as a main dish with rice (92% of households harvest insects for this reason), and only sometimes as snacks (7%). Moreover, when asked whether they would choose a plate of insects over a plate of meat, if offered both for free, 52% responded that they would choose the latter, 37% would prefer insects over meat, and 11% like meat and insects equally. Around three quarters of households would like to consume more insects and are, therefore, interested in starting to rear them.

In total, 82% of households would like to collect more insects. The reasons why they do not harvest more refer much more to a lack of insects ("they are rare", 52%; "rare because of climate and habitat change", 27% of answers) than to labour constraints ("too busy to catch more", 12%). Moreover, 81% of interviewees reported a diminishing trend of harvested insect quantities in recent years.

3.4. Seasonality of Harvesting and Consumption

Many edible insect species are only seasonally available [18]. However, different species appear at different times of the year, and there are some insects which are available year-round. It is not clear how seasonal availability translates into seasonal patterns of consumption. As Figure 1 shows, however, there is a clear pattern in Sandrandahy: the harvesting of insects is highest between October and December. (As explained above, we did not collect data on monthly consumption of insects but, as the households consume nearly all insects immediately and hardly any insects are bought, sold, or preserved, seasonal consumption should follow seasonal harvesting patterns.) Some 82% of the total amount is collected in these three months, which are, at the same time, the start of the lean season. This is explained by the fact that the most commonly consumed species—adult scarab beetles, but also cicada—only appear in these months, whereas other species that are available year-round, such as locusts, crickets, and the larvae of Coleoptera, are not collected much.

For comparison, Figure 2 shows the monthly amounts (in kg) of beef, pork, chicken, and fish consumed by households over the year, as well as insects. The consumption of meat fluctuates only slightly throughout the year (Coefficient of Variation (CV) of 15%), whereas that of (fresh and dried) fish varies more (CV of 32%), both together reaching their highest levels between April and June, diminishing afterwards to reach their lowest levels in October and November, exactly the months when insect harvesting is at its peak. Between October and December, insects constitute 44% of the total amount of meat, fish, and insects consumed, meaning that edible insects (over-)compensate for the lower levels of meat and fish consumption. Furthermore, due to insect consumption, this is the period with the highest total consumption of meat, fish, and insects.

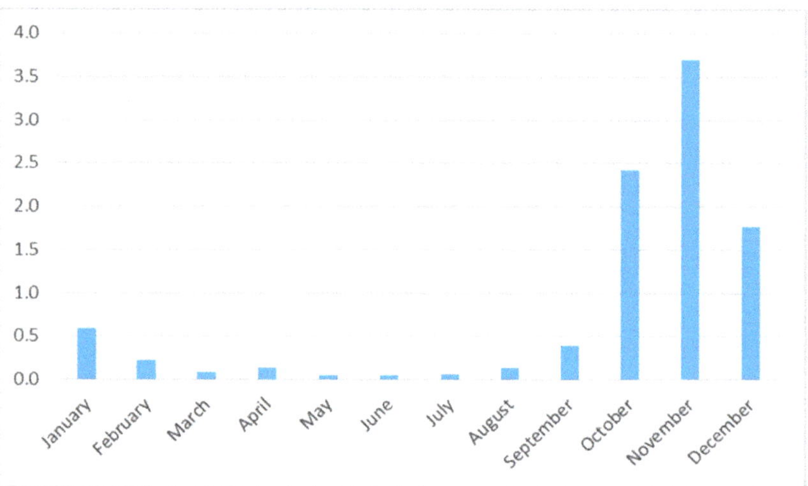

Figure 1. Monthly harvesting amounts of edible insects, in kg per household. Source: own calculations based on survey data.

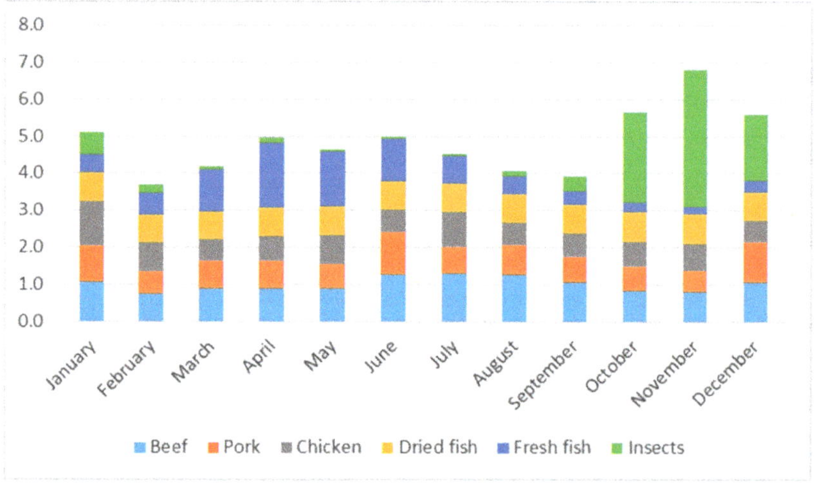

Figure 2. Monthly amounts of meat and fish consumed and edible insects harvested, in kg per household. Source: own calculations based on survey data.

3.5. Protein Intake

The quantities of meat, fish, and insects shown in Figure 2, and of protein-rich crops were converted into amounts of protein consumed by taking the average protein content of the different sources (see Tables A1 and A2 in the Appendix A). In general, animal protein sources are limited: an average household of 5.5 members ingests 69 kg of animal products (including insects) per year, containing 9.9 kg of proteins (SD 6.2 kg), of which 1.8 kg comes from insects (SD 2.0 kg). This equals an average per capita animal protein consumption of around 5 g/day (not differentiating between adults and children, male and female), lower than the average national supply of 9 g/day [19]. Besides animal products, protein-rich crops, mainly legumes (beans, bambara nuts, peas), are consumed. The protein coming from these plants is 8.3 kg per household, not much lower than that of animal products. Figure 3 shows the different sources of proteins in the region. Insects are responsible for nearly 10% of protein consumption of the rural population, in the same range as pork and beef.

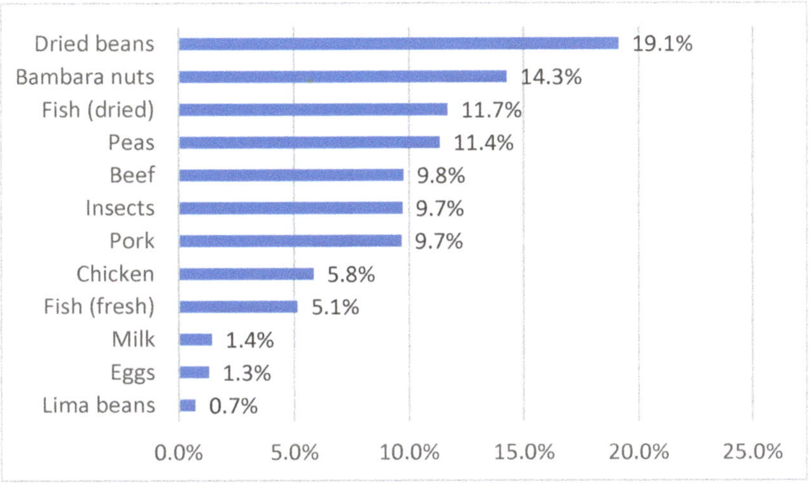

Figure 3. Average protein intake of different sources, in % of total. Source: own calculations based on survey data and data from Tables A1 and A2 in the Appendix A.

However, probably the most important protein source is rice (not considering that rice is not the best protein source, as it contains less essential amino acids), which is eaten on a daily basis. To illustrate, paddy production for personal consumption, which is on average 737 kg per household (we only collected production data, but not consumption data of rice, corn, and other cereals), contains around 44 kg of protein per year per family, compared to just 18 kg coming from animal products and legumes. Per capita protein supply from animal products, legumes, and rice together is around 31 g/day, lower than the national average of 43 g [19].

3.6. Cost of Different Protein Sources

It has been assumed that insects are a cheaper source of macro- and micronutrients than meat [3]. For our study region, Figure 4 demonstrates that meat, as well as milk and eggs, are indeed a relatively expensive source of protein: 100 g of their protein costs between MGA 6000 and 8000 (EUR 1.50 to 2.00). Fresh fish is the most expensive source (around MGA 9000 or EUR 2.25 per 100 g), whereas dried fish protein is much cheaper (nearly MGA 3400 or EUR 0.84 per 100 g). The calculated protein price of insects varies according to the species (as mentioned above, we only have sufficient price data for two types of insects). Whereas protein from adult scarab beetles is relatively cheap (around MGA 2500 or EUR 0.63 per 100 g of protein), that from silkworm chrysalis is at MGA 5000 (EUR 1.25), twice as expensive. In general, plant-based proteins are much cheaper (around MGA 1000 to 2000 or EUR 0.50 to 1.00/100 g) compared to animal protein sources.

As almost all insects are harvested in the wild by family members, there are hardly any cash expenses involved. The costs inferred are only the opportunity costs of harvesting them. The mean time to collect one kilogram of insects is, according to our data, 6.6 h; if this time could be used for other productive purposes instead, this would be an opportunity cost. This might differ from household to household but, assuming an opportunity cost that equals the agricultural minimum wage of MGA 675/h (https://www.minimum-wage. org/international/madagascar, accessed on 3 June 2021), collecting one kilogram of insects would have an opportunity cost of around MGA 4500, or approximately MGA 2250 (EUR 0.56) per 100 g of protein (considering a weighted average of 20 g protein/100 g of fresh insect weight), a bit lower than the price of protein from adult beetles calculated above.

However, as more than two thirds (68%) of the harvesting is done by children and youths (5–15 years old), the opportunity costs are probably lower (assuming that insect collection is not done at the expense of school attendance).

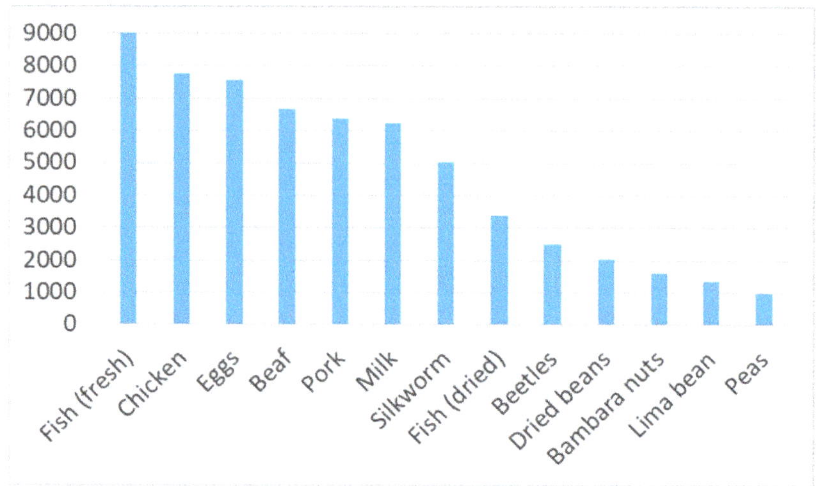

Figure 4. Protein cost of different foods, in MGA (at the time of the survey, EUR 1 = MGA 4000 (www.oanda.com), accessed on 5 May 2021) per 100 g of protein. Source: own calculations based on survey and data from Tables A1 and A2 in the Appendix A.

3.7. Micronutrient Intake

Many authors have been highlighting the contribution of entomophagy to the protein intake in protein-scarce food environments. However, edible insects' contribution to micronutrient intake might be even more important in some cases [35]. Especially in low-income countries, many people, especially children and women of reproductive age, suffer from iron and zinc deficiencies, which represent two of the most severe micronutrient deficiencies globally [36]. Adult scarab beetles (*Holotrichia* sp.), the far most consumed insect in our project region (*Voangory*), have an iron and zinc content of 9.1 mg and 8.8 mg/100 g edible portion, respectively (Köhler et al. 2019). As the daily required intake of children, women of reproductive age, and men is 10 mg, 18 mg, and 8 mg/day for iron, and 5 mg, 11 mg, and 8 mg/day for zinc [37], respectively, a family consisting of one couple with three children would need 34 mg of zinc and 56 mg of iron per day. The actual intake of 5.92 kg beetles per family would only allow for around 10 (iron) to 15 (zinc) days of the total required intake. As these beetles are consumed only seasonally (1–3 months), at least over that time, they can contribute substantially to the daily required intake of these important micronutrients

4. Hypotheses

4.1. Poorer Households Consume Significantly More Insect Proteins Compared to Richer Households

Given that insects are a cheaper source of protein compared to other animals, to the extent that they are almost such a thing as a "free lunch" [38], it can be supposed that poorer people will rely more on this protein source than richer ones. We used cash income as a proxy for poverty. As few people buy insects, we do not expect any direct positive effect of income on insect protein consumption, but one on the consumption of other animal protein sources (meat, fish, milk, eggs), which could then indirectly lead to lower insect consumption of richer households, in relative and/or absolute terms. Table 2 shows the correlation between the variables: no relationship could be established between insect protein consumption (in kg per capita per year) and per capita cash income, but a relatively strong correlation ($r = 0.546$ **) between income and other animal protein consumption. This means that richer households eat more meat and fish, but not less insects compared to poorer households. Insect and other animal protein consumption are not significantly correlated. It seems that households do not ingest less insect proteins only because other animal protein sources are more available, or, in other words, people with less income and less meat and fish consumption do not compensate this by higher (absolute) insect consumption. However, as poorer households have less access to meat and fish protein, the share of insects in their protein consumption tends to be higher compared to higher cash income households ($r = -0.177$ **). Finally, the more proteins from meat and fish are consumed, the lower the share of insect protein ($r = -0.288$ **).

Table 2. Correlation between insect and animal (meat, fish, milk, eggs) protein consumption, income, and insect relative to animal protein consumption, per capita.

Pearson Correlation Coefficient			
	Cash Income	**Insect Protein**	**Relative Insect Protein**
Insect protein	−0.061		
Relative insect protein	−0.177 **	0.634 **	
Animal protein	0.546 **	0.098	−0.288 **

Source: own calculations based on survey data; ** significant at 0.01 level.

4.2. Significant Differences in Protein Consumption Patterns between Villages, and between Rural and Urban Sites Exist

Different market access for animal products and different occurrence of insects for wild harvesting might lead to locally diverse protein consumption patterns in the 12 villages of our sample. Still, we found few differences between the places concerning per capita insect protein consumption: ANOVA post hoc tests (alpha = 0.05) only showed two different groups of villages, seven out of the twelve villages being in both groups. Concerning animal protein consumption, only two homogeneous groups could be identified: the (semi-)urban fokontany of Sandrandahy, with a higher consumption level, and the other villages. We did not find other patterns of low versus high consumption: some villages have below total average animal protein consumption and below total average insect protein consumption (4 out of 12). There are others with lower animal, but higher insect protein consumption (3), there are those with higher animal, but lower insect protein consumption (4), and, finally, one village with higher than average consumption of both protein sources. Consequently, there is no significant correlation between average animal protein and average insect protein consumption per village ($r = -0.287, p = 0.365$).

4.3. Insect Protein Consumption Is Positively Related to Land Size and Harvesting Time

As insects are mainly collected in the wild, bigger households might be able to collect and consume more insects than smaller households. Table 3 shows that insect protein consumption per capita is significantly correlated with the amount of time spent on insect harvesting, but not with household size. However, household size and number of hours spent on harvesting are positively correlated. This means that larger households have more labour available to catch more insects. However, this does not translate into higher per capita consumption in larger households because of the counteracting effect of the greater number of consumers in these households.

Since insects are collected in the fields, larger farms should have more space to search for wild insects and, hence, could harvest and consume more. Table 3 shows that farm size is indeed positively correlated with insect protein consumption, and correlates significantly with time spent on harvesting. It seems that, if more land is available, this opportunity is used by spending more time on harvesting insects, which allows for higher consumption.

Table 3. Correlation between insect protein consumption per capita, household size, farm size, and hours harvested per household.

Pearson Correlation Coefficient			
	Household Size	**Farm Size**	**Insect Protein**
Insect protein	−0.073	0.144 *	
Hours harvested	0.316 **	0.190 **	0.369 **

Source: own calculations based on survey data; * significant at 0.05, ** significant at 0.01 level.

Finally, we tested all three hypotheses together using a double-logarithmic regression model, with the dependent variable being protein per capita consumption. One influential case (Cook's d) was excluded. The model explains 54% of variances, which is a good fit (see Table 4).

Table 4. Multiple linear regression model summary; method: standard; one influential case excluded (n = 215), dependent variable: log of insect consumption per capita in kg.

R	R Square	Adjusted R Square	Std. Error of the Estimate	F	Sig.	Durbin-Watson
0.740	0.548	0.535	1.310	41.968	0.000	1.910

Table 5 presents the coefficients of the logarithmic variables. Because of heteroscedasticity problems (Breusch-Pagan test statistics: chi-square = 210.7, p = 0.000) and non-normal distribution of residuals, we estimated the parameters using robust standard errors (HC3 method). Only two variables have significant p-values: animal protein consumption (p = 0.003) and harvesting time (p = 0.000), whereas cash income and household size, but also farm size (which has a weak but significant correlation coefficient, see Table 3), have nonsignificant p-values. The variable harvesting time shows the highest partial Eta-squared of 18, much higher than that of animal protein (0.04), meaning that the former variable has highest effect size and explains most of the regression results. For example, if we run a regression only including harvesting time as an independent variable, R square is still as high as 506 (F = 218.015, p = 0.000), whereas, if we run the regression with all other variables, excluding harvesting time, R square is merely 0.073 (F = 3.281, p = 0.007). It seems that the most important determinant of insect consumption is the time spent on collecting them, confirming part of hypothesis three.

Table 5. Model coefficients.

Parameter	Coefficient B	Std. Error	T	Sig.	Part. Eta Squared
Constant	−3.527	1.330	−2.653	0.009	0.033
Log_HH size	−0.185	0.288	−0.642	0.522	0.002
Log_Animal protein	0.421	0.141	2.979	0.003	0.041
Log_Income capita	0.064	0.093	0.687	0.493	0.002
Log_Farm size	−0.020	0.106	−0.188	0.851	0.000
Rural/urban	0.298	0.576	0.517	0.606	0.001
Log_Harvest. time	0.480	0.070	6.822	0.000	0.183

Dependent variable: Log_Insectprotein.

5. Discussion

In our research area, entomophagy is widespread, with 95% of households consuming insects. This high prevalence of entomophagy is common in many African countries. For example, Anankware et al. [39] stated that more than 80% of interviewees in Ghana consume edible insects as a protein source. However, as explained above, not much quantitative evidence on insects as a protein source exists. In the commune of Sandrandahy, insects contribute nearly 10% to protein intake derived from animals and legumes, comparable with the proportions of beef or pork. Entomophagy as an important element of protein intake has only been documented in a few other studies. For example, Roulon-Doko, 1998, cited in [8], documented that insects consumed by the indigenous Gbaya people in the Democratic Republic of the Congo contribute to 15% of total protein intake.

However, the amounts of insects consumed and, therefore, of proteins coming from insects vary widely between households in our study. One explanation could be that poorer households consume more insects, substituting meat or fish proteins. However, in contrast to meat and fish, of which poorer households consume less, insect protein consumption is not correlated with household income level. This finding is consistent with the results of Manditsera et al. [6], who did not find a significant association between income and insect consumption in rural areas of Zimbabwe, but contrasts with the outcome of the study by Kisaka [40] for rural Kenyan households, where increased income leads to higher consumption of winged termites. For urban households, Kisaka [40] reports lower termite consumption, possibly because of better access to other protein sources, such as meat. In our study, meat and fish consumption is not correlated with edible insect consumption, meaning that households that can afford to buy more meat and fish continue to consume insects. This may be because total protein intake is still low, even for richer households, which continue using insects as a complement to their diets. In addition, insects are considered a free and easily accessible food, which everyone still uses, independently of the income situation. Finally, it might just be because most people, richer as well as poorer people, find insects tasty. Hence, it seems that insects are not viewed as an inferior good in relation to meat and fish, and are not substituted when income increases, as repeatedly stressed by various authors and summarised in the review by Meyer-Rochow et al. [17].

Even if insects are cheaper than meat and fish, insects are hardly bought by the rural population, but almost exclusively harvested in the wild. Hence, households' collection of wild insects seems to be key in explaining consumption patterns in rural areas. The time dedicated to harvesting is strongly correlated with amounts of insect proteins consumed and explains most of the regression results. This seems obvious in a situation where consumption depends almost entirely on wild harvesting. Especially in rural areas, access to insects in the wild greatly influences their availability for consumption [6]. However, the variable "harvesting time" was not considered in any other study we have found. On average, 34 h per month are dedicated to harvesting insects during season, but this varies widely (SD: 37 h). The question then might be: what influences harvesting time? Or, in other words, why do some households spend more time on harvesting than others?

We can only speculate on this: it might be (limited) labour availability, but also natural limitations. As seen above, most households would be willing to collect more insects if possible. The limitations mentioned refer more to lack of insects due to loss of habitat, especially in recent years, and less to labour constraints. Overexploitation of insects and changes in habitat (i.e., deforestation) have also been reported as a threat to entomophagy in other studies (summarised by van Huis [41]). Another explanation could be the interest and willingness of household members to collect insects. As children are the main collectors (68% of harvesting time), a fact which is confirmed by another study in Madagascar [12], it would be necessary to know what influences their harvesting activities. For example, school attendance might influence gathering time in communities where insects are mainly collected by children [42]. Moreover, besides the time spent, the amounts harvested per hour (i.e., the productivity) also differ greatly between households, a fact that might be explained by the diverse knowledge and skills (where to find and how to catch insects) of the gatherers, i.e., mainly of children.

Most insect collection is done seasonally, in up to three months of the year, and concentrated on adult scarab beetles (*Voangory*). On the one hand, this limits consumption opportunities. As different species normally appear at different times of the year, diversification could extend the availability of edible insects during the year. On the other hand, the main harvesting season for insects in Sandrandahy coincides with the period when other protein sources, such as fish, are rare, so that insects can compensate for this scarcity, contributing nearly half (44%) of meat, fish, and insect consumption in this period. A similar fact was found in the Lake Tumba region in Congo (at the time called Zaire), where Pagezy 1975, cited in [3], documented that, in some months of the year, when the occurrence of caterpillars is high, those of fish and game are low, and vice versa. High percentages (60%) of insects as an animal protein source during two months of the rainy season were also estimated by Paoletti et al. [43] for an Amerindian group of the Amazon. Muafor et al. [44] report that adult beetles consumed in Cameroon are only seasonally available, but constitute an alternative protein source to meat and fish in the months when they occur. This means that insect consumption can smoothen protein intake of rural households during the year. Other studies have also stressed the importance of some insect species, such as caterpillars, for seasonal food security [45].

Finally, some of the limitations of our study should be mentioned. Firstly, as a case study design with a strong regional focus was chosen, the results cannot be transferred to other regions of the country especially when considering that, in Madagascar, remarkable differences in insect consumption between different regions exist [23]. Second, an exact quantification of the consumption of edible insects is difficult due to the frequent recall bias [27], especially if the consumption quantities are recorded over a longer period of time. In addition, insects are mainly harvested by children, which makes it difficult for respondents to have an overview of how much is collected. As mentioned above (Section 2.3), we tried to check this by comparing the quantities harvested with the quantities consumed, both of which were asked independently in the questionnaire. Even if this comparison showed only minor deviations, we cannot rule out the possibility that both amounts are biased. Moreover, the frequencies of the other food items may have been affected by the recall bias. Third, we focused only on the protein content of insects (and some minerals found in *Voangory*), but did not consider, for example, fatty acids or vitamins. Moreover, we used data of protein content from the literature, which may differ for the insects found in the research area. Nutritional values depend not only on the species, but also on their diets and, hence, on their habitats, and on the processing methods [17]. Therefore, nutritional data on local insects and the influence of processing practices are required (and will be provided in the future by ProciNut project) in order to better assess their specific nutritional values. Fourth, the food safety aspect was not taken into account in our research. It is known that edible insects can be contaminated by micro-organisms, but also by pesticides. Even if contamination of microbes is reduced or eliminated by processing methods, such as boiling or deep-frying [17], further research is needed to ensure food safety of local insect consumption.

6. Conclusions

Apart from the general recognition of edible insects as an important protein source in developing countries [3], not much quantitative information on their contribution to food security is available. This study tried to partly fill this research gap by examining insect consumption in one rural commune in the highlands of Madagascar. Our results might be different in other rural areas. However, given the widespread tradition of entomophagy in this country [23], it can be concluded that entomophagy is a relevant and cheap local food source—a "free lunch". This is especially the case in the lean season, when other protein sources are scarce. However, the seasonality of insect occurrence also limits the possible amounts for consumption purposes. Moreover, due to climate and habitat changes, insects have become rare. The decreased occurrence and limited availability of insects, and the fact that they are liked as much as meat by the local population, represents an incentive for starting to rear them. However, the type of insect that is most consumed in the region, the adult scarab beetle, is difficult to rear because of its long life cycle and underground life when in the larval stage. Commonly reared species, such as crickets, are not consumed much yet by the local population, but might be accepted by consumers when offered to them in other forms, for example, as cricket flour used in buns [46]. Given local demand and appropriate incentives, such as extension services and knowledge transfer, as well as start-up aid for investments, the raising of insects could become an interesting business for smallholders, as seen in other countries, such as Thailand [47]. This could allow food and nutrition security and the tradition of entomophagy in poor rural areas of Madagascar to be strengthened.

Author Contributions: Conceptualization, J.D. and C.R.; methodology, J.D.; software, J.D. and C.R.; validation, J.D. and C.R.; formal analysis, J.D.; investigation, J.D. and C.R.; resources, J.D. and C.R.; data curation, J.D.; writing—original draft preparation, J.D.; writing—review and editing, J.D. and C.R.; visualization, J.D.; supervision, J.D.; project administration, J.D. and C.R.; funding acquisition, J.D. All authors have read and agreed to the published version of the manuscript.

Funding: This article has been written within the ProciNut project at the Center for Development Research, University of Bonn, Germany. ProciNut is financially supported by the German Federal Ministry of Food and Agriculture (BMEL) based on the decision of the Parliament of the Federal Republic of Germany through the Federal Office for Agriculture and Food (BLE), grant number FKZ: 2816PROC06.

Institutional Review Board Statement: The study was conducted according to the guidelines of the Declaration of Helsinki, and approved by the Ethics Committee of the Center for Development Research (ZEF) of the University of Bonn, Germany (date of approval: 2 June 2021).

Informed Consent Statement: Informed consent was obtained from all subjects involved in the study.

Data Availability Statement: Data can be made available upon request.

Acknowledgments: The authors gratefully acknowledge the contributions of our team members Anne Meysing, Sebastian Forneck, Hery Andriamazaoro, Narilala Randrianarison, Andrianantenaina Razafindrakotomamonjy, and Harifetra Andrianintsoa, who organised the field study very well under difficult conditions. Without their commitment, this study would not have been possible. We are also very grateful to the enumerators Saholiniaina Rasoanaivo, Irené Ratojonirainy, Yves Ralison, Tsiory Rakotobe, José Rasolofonirina, and Zo Randrianirisoa, who worked very hard and carried out thoroughly as many interviews as possible in the limited time available. The authors further thank the households that participated in the survey for their patience and time. We are also grateful for the valuable comments from Simone Kathrin Kriesemer, Lukas Kornher, and Béatrice Knerr and from the anonymous reviewers.

Conflicts of Interest: The authors declare that they have no conflict of interest.

Appendix A

Table A1. Protein content of different insects found in the study area.

Malagasy Name	English Name	Family	Order	DRY MATTER BASIS	FRESH MATTER BASIS		DATA USED	References/Observations
				Protein (g/100 g EP)	Protein (g/100 g EP)	Moisture Content (%)	Protein (g/100 g EP)	
Voanosy (Adult)	Weevils	CURCULIONIDAE	COLEOPTERA				21.2	Data used from *Voangory*
Voangory (Adult)	Scarab beetles	SCARABAEIDAE	COLEOPTERA		28.9 13.4	74.1	21.2	Mean of Köhler et al. [16] and Yhoung-Aree & Viwatpanich [48]
Sakivy (Larvae)	Scarab beetles Weevils	SCARABAEIDAE CURCULIONIDAE	COLEOPTERA	49.2			14.8	Gosh et al. [15]: mean of two species; moisture content assumed: 70%
Abado (Pupae)	Scarab beetles Weevils	SCARABAEIDAE CURCULIONIDAE	COLEOPTERA				14.8	Data used from *Sakivy*
Tsikovoka (Adult)	Water beetles	DYTISCIDAE	COLEOPTERA		25.1 21.0	49.1 61.2	23.0	Mean of Shantibala et al. [49] and Yhoung-Aree et al. [28]
Jorery (Adult)	Cicadas	CICADIDAE	HEMIPTERA	47.2			14.2	Raksakantong et al. [50]; moisture content assumed: 70%
Valala (Adult)	Grasshoppers Locusts	ACRIDIDAE OEDIPODINAE	ORTHOPTERA		14.3 20.1	76.7 69.0	17.2	Mean of Yhoung-Aree & Viwatpanich [48] and Oonincx & van der Poel [51]
Akitra (Adult)	Crickets	GRYLLIDAE	ORTHOPTERA		14.7	70.8	14.7	Yhoung-Aree & Viwatpanich [48]; mean for different crickets
Zana-dandy (Larvae)	Domesticated silkworms	BOMBYCIDAE	LEPIDOPTERA		18.9 26.6 12.2	69.9 75.3	19.2	Mean of Frye & Calvert [52] and Köhler et al. [16] and Yhoung-Aree & Viwatpanich [48]
Landibe (Larvae)	Wild silkworms	LASIOCAMPIDAE	LEPIDOPTERA		14.7	70.8	19.2	Data used from *Zana-dandy*

Sources: Taxonomy by Andrianantenaina Razafindrakotomamonjy (ProciNut project); own calculations based on given references; EP: edible portion.

Table A2. Protein content of different animal-based and plant-based foods. Composition (pour 100 g de partie comestible/100 g edible parts).

Aliments/Foods	Protein (g)
Animal proteins	
Viande de boeuf/Beef	14.60
Viande de porc/Pork meat	18.70
Viande de poulet/Poultry meat	12.30
Poissons frais/Fresh fish	10.30
Poissons seches/Dried fish	22.90
Lait frais/Fresh milk	3.10
Oeuf/Eggs	10.30
Plant proteins	
Tsaramaso mainty (sp vulgaris)/Dried beans	14.13
Pois du cap/Peas	24.05
Voanjobory/Bambara nuts	17.21
Kabaro fotsy (sp lunatus)/Lima beans	34.88
Kabaro sadamena (sp lunatus)	25.90
Kabaro mena (sp lunatus)	17.45
Kabaro mainty (sp lunatus)	18.93
Kabaro maramainty (sp lunatus)	20.81
Kabaro/Lima beans (average)	23.59
Arachide nature/Peanuts	29.91
Rice	
Paddy	6.00

Source: Organisation Nationale de Nutrition (ONN, Madagascar), without date.

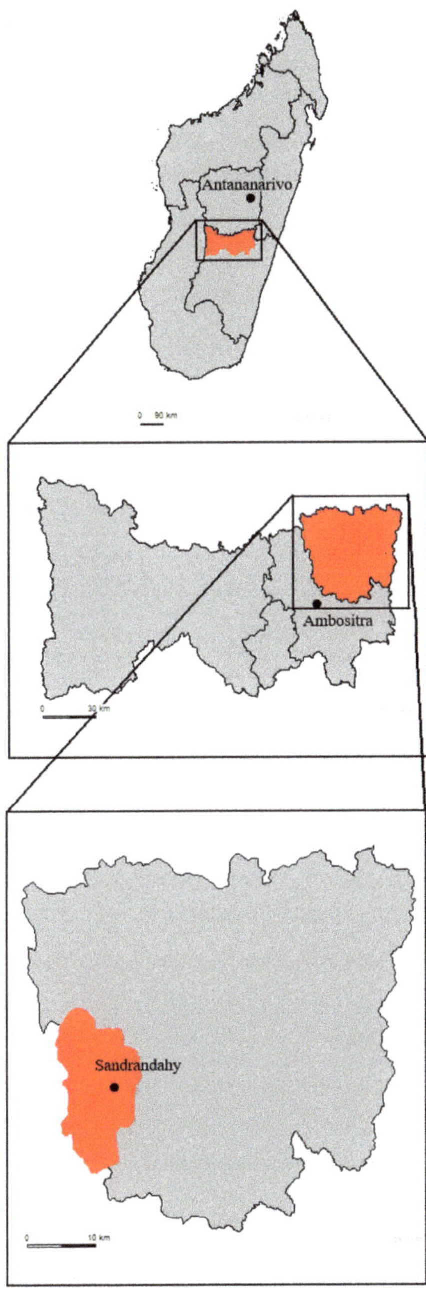

Figure A1. Location of Sandrandahy Commune within the district of Fandriana, the district within the region Amoron'i Mania, and the region within Madagascar. Source: ProciNut project by Sebastian Forneck; based on maps from Global Administrative Areas (2018). GADM database of Global Administrative Areas, version 3.6. [online] URL: www.gadm.org (accessed on 20 May 2020).

References

1. Meyer-Rochow, V.B. Can insects help to ease the problem of world food shortage? *Search* **1975**, *6*, 261–262.
2. Chavunduka, D.M. Insects as a source of protein to the African. *Rhod. Sci. News* **1976**, *9*, 217–220.
3. FAO (Food and Agriculture Organization of the United Nations). *Edible Insects: Future Prospects for Food and Feed Security*; van Huis, A., Van Itterbeeck, J., Klunder, H., Mertens, E., Halloran, A., Muir, G., Vantomme, P., Eds.; FAO: Rome, Italy, 2013.
4. Gahukar, R.T. Entomophagy and human food security. *Int. J. Trop. Insect Sci.* **2011**, *31*, 129–144. [CrossRef]
5. Obopile, M.; Seeletso, T.G. Eat or not eat: An analysis of the status of entomophagy in Botswana. *Food Secur.* **2013**, *5*, 817–824. [CrossRef]
6. Manditsera, F.A.; Lakemond, C.M.M.; Fogliano, V.; Zvidzai, C.J.; Luning, P.A. Consumption patterns of edible insects in rural and urban areas of Zimbabwe: Taste, nutritional value and availability are key elements for keeping the insect eating habit. *Food Secur.* **2018**, *10*, 561–570. [CrossRef]
7. Van Huis, A. Potential of Insects as Food and Feed in Assuring Food Security. *Annu. Rev. Entomol.* **2013**, *58*, 563–583. [CrossRef] [PubMed]
8. Kelemu, S.; Niassy, S.; Torto, B.; Fiaboe, K.; Affognon, H.; Tonnang, H.; Maniania, N.K.; Ekesi, S. African edible insects for food and feed: Inventory, diversity, commonalities and contribution to food security. *J. Insects Food Feed* **2015**, *1*, 103–119. [CrossRef]
9. Cox, S.; Payne, C.; Badolo, A.; Attenborough, R.; Milbank, C. The nutritional role of insects as food: A case study of 'chitoumou' (*Cirina butyrospermi*), an edible caterpillar in rural Burkina Faso. *J. Insects Food Feed* **2018**, *6*, 69–80. [CrossRef]
10. Ghosh, S.; Jung, C.; Meyer-Rochow, V.B.; Dekebo, A. Perception of entomophagy by residents of Korea and Ethiopia revealed through structured questionnaire. *J. Insects Food Feed* **2020**, *6*, 59–64. [CrossRef]
11. Rumpold, B.A.; Schlüter, O.K. Nutritional composition and safety aspects of edible insects. *Mol. Nutr. Food Res.* **2013**, *57*, 802–823. [CrossRef]
12. Van Itterbeeck, J.; Rakotomalala Andrianavalona, I.N.; Rajemison, F.I.; Rakotondrasoa, J.F.; Ralantoarinaivo, V.R.; Hugel, S.; Fisher, B.L. Diversity and Use of Edible Grasshoppers, Locusts, Crickets, and Katydids (*Orthoptera*) in Madagascar. *Foods* **2019**, *8*, 666. [CrossRef]
13. Poelaert, C.; Francis, F.; Alabi, T.; Megido, R.C.; Crahay, B.; Bindelle, J.; Beckers, Y. Protein value of two insects, subjected to various heat treatments, using growing rats and the protein digestibility-corrected amino acid score. *J. Insects Food Feed* **2018**, *4*, 77–87. [CrossRef]
14. Kouřimská, L.; Adámková, A. Nutritional and sensory quality of edible insects. *NFS J.* **2016**, *4*, 22–26. [CrossRef]
15. Ghosh, S.; Lee, S.-M.; Jung, C.; Meyer-Rochow, V.B. Nutritional composition of five commercial edible insects in South Korea. *J. Asia-Pac. Entomol.* **2017**, *20*, 686–694. [CrossRef]
16. Köhler, R.; Kariuki, L.; Lambert, C.; Biesalski, H.K. Protein, amino acid and mineral composition of some edible insects from Thailand. *J. Asia-Pac. Entomol.* **2019**, *22*, 372–378. [CrossRef]
17. Meyer-Rochow, V.B.; Gahukar, R.T.; Ghosh, S.; Jung, C. Chemical Composition, Nutrient Quality and Acceptability of Edible Insects Are Affected by Species, Developmental Stage, Gender, Diet, and Processing Method. *Foods* **2021**, *10*, 1036. [CrossRef]
18. Van Huis, A. Insects as Food in Sub-Saharan Africa. *Insect Sci. Appl.* **2003**, *23*, 163–185. [CrossRef]
19. FAOSTAT. Statistical Division of the UN Food and Agriculture Organization. 2021. Available online: https://faostat.fao.org/ (accessed on 5 May 2021).
20. FAO/WHO/UNU. *Technical Report Series 935: Protein and Amino Acid Requirements in Human Nutrition*; WHO Press: Geneva, Switzerland, 2007.
21. Badjeck, B.; Ibrahima, N.C.; Slaviero, F. *Mission FAO/PAM d'Evaluation de la Sécurité Alimentaire à Madagascar. Rapport aux Food and Agriculture Organization et Programme Alimentaire Mondial*; FAO: Rome, Italy, 2013.
22. INSTAT (Institut National de la Statistique); Ministère de l'Économie et de l'Industrie; ICF Macro. *Enquête Démographique et de Santé Madagascar 2008–2009*; INSTAT et ICF Macro: Antananarivo, Madagascar, 2010.
23. Randrianandrasana, M.; Berenbaum, M.R. Edible Non-Crustacean Arthropods in Rural Communities of Madagascar. *J. Ethnobiol.* **2015**, *35*, 354–383. [CrossRef]
24. Agbidye, F.S.; Ofuya, T.I.; Akindele, S.O. Some edible insect species consumed by the people of Benue State, Nigeria. *Pak. J. Nutr.* **2009**, *8*, 946–950. [CrossRef]
25. Bomolo, O.; Niassy, S.; Chocha, A.; Longanza, B.; Bugeme, D.M.; Ekesi, S.; Tanga, C.M. Ecological diversity of edible insects and their potential contribution to household food security in Haut-Katanga Province, Democratic Republic of Congo. *Afr. J. Ecol.* **2017**, *55*, 640–653. [CrossRef]
26. Roos, N. Insects and Human Nutrition. In *Edible Insects in Sustainable Food Systems*; Halloran, A., Flore, R., Vantomme, P., Roos, N., Eds.; Springer: Cham, Switzerland, 2018; pp. 83–92.
27. Barennes, H.; Phimmasane, M.; Rajaonarivo, C. Insect Consumption to Address Undernutrition. A National Survey on the Prevalence of Insect Consumption among Adults and Vendors in Laos. *PLoS ONE* **2015**, *10*, e0136458. [CrossRef] [PubMed]
28. Yhoung-Aree, J.; Puwastein, P.; Attig, G.A. Edible insects in Thailand: An unconventional protein source? *Ecol. Food Nutr.* **1997**, *36*, 133–149. [CrossRef]
29. Acuña, A.M.; Caso, L.; Aliphat, M.M.; Vergara, C.H. Edible insects as part of the traditional food system of the Popoloca town of Los Reyes Metzontla, Mexico. *J. Ethnobiol.* **2011**, *31*, 150–169. [CrossRef]

30. N'Gasse, G.; Moussa, J.-B.; Mapunzu, P.M.; Balinga, M.P. *Contribution des Chenilles/Larves Comestibles à la Reduction de L'Insécurité Alimentaire en République Centrafricaine (RCA)*; Food and Agriculture Organization of the United Nations: Rome, Italy, 2004. Available online: http://www.fao.org/3/j3463f/j3463f05.htm (accessed on 5 May 2021).

31. ONE (Office National de l'Environnement). *Tableau de Bord Environnemental—Région Amorin'i Mania*; ONE: Antananarivo, Madagascar, 2007.

32. MEP (Ministère de l'Economie et de la Planification). *Région Amoron'I Mania—Monographie Régionale*; MEP: Antananarivo, Madagascar, 2015.

33. FAO (Food and Agriculture Organization of the United Nations); Programme Alimentaire Mondiale (PAM). *Rapport Spécial, Mission FAO/PAM d'Evaluation des Récoltes et Sécurité Alimentaire à Madagascar*; FAO: Rome, Italy, 2017.

34. Magnani, R. *Sampling Guide*; FHI 360/FANTA: Washington, DC, USA, 1999.

35. Michaelsen, K.F.; Hoppe, C.; Roos, N.; Kaestel, P.; Stougaard, M.; Lauritzen, L.; Mølgaard, C.; Girma, T.; Friis, H. Choice of foods and ingredients for moderately malnourished children 6 months to 5 years of age. *Food Nutr. Bull.* **2009**, *30* (Suppl. S3), 343–404. [CrossRef]

36. Müller, O.; Krawinkel, M. Malnutrition and health in developing countries. *CMAJ* **2005**, *173*, 279–286. [CrossRef] [PubMed]

37. Oria, M.; Harrison, M.; Stallings, V.A. (Eds.) *Dietary Reference Intakes for Sodium and Potassium. National Academies of Sciences, Engineering, and Medicine; Health and Medicine Division; Food and Nutrition Board; Committee to Review the Dietary Reference Intakes for Sodium and Potassium*; National Academies Press: Washington, DC, USA, 2019.

38. Dürr, J.; Andriamazaoro, H.; Nischalke, S.; Preteseille, N.; Rabenjanahary, A.; Randrianarison, N.; Ratompoarison, C.; Razafindrakotomamonjy, A.; Straub, P.; Wagler, I. "It is edible, so we eat it": Insect supply and consumption in the central highlands of Madagascar. *Int. J. Trop. Insect Sci.* **2020**, *40*, 167–179. [CrossRef]

39. Anankware, P.J.; Osekre, E.A.; Obeng-Ofori, D.; Khamala, C.M. Factors that affect entomophagical practices in Ghana. *J. Insects Food Feed* **2017**, *3*, 33–41. [CrossRef]

40. Kisaka, C. Evaluation of Consumers Acceptance and Pricing of Edible Winged Termites (*Macrotermes subhylanus*) in Kimilili Sub-County, Kenya. Master's Thesis, Egerton University, Njoro, Kenya, 2018; p. 96.

41. Van Huis, A. Insects as Human Food. In *Ethnozoology. Animals in Our Lives*; Nóbrega Alves, R.R., Albuquerque, U.P., Eds.; Chapter 11; Academic Press: Cambridge, MA, USA, 2018; pp. 195–213.

42. Riggi, L.G.; Veronesi, M.; Goergen, G.; MacFarlane, C.; Verspoor, R.L. Observations of entomophagy across Benin—Practices and potentials. *Food Secur.* **2016**, *8*, 139–149. [CrossRef]

43. Paoletti, M.G.; Dufour, D.L.; Cerda, H.; Torres, F.; Pizzoferrato, L.; Pimentel, D. The importance of leaf- and litter-feeding invertebrates as sources of animal protein for the Amazonian Amerindians. *Proc. R. Soc. Lond.* **2000**, *267*, 2247–2252. [CrossRef] [PubMed]

44. Muafor, F.J.; Levang, P.; Le Gall, P. A crispy delicacy: Augosoma Beetle as alternative source of protein in East Cameroon. *Int. J. Biodivers.* **2014**, 214071. [CrossRef]

45. Payne, C.L.R.; Badolo, A.; Cox, S.; Sagnon, B.; Dobermann, D.; Milbank, C.; Scarborough, P.; Sanon, A.; Bationo, F.; Balmford, A. The contribution of 'chitoumou', the edible caterpillar Cirina butyrospermi, to the food security of smallholder farmers in southwestern Burkina Faso. *Food Secur.* **2020**, *12*, 221–234. [CrossRef]

46. Alemu, M.H.; Olsen, S.B.; Vedel, S.E.; Kinyuru, J.N.; Pambo, K.O. Can insects increase food security in developing countries? An analysis of Kenyan consumer preferences and demand for cricket flour buns. *Food Secur.* **2017**, *9*, 471–484. [CrossRef]

47. Hanboonsong, Y.; Jamjanya, T.; Durst, P.B. *Six-Legged Livestock: Edible Insect Farming, Collecting and Marketing in Thailand*; FAO: Bangkok, Thailand, 2013.

48. Yhoung-Aree, J.; Viwatpanich, K. Edible insects in the Laos PDR, Myanmar, Thailand, and Vietnam. In *Ecological Implications of Minilivestock*; Paoletti, M.G., Ed.; Science Pub.: Enfield, NH, USA, 2005; pp. 415–440.

49. Shantibala, T.; Lokeshwari, R.K.; Debaraj, H. Nutritional and antinutritional composition of the five species of aquatic edible insects consumed in Manipur, India. *J. Insect Sci.* **2014**, *14*, 14. [CrossRef] [PubMed]

50. Raksakantong, P.; Meeso, N.; Kubola, J.; Siriamornpun, S. Fatty acids and proximate composition of eight Thai edible terricolous insects. *Food Res. Int.* **2010**, *43*, 350–355. [CrossRef]

51. Oonincx, D.G.A.B.; van der Poel, A.F.B. Effects of Diet on the Chemical Composition of Migratory Locusts (*Locusta migratoria*). *Zoo Biol.* **2011**, *30*, 9–16. [CrossRef]

52. Frye, F.; Calvert, C. Preliminary Information on the Nutritional Content of Mulberry Silk Moth (*Bombyx mori*) Larvae. *J. Zoo Wildl. Med.* **1989**, *20*, 73–75.

Review

Edible Aquatic Insects: Diversities, Nutrition, and Safety

Min Zhao [1], Cheng-Ye Wang [1], Long Sun [1], Zhao He [1], Pan-Li Yang [1], Huai-Jian Liao [2,*] and Ying Feng [1,*]

[1] Key Laboratory of Breeding and Utilization of Resource Insects of National Forestry and Grassland Administration, Research Institute of Resource Insects, Chinese Academy of Forestry, Kunming 650224, China; mzhao@caf.ac.cn (M.Z.); cywang11@126.com (C.-Y.W.); sunlong1818@126.com (L.S.); hezhao@caf.ac.cn (Z.H.); yangpl1845@163.com (P.-L.Y.)

[2] Institute of Leisure Agriculture, Jiangsu Academy of Agricultural Sciences, Nanjing 210014, China

*   Correspondence: lhj@jaas.ac.cn (H.-J.L.); fengying@caf.ac.cn (Y.F.); Tel.: +86-871-6386-0020 (Y.F.)

Abstract: Edible insects have great potential to be human food; among them, aquatic insects have unique characteristics and deserve special attention. Before consuming these insects, the nutrition and food safety should always be considered. In this review, we summarized the species diversity, nutrition composition, and food safety of edible aquatic insects, and also compared their distinguished characteristics with those of terrestrial insects. Generally, in contrast with the role of plant feeders that most terrestrial edible insect species play, most aquatic edible insects are carnivorous animals. Besides the differences in physiology and metabolism, there are differences in fat, fatty acid, limiting/flavor amino acid, and mineral element contents between terrestrial and aquatic insects. Furthermore, heavy metal, pesticide residue, and uric acid composition, concerning food safety, are also discussed. Combined with the nutritional characteristics of aquatic insects, it is not recommended to eat the wild resources on a large scale. For the aquatic insects with large consumption, it is better to realize the standardized cultivation before they can be safely eaten.

Keywords: entomophagy; terrestrial insects; health benefits; contaminant; selenium; insect farming

Citation: Zhao, M.; Wang, C.-Y.; Sun, L.; He, Z.; Yang, P.-L.; Liao, H.-J.; Feng, Y. Edible Aquatic Insects: Diversities, Nutrition, and Safety. *Foods* **2021**, *10*, 3033. https://doi.org/10.3390/foods10123033

Academic Editors: Victor Benno Meyer-Rochow and Chuleui Jung

Received: 29 October 2021
Accepted: 2 December 2021
Published: 6 December 2021

Publisher's Note: MDPI stays neutral with regard to jurisdictional claims in published maps and institutional affiliations.

1. Introduction

Human beings have a very long and rich history of entomophagy, especially in Africa, Asia, and Latin America [1–4]. There are plentiful and colorful customs and cultures of entomophagy in these areas. To some ethnic groups, such as Mazahua and Nahuas in Mexico [5], Gelao in China [2], and Kiriwinians in Papua New Guinea [6], entomophagy not only means an indigenous practice of the consumption of insects, but is also imbibed into the ethnic culture and traditional ecological knowledge [7,8]. Entomophagy has decreased significantly with the development of modern agriculture; however, it still exists and plays important roles in the lives of people in underdeveloped areas. In recent years, considering their outstanding source of nutrition [9], low levels of greenhouse gas emissions [10], limited agricultural land being required [11], and potential socio-economic benefits, edible insects have been considered as valuable and sustainable alternative nutrition sources for food security [12]. The topics of "edible insects as food candidates" or "insects as an alternative protein source" have received a huge amount of attention [3,13]. Up until now, great progress in those topics has been achieved due to the encouragement of the greater use of insects in our diets by the Food and Agriculture Organization of the United Nations [14], as well as technological advances in research on the nutrition, safety and farming of insects.

Aquatic insects are composed of around 76,000 species, found in a wide range of aquatic (and semiaquatic) habitats, from springs, ponds and lakes to large rivers [15,16]. As edible insects, edible aquatic insects have a long history of utilization [5] too. These insects could have great potential to service humans with nutrition and health benefits; for example, the captured biomass of aquatic insects in Zaire (Central Africa) has already reached a very high value, with 16 tons/year in 1989 [17]. However, as compared to edible terrestrial insects, such as the black soldier fly, mealworm, cricket, and grasshopper, fewer aquatic insects appear in the market and daily life, and the same situation exists in terms of research. Insects inhabit a large variety of environments with different feeding habits; the majority of terrestrial insects are phytophagous, while the majority of aquatic insects are carnivores [18]. What is the difference or potentiality between the use of edible aquatic insects and terrestrial insects? In this review, the characteristics of aquatic insects were summarized to provide information to better develop and utilize this kind of resource. Here, 'aquatic insects' refers to aquatic and semi-aquatic insects in or from freshwater.

2. Aquatic Insects and Its Resource as Food and Feed

Aquatic insects have very rich species diversity, though aquatic insects represent only 10% of the insect species and only include 12 orders [19,20], and they share some of the same orders with terrestrial insects taxonomically. The relevant biology, natural habitats, and comparisons of aquatic insect orders have been well summarized by D. Dudley Williams and Siân S. Williams [19]. Six of the 12 orders of aquatic insects are likely to contain candidate species for food and feed [19,21,22]. They are Coleoptera (beetles), Diptera (true flies), Ephemeroptera (mayflies), Hemiptera (true bugs), Odonata (dragonflies/damselflies), and Trichoptera (caddisflies). Judging from the research and utilization reports from Southwest China and Japan, the order Megaloptera has the potential to be a candidate species, because it has been used as food and folk medicine for a long time, with remarkable economic value [23,24]. The insects are widely distributed all around the world, and its breeding technology of some species has gradually matured [25,26].

To update the list of edible aquatic insects, the methods of Macadam and Stockan [22] have been followed. Two main sources, Jongema (2017) [27] and Mitsuhashi (2016) [28], and other sources (Supplementary Table S1), were screened for the list. The list contains 329 species belonging to 153 genera, and 51 families coming from 46 countries (Table 1). Given that over 2000 insect species are eaten, edible aquatic insects account for about 15% of the total number. The species belong to eight orders, namely, Coleoptera, Odonata, Hemiptera, Diptera, Trichoptera, Megaloptera, Ephemeroptera, and Plecoptera, ordered from high to low. Among them, Coleoptera, Odonata, and Hemiptera contribute over 3/4 of the number of species, and they are all predatory. The naiad/larva is the main edible stage of aquatic insects. Most species are consumed in Mexico, Japan, China, Thailand, India, and Venezuela.

Table 1. The species information of edible aquatic insects.

Order	Number	Feeding Habits	Edible Stage	Mainly Edible Family—Genus (Edible Species Number)	Edible Country (Edible Species Number)
Ephemeroptera	11	phytophagous	naiad, adult	Ephemeridae—Ephemera (3), Baetidae—Cloeon (2)	India (3), Mexico (2), Malawi (2), Japan (2), China, Papua New Guinea, Kenya, Malawi, Tanzania, Uganda
Odonata	72	predatism	naiad, adult	Libellulidae—Sympetrum (8), Libellulidae—Orthetrum (7), Aeschnidae—Rhionaeschna (4), Libellulidae—Neurothemis (3), Libellulidae—Trithemis (3), Aeschnidae—Anax (3),	India (14), China (13), Thailand (13), Indonesia (12), Venezuela (10), Japan (10), Ecuador (4), Mexico (3), Laos (3), Madagascar (2), Myanmar (2), Vietnam (2), USA, Italy, South Korea, D.R.Congo
Plecoptera	11	omnivorous	naiad	Pteronarcyidae—Pteronarcys (4)	Japan (6), USA (4), India (2)

Table 1. *Cont.*

Order	Number	Feeding Habits	Edible Stage	Mainly Edible Family—Genus (Edible Species Number)	Edible Country (Edible Species Number)
Hemiptera	68	predatism	egg, naiad, adult	Nepidae—Laccotrephes (7), Belostomatidae—Lethocerus (6), Nepidae—Ranatra (5) Belostomatidae—Abedus (3), Belostomatidae—Diplonychus (3), Belostomatidae—Sphaerodema (3), Corixidae—Corisella (3), Corixidae—Graptocorixa (3)	Mexico (17), Thailand (17), China (8), Japan (8), India (7), Venezuela (4), Madagascar (4), Laos (4), USA (3), Myanmar (3), Vietnam (2), D.R. Congo, Cameroon, Zambezian region, Malaysia, Mali, Singapore, Sri Lanka, Republic of Congo, Togo, Malawi, South Korea
Megaloptera	15	predatism	larva	Corydalidae—Acanthacorydalis (6), Corydalidae—Corydalus (4)	China (9), Japan (2), Peru (2), Mexico, Colombia, Venezuela
Coleoptera	108	predatism	larva, pupa, adult	Dytiscidae—Cybister (32), Hydrophilidae—Hydrophilus (14), Dytiscidae—Dytiscus (6), Dytiscidae—Rhantus (6), Dytiscidae—Laccophilus (5), Hydrophilidae—Tropisternus (5), Gyrinidae—Gyrinus (4)	Mexico (35), China (26), Japan (22), Thailand (21), India (11), Madagascar (10), Laos (9), Vietnam (7), Myanmar (6), Senegal (4), North Korea (3), Sri Lanka(3), Benin (3), Malaysia (3), Togo (3), Chile (3), Sabah (2), Cambodia (2), South Korea (2), USA (2), Indonesia (2), Cameroon (2), Turkey (2), Peru (2), Australia (2), Sierra Leone, D.R.Congo, Gabon, Panama
Diptera	27	saprophagous	larva, pupa	Chaoboridae—Chaoborus (4), Tipulidae—Tipula (4), Ephydridae—Ephydra (3), Simuliidae—Simulium (3)	USA (7), Uganda (6), Mexico (5), Venezuela (2), Japan (2), Kenya (2), Tanzania (2), Brazil, China, Colombia, Malawi, N. Am., Nearctic, Sri Lanka, Thailand
Trichoptera	17	phytophagous	larva	Stenopsychidae—Stenopsyche (3)	Japan (12), Venezuela (4), Mexico (2), Colombia

Noticeably, the number of species of edible aquatic insects might be underestimated [22] because the mainly edible stage of aquatic insects—naiads/larva—is morphologically undistinguishable most of the time. More and more species of edible aquatic insects will be identified with the help of molecular biology techniques [29].

3. Nutritional and Health Benefits of Edible Aquatic Insects

3.1. Protein Content and Amino Acid Composition of Aquatic Insects

Aquatic insects have an average protein content of 59.55% (Table 2), and this value is higher than that of conventional animal meats [30–32]. Furthermore, investigations identified that proteins from aquatic insects not only contain 45.93–62.01% essential amino acids (Table 2), but also have a good balance of different kinds of amino acids [33,34]. Meanwhile, the ratio of essential amino acids in aquatic insect proteins is close to human proteins, indicating a high nutritional value of aquatic insects [35].

Different edible insects have very different amino acid patterns [14,36,37]; we summarized the first limiting amino acid and highest amino acids in aquatic insects and compared these to terrestrial insects (Table 2). In aquatic insects, the average essential amino acid content is 51.60%, and the most abundant essential amino acid is Leu (7.8% on average). The limiting amino acids were Met + Cys (2.3% on average) and Trp (1.2% on average). The content of Glu (11.30% on average) is the highest among all the amino acids in aquatic insects. Jiang et al. compared the composition of the essential amino acids of dragonfly larvae in different regions and found that the difference is little, even between different species [38].

More attention has been paid to the amino acid content of terrestrial insects (Table 2) [32,37]. In terrestrial insects, the average essential amino acid content is 49.49%, and the most abundant essential amino acid is Leu (7.35% on average). The limiting amino acids were Thr (4.20% on average) and Trp (3.61% on average). The content of Glu (15.59% on average) is the highest among all the amino acids in terrestrial insects.

Although some studies have shown that the contents of some amino acids are different between aquatic and terrestrial insects, such as tryptophan [39], the difference is not significant judging from the average data, and this indicates that aquatic insects and terrestrial insects have similar amino acid compositions (Table 2). For example, Lys is present in both terrestrial edible insects and aquatic edible insects, with a rich content [38]. However, this essential amino acid is usually lacking in cereal protein, so both aquatic insects and terrestrial insects complement cereal protein in nutrition.

3.2. Characteristics of Fatty Acids in Aquatic Insects

As the second most abundant nutritional ingredient (only behind the protein content), fatty acids always play a crucial role in the growth and development of insects [40]. It is strongly believed that aquatic insects are also a rich source of saturated fatty acids (SFAs), monounsaturated fatty acids (MUFAs), and polyunsaturated fatty acids (PUFAs), especially PUFAs, such as linoleic acid (18:2), linolenic acid (18:3), arachidonic acid (20:4, AA), and eicosapentaenoic acid (20:5, EPA) [41,42]. In general, linoleic acid, linolenic acid, AA, and EPA belong to the omega-6 or omega-3 fatty acid family, which are beneficial to the health of human beings and must be obtained from the diet [43]. Table 3 presents some fatty acid compositions of selected aquatic and terrestrial insects. Noticeably, much research is carried out on the fatty acid composition of edible terrestrial insects. In contrast, there are much fewer data of the fatty acid composition of edible aquatic insects. For edible aquatic insects, SFAs are dominated by palmitic acid (16:0) and, to a lesser extent, by stearic acid (18:0), and oleic acid (18:1) is the most abundant composition among MUFAs, which is generally similar to the reported terrestrial species. In addition, edible terrestrial insects have a higher content of linoleic acid (18:2), with the exception of a few species, in comparison to aquatic groups. The great difference between aquatic insects and terrestrial insects is that the former are significantly enriched in PUFAs that contain four or more double binds, mainly in terms of AA and EPA [44–46]. The content of unsaturated fatty acids (UFA), palmitic acid, and oleic acid in the oil of dragonfly naiads is high, and the oil contains 1.23–7.05% odd carbon fatty acids (OCFA), which have the characteristics of general insect oil. Meanwhile, the oil of dragonflies contains some long-chain polyunsaturated fatty acids, including AA, EPA, and docosahexaenoic acids (DHA), which is similar to freshwater fish, and the similarity might result from the aquatic life stage of dragonfly naiads [47]. The presence of relatively substantial amounts of these PUFAs was possibly associated with membrane fluidity, because PUFAs have a lower melting point than SFAs or MUFAs, which helps aquatic insects to better adapt to the cold-water environment. When these aquatic insects were transferred to land and became adults, they were expected to have less need for cold endurance, and could be found to have low levels of ARA and EPA, or even undetectable levels [48].

3.3. Characteristics of Mineral Elements in Aquatic Insects

Minerals are particularly rich in insect-based foods [49,50], which indicate that insect-based foods are excellent mineral providers [51]; for example, both aquatic and terrestrial insects have higher contents of calcium, iron, and zinc compared with common meat [32].

When edible insects are divided into aquatic insects and terrestrial insects, the difference in their mineral element contents shows great heterogeneity in different studies. Sometimes, aquatic and terrestrial insects have similar elemental compositions [52]; for example, they have almost the same concentration of zinc [30,53]. However, for calcium and iron contents, studies have shown that there is a considerable gap between aquatic and terrestrial insects. A much higher calcium content was identified in aquatic insects (24.3–96 mg/100 g) [30] than in different terrestrial insects (0.0012–0.126 mg/100 g) [54]. The iron contents in aquatic insects (e.g., *Lethocerus indicus* 410 mg/100 g, *Hydrophilus olivaceous* 461 mg/100 g) [30] are significantly higher than in terrestrial insects (*Bombyx mori* 1.8 mg/100 g [55], *Cirina forda* 5.34 mg/100 g) [56].

Dragonflies are usually consumed as a dish in Southwest China, and their selenium content has been a particular concern [57]. In particular, the selenium content of common edible insects in China has been analyzed, and it was found that the average selenium content of terrestrial insects was 0.15 mg/kg, and that of aquatic insects was 0.31 mg/kg, which was two-fold higher than that in terrestrial insects [58].

To better compare the mineral element contents of aquatic and terrestrial insects as a whole, we summarized the insect species with relatively complete data, and divided them into aquatic and terrestrial insects. The comparisons of the average mineral element content between aquatic and terrestrial insects are obviously different from the comparison results from several separate literatures. In general, the contents of iron and sodium in aquatic insects were significantly higher than those in terrestrial insects, while the contents of magnesium and potassium were significantly lower than those in terrestrial insects (Table 4).

Table 2. Amino acid composition of edible aquatic insects and selected edible terrestrial insects.

Order	Species	Edvelopmental Stage	Protein (%)	Amino Acid Composition (% of Total Amino Acids or Protein)																		Total Amino Acids (g/100 g DM)	Reference
				Val	Ile	Leu	Lys	Tyr	Thr	Phe	Trp	His	Met + Cys	Total EAA [++]	Arg	Asp	Ser	Glu	Gly	Ala	Pro		
Edible aquatic insects																							
Odonata	*Epophthalmia elegans*	L	65.23	9.96	2.78	6.43	5.62	7.06	3.84	9.26	0.58	3.74	1.63	50.90	11.79	6.91	3.94	10.92	4.42	5.97	5.15	60.16	[57]
	Anax parthenope	L	65.76	10.04	3.20	6.96	5.72	6.96	3.87	8.82	0.63	3.20	1.13	50.55	11.91	7.04	4.09	10.84	4.32	5.93	5.33	53.99	
	Ictinogomphus rapax	L	62.37	10.08	3.54	7.05	4.62	7.77	4.24	12.12	0.50	3.09	1.44	54.45	9.92	6.99	3.97	8.30	4.84	6.85	4.67	55.63	
	Sinictinogomphus clavatus	L	63.64	9.55	3.51	7.19	5.42	7.64	4.32	10.51	0.53	2.89	1.66	53.23	9.95	7.74	4.00	9.32	4.91	5.98	4.87	52.98	
	Pantala flavescens	L	65.18	9.78	3.30	7.20	5.97	6.48	4.06	8.61	0.65	2.87	1.32	50.26	12.47	7.38	4.07	10.73	4.64	6.00	4.45	58.16	
	Orthetrum pruinosum	L	71.53	9.63	3.33	7.11	5.85	6.43	3.86	8.68	0.46	2.78	1.90	50.01	12.00	7.38	4.02	11.11	4.39	5.81	5.28	54.73	
	Crocothemis servilia	L	65.45	6.12	3.91	7.45	7.93	6.23	5.25	2.88	2.20	5.45	3.77	51.18	8.04	8.47	4.26	10.99	5.03	7.31	4.72	51.70	
	Gomphus cuneatus	L	64.64	6.59	7.33	3.96	6.33	7.17	4.79	3.37	0.67	6.93	4.07	51.21	4.86	6.34	4.32	14.40	5.46	7.85	5.61	50.26	[59]
	Lestes praemorsus	L	46.37	6.01	6.96	4.16	8.37	7.26	4.98	3.22	5.23	6.54	2.72	55.44	8.54	6.41	4.20	13.36	4.45	7.26	5.23	36.1	
Ephemeroptera	*Ephermeterella jianghongensis*	L	66.26	5.75	5.29	8.51	5.51	6.00	4.88	3.27	-	3.33	3.39	45.93	5.75	8.71	4.55	15.21	4.96	9.15	5.74	65.54	[60]
Coleptera	*Cybister japonicus*	L	57.34	6.33	14.18	11.96	5.14	1.06	3.95	4.53	-	4.32	2.76	54.23	6.45	8.44	4.47	8.62	8.14	7.12	2.53	47.89	[61]
	Dytiscus dauricus	L	57.97	6.50	12.06	11.82	5.91	1.89	4.43	3.61	-	3.75	3.12	53.10	5.72	8.88	4.92	9.11	7.76	8.43	2.07	48.74	
	Hydrophilus acminatus	L	56.41	6.12	11.24	12.16	7.08	1.19	4.14	3.28	-	3.30	2.74	51.25	4.87	10.01	4.76	8.98	8.04	7.25	2.74	47.86	
	H. acminatus	L	20.37	5.76	4.38	7.59	6.89	5.11	4.26	4.33	-	6.42	2.74	47.48	5.76	9.56	3.56	9.25	8.08	10.38	5.93	42.69	[62]
Megaloptera	*Acanthacorydalis orientalis*	L	56.56	5.63	5.61	6.96	6.25	5.76	4.88	4.39	-	4.18	2.89	46.54	6.75	10.13	4.11	17.13	5.14	5.74	4.46	53.31	[60]
	Acanthacory dalisasiatice	A	-	7.58	5.58	9.00	7.08	9.81	4.13	10.61	-	4.02	4.21	62.01	3.90	11.73	7.69	-	-	14.34	-	52.01	[63]
	Neochauliodes sparsus	L	67.69	6.35	4.75	7.41	7.43	6.10	4.55	4.21	0.70	4.27	3.82	49.59	7.23	9.09	4.25	12.73	4.80	7.41	4.91	56.02	[64]
Edible terrestrial insects																							
Hymenoptera	*Polybia occidentalis nigratella*	B	61.00	5.90	4.50	7.80	7.40	5.60	4.00	3.30	0.70	3.00	5.00	47.20	5.70	8.40	4.50	12.90	7.10	6.50	6.30	-	[65]
	Polybia parvulina	B	61.00	6.10	4.70	7.80	7.30	5.90	4.10	3.40	0.70	3.40	5.30	48.70	5.70	7.80	4.40	13.30	7.20	6.40	6.50	-	
	Vespa velutina	B	-	6.10	5.50	8.70	6.10	6.60	4.20	4.20	-	4.20	2.40	47.00	4.50	6.30	6.30	20.10	6.30	5.50	6.10	37.90	[66]
	V. mandarinia	B	-	6.30	5.70	8.70	6.30	7.30	4.30	4.30	-	4.30	2.70	48.90	2.20	6.50	6.50	21.20	6.30	5.40	5.70	36.80	
	V. basalis	B	-	5.70	5.30	8.50	6.80	7.10	4.30	4.30	-	4.30	1.40	46.60	4.30	6.40	6.40	22.10	5.70	5.00	5.70	28.10	
Coleoptera	*Allomyrina dichotoma*	L	54.18	5.58	4.35	6.40	4.97	7.73	3.84	3.59	-	4.82	8.92	50.21	5.29	5.46	5.95	17.83	5.70	4.51	5.05	48.74	[32]
	Protaetia brevitarsis	L	44.23	6.36	4.14	5.90	4.47	8.43	3.96	4.14	-	4.65	7.51	49.54	5.34	5.77	6.51	14.15	5.72	6.23	6.72	39.16	
	Tenebrio molitor	L	53.22	6.61	4.45	7.57	4.52	7.75	4.11	3.96	-	6.29	7.10	52.36	5.01	6.20	4.94	12.99	5.87	8.90	3.73	44.50	
Orthoptera	*Teleogryllus emma*	A	55.65	5.85	4.30	7.93	5.23	5.23	3.84	3.58	-	4.82	7.63	48.41	7.43	7.71	5.91	13.03	5.09	9.19	3.24	49.95	
	Gryllus bimaculatus	A	58.32	5.94	4.01	7.38	4.50	5.07	3.72	3.40	-	4.64	9.98	48.63	6.69	6.69	5.07	11.87	6.17	10.48	3.70	53.83	
Lepidoptera	*Antheraea pernyi*	P	71.9	6.63	7.95	3.24	4.54	2.06	4.64	8.10	4.05	2.94	1.62	45.77	4.12	6.41	4.64	12.74	4.42	6.26	12.22	-	[67]
	Bombyx mori	P	-	5.60	5.70	8.30	7.50	5.40	5.40	5.10	9.00	2.50	6.00	60.50	6.80	10.90	4.70	14.90	4.60	5.50	4.00	-	

Note: L = larva, P = pupa, A = adult, B = brood; DM = dry matter; "-" = not determined or not estimated; [++] EAA: essential amino acids; we include essential amino acids (Val, Ile, Leu, Lys, Thr, Trp, Phe, His, Met) and two conditional essential amino acids (Tyr, Cys).

Table 3. Fatty acid composition of edible aquatic insects and selected edible terrestrial insects.

Order	Species	Developmental Stage	Lipid %	Fatty Acid Composition (% of Total Fatty Acids)										Reference
				C14:0	C16:0	C18:0	SFA	C18:1	MUFA	C18:2	C20:4	C20:5	PUFA	
Aquatic insects														
Odonata	Epophthalmia elegans	L	9.14	1.67	21.88	7.85	33.73	16.97	32.84	4.46	5.02	6.11	21.05	[47]
	Anax parthenope julius	L	11.06	1.18	24.41	7.00	34.68	17.65	27.85	7.01	3.73	3.08	27.85	
	Ictinogomphus rapax	L	10.59	2.07	19.30	6.98	30.11	19.01	35.75	7.49	1.89	5.08	20.33	
	Pantala flavescens	L	10.4	0.75	21.6	8.55	32.3	11.98	33.07	9.59	1.91	9.44	27.51	
	Inictinogomphus clavatus	L	11.9	0.89	24.61	7.74	34.43	20.58	42.28	6.66	1.24	3.87	17.75	
	Orthetrum pruinosum neglectum	L	5.72	2.34	17.57	7.65	34.97	6.85	31.53	5.77	6.70	7.83	26.70	
Diptera	Stictochironomus pictulus	L	-	4.70	16.10	6.20	34.00	11.00	49.50	6.70	1.00	3.60	14.30	[68]
	Anopheles albimanus	L	-	1.94	28.03	7.76	44.50	22.26	31.13	7.59	3.07	4.19	24.40	
	A. vestitipennis	L	-	1.86	21.82	6.62	37.39	22.06	36.16	12.28	2.85	2.47	26.43	[69]
	A. darlingi	L	-	1.26	25.42	6.64	39.53	22.44	30.76	18.66	2.46	2.39	29.69	
Coleoptera	Cybister japonicus	A	27.66	3.41	12.01	5.18	27.56	35.61	49.94	9.82	3.55	3.96	22.52	[61]
	Dytiscus danmcus	A	27.56	2.86	21.63	2.45	35.1	29.94	46.95	6.53	3.54	4.08	19.46	
	Hydrophilus aoninatus	A	31.86	9.16	19.09	-	65.2	1.83	3.81	1.98	-	-	30.92	
Terrestrial insects														
Orthoptera	Gryllus bimaculatus	L	28.90	-	25.44	8.74	34.67	25.86	26.54	37.05	-	-	38.79	[70]
	Ruspolia differens	A	48.20	0.90	31.50	5.50	38.30	24.60	26.60	31.20	-	-	34.40	[53]
Coleoptera	Tenebrio molitor	L	31.97	4.45	21.33	7.92	33.70	35.83	37.80	22.83	0	0	22.94	[71]
Hymenoptera	Apis mellifera	L	4.90	2.40	37.30	11.80	51.80	47.50	48.20	0	-	-	0	[72]
	Vespa mandarinia	B	20.20	2.50	21.30	5.00	30.70	27.70	29.20	33.70	-	-	40.10	
Lepidoptera	Bombyx mori	P	32.20	0.10	24.20	4.50	24.30	26.00	27.70	7.30	-	-	36.30	[73]

Note: L = larva, P = pupa, A = adult; "-" = not determined or not estimated; oil. SFA = saturated fatty acids, MUFA = monounsaturated fatty acids, PUFA = polyunsaturated fatty acids.

Table 4. Mineral content [mg/kg, DM] of selected edible aquatic insects and edible terrestrial insects.

Species	Developmental Stage	Ca	Mg	K	Na	Fe	P	Mn	Cu	Zn	Se	References
Edible aquatic insects												
Anax parthenope	L	124.960	116.900	1591.900	1339.760	158.210	-	6.790	4.180	74.770	0.193	[57,58]
Epophthalmia elegans	L	90.110	101.800	1350.790	1372.490	22.640	-	12.050	2.460	40.410	0.223	[57,58]
Crocothemes servillia	L	865.000	370.000	2680.000	14,100.000	113.000	-	-	19.000	93.000	-	[30]
Lethocerus indicus	L & A	960.000	703.300	1700.000	8550.000	4100.000	-	-	11.000	295.000	-	[30]
Laccotrephes maculatus	L & A	665.000	460.000	5500.000	15,000.000	250.000	-	-	137.000	231.500	-	[30]
Cybister tripunctatus	A	277.000	336.000	6430.000	3050.000	73.000	-	-	51.000	57.500	-	[30]
C. japonicus	A	3602.820	774.520	6722.650	2251.760	148.160	5809.000	8.700	29.370	93.990	0.360	[58,74]
Hydrophilus olivaceous	A	243.000	990.000	3900.000	8160.000	4610.000	-	-	17.000	118.000	-	[30]
Hydrous acuminatus	A	106.700	109.200	1807.700	-	83.200	1905.100	-	-	29.000	-	[62]
Edible terrestrial insects												
Gryllus bimaculatus	A	1660.850	1073.750	8607.500	3649.450	81.800	11,696.000	66.300	36.250	232.650	0.490	[32,70]
Acheta domesticus	L & A	1261.150	1040.550	13,318.700	5122.900	77.650	10,291.150	38.100	21.200	257.400	0.500	[75]
Teleogryllus emma	A	1935.400	1524.800	8955.000	2782.300	107.500	10,854.000	58.600	21.900	184.700	-	[32]
Tenebrio molitor	L	504.800	2450.800	8212.375	1047.125	98.393	8282.233	14.170	18.138	116.660	0.377	[32,75,76]
Zophobas morio	L	420.400	1182.900	7505.900	1128.300	39.200	5629.500	10.200	8.600	72.900	0.300	[75]
Protaetia brevitarsis	L	2585.600	3276.000	20,014.000	2116.000	162.000	11,404.000	58.900	18.200	118.900	-	[32]
Allomyrina dichotoma	L	1234.000	2835.600	12,491.000	1483.800	142.600	8606.900	86.400	14.300	102.600	0.064	[32,58]
Anoplophora chinensis	L	269.300	1881.000	5647.000	92.850	131.280		35.020	8.980	223.640	0.050	[76]
Galleria mellonella	L	585.500	761.400	5325.300	397.600	50.400	4698.800	3.100	9.200	61.200	0.300	[75]
Bombyx mori	L & P	1023.100	2878.600	18,265.900	2745.700	95.400	13,699.400	24.900	20.800	177.500	0.575	[58,75]
Antheraea pernyi	P	234.000	707.000	4020.000	57.400	13.400	-	2.200	2.900	30.600	0.210	[77]
Vespa velutina	L & P	388.000	639.000	7516.000	104.000	100.000	5612.000	6.000	22.000	72.000	-	[66]
Polyrhachis vicina	A	785.500	664.500	-	-	858.500	4028.500	291.000	21.500	147.500	0.335	[78]

Note: L = larva, P = pupa, A = adult; "-" = not determined or not estimated; Mineral compositions [mg/kg] of calcium (Ca), magnesium (Mg), potassium (K), sodium (Na), iron (Fe), phosphorus (P), manganese (Mn), copper (Cu), zinc (Zn), and selenium (Se) were analyzed based on dry matter (DM); if different values were involved in multiple literatures, the mean value was listed.

Environmental factors seem to be the main factor determining the mineral element contents of aquatic insects. It is reported that, despite their body size, dragonflies had good ability to absorb environmental metallic elements (Hg, As, Pb, Cr, Cu, Cd, Ni, Se, Al, and Au) [79], thus they are good ambient heavy metal content indicators.

3.4. Chitin and Chitosan

Chitin is a significant biopolymer [80], and chitosan is formed by the deacetylation of chitin. They could be used in food, biomedical and cosmetic industries, and for wastewater treatment and textiles. Up to date, crustacean shells from the marine food industry provide us with the chief commercial sources of chitin and chitosan [80,81]. Due to the COVID-19 pandemic, biopolymer materials have been increasingly in demand, and the market for chitin and chitosan is growing steadily [81].

Chitin is found throughout the exoskeletons of most insects. In recent years, some terrestrial insect species have been investigated as alternative chitin sources [82–84], since insects were considered as a potential resources of food. As compared to the existing sources, the extraction of chitin and chitosan from insects is simple, requires less chemical consumption and time, and they can be extracted in a higher yield [81]. Moreover, insect-derived chitin and chitosan have numerous biological effects, such as antioxidant and antibacterial activities with substantial rheological properties [82,83]. In Table 5, the percentages of chitin and chitosan from terrestrial insects are species specific and are in the ranges of 2.59–36.80% and 16.00–96.35%, respectively. However, for aquatic insects, far fewer literature studies are reported. The contents of chitin in aquatic insects have almost no differences with terrestrial insects; the former may also be used as alternative chitin sources [85]. More extensive and comparative research of chitin and chitosan between aquatic insects and terrestrial insects is needed.

Table 5. Chitin and chitosan from aquatic insect and selected edible terrestrial insects.

Order	Species	Developmental Stage	Yield of Chitin (%)	Yield of Chitosan (%)	Reference
Aquatic insects					
Coleoptera	*Agabus bipustulatus*	-	14.00–15.00	71.00	
	Hydrophilus piceus	-	19.00–20.00	74.00	
Odonata	*Anax imperator*	L	11.00–12.00	67.00	[85]
Hemiptera	*Notonecta glauca*	-	10.00–11.00	69.00	
	Ranatra linearis	-	15.00–16.00	70.00	
Terrestrial insects					
Lepidoptera	*Bombyx mori*	P	2.59–4.23	73.00–96.35	[86]
Coleoptera	*Catharsius molossus*	A	24.00	70.83	[87]
Orthoptera	*Pterophylla beltrani*	-	11.80	58.80	[88]
	Brachytrupes portentosus	-	4.30–7.10	55.81–81.69	[89]
	Calliptamus barbaru	A	20.50	74.00–75.00	[90]
	Oedaleus decorus	A	16.50	75.00–76.00	
Hymenoptera	*Apsis mellifera*	A	19.00–36.80	16.00–30.00	[91]
Diptera	*Musca domestica*	P	8.02	73.19	[92]
	Hermetia illucens	L	7.00	32.00	[93]
	Drosophila melanogaster	A	7.85	70.91	[94]
Blattodea	*Periplaneta americana*	-	12.17	59.82	[95]

Note: L = larva, P = pupa, A = adult, "-" = not sure; the yield of chitosan is calculated from chitin, or degree of deacetylation.

3.5. Active Substances and Healthcare

Edible insects are considered to have superior health benefits, due to their high quantities of nutrients, such as essential amino acids, omega-3 and omega-6 fatty acids, vitamin B12, iron, and zinc [2,72,96]. In addition to the health benefits of edible insect nutrients, the rich active substances in edible aquatic insects have also attracted attention [97]. Edible aquatic insects, such as dragonflies, water strider, and whirligig beetle, have been used in healthcare or for treating human diseases since the ancient times [98–100], especially in the countries of East Asia [101], e.g., China, Japan, and South Korea. The theory of traditional Chinese medicine is that the larva of aquatic insects has the effects of boosting the kidney, nourishing essence, moisturizing the lungs, and relieving coughs [102]. They can be used alone or in combination with other materials for medical applications; for example, the whole body of a dried adult dragonfly can be used to treat impotence and nocturnal emission, sore throat, and whooping cough [103]. It is believed that *Cybister tripunctatus*, or *C. japonicus*, can reinforce kidney function and invigorate the circulation of blood in human beings [102]. These traditional practices of edible aquatic insects also provide ideas for the exploitation of modern drugs [104]; for example, the methanol extract of *C. tripunctatus* displays strong antioxidant activity at a concentration of 110 μg/mL [30]. It is reported that the water and liposoluble extracts of the giant water bugs *L. indicus* have negligible values of antioxidant capacity compared to the extracts of terrestrial insects, such as grasshoppers, silkworm, and crickets, in vitro [105].

The bioactive ingredients in insects have different sources depending on the species [106–108]. Some natural toxins are produced by insects in their special organs, such as bee venom and cantharidin [109,110]. The giant water bug *L. indicus*, which is consumed as both a medicinal and edible insect [111], could produce complex chemicals in its odorous gland [112,113]. On the other hand, some other components from insects are enriched and accumulated from the plants, fungi, and algae that they feed on. Phytophagous insects have evolved unique systems to keep the secondary metabolites taken up from the plant in their bodies [114] and utilize them for escaping from predators [115,116], and these accumulated secondary metabolites have a certain value in drug development [117,118]. However, exploration of the source of functional substances of aquatic insects is still very limited. This is an aspect worth exploring further.

4. Safety in Utilization of Edible Aquatic Insects

4.1. Contaminant

Aquatic insects, as environmental indicators, have attracted extensive attention for a long time, and their related research is very rich. A water body is an open environment, which readily gathers various substances through water flow, soil, etc. Due to the close relationship between aquatic organisms and water, aquatic insects easily accumulate various pollutants, including heavy metals, pathogens, pesticides, and so on. These contaminants may enter the human body through the food chain, which will cause harm to human health. We should draw more attention to them. This part mainly indicates the possible risks of the utilization of aquatic insects, but the relevant processes and pollution mechanisms are beyond the scope of this review.

Heavy metals. Heavy metal pollution in the freshwater ecosystem poses a great threat to the growth of aquatic insects [119,120]; therefore, there is a growing concern that it will eventually affect the health of humans [121,122]. Up to date, over 33 metallic elements of aquatic edible insects have been detected, and heavy metals, such as Hg, Pb, Cd, and Cr, have attracted serious concern [123]. Most of the metals in nature will enter the water, and the aquatic ecosystem plays an important role in the transfer and circulation of metals. Insects could absorb metal elements and accumulate a higher concentration of them than the environment [124]. The degree of metal element absorption by aquatic insects is related to environmental factors; for example, aquatic insects collected near wastewater treatment plants, or near mines, may show higher metallic element concentrations when compared with other sites [125,126]. Meanwhile, the mercury content in carnivorous insects was generally higher than that in herbivorous insects, and the mercury content in aquatic insects was much higher than that in terrestrial insects (*t*-test, $p < 0.01$), with the investigation of 42 insect species from Yunnan, China [58]. Many studies [121,127–132] have indicated that the larvae of aquatic insects accumulate metals in aquatic ecological environments and retain them until the adult stage. After being preyed on by bats, spiders, birds, fish, and so on, they become the connector of the aquatic and terrestrial food chains, and, thus, bring metal elements into the terrestrial food chain. This suggests that there are the same risks in the process of human consumption of aquatic insects.

Pathogen. Aquatic insects are important vectors (e.g., malaria) of environmental pathogens. Mainly originating from the Japanese encephalitis (JE) virus, the viral encephalitis in Southeast Asia was detected in *Culex gelidus* in 1976 [133]. Aquatic insects are possible vectors of *Mycobacterium ulcerans*, which causes chronic skin ulcers in tropical countries [134]. Two conditioned pathogens, *Lelliottia amnigena* and *Citrobacter freundii*, were detected from edible aquatic insects of the genus *Cybister* [135].

Pesticide. Pesticide residues in food can cause damage to the human nervous and reproductive systems, interfere with the normal operation of the human immune or endocrine system, and even induce cancer. Pesticides could enter aquatic insects through contact or feeding [136]. The impact of pesticides reaching streams and potentially harming the aquatic juvenile stages has drawn extensive concerns when investigating the effects of insecticides on freshwater insects [137,138]. There are few studies on the effects of pesticide residues in insects on human health.

4.2. Purine Derivatives and Uric Acid

Uric acid serves as an antioxidant, and is important for protecting human blood vessels. It is metabolized from purines [139], which are important nucleic acid components in all organisms. However, increasing numbers of people are suffering from hyperuricemia, gout, and other disease caused by frequent and high intake of purine-rich and protein-rich foods, which enhances serum uric acid levels. An important way to treat people with hyperuricemia or gout is dietary restriction of purine-rich foods [140,141].

The concentration of purine derivative and uric acid in edible insects varies, primarily with species and gender (in some cases) [140,142,143]. Though no common characteristics of purine and uric acid contents in different edible insects or at different life cycle stages of the same insect were indicated, much lower uric acid levels are detected in aquatic insects than in terrestrial insects [143]. A possible reason for this could be the extremely moist environments that aquatic insects live in, where they can excrete large quantities of ammonia. However, for most terrestrial insects, loss by excretion has to be minimized because water conservation is essential throughout their life cycles. Hence, more than 80% of the total nitrogenous materials excreted from most terrestrial insects are uric acid [144].

4.3. Allergy

Seafood is an important origin of allergies, and, thus, has attracted attention as a food safety issue. Similarly, most known edible insect allergens have cross-reactivity with homologous proteins in shellfish [145]; therefore, enough attention should be paid to the insect allergy too. At present, all the reported cases of insect allergies are from terrestrial insects, such as silkworm, mealworm, caterpillars, wasps, grasshoppers, cicada, and bees [146].

5. Discussion
5.1. Characteristics of Nutrition in Aquatic Insects

There is much less available information on aquatic insects, especially on their nutritional value, when compared to terrestrial insects. In this paper, we summarize some results of the research on aquatic insect nutrition. It should be noted that, for the nutritional composition, there must be variations, even in the same species, attributed to some internal (i.e., feed, developmental and physiological state of the samples analyzed) and external (i.e., season, geographic location, techniques employed, etc.) factors [30,147–149].

Both aquatic insects and terrestrial insects have high protein contents, with some differences in amino acid compositions, kinds of restricted amino acids, and the dominant amino acids. It should be encouraged that proteins from different sources (e.g., aquatic insects and terrestrial insects) are used simultaneously, so as to complement each other and improve the biological value of insect proteins.

A remarkable feature of aquatic insects is the rich, highly unsaturated fatty acid (HUFA) content, which is much higher than in terrestrial insects, especially the EPA content. This big difference might be explained by their food webs. The food of terrestrial insects, such as vascular plants, usually only contains the HUFA precursor α-Linolenic acid (ALA) [150], while aquatic insects feed on freshwater algae and diatoms, which are particularly rich in HUFA and EPA [151].

The living environment is a main factor affecting the selenium content in insects. The reason for the significant effect of aquatic habitat on the selenium content of insects may be related to the characteristics of selenium. Selenium in soil can be leached into the aquatic ecosystem; therefore, aquatic organisms are more easily exposed to selenium than terrestrial organisms [152]. Aquatic organisms usually have a larger selenoprotein group, while terrestrial organisms have a significantly smaller selenoprotein group. It is speculated that terrestrial organisms gradually lost their selenoprotein-coding genes, or replaced the Sec with cysteine in the original selenoprotein-coding genes [153,154]. This may help explain why the selenium content of aquatic insects is higher than that of terrestrial insects [58].

5.2. Edible Aquatic Insects Resources and Farming

In order to protect wild resources and control the quantity and quality of products, it is very necessary to cultivate insects when we promote insect consumption. Edible insect farming has already taken off all around the world. The farming technology of some terrestrial insect food or feed species, such as mealworm, black soldier flies, crickets, and houseflies, has largely advanced, and their yields are very high and sustainable nowadays [21]. However, there is little information about rearing aquatic insects, especially edible aquatic insects.

Aquatic insects should be cultivated on a large scale for the following reasons. Firstly, excessive collection of natural resources will lead to a reduction in aquatic insect resources and regional extinction of species. The majority of edible insects commercially consumed are mainly collected from nature [23]. Due to the lack of rearing technology and boom of traditional customs, collecting edible aquatic insects from the wild is much more prevalent than collecting terrestrial insects. However, the overexploitation of several species in Mexico has led to a subsequent decrease in the population [155], and the uncontrolled harvesting of wild populations may prove unsustainable [22]. Fortunately, the production patterns of edible insects in Thailand and Lao People's Democratic Republic have changed to semi-domestication and insect farming, besides the traditional harvesting of insects from wild habitats [156]. Secondly, the artificial cultivation of insects could avoid the disturbance from pollutants. As mentioned previously, a water body is an open environment, which is easily disturbed by pollutants, affecting the population growth and consumption safety of aquatic edible insects. Thirdly, artificial cultivation can improve the nutritional quality of aquatic insects through feeding technology [157], such as to attain a beneficial n-6/n-3 ratio, or to better meet the nutritional demands of the consumer. Finally, population outbreaks do not theoretically easily happen in aquatic insects [158]; therefore, it is necessary to expand the population by artificial promotion.

In addition, the development of aquatic edible insect cultivation technology can be supported by several aspects of experience. Historically, there have been some practices of raising insects all over the world. Semi-domestication also has the potential to promote the yield of insects greatly [22]. A famous example of semi-domestication from Mexico showed that eggs of semi-cultivated aquatic Hemiptera were used by local people, through water management and to provide egg laying sites [1]. In modern times, the breeding of aquatic insects is not a blank, but has accumulated technologies and experience in species from several orders [159–161]. Moreover, the successful breeding of terrestrial insects (e.g., crickets, mealworms, and black solider flies) can provide a reference for the breeding of aquatic insects [19]. In recent years, some species from Hemiptera, Coleoptera, and Megaloptera have had some success in farming, e.g., *L. indicus* in Thailand, and predacious diving beetle [162,163] and hellgrammites [25,26] in China.

Nevertheless, the aquaculture of aquatic insects still needs to face the problems of high input costs, such as water quality, energy, and feed [18,164]. It is suggested that different types of aquatic insect breeding technologies should be developed and utilized, according to market demand, product value, regional conditions, and other factors, and regional planning should be performed for aquatic insect breeding. For example, for the species with high market recognition and health care function, priority should be given to the development of automatic three-dimensional breeding, and most aquatic edible insects should be cultivated with semi-artificial technology, or artificial promotion and management technology. Similarly, a large area and protected water source can harvest a large number of aquatic insect products. On the contrary, such appropriate conditions can guide more people to breed in this area, so as to improve the technical maturity. Of course, the cultivation of aquatic insects will also help to prevent environmental pollution; for example, in order to harvest aquatic insects raised in the farmland system, excessive pesticides and fertilizers cannot be applied, which reduces the pressure of environmental pollution.

5.3. Enrich the Use of Aquatic Insects

Edible aquatic insects also have the potential to serve human health and feed domestic animals. The utilization of aquatic insects is small in scale, but the prospects are obvious, and the development of new uses can increase its attraction. The active substances of aquatic insects are specific products in aquatic environments, which are worthy of further study.

It is a valuable and innovative proposal to use edible aquatic insects as feed, and some species, such as mosquito larvae, could provide a convenient source of feed [19,165]. Mosquitos have the distinct biological characteristics of a short lifecycle, high occurrence densities, and wide distribution. In one case, the larvae of chironomid from chicken manure were used for fish food, both in local areas and North America [159]. Aquatic insects with high HUFA are being designed to be aquafeeds and farmed marine fishes, for meeting the demand of the HUFA of fishes [19,166].

6. Conclusions

Edible insects are promising safe food that have high potential to be more sustainable and more equitable at the global scale. Resources of insects, both from terrestrial and aquatic environments, should be encouraged. Although the uses of aquatic insects as food, as feed, and in healthcare were ignored to a certain extent, edible aquatic insects have great potential for utilization.

There has been the discovery of a few characteristics of edible aquatic insects with limited documents, compared with terrestrial insects. Two sources of edible insects have a high content of protein, calcium, iron, and zinc, and similar contents of chitin. There is no doubt that they are all excellent biological resources. The living environment seems to be the main factor for the origin of the difference between aquatic and terrestrial insects. A water body is an open environment; therefore, aquatic organisms are more easily exposed to various substances. Compared with edible terrestrial insects, aquatic insects usually have different limiting amino acids and a higher content of delicious amino acids, and contain highly unsaturated omega-3 fatty acids and some long-chain polyunsaturated fatty acids, lower uric acid, much higher amounts of calcium and selenium, and a larger selenoprotein group, and face higher risk of exposure to contaminants (e.g., heavy metal, pesticides). More studies investigating the distinct differences between aquatic and terrestrial insects are needed for better conservation and utilization of these insects.

Supplementary Materials: The following are available online at https://www.mdpi.com/article/10.3390/foods10123033/s1, Table S1: full list of named edible aquatic insects.

Author Contributions: M.Z., C.-Y.W., H.-J.L. and Y.F. conceptualized the manuscript. M.Z., C.-Y.W., L.S., Z.H. and P.-L.Y. drafted the manuscript, M.Z., C.-Y.W., H.-J.L. and Y.F. edited the manuscript. All authors have read and agreed to the published version of the manuscript.

Funding: This research was funded by Lancang-Mekong Cooperation Special Fund Projects (JiaoWaiSiYa [2020]619), Jiangsu Agricultural Science and Technology Innovation Fund (no. CX(19)2036), Cooperation Project of Zhejiang Province and CAF (no. 2019SY02), and the Fundamental Research Funds for the Central Non-profit Research Institution of CAF (no. CAFYBB2018GB001).

Institutional Review Board Statement: Not applicable.

Informed Consent Statement: Not applicable.

Data Availability Statement: Not applicable.

Acknowledgments: The authors would like to acknowledge Yu-Xuan Long (student of Southwest Forestry University, China) for checking information in the Supplementary Materials, and Wei-feng Ding for providing convenience for literature retrieval. We also sincerely thank three anonymous reviewers for their critical comments regarding the manuscript.

Conflicts of Interest: The authors declare that the research was conducted in the absence of any commercial or financial relationship that could be construed as a potential conflict of interest.

References

1. Van Itterbeeck, J.; van Huis, A. Environmental manipulation for edible insect procurement: A historical perspective. *J. Ethnobiol. Ethnomed.* **2012**, *8*, 3. [CrossRef] [PubMed]
2. Feng, Y.; Chen, X.-M.; Zhao, M.; He, Z.; Sun, L.; Wang, C.-Y.; Ding, W.-F. Edible insects in China: Utilization and prospects. *Insect Sci.* **2017**, *25*, 184–198. [CrossRef]
3. Kim, T.-K.; Yong, H.I.; Kim, Y.-B.; Kim, H.-W.; Choi, Y.-S. Edible Insects as a Protein Source: A Review of Public Perception, Processing Technology, and Research Trends. *Food Sci. Anim. Resour.* **2019**, *39*, 521–540. [CrossRef] [PubMed]
4. Raheem, D.; Carrascosa, C.; Oluwole, O.B.; Nieuwland, M.; Saraiva, A.; Millán, R.; Raposo, A. Traditional consumption of and rearing edible insects in Africa, Asia and Europe. *Crit. Rev. Food Sci. Nutr.* **2018**, *59*, 2169–2188. [CrossRef] [PubMed]
5. Ramos-Elorduy, J. Anthropo-entomophagy: Cultures, evolution and sustainability. *Entomol. Res.* **2009**, *39*, 271–288. [CrossRef]
6. Meyer-Rochow, V.B. Edible insects in three different ethnic groups of Papua and New Guinea. *Am. J. Clin. Nutr.* **1973**, *26*, 673–677. [CrossRef] [PubMed]
7. Séré, A.; Bougma, A.; Ouilly, J.T.; Traoré, M.; Sangaré, H.; Lykke, A.M.; Ouédraogo, A.; Gnankiné, O.; Bassolé, I.H.N. Traditional knowledge regarding edible insects in Burkina Faso. *J. Ethnobiol. Ethnomed.* **2018**, *14*, 59. [CrossRef] [PubMed]
8. Hlongwane, Z.T.; Slotow, R.; Munyai, T.C. Indigenous Knowledge about Consumption of Edible Insects in South Africa. *Insects* **2020**, *12*, 22. [CrossRef]
9. Mwangi, M.N.; Oonincx, D.G.A.B.; Stouten, T.; Veenenbos, M.; Melse-Boonstra, A.; Dicke, M.; Van Loon, J.J.A. Insects as sources of iron and zinc in human nutrition. *Nutr. Res. Rev.* **2018**, *31*, 248–255. [CrossRef] [PubMed]
10. Oonincx, D.G.A.B.; van Itterbeeck, J.; Heetkamp, M.J.W.; Brand, H.V.D.; van Loon, J.J.A.; van Huis, A. An Exploration on Greenhouse Gas and Ammonia Production by Insect Species Suitable for Animal or Human Consumption. *PLoS ONE* **2010**, *5*, e14445. [CrossRef] [PubMed]
11. Gahukar, R.T. Edible Insects Farming: Efficiency and Impact on Family Livelihood, Food Security, and Environment Compared With Livestock and Crops. In *Insects Sust Food Ingredients*; Dossey, A.T., Morales-Ramos, J.A., Rojas, M.G., Eds.; Elsevier: Amsterdam, The Netherlands, 2016; pp. 85–111. [CrossRef]
12. Hlongwane, Z.T.; Slotow, R.; Munyai, T.C. Nutritional Composition of Edible Insects Consumed in Africa: A Systematic Review. *Nutrients* **2020**, *12*, 2786. [CrossRef] [PubMed]
13. Patel, S.; Suleria, H.A.R.; Rauf, A. Edible insects as innovative foods: Nutritional and functional assessments. *Trends Food Sci. Technol.* **2019**, *86*, 352–359. [CrossRef]
14. Huis, A.v.; Itterbeeck, J.V.; Klunder, H.; Mertens, E.; Halloran, A.; Muir, G.; Vantomme, P. *Edible Insects: Future Prospects for Food and Feed Security*; FAO: Rome, Italy, 2013.
15. Koroiva, R.; Pepinelli, M. Distribution and Habitats of Aquatic Insects. In *Aquatic Insects: Behavior and Ecology*; Del-Claro, K., Guillermo, R., Eds.; Springer International Publishing: Cham, Switzerland, 2019; pp. 11–33. [CrossRef]
16. Williams, D.D.; Feltmate, B.W. *Aquatic Insects*; Blackburn Press: Caldwell, NJ, USA, 2017; p. 372.
17. Kitsa, K. Contribution des insectes comestibles a l'amélioration de la ration alimentaire au Kasai Occidental. *Zaire-Afrique* **1989**, *239*, 511–519.
18. Yen, A. Insects as food and feed in the Asia Pacific region: Current perspectives and future directions. *J. Insects Food Feed* **2015**, *1*, 33–55. [CrossRef]
19. Williams, D.D.; Williams, S.S. Aquatic Insects and their Potential to Contribute to the Diet of the Globally Expanding Human Population. *Insects* **2017**, *8*, 72. [CrossRef]

20. Hotaling, S.; Kelley, J.; Frandsen, P. Aquatic Insects Are Dramatically Underrepresented in Genomic Research. *Insects* **2020**, *11*, 601. [CrossRef] [PubMed]
21. Williams, D.; Williams, S.; van Huis, A. Can we farm aquatic insects for human food or livestock feed? *J. Insects Food Feed* **2021**, *7*, 121–127. [CrossRef]
22. Macadam, C.; Stockan, J. The diversity of aquatic insects used as human food. *J. Insects Food Feed* **2017**, *3*, 203–209. [CrossRef]
23. Feng, Y.; Zhao, M.; Ding, W.; Chen, X. Overview of edible insect resources and common species utilisation in China. *J. Insects Food Feed* **2020**, *6*, 13–25. [CrossRef]
24. Tanaka, R.; Oda, M. Cyclic Dipeptide of D-ornithine obtained from the dobsonfly, *Protohermes grandis* Thunberg. *Biosci. Biotechnol. Biochem.* **2009**, *73*, 1669–1670. [CrossRef] [PubMed]
25. Cao, C. Rearing hellgrammites for food and medicine in China. *J. Insects Food Feed* **2016**, *2*, 263–267. [CrossRef]
26. Shi, Z.X. A research on artificial culture of climbing-sand worms-panxi special aquatic organisms based on computer control. *J. Xichang Coll. (Nat. Sci. Ed.)* **2008**, *22*, 72–75.
27. Jongema, Y. List of Edible Insect Species of the World. Available online: https://www.wur.nl/en/Research-Results/Chair-groups/Plant-Sciences/Laboratory-of-Entomology/Edible-insects/Worldwide-species-list.htm (accessed on 15 June 2021).
28. Mitsuhashi, J. *Edible Insects of the World*; CRC Press: Boca Raton, FL, USA, 2016; pp. 1–296. [CrossRef]
29. Wang, C.Y.; Zhao, M.; Wang, J.D.; Jiang, Y.Y.; He, Z.; Feng, Y. Molecular identification of a new species of edible dragonfly. *Biot. Resour.* **2018**, *40*, 164–169. [CrossRef]
30. Shantibala, T.; Lokeshwari, R.K.; Debaraj, H. Nutritional and antinutritional composition of the five species of aquatic edible insects consumed in Manipur, India. *J. Insect Sci.* **2014**, *14*, 14. [CrossRef] [PubMed]
31. Nurhasan, M.; Maehre, H.K.; Malde, M.K.; Stormo, S.K.; Halwart, M.; James, D.; Elvevoll, E.O. Nutritional composition of aquatic species in Laotian rice field ecosystems. *J. Food Compos. Anal.* **2010**, *23*, 205–213. [CrossRef]
32. Ghosh, S.; Lee, S.-M.; Jung, C.; Meyer-Rochow, V.B. Nutritional composition of five commercial edible insects in South Korea. *J. Asia-Pac. Entomol.* **2017**, *20*, 686–694. [CrossRef]
33. Bergeron, D.; Bushway, R.J.; Roberts, F.L.; Kornfield, I.; Okedi, J.; Bushway, A.A. The nutrient composition of an insect flour sample from Lake Victoria, Uganda. *J. Food Compos. Anal.* **1988**, *1*, 371–377. [CrossRef]
34. Okedi, J. Chemical evaluation of Lake Victoria lakefly as nutrient source in animal feeds. *Int. J. Trop. Insect Sci.* **1992**, *13*, 373–376. [CrossRef]
35. FAO; WHO; UNU. *Protein and Amino Acid Requirements in Human Nutrition: Report of a Joint WHO/FAO/UNU Expert Consultation*; World Health Organization: Geneva, Switzerland, 2007.
36. Bukkens, S. The nutritional value of edible insects. *Ecol. Food Nutr.* **1997**, *36*, 287–319. [CrossRef]
37. Zielińska, E.; Baraniak, B.; Karaś, M.; Rybczyńska-Tkaczyk, K.; Jakubczyk, A. Selected species of edible insects as a source of nutrient composition. *Food Res. Int.* **2015**, *77*, 460–466. [CrossRef]
38. Ekpo, K.E. Effect of processing on the protein quality of four popular insects consumed in Southern Nigeria. *Arch. Appl. Sci. Res.* **2011**, *3*, 307–326.
39. De Guevara, O.L.; Padilla, P.; García, L.; Pino, J.M.; Ramos-Elorduy, J. Amino acid determination in some edible Mexican insects. *Amino Acids* **1995**, *9*, 161–173. [CrossRef]
40. Aguilar, J.G.d.S. An overview of lipids from insects. *Biocatal. Agric. Biotechnol.* **2021**, *33*, 101967. [CrossRef]
41. Hixson, S.M.; Sharma, B.; Kainz, M.J.; Wacker, A.; Arts, M. Production, distribution, and abundance of long-chain omega-3 polyunsaturated fatty acids: A fundamental dichotomy between freshwater and terrestrial ecosystems. *Environ. Rev.* **2015**, *23*, 414–424. [CrossRef]
42. Bell, J.; Ghioni, C.; Sargent, J.R. Fatty acid compositions of 10 freshwater invertebrates which are natural food organisms of Atlantic salmon parr (*Salmo salar*): A comparison with commercial diets. *Aquaculture* **1994**, *128*, 301–313. [CrossRef]
43. Béligon, V.; Christophe, G.; Fontanille, P.; Larroche, C. Microbial lipids as potential source to food supplements. *Curr. Opin. Food Sci.* **2016**, *7*, 35–42. [CrossRef]
44. Twining, C.W.; Brenna, J.T.; Lawrence, P.; Winkler, D.W.; Flecker, A.S.; Hairston, N.G., Jr. Aquatic and terrestrial resources are not nutritionally reciprocal for consumers. *Funct. Ecol.* **2019**, *33*, 2042–2052. [CrossRef]
45. Popova, O.N.; Haritonov, A.Y.; Sushchik, N.N.; Makhutova, O.N.; Kalachova, G.S.; Kolmakova, A.A.; Gladyshev, M.I. Export of aquatic productivity, including highly unsaturated fatty acids, to terrestrial ecosystems via Odonata. *Sci. Total Environ.* **2017**, *581–582*, 40–48. [CrossRef] [PubMed]
46. Twining, C.W.; Shipley, J.R.; Winkler, D.W. Aquatic insects rich in omega-3 fatty acids drive breeding success in a widespread bird. *Ecol. Lett.* **2018**, *21*, 1812–1820. [CrossRef]
47. Jiang, Y.Y.; He, Z.; Zhao, M.; Wang, C.Y.; Sun, L.; Feng, Y. Oil content and fatty acid composition of six kinds of common edible dragonfly naiads. *China Oils Fats* **2017**, *42*, 135–139.
48. Hanson, B. Lipid Content and Fatty Acid Composition of Aquatic Insects: Dietary Influence and Aquatic Adaptation. Ph.D. Thesis, Oregon State University, Corvallis, OR, USA, 1983.
49. Cerritos, R. Insects as food: An ecological, social and economical approach. *CAB Rev. Perspect. Agric. Vet. Sci. Nutr. Nat. Resour.* **2009**, *4*, 1–10. [CrossRef]
50. Banjo, A.D.; Lawal, O.A.; Songonuga, E. The nutritional value of fourteen species of edible insects in southwestern Nigeria. *Afr. J. Biotechnol.* **2006**, *5*, 298–301.

51. Kinyuru, J.N.; Mogendi, J.B.; Riwa, C.A.; Ndung'u, N.W. Edible insects-A novel source of essential nutrients for human diet: Learning from traditional knowledge. *Anim. Front.* **2015**, *5*, 14–19. [CrossRef]

52. Elser, J.J.; Fagan, W.F.; Denno, R.F.; Dobberfuhl, D.R.; Folarin, A.; Huberty, A.F.; Interlandi, S.J.; Kilham, S.S.; McCauley, E.; Schulz, K.; et al. Nutritional constraints in terrestrial and freshwater food webs. *Nature* **2000**, *408*, 578–580. [CrossRef] [PubMed]

53. Kinyuru, J.N.; Kenji, G.M.; Muhoho, S.N.; Ayieko, M. Nutritional potential of longhorn grasshopper (*Ruspolia differens*) consumed in Siaya District, Kenya. *J. Agric. Sci. Technol.* **2009**, *12*, 32–46.

54. Adeduntan, S.A. Nutritonal and Antinutritional Characteristics of Some Insects Foragaing in Akure Forest Reserve Ondo State, Nigeria. *J. Food Technol.* **2005**, *3*, 563–567.

55. Dunkei, F.V. Nutritional value of various insects per 100 grams. *Food Insect Newsl.* **1996**, *9*, 1–8.

56. Omotoso, O. Nutritional quality, functional properties and anti-nutrient compositions of the larva of *Cirina forda* (Westwood) (Lepidoptera: Saturniidae). *J. Zhejiang Univ. Sci. B* **2006**, *7*, 51–55. [CrossRef]

57. Jiang, Y.Y.; Zhao, M.; He, Z.; Wang, C.Y.; Sun, L.; Feng, Y. Nutrition composition and evaluation of six edible dragonfly naiads. *Biot. Resour.* **2017**, *39*, 352–359. [CrossRef]

58. Wang, J.D.; Wang, C.Y.; Zhao, M.; He, Z.; Sun, L.; Feng, Y. Contents of Mercury and Selenium in Common Edible and Medicinal Insects in Yunnan and Their Correlation Analysis. *J. Yunnan Agric. Univ. Nat. Sci.* **2019**, *34*, 1033–1040.

59. Feng, Y.; Chen, X.M.; Wang, S.Y.; Ye, S.D.; Chen, Y. Three Edible Odonata Species and Their Nutritive Value. *For. Res.* **2001**, *14*, 421–424.

60. Feng, Y.; Chen, X.M.; Zhao, M. *Edible Insects of China*; Science Press: Beijing, China, 2016.

61. Ma, Y.K.; Zheng, L.J.; Sun, Z.W.; Fu, B.R. Analysis of nutrient components of three species of aquatic beetles. *Acta Nutr. Sin.* **2002**, *24*, 90–92.

62. Guo, L.Z.; Wang, R.L.; Liang, A.P.; Pan, S.M.; Chen, S.H. Analysis and evaluation of nutritional components of *Hydrous acuminatus*. *Entomol. Knowl.* **2003**, *40*, 366–368.

63. Shi, Y.C.; Ren, Y.; Liu, H.-B.; Gu, J.; Cao, C.Q. Analysis of Main Nutritional Componets in Hellgrammites and Its Anti-Diuretic Effect. *Sci. Technol. Food Ind.* **2019**, *40*, 87–92.

64. Wang, F.B.; Liu, Y.S. Analysis and evaluation of resource components of *Neochauliodes sparsus* larvae. *Chin. J. Appl. Entomol.* **2011**, *48*, 147–151.

65. Ramos-Elorduy, J.; Moreno, J.M.P.; Prado, E.E.; Perez, M.A.; Otero, J.L.; de Guevara, O.L. Nutritional Value of Edible Insects from the State of Oaxaca, Mexico. *J. Food Compos. Anal.* **1997**, *10*, 142–157. [CrossRef]

66. Ghosh, S.; Namin, S.; Meyer-Rochow, V.; Jung, C. Chemical Composition and Nutritional Value of Different Species of *Vespa* Hornets. *Foods* **2021**, *10*, 418. [CrossRef]

67. Zhou, J.; Han, D. Proximate, amino acid and mineral composition of pupae of the silkworm *Antheraea pernyi* in China. *J. Food Compos. Anal.* **2006**, *19*, 850–853. [CrossRef]

68. Kiyashko, S.I.; Imbs, A.B.; Narita, T.; Svetashev, V.I.; Wada, E. Fatty acid composition of aquatic insect larvae *Stictochironomus pictulus* (Diptera: Chironomidae): Evidence of feeding upon methanotrophic bacteria. *Comp. Biochem. Physiol. Part B Biochem. Mol. Biol.* **2004**, *139*, 705–711. [CrossRef]

69. Komínková, D.; Rejmánková, E.; Grieco, J.; Achee, N. Fatty acids in anopheline mosquito larvae and their habitats. *J. Vector Ecol.* **2012**, *37*, 382–395. [CrossRef]

70. He, Z.; Sun, L.; Wang, C.Y.; Feng, Y.; Zhao, M. Nutritional composition analysis and evaluation of the two-spotted cricket *Gryllus bimaculatus* (Orthoptera: Gryllidae). *Biotic Resources* **2021**, *43*, 303–308.

71. Paul, A.; Frederich, M.; Megido, R.C.; Alabi, T.; Malik, P.; Uyttenbroeck, R.; Francis, F.; Blecker, C.; Haubruge, E.; Lognay, G. Insect fatty acids: A comparison of lipids from three Orthopterans and *Tenebrio molitor* L. larvae. *J. Asia Pac. Entomol.* **2017**, *20*, 337–340. [CrossRef]

72. Meyer-Rochow, V.; Gahukar, R.; Ghosh, S.; Jung, C. Chemical Composition, Nutrient Quality and Acceptability of Edible Insects Are Affected by Species, Developmental Stage, Gender, Diet, and Processing Method. *Foods* **2021**, *10*, 1036. [CrossRef] [PubMed]

73. Tomotake, H.; Katagiri, M.; Yamato, M. Silkworm Pupae (Bombyx mori) Are New Sources of High Quality Protein and Lipid. *J. Nutr. Sci. Vitaminol.* **2010**, *56*, 446–448. [CrossRef]

74. Weng, Y.C. A preliminary study on the nutritional components of *Cybister japonicus* sharp. *Fisheries Science & Technology of Guangxi* **1998**, *3*, 16–20.

75. Rumpold, B.A.; Schlüter, O.K. Nutritional composition and safety aspects of edible insects. *Mol. Nutr. Food Res.* **2013**, *57*, 802–823. [CrossRef]

76. Wu, R.A.; Ding, Q.; Yin, L.; Chi, X.; Sun, N.; Chen, R.A.W.; Luo, L.; Ma, H.; Li, Z. Comparison of the nutritional value of mysore thorn borer (*Anoplophora chinensis*) and mealworm larva (*Tenebrio molitor*): Amino acid, fatty acid, and element profiles. *Food Chem.* **2020**, *323*, 126818. [CrossRef] [PubMed]

77. Yue, D.M.; Li, S.Y.; Zhang, J.; Wei, Q.; Zhao, M.N.; Wang, L.M. Analysis on Nutriential Content and Amino Acid Composition of *Antheraea yamamai* Pupa. *Science of Sericulture* **2017**, *43*, 479–485.

78. Bhulaidok, S.; Sihamala, O.; Shen, L. Nutritional and fatty acid profiles of sun-dried edible black ants (*Polyrhachis vicina* Roger). *Maejo Int. J. Sci. Technol.* **2010**, *4*, 101–112.

79. Lesch, V.; Bouwman, H. Adult dragonflies are indicators of environmental metallic elements. *Chemosphere* **2018**, *209*, 654–665. [CrossRef]

80. Abidin, N.Z.; Kormin, F.; Abidin, N.Z.; Anuar, N.M.; Abu Bakar, M. The Potential of Insects as Alternative Sources of Chitin: An Overview on the Chemical Method of Extraction from Various Sources. *Int. J. Mol. Sci.* **2020**, *21*, 4978. [CrossRef] [PubMed]

81. Mohan, K.; Ganesan, A.R.; Muralisankar, T.; Jayakumar, R.; Sathishkumar, P.; Uthayakumar, V.; Chandirasekar, R.; Revathi, N. Recent insights into the extraction, characterization, and bioactivities of chitin and chitosan from insects. *Trends Food Sci. Technol.* **2020**, *105*, 17–42. [CrossRef] [PubMed]

82. Luo, Q.; Wang, Y.; Han, Q.; Ji, L.; Zhang, H.; Fei, Z.; Wang, Y. Comparison of the physicochemical, rheological, and morphologic properties of chitosan from four insects. *Carbohydr. Polym.* **2019**, *209*, 266–275. [CrossRef] [PubMed]

83. Kaya, M.; Baublys, V.; Can, E.; Šatkauskienė, I.; Bitim, B.; Tubelytė, V.; Baran, T. Comparison of physicochemical properties of chitins isolated from an insect (*Melolontha melolontha*) and a crustacean species (*Oniscus asellus*). *Zoomorphology* **2014**, *133*, 285–293. [CrossRef]

84. Soon, C.Y.; Tee, Y.B.; Tan, C.H.; Rosnita, A.T.; Khalina, A. Extraction and physicochemical characterization of chitin and chitosan from Zophobas morio larvae in varying sodium hydroxide concentration. *Int. J. Biol. Macromol.* **2018**, *108*, 135–142. [CrossRef]

85. Kaya, M.; Baran, T.; Mentes, A.; Asaroglu, M.; Sezen, G.; Tozak, K.O. Extraction and Characterization of α-Chitin and Chitosan from Six Different Aquatic Invertebrates. *Food Biophys.* **2014**, *9*, 145–157. [CrossRef]

86. Paulino, A.; Simionato, J.I.; Garcia, J.C.; Nozaki, J. Characterization of chitosan and chitin produced from silkworm crysalides. *Carbohydr. Polym.* **2006**, *64*, 98–103. [CrossRef]

87. Ma, J.; Xin, C.; Tan, C. Preparation, physicochemical and pharmaceutical characterization of chitosan from *Catharsius molossus* residue. *Int. J. Biol. Macromol.* **2015**, *80*, 547–556. [CrossRef]

88. Torres, J.; Garcia, S.R.S.; Lara-Villalón, M.; Ávila, G.C.G.M.; Mora-Olivo, A.; Reyes-Soria, F.A. Evaluation of Biochemical Components from *Pterophylla beltrani* (Bolivar & Bolivar) (Orthoptera: Tettigoniidae): A Forest Pest from Northeastern Mexico. *Southwest. Entomol.* **2015**, *40*, 741–751. [CrossRef]

89. Ibitoye, B.E.; Idris, L.H.; Noor, M.M.; Meng, G.Y.; Bakar, M.A.; Jimoh, A.A. Extraction and physicochemical characterization of chitin and chitosan isolated from House Cricket. *Biomed. Mater.* **2018**, *13*, 025009. [CrossRef]

90. Kaya, M.; Baran, T.; Asan-Ozusaglam, M.; Cakmak, Y.S.; Tozak, K.O.; Mol, A.; Menteş, A.; Sezen, G. Extraction and characterization of chitin and chitosan with antimicrobial and antioxidant activities from cosmopolitan Orthoptera species (Insecta). *Biotechnol. Bioprocess Eng.* **2015**, *20*, 168–179. [CrossRef]

91. Nemtsev, S.V.; Zueva, O.Y.; Khismatullin, M.R.; Albulov, A.I.; Varlamov, V.P. Isolation of Chitin and Chitosan from Honeybees. *Appl. Biochem. Microbiol.* **2004**, *40*, 39–43. [CrossRef]

92. Kim, M.-W.; Han, Y.S.; Jo, Y.H.; Choi, M.H.; Kang, S.H.; Kim, S.-A.; Jung, W.-J. Extraction of chitin and chitosan from housefly, *Musca domestica*, pupa shells. *Entomol. Res.* **2016**, *46*, 324–328. [CrossRef]

93. Khayrova, A.; Lopatin, S.; Varlamov, V. Black Soldier Fly Hermetia illucens as a Novel Source of Chitin and Chitosan. *Int. J. Sci.* **2019**, *8*, 81–86. [CrossRef]

94. Kaya, M.; Akyuz, B.; Bulut, E.; Sargin, I.; Eroglu, F.; Tan, G. Chitosan nanofiber production from Drosophila by electrospinning. *Int. J. Biol. Macromol.* **2016**, *92*, 49–55. [CrossRef] [PubMed]

95. Kim, M.-W.; Song, Y.-S.; Seo, D.-J.; Han, Y.S.; Jo, Y.H.; Noh, M.Y.; Yang, Y.C.; Park, Y.-K.; Kim, S.-A.; Choi, C.; et al. Extraction of Chitin and Chitosan from the Exoskeleton of the Cockroach (*Periplaneta americana* L.). *J. Chitin Chitosan* **2017**, *22*, 76–81. [CrossRef]

96. Nowakowski, A.C.; Miller, A.C.; Miller, M.E.; Xiao, H.; Wu, X. Potential health benefits of edible insects. *Crit. Rev. Food Sci. Nutr.* **2021**, 1–10. [CrossRef]

97. D'Antonio, V.; Serafini, M.; Battista, N. Dietary Modulation of Oxidative Stress From Edible Insects: A Mini-Review. *Front. Nutr.* **2021**, *8*, 642551. [CrossRef]

98. Chakravorty, J.; Ghosh, S.; Meyer-Rochow, V.B. Practices of entomophagy and entomotherapy by members of the Nyishi and Galo tribes, two ethnic groups of the state of Arunachal Pradesh (North-East India). *J. Ethnobiol. Ethnomed.* **2011**, *7*, 5. [CrossRef]

99. Meyer-Rochow, V.B. Therapeutic arthropods and other, largely terrestrial, folk-medicinally important invertebrates: A comparative survey and review. *J. Ethnobiol. Ethnomed.* **2017**, *13*, 9. [CrossRef] [PubMed]

100. Yang, L.F.; Tian, L.X. A brief history of aquatic insect research in China. *Entomol. Knowl.* **1994**, *31*, 308–311.

101. Feng, Y.; Zhao, M.; He, Z.; Chen, Z.; Sun, L. Research and utilization of medicinal insects in China. *Entomol. Res.* **2009**, *39*, 313–316. [CrossRef]

102. Zhu, L.C. *Applications of Insect Medicine*; People's Medical Publishing House: Beijing, China, 2012.

103. Jiang, Y.X. Sympetrum infuscatum as a medicinal dragonfly species in Heilongjiang. *Forest By-Prod. Spec. China* **2004**, *4*, 29–30.

104. Dossey, A.T. Insects and their chemical weaponry: New potential for drug discovery. *Nat. Prod. Rep.* **2010**, *27*, 1737–1757. [CrossRef] [PubMed]

105. Di Mattia, C.; Battista, N.; Sacchetti, G.; Serafini, M. Antioxidant Activities in vitro of Water and Liposoluble Extracts Obtained by Different Species of Edible Insects and Invertebrates. *Front. Nutr.* **2019**, *6*, 106. [CrossRef] [PubMed]

106. Zielińska, E.; Baraniak, B.; Karaś, M. Antioxidant and Anti-Inflammatory Activities of Hydrolysates and Peptide Fractions Obtained by Enzymatic Hydrolysis of Selected Heat-Treated Edible Insects. *Nutrients* **2017**, *9*, 970. [CrossRef]

107. Rinehart, K.L.; Holt, T.G.; Fregeau, N.L.; Keifer, P.A.; Wilson, G.R.; Perun, T.J., Jr.; Sakai, R.; Thompson, A.G.; Stroh, J.G.; Shield, L.S.; et al. Bioactive Compounds from Aquatic and Terrestrial Sources. *J. Nat. Prod.* **1990**, *53*, 771–792. [CrossRef] [PubMed]

108. Feng, Y.; Chen, X.M.; Ma, Y.; He, Z. Experimental study on immunomodulation of white wax scale (*Ericerus pela* Chavannes). *For. Res.* **2006**, *19*, 221–224.

109. Bridges, A.R.; Owen, M.D. The morphology of the honey bee (*Apis mellifera* L.) venom gland and reservoir. *J. Morphol.* **1984**, *181*, 69–86. [CrossRef]

110. Jiang, M.; Lü, S.; Zhang, Y. The Potential Organ Involved in Cantharidin Biosynthesis in *Epicauta chinensis* Laporte (Coleoptera: Meloidae). *J. Insect Sci.* **2017**, *17*, 52. [CrossRef]

111. Mozhui, L.; Kakati, L.N.; Meyer-Rochow, V.B. Entomotherapy: A study of medicinal insects of seven ethnic groups in Nagaland, North-East India. *J. Ethnobiol. Ethnomed.* **2021**, *17*, 17. [CrossRef]

112. Devakul, V.; Maarse, H. A second compound in the odorous gland liquid of the giant water bug *Lethocerus indicus* (Lep. and Serv.). *Anal. Biochem.* **1964**, *7*, 269–274. [CrossRef]

113. Kiatbenjakul, P.; Intarapichet, K.-O.; Cadwallader, K.R. Characterization of potent odorants in male giant water bug (*Lethocerus indicus* Lep. and Serv.), an important edible insect of Southeast Asia. *Food Chem.* **2015**, *168*, 639–647. [CrossRef] [PubMed]

114. Yang, Z.-L.; Nour-Eldin, H.H.; Hänniger, S.; Reichelt, M.; Crocoll, C.; Seitz, F.; Vogel, H.; Beran, F. Sugar transporters enable a leaf beetle to accumulate plant defense compounds. *Nat. Commun.* **2021**, *12*, 2658. [CrossRef]

115. Kazana, E.; Pope, T.W.; Tibbles, L.; Bridges, M.; Pickett, J.A.; Bones, A.; Powell, G.; Rossiter, J.T. The cabbage aphid: A walking mustard oil bomb. *Proc. R. Soc. B Boil. Sci.* **2007**, *274*, 2271–2277. [CrossRef] [PubMed]

116. Kumar, P.; Pandit, S.S.; Steppuhn, A.; Baldwin, I.T. Natural history-driven, plant-mediated RNAi-based study reveals CYP6B46's role in a nicotine-mediated antipredator herbivore defense. *Proc. Natl. Acad. Sci. USA* **2013**, *111*, 1245–1252. [CrossRef]

117. Heidari, H.; Azizi, Y.; Maleki-Ravasan, N.; Tahghighi, A.; Khalaj, A.; Pourhamzeh, M. Nature's gifts to medicine: The metabolic effects of extracts from cocoons of *Larinus hedenborgi* (Coleoptera: Curculionidae) and their host plant *Echinops cephalotes* (Asteraceae) in diabetic rats. *J. Ethnopharmacol.* **2021**, *284*, 114762. [CrossRef] [PubMed]

118. Nakane, W.; Nakamura, H.; Nakazato, T.; Kaminaga, N.; Nakano, M.; Sakamoto, T.; Nishiko, M.; Bono, H.; Ogiwara, I.; Kitano, Y.; et al. Construction of TUATinsecta database that integrated plant and insect database for screening phytophagous insect metabolic products with medicinal potential. *Sci. Rep.* **2020**, *10*, 17509. [CrossRef]

119. Lidman, J.; Jonsson, M.; Berglund, M. The effect of lead (Pb) and zinc (Zn) contamination on aquatic insect community composition and metamorphosis. *Sci. Total Environ.* **2020**, *734*, 139406. [CrossRef]

120. Scheibener, S.; Conley, J.M.; Buchwalter, D. Sulfate transport kinetics and toxicity are modulated by sodium in aquatic insects. *Aquat. Toxicol.* **2017**, *190*, 62–69. [CrossRef]

121. Neff, E.; Dharmarajan, G. The direct and indirect effects of copper on vector-borne disease dynamics. *Environ. Pollut.* **2020**, *269*, 116213. [CrossRef] [PubMed]

122. Chaves-Ulloa, R.; Taylor, B.W.; Broadley, H.; Cottingham, K.L.; Baer, N.A.; Weathers, K.C.; Ewing, H.A.; Chen, C.Y. Dissolved organic carbon modulates mercury concentrations in insect subsidies from streams to terrestrial consumers. *Ecol. Appl.* **2016**, *26*, 1771–1784. [CrossRef] [PubMed]

123. Hare, L. Aquatic Insects and Trace Metals: Bioavailability, Bioaccumulation, and Toxicity. *Crit. Rev. Toxicol.* **1992**, *22*, 327–369. [CrossRef] [PubMed]

124. Aydoğan, Z.; Şişman, T.; Incekara, Ü.; Gürol, A. Heavy metal accumulation in some aquatic insects (Coleoptera: Hydrophilidae) and tissues of Chondrostoma regium (Heckel, 1843) relevant to their concentration in water and sediments from Karasu River, Erzurum, Turkey. *Environ. Sci. Pollut. Res.* **2017**, *24*, 9566–9574. [CrossRef] [PubMed]

125. Cain, D.J.; Luoma, S.N.; Wallace, W.G. Linking metal bioaccumulation of aquatic insects to their distribution patterns in a mining-impacted river. *Environ. Toxicol. Chem.* **2004**, *23*, 1463–1473. [CrossRef]

126. Eisler, R. Arsenic hazards to humans, plants, and animals from gold mining. *Rev. Environ. Contam. Toxicol.* **2004**, *180*, 133–165. [CrossRef]

127. Hall, B.D.; Rosenberg, D.M.; Wiens, A.P. Methyl mercury in aquatic insects from an experimental reservoir. *Can. J. Fish. Aquat. Sci.* **1998**, *55*, 2036–2047. [CrossRef]

128. Cristol, D.A.; Brasso, R.L.; Condon, A.M.; Fovargue, R.E.; Friedman, S.L.; Hallinger, K.K.; Monroe, A.P.; White, A.E. The Movement of Aquatic Mercury Through Terrestrial Food Webs. *Science* **2008**, *320*, 335. [CrossRef] [PubMed]

129. Raikow, D.F.; Walters, D.M.; Fritz, K.M.; Mills, M.A. The distance that contaminated aquatic subsidies extend into lake riparian zones. *Ecol. Appl.* **2011**, *21*, 983–990. [CrossRef] [PubMed]

130. Kraus, J.M.; Schmidt, T.; Walters, D.M.; Wanty, R.B.; Zuellig, R.E.; Wolf, R.E. Cross-ecosystem impacts of stream pollution reduce resource and contaminant flux to riparian food webs. *Ecol. Appl.* **2014**, *24*, 235–243. [CrossRef]

131. Wang, J.D.; Wang, C.Y.; Zhao, M.; Feng, Y. Mercury absorption, enrichment and water-land transfer by aquatic insects. *Biot. Resour.* **2018**, *40*, 507–511. [CrossRef]

132. Cetinić, K.A.; Previšić, A.; Rožman, M. Holo- and hemimetabolism of aquatic insects: Implications for a differential cross-ecosystem flux of metals. *Environ. Pollut.* **2021**, *277*, 116798. [CrossRef]

133. Abu Hassan, A.; Dieng, H.; Satho, T.; Boots, M.; Al Sariy, J.S. Breeding patterns of the JE vector Culex gelidus and its insect predators in rice cultivation areas of northern peninsular Malaysia. *Trop. Biomed.* **2010**, *27*, 404–416. [PubMed]

134. Marsollier, L.; Robert, R.; Aubry, J.; André, J.-P.S.; Kouakou, H.; Legras, P.; Manceau, A.-L.; Mahaza, C.; Carbonnelle, B. Aquatic Insects as a Vector for Mycobacterium ulcerans. *Appl. Environ. Microbiol.* **2002**, *68*, 4623–4628. [CrossRef] [PubMed]

135. Bektaş, M.; Orhan, F.; Erman, Ö.K.; Barış, Ö. Bacterial microbiota on digestive structure of Cybister lateralimarginalis torquatus (Fischer von Waldheim, 1829) (Dytiscidae: Coleoptera). *Arch. Microbiol.* **2020**, *203*, 635–641. [CrossRef] [PubMed]

136. Bruus, M.; Rasmussen, J.J.; Strandberg, M.; Strandberg, B.; Sørensen, P.B.; Larsen, S.E.; Kjær, C.; Lorenz, S.; Wiberg-Larsen, P. Terrestrial adult stages of freshwater insects are sensitive to insecticides. *Chemosphere* **2019**, *239*, 124799. [CrossRef]
137. Maloney, E.; Taillebois, E.; Gilles, N.; Morrissey, C.; Liber, K.; Servent, D.; Thany, S. Binding properties to nicotinic acetylcholine receptors can explain differential toxicity of neonicotinoid insecticides in Chironomidae. *Aquat. Toxicol.* **2020**, *230*, 105701. [CrossRef] [PubMed]
138. Jinguji, H.; Ohtsu, K.; Ueda, T.; Goka, K. Effects of short-term, sublethal fipronil and its metabolite on dragonfly feeding activity. *PLoS ONE* **2018**, *13*, e0200299. [CrossRef]
139. Fukuuchi, T.; Yasuda, M.; Inazawa, K.; Ota, T.; Yamaoka, N.; Mawatari, K.-I.; Nakagomi, K.; Kaneko, K. A Simple HPLC Method for Determining the Purine Content of Beer and Beer-like Alcoholic Beverages. *Anal. Sci.* **2013**, *29*, 511–517. [CrossRef] [PubMed]
140. Bednářová, M.; Borkovcová, M.; Komprda, T. Purine derivate content and amino acid profile in larval stages of three edible insects. *J. Sci. Food Agric.* **2013**, *94*, 71–76. [CrossRef]
141. Kaneko, K.; Aoyagi, Y.; Fukuuchi, T.; Inazawa, K.; Yamaoka, N. Total Purine and Purine Base Content of Common Foodstuffs for Facilitating Nutritional Therapy for Gout and Hyperuricemia. *Biol. Pharm. Bull.* **2014**, *37*, 709–721. [CrossRef]
142. Sabolová, M.; Kulma, M.; Kouřimská, L. Sex-dependent differences in purine and uric acid contents of selected edible insects. *J. Food Compos. Anal.* **2020**, *96*, 103746. [CrossRef]
143. He, Z.; Zhao, M.; Wang, C.; Sun, L.; Jiang, Y.; Feng, Y. Purine and uric acid contents of common edible insects in Southwest China. *J. Insects Food Feed* **2019**, *5*, 293–299. [CrossRef]
144. Dow, J.A.T. Excretion and salt and water regulation. In *The Insects: Structure and Function*, 5th ed.; Douglas, A.E., Chapman, R.F., Simpson, S.J., Eds.; Cambridge University Press: Cambridge, UK, 2012; pp. 546–587. [CrossRef]
145. Jeong, K.Y.; Park, J.-W. Insect Allergens on the Dining Table. *Curr. Protein Pept. Sci.* **2020**, *21*, 159–169. [CrossRef] [PubMed]
146. De Gier, S.; Verhoeckx, K. Insect (food) allergy and allergens. *Mol. Immunol.* **2018**, *100*, 82–106. [CrossRef]
147. Payne, C.L.R.; Scarborough, P.; Rayner, M.; Nonaka, K. Are edible insects more or less 'healthy' than commonly consumed meats? A comparison using two nutrient profiling models developed to combat over- and undernutrition. *Eur. J. Clin. Nutr.* **2015**, *70*, 285–291. [CrossRef] [PubMed]
148. Kolakowska, A.; Olley, J.N.; Dunstan, G.A. Fish lipids. In *Chemical and Functional Properties of Food Lipids*; Sikorski, Z.E., Kolakowska, A., Eds.; CRC Press: New Work, NY, USA, 2002; pp. 1–388.
149. Wang, D.; Zhai, S.-W.; Zhang, C.-X.; Zhang, Q.; Chen, H. Nutrition value of the Chinese grasshopper *Acrida cinerea* (Thunberg) for broilers. *Anim. Feed Sci. Technol.* **2007**, *135*, 66–74. [CrossRef]
150. Twining, C.W.; Brenna, J.T.; Hairston, N.G.; Flecker, A.S. Highly unsaturated fatty acids in nature: What we know and what we need to learn. *Oikos* **2015**, *125*, 749–760. [CrossRef]
151. Honeyfield, D.C.; Maloney, K.O. Seasonal patterns in stream periphyton fatty acids and community benthic algal composition in six high-quality headwater streams. *Hydrobiologia* **2014**, *744*, 35–47. [CrossRef]
152. Dai, Z.H.; Liao, B.H.; Wang, Z.H.; Wang, X.J.; Liu, Y.X. A preliminary study on leaching of selenium in the soils of china. *J. Environ. Sci.* **1995**, 338–345.
153. Lobanov, A.V.; Fomenko, D.E.; Zhang, Y.; Sengupta, A.; Hatfield, D.L.; Gladyshev, V.N. Evolutionary dynamics of eukaryotic selenoproteomes: Large selenoproteomes may associate with aquatic life and small with terrestrial life. *Genome Biol.* **2007**, *8*, R198. [CrossRef] [PubMed]
154. Lobanov, A.V.; Hatfield, D.L.; Gladyshev, V.N. Selenoproteinless animals: Selenophosphate synthetase SPS1 functions in a pathway unrelated to selenocysteine biosynthesis. *Protein Sci.* **2007**, *17*, 176–182. [CrossRef] [PubMed]
155. Ramos-Elorduy, J. Threatened edible insects in Hidalgo, Mexico and some measures to preserve them. *J. Ethnobiol. Ethnomed.* **2006**, *2*, 51. [CrossRef]
156. Durst, P.; Hanboonsong, Y. Small-scale production of edible insects for enhanced food security and rural livelihoods: Experience from Thailand and Lao People's Democratic Republic. *J. Insects Food Feed* **2015**, *1*, 25–31. [CrossRef]
157. Oonincx, D.G.A.B.; van der Poel, A.F.B. Effects of diet on the chemical composition of migratory locusts (*Locusta migratoria*). *Zoo Biol.* **2010**, *30*, 9–16. [CrossRef] [PubMed]
158. Lancaster, J.; Downes, B.J. Aquatic versus terrestrial insects: Real or presumed differences in population dynamics? *Insects* **2018**, *9*, 157. [CrossRef] [PubMed]
159. Shaw, P.-C.; Mark, K.-K. Chironomid farming—A means of recycling farm manure and potentially reducing water pollution in Hong Kong. *Aquaculture* **1980**, *21*, 155–163. [CrossRef]
160. Yokoyama, A.; Hamaguchi, K.; Ohtsu, K.; Ishihara, S.; Kobara, Y.; Horio, T.; Endo, S. A method for mass-rearing caddisfly, *Cheumatopsyche brevilineata* (Iwata) (Trichoptera: Hydropsychidae), as a new test organism for assessing the impact of insecticides on riverine insects. *Appl. Entomol. Zool.* **2009**, *44*, 195–201. [CrossRef]
161. Locklin, J.L.; Huckabee, J.S.; Gering, E.J. A Method for Rearing Large Quantities of the Damselfly, *Ischnura ramburii* (Odonata: Coenagrionidae), in the Laboratory. *Fla. Entomol.* **2012**, *95*, 273–277. [CrossRef]
162. Wang, X.L.; Wang, T.F.; Su, X.D.; Niu, D.M. Study on the techniques of artificial breeding predacious diving beetle. *J. Baicheng Norm. Univ.* **2019**, *33*, 1–4.
163. Wen, Y.C.; Zhang, T.L. A tentative researche to the large aquatic insect in two classes-predacious diving beetles (Dytiscidae) and giant water burs *Lethocerus indicus* (Lepeletier et Serville). *J. Zhanjiang Ocean. Univ.* **1993**, *13*, 22–27.

164. Van Huis, A.; Oonincx, D.G.A.B. The environmental sustainability of insects as food and feed. A review. *Agron. Sustain. Dev.* **2017**, *37*, 43. [CrossRef]
165. Bektas, M.; Guler, O. Usage of edible aquatic insects for feed rations of poultry animals. *Int. J. Sci. Technol. Res.* **2019**, *5*, 70–80. [CrossRef]
166. Henry, M.; Gasco, L.; Piccolo, G.; Fountoulaki, E. Review on the use of insects in the diet of farmed fish: Past and future. *Anim. Feed Sci. Technol.* **2015**, *203*, 1–22. [CrossRef]

 *foods*

Article

Evaluation of the Quality Characteristics and Development of a Puffed-Rice Snack Enriched with Honeybee (*Apis mellifera* L.) Drone Pupae Powder

Woo-Hee Cho [1,†], Jung-Min Park [2,†], Eun-Ji Kim [2], Md. Mohibbullah [1,3] and Jae-Suk Choi [1,4,*]

[1] Seafood Research Center, IACF, Silla University, 606, Advanced Seafood Processing Complex, Wonyang-ro, Amnam-dong, Seo-gu, Busan 49277, Korea; ftrnd3@silla.ac.kr (W.-H.C.); mmohib.fpht@sau.edu.bd (M.M.)
[2] With Food Corp., 204, Busan Inno-Biz Center, Mandeok 3-ro 16beon-gil, Buk-gu, Busan 46570, Korea; pjm_kdessert@naver.com (J.-M.P.); 1abmiracle@naver.com (E.-J.K.)
[3] Department of Fishing and Post Harvest Technology, Sher-e-Bangla Agricultural University, Sher-e-Bangla Nagar, Dhaka 1207, Bangladesh
[4] Department of Food Biotechnology, Silla University, 140, Baegyang-daero 700beon-gil, Sasang-gu, Busan 46958, Korea
* Correspondence: jsc1008@silla.ac.kr; Tel.: +82-51-248-7789
† These authors contributed equally to this work.

Abstract: Edible insect ingredients have gained importance as environmental-friendly energy sources world-wide; the honeybee (*Apis mellifera* L.) drone pupae has gained prominence as a nutritional material. In this study, bee drone pupae were processed under different heating and drying conditions and incorporated into a puffed-rice snack with honey. The sensory, physicochemical, nutritional and microbial qualities of drone pupae powders were tested. The deep-fried and hot-air dried powder was selected; the values of 5.54% (powder) and 2.13% (honey) were obtained on optimization with honey by response surface methodology. Subsequently, the puffed-rice snack product enriched with drone pupae powder was stored at different temperatures for 180 days. The prepared product showed a higher content of proteins, fats, amino acids, and fatty acids compared to the control. The high content of a few minerals were maintained in the processed powder and the product, whereas heavy metals were not detected. The storage test indicated acceptable sensory qualities and safety results, considering important quality parameters. Thus, drone pupae powder and the developed product can be consumed as nutritional food materials; the quality characteristics can be improved through optimal processing.

Keywords: puffed-rice snack; drone pupae; quality characteristics; *Apis mellifera* L.; nutritional profile

Citation: Cho, W.-H.; Park, J.-M.; Kim, E.-J.; Mohibbullah, M.; Choi, J.-S. Evaluation of the Quality Characteristics and Development of a Puffed-Rice Snack Enriched with Honeybee (*Apis mellifera* L.) Drone Pupae Powder. *Foods* **2022**, *11*, 1599. https://doi.org/10.3390/foods11111599

Academic Editors: Chuleui Jung and Victor Benno Meyer-Rochow

Received: 3 May 2022
Accepted: 27 May 2022
Published: 28 May 2022

Publisher's Note: MDPI stays neutral with regard to jurisdictional claims in published maps and institutional affiliations.

1. Introduction

Honeybee (*Apis mellifera* L.) drone pupae are rich sources of protein and other essential nutrients (such as carbohydrates, fats, amino acids, minerals, and vitamins) and can be a valuable food ingredient [1]. Bee drone pupae are used as nutritious and high-protein food in Korea, Japan, China, the United States, and many European countries, and can be added to various dishes (such as, soups, confectioneries, and baked goods) [2]. To popularize drone pupae powder as a food material, the RDA (Rural Development Administration) of South Korea has selected 11 drone pupae powder-added food items (including cookies, sesame tapioca bread, chocolates, and shakes) and published their recipe book [3]. Furthermore, a jelly-type energy supplement prepared using drone pupae powder is already in existence, while several studies confirm the application of ground insects in familiar edible items (such as, bread, biscuits, and so on) [4,5]. Additionally, it is used as a raw material for chocolate, confectionery, alcoholic beverages, and functional foods [2,6,7].

The Food and Agriculture Organization (FAO) has published reports on the nutritional value, collection, storage, quality control, and use of bee larvae, pupae, and adult insects in food, medicine, and cosmetics (Krell, 1996) [8]. Honeybee (*Apis mellifera* L.) drone pupae were registered as a food ingredient in Korea by the MFDS (Ministry of Food Drug Safety) in July 2020 [9].

To promote the human consumption of honeybee drone pupae, in addition to the development of novel promising functional foods and pharmaceuticals, the direct application of insect powder in food is under consideration. Various snack products enriched with insects, such as, grasshoppers (*Sphenarium purpurascens* Ch.) [10], migratory locusts (*Locusta migratoria*) [11], lesser mealworms (*Alphitobius diaperinus*) [12] and yellow mealworms (*Tenebrio molitor*) have been developed recently [13]. These insects, being rich in protein, increase the protein content of grain-based snacks. There are no detailed reports on grain-based snacks enriched with honeybee (*Apis mellifera* L.) drone pupae.

Grain-based snacks incorporating insects mainly use wheat [13], maize (*Zea mays* L.) [10], and chickpea [11]. Wheat, a protein-rich resource, is widely used in the food industry because of its low price, convenience of procurement, easy processibility, etc. However, gluten-free alternatives are required for people with wheat-gluten allergies [14]. Rice is rapidly gaining popularity as an alternative to wheat; soft, low-allergenic, and easily digestible puffed-rice snacks are particularly popular [15].

In this study, the best drone pupae processing method was identified, a puffed-rice snack enriched with honeybee (*Apis mellifera* L.) drone pupae powder was developed, and its quality characteristics were evaluated after a storage test.

2. Materials and Methods

2.1. Materials

Honeybee (*Apis mellifera* L.) drone pupae (20 days old) were purchased from the Cho-Won beekeeping farm (Pohang, Gyeongsangbuk-do, Korea). The honeycomb containing the collected bee drone pupae was rapidly frozen using a deep-freezer (DF3503S, ilShinBioBase Co., Ltd., Dongducheon, Gyeonggi-do, Korea) and stored at −20 °C until experimentation. Puffed rice (Woori food Co., Anseong, Gyeonggi-do, Korea) was used for the development of the snack product (prepared at about 250 °C). Soybean oil, white sugar, and starch syrup were purchased from the CJ Cheiljedang Corporation (Seoul, Korea). Organic solvents and indicators of the highest grade were used for the physicochemical analysis (Sigma-Aldrich Co. LLC., St. Louis, MO, USA).

2.2. Drone Pupae Preparation and Pre-Treatment for Processing to Powder

The processing flow shown in Figure 1 was used to process drone pupae powder and develop the rice snack. The frozen drone pupae were individually picked up from the hive and thawed at room temperature (RT) within 10 min, immediately after being taken out of the freezer. After the removal of water with a paper towel (Kimtech Science Wipers, Yuhan-Kimberly, Seoul, Korea), the thawed drone pupae were used in the next process.

Heating-drying was used for processing to powder form. Three groups of samples underwent deep-frying with oil (150 °C, 12 min), stir-frying without oil (180 °C, 3 min), or non-heating. The heated sample groups were treated by hot-air drying and freeze-drying. A hot-air drying machine (P-IOV864, Labmate, Gumi, Gyeongsangbuk-do, Korea) was used at 70 °C for 14 h, while a freeze-dryer (Lyoph-prido 20R, ilShinBioBase Co., Ltd., Dongducheon, Gyeonggi-do, Korea) was used for two days until the sample dried up to about 4.5% (*w/w*) of its moisture content.

The dried drone pupae were pulverized in a grinder (HMF-3100S, Hanil Electric Co., Ltd., Seoul, Korea) for about 10 min and then filtered using a 7 mm sieve. The prepared drone pupae powder was sealed in a zipper pack and stored in a freezer (DF3503S, ilShinBioBase Co., Ltd. Dongducheon, Gyeonggi-do, Korea) at −35 °C or less until further experimentation.

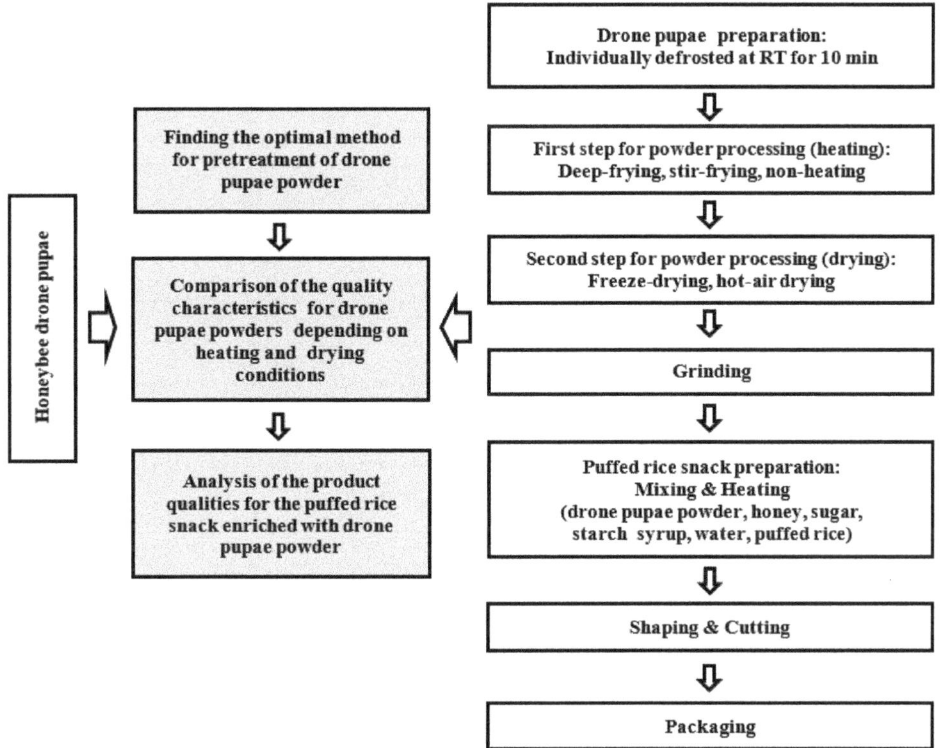

Figure 1. Schematic design of the drone pupae powder and the developed product research (RT; room temperature; 15–20 °C).

2.3. Development of the Puffed-Rice Snack Enriched with Drone Pupae Powder

To prepare the syrup base, 50 g of starch syrup, sugar, and water were mixed together in a pot and boiled over low heat until the weight of the syrup was reduced to 100 g. After turning off the heat, puffed rice (50 g) was added and stirred with the syrup base to ensure smooth mixing. Subsequently, drone pupae powder and honey were mixed into the puffed-rice mixture in different concentrations according to the central composite design (CCD) and response surface methodology (RSM). Finally, the puffed-rice mixture was placed in a rectangular stainless-steel tray, flattened by a rolling stick to a 2.5 cm-thick sheet, and cooled (at RT) for 10 min. The puffed-rice snack enriched with drone pupae powder was cut into 3.5 cm (H) × 3.5 cm (W) × 2.5 cm (T) pieces and stored in a zipper bag at RT.

2.4. Optimization of the Mixing Conditions for Drone Pupae Powder and Honey with Puffed Rice

Based on the descriptive sensory test, the mixing conditions (%, w/v) for drone pupae powder and honey with puffed rice were optimized using the response surface methodology (RSM). Table 1 shows the mixing conditions of the prepared materials in terms of the central composition using different variables and code values (-1.414, -1, 0, $+1$, $+1.414$). The concentrations of drone pupae powder (X_1) and honey (X_2) with a base rice snack were set as independent variables (factors), while the descriptive sensory terms for nutty aroma (Y_1) and sweetness taste (Y_2) were applied as dependent variables (responses) using a 10-point strong level scale. With this experimental design, a comparison using the descriptive sensory analysis was tested with a series of snack samples. MINITAB 18

(Minitab Inc., State College, PA, USA) was used for the model calculation of each response, using the following Equation (1):

$$Y = \beta_0 + \beta_1 X_1 + \beta_2 X_2 + \beta_3 X_1^2 + \beta_4 X_2^2 + \beta_5 X_1 X_2 \tag{1}$$

Table 1. Factors and coded levels for the central composite design of processing the puffed-rice snack.

Independent Variables	Symbol	Unit	Range Level				
			−1.414	−1	0	+1	+1.414
Drone pupae powder	X_1	%	2.172	3.0	5.0	7.0	7.828
Honey	X_2	%	0.586	1.0	2.0	3.0	4.414

The formula contains the coefficients for regression β_0 (intercept), $\beta_{1,2}$ (linear), $\beta_{3,4}$ (quadratic), and β_5 (interaction). The final model was reduced in terms of significance (R^2) by eliminating the insignificant terms. To optimize with these models, Minitab ver. 19.0 software (Minitab Inc, State College, PA, USA) was used for an analysis of variance (ANOVA) and experimental validation of the computed conditions.

2.5. Yield

The yield of the drone pupae powder was calculated by dividing the weight of the dried powder by that of the control drone pupae (frozen; 700 g). Weights were recorded on an electronic scale (WH-1A, Wessglobal, Co., Ltd., Seoul, Korea) after drying the pretreated drone pupae powders in an oven (VS-1202D3, Vision Scientific Co., Ltd., Bucheon, Korea).

2.6. Color

The method of Kang et al. [16] was used to measure the color values of the drone pupae powders, using a color difference meter (CM-700d Spectrophotometer; Konica Minolta Sensing Inc., Tokyo, Japan). Hunter system values, including L (lightness), a (redness), and b (yellowness) were monitored using SpectraMagic software (version 2.11; Minolta Cyber Chrome Inc. Osaka, Japan). The overall color difference (ΔE) was made between treatment groups of drone pupae and calculated using the following formula: $\Delta E = ((\Delta L)^2 + (\Delta a)^2 + (\Delta b)^2)^{1/2}$.

2.7. Odor Intensity

The odor intensity was measured according to the method of Kang et al. [17]. Each prepared drone pupae powder (5 g) was placed in a 50 mL conical tube (SPL Life Science Co., Ltd., Pocheon, Gyeonggi-do, Korea) with a control group (the frozen drone pupae). The suction port of the odor-concentration meter (XP-329R, New Cosmos Electric Co., Ltd., Osaka, Japan) was inserted into the conical tube and sealed with parafilm to enable odor retention. Subsequently, the odor intensity of each drone pupae powder was measured; the measurement mode of the odor concentration meter was set to batch, with a unit of odor intensity (VCI; Volatile Component Intensity).

2.8. pH Value

A pH meter (ST3100, Ohaus, Parsippany, NJ, USA) was used to measure the pH after adding 25 mL of distilled water to 5 g of the sample and mixing. Each sample was independently measured three times and expressed as an average pH value.

2.9. Acid Value

The acid value was analyzed according to the Food Code [18]. Extracted lipid (5 g) from the sample was taken in an Erlenmeyer flask (1000 mL), followed by the addition of 100 mL of ethanol and ether-mixed solution (1:2, v/v). After adding 1% phenolphthalein solution as indicator, the solution was titrated using 0.1 N ethanolic potassium hydroxide until a pale red color persisted for 30 s.

2.10. Volatile Basic Nitrogen (VBN)

The volatile basic nitrogen (VBN) content was measured by the micro-diffusion method using the Conway unit mentioned in the Food Code (MFDS, 2021) [18].

2.11. Thiobarbituric Acid Reactive Substance (TBARS)

Thiobarbituric acid reactive substance (TBARS) values were measured by the method of Buege and Aust [19]. After adding 12.5 mL of 20% trichloroacetic acid (TCA) in 2 M phosphoric acid to 5 g of each sample, and homogenization at 1000 rpm for 1 min using a homogenizer (SHG-15D, SciLab Co., Ltd., Seoul, Korea), the volumes were adjusted to 25 mL each using distilled water. The homogenates were centrifuged at $1600 \times g$ for 15 min at 4 °C using a centrifuge (Labogene 416, Gyrozen Co., Ltd., Gimpo-si, Korea). Subsequently, the supernatants were filtered with filter paper (70 mm, Toyo Roshi Kaisha Ltd., Tokyo, Japan); 2 mL of each filtrate was mixed with 0.005 M thiobarbituric acid (TBA) solution, heated to 95 °C for 30 min in a water bath (JSWB-22TL, JS Research Inc., Gongju-si, Korea), and cooled to room temperature. The quantification of TBARS involved measuring absorbance at 530 nm using a microplate reader (SPECTRO star-nano, BMG Labtech, Ortenberg, Germany).

2.12. Sensory Evaluation

The sensory evaluation was performed for different purposes by two different methods: the optimizing process (descriptive analysis) and controlling quality (affective analysis). To establish optimal conditions, eight trained panelists (from Woori Corp, Pyongtaek, Korea), equipped to provide detailed information about bee-processing products participated in the descriptive tests. Each descriptive term (darkness color, brownness color, nutty aroma, sweetness taste, burnt taste, bitter taste, sticky texture, rough texture) was evaluated by a 10-point strong level scale according to the glossary and standard scores shown in Table 2. After the storage test, a 9-point hedonic scale (1: I hate it very much, 9: I like it very much) evaluation (appearance, odor, taste, texture, overall acceptance) was carried out by 21 semi-trained panelists belonging to the Silla University Industry-Academic Cooperation Foundation.

The former method is in accordance with a standard descriptive analysis in the International Organization for Standardization (ISO 13299: 2016) [20], while the latter follows the guidelines of the shelf-life test in the MFDS. These sensory tests have been approved by the Institutional Review Board of Silla University (IRB approval number; 1041449-202108-HR-002).

2.13. Total Amino-Acid Content

The total amino-acid content was measured according to a method (protocol 994.12) of the AOAC (Association of Official Analytical Chemists) [21]. One gram of the sample was placed in a test tube and mixed with 15 mL of 6 N HCl solution. Subsequently, the test sample was sealed under reduced pressure and acid hydrolyzed at 110 °C for 24 h in a drying oven (VS-1202D3, Vision Scientific Co., Ltd., Bucheon, Korea). The decomposed solution was filtered with a glass filter, and the filtrate was concentrated under reduced pressure at 55 °C to completely evaporate HCl and water using a rotary vacuum evaporator (EYELA N-1000, Riakikai Co., Ltd., Tokyo, Japan). The concentrated sample was aliquoted with a sodium citrate buffer (pH 2.2) and filtered by a 0.45 μm-thick membrane filter. The total amino-acid content was measured using an automatic amino acid analyzer (Biochrom 20, Pharmacia Biotech Ltd., Cambridge, UK).

Table 2. The glossary of standards and their scores, defining the quality characteristics used in the sensory analysis.

Sensory Quality	Descriptive Terms	Definition	Standard and Scores
Appearance	Darkness color	Low illumination and absorbs light, such as black and brown.	*White sugar = 0, Grey oyster mushroom = 4, Black sesame = 10*
	Brownness color	Orange of low brightness and saturation.	*Curry powder = 3, Chocolate = 6, Americano coffee = 10*
Aroma	Nutty aroma	Containing, smelling of, or similar to nuts.	*Water = 0, Cheese powder = 6, Parched cereal powder = 8, Sesame oil = 10*
Taste	Sweetness taste [1]	Perceived when eating foods rich in sugars	*Whipped cream = 5, Chocolate = 8, Honey = 10*
	Burnt taste	Overwhelmingly bitter and unpleasantly overshadowed by acridness.	*Water = 0, Grilled meat surface = 5, Over extracted-espresso = 10*
	Bitter taste	Sharp, pungent, or disagreeable flavor.	*Lettuce = 3, Grapefruit = 7, Ginseng = 10*
Texture	Sticky texture	Tending to hold like glue.	*Tofu (okara) powder = 5, Ketchup = 7, Caramel = 10*
	Rough texture	Uneven surface and not smooth.	*Flour = 1, White sugar = 6, Bread crumbs = 10*

[1] Sweetness taste was used in the optimization of drone pupae powder and honey on processing the puffed-rice snack. The standards written in Italic font were used for the targeted scores in RSM.

2.14. Fatty-Acid Composition

The fatty-acid composition was analyzed by the following protocol. First, lipids were extracted from the samples by the method of Bligh and Dyer [22]. Subsequently, after methyl esterification of fatty acids according to a method of the AOCS (American Oil Chemists' Society) [23,24], the fatty-acid composition was analyzed using a gas chromatography instrument (Shimadzu GC-2010; carrier gas, He; detector, FID) equipped with a capillary column (Supelcowax-10 fused silica wall-coated open tubular column, 30 m × 0.25 mm Id, Supelco Japan Ltd. Tokyo, Japan). The injector and detector (FID) temperatures were set to 250 °C, respectively, while the column temperature was increased to 230 °C and maintained for 15 min. He (1.0 kg/cm^2) was used as the carrier gas with a 1:50 split ratio. The fatty-acid composition was determined by comparing the recorded retention times with those of standard fatty acids (Applied Science Lab. Co., Baldwin Park, CA, USA).

2.15. Analysis of Nutrition and Minerals

According to the standard test methods (protocol 925.09, 923.03, 979.09, 962.09, and 923.05) of AOAC [25], 14 nutrients (calories, sodium, carbohydrate, sugar, dietary fiber, crude fat, trans fat, saturated fat, cholesterol, crude protein, vitamin D, potassium, iron, and calcium) of the puffed-rice snack enriched with drone pupae powder were analyzed. The minerals (Na, Ca, K, Fe, P, Mg, Zn, and Cu) of the selected drone pupae powder were analyzed using a method of the Food Code (MFDS, 2021) using inductively coupled plasma (ICP; Optima 5300 DV; Perkin Elmer, Waltham, MA, USA) [18].

2.16. Heavy Metals

Four heavy metals (lead, cadmium, mercury, and arsenic) were analyzed according to a method of the Food Code [26]. Each sample (10 g) was heated in a microwave oven (MARS 6, CEM Corporation, Matthews, NC, USA) at 450 °C until completely carbonized, and then homogenized using a homogenizer (SHG-15D, SciLab Co., Ltd., Seoul, Korea). Subsequently, they were dissolved in 5 N nitric acid solutions to obtain at least 20 mL of experimental solution for each sample. The concentrations (mg/kg) of lead, cadmium, and arsenic were measured by inductively coupled plasma (ICP; Optima 5300 DV; Perkin Elmer, Waltham, MA, USA) in triplicates, whereas 1 g of the sample was analyzed using a mercury analyzer (MA-2; Nippon Instruments Corporation, Tokyo, Japan) to determine the mercury concentration.

2.17. Microbial Analysis

The total bacteria count was measured using methods (protocol 990.12 and 991.14) of the AOAC (2002) [27]. The sample (10 g) was collected in a sterile pack (3M Co., Ltd., Saint Paul, MN, USA), mixed with 900 mL of 0.85% NaCl sterilized saline, and inoculated on a 3M dry film after ten-times dilution. The number of colonies were counted after culturing at 35 ± 1 °C for 24–48 h; the valid number of colonies (showing a red dot with gas) were counted to detect *E. coli* and the total coliform group.

2.18. Moisture Content

The moisture content of the samples were measured using a method of the Food Code [18]. Each sample (5 g) was dried at 105 °C for 24 h using a drying oven, cooled in a desiccator for 30 min, and weighed. The moisture content (w/w % of wet basis) was calculated using the following equation: Moisture (%) = ((g of initial weight)—(g of final weight))/(g of initial weight) × 100.

2.19. Statistical Analysis

All experiments (except the optimization by response surface methodology) were performed thrice and the measured values were statistically analyzed by one-way ANOVA along with *t*-tests using the SPSS version 23.0 (IBM Corp., Armonk, NY, USA). The significant difference was evaluated at *p*-values < 0.05.

3. Results and Discussion

3.1. Establishment of Optimal Processing Conditions for Drone Pupae Powder

To develop the puffed-rice snack enriched with drone pupae powder, the heating and drying conditions were first tested for processing from drone pupae to powder. Through a comparison of the quality characteristics of the treated drone pupae powders, the optimal processing conditions for their pre-treatment were determined. Here, different heating methods (stir-frying without oil, deep-frying with oil, and unheated condition) were used, followed by drying (hot-air drying, freeze-drying), for three groups of drone pupae powder. Finally, six groups of heated and dried drone pupae were pulverized, and the different quality characteristics (sensory characteristics, yield, color, odor intensity, acid value, pH, TBARS, and VBN) were evaluated.

3.1.1. Sensory Evaluation Results of Drone Pupae Powders Using the Descriptive Test

Table 3 shows the results of the descriptive analysis for color, aroma, taste, and texture of the drone pupae powders. The color parameter was evaluated considering darkness and brownness. The highest darkness levels were scored by the DFD (8.00) and DHD (7.38) groups, with an insignificant difference (*p* < 0.05); statistically, the DHD (8.88) and DFD (8.13) groups exhibited the highest brownness values, while the UFD (1.50) and SFD (2.38) groups exhibited the lowest values. Both DHD (7.38) and DFD (8.00) exhibited the strongest points for nutty-aroma, with an insignificant difference. On heating protein compounds in food to the temperature range of 130–140 °C, different colors and flavors are obtained by the Maillard reaction, which cause browning and impart a nutty smell to food (Lin et al., 2009) [28].

In this study, the sensory panelists also scored on the burnt and bitter taste to account for any changes in surface and chemical conditions due to the heating treatments. The UHD (2.63), UFD (2.13) and SFD (2.38) groups showed significantly lower burnt tastes than the other three powders (*p* < 0.05). The highest bitterness (taste) was observed in the UFD (4.38) and SFD (4.25) groups, which were significantly higher than the other powders. Each powder, treated by a different combination of frying (deep frying, stir-frying, control) and drying (freeze-drying, hot-air drying) conditions, exhibited different texture characteristics (sticky, rough). The DHD (6.25) and DFD (5.13) groups showed the highest values for sticky texture. Thus, cooking oil (used for deep-frying) considerably influenced sticky texture, while the rough texture was also attributed to deep-frying ($p < 0.05$).

Table 3. Sensory evaluation results by the descriptive test on drone pupae powders processed using different methods.

Descriptive Terms	UHD [1]	SHD [2]	DHD [3]	UFD [4]	SFD [5]	DFD [6]
Darkness color	6.25 ± 0.46 [b]	5.75 ± 0.46 [b]	7.38 ± 0.74 [a]	2.75 ± 0.46 [c]	3.00 ± 0.76 [c]	8.00 ± 0.76 [a]
Brownness color	5.63 ± 0.74 [b]	3.50 ± 0.53 [c]	8.88 ± 0.64 [a]	1.50 ± 0.74 [d]	2.38 ± 0.52 [d]	8.13 ± 0.99 [a]
Nutty aroma	6.13 ± 0.64 [b]	6.25 ± 0.46 [b]	7.38 ± 0.52 [a]	4.75 ± 0.89 [c]	5.63 ± 0.52 [bc]	8.00 ± 0.76 [a]
Burnt taste	2.63 ± 0.92 [b]	3.88 ± 0.35 [a]	4.50 ± 0.76 [a]	2.13 ± 0.35 [b]	2.38 ± 0.52 [b]	4.50 ± 0.76 [a]
Bitter taste	2.13 ± 0.64 [b]	2.00 ± 0.93 [b]	2.63 ± 0.92 [b]	4.38 ± 0.74 [a]	4.25 ± 0.71 [a]	2.63 ± 0.52 [b]
Sticky texture	2.63 ± 0.52 [c]	2.75 ± 0.71 [c]	6.25 ± 0.89 [a]	1.25 ± 0.46 [d]	1.13 ± 0.35 [d]	5.13 ± 0.64 [b]
Rough texture	2.50 ± 0.53 [b]	2.63 ± 0.92 [b]	5.25 ± 0.71 [a]	2.75 ± 1.04 [b]	3.63 ± 0.52 [b]	6.00 ± 0.76 [a]

Values are mean ± standard deviation. Different letters (a–d) in each column indicate significant differences among means by the Tukey's test ($p < 0.05$). [1] Unheated and hot air-dried drone pupae powder. [2] Stir-fried and hot air-dried drone pupae powder. [3] Deep fried and hot air-dried drone pupae powder. [4] Unheated and freeze-dried drone pupae powder. [5] Stir-fried and freeze-dried drone pupae powder. [6] Deep fried and freeze-dried drone pupae powder.

3.1.2. Yields of Drone Pupae Powders

The yield values for the drone pupae, considering the heating and drying method, are shown in Table 4. The post pre-treatment weight and corresponding yields for all the powders were: DFD (222 g; 31.71%), DHD (206 g; 29.43%), UFD (172 g; 24.57%), SFD (167 g; 23.86%), UHD (156 g; 22.29%), and SHD (150 g; 21.43%). Considering the statistical differences, DFD (31.71%) and UHD (22.29%) exhibited the highest and lowest yields, respectively ($p < 0.05$). Overall, the deep-fried samples exhibited higher yields than the other groups; this could be attributed to the presence of some frying oil portion and its degraded polar compounds absorbed by food during deep-frying [29].

Table 4. Yield results of the drone pupae powders processed using different methods.

Powders	Control Drone Pupae (Frozen)	UHD	SHD	DHD	UFD	SFD	DFD
Weight (g)	700	156 ± 2.65 [d]	150 ± 3.61 [d]	206 ± 2.65 [b]	172 ± 2.00 [c]	167 ± 1.00 [c]	222 ± 2.65 [a]
Yield (w/w %)	-	22.29 ± 0.38 [d]	21.43 ± 0.52 [d]	29.43 ± 0.38 [b]	24.57 ± 0.29 [c]	23.86 ± 0.14 [c]	31.71 ± 0.38 [a]

Values are mean ± SD. Different letters (a–d) in each row indicate significant differences among the means by the Tukey's test ($p < 0.05$). Dash (-) indicates not performed.

3.1.3. Instrumental Sensory Characteristic Results of Drone Pupae Powders

The color values (L, a, b, and ΔE) and odor-intensity values of the instrumental sensory characteristics of variously treated drone pupae powders are shown in Table 5, while the actual sample pictures of the powders are shown in Figure 2. All the measured color values showed significant differences with each other ($p < 0.05$). After the control drone pupae (56.84), the UFD group exhibited the highest L value (41.28). Additionally, the ΔE (21.87) for UFD was closest to that of the control (32.92), indicating high similarity with the original color. It has been reported in a previous study (Hwang and Kim, 2020) [30] that a similarly treated (unheated and freeze-dried) silkworm powder shows a higher L value than other differently treated powder samples. The a-values (indicating red color) for the freeze-dried drone pupae powders (UFD, 2.26; SFD, 2.96; DFD, 5.82) were lower than those of the hot-air dried powders (UHD, 8.83; SHD, 7.16; DHD, 8.21), regardless of the heating conditions (unheated, stir-frying, deep-frying).

Table 5. The color and odor intensity of drone pupae powders processed using different treatment methods.

Powders	Color				Odor Intensity (VCI)
	L	a	b	ΔE	
Control drone pupae	56.84 ± 0.67 [a]	3.31 ± 0.11 [e]	25.55 ± 0.45 [a]	32.92 ± 0.73 [a]	349.67 ± 2.5 [d]
UHD	27.36 ± 0.22 [d]	8.83 ± 0.08 [a]	19.37 ± 0.10 [b]	17.76 ± 0.12 [c]	382.33 ± 1.53 [b]
SHD	26.42 ± 0.09 [e]	7.16 ± 0.02 [c]	17.63 ± 0.04 [c]	16.33 ± 0.04 [e]	365.67 ± 1.53 [c]
DHD	15.09 ± 0.05 [g]	8.21 ± 0.06 [b]	12.03 ± 0.06 [e]	12.91 ± 0.03 [g]	376.00 ± 2.00 [bc]
UFD	41.28 ± 0.03 [b]	2.26 ± 0.01 [g]	12.02 ± 0.01 [e]	21.87 ± 0.02 [b]	453.67 ± 3.06 [a]
SFD	33.62 ± 0.02 [c]	2.96 ± 0.02 [f]	12.88 ± 0.02 [d]	16.90 ± 0.01 [d]	336.00 ± 8.72 [e]
DFD	21.07 ± 0.07 [f]	5.82 ± 0.04 [d]	11.41 ± 0.04 [f]	13.19 ± 0.14 [f]	302.67 ± 0.58 [f]

Values are mean ± SD. Different letters (a–g) in each column indicate significant differences among the means by the Tukey's test ($p < 0.05$). VCI indicates the volatile component intensity.

Figure 2. Sample groups of drone pupae (*Apis mellifera* L.) powder processed under different conditions: (**i**) Control drone pupae; (**ii**) UHD; (**iii**) SHD; (**iv**) DHD; (**v**) UFD; (**vi**) SFD; (**vii**) DFD.

The results were consistent with the colors observed in actual photographs; UHD, SHD, and DHD exhibited some red color while UFD, SFD and DFD appeared mainly achromatic (white, grey, and black). The appearance of non-achromatic colors (such as brown or red) with high a-values could be attributed to melanoidin, chemically derived by the Maillard reaction. Similar to the trend of a-value distribution, the b-values (indicating yellowness) of the hot-air dried samples (UHD, 19.37; SHD, 17.63; DHD, 12.03) were higher than those of the freeze-dried ones (UFD, 12.02; DFD, 11.41), except for SFD (12.88).

The treated drone pupae powders exhibited odor intensities in the range of 300–400 VCI, except UFD (453.67). The unheated powders (UHD, UFD) showed the highest odor-intensity values (382.33 and 453.67, respectively) in their groups. Thus, the additional heat treatments of stir-frying and deep-frying could have a higher effect on decreasing odor intensity levels than non-heating.

3.1.4. Physicochemical Quality Characteristics of Drone Pupae Powders

Figure 3 shows the physicochemical quality characteristics (acid value, pH, VBN, TBARS) of the drone pupae powders and the control (freeze-thawed raw drone pupae), depending on the combination of the three heating conditions and two drying methods used for processing.

As shown in Figure 3a, the acid values of UHD (3.79 mg/g), SHD (3.59 mg/g), DHD (2.57 mg/g), UFD (4.08 mg/g), SFD (3.9 mg/g), and DFD (3.1 mg/g) were higher than that of the control drone pupae (2.29 mg/g). Herein, the control (freeze-thawed raw drone pupae) was measured using the raw drone pupae (2.92 mg/g) according to a previously published study [31]. The high content of unsaturated fatty acids in the drone pupae indicated that the heating and drying processes affected lipid oxidation. However, the acid values of the six drone pupae powders were within the acceptable limit (5.0 mg/g) for processed insect foods in the Food Code [18], indicating their applicability as food material. Furthermore, samples that underwent less oil treatment exhibited higher acid values, regardless of the drying method. However, no studies on drone pupae processed by different frying methods have been reported so far. Thus, as reported in a study by Andrikopoulos et al. [32] on potato processing, pan-frying was assumed to increase the acid value to a greater extent than deep-frying.

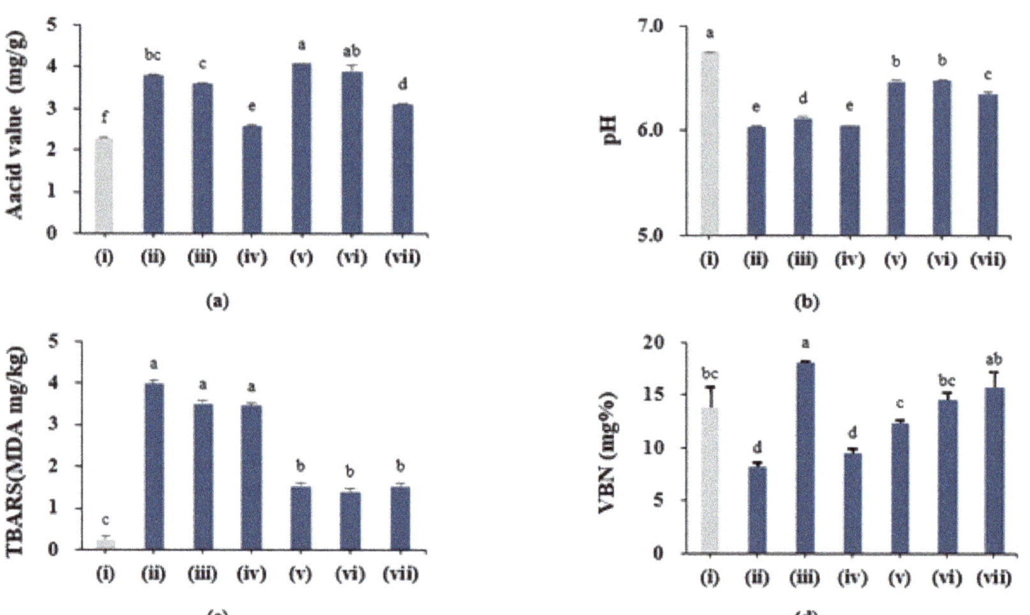

Figure 3. The chemical quality parameters ((**a**), acid value; (**b**), pH; (**c**), TBARS; (**d**), VBN) of the drone pupae powders depending on the pretreatment conditions. (i) The control drone pupae; (ii) UHD; (iii) SHD; (iv) DHD; (v) UFD; (vi) SFD; (vii) DFD. Values are mean ± SD. Different letters (a–f) in each column indicate significant differences among the means by the Tukey's test ($p < 0.05$).

The hydrogen-ion concentrations (pH values) for the six drone pupae powders are shown in Figure 3b; all the powders exhibited lower values (6.04–6.48) than that of the control drone pupae (6.75). Overall, the average pH value of the hot air-dried powders (UHD, SHD, DHD; 6.07) were lower than that of the freeze-dried powders (UFD, SFD, DFD; 6.43), and the difference between the two groups was statistically significant ($p < 0.05$). Here, this could be due to a difference in the drying temperature of the two groups; hot-air drying increased the acidity levels to a greater extent.

The TBARS value is an indicator of the degree of lipid oxidation, showing the intensity of malonaldehyde (MDA) produced by the oxidation of fat and thio-barbituric acid [33]. As shown in Figure 3c, the levels of MDA for UHD (3.99), SHD (3.50), and DHD (3.45) were higher than those for UFD (1.54), SFD (1.41), and DFD (1.55), respectively, with significant differences ($p < 0.05$). Thus, the freeze-drying treatment exhibited better stability, as indicated by lipid oxidation, than the hot-air drying treatment, in the processing of drone pupae powder. In the present study, the moisture levels of the hot-air dried powders were remarkably decreased due to their long drying time (14 h) at 70 °C, which could increase the MDA levels and make the lipids unstable [34]. Al-Kahtani et al. [35] have reported that the acceptable limit of TBARS for meat products is 3.0 MDA mg/kg, and lipids do not undergo significant oxidation under this level. Although three hot-air dried drone pupae powders (UHD, SHD, DHD) showed higher TBARS values than the suggested limit, they did not exhibit a bitter taste or unnatural odor due to lipid oxidation, as reported by the sensory panelists, indicating edible-food quality.

Volatile basic nitrogen (VBN) is a general term for volatile amines (such as, ammonia, nitrogen, and trimethylamine) and is an indicator of the decomposition quality of proteins [36]. As the drone pupae has a high protein content, the level of decomposition could be an indicator of its protein-content quality. Figure 3d shows the VBN values of six drone pupae powders with the control. The highest and lowest VBN values were measured in SHD (18.08 mg/100 g) and UHD (8.17 mg%), respectively. Additionally, due to the high value for SHD, the variation of VBN did not exhibit a regular pattern. Fresh meat exhibits VBN values in the range of 5–10 mg%, while a value higher than 30 mg% indicates spoilage, according to Jin et al. [37]. Notably, these recommendations are applicable for general food types, and a different specific standard should be used while considering insect-based foods. No standard for insect-based foods and materials has been suggested to date; however, the VBN values of the drone pupae powders were within the general-food acceptable limits.

In this study, the acid and VBN values were used to determine the optimal heating and drying conditions for processing drone pupae to powder, as they indicate lipid oxidation and protein-content quality. DHD showed the lowest acid value (2.57 mg/g) and VBN level (9.57 mg%). Therefore, the optimal heating and drying conditions for processing drone pupae were selected to be deep-frying and hot-air drying (DHD), and further experiments were conducted using the DHD powder.

3.2. Analysis of Microbial Quality Characteristics and Moisture Content of the Optimal Drone Pupae Powder

Table 6 summarizes the total bacteria count (TBC), *E. coli*, total coliform group (TCG), and moisture content of the optimal drone pupae powder (DHD) and the control drone pupae. For both samples, the level of TBC did not exceed the recommended value (5 log CFU/g) suggested in the Food Code, while *E. coli* and TCG were not detected. The moisture content of the control material was 51.9%, which was about 11 times higher than that of the optimal drone pupae powder (4.54%) ($p < 0.05$). Kim et al. [38] have reported that the freeze-dried powder of drone pupae (16–20 days old) shows 8.79% moisture content; this is higher than the value obtained for DHD here. This could be attributed to the higher effectiveness of the freeze-drying method compared to hot-air drying; furthermore, deep-frying increases the evaporation rate. The moisture content of the final product (the puffed-rice snack enriched with drone pupae powder) and control snack were 10% and 6.3%, respectively, as indicated by further experimentation.

Table 6. The microbial qualities and moisture content of the control drone pupae and optimal drone pupae powder.

Samples	Total Bacteria Count (Log CFU/g)	*E. coli* and (CFU/g)	Total Coliform Group (CFU/g)	Moisture Content (*w/w* % Wet Basis)
Control drone pupae	2.56 ± 0.29 [a]	-	-	51.9 ± 0.10 [a]
Drone pupae powder (DHD)	2.30 ± 0.01 [a]	-	-	4.54 ± 0.02 [b]

Values are mean ± SD. Different letters (a,b) in each column indicate significant differences among the means by the *t*-test ($p < 0.05$). Dash (-) indicates not identified.

3.3. The Nutritional Composition Results of the Optimal Drone Pupae Powder

3.3.1. Amino Acids

The amino-acid composition of the optimal drone pupae powder (DHD) is summarized in Table 7, along with the recommended daily requirement of essential amino acids according to the World Health Organization (WHO) and the MOHW (Korea). Considering essential amino acids (EAAs), with 15.65 g/100 g of the total EAA, aromatic amino acids (AAAs; tyrosine and phenylalanine) (3.08 g/100 g) were mainly present, followed by leucine (2.89 g/100 g), lysine (2.19 g/100 g), and valine (2.14 g/100 g). The total non-essential amino acids (NAAs) accounted for 18.07 g/100 g (a value that is higher than the sum of the EAAs); glutamic acid (5.43 g/100 g) and aspartate (3.40 g/100 g) were predominantly present.

Table 7. The amino-acid composition of the optimal drone pupae powder.

Amino Acids	The Experimental Sample (g/100 g)			Recommended Daily Requirement	
	Amount	% WHO	% MOHW	WHO [1] (g/70 kg Body Weight)	MOHW [2] (g of 19–29 Ages/Male)
Histidine	0.92	87.6	92.0	1.05	1.0
Isoleucine	1.86	88.6	143.1	2.10	1.3
Leucine	2.89	70.0	96.3	4.13	3.0
Lysine	2.19	69.5	73.0	3.15	3.0
SAA [3]	1.04	67.5	80.0	1.54	1.3
AAA [4]	3.08	115.8	110.0	2.66	2.8
Threonine	1.28	79.5	91.4	1.61	1.4
Tryptophan	0.25	6.0	83.3	4.20	0.3
Valine	2.14	78.4	164.6	2.73	1.3
∑ EAA [5]	15.65	67.5	101.6	23.17	15.4
Aspartate	3.40				
Serine	1.18				
Glutamic	5.43				
Proline	2.20				
Glycine	2.06				
Alanine	2.08				
Arginine	1.72				
∑ NAA [6]	18.07				

[1] According to the Protein and Amino Acid Requirements in Human Nutrition (WHO, 2007). [2] According to the Dietary Reference Intakes for Koreans (Ministry of Health and Welfare, MOHW; 2020). [3] SAA: sulfur amino acids (cysteine, methionine). [4] AAA: aromatic amino acids (tyrosine, phenylalanine). [5] EAA: essential amino acids. [6] NAA: nonessential amino acids.

Thus, the EAA value for the optimal drone pupae powder accounted for 67.5% and 101.6% of the recommended daily requirement according to WHO and MOHW, respectively. The percentages of BCAA (branched-chain amino acid), such as, isoleucine, leucine, and valine, were significantly higher while considering the recommendations of the MOHW than those of WHO. Overall, the optimal drone pupae powder exhibited a lower amino-acid content than freeze-dried drone pupae powders reported in previous studies [39,40], while it exhibited similar values as an oven-dried sample reported in a previous study [41].

3.3.2. Fatty Acids

Table 8 summarizes the fatty-acid composition of the optimal drone pupae powder (DHD) according to a percentage of total fatty acids. Among the saturated fatty acids (32.65%, *w/w*), palmitic acid (24.22%, *w/w*) occurred predominantly, while oleic acid (32.19% *w/w*) and linoleic acid (31.14%, *w/w*) mainly accounted for the total unsaturated fatty acid (UFA) content of 67.34% (*w/w*). These results are different from those reported in previous studies [39,41], wherein SFA exhibits a higher percentage than UFA. In this study, the tested drone pupae powder was prepared using deep-frying, and could be affected by the high linoleic acid in soybean oil [42]. According to the NNR (Nordic Nutrition Recommendations, 2012) [43], the UFA percentage is recommended to be above two thirds of the total fatty-acid content; the optimal drone pupae powder used here met this criterion.

Table 8. Fatty-acid compositions of the optimal drone pupae powder and other references for bee powder (% of total fatty acids).

Fatty Acids	The Experimental Sample	Ghosh et al. [41]	Kim et al. [39]
	Deep Fried and Hot Air-Dried	Oven-Dried	Freeze-Dried (−70~85 °C, 72 h)
	20 Days Old	21 Days within	16–20 Days Old
Caprylic acid	0.02	-	-
Capric acid	0.01	-	-
Lauric acid	0.09	0.4	0.64
Myristic acid	1.21	2.9	4.64
Pentadecanoic acid	0.03	-	-
Palmitic acid	24.22	35.1	35.49
Magaric acid	0.05	-	-
Stearic acid	6.79	12.6	14.46
Arachidic acid	0.21	-	0.74
Heneicosylic acid	-	-	-
Behenic acid	-	-	0.22
Tricosanoic acid	-	-	2.17
Lignoceric acid	0.02	-	1.26
∑ SFA	32.65	51.1	59.62
Myristoleic acid	0.02	-	-
Pentadecenoic acid	-	-	-
Palmitoleic acid	0.35	0.6	1.13
Magaoleic acid	0.05		-
Oleic acid	32.19	47.6	35.91
Eicosenoic acid	0.09	0.8	0.17
Eicosadienoic acid	0.01	-	-
Erucic acid	0.01	-	-
∑ MUFA	32.72	48.9	37.20
Linoleic acid	31.14	-	1.14
γ-Linolenic acid	-	-	-
Dihomo γ-Linolenic acid	0.02	-	-
Arachidonic acid	0.04	-	-
Docosadienoic acid	-	-	-
∑ n-6	31.2	-	1.14
Linolenic acid	3.34	-	2.04
Eicosatrienoic acid	-	-	-
Eicosapentaenoic acid	0.08	-	-
Docosapentaenoic acid	-	-	-
Docosahexaenoic acid	-	-	-
∑ n-3	3.42	-	2.04
∑ PUFA	34.62	-	3.18
∑ UFA	67.34	48.9	40.38
Total fatty acid (%)	100.0	100.0	100.0

Dash (-) means not measured/suggested.

3.3.3. Minerals and Heavy Metals

Table 9 summarizes the results of mineral and heavy-metal analysis for the optimal drone pupae powder (DHD) according to the recommendations of the MFDS and international standards. The drone pupae powder showed the following values for the eight tested minerals: sodium (41.90 mg), calcium (0.03 g), potassium (1.04 g), iron (5 mg), phosphorus (0.57 g), magnesium (0.07 g), zinc (4.44 mg), and copper (1.19 mg). According to Kim et al. [44], these minerals are not obtained by themselves and are stored in honey and pollen.

Table 9. Analysis of the mineral and heavy-metal composition of the optimal drone pupae powder.

Minerals and Heavy Metals	Unit	Result	Amount % and Recommendations				Remark
			%	MFDS	%	International	
Sodium	mg/100 g	41.90	2.1	<2000 [1]	1.8	<2300 [2]	Following
Calcium	g/100 g	0.03	4.3	>0.7 [1]	2.3	>1.3 [2]	the standard
Potassium	g/100 g	1.04	29.4	>3.5 [1]	22.1	>4.7 [2]	of nutrition
Iron	mg/100 g	5	41.67	>12 [1]	62.5	>8 [2]	facts
Phosphorus	g/100 g	0.57	81.4	>0.7 [3]	81.4	>0.7 [4]	Following
Magnesium	g/100 g	0.07	19.4	>0.36 [3]	16.7	>0.42 [4]	the standard
Zinc	mg/100 g	4.44	44.4	>10 [3]	40.4	>11 [4]	of minerals
Copper	mg/100 g	1.19	12.5	<9.5 [3]	11.9	<10 [4]	
Lead	mg/kg	0.01	10	<0.1 [5]	10	<0.1 [6]	Following
Cadmium	mg/kg	0.01	10	<0.1 [5]	20	<0.05 [7]	the standard
Mercury	mg/kg	<0.01	-	-	<2	<0.5 [8]	of heavy
Arsenic	mg/kg	0.01	10	<0.1 [5]	10	<0.1 [6]	metals

[1] According to Nutrition Facts Labeling Requirements, US Food and Drug Administration. Industry Resources on the Changes to the Nutrition Facts Label. [2] According to Article 6 (Nutrition Facts Label) in Act on Labeling and Advertising of Foods (MFDS, 2020). [3] According to the Dietary Reference Intakes for Koreans (Ministry of Health and Welfare, MOHW; 2020). [4] According to the Dietary Reference Intakes (FDA, 2020). [5] According to the Food material recognition of MFDS (No. 2020-5). [6] According to the recommendation for metal amount in food (FDA, 2020). [7] According to the maximum limit for cadmium (European Community, 2005). [8] According to the maximum limit for mercury (Health Canada, 2019).

In this study, four minerals (sodium, calcium, potassium, iron) were evaluated by standard nutrition facts while general mineral compounds were used to estimate the other four (phosphorus, magnesium, zinc, copper). Here, phosphorus exhibited the highest percentage (0.57 g; 81.4%) according to the recommended amount (>0.7 g) by the MFDS and international standards. The amounts of heavy metals were within the acceptable limits according to the MFDS and international recommendations: lead (0.01 mg), cadmium (0.01 mg), mercury (0.01 mg or less), and arsenic (0.01 mg).

3.4. Processing Optimization of the Puffed-Rice Snack Product Using Optimal Drone Pupae Powder and Honey by RSM

For the central composite design (CCD) of response surface methodology, the expected concentration (zero point in Table 1) of the optimal drone pupae powder was selected as the result of the preliminary experiment in the middle of study; this confirmed the acceptable color, smell, and taste depending on the mixed concentrations (1, 2, 3, 4, 5, 6, 7, 8, 9, and 10%) of drone pupae powder with the product, as shown in Figure 4. Subsequently, 5% of drone pupae powder was determined as the zero point for CCD, followed by optimization.

Figure 4. The puffed-rice snacks with different concentrations (*w/w*%) of drone pupae powder: (**a**) Control; (**b**) 1%; (**c**) 2%; (**d**) 3%; (**e**) 4%; (**f**) 5%; (**g**) 6%; (**h**) 7%; (**i**) 8%; (**j**) 9%; (**k**) 10%.

Table 10 summarizes the experimental results for the responses Y_1 and Y_2 for the optimization of the puffed-rice snack prepared by processing with different concentrations of drone pupae powder (X_1) and honey (X_2). According to these responses, the regression formulae for the response surface model were computed as shown:

$$Y_1 \text{ (nutty aroma)} = 7.963 - 0.487\,X_1 - 0639\,X_2 - 0.169\,X_1\,X_1 - 0.200\,X_2\,X_2$$

$$Y_2 \text{ (sweetness taste)} = 8.003 - 0.3018\,X_1 + 0.5711\,X_2 + 0.001\,X_1\,X_1 - 0.89\,X_2\,X_2$$

Table 10. Experimental results for central composite design.

Run No.	Coded Values		Actual Values		Responses	
	X_1	X_2	Drone Pupae Powder (X_1)	Honey (X_2)	Nutty Aroma (Y_1)	Sweetness Taste (Y_2)
1	−1	−1	3	1	8.00	7.88
2	+1	−1	7	1	9.13	7.13
3	−1	+1	3	3	6.13	8.50
4	+1	+1	7	3	7.13	8.25
5	−1.414	0	2.17	2	7.00	8.38
6	+1.414	0	7.83	2	8.25	7.38
7	0	−1.414	5	0.59	8.00	6.50
8	0	+1.414	5	4.41	7.13	8.50
9	0	0	5	2	8.13	8.00
10	0	0	5	2	7.88	8.13
11	0	0	5	2	7.88	7.88

The statistical correlations for both the models (Y_1 and Y_2) were satisfied with the standard *p*-values (<0.05), and they showed high R^2 values, as shown in Table 11. Lack of fits for the respective models were insignificant (showing *p*-values higher than 0.05), indicating that the models were organized using suitable correlations. Therefore, the exported models could be used to derive optimal conditions for the targeted response qualities (which are written in bold in Table 1).

Table 11. Analysis of variance for responses on the optimization of mixing conditions for drone pupae powder and honey with a puffed-rice snack: X_1, drone pupae powder; X_2, honey; Y_1, nutty aroma; Y_2, sweetness taste.

Responses	R^2	Lack of Fit (>0.05)	*p*-Values (<0.05)			
			Model	Linear (X_1, X_2)	Quadratic (X_1X_1, X_2X_2)	Interaction (X_1X_2)
Y_1 (nutty aroma)	0.854	0.087	>0.011	0.004, 0.013	>0.526, 0.821	-
Y_2 (sweetness taste)	0.902	0.155	>0.001	0.015, 0.001	>0.970, 0.165	-

Dash (-) means not identified.

As shown in Figure 5a,b, the nutty aroma (Y1) increased on increasing the percentage of drone pupae powder (X_1), while mixing a higher concentration of honey (X_2) increased the sweetness taste (Y_2). Normally, honey used for food processing has a high effect on sweetness due to its high level of fructose (about 75–80%), which is a strong sweetener for human beings (Aparna and Rajalakshmi, 2009) [45]. Furthermore, in this study, adding high concentrations of insect powder (such as drone pupae) in food increased the nutty scent, as the preparatory processes using heat caused the Maillard reaction (Jensen et al., 2016) [46]. The sensory panelists reported a strong nutty aroma and sweet flavor in the puffed-rice snack with a high percentage of drone pupae powder and honey.

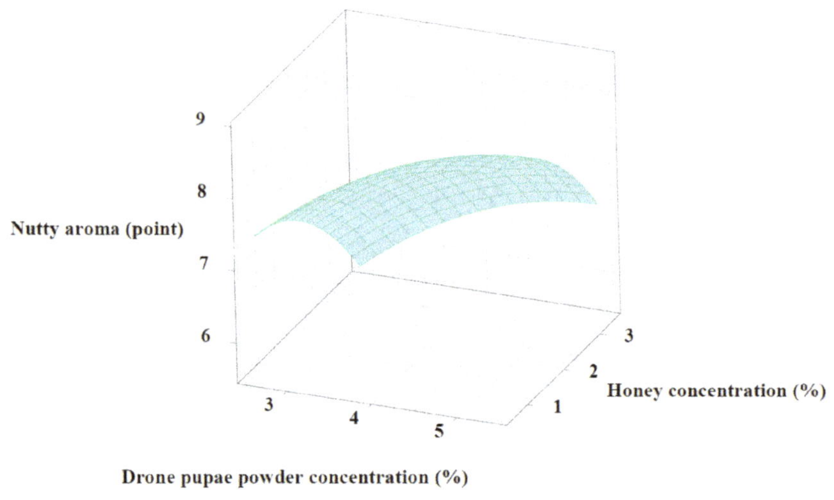

(a)

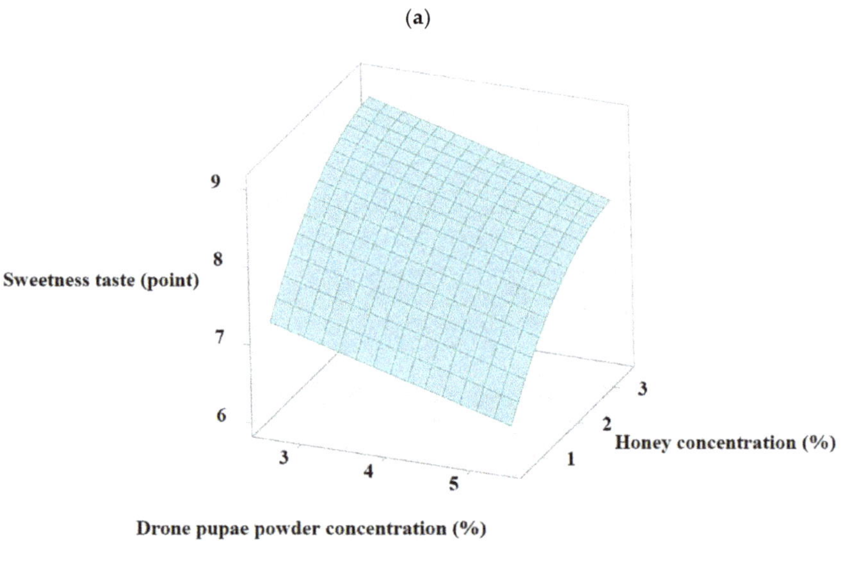

(b)

Figure 5. Three-dimensional response surface plots of the puffed-rice snack with respect to drone pupae powder and honey concentrations: (**a**) Nutty aroma; (**b**) Sweetness taste.

Table 12 summarizes the optimal mixing condition for drone pupae powder and honey with the puffed-rice snack, with the expected response values from references, along with the validation results. According to Table 2, the targeted nutty aroma was the level of parched cereal powder (8 point), while chocolate (8 point) was set as the goal score for sweetness taste. Based on model optimization, the predicted responses were 7.992 (nutty aroma) and 7.997 (sweetness taste) on using 5.54% (drone pupae powder) and 2.13% (honey), showing good desirability values. In accordance with the RSM process, these conditions were practically validated with additional experimentation, whereby nutty aroma and sweetness taste were rated at 8.19 and 8.43, respectively (almost equivalent to the predicted values). Thus, this condition was determined as the final mixing concentration and was applied in the subsequent steps.

Table 12. Optimization of the mixing conditions for drone pupae powder and honey with a puffed-rice snack by RSM.

Responses	Optimal Conditions		Predicted Values	Experimental Values	Desirability
	X_1 (%)	X_2 (%)			
Y_1 (nutty aroma)	+0.271 (5.54%)	+0.129 (2.13%)	7.992	8.19 ± 0.40	0.9714
Y_2 (sweetness taste)			7.997	8.43 ± 0.51	0.9183

3.5. Product Quality Characteristics of the Puffed-Rice Snack Enriched with Drone Pupae Powder

3.5.1. Nutrition

Table 13 shows the contents of 14 nutrients in the control and drone-pupae-powder enriched puffed-rice snack product as percentages of daily intake recommendations according to the FDA (USA) and MFDS (Korea). The protein (4.45 g) and fat (4.60 g) content of the puffed-rice snack enriched with drone pupae powder was higher than those of the control (2.36 g and 1.06 g, respectively). Thus, the addition of drone pupae powder to food increased its protein and fat content; this is in agreement with a previous study by Biró et al. [47] in which the insect-enriched snack shows increasing protein and fat values with increasing insect concentration. Furthermore, the developed snack product exhibited a high cholesterol value (17.16 mg) due to its high fat content. Considering the suggested daily intake recommendations, the developed product possessed high percentages (76.4% and 38.2%) of sugar content (considering the limits according to the FDA and MFDS), while it contained 49.8% (FDA) and 74.8% (MFDS) of iron.

Table 13. Fourteen nutrients in the control and drone-pupae-powder enriched puffed-rice snack.

Items	The puffed Rice Snack						Daily Values	
	Control			Enriched with Drone Pupae Powder				
	Amount	% FDA	% MFDS	Amount	% FDA	% MFDS	FDA [1]	MFDS [2]
Calories (cal)	375.14	-	-	389.9	-	-	-	-
Sodium (mg)	15.22	0.7	0.8	12.87	0.6	0.6	<2300	<2000
Carbohydrate (g)	89.04	32.4	27.5	82.65	30.1	25.5	>275	>324
Sugar (g)	44.28	88.6	44.3	38.18	76.4	38.2	<50	<100
Dietary fiber (g)	1.33	4.8	5.3	1.44	5.1	5.8	>28	>25
Crude fat (g)	1.06	1.4	2.0	4.60	5.9	8.5	<78	<54
Trans fat (g)	-	-	-	-	-	-	<2	-
Saturated fat (g)	0.34	1.7	2.3	1.44	7.2	9.6	<20	<15
Cholesterol (mg)	-	-	-	17.16	5.7	5.7	<300	<300
Crude protein (g)	2.36	4.7	4.3	4.45	8.9	8.1	>50	>55
Vitamin D (μg)	-	-	-	-	-	-	>20	>10
Potassium (g)	0.04	0.9	1.1	0.10	2.1	2.9	>4.7	>3.5
Iron (mg)	8.54	47.4	71.2	8.97	49.8	74.8	>18	>12
Calcium (g)	0.02	1.5	2.9	0.01	0.8	1.4	>1.3	>0.7

[1] According to the Nutrition Facts Labeling Requirements, US Food and Drug Administration. [2] According to Article 6 (Nutrition Facts Label) in the Act on Labeling and Advertising of Foods (MFDS, 2020). Dash (-) means not identified.

3.5.2. Amino-Acid Composition

The amino-acid composition of the puffed-rice snack enriched with drone pupae powder is summarized in Table 14, along with that of the control snack. The amino-acid contents of the developed product were higher than those of the control snack. In particular, the levels of isoleucine (0.19 g), lysine (0.16 g), valine (0.25 g), and glycine (0.22 g) for the developed product were approximately two times that of the control snack. These differences between the developed product and control were reasonable, as the drone pupae powder contained high levels of amino acids (Table 7). Kim et al. [44] have reported that bee drone pupae powder contains substantially high amounts of EAA, such as, isoleucine, lysine, and valine.

Table 14. The total amino-acid composition of the control and drone-pupae-powder enriched puffed-rice snack.

| Amino Acids | The Puffed Rice Snack (g/100 g) | | | | | |
| | Control | | | Enriched with Drone Pupae Powder | | |
	Amount	% WHO	% MOHW	Amount	% WHO	% MOHW
Histidine	0.06	5.7	6.0	0.10	9.5	10.0
Isoleucine	0.08	3.8	6.2	0.19	9.0	14.6
Leucine	0.20	4.8	6.7	0.35	8.5	11.7
Lysine	0.07	2.2	2.3	0.16	5.1	5.3
SAA	0.11	7.1	8.5	0.15	9.7	11.5
AAA	0.22	8.3	7.9	0.39	14.7	13.9
Threonine	0.09	5.6	6.4	0.16	9.9	11.4
Tryptophan	0.02	0.5	6.7	0.03	0.7	10.0
Valine	0.12	4.4	9.2	0.25	9.2	19.2
$\sum$ EAA	0.97	4.2	6.3	1.78	7.7	11.6
Aspartate	0.22			0.40		
Serine	0.14			0.21		
Glutamic	0.47			0.75		
Proline	0.14			0.25		
Glycine	0.11			0.22		
Alanine	0.15			0.26		
Arginine	0.19			0.28		
$\sum$ NAA	1.42			2.37		

According to the same study [44], bee drone pupae powder contains high levels of glutamic acid; this is in agreement with the high value of glutamic acid recorded in the puffed-rice snack enriched with drone pupae powder in the present study. Comparing the results to daily intake requirements of WHO and the MOHW, the developed product showed higher percentages (7.7% and 11.6% in total) of EAA than the control product (4.2% and 6.3% in total).

3.5.3. Fatty-Acid Composition

Table 15 summarizes the fatty-acid compositions (%) of the control and drone-pupae-powder enriched puffed-rice snack. Overall, the developed product exhibited a higher percentage (68.57%) of UFA than the control (67.83%) due to the high UFA content of the powder. Furthermore, the percentage of stearic acid (among the SFAs) was slightly higher in the developed product (6.47%) compared to the powder before processing (2.97%). According to a report of the RDA (2021) [48], drone pupae contains abundant phospholipids, from which stearic acid can be obtained by heating. In this study, the powder was mixed with the other ingredients (syrup and puffed rice) in the heated pot immediately after heating; thus, the mixing step could involve the aforementioned phenomenon.

Considering the UFA group, in the developed snack compared to the control, the content of oleic acid and linolenic acid increased from 31.1% to 35.01% and 1.39% to 3.05%, respectively, whereas the content of linoleic acid decreased from 34.29% to 29.69%. The drone pupae powder used for developing the final product was prepared using the deep-frying method with cooking oil; according to a study by Choi and Gil [49], deep-fried foods exhibit a long heating time and high oleic acid and low linoleic acid contents, regardless of the oil type used.

According to Ney [50], the rise of cholesterol levels by fatty acids is largely due to saturated fatty acids with 12, 14, and 16 carbons. In this study, these fatty acids did not increase, while there were high percentages of C18 fatty acids in the developed product. Overall, the puffed-rice snack enriched with drone pupae powder possessed a higher ratio of n-3 to n-6 fatty acids (n-3/n-6 value of 10.8%) than the control (4.5%). Hernandez and Hosokawa [51] have reported that, based on WHO recommendations, the n-3/n-6 value should be 1:5–1:10; thus, the product developed in this study exhibited a high nutritional value.

Table 15. Fatty-acid compositions of the control and drone-pupae-powder enriched puffed-rice snack (% of total fatty acids).

Fatty Acids	Shorthand	The Puffed Rice Snack	
		% Control	% Enriched with Drone Pupae Powder
Caprylic acid	C8:0	0.15	0.03
Capric acid	C10:0	0.08	0.01
Lauric acid	C12:0	0.25	0.10
Myristic acid	C14:0	1.14	1.15
Pentadecanoic acid	C15:0	0.18	0.02
Palmitic acid	C16:0	26.92	23.24
Magaric acid	C17:0	0.17	0.09
Stearic acid	C18:0	2.97	6.47
Arachidic acid	C20:0	0.22	0.28
Heneicosylic acid	C21:0	-	-
Behenic acid	C22:0	-	-
Lignoceric acid	C24:0	0.07	0.04
$\sum$ SFA		32.15	31.43
Myristoleic acid	C14:1	0.06	0.02
Pentadecenoic acid	C15:1	0.05	0.01
Palmitoleic acid	C16:1	0.43	0.35
Magaoleic acid	C17:1	0.09	0.03
Oleic acid	C18:1	31.10	35.01
Eicosenoic acid	C20:1	0.15	0.14
Eicosadienoic acid	C20:2	-	0.02
Erucic acid	C22:1	0.09	0.02
$\sum$ MUFA		31.97	35.6
Linoleic acid	C18:2 n-6	34.29	29.69
γ-Linolenic acid	C18:3 n-6	0.03	0.01
Dihomo γ-Linolenic acid	C20:3 n-6	-	0.03
Arachidonic acid	C20:4 n-6	-	0.03
$\sum$ n-6		34.32	29.76
Linolenic acid	C18:3 n-3	1.39	3.05
Eicosatrienoic acid	C20:3 n-3	-	-
Eicosapentaenoic acid (EPA)	C20:5 n-3	0.10	0.13
Docosapentaenoic acid (DPA)	C22:5 n-3	-	-
Docosahexaenoic acid (DHA)	C22:6 n-3	0.05	0.03
$\sum$ n-3		1.54	3.21
n-3/n-6 (%)		4.5	10.8
$\sum$ PUFA		35.86	32.97
$\sum$ UFA		67.83	68.57
Total fatty acid (%)		100	100

3.5.4. Minerals and Heavy Metals

Table 16 summarizes the mineral and heavy-metal contents of the control and drone-pupae-powder enriched puffed-rice snack. The results are shown as percentages of the international (Canada, USA, Europe) and MFDS (Korea) recommendations. The developed product containing drone pupae powder exhibited higher values (0.08, 0.02, and 0.35 mg) of phosphorus, magnesium, and copper compared to the control (a normal puffed-rice snack) (0.04, 0.01, and 0.23 mg, respectively), while the amount of zinc was lower in the developed product (1.62 mg) compared to the control (1.92 mg). Here, the drone-pupae-powder enriched product exhibited a very low amount of zinc (Table 9); this was investigated further.

Table 16. The mineral and heavy-metal composition of the control and drone-pupae-powder enriched puffed-rice snack.

Minerals and Heavy Metals	Unit	The Puffed Rice Snack					
		Control			Enriched with Drone Pupae Powder		
		Amount	% MFDS	% Int'l	Amount	% MFDS	% Int'l
Phosphorus	g/100 g	0.04	6.7	6.7	0.08	10.9	10.9
Magnesium	g/100 g	0.01	2.7	2.3	0.02	4.1	3.5
Zinc	mg/100 g	1.92	19.2	17.5	1.62	16.2	14.7
Copper	mg/100 g	0.23	2.4	2.3	0.35	3.7	3.5
Lead	mg/kg	0.01	10	10	0.01	10	10
Cadmium	mg/kg	0.01	10	20	0.01	10	20
Mercury	mg/kg	<0.01	-	<2	<0.01	-	<2
Arsenic	mg/kg	0.01	10	10	0.01	10	10

Considering recommended amounts, magnesium and copper occurred under 10%, whereas phosphorus and zinc occurred in the range of 10–20%, regardless of sample group and recommendation source. The heavy metals exhibited the following values and percentages according to recommendations (MFDS; international): lead (0.01 mg; 10%; 10%), cadmium (0.01 mg; 10%; 20%), mercury (<0.01 mg; -; <2%), and arsenic (0.01 mg; 10%; 10%). Thus, the developed product is a safe and edible insect-based food.

3.6. Effects of Different Storage Conditions on the Quality of the Puffed-Rice Snack Enriched with Drone Pupae Powder

Figure 6 shows the changes in pH, TBARS, and VBN of the developed product for 180 days under different storage temperatures (15 °C, 25 °C, and 35 °C). As shown in Figure 6a, the storage conditions did not significantly change the pH values (6.26, 6.38, and 6.32 at 15 °C, 25 °C, and 35 °C, respectively) compared to the initial value (6.15), they were just slightly increased. All of the pH values were slightly acidic (5.0–6.5), similar to a previous study [52] reporting a rice-cracker product stored at different temperatures.

The TBARS values are shown in Figure 6b; it was 1.924 MDA mg/kg before storage, and significantly increased after 180 days at all storage temperatures. The final values were 4.287, 4.533, and 4.513 MDA mg/kg at 15 °C, 25 °C, and 35 °C, respectively. Thus, storage at 15 °C inhibit lipid oxidation to a greater extent than that at higher temperatures (such as, 25 °C and 35 °C). According to a study by Son et al. [53], insects with high fat content (such as bee drone pupae) show slightly increased TBARS values (lower than twice the initial value) after storage at fresh temperatures.

Figure 6c shows the VBN values for the storage period. After 180 days, the developed rice snack (using drone pupae powder) exhibited significantly higher VBN values at all temperatures (10.62 mg% at 15 °C, 10.50 mg% at 25 °C, and 10.97 mg% at 35 °C) compared to the initial value (4.8 mg%) ($p < 0.05$), and the values were all lower than 20 mg%, indicating acceptable VBN quality according to the MFDS. The drone pupae powder added to the developed product caused the change in VBN values during the storage period, similar to the increments in protein and amino-acid content (Tables 13 and 14) discussed previously.

To assess microbial quality, the levels of TBC, *E. coli*, and TCG of the final product during storage have been summarized in Table 17. The TBC values did not exhibit a significant change ($p < 0.05$), whereas *E. coli* and TCG were not detected. The TBC values were below the recommended limit (5 log CFU/g) of the MFDS. This could be attributed to the low moisture content and sealed condition of the package, with low water activation conditions due to low relative humidity.

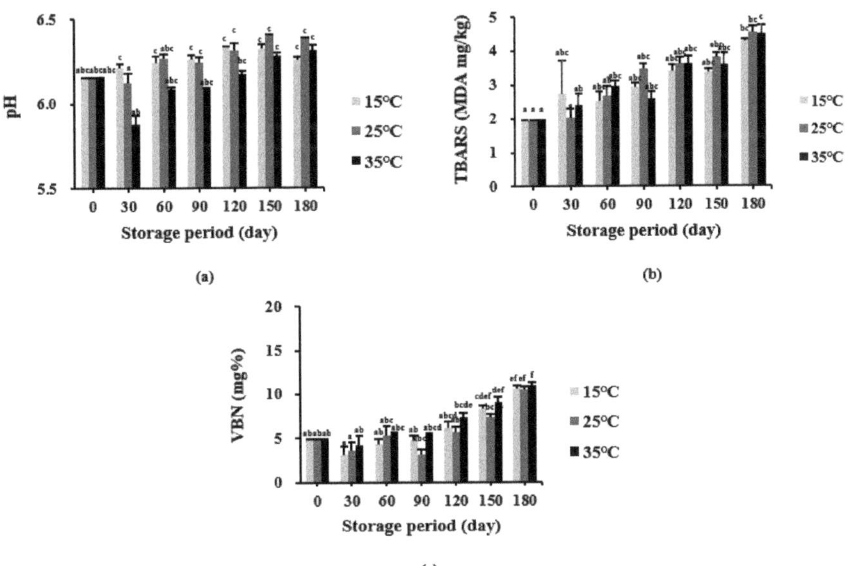

Figure 6. Chemical quality parameters of the puffed-rice snack enriched with drone pupae powder at different storage temperatures (15 °C, 25 °C, and 35 °C) for 180 days: (**a**) pH values; (**b**) TBARS values; (**c**) VBN values. Values are mean ± SD. Different letters (a–f) in the respective columns indicate significant differences among the means by the Tukey's test ($p < 0.05$).

Table 17. Changes in the total bacteria count (TBC), *E. coli*, and total coliform group (TCG) in the puffed-rice snack enriched with drone pupae powder at different storage temperatures (15 °C, 25 °C, 35 °C) for 180 days.

Temperature	Day	Total Bacteria Count (Log CFU/g)	*E. coli* and Total Coliform Group (CFU/g)
15 °C	0	1.57 ± 0.09 [a]	-
	30	1.29 ± 0.21 [a]	-
	60	1.37 ± 0.15 [a]	-
	90	1.22 ± 0.20 [a]	-
	120	1.39 ± 0.07 [a]	-
	150	1.25 ± 0.10 [a]	-
	180	1.41 ± 0.19 [a]	-
25 °C	0	1.57 ± 0.09 [a]	-
	30	1.35 ± 0.13 [a]	-
	60	1.22 ± 0.20 [a]	-
	90	1.37 ± 0.05 [a]	-
	120	1.29 ± 0.09 [a]	-
	150	1.10 ± 0.14 [a]	-
	180	1.28 ± 0.14 [a]	-
35 °C	0	1.57 ± 0.09 [a]	-
	30	1.53 ± 0.10 [a]	-
	60	1.29 ± 0.09 [a]	-
	90	1.16 ± 0.12 [a]	-
	120	1.26 ± 0.20 [a]	-
	150	1.29 ± 0.21 [a]	-
	180	1.10 ± 0.14 [a]	-

Values are mean ± SD. The small letter (a) in a column indicates the significant difference among means by the Tukey's test ($p < 0.05$). Dash (-) indicates not detected.

Figure 7 shows the sensory evaluation results of the puffed-rice snack (enriched with drone pupae powder) stored at different temperatures for 180 days by the hedonic scale method. The developed product maintained scores above 7-points during the storage period for all the sensory items, regardless of the storage temperature. Figure 7a shows the scores on appearance that changed from 8.48 (0 day) to 7.33, 7.76, and 7.33 at 15 °C, 25 °C, and 35 °C (180 days), respectively, with significant differences ($p < 0.05$). Interestingly, the appearance scores at each temperature rapidly decreased after 150 days. According to the sensory panelists, the final samples looked a bit yellowish and non-fresh after 180 days. The odor scores are shown in Figure 7b; similar to the appearance results, the tested products showed significantly lower points for odor than the initial state after 180 days under all storage conditions. As shown in Figure 6b, the samples stored for 180 days showed TBARS values above 4 MDA mg/kg; thus, the sensory qualities could be affected by lipid oxidation. However, the taste, texture, and overall acceptance scores did not change significantly after storage for 180 days at different temperatures, as shown in Figure 7c–e.

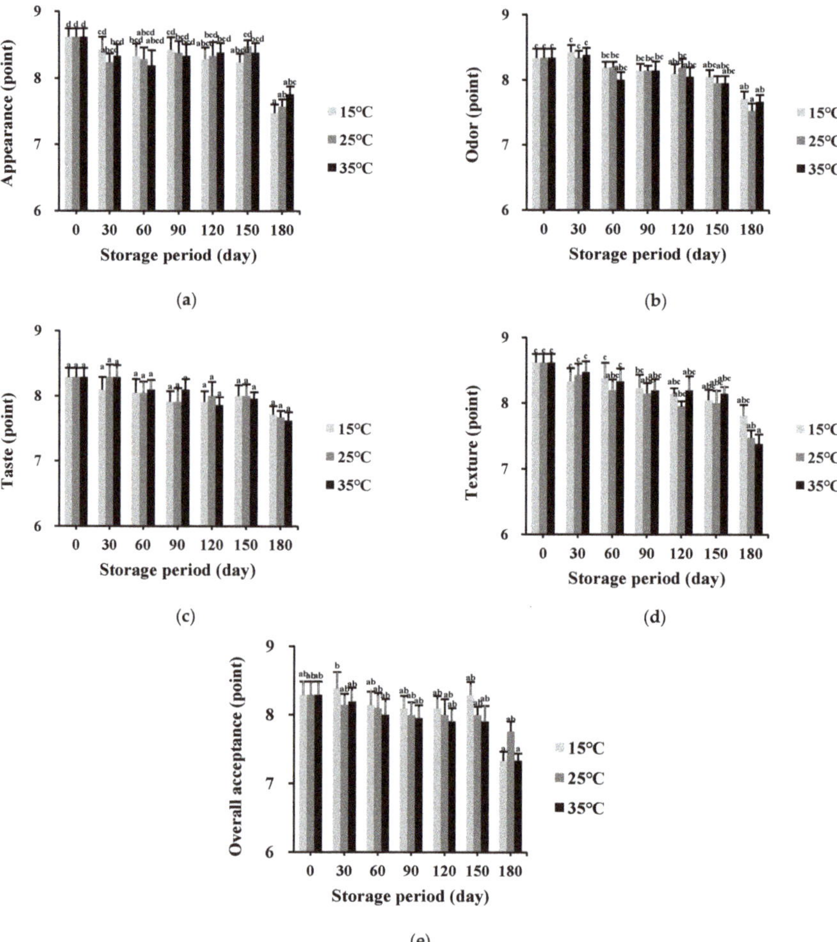

Figure 7. Changes in sensory evaluation of the puffed-rice snack enriched with drone pupae powder at different storage temperatures (15 °C, 25 °C, and 35 °C) for 180 days: (**a**) Appearance; (**b**) Odor; (**c**) Taste; (**d**) Texture; (**e**) Overall acceptance. The values are mean ± SD. Different letters (a–c) in the columns indicate significant differences among the means by the Tukey's test ($p < 0.05$).

4. Conclusions

In this study, the best processing method for honeybee drone pupae powder and the development of a puffed-rice snack product using the powder has been described. The combination of deep-frying and hot-air drying was confirmed to be the best method for processing drone pupae to powder, considering various quality characteristics. Subsequently, a puffed-rice snack product enriched with the optimal powder (DHD) was prepared and its nutritional qualities and storage effect under different conditions were tested. The drone pupae powder exhibiting the best acid value and TBARS was selected among the different processing groups; it showed good sensory qualities (high nutty aroma and low bitter taste) and nutritional values considering institutional recommendations (high EAA and UFA). The drone pupae powder (5.54%) and honey (2.13%) were optimized using response surface methodology by adjusting the mixing concentrations with the puffed-rice snack to obtain the targeted nutty aroma and sweetness taste. The developed puffed-rice snack enriched with drone pupae powder exhibited higher values of nutrients (including proteins, fats, amino acids, and fatty acids) compared to the control rice snack. Additionally, the mineral content of the selected drone pupae powder and snack product were examined; heavy metals were not detected. In the storage test, the final product exhibited safe quality values considering important parameters (such as VBN and bacterial levels), along with a stable sensory likeness for 180 days. Therefore, the processed drone pupae powder is suitable to be used as an edible food ingredient. Furthermore, both the powder and the powder-enriched rice snack product exhibit high nutritional value and are safe for consumption.

Author Contributions: Conceptualization, E.-J.K. and J.-S.C.; methodology, W.-H.C. and J.-S.C.; software, W.-H.C. and J.-S.C.; validation, W.-H.C. and J.-S.C.; formal analysis, W.-H.C. and J.-M.P.; investigation, W.-H.C. and J.-M.P.; resources, E.-J.K.; data curation, J.-S.C.; writing—original draft preparation, W.-H.C. and J.-M.P.; writing—review and editing, M.M. and J.-S.C.; visualization, W.-H.C. and J.-S.C.; supervision, J.-S.C.; project administration, E.-J.K. and J.-S.C.; funding acquisition, E.-J.K. and J.-S.C. All authors have read and agreed to the published version of the manuscript.

Funding: This research was supported by the Ministry of Trade, Industry, and Energy (MOTIE), Korea Institute for Advancement of Technology (KIAT) For Regional Program Evaluation (IRPE) through the Encouragement Program for The Social and Economic Innovation Growth (project number P0017633).

Institutional Review Board Statement: The study was conducted according to the guidelines of the Declaration of Helsinki, and approved by the Institutional Review Board of Silla University (protocol code 1041449-202108-HR-002).

Data Availability Statement: Data supporting the reported results are available upon request.

Acknowledgments: We would like to thank to S.-K.P. for his technical support.

Conflicts of Interest: The authors and With Food Corp. declare that they have no conflict of interest.

References

1. Choi, J.S. Nutrition, Safety, Health Functional Effects, and Availability of Honeybee (*Apis mellifera* L.) Drone Pupae. *Insects* **2021**, *12*, 771. [CrossRef] [PubMed]
2. Evans, J.; Müller, A.; Jensen, A.B.; Dahle, B.; Flore, R.; Eilenberg, J.; Frøst, M.B. A descriptive sensory analysis of honeybee drone brood from Denmark and Norway. *J. Insects Food Feed.* **2016**, *2*, 277–283. [CrossRef]
3. Han, S.M.; Moon, H.J.; Woo, S.O.; Bang, K.W.; Kim, S.G.; Kim, H.Y.; Choi, H.M.; Lee, M.Y. *Bee Pupa Cookbook Nutritious, Bee Pupa Cooking*, 1st ed.; Rural Development Administration: Wanju-gun, Korea, 2019.
4. Kulma, M.; Tůmová, V.; Fialová, A.; Kouřimská, L. Insect consumption in the Czech Republic: What the eye does not see, the heart does not grieve over. *J. Insects Food Feed.* **2020**, *6*, 525–535. [CrossRef]
5. Wilkinson, K.; Muhlhausler, B.; Motley, C.; Crump, A.; Bray, H.; Ankeny, R. Australian consumers' awareness and acceptance of insects as food. *Insects* **2018**, *9*, 44. [CrossRef] [PubMed]
6. Winston, M.L. *The Biology of the Honey Bee*; Harvard University Press: Cambridge, MA, USA, 1991.
7. Ulmer, M.; Smetana, S.; Heinz, V. Utilizing honeybee drone brood as a protein source for food products: Life cycle assessment of apiculture in Germany. *Resour. Conserv. Recy.* **2020**, *154*, 104576. [CrossRef]

8. Krell, R. *Value-Added Products from Beekeeping*; Food and Agriculture Organization of the United Nations: Rome, Italy, 1996; Volume 124, pp. 12–28.

9. Ministry of Food and Drug Safety. Status of Temporary Recognition of Standards and Specifications for Food Ingredients. 44. Honeybee Drone Pupae (*Apis mellifera* L.). Available online: https://www.foodsafetykorea.go.kr/portal/board/boardDetail.do (accessed on 3 November 2021).

10. Ramírez-Rivera, E.J.; Hernández-Santos, B.; Juárez-Barrientos, J.M.; Torruco-Uco, J.G.; Ramírez-Figueroa, E.; Rodríguez-Miranda, J. Effects of formulation and process conditions on chemical composition, color parameters, and acceptability of extruded insect-rich snack. *J. Food Processing Preserv.* **2021**, *45*, e15499. [CrossRef]

11. García-Gutiérrez, N.; Mellado-Carretero, J.; Bengoa, C.; Salvador, A.; Sanz, T.; Wang, J.; Ferrando, M.; Güell, C.; Lamo-Castellví, S.D. ATR-FTIR Spectroscopy Combined with Multivariate Analysis Successfully Discriminates Raw Doughs and Baked 3D-Printed Snacks Enriched with Edible Insect Powder. *Foods* **2021**, *10*, 1806. [CrossRef]

12. Roncolini, A.; Milanović, V.; Aquilanti, L.; Cardinali, F.; Garofalo, C.; Sabbatini, R.; Clementi, F.; Belleggia, L.; Pasquini, M.; Mozzon, M.; et al. Lesser mealworm (*Alphitobius diaperinus*) powder as a novel baking ingredient for manufacturing high-protein, mineral-dense snacks. *Food Res. Int.* **2020**, *131*, 109031. [CrossRef]

13. Severini, C.; Azzollini, D.; Albenzio, M.; Derossi, A. On printability, quality and nutritional properties of 3D printed cereal based snacks enriched with edible insects. *Food Res. Int.* **2018**, *106*, 666–676. [CrossRef]

14. Gallagher, E. *Gluten-Free Food Science and Technology*; John Wiley & Sons: Hoboken, NJ, USA, 2009.

15. We, G.J.; Lee, I.A.; Cho, Y.S.; Yoon, M.R.; Shin, M.S.; Ko, S.H. Development of rice flour-based puffing snack for early childhood. *Food Eng. Prog.* **2010**, *14*, 322–327.

16. Kang, S.M.; Maeng, A.R.; Seong, P.N.; Kim, J.H.; Cho, S.; Kim, Y.; Choi, Y.S. Effect of drone pupa meal added as replacement of sodium nitrite and vitamin C on physico-chemical quality characteristics of emulsion-type sausage. *Korean J. Food Nutr.* **2018**, *31*, 802–810.

17. Kang, S.I.; Kim, K.H.; Lee, J.K.; Kim, Y.J.; Park, S.J.; Kim, M.W.; Choi, B.D.; Kim, D.; Kim, J.S. Comparison of the food quality of freshwater rainbow trout *Oncorhynchus mykiss* cultured in different regions. *Korean J. Fish. Aquat. Sci.* **2014**, *47*, 103–113. [CrossRef]

18. Ministry of Food and Drug Safety (MFDS). *8th General Analysis Method. Food Code (Sik-Poom-Gong-Jeon)*; MFDS: Cheongju, Korea, 2021. Available online: https://www.foodsafetykorea.go.kr/foodcode/01_03.jsp?idx=11031 (accessed on 10 November 2021).

19. Buege, J.A.; Aust, S.D. MDA levels in plasma and TBARs. *Methods Enzym.* **1978**, *12*, 302–310.

20. *ISO1329*; Sensory Analysis—Methodology—General Guidance for Establishing a Sensory Profile. International Organization for Standardization: Geneva, Switzerland, 2016.

21. AOAC International. *International Official Methods of Analysis Official Methods 994.12*, 17th ed.; Association of Official Analytical Chemists: Washington, DC, USA, 2000.

22. Bligh, E.G.; Dyer, W.J. The diet of amphioxus in subtropical Hong Kong as indicated by fatty acid and stable isotopic analyses. *Can. J. Biochem. Physiol.* **1959**, *37*, 911–917. [CrossRef]

23. Official Methods and Recommended Practices of the AOCS. *AOCS Official Method Ce 1c-89*, 7th ed.; American Oil Chemists' Society: Urbana, IL, USA, 2006.

24. American Oil Chemists' Society. *AOCS Official Method Cd 14c-94*; American Oil Chemists' Society: Urbana, IL, USA, 2006.

25. AOAC International. *International Official Methods of Analysis Official Methods 925.09, 923.03, 979.09, 962.09, and 923.05*, 17th ed.; Association of Official Analytical Chemists: Washington, DC, USA, 2000.

26. AOAC International. Method of Analysis–Official methods 990.12. In *International Official Methods of Analysis Official Methods*, 18th ed.; Association of Official Analytical Chemists: Arlington, VA, USA, 2002.

27. AOAC International. Method of Analysis–Official methods 991.14. In *International Official Methods of Analysis Official Methods*, 18th ed.; Association of Official Analytical Chemists: Arlington, VA, USA, 2002.

28. Lin, C.Y.; Lin, Y.C.; Kuo, H.K.; Hwang, J.J.; Lin, J.L.; Chen, P.C.; Lin, L.Y. Association among acrylamide, blood insulin, and insulin resistance in adults. *Diabetes Care* **2009**, *32*, 2206–2211. [CrossRef]

29. Suderman, D.R.; WIKER, J.; Cunningham, F.E. Factors affecting adhesion of coating to poultry skin: Effects of various protein and gum sources in the coating composition. *J. Food Sci.* **1981**, *46*, 1010–1011. [CrossRef]

30. Hwang, S.Y.; Kim, G.S. A Study on Quality Characteristics of Silkworm (*Bombyx mori* L.) by various pretreatment methods. *Culin. Sci. Hosp. Res.* **2020**, *26*, 149–161.

31. Pyo, S.J.; Jung, C.E.; Sohn, H.Y. Platelet aggregatory and antidiabetic activities of larvae, pupae, and adult of honeybee drone (*Apis mellifera*). *J. Apic.* **2020**, *35*, 41–48.

32. Andrikopoulos, N.K.; Dedoussis, G.V.; Falirea, A.; Kalogeropoulos, N.; Hatzinikola, H.S. Deterioration of natural antioxidant species of vegetable edible oils during the domestic deep-frying and pan-frying of potatoes. *Int. J. Food Sci. Nutr.* **2002**, *53*, 351–363. [CrossRef]

33. Guillen-Sans, R.; Guzman-Chozas, M. The thiobarbituric acid (TBA) reaction in foods: A review. *Crit. Rev. Food Sci. Nutr.* **1998**, *38*, 315–350. [CrossRef]

34. Koh, H.Y.; Kowon, Y.J. Effect of storage temperature and humidity on water absorption and rancidity of peanuts. *J. Korean Soc. Food Nutr.* **1989**, *18*, 216–222.

35. AL-KAHTANI, H.A.; ABU-TARBOUSH, H.M.; BAJABER, A.S.; ATIA, M.; ABOU-ARAB, A.A.; EL-MOJADDIDI, M.A. Chemical changes after irradiation and post-irradiation storage in tilapia and Spanish mackerel. *J. Food Sci.* **1996**, *61*, 729–733. [CrossRef]

36. Santos, M.S. Biogenic amines: Their importance in foods. *Int. J. Food Microbiol.* **1996**, *29*, 213–231. [CrossRef]
37. Jin, S.K.; Kim, I.S.; Hah, K.H.; Hur, S.J.; Lyou, H.J.; Park, K.H.; Bae, D.S. Changes of qualities in aerobic packed ripening pork using a Korea traditional seasoning during storage. *J. Anim. Sci. Technol.* **2005**, *47*, 73–82.
38. Kim, J.E.; Kim, S.G.; Kang, S.J.; Kim, Y.H. *Physicochemical Properties of the Bee Pupa and Main Pollen, Symposium for the Development of the Beekeeping Industry, Korea, 20–21 October 2016*; The Apicultural Society of Korea: Suwonsi, Korea, 2016; p. 142.
39. Kim, J.E.; Kim, D.I.; Koo, H.Y.; Kim, H.J.; Kim, S.Y.; Lee, Y.B.; Kim, J.S.; Kim, H.H.; Moon, J.H.; Choi, Y.S. Analysis of Nutritional Compounds and Antioxidant Effect of Freeze-Dried powder of the Honey Bee (*Apis mellifera* L.) Drone (Pupal stage). *Korean J. Appl. Entomol.* **2020**, *59*, 265–275.
40. Ghosh, S.; Sohn, H.Y.; Pyo, S.J.; Jensen, A.B.; Meyer-Rochow, V.B.; Jung, C. Nutritional composition of *Apis mellifera* drones from Korea and Denmark as a potential sustainable alternative food source: Comparison between developmental stages. *Foods* **2020**, *9*, 389. [CrossRef]
41. Ghosh, S.; Jung, C.; Meyer-Rochow, V.B. Nutritional value and chemical composition of larvae, pupae, and adults of worker honey bee, *Apis mellifera ligustica* as a sustainable food source. *J. Asia-Pac. Entomol.* **2016**, *19*, 487–495. [CrossRef]
42. Lee, J.S.; Cho, S.Y.; Choi, S.G. *Application of Soybean Oil in the Food Industry, Food Industry and Nutrition, Korea, December 1996, Busan*; The Korean Society of Food Science and Nutrition: Daejeon, Korea, 1996; pp. 27–36.
43. Recommended Intakes of Macronutrients (Nordic Nutrition Recommendations 2012). Available online: http://norden.diva-portal.org/smash/get/diva2:704251/FULLTEXT01.pdf (accessed on 6 November 2021).
44. Kim, S.G.; Woo, S.O.; Bang, K.W.; Jang, H.R.; Han, S.M. Chemical composition of drone pupa of *Apis mellifera* and its nutritional evaluation. *Korean J. Apic.* **2018**, *33*, 17–23.
45. Aparna, A.R.; Rajalakshmi, D. Honey—its characteristics, sensory aspects, and applications. *Food Rev. Int.* **1999**, *15*, 455–471. [CrossRef]
46. Jensen, A.B.; Evans, J.; Jonas-Levi, A.; Benjamin, O.; Martinez, I.; Dahle, B.; Roos, N.; Lecocq, A.; Foley, K. Standard methods for *Apis mellifera* brood as human food. *J. Apic. Res.* **2019**, *58*, 1–28. [CrossRef]
47. Biró, B.; Sipos, M.A.; Kovács, A.; Badak-Kerti, K.; Pásztor-Huszár, K.; Gere, A. Cricket-Enriched Oat Biscuit: Technological Analysis and Sensory Evaluation. *Foods* **2020**, *9*, 1561. [CrossRef] [PubMed]
48. Monthly Agricultural Technology (Nongsaro Agricultural Technology Portal in Rural Development Administration). Available online: http://www.nongsaro.go.kr/portal/ps/psv/psvr/psvre/curationDtl.ps?menuId=PS03352&srchCurationNo=1689&totalSearchYn=Y (accessed on 6 November 2021).
49. Choi, E.S.; Gil, B.I. Effects of Thermooxidation of Soybean Oil in Association with Fried Foods on Quantity Food Production. *J. East Asian Soc. Diet. Life* **2011**, *21*, 723–730.
50. Ney, D.M. Potential for enhancing the nutritional properties of milk fat. *J. Dairy Sci.* **1991**, *74*, 4002–4012. [CrossRef]
51. Hernandez, E.; Hosokawa, M. (Eds.) *Omega-3 Oils: Applications in Functional Foods*; Elsevier: Amsterdam, The Netherlands, 2015.
52. Gunaratne, T.M.; Gunaratne, N.M.; Navaratne, S.B. Selection of best packaging method to extend the shelf life of rice crackers. *Int. J. Sci. Eng. Res.* **2015**, *6*, 638–645.
53. Son, Y.J.; Ahn, W.; Kim, S.H.; Park, H.N.; Choi, S.Y.; Lee, D.G.; Kim, A.N.; Hwang, I.K. Study on the oxidative and microbial stabilities of four edible insects during cold storage after sacrificing with blanching methods. *Korean J. Food Nutr.* **2016**, *29*, 849–859. [CrossRef]

 *foods*

Article

Nutritional Quality of Four Novel Porridge Products Blended with Edible Cricket (*Scapsipedus icipe*) Meal for Food

Nelly C. Maiyo [1,2], Fathiya M. Khamis [1], Michael W. Okoth [2], George O. Abong [2], Sevgan Subramanian [1], James P. Egonyu [1], Cheseto Xavier [1], Sunday Ekesi [1], Evanson R. Omuse [1], Dorothy Nakimbugwe [3], Geoffrey Ssepuuya [3], Changeh J. Ghemoh [4] and Chrysantus M. Tanga [1,*]

[1] International Centre of Insect Physiology and Ecology (icipe), P.O. Box 30772, Nairobi 00100, Kenya; nmaiyo@icipe.org (N.C.M.); fkhamis@icipe.org (F.M.K.); ssubramania@icipe.org (S.S.); pegonyu@icipe.org (J.P.E.); cxavier@icipe.org (C.X.); sekesi@icipe.org (S.E.); evansonomuse12@gmail.com (E.R.O.)

[2] Department of Food Science, Nutrition and Technology, University of Nairobi, P.O. Box 30197, Nairobi 00100, Kenya; mwokoth@uonbi.ac.ke (M.W.O.); ooko.george@uonbi.ac.ke (G.O.A.)

[3] Department of Food Technology and Nutrition, School of Food Technology, Nutrition and Bioengineering, Makerere University, Kampala P.O. Box 7062, Uganda; dnakimbugwe@gmail.com (D.N.); gksepuya@gmail.com (G.S.)

[4] Centre for African Bio-Entrepreneurship (CABE), P.O. Box 25535, Nairobi 00603, Kenya; janiceghemoh@yahoo.com

* Correspondence: ctanga@icipe.org; Tel.: +254-702-729-931

Citation: Maiyo, N.C.; Khamis, F.M.; Okoth, M.W.; Abong, G.O.; Subramanian, S.; Egonyu, J.P.; Xavier, C.; Ekesi, S.; Omuse, E.R.; Nakimbugwe, D.; et al. Nutritional Quality of Four Novel Porridge Products Blended with Edible Cricket (*Scapsipedus icipe*) Meal for Food. *Foods* 2022, 11, 1047. https://doi.org/10.3390/foods11071047

Academic Editors: Victor Benno Meyer-Rochow and Chuleui Jung

Received: 25 February 2022
Accepted: 23 March 2022
Published: 5 April 2022

Publisher's Note: MDPI stays neutral with regard to jurisdictional claims in published maps and institutional affiliations.

Abstract: Currently, no data exist on the utilization of the newly described cricket species (*Scapsipedus icipe*) meal as additive in food products, though they have high protein (57%) with 88% total digestibility as well as a variety of essential amino acids. This article presents the first report on the effects of processing techniques and the inclusion of cricket meal (CM) on the nutrient and antinutrient properties of four porridge products compared to a popularly consumed commercial porridge flour (CPF). Porridge enriched with CM had significantly higher protein (2-folds), crude fat (3.4–4-folds), and energy (1.1–1.2-folds) levels than the CPF. Fermented cereal porridge fortified with CM had all three types of omega-3 fatty acids compared to the others. The vitamin content across the different porridge products varied considerably. Germinated cereal porridge with CM had significantly higher iron content (19.5 mg/100 g). Zinc levels ranged from 3.1–3.7 mg/100 g across the various treatments. Total flavonoid content varied significantly in the different porridge products. The phytic acid degradation in germinated and fermented porridge products with CM was 67% and 33%, respectively. Thus, the fortification of porridge products with cricket and indigenous vegetable grain powder could be considered an appropriate preventive approach against malnutrition and to reduce incidences in many low-and middle-income countries.

Keywords: underutilized food resources; grain amaranth; edible cricket meal; complementary porridge; nutrient quality; malnutrition and food security

1. Introduction

The malnutrition of children in sub–Saharan African (SSA) countries is a serious health concern [1,2]. Protein-energy malnutrition (PEM) and deficiencies of micronutrients including iron, zinc, and vitamin A are the most common forms of malnutrition reported in these countries [3,4]. Kenya is amongst the 20 countries accounting for 80% of the world's malnourished children [5] where stunting, wasting, and underweight in children below five years have been estimated at 26%, 4%, and 11%, respectively [6]. Malnutrition in infancy and early childhood affects physical growth and cognitive behaviour, leading to delays in mental and motor development, as well as increased morbidity and mortality [7]. Low socio-economic status is considered the key underlying cause where children lack food or survive on diets of low nutritional quality [8].

The African continent is blessed with a rich diversity of food crops, most of which have received little or no attention in terms of research and development of policy frameworks that could promote their effective commercial and industrial utilization. Grain amaranth (*Amaranthus* spp.) and finger millets are some of such neglected and underutilized species that could be used to produce porridge products to serve as important traditional beverages and complementary food for adults and children, respectively, of all ages in Africa [9,10]. Despite these cereals being rich in carbohydrates, their energy and nutrient densities are extremely low [11], partly due to the presence of anti-nutritional factors that restricts the bioavailability of essential nutrients [12]. Finger millet grain contains high amounts of proteins and minerals compared to other staple cereals [13] and is the most preferred grain in composite flour for making porridge [1]. Amaranth (*Amaranthus* spp.) is an indigenous African leafy vegetable grown in at least fifty tropical countries and consumed by several million people [2,14] for many nutritional reasons. Farmers in sub-Saharan Africa cultivate amaranth either for its leaves or for its grain [14]. The leaves are rich in vitamin C and pro-vitamin A as well as in iron, zinc, and calcium [15]. The grains are also rich in quality protein, lysine, and calcium and are consumed directly or used to fortify maize flour [16,17]. Amaranth grain contains bioactive compounds with health promoting effects [18]. Studies have shown that regular consumption of amaranths has the potential [18] to reduce cholesterol levels [19], benefit people suffering from hypertension and cardiovascular disease [20,21], improve liver functions [22,23], and prevent cancers [24]. The ability health benefits of amaranth are due to the bioactive compounds such as protocatechuic, hydroxybenzoic, caffeic and ferulic acids, rutine, nicotiflorin, and isoquercetin present in it [25]. While amaranth is being widely cultivated in Africa as a vegetable, its grains have been documented to contain excellent nutritional properties and have been milled and blended with flours of other cereal staples to improve their overall acceptability [26]. However, these grains contain a high content of anti-nutrients, mostly in the form of phytic acid [27,28]. These anti-nutrients can be reduced to improve the nutritional quality through traditional food processing methods, including soaking, germination, fermentation, roasting, and milling [13].

The use of edible insect meal to fortify porridge products has received limited research attention, even though it has been postulated to contain high-quality nutrients, which are easily digestible and more bio-available than those available from plant and animal food sources [29,30]. This will contribute to the Sustainable Development Goal (SDG) 2 of zero hunger championed by the United Nations, which aims at improving food and nutritional security [29]. Food-to-food fortification of complementary foods with nutrient-rich ingredients like edible insects that are rich in protein, amino acids, fatty acids, vitamins, and minerals (e.g., zinc, iron, and calcium) [30–33] can help in the attainment of this goal. Thus, cricket consumption could be an immediate solution to many of the nutrient deficiency issues and could be both an inexpensive and effective option. The use of naturally available food resources such as insects like crickets, would easily be favoured by policy makers in view of their sustainability [34–36]. Murugu et al. [37] have demonstrated that the newly described cricket *Scapsipedus icipe* Hugel and Tanga is significantly rich in protein (57%), fat (36%), amino acids, minerals, and vitamins. However, no information exists on their inclusion into cereal-based human food products, although there are promising indications that they can significantly improve the nutritional quality of food products.

Furthermore, edible insects are commonly processed using different techniques (frying, drying, roasting, smoking, boiling, toasting, and steaming) to ensure microbial safety, increase shelf-life, and improve on the sensory appeal [4,7,37,38]. The utilization of appropriate processing methods is critical to mitigate spoilage [5]. Several studies have shown that processing methods can cause significant losses and degradation of essential nutrients besides enhancing their levels, digestibility, and bioactivity of the others [3,6,7], either through solubilization, leakages, and intra- and inter-biochemical reactions. Ssepuuya et al. [8] reported the effects of two thermal processes on the nutritional composition, colour, and aroma compounds of *R. differens*. Furthermore, investigation by Nyangena et al. [4] apprised the effects of toasting, boiling, and drying techniques on proximate composition and microbial quality of *R. differens* and other edible insects only. None of these studies looked at the potential effects of processing techniques and mixing of amaranth integrated with cricket meal on compounded human food products. This study sought to bridge this gap by comparing the effects of four common food processing methods of mixing of amaranth, cricket meal, and finger millets into porridge products on the nutrient (proximate composition, fatty acids, and minerals) and antinutrient composition as well as bioactive compounds (total flavonoid content and vitamins). This detailed report on the effect of the different processing methods applicable to cereal-based porridge food products with cricket meal is inevitable, especially with recent findings revealing the significance of edible insects and amaranth in the fight against malnutrition and food insecurity [9]. The campaign for consumption of insects, grains, and vegetables for the maintenance of good health has become ever more popular globally [29,30,39].

2. Materials and Methods

2.1. Raw Materials

Grain amaranth (*Amaranthus cruentus*) and finger millet (*Eleusine coracan* L.) were purchased from a local market in Nairobi, Kenya. One of the most traded brands of commercial porridge flour (CPF) (FAMiLA® pure WIMBI porridge, Unga Limited, Nairobi, Kenya) was purchased from a supermarket in the Kenyan capital, Nairobi. The unique formulation of FAMiLA® consists of pure finger-millet flour fortified with minerals (calcium), carbohydrates, and proteins (particularly lysine). This product was chosen because it is popular and consumed by adults and children of all ages. On the other hand, edible crickets (*Scapsipedus icipe*) were obtained from the Animal Rearing and Containment Unit (ARCU) at the International Centre of Insect Physiology and Ecology (*icipe*), Nairobi, Kenya.

2.2. Preparation of Raw Materials

2.2.1. Preparation of Crickets

Frozen cricket samples were allowed to thaw overnight at a 5 °C refrigerated temperature. The samples were then washed thrice in fresh, clean tap water at 18 °C to remove dirt, drained, and appropriately heat-treated in hot water at 100 °C for 5 min for efficient sterilization. Thereafter, the crickets were oven-dried (WTB binder, Tuttlingen, Germany) at 60 °C for 24 h, milled using an electric grinder [Medical Research Council (MRC) laboratory grinder, London, UK], and sieved through a 0.595-mm aperture sieving mesh. The resulting powder was packaged in a sterile zip-lock polyethylene bag and stored at 4 °C for subsequent composite flour formulation.

2.2.2. Processing of Finger Millet and Amaranth Grains

Finger millet and amaranth grains were each divided into a batch of 500 g in four replications. The first batch was germinated according to the method described by Onyango et al. [40]. Briefly, the grains were cleaned and steeped in tap water for 24 h at 25 °C and germinated for 72 h at 25 °C in the dark, while being moistened and turned at 12–hour intervals. Germination was halted by oven-drying at 50 °C for 12 h. The dried germinated grains were then removed and milled using an electric grinder (MRC laboratory grinder, London, UK). The second batch was roasted according to the modified method [41]. The grains were spread in a uniform, thin layer on an oven tray and roasted in a pre-heated oven at 120 °C for 20 min. The grains were allowed to cool to room temperature before milling. The third batch was milled and fermented according to the method described by Onyango et al. [10] with slight modifications. The composite flour was mixed with tap water that had been boiled, then cooled to 45 °C, at a ratio of 2:3 (flour to water). The slurries were fermented spontaneously in round-bottomed flasks placed in a water bath (Blue M Electric Company, IL 60406, USA) at 45 °C. After 24 h the fermented slurries were inoculated into freshly prepared slurries before fermenting at 45 °C for 24 h. The samples were then spread on trays and dried in an oven at 50 °C for 24 h and ground in an electric grinder (MRC laboratory grinder,

London, UK). The fourth batch was unprocessed grains treated according to farmers' field practices of open sun drying and storage, using appropriate technology to ensure availability throughout the year [42]. Thereafter, the grains were milled and prepared for further processing into porridge products. The fifth batch was considered as the "control treatment" and consisted of the CPF, which had undergone approved commercial standards of processing. All the samples were ensured to have a moisture content of <12.0%.

2.3. Porridge Flour Formulations

Composite flour formulations were developed and optimized using the Linear Programming of the Excel solver 2010 version, based on the minimum recommended dietary allowance (RDA) for protein (13 g), energy (500 kcal), calcium (500 mg), iron (7 mg), and zinc (4.1 mg) for children aged 1–3 years [11,12]. Porridge formulations were prepared by mixing flours of finger millet, amaranth, and cricket, at a ratio of 60:30:10. The porridge mix samples were coded as follows: commercial porridge flour (CPF), fermented finger millet − amaranth + cricket (FFM–AC), germinated finger millet − amaranth + cricket (GFM–AC), roasted finger millet − amaranth + cricket (RFM–AC), and unprocessed sundried finger millet − amaranth + cricket (UFM–AC). These new products supplemented with cricket meal and amaranth were compared with a commercial porridge product widely used by millions of households in East Africa.

2.4. Analysis of Proximate Composition

Proximate analysis for the flour was determined following the Association of Official Analytical Chemists (AOAC) methods [13] in three replications. Moisture content was determined in an air oven adjusted to 105 °C (method 925.10). The Kjeldahl method (method 978.04) was used to assess the crude protein (N × 6.25). Fat content was extracted using petroleum ether in a Soxhlet extractor (method 930.09). Ash content was determined by gravimetries (method 930.05). Fibre was determined by acid digestion and loss of ignition (method 930.10). Carbohydrate contents were determined as the difference (CHO% = [100−protein %−fat %−crude fibre %−ash %]). The energy value was computed by multiplying the Atwater factors of 4, 9, and 4 with protein %, fat % and carbohydrate % contents, respectively.

2.5. Determination of Mineral Composition

Mineral composition was established in three replications by the ICP OES quantitation method as follows. Exactly 0.5 g of each sample was digested with concentrated HNO_3 (8 mL) and 30% H_2O_2 (2 mL) and left overnight in a fume chamber. Samples were then digested in a temperature-controlled block digester (Model TE007—A, TECNAL, São Paulo, SP, Brazil) following these conditions: 75 °C for 30 min, 120 °C for 20 min, 180 °C for 20 min, and 200 °C for 10 min. Resulting solutions were cooled and transferred to 25 mL falcon tubes and diluted with 2% nitric acid. Mineral compositions were assessed using an inductively coupled plasma optical emission spectrometer (ICP OES) (Model Optima 2100 DV Perkin Elmer, MA, USA) and analysed using WinLab 32 software (Perkin Elmer, Poway, CA 92064, USA). The following operational conditions were used: radiofrequency power (1.45 kW), auxiliary gas flow rate (1.5 L min^{-1}), plasma gas flow rate (15.0 L min^{-1}), nebuliser gas flow rate (0.7 L min^{-1}), sample flow rate (1.5 L min^{-1}), source equilibrium time (10 s), and delay time (10 s). Signal intensity measurements of the analytes in all samples solutions were carried out at wavelengths (nm) as follows: Mg: 285.213, Fe: 259.939, Mn: 257.61, Ca: 317.933, P: 213.617, Mo: 202.031, K: 766.49, Al: 396.153, Cu: 224.7, Co: 228.616, and Zn: 213.857. ICP OES quantification was carried out using a multi-element standard solution (TraceCERT) CatNo.43843 (Sigma-Aldrich, Saint Louis, MO, USA). The calibration standards were prepared by titrating the standard solution in 2% (*v*/*v*) nitric acid to obtain the working ranges required (400–4000 µg L $^{-1}$). The correlation coefficient obtained was ≥0.999. Calibration was performed using WinLab 32 software (Perkin Elmer, Poway, CA 92064, USA).

2.6. Assessment of Fatty Acids

2.6.1. Folch Extraction Method

Oil extraction from the formulated porridge flours was achieved based on the modified method [43]. In three replications, 5 g each of sample were diluted with 10 mL of 2:1 dichloromethane (DCM) and methanol (MeOH). The mixtures were vortexed for 1 min, sonicated for 10 min, and centrifuged at 4200 rpm for 10 min. The supernatants were filtered using grade 1 and 90 mm (diameter) Whatman filter paper into a 250 mL round bottomed flask and evaporated in vacuo to yield approximately 200 mg of oil.

2.6.2. Fatty Acid Determination

Compositions of fatty acid (FA) in the oil extract from formulated porridge flours were examined as fatty acid methyl esters (FAMEs), according to the modified method [27]. Approximately 1 mL of sodium methoxide solution (100 mg/mL) was added to 100 mg of recovered oil extract, vortexed for 1 min, followed by sonication for 10 min. The resulting mixture was placed in a water bath (70 °C) for 1 h and the reaction halted by adding of 100 µL deionized water, followed by vortexing for another 1 min. To extract the FAMEs, 1 mL of gas chromatography (GC)-grade hexane (Sigma-Aldrich, St. Louis, MO, USA) was added to the mixture, followed by 20 min centrifugation at 14,000 rpm. The resulting hexane layer (upper) was dried over anhydrous Na_2SO_4, and analysis was performed using an Agilent GC–MS on a 7890A GC, connected to a 5975 C mass selective detector (Agilent Technologies Inc., Santa Clara, CA, USA). GC was fitted with a (5%–phenyl)–methylpolysiloxane (HP5 MS) low bleed capillary column (30 m × 0.25 mm i.d., 0.25 µm (J&W, Folsom, CA, USA). The sample injection volume was 1 µL. Helium acted as the carrier gas at flow rates of 1.25 mL/min. The oven temperature was programmed between 35 to 285 °C, with the initial and final temperature kept for 5 and 20.4 min, respectively, with a rising rate of 10 °C minute^{-1}. An ion trap mass selective detector was maintained at the ion source temperature of 230 °C and quad temperature of 180 °C. The mass detector was run in electron impact (EI) mode (70 eV). Fragment ions were determined in the full scan mode at over 40–550 m/z mass range. The filament was delayed for 3.3 min. Authentic standard methyl octadecenoate (0.2–125 ng/µL) serial dilutions prepared from octadecanoic acid ($\geq$95% purity) (Sigma-Aldrich, St. Louis, MO, USA) were analysed using GC-MS in full scan mode. GC-MS generated a linear calibration graph of peak area versus concentration with the equation; [$y = 5E + 0.7x + 2E + 0.7$] with R^2 of 0.9997. This regression equation was used for the external quantification of the different FA and sterols from the samples.

A Hewlett-Packard (HP Z220 SFF intel xeon) workstation with ChemStation B.02.02. data acquisition software (Palo Alto, CA, USA) was used to control the operation of GC-MS. ChemStation integration parameters were calibrated as follows: initial threshold = 3, initial peak width = 0.010, initial area reject = 1, and shoulder detection = on. These calibrations yielded the mass spectrum for the peaks. The identification of compounds were based on the comparison of the generated mass spectra and retention times with those of authentic standards and reference spectra in library-MS databases: National Institute of Standards and Technology (NIST) 05, 08, and 11. The determination of the FAMES in all the flour samples was carried out in triplicates.

2.7. Determination of Water-Soluble Vitamins

The determination of water-soluble vitamins was carried out according to Thermo Fisher Scientific [28] (Waltham, MA, USA). For sample preparation, 100 mg of flour sample was mixed with 25 mL of distilled water in a 50 mL falcon tube. The mixture was ultra-sonicated for 15 min and solution filtered through 0.2-µm filters into UPLC vials. The chromatographic analysis was performed on a Liquid chromatographic system with a diode array detector (LC-30AC with Nexera column oven CTO-30A, Shimadzu, Tokyo, Japan). A Phenomenex C18 Column Synergi, 100 × 3.00 mm, 2.6 µm polar (Phenomenex, Torrance, CA, USA) at 30 °C, was used. The mobile phase consisted of two phases: mobile phase A: 25 mM phosphate buffer; mobile phase B: 7:3 *v/v* Acetonitrile-Mobile phase A. The total run time was 12 min with a flow rate of 0.4 mL min^{-1}. Stock solutions of 1.0 mg/mL were prepared by dissolving the individual water-soluble vitamin standards in distilled water except for Vitamin B_2 (in 5 mM potassium hydroxide) and Vitamin B_9 (in 20 mM potassium hydrogen carbonate). Four calibration standards at a concentration of 2, 5, 10 and 15 µg/mL were prepared from the mixed stock solutions. The retention times (mins) for the vitamins were as follows: Vitamin C—1.596, Vitamin B_1—1.922, Nicotinic acid—2.228, Vitamin B_6—3.496, Nicotinamide—5.050, Vitamin B_5—6.772, Vitamin B_9—8.236, Vitamin B_{12}—8.936, and Vitamin B_2—9.224. R^2 was ≥ 0.996. All determinations were carried out in triplicates.

2.8. Determination of Fat-Soluble Vitamins

Fat-soluble vitamins was determined according to a method described by Bhatnagar et al. [29]. Briefly, 6 mL of ethanol with 0.1% (BHT)) was added to 500 mg of the flour sample and homogenized. To the resulting mixture, 120 µL of potassium hydroxide 80% (w/v) was added and vortexed for 1 min followed by incubation at 80 °C for 5 min. Cooling was carried out by placing the test tubes in ice, and 4 mL of deionized water was added to the mixture to enhance phase separation, followed by vortexing for 1 min. For extraction, 5 mL HPLC-grade hexane (Sigma-Aldrich, St. Louis, MO, USA) was added to the mixture, followed by a 5 min centrifugation at 3000 rpm. The resulting hexane layer (upper) was transferred into a separate test tube, and pellet was re-extracted twice more using hexane and the upper phases collected and pooled. Drying was carried out under nitrogen gas flow, and the residue reconstituted in 1 mL of methanol: tetrahydrofuran (85:15 v/v), vortexed and sonicated for 30 s and filtered into HPLC sample vials. Analysis was performed using reverse-phase HPLC (Shimadzu, Tokyo, Japan) linked to SPD-M2A detector. The UPLC was fitted with a YMC C30, carotenoid column (3 µm, 150X3.0 mm, YMC Wilmington, NC, USA). The mobile phase consisted of two phases: A: methanol/tert-butyl methyl ether/water (85:12:3, $v/v/v$, with 1.5% ammonium acetate in the water) and B: methanol/tert-butyl methyl ether/water (8:90:2, $v/v/v$, with 1% ammonium acetate in the water). The injection volume was 10 µL with a total flow rate of 0.4 mL/min. The retention times for retinol and α- and γ-tocopherol were 2.74, 5.40, and 6.29 min, respectively. Compounds presenting the eluting sample were monitored at 290 nm. Peaks were identified by their retention time, and absorption spectra were compared to those of known standards (Sigma Chemicals). Sample concentrations were calculated by comparing peak area of samples to peak area of the standards.

2.9. Assessment of Phytic Acid, Tannins, and Flavonoids

Phytic acid contents were determined using a K–PHYT Phytic Acid (Phytate)/Total Phosphorus kit (Megazyme Int. Ireland Ltd., Wicklow, Ireland) following the manufacturer's instructions [30]. A flour sample (1 g) was added to 20 mL of 0.8 M HCl and mixed by shaking at room temperature overnight, and 1.5 mL of the resulting extract was centrifuged at 13,000 rpm for 10 min. The supernatant (0.5 mL) was neutralized with 0.5 mL of 0.8 M NaOH and used for the enzymatic dephosphorylation reaction and the subsequent total phosphorous and inorganic phosphate quantification. In parallel, 0.5 mL of the neutralized sample were used to quantify the inorganic phosphate of the sample. After the enzymatic treatment, the total phosphorous and total inorganic phosphate of the samples were used for colorimetric development to estimate the phosphorous content. This kit measures the inorganic phosphate released from the extracted flour sample after treatment with phytase and alkaline phosphatase. The free inorganic phosphate content was estimated from samples not treated with phytase. Experiments were performed in triplicate.

Total tannin content was determined by Folin–Denis method, using a microplate reader as outlined by Saxena et al. [31]. The results were expressed in mg tannic acid equivalents (TAE)/100 g of sample. The total flavonoid content was analysed using the Aluminium Chloride colorimetric [32]. The results were expressed in mg catechin equivalents (CEQ)/100 g of dry sample. All determinations were carried out in triplicates.

A flowchart of the various processes involved throughout the development of the novel porridge products and nutrient profiling are summarized in Figure 1 below.

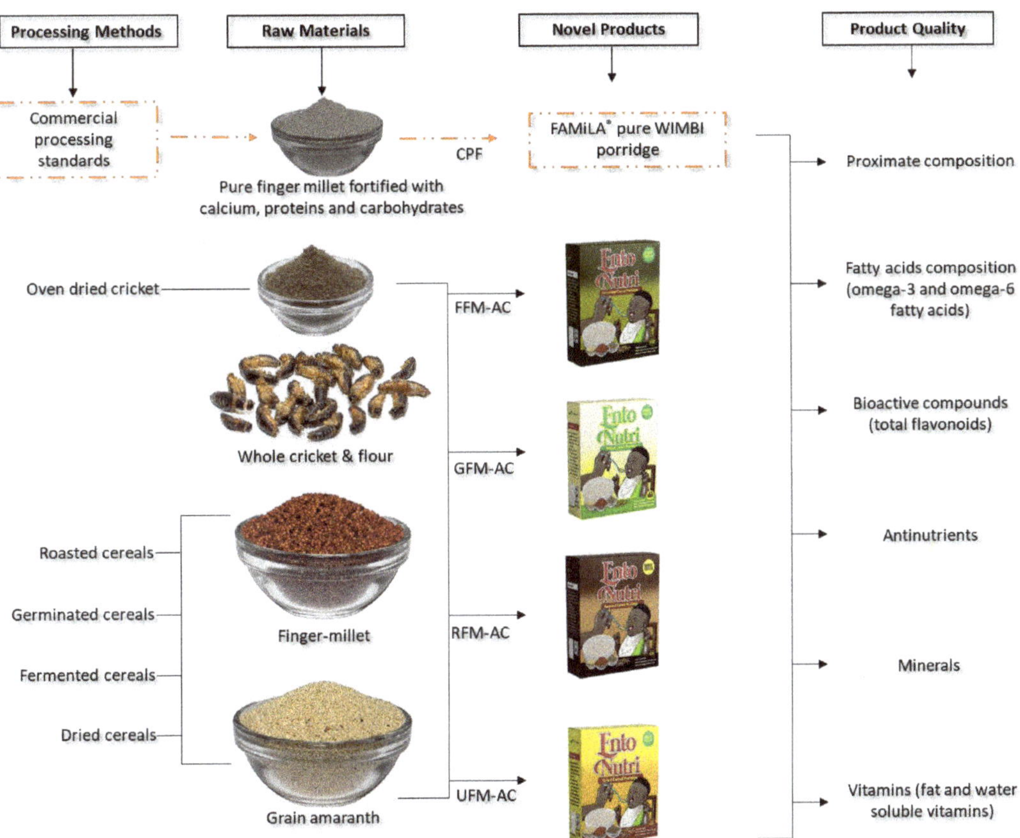

Figure 1. Flow diagram showing the raw materials (grain amaranth, finger-millet, and cricket) and the formulated porridge products [CPF = commercial porridge flour; FFM–AC = fermented finger millet − amaranth + cricket; GFM–AC = germinated finger millet − amaranth + cricket; RFM–AC = roasted finger millet − amaranth + cricket; and UFM–AC = unprocessed finger millet − amaranth + cricket.

2.10. Determination of Molar Ratios

Mineral bioavailability (zinc, iron, and calcium) was expressed as a phytate/mineral molar ratio [33]. Moles of phytic acid were calculated by dividing the recorded value of phytic acid with its atomic weight (660), while the moles of minerals (zinc, iron, and calcium) were determined by dividing the recorded values by the individual molecular weight of the respective compounds (i.e., Zn = 65, Fe = 56, and Ca = 40).

2.11. Statistical Analysis

Datasets of proximate compositions, mineral contents, fatty acids (FAs), and anti-nutritional data were subjected to analysis of variance (ANOVA). The means were separated using the HSD Tukey tests. Statistical analysis was performed using R Studio software version 1.3.1093-1(Boston, MA, USA) [34].

3. Results

3.1. Proximate Composition

The results of the proximate composition and energy values of the formulated porridge products, which consisted of grain amaranth, finger-millet, and cricket, are presented in Table 1. Samples varied significantly ($p < 0.05$) in their proximate values, except on ash content. All products enriched with cricket had significantly ($p < 0.05$) higher protein (2-fold), fat (3.4–4-fold), and energy content (1.1–1.2-fold) when compared to the commercial porridge flour. The germinated sample presented the highest protein content (16.12 g/100 g) followed by the unprocessed sample (15.9 g/100 g), whereas the fat content of the fermented product (7.2 g/100 g) was significantly ($p < 0.05$) lower than that of other formulated products (8.1 g–8.31 g/100 g), but higher when compared to the commercial porridge flour (2.4 g/100 g). The highest crude fibre content was recorded in the germinated sample (5.3 mg/100 g), whereas the lowest was recorded in the fermented sample (3.3 mg/100 g).

Table 1. Proximate and energy values of porridge flour on a dry weight basis.

Products	Proximate Composition						
	Moisture (%)	Ash (g/100 g)	Fiber (g/100 g)	Protein (g/100 g)	Fat (g/100 g)	CHO (g/100 g)	Energy (kcal/100 g)
CPF	11.76 ± 0.25 [e]	2.57 ± 0.04 [ab]	4.79 ± 0.28 [c]	8.55 ± 0.16 [a]	2.14 ± 0.16 [a]	81.94 ± 0.26 [d]	381.28 ± 0.35 [a]
FFM–AC	4.84 ± 0.25 [b]	2.22 ± 0.15 [a]	3.33 ± 0.29 [a]	15.34 ± 0.17 [b]	7.22 ± 0.06 [b]	71.88 ± 0.29 [c]	413.92 ± 0.80 [c]
GFM–AC	5.92 ± 0.42 [c]	2.88 ± 0.48 [b]	5.34 ± 0.02 [d]	16.12 ± 0.15 [d]	8.19 ± 0.09 [c]	67.46 ± 0.39 [a]	408.12 ± 2.27 [b]
RFM–AC	3.02 ± 0.22 [a]	2.76 ± 0.05 [ab]	3.88 ± 0.04 [b]	15.54 ± 0.16 [bc]	8.08 ± 0.28 [c]	69.74 ± 0.13 [b]	413.83 ± 1.44 [c]
UFM–AC	7.44 ± 0.26 [d]	2.83 ± 0.09 [ab]	4.64 ± 0.11 [c]	15.90 ± 0.28 [cd]	8.31 ± 0.38 [c]	68.31 ± 0.71 [a]	411.68 ± 1.71 [bc]

Products: CPF = commercial porridge flour; FFM–AC = fermented finger millet − amaranth + cricket; GFM–AC = germinated finger millet − amaranth + cricket; RFM–AC = roasted finger millet − amaranth + cricket; UFM–AC = unprocessed finger millet − amaranth + cricket. Values are mean (± standard deviation). Moisture content not based on dry weight. Mean values with different superscript letters within columns are significantly different at $p < 0.05$ according to the Tukey test.

3.2. Fatty Acids

The results of FAMEs in different porridge flour samples are presented in Table 2. Of the 44 FAs detected, the proportion of saturated fatty acids (SFAs), monounsaturated fatty acids (MUFAs), and polyunsaturated fatty acids (PUFAs) present in porridge flour samples were 23%, 54%, and 23%, respectively. More FAMEs (3 to 4-fold) were detected in the cricket enriched porridge flours than in the commercial porridge flour. Additionally, the proportion of each group of FAMEs (SFA, MUFA, and PUFA) across the different porridge products are illustrated in Figure 2. Of the 24 SFAs detected, Methyl Hexadecanoate (palmitic acid) contributed the highest proportion, followed by Methyl Octadecanoate (stearic acid), across the flour samples. In addition, Methyl 9E-Octadecenoate (oleic acid) was the predominant MUFA, whereas Methyl 9Z,12Z-Octadecadienoate (linoleic acid, LA) accounted for the highest proportion of the PUFAS.

Table 2. Compositions of fatty acids (μg/g of oil) of porridge flour samples analysed using gas chromatography coupled with mass spectrometry (GC-MS).

Peak No.	tR (min)	Compound Name	ω-n(Δn)	CPF	FFM-AC	GFM-AC	RFM-AC	UFM-AC
1	18.96	Methyl Dodecanoate	C12:0	-	0.67 ± 0.09	0.27 ± 0.01	0.58 ± 0.05	0.35 ± 0.02
2	19.72	Methyl 11-Methyldodecanoate	Iso-methyl-C12:0	-	0.05 ± 0.01	0.06 ± 0.00	-	
3	20.12	Methyl Tridecanoate	C13:0	-	0.12 ± 0.01	0.06 ± 0.00	0.09 ± 0.01	0.66 ± 0.00
4	20.82	Methyl 12-Methyltridecanoate	Iso-methyl-C13:0	-	0.21 ± 0.02	0.12 ± 0.00	0.07 ± 0.01	0.13 ± 0.01
5	21.22	Methyl Tetradecanoate	C14:0	0.85 ± 0.04	14.72 ± 0.84	6.09 ± 0.10	12.04 ± 1.21	8.26 ± 0.97
6	21.78	Methyl 4-Methyldodecanoate	Iso-methyl-C12:0	-	0.72 ± 0.07	0.44 ± 0.00	-	-
7	22.00	Methyl 13-Methyltetradecanoate	Iso-methyl-C14:0	-	3.64 ± 0.21	1.98 ± 0.01	1.42 ± 0.17	0.89 ± 0.17
8	22.00	Methyl 12-Methyltetradecanoate	Iso-methyl-C14:0	-	0.85 ± 0.04	0.54 ± 0.00	1.84 ± 0.10	0.29 ± 0.19
9	22.29	Methyl Pentadecanoate	C15:0	0.38 ± 0.04	3.79 ± 0.28	1.61 ± 0.00	3.20 ± 0.48	2.93 ± 0.56
10	22.74	Methyl 5,9,13-Trimethyltetradecanoate	Iso-trimethyl-C14:0	-	0.45 ± 0.05	0.00 ± 0.00	-	-
11	22.94	Methyl 14-Methylpentadecanoate	Iso-methyl-C15:0	-	1.22 ± 0.10	0.85 ± 0.01	1.24 ± 0.07	0.94 ± 0.06
12	23.37	Methyl Hexadecanoate	C16:0	32.22 ± 1.37	200.94 ± 6.55	255.32 ± 7.40	559.63 ± 50.55	319.22 ± 28.46
13	23.93	Methyl 15-Methylhexadecanoate	Iso-methyl-C16:0	-	3.89 ± 0.38	2.90 ± 0.06	3.09 ± 0.42	3.15 ± 0.39
14	24.02	Methyl 14-Methylhexadecanoate	Iso-methyl-C16:0	1.30 ± 0.13	11.82 ± 0.94	5.85 ± 0.12	12.80 ± 1.11	6.23 ± 0.88
15	24.29	Methyl Heptadecanoate	C17:0	1.11 ± 0.22	9.92 ± 0.71	3.08 ± 0.05	7.40 ± 0.68	4.58 ± 0.91
16	24.69	Methyl 14-Methylheptadecanoate	Iso-methyl-C17:0	-	1.40 ± 0.18	1.43 ± 0.02	2.99 ± 0.24	1.66 ± 0.20
17	25.25	Methyl Octadecanoate	C18:0	13.02 ± 0.51	170.54 ± 2.98	78.14 ± 1.41	36.71 ± 2.68	114.00 ± 23.09
18	26.12	Methyl Nonadecanoate	C19:0	-	2.08 ± 0.14	1.46 ± 0.05	1.92 ± 0.06	1.44 ± 0.30
19	26.98	Methyl Eicosanoate	C20:0	3.05 ± 0.06	13.07 ± 0.15	11.07 ± 0.00	26.99 ± 1.49	16.48 ± 2.30
20	27.58	Methyl 18-Methyleicosanoate	Iso-methyl-C20:0	-	4.90 ± 0.78	3.45 ± 0.00	5.11 ± 0.76	2.80 ± 0.17
21	27.80	Methyl Heneicosanoate	C21:0	-	2.30 ± 0.23	3.34 ± 0.30	3.75 ± 0.66	1.79 ± 0.10
22	28.59	Methyl Docosanoate	C22:0	2.47 ± 0.02	5.35 ± 0.13	6.30 ± 0.13	9.01 ± 0.45	6.92 ± 0.85
23	29.37	Methyl Tricosanoate	C23:0	-	3.95 ± 0.09	3.93 ± 0.01	4.63 ± 0.39	2.89 ± 0.21
24	30.13	Methyl Tetracosanoate	C24:0	-	9.45 ± 0.54	6.07 ± 0.04	10.98 ± 0.48	4.43 ± 1.23
		∑ SFA						
25	20.95	Methyl 11Z-Tetradecenoate	C14:1 (n-3)	-	0.52 ± 0.01	-	0.93 ± 0.09	-
26	21.08	Methyl 9Z-Tetradecenoate	C14:1 (n-3)	-	0.89 ± 0.09	-	0.67 ± 0.07	-
27	23.12	Methyl 9Z-Hexadecenoate	C16:1 (n-7)	2.24 ± 0.26	55.80 ± 1.46	18.50 ± 0.16	67.70 ± 1.84	31.80 ± 2.43
28	24.09	Methyl 10Z-Heptadecenoate	C17:1 (n-7)	-	7.42 ± 1.10	1.37 ± 0.03	1.75 ± 0.16	0.81 ± 0.08
29	25.00	Methyl 11-Octadecenoate	C18:1 (n-9)	-	3.69 ± 0.18	3.48 ± 0.11	2.83 ± 0.19	3.72 ± 0.12
30	25.07	Methyl 9E-Octadecenoate	C18:1 (n-9)	111.68 ± 6.88	729.29 ± 41.54	630.70 ± 12.89	1149.35 ± 93.82	332.62 ± 40.54
31	25.88	Methyl 10-Nonadecenoate	C19:1 (n-9)	-	2.69 ± 0.14	3.81 ± 0.17	8.99 ± 0.45	4.55 ± 0.46
32	26.77	Methyl 11Z-Eicosenoate	C20:1 (n-9)	2.94 ± 0.14	8.63 ± 0.55	8.52 ± 0.16	19.96 ± 0.57	11.31 ± 2.20
33	28.41	Methyl 11-Docosenoate	C22:1 (n-11)	-	4.23 ± 0.20	-	3.27 ± 0.30	-
34	29.95	Methyl 15Z-Tetracosenoate	C24:1 (n-9)	-	3.31 ± 0.16	-	-	-
		∑ MUFA						
35	24.74	Methyl 9Z,12Z-Octadecadienoate	C18:2 (n-6)	-	74.70 ± 4.02	38.38 ± 2.09	16.69 ± 1.10	22.62 ± 2.07
36	24.76	methyl 6Z,9Z,12Z-Octadecatrienoate	C18:3 (n-3)	-	4.60 ± 0.43	3.48 ± 0.40	1.67 ± 0.15	2.80 ± 0.03
37	25.41	Methyl 7,12-Octadecadienoate	C18:2 (n-7)	-	8.87 ± 0.66	11.07 ± 0.94	6.55 ± 1.96	5.22 ± 0.27
38	25.79	Methyl 9Z,11E-Octadecadienoate	C18:2 (n-7)	-	2.39 ± 0.09	2.60 ± 0.42	1.62 ± 0.21	1.35 ± 0.01

Table 2. *Cont.*

Peak No.	tR (min)	Compound Name	ω-n(Δn)	CPF	FFM-AC	GFM-AC	RFM-AC	UFM-AC
39	26.24	Methyl 9Z,11E,13E-Octadecatrienoate (α-ESA)	C18:3 (n-3)	-	1.03 ± 0.04	0.92 ± 0.10	0.57 ± 0.00	0.74 ± 0.03
40	26.26	Methyl 9Z,12Z,15Z-Octadecatrienoate (ALA)	C18:3 (n-3)	-	7.01 ± 0.23	7.10 ± 0.61	3.79 ± 0.01	5.07 ± 0.02
41	26.44	Methyl 5Z,8Z,11Z,14Z-Eicosatetraenoate (AA)	C20:4 (n-6)	-	1.55 ± 0.09	2.76 ± 0.06	1.89 ± 0.04	1.95 ± 0.19
42	26.50	Methyl 5Z,8Z,11Z,14Z,17Z-Eicosapentaenoate (EPA)	C20:5 (n-3)	-	13.07 ± 0.77	4.44 ± 0.02	-	-
43	26.64	Methyl 8,11,14,17-Eicosatetraenoate (AA)	C20:4 (n-6)	-	2.040 ± 0.10	-	-	-
44	28.07	Methyl 4Z,7Z,10Z,13Z,16Z,19Z-Docosahexaenoate (DHA)	C22:6 (n-3)	-	2.40 ± 0.34	-	-	-
		∑ PUFA		-	117.66 ± 3.17	70.76 ± 1.30	32.68 ± 1.72	39.73 ± 2.16
		∑ n-6 PUFA		-	89.55 ± 4.42	54.82 ± 0.92	26.65 ± 1.68	31.13 ± 2.15
		∑ n-3 PUFA		-	28.11 ± 0.75	15.94 ± 0.99	6.03 ± 1.15	8.60 ± 0.01
		∑ n-6/n-3		-	3.2	3.4	4.4	3.6
		∑ ALA + EPA + DHA		-	27.08 ± 0.70	15.02 ± 1.00	5.47 ± 1.15	7.87 ± 0.04

(tR Retention time, Mean ± standard deviation). SFA = saturated fatty acids, MUFA = monounsaturated fatty acids, PUFA = polyunsaturated fatty acids, ALA = α -Linolenic acid, EPA = Eicosapentaenoic acid, DHA = Docosapentaenoic acid, α-ESA = alpha Eleostearic acid.

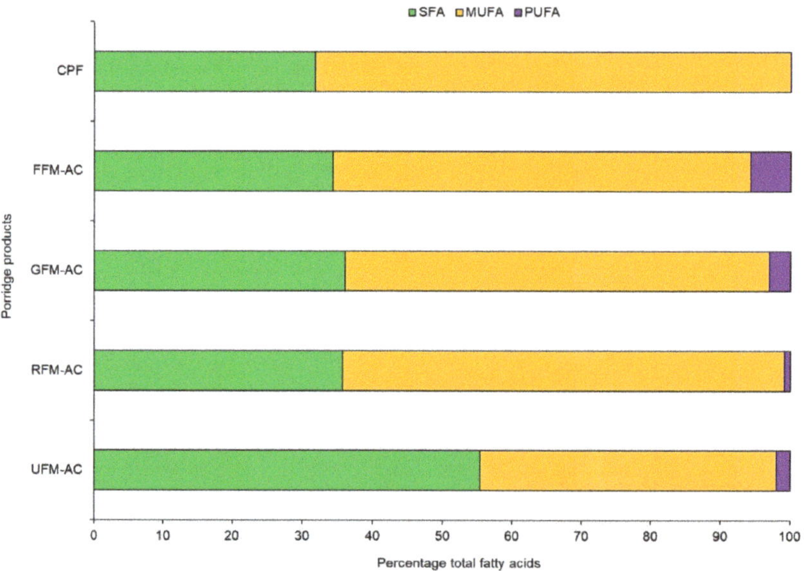

Figure 2.]Total fatty acid composition in different porridge flour samples. SFA = Saturated fatty acid; MUFA = monounsaturated fatty acid; PUFA = polyunsaturated fatty acid (PUFA); FFM–AC = fermented finger millet − amaranth + cricket; RFM–AC = roasted finger millet − amaranth + cricket; GFM–AC = germinated finger millet − amaranth + cricket; UFM–AC = unprocessed finger millet − amaranth + cricket; CPF = commercial porridge flour.

The proportion of omega-3 fatty acids, namely α-linolenic acid (ALA), Eicosapentaenoic acid (EPA), Alpha Eleostearic acid (α-ESA), and Docosapentaenoic acid (DHA) differed considerably across the flour samples. α-linolenic acid (ALA) was detected in all the cricket enriched samples, while EPA was present in the fermented and germinated sample. DHA was only present in the fermented flour. On the effect of processing, PUFAs increased significantly ($p < 0.05$) by 30% during fermentation and decreased by 3% during roasting. Additionally, the roasting process caused a significant ($p < 0.05$) increase in both MUFAs and SFAs by 27% and 10%, respectively. Germination caused a slight increase in 9% and 12% in both MUFAs and PUFAs, respectively. The ratio of omega-6 to omega-3 varied significantly among the different flour samples and ranged from 3.2 to 4.4.

3.3. Vitamin Content

The results show significant variations in the vitamin content (Table 3). Cricket enriched formulations had a significantly ($p < 0.05$) higher content of vitamin B_{12}, vitamin B_5, B_6, nicotinamide, and thiamine. Thiamine content ranged from 4.3–39.5 mg/100 g while vitamin B_{12} was in the range of 3.2–37.7 mg/100 g. Processing methods had a significant ($p < 0.05$) effect on the vitamin levels. Germinated and fermented samples had enhanced levels of vitamin C, nicotinamide, vitamin B_5, folate, and vitamin B_6. However, there was a significant ($p < 0.05$) decrease to undetectable levels in the contents of nicotinic acid and thiamine during fermentation and germination, respectively. Vitamin B_{12} showed significant reductions in all the processed porridge flours, while riboflavin content was not affected by processing. The roasting process had a negligible effect on vitamins, except for significant ($p < 0.05$) reductions in nicotinic acid and vitamin B_{12}, and a slight reduction in vitamin C. The retinol and α- and γ-tocopherol content differed significantly ($p < 0.05$) in different porridge flour samples. Retinol content increased slightly during fermentation, while germination process decreased the levels of retinol and increased the α-tocopherol content. However, the roasting process did not affect the levels of α- and γ-tocopherol.

Table 3. Vitamin content (mg/100 g) in porridge flour products.

Vitamins	Porridge Products				
	CPF	**FFM–AC**	**GFM–AC**	**RFM–AC**	**UFM–AC**
Vitamin C	149.6 ± 2.1 [c]	146.5 ± 2.8 [c]	72.0 ± 6.5 [b]	55.2 ± 2.9 [a]	58.0 ± 5.1 [a]
Thiamine (B_1)	–	39.5 ± 3.0 [c]	–	4.3 ± 0.2 [b]	5.9 ± 0.4 [b]
Nicotinic acid (B_3)	27.7 ± 2.9 [e]	–	19.5 ± 1.2 [d]	6.1 ± 0.5 [b]	10.6 ± 0.7 [c]
Pyridoxine (B_6)	0.5 ± 0.1 [a]	10.8 ± 1.1 [c]	6.0 ± 0.3 [b]	–	–
Nicotinamide	3.0 ± 0.3 [a]	47.9 ± 2.1 [d]	33.9 ± 1.1 [c]	7.1 ± 0.2 [b]	8.8 ± 0.4 [b]
Pantothenic acid (B_5)	26.4 ± 2.5 [a]	423.3 ± 3.4 [d]	453.8 ± 44.5 [d]	314.4 ± 18.4 [c]	209.6 ± 3.8 [b]
Folate (B_9)	28.6 ± 2.6 [a]	38.8 ± 1.5 [b]	42.4 ± 3.7 [b]	41.8 ± 0.3 [b]	29.3 ± 1.8 [a]
Cyanocobalamin (B_{12})	3.2 ± 0.3 [a]	21.9 ± 1.4 [c]	13.7 ± 4.0 [b]	12.4 ± 0.9 [b]	37.7 ± 4.0 [d]
Riboflavin (B_2)	74.2 ± 8.2 [b]	41.6 ± 1.0 [a]	34.8 ± 2.9 [a]	45.5 ± 2.4 [a]	41.6 ± 5.0 [b]
Retinol	0.55 ± 0.03 [c]	0.54 ± 0.08 [c]	0.07 ± 0.01 [a]	0.29 ± 0.01 [ab]	0.38 ± 0.19 [bc]
γ-Tocopherols	0.88 ± 0.04 [c]	0.17 ± 0.01 [a]	0.19 ± 0.01 [a]	0.54 ± 0.00 [b]	0.52 ± 0.02 [b]
α-Tocopherols	0.46 ± 0.06 [a]	0.35 ± 0.09 [a]	1.48 ± 0.03 [c]	0.76 ± 0.08 [b]	0.83 ± 0.03 [b]

Porridge product: FFM–AC = fermented finger millet − amaranth + cricket; RFM–AC = roasted finger millet − amaranth + cricket; GFM–AC = germinated finger millet − amaranth + cricket; UFM–AC = unprocessed finger millet − amaranth + cricket; CPF = commercial porridge flour. Values are mean ($\pm$ standard deviation). Mean values with different superscript letters within rows are significantly different at $p < 0.05$, according to the Tukey test.

3.4. Mineral Content and Molar Ratios

The mineral content varied significantly across the porridge flour products as shown in Table 4. The iron content of the formulated flours ranged between 8.59–19.48 mg/100 g with the germinated sample having the highest content (19.48 mg/100 g). Zinc content was in the range of 3.08–3.70 mg/100 g, while the range obtained for calcium was from 234.87 mg–278.61 mg/100 g. The calcium content of the commercial porridge flour was significantly ($p < 0.05$) higher at 312.69 mg/100 g when compared to the formulated samples, while the Zn content was significantly lower (1.86 mg/100 g). The fermented sample had significantly ($p < 0.05$) lower levels of magnesium and phosphorous, but the levels of calcium, copper, iron, and zinc did not vary when compared with the unprocessed formulation. Calcium and iron content in the germinated sample were higher than in other formulations. The roasting process did not significantly ($p > 0.05$) affect the mineral content.

Table 4. Mineral content and bioavailability of porridge flour formulations.

Minerals	Porridge Products				
	CPF	**FFM–AC**	**GFM–AC**	**RFM–AC**	**UFM–AC**
Mg (mg/100 g)	145.08 ± 0.25 [a]	169.57 ± 8.03 [b]	207.82 ± 11.93 [c]	210.84 ± 1.45 [c]	203.09 ± 5.14 [c]
Fe (mg/100 g)	9.86 ± 2.08 [a]	8.56 ± 1.45 [a]	19.48 ± 6.69 [b]	9.18 ± 1.18 [a]	9.55 ± 2.42 [a]
Ca (mg/100 g)	312.69 ± 0.57 [c]	234.87 ± 17.60 [a]	278.61 ± 17.90 [b]	257.69 ± 1.37 [ab]	244.69 ± 4.94 [a]
Zn (mg/100 g)	1.86 ± 0.04 [a]	3.23 ± 0.28 [b]	3.71 ± 0.18 [b]	3.39 ± 0.31 [b]	3.08 ± 0.16 [b]
P (mg/100 g)	221.63 ± 5.57 [a]	372.71 ± 19.14 [b]	469.28 ± 9.55 [c]	458.70 ± 3.76 [c]	476.72 ± 17.46 [c]
Mn (mg/100 g)	22.44 ± 0.20 [c]	10.92 ± 0.63 [b]	9.32 ± 0.67 [a]	9.45 ± 0.15 [a]	8.64 ± 0.04 [a]
Cu (μg/100 g)	477.42 ± 1.88 [a]	728.78 ± 15.37 [b]	787.20 ± 42.28 [b]	724.94 ± 22.80 [b]	736.28 ± 14.47 [b]
Molar ratios (Bioavailability)					
Phy:Fe	1.46 ± 0.05 [a]	5.69 ± 1.38 [b]	1.31 ± 0.42 [a]	9.63 ± 1.41 [c]	7.76 ± 1.90 [bc]
Phy:Zn	9.03 ± 1.97 [a]	17.27 ± 1.40 [b]	7.43 ± 0.64 [a]	30.48 ± 5.39 [c]	26.90 ± 2.04 [c]
Phy:Ca	0.03 ± 0.01 [a]	0.14 ± 0.00 [c]	0.06 ± 0.01 [a]	0.24 ± 0.01 [e]	0.21 ± 0.00 [d]

Porridge flour product: FFM–AC = fermented finger millet − amaranth + cricket; RFM–AC = roasted finger millet − amaranth + cricket; GFM–AC = germinated finger millet − amaranth + cricket meal; UFM–AC = unprocessed finger millet − amaranth + cricket; CPF = commercial porridge flour. Values are mean (± standard deviation). Mean values with different superscripts within rows are significantly different at $p < 0.05$ according to the Tukey test.

Processing methods had a significant ($p < 0.05$) influence on the phytate/mineral molar ratios (Table 4). The critical values for the Phy:Fe, Phy:Zn, and Phy:Ca ratios were 1.0, 15.0 and 0.24, respectively [35]. Germinated sample had the lowest ratios for Phy:Fe (1.31) and Phy:Zn (7.43), whereas commercial flour had the lowest ratio for Phy:Ca (0.03). Phy:Ca ratios recorded in all the flours were below the limit threshold, except for the roasted sample (0.24) which had the highest values of all molar ratios.

3.5. Phytic Acid, Tannins, and Flavonoids

The phytic acid, tannin, and flavonoid contents of the flour samples are summarized in Figure 3. The phytic acid content of the roasted sample was significantly ($p < 0.05$) higher (1029 mg/100 g), followed by the unprocessed sample (837 mg/100 g), while the germinated sample had the lowest (279 mg/100 g) among the processed formulations. The highest ($p < 0.05$) tannin content (683 mg/100 g) was recorded in the fermented sample and the lowest content (381 mg/100 g) was detected in the germinated sample. The flavonoid content of the porridge flours was in the range of 60 mg–166 mg/100 g, with the highest value (166 mg/100 g) recorded for the commercial porridge flour.

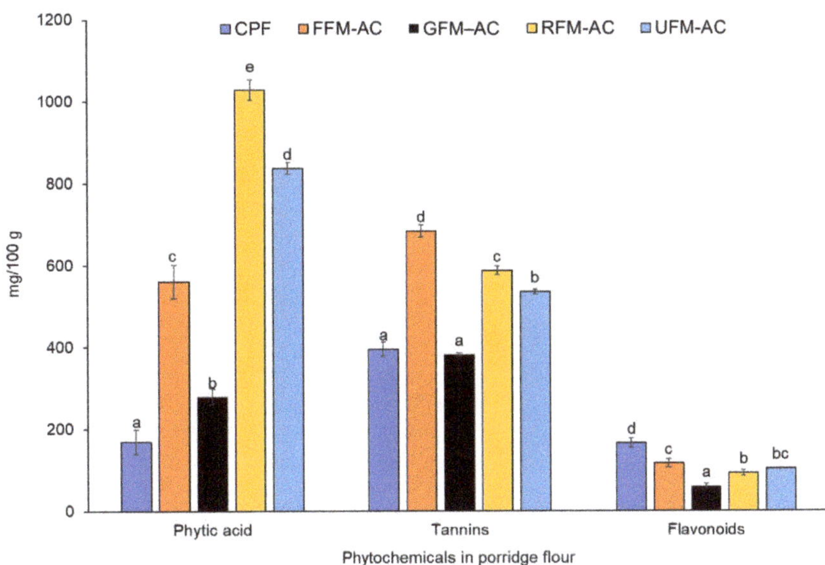

Figure 3. Bar graphs represent mean and error bar standard deviation of phytochemical content in the flour samples. Different letters above the error bar indicate significant differences in phytochemical content. CPF = commercial porridge flour; FFM–AC = fermented finger millet − amaranth + cricket; GFM–AC = germinated finger millet − amaranth + cricket; RFM–AC = roasted finger millet − amaranth + cricket; and UFM–AC = unprocessed finger millet − amaranth + cricket. mg tannic acid equivalents (TAE)/100 g of sample, mg catechin equivalents (CEQ)/100 g of dry sample.

4. Discussion

In low- and middle-income countries, child malnutrition contributes to about 45% of under-five child mortality, and this portends great danger to Africa's growth and development. One third of child deaths in Africa are attributable largely to protein energy malnutrition and micronutrient deficiencies [44,45], which could be solved by exploring underutilized nutritious crop and animal species of African origin. Traditional cereal-based porridge preparation has been considered as one of the major causes of protein energy malnutrition in developing countries [35]. The traditional complementary foods are usually low in protein and energy density, while containing high anti-nutrient content [36]. This study shows that formulating porridge products with underutilized edible cricket and amaranth meal, using household techniques to increase the protein, energy, vitamin, fatty acids, and mineral content in the porridge flour, could be the best substitute for traditional weaning flours. The recommended daily intake of protein in children aged 1–3 years is 13 g/day [12], and our study showed that the protein content in cricket and amaranth enriched flour ranged between 15.34–16.12 g/100 g (dwb). The observed protein content in cricket and amaranth enriched porridge flour is higher than that reported by Agbemafle et al. [46] in complementary food with orange-fleshed sweet potato, enriched with palm weevil larvae. The high protein in our formulated porridge products may be accredited to the addition of cricket and amaranth powder, which are known to be highly rich in protein [47].

The crude protein of the various porridge products fortified with cricket and amaranth meal varied slightly. This variation could be associated with the amino acids being metabolized into ammonia and other volatile flavour compounds during the various process methods, especially during fermentation [48]. A similar trend was reported during the fermentation of pearl millet by Osman [49]. Conversely, there was a slight increase in the protein content of the germinated product, which could be attributed to the mobilization of storage nitrogen and the synthesis of enzymatic proteins by the sprouting seeds during germination [50].

Fat content is an important factor in influencing the energy density of foods. It also provides essential FAs, improves the absorption of fat-soluble vitamins, and the sensory quality of food [51]. The fat content of formulated porridge products with cricket and amaranth meal was higher than that of commercial porridge products. Fermentation process caused a 13% reduction in fat content, indicating a possible active utilization of fats by microorganisms. These findings are in agreement with the observations by Assohoun et al. [52] who reported a reduction in fat content in fermented maize porridge products. The fiber content was increased during germination but reduced during fermentation and roasting. The increase in fibre content during germination could be attributed to the utilization of sugars by spouting seeds [53]. According to the Codex Alumentarius [54], the energy requirement for complementary foods is 400 kcal/100 g on dry weight basis. The energy contents in all the processed and formulated porridge products ranged from 408.12 to 413.92 kcal/100 g, which is higher compared to the commercial porridge flour (381.28 kcal/100 g). This implies that the new porridge products can sufficiently and adequately meet everyone's recommended energy allowance and nutrient needs.

Fatty acids, especially PUFAs, have enormous health benefits, including preventing degenerative ailments and promoting brain development and proper immune functioning [55,56]. The intake of omega-3 FAs during pregnancy and lactation reduces mortality and allergies in infants and improves their cognitive functions [57,58]. The formulations varied in the composition of FA with the fermented porridge products integrated with cricket and amaranth meal having the highest proportion of PUFAs, including the omega-3 FAs. The ratios of omega-6 to omega-3 FAs were lower than those reported by Paucean et al. [59] on wheat-lentil composite flour and Cheseto et al. [27] on insect oils and cookies. The fermentation process caused an increase in the levels of PUFAs and MUFAs and a decrease in the SFAs. Two omega-3 FAs (ALA + EPA) were detected in all the newly formulated flour products, but DHA was only present in the fermented porridge product, an indication of biosynthesis of DHA during fermentation. The increase in PUFAs, particularly omega-3 FAs, during fermentation could be attributed to microbial synthesis [60]. In this study, roasting increased the amounts of SFAs and oleic acid but decreased the amounts of unsaturated linoleic acid and omega-3 FAs. Studies indicate that the rate of FA oxidation increases as the number of double bonds increases, hence PUFAs will isomerize at a higher rate than MUFAs (oleic acid) [61], which partly explains the increase and decrease of oleic and linoleic acids, respectively, during roasting. Germination caused a 9% decrease in MUFAs and 12% increase in the levels of PUFAs. The resulted in an increase in PUFAs during germination, which is similar to the findings reported by Mariod et al. [62] on germinated black cumin seeds.

The formulated porridge products had increased amounts of essential minerals including zinc and iron. The iron content in the formulated products (8.6–19.5 mg/100 g) met the 7 mg/day recommended daily allowance (RDA) for iron in young children aged 1–3 years. Zinc content was in the range of 3.1–3.7 mg/100 g, which contributes about 75–90% RDA for zinc (4.1 mg/day) in young children. All formulated porridge products were high in calcium content (234.9–312.8 mg/100 g), although the levels were inadequate for providing the RDA (500 mg/day) for all the target age groups [11]. Fermentation processing only reduced the concentrations of Mg and P, while increasing that of Mn. reduction in mineral content during fermentation, has also been observed in previous studies involving cowpea porridge flour [63]. However, germination has been reported to increase the levels of Ca and Fe, which is consistent with previously reported studies by Tizazu et al. [64]. The increase in mineral content could be attributed to losses of water-soluble compounds during soaking.

Vitamins are organic compounds that are required in small amounts for the maintenance of normal health and biological reactions in the cells [65]. In this study, the fermentation process caused an increase in vitamin C, thiamine (B$_1$), pyridoxine (B$_6$), pantothenic acid (B$_5$), folate (B$_9$), and nicotinamide. The increase in water-soluble vitamins, particularly of the B-group, during fermentation could be as a result of microbial synthesis [66]. The results are in agreement with Kaprasob et al. [67] who reported an increase in B-group vitamins during fermentation of cashew apples, using probiotic strains of lactic acid bacteria (LAB). However, fermentation reduced nicotinic acid and vitamin B$_{12}$ in the various porridge products, and the losses might be attributed to utilization by LAB in their metabolic biosynthesis necessary for growth [68]. Germination on the other hand, enhanced the concentrations of vitamin C, nicotinic acid, vitamin B$_6$, B$_5$, and B$_9$. The increase could be due to the synthesis of these vitamins by the germinating seeds [69]. The increase in vitamin C during germination has been reported by other researchers and is attributed to the enzymatic hydrolysis of starch by amylases, leading to an increase in the bioavailability of glucose for biosynthesis of vitamin C [69,70]. However, the loss of thiamine to undetectable levels was recorded in the germinated porridge products; this might be due to leaching in the spouting medium [71]. The slight effect of roasting on riboflavin and nicotinamide in porridge flours suggests high thermal stability of B group vitamins [72]. The germination process increased the α-tocopherol content, and these results are comparable to those reported by Young et al. [73] on germinated rough rice seeds. The roasting process, however, did not affect the levels of tocopherols; this could be due to the fact that tocopherols are resistant to thermal degradation [74]. Contrarily, Stuetz et al. [75] reported the loss of tocopherols in roasted nuts compared to raw nuts; this variation might be explainable by the different roasting conditions used in the study.

Phytic acid is the principal storage form of phosphorous in plant seeds [76] and is known to reduce the bioavailability of dietary Zn, Fe, and Ca in humans and monogastric animals, owing to their chelating properties [77]. Both fermented and germinated porridge products had reduced levels of phytic acid when compared to the unprocessed sample. The reduction in phytic acid in the fermented product may be attributed to the effect of microbial phytase and the low pH which favours the activity of endogenous cereal phytase during fermentation [78]. Phytic acid degradation during the germination process could be due to increased activity of enzyme phytase in spouting grains which hydrolyse phytic acid into lower inositol phosphates [79]. Germination decreased phytic acid by 67% as compared to fermentation at 33%. The effect of germination on phytic acid levels could be attributed to a combination of processes such as phytase activity and the leaching of phytate ions into water during soaking before the germination of grains. Minor degradation of phytic acid during fermentation could also be attributed to the high phytic acid content and presence of endogenous phytase inhibitory compounds, such as tannins in both finger millet and amaranth grains [80]. Fermentation is more effective when carried out in grains with low anti-nutrients levels [81]. Contrarily, the increase in phytic acid during roasting could be as a result of an increase in lower phosphorylated inositol phosphates, which in turn increased the total phytic acid content. Similar results were observed by Frontela et al. [82]. However, our study did not differentiate between the classes of phytates but rather focused on the total phytic acid content. Moreover, the roasting temperature (120 °C) restricted endogenous phytase enzyme activity whose optimum temperature varies from 35 to 80 °C [83]. The low levels of phytic acid in the commercial porridge flour could be due to the use of exogenous phytate degrading enzymes during the industrial processing of the flour. The complete degradation of phytic acid in the four formulated porridge products is feasible with the use of additional exogenous phytases [83].

The phytate/mineral molar ratio is an indicator of mineral bioavailability in plant–based foods [33]. Flour samples from germinated grains had good bioavailability of Zn, while all porridge products, except for the roasted sample, had good bioavailability of Ca. The increased bioavailability of minerals corresponds with the reduction in phytic acid. The commercial porridge flour had good bioavailability of minerals except for Fe, and this could be attributed to low levels of phytic acid as well as the addition of extra minerals during fortification (as stated in the label).

Flavonoids exhibit powerful antioxidative properties, which makes them significant in the prevention of degenerative diseases [84]. The total flavonoid contents in our products ranged between 59.8–165.5 mg/100 g. Flavonoid content increased during fermentation, and this is consistent with the observations by Adetuyi and Ibrahim [85], who reported an increase in total flavonoid content during the fermentation of okra seeds. The increase in flavonoid content could be attributed to the release of simple phenolic compounds during the acid and microbial hydrolysis of complex phenolic compounds during fermentation [86]. A reduction in flavonoid content was noted during germination (42%) and roasting (10%). The reduction during roasting could be due to flavonoids' sensitivity to high temperatures [87]. The results on the reduction of flavonoids during roasting are similar to the findings of other authors [88].

Tannins are regarded as antinutrients because of their ability to inhibit digestive enzymes, lowering the digestibility of most nutrients, particularly proteins [89]. Tannin content reduced during germination is probably due to the complexing of tannins with seed proteins and metabolic enzymes [90] and leaching into the sprouting medium [91]. Significant increase in tannin content observed during fermentation might be due to the hydrolysis of various components, including condensed tannins by catabolic enzymes. Similar findings were reported by Osman [49] during the fermentation of pearl millet. However, reductions in tannin content during fermentation have also been reported [92].

5. Conclusions

The projected growth in global population with the attendant increase in food demands, especially in the African region, calls for focusing and devoting more attention to the underutilized crop and animal species, which have great potential to influence and improve food security for African nations and promote sustainable rural growth and development. Grain amaranth and edible insects are some of such neglected and underutilized species, despite their great inherent health-promoting components, which are good for human applications and uses. The demand for indigenous vegetable/grains and cricket-based products have risen dramatically in recent years because of the rapidly expanding human population. Our findings have also demonstrated the importance of integrating cricket and amaranth grain meal into porridge products to produce nutritious food sufficient for meeting the recommended daily allowance of the most vulnerable segments of the population, if processed properly. The comparative analysis of various processing protocols would empower communities to develop acceptable composite porridge products that are appealing to the consumers and increase demand. The four porridge products developed in this study are higher in energy density and nutritional value, as evidenced by the improved fat, protein, vitamin, and mineral contents in the various products, compared to the widely consumed commercial porridge product. Although complete degradation of phytic acid was not achieved, this study demonstrated that germination and fermentation techniques played a key role in improving the bioavailability of nutrients in the porridge products. Additional, innovative traditional processing technologies to increase phytic acid degradation are warranted. From a nutritional point of view, further studies on protein digestibility and utilization, the bioavailability of essential mineral and trace elements as well as other factors such as pH, the concentration of enhancers, and inhibitors (dietary fibre and polyphenols) would be crucial, particularly for cricket-based food products.

Author Contributions: Conceptualization, N.C.M., F.M.K. and C.M.T.; methodology, N.C.M., F.M.K., C.X., G.O.A. and C.M.T.; formal analysis, N.C.M., C.M.T. and E.R.O.; investigation, N.C.M. and C.M.T.; re-sources, S.E., F.M.K. and C.M.T.; data curation, N.C.M. and E.R.O.; writing—original draft preparation, N.C.M.; writing–review and editing, N.C.M., F.M.K., M.W.O., G.O.A., C.X., S.S., J.P.E., E.R.O., D.N., G.S., C.J.G. and C.M.T.; visualization, N.C.M., C.X., E.R.O. and C.M.T.; supervision, F.M.K., M.W.O., G.O.A. and C.M.T.; project administration, C.M.T.; funding acquisition, C.M.T. All authors have read and agreed to the published version of the manuscript.

Funding: The authors gratefully acknowledge the financial support for this research by the following organizations and agencies: BioInnovate Africa Programme (INSBIZ—Contribution ID No. 51050076); the Curt Bergfors Foundation Food Planet Prize Award; Bill & Melinda Gates Foundation (INV-032416); Canadian International Development Research Centre (IDRC); the Australian Centre for International Agricultural Research (ACIAR) (INSFEED—Phase 2: Cultivate Grant No: 108866–001); Norwegian Agency for Development Cooperation, the section for research, innovation, and higher education grant number RAF–3058 KEN–18/0005 (CAP–Africa); the Swedish International Development Cooperation Agency (Sida); the Swiss Agency for Development and Cooperation (SDC); the Federal Democratic Republic of Ethiopia; and the Government of the Republic of Kenya. The funders had no role in study design, data collection and analysis, decision to publish, or preparation of the manuscript. The views expressed herein do not necessarily reflect the official opinion of the donors.

Institutional Review Board Statement: Not applicable.

Informed Consent Statement: Not applicable.

Data Availability Statement: Data are contained within the article.

Conflicts of Interest: The authors declare no conflict of interest.

References

1. Wanjala, G.; Onyango, A.; Makayoto, M.; Onyango, C. Indigenous Technical Knowledge and Formulations of Thick (Ugali) and Thin (Uji) Porridges Consumed in Kenya. *Afr. J. Food Sci.* **2016**, *10*, 385–396. [CrossRef]
2. Békés, F.; Schoenlechner, R.; Tömösközi, S. Ancient wheats and pseudocereals for possible use in cereal-grain dietary intolerances. In *Cereal Grains*; Wrigley, C., Ed.; Woodhead Publishing: Amsterdam, The Netherlands, 2017; pp. 353–389. [CrossRef]
3. Liceaga, A.M. Processing insects for use in the food and feed industry. *Curr. Opin. Insect. Sci.* **2021**, *48*, 32–36. [CrossRef] [PubMed]
4. Nyangena, D.N.; Mutungi, C.; Imathiu, S.; Kinyuru, J.; Affognon, H.; Ekesi, S.; Nakimbugwe, D.; Fiaboe, K.K. Effects of Traditional Processing Techniques on the Nutritional and Microbiological Quality of Four Edible Insect Species Used for Food and Feed in East Africa. *Foods* **2020**, *9*, 574. [CrossRef] [PubMed]
5. Fombong, F.T.; Van Der Borght, M.; Vanden Broeck, J. Influence of Freeze-Drying and Oven-Drying Post Blanching on the Nutrient Composition of the Edible Insect *Ruspolia Differens*. *Insects* **2017**, *8*, 102. [CrossRef]
6. Dobermann, D.; Field, L.M.; Michaelson, L.V. Impact of Heat Processing on the Nutritional Content of Gryllus Bimaculatus (Black Cricket). *Nutr. Bull.* **2019**, *44*, 116–122. [CrossRef]
7. Mutungi, C.; Irungu, F.G.; Nduko, J.; Mutua, F.; Affognon, H.; Nakimbugwe, D.; Ekesi, S.; Fiaboe, K.K.M. Postharvest Processes of Edible Insects in Africa: A Review of Processing Methods, and the Implications for Nutrition, Safety and New Products Development. *Crit. Rev. Food Sci. Nutr.* **2019**, *59*, 276–298. [CrossRef]
8. Ssepuuya, G.; Nakimbugwe, D.; De Winne, A.; Smets, R.; Claes, J.; Van Der Borght, M. Effect of Heat Processing on the Nutrient Composition, Colour, and Volatile Odour Compounds of the Long-Horned Grasshopper *Ruspolia differens* Serville. *Food Res. Int.* **2020**, *129*, 108831. [CrossRef]
9. Mmari, M.W.; Kinyuru, J.N.; Laswai, H.S.; Okoth, J.K. Traditions, Beliefs and Indigenous Technologies in Connection with the Edible Longhorn Grasshopper *Ruspolia differens* (Serville 1838) in Tanzania. *J. Ethnobiol. Ethnomed.* **2017**, *13*, 60. [CrossRef]
10. Onyango, C.; Okoth, M.W.; Mbugua, S.K. Effect of Drying Lactic Fermented *Uji* (an East African Sour Porridge) on Some Carboxylic Acids. *J. Sci. Food Agric.* **2000**, *80*, 1854–1858. [CrossRef]
11. WHO; FAO. *WHO | Vitamin and Mineral Requirements in Human Nutrition. World Health Organization WHO/FAO (2004)*, 2nd ed.; WHO: Geneva, Switzerland, 2004. Available online: https://apps.who.int/iris/handle/10665/42716 (accessed on 13 February 2022).
12. WHO; FAO; UNU. *Joint FAO/WHO/UNU Expert Consultation on Protein and Amino Acid Requirements in Human Nutrition (2002: Geneva, Switzerland)*; Food and Agriculture Organization of the United Nations, World Health Organization & United Nations University: Geneva, Switzerland, 2007.
13. AOAC. *Official Methods of Analysis of AOAC International*, 21st ed.; Latimer, G.W.J., Ed.; AOAC International: New York, NY, USA, 2019.
14. Ochieng, J.; Schreinemachers, P.; Ogada, M.; Dinssa, F.F.; Barnos, W.; Mndiga, H. Adoption of improved amaranth varieties and good agricultural practices in East Africa. *Land Use Policy* **2019**, *83*, 187–194. [CrossRef]
15. Yang, R.Y.; Keding, G.B. *Nutritional Contributions of Important African Indigenous Vegetables. African Indigenous Vegetables in Urban Agriculture*; Earthscan: London, UK, 2009; pp. 105–143.
16. Petr, J.; Michalik, I.; Tlaskalova, H.; Capouchova, I.; Famera, O.; Urminska, D.; Tukova, L.; Knoblochova, H. Extension of the spectra of plant products for the diet in celiac disease. *Czech J. Food Sci.* **2003**, *21*, 59–70. [CrossRef]
17. Macharia-Mutie, C.W.; van de Wiel, A.M.; Moreno-Londono, A.M.; Mwangi, A.M.; Brouwer, I.D. Sensory acceptability and factors predicting the consumption of grain amaranth in Kenya. *Ecol. Food Nutr.* **2011**, *50*, 375–392. [CrossRef] [PubMed]

18. Karamać, M.; Gai, F.; Longato, E.; Meineri, G.; Janiak, M.A.; Amarowicz, R.; Peiretti, P.G. Antioxidant Activity and Phenolic Composition of Amaranth (*Amaranthus caudatus*) during Plant Growth. *Antioxidants* **2019**, *8*, 173. [CrossRef] [PubMed]

19. Chavez-Jauregui, R.N.; Silva, M.E.M.P.; Arěas, J.A.G. Extrusion cooking process for amaranth. *J. Food Sci.* **2000**, *65*, 1009–1015. [CrossRef]

20. Olaniyi, J.O. Evaluation of yield and quality performance of grain amaranth varieties in the southwestern Nigeria. *Res. J. Agron.* **2007**, *1*, 42–45.

21. Kolawole, E.L.; Sarah, O.A. Growth and yield performance of Amaranthus cruentus influenced by planting density and poultry manure application. *Not. Bot. Horti Agrobot. Cluj-Napoca.* **2009**, *37*, 195–199.

22. Kim, H.K.; Kim, M.J.; Cho, H.Y.; Kim, E.-K.; Shin, D.H. Antioxidative and anti-diabetic effects of amaranth (*Amaranthus esculantus*) in streptozotocin-induced diabetic rats. *Cell Biochem. Funct.* **2006**, *24*, 195–214. [CrossRef]

23. Kim, H.K.; Kim, M.J.; Shin, D.H. Improvement of lipid profile by amaranth (*Amaranthus esculantus*) supplementation in streptozotocin-induced diabetic rats. *Ann. Nutr. Metab.* **2006**, *50*, 277–281. [CrossRef]

24. Bario, D.A.; Añón, M.C. Potential antitumor properties of a protein isolate obtained from the seeds of *Amaranthus mantegazzianus*. *Eur. J. Nutr.* **2010**, *49*, 73–82. [CrossRef]

25. Aderibigbe, O.R.; Ezekiel, O.O.; Owolade, S.O.; Korese, J.K.; Sturm, B.; Hensel, O. Exploring the potentials of underutilized grain amaranth (*Amaranthus* spp.) along the value chain for food and nutrition security: A review. *Crit. Rev. Food Sci. Nutr.* **2022**, *62*, 656–669. [CrossRef]

26. Okoth, J.K.; Ochola, S.; Gikonyo, N.K.; and Makokha, A.O. Efficacy of amaranth sorghum grains porridge in rehabilitating moderately acute malnourished children in a low-resource setting in Kenya: A randomized controlled trial. *Integr. Food Nutr. Metal.* **2017**, *4*, 1–7. [CrossRef]

27. Cheseto, X.; Baleba, S.B.S.; Tanga, C.M.; Kelemu, S.; Torto, B. Chemistry and Sensory Characterization of a Bakery Product Prepared with Oils from African Edible Insects. *Foods* **2020**, *9*, 800. [CrossRef] [PubMed]

28. Thermo Fisher Scientific. *Determination of Water–and Fat–Soluble Vitamins by HPLC, Knowledge Creation Diffusion Utilization*; Thermo Fisher Scientific: Waltham, MA, USA, 2010; pp. 1–23.

29. Bhatnagar-Panwar, M.; Bhatnagar-Mathur, P.; VijayAnand Bhaaskarla, V.; Reddy Dumbala, S.; Sharma, K.K. Rapid, Accurate and Routine HPLC Method for Large-Scale Screening of pro-Vitamin A Carotenoids in Oilseeds. *J. Plant Biochem. Biotechnol.* **2015**, *24*, 84–92. [CrossRef]

30. Megazyme. *Megazyme—Phytic Acid (Phytate)/Total Phosphoru Assay Kit Procedure*; Megazyme: Wicklow, Ireland, 2017; Volume 17.

31. Saxena, V.; Mishra, G.; Saxena, A.; Vishwakarma, K.K. A Comparative Study on Quantitative Estimation of Tannins in Terminalia Chebula, Terminalia Belerica, Terminalia Arjuna and Saraca Indica Using Spectrophotometer. *Asian J. Pharm. Clin. Res.* **2013**, *6* (Suppl. 3), 148–149.

32. Zhishen, J.; Mengcheng, T.; Jianming, W. The Determination of Flavonoid Contents in Mulberry and Their Scavenging Effects on Superoxide Radicals. *Food Chem.* **1999**, *64*, 555–559. [CrossRef]

33. Norhaizan, M.E.; Nor Faizadatul Ain, A.W. Determination of Phytate, Iron, Zinc, Calcium Contents and Their Molar Ratios in Commonly Consumed Raw and Prepared Food in Malaysia. *Malays. J. Nutr.* **2009**, *15*, 213–222.

34. R Core Team. *R: A Language and Environment for Statistical Computing. R Foundation for Statistical Computing*; R Core Team: Vienna, Austria, 2020.

35. Anigo, K.; Ameh, D.; Ibrahim, S.; Danbauchi, S. Nutrient Composition of Complementary Food Gruels Formulated from Malted Cereals, Soybeans and Groundnut for Use in North-Western Nigeria. *Afr. J. Food Sci.* **2010**, *4*, 65–72.

36. Dewey, K.G. The Challenge of Meeting Nutrient Needs of Infants and Young Children during the Period of Complementary Feeding: An Evolutionary Perspective. *J. Nutr.* **2013**, *143*, 2050–2054. [CrossRef]

37. Murugu, D.K.; Onyango, A.N.; Ndiritu, A.K.; Osuga, I.M.; Xavier, C.; Nakimbugwe, D.; Tanga, C.M. From Farm to Fork: Crickets as Alternative Source of Protein, Minerals, and Vitamins. *Front. Nutr.* **2021**, *505*, 704002. [CrossRef]

38. Melgar-Lalanne, G.; Hern'andez-´Alvarez, A.; Salinas-Castro, A. Edible insects processing: Traditional and innovative technologies. *Compr. Rev. Food Sci. Food Saf.* **2019**, *18*, 1166–1191. [CrossRef]

39. Lillioja, S.; Neal, A.L.; Tapsell, L.; Jacobs, D.R. Whole grains, type 2 diabetes, coronary heart disease, and hypertension: Links to the aleurone preferred over indigestible fiber. *BioFactors* **2013**, *39*, 242–258. [CrossRef] [PubMed]

40. Onyango, C.A.; Ochanda, S.O.; Mwasaru, M.A.; Ochieng, J.K.; Mathooko, F.M.; Kinyuru, J.N. Effects of Malting and Fermentation on Anti-Nutrient Reduction and Protein Digestibility of Red Sorghum, White Sorghum and Pearl Millet. *J. Food Res.* **2013**, *2*, 41. [CrossRef]

41. Min, Z.; Chen, H.; Li, J.; Pei, Y.; Liang, Y. Antioxidant properties of tartary buckwheat extracts as affected by different thermal processing methods. *LWT Food Sci. Technol.* **2010**, *43*, 181–185.

42. Masarirambi, M.; Mavuso, V.; Songwe, V.; Nkambule, T.; and Mhazo, N. Indigenous Post-Harvest Handling and Processing of Traditional Vegetables in Swaziland: A Review. *Afr. J. Agric. Res.* **2010**, *5*, 3333–3341.

43. Folch, J.; Lees, M.; Sloane Stanley, G.H. A Simple Method for the Isolation and Purification of Total Lipides from Animal Tissues. *J. Biol. Chem.* **1957**, *226*, 497–509. [CrossRef]

44. Luchuo, E.B.; Paschal, K.A.; Geraldine, N.; Kindong, N.P.; Nsah, Y.S.; Tanjeko, A.T. Malnutrition in Sub—Saharan Africa: Burden, causes and prospects. *Pan Afr Med. J.* **2013**, *15*, 1–20.

45. Branca, F.; Demaio, A.; Udomkesmalee, E.; Baker, P.; Aguayo, V.M.; Barquera, S.; Dain, K.; Keir, L.; Lartey, A.; Mugambi, G. Dynamics of the double burden of malnutrition and the changing nutrition reality. *Lancet* **2020**, *395*, 65–74.

46. Agbemafle, I.; Hadz, D.; Amagloh, F.K.; Zotor, F.B.; Reddy, M.B. Orange-Fleshed Sweet Potato and Edible Insects. *Foods* **2020**, *9*, 1225. [CrossRef]

47. Magara, H.J.O.; Niassy, S.; Ayieko, M.A.; Mukundamago, M.; Egonyu, J.P.; Tanga, C.M.; Kimathi, E.K.; Ongere, J.O.; Fiaboe, K.K.M.; Hugel, S.; et al. Edible Crickets (Orthoptera) Around the World: Distribution, Nutritional Value, and Other Benefits—A Review. *Front. Nutr.* **2021**, *7*, 257. [CrossRef]

48. Pranoto, Y.; Anggrahini, S.; Efendi, Z. Effect of Natural and Lactobacillus Plantarum Fermentation on In-Vitro Protein and Starch Digestibilities of Sorghum Flour. *Food Biosci.* **2013**, *2*, 46–52. [CrossRef]

49. Osman, M.A. Effect of Traditional Fermentation Process on the Nutrient and Antinutrient Contents of Pearl Millet during Preparation of Lohoh. *J. Saudi Soc. Agric. Sci.* **2011**, *10*, 1–6. [CrossRef]

50. Nnam, N.M. Chemical Evaluation of Multimixes Formulated from Some Local Staples for Use as Complementary Foods in Nigeria. *Plant Foods Hum. Nutr.* **2000**, *55*, 255–263. [CrossRef] [PubMed]

51. Onabanjo, O.O.; Oguntona, C.R.B.; Maziya-Dixon, B.; Olayiwola, I.O.; Oguntona, E.B.; Dixon, A.G.O. Nutritional Evaluation of Four Optimized Cassava-Based Complementary Foods. *Afr. J. Food Sci.* **2008**, *2*, 136–142.

52. Assohoun, M.C.N.; Djeni, T.N.; Koussémon-Camara, M.; Brou, K. Effect of Fermentation Process on Nutritional Composition and Aflatoxins Concentration of Doklu, a Fermented Maize Based Food. *Food Nutr. Sci.* **2013**, *4*, 1120–1127. [CrossRef]

53. Ikenebomah, M.; Kok, J.R.; Ingram, J. Processing and Fermentation of the African Locust Bean (Parkia Felocoidea) Can. *Int. J. Food Sci. Technol.* **2019**, *17*, 48–50.

54. Codex Alimentarius. *Guidelines for Development of Supplementary Foods for Older Infants and Children*; FAO: Rome, Italy, 1991.

55. Islam, M.S.; Castellucci, C.; Fiorini, R.; Greco, S.; Gagliardi, R.; Zannotti, A.; Giannubilo, S.R.; Ciavattini, A.; Frega, N.G.; Pacetti, D.; et al. Omega-3 Fatty Acids Modulate the Lipid Profile, Membrane Architecture, and Gene Expression of Leiomyoma Cells. *J. Cell. Physiol.* **2018**, *233*, 7143–7156. [CrossRef]

56. Martínez Andrade, K.A.; Lauritano, C.; Romano, G.; Ianora, A. Marine Microalgae with Anti-Cancer Properties. *Mar. Drugs* **2018**, *16*, 165. [CrossRef]

57. Stark, K.D.; Van Elswyk, M.E.; Higgins, M.R.; Weatherford, C.A.; Salem, N. Global Survey of the Omega-3 Fatty Acids, Docosahexaenoic Acid and Eicosapentaenoic Acid in the Blood Stream of Healthy Adults. *Prog. Lipid Res.* **2016**, *63*, 132–152. [CrossRef]

58. Gunaratne, A.W.; Makrides, M.; Collins, C.T. Maternal Prenatal and/or Postnatal n-3 Fish Oil Supplementation for Preventing Allergies in Early Childhood. *Cochrane Database Syst. Rev.* **2012**, *9*, 1–23. [CrossRef]

59. Paucean, A.; Moldovan, O.P.; Mureşan, V.; Socaci, S.A.; Dulf, F.V.; Alexa, E.; Man, S.M.; Mureşan, A.E.; Muste, S. Folic Acid, Minerals, Amino-Acids, Fatty Acids and Volatile Compounds of Green and Red Lentils. Folic Acid Content Optimization in Wheat-Lentils Composite Flours. *Chem. Cent. J.* **2018**, *12*, 88. [CrossRef]

60. Kannan, N.; Rao, A.S.; Nair, A. Microbial Production of Omega-3 Fatty Acids: An Overview. *J. Appl. Microbiol.* **2021**, *131*, 2114–2130. [CrossRef] [PubMed]

61. Yoshida, H.; Abe, S.; Hirakawa, Y.; Takagi, S. Roasting Effects on Fatty Acid Distributions of Triacylglycerols and Phospholipids in Sesame (Sesamum Indicum) Seeds. *J. Sci. Food Agric.* **2001**, *81*, 620–626. [CrossRef]

62. Mariod, A.A.; Edris, Y.A.; Cheng, S.F.; Abdelwahab, S.I. Effect of Germination Periods and Conditions on Chemical Composition, Fatty Acids and Amino Acids of Two Black Cumin Seeds. *Acta Sci. Pol. Technol. Aliment.* **2012**, *11*, 401–410.

63. Difo, H.V.; Onyike, E.; Ameh, D.A.; Ndidi, U.S.; Njoku, G.C. Chemical Changes during Open and Controlled Fermentation of Cowpea (Vigna Unguiculata) Flour. *Int. J. Food Nutr. Saf.* **2014**, *5*, 1–10.

64. Tizazu, S.; Urga, K.; Abuye, C.; Retta, N. Improvement of Energy and Nutrient Density of Sorghumbased Complementary Foods Using Germination. *Afr. J. Food Agric. Nutr. Dev.* **2010**, *10*, 8. [CrossRef]

65. Walther, B.; Schmid, A. *Effect of Fermentation on Vitamin Content in Food*; Elsevier Inc.: Amsterdam, The Netherlands, 2017. [CrossRef]

66. Barrios-González, J. Solid-State Fermentation: Physiology of Solid Medium, Its Molecular Basis and Applications. *Process Biochem.* **2012**, *47*, 175–185. [CrossRef]

67. Kaprasob, R.; Kerdchoechuen, O.; Laohakunjit, N.; Somboonpanyakul, P. B Vitamins and Prebiotic Fructooligosaccharides of Cashew Apple Fermented with Probiotic Strains Lactobacillus Spp., Leuconostoc Mesenteroides and Bifidobacterium Longum. *Process Biochem.* **2018**, *70*, 9–19. [CrossRef]

68. Tabaszewska, M.; Gabor, A.; Jaworska, G.; Drożdż, I. Effect of Fermentation and Storage on the Nutritional Value and Contents of Biologically-Active Compounds in Lacto-Fermented White Asparagus (*Asparagus Officinalis* L.). *LWT Food Sci. Technol.* **2018**, *92*, 67–72. [CrossRef]

69. Zilic, S.; Delic, N.; Basic, Z.; Ignjatovic-Micic, D.; Jankovic, M.; Vancetovic, J. Effects of Alkaline Cooking and Sprouting on Bioactive Compounds, Their Bioavailability and Relation to Antioxidant Capacity of Maize Flour. *J. Food Nutr. Res.* **2015**, *54*, 155–164.

70. Huang, X.; Cai, W.; Xu, B. Kinetic Changes of Nutrients and Antioxidant Capacities of Germinated Soybean (Glycine Max l.) and Mung Bean (*Vigna Radiata* L.) with Germination Time. *Food Chem.* **2014**, *143*, 268–276. [CrossRef]

71. Nkhata, S.G.; Ayua, E.; Kamau, E.H.; Shingiro, J.B. Fermentation and Germination Improve Nutritional Value of Cereals and Legumes through Activation of Endogenous Enzymes. *Food Sci. Nutr.* **2018**, *6*, 2446–2458. [CrossRef] [PubMed]
72. Fuliaş, A.; Vlase, G.; Vlase, T.; Oneţiu, D.; Doca, N.; Ledeţi, I. Thermal Degradation of B-Group Vitamins: B1, B2 and B6: Kinetic Study. *J. Therm. Anal. Calorim.* **2014**, *118*, 1033–1038. [CrossRef]
73. Young, H.; Guk, I.; Myoung, T.; Sik, K.; Sik, D.; Hyun, J.; Joong, D.; Lee, J.; Ri, Y.; Sang, H. Chemical and Functional Components in Different Parts of Rough Rice (*Oryza Sativa* L.) before and after Germination. *Food Chem.* **2012**, *134*, 288–293. [CrossRef]
74. Alamprese, C.; Ratti, S.; Rossi, M. Effects of Roasting Conditions on Hazelnut Characteristics in a Two-Step Process. *J. Food Eng.* **2009**, *95*, 272–279. [CrossRef]
75. Stuetz, W.; Schlörmann, W.; Glei, M. B-Vitamins, Carotenoids and Tocopherols in Nuts. *Food Chem.* **2016**, *221*, 222–227. [CrossRef]
76. Zhang, Y.Y.; Stockmann, R.; Ng, K.; Ajlouni, S. Revisiting Phytate-Element Interactions: Implications for Iron, Zinc and Calcium Bioavailability, with Emphasis on Legumes. *Crit. Rev. Food Sci. Nutr.* **2020**, *62*, 1696–1712. [CrossRef]
77. Konietzny, U.; Greiner, R. PHYTIC ACID | Properties and Determination. *Am. J. Med. Sci* **2003**, *317*, 370–376.
78. Castro-Alba, V.; Lazarte, C.E.; Perez-Rea, D.; Carlsson, N.G.; Almgren, A.; Bergenståhl, B.; Granfeldt, Y. Fermentation of Pseudocereals Quinoa, Canihua, and Amaranth to Improve Mineral Accessibility through Degradation of Phytate. *J. Sci. Food Agric.* **2019**, *99*, 5239–5248. [CrossRef]
79. Inyang, C.U.; Zakari, U.M. Effect of Germination and Fermentation of Pearl Millet on Proximate Chemical and Sensory Properties of Instant "Fura"—A Nigerian Cereal Food. *Pak. J. Nutr.* **2008**, *7*, 9–12. [CrossRef]
80. García-Mantrana, I.; Monedero, V.; Haros, M. Application of Phytases from Bifidobacteria in the Development of Cereal-Based Products with Amaranth. *Eur. Food Res. Technol.* **2014**, *238*, 853–862. [CrossRef]
81. Deshpande, S.S.; Salunke, D.K. Grain legumes, seeds and nuts: Rationale for fermentation. Fermented grains legumes, seeds and nuts: A global perspective. *FAO Agric. Serv. Bull.* **2002**, *142*, 1–32.
82. Frontela, C.; García-Alonso, F.J.; Ros, G.; Martínez, C. Phytic Acid and Inositol Phosphates in Raw Flours and Infant Cereals: The Effect of Processing. *J. Food Compos. Anal.* **2008**, *21*, 343–350. [CrossRef]
83. Greiner, R.; Konietzny, U. Phytase for Food Application Phytase for Food Application. *J. Food Technol. Biotechnol.* **2006**, *44*, 125–140.
84. Panche, A.N.; Diwan, A.D.; Chandra, S.R. Flavonoids: An Overview. *J. Nutr. Sci.* **2016**, *5*, 1–15. [CrossRef] [PubMed]
85. Adetuyi, F.O.; Ibrahim, T.A. Effect of Fermentation Time on the Phenolic, Flavonoid and Vitamin C Contents and Antioxidant Activities of Okra (Abelmoschus Esculentus) Seeds. *Niger. Food J.* **2014**, *32*, 128–137. [CrossRef]
86. Hur, S.J.; Lee, S.Y.; Kim, Y.C.; Choi, I.; Kim, G.B. Effect of Fermentation on the Antioxidant Activity in Plant-Based Foods. *Food Chem.* **2014**, *160*, 346–356. [CrossRef] [PubMed]
87. Chaaban, H.; Ioannou, I.; Chebil, L.; Slimane, M.; Gérardin, C.; Paris, C.; Charbonnel, C.; Chekir, L.; Ghoul, M. Effect of Heat Processing on Thermal Stability and Antioxidant Activity of Six Flavonoids. *J. Food Process. Preserv.* **2017**, *41*, 1–12. [CrossRef]
88. Modgil, R.; Sood, P. Effect of Roasting and Germination on Carbohydrates and Anti-Nutritional Constituents of Indigenous and Exotic Cultivars of Pseudo-Cereal (Chenopodium). *J. Life Sci.* **2017**, *9*, 64–70. [CrossRef]
89. Ali, M.A.M.; Tinay, A.H.; Tinay, E.l.; Abdalla, A.H. Effect of Fermentation on the in Vitro Protein Digestibility of Pearl Millet. *Food Chem.* **2003**, *80*, 51–54. [CrossRef]
90. Shimelis, E.A.; Rakshit, S.K. Effect of Processing on Antinutrients and in Vitro Protein Digestibility of Kidney Bean (*Phaseolus Vulgaris* L.) Varieties Grown in East Africa. *Food Chem.* **2007**, *103*, 161–172. [CrossRef]
91. Kunyanga, C.N.; Imungi, J.K.; Okoth, M.; Momanyi, C.; Biesalski, H.K.; Vadivel, V. Antioxidant and Antidiabetic Properties of Condensed Tannins in Acetonic Extract of Selected Raw and Processed Indigenous Food Ingredients from Kenya. *J. Food Sci.* **2011**, *76*, C560–C567. [CrossRef] [PubMed]
92. Abdelhaleem, H.; Tinay, A.H.; El Mustafa, A.I.; Babiker, E.E. Effect of Fermentation, Malt-Pretreatment and Cooking on Antinutritional Factors and Protein Digestibility of Sorghum Cultivars. *Pak. J. Nutr.* **2008**, *7*, 335–341.

Article

Proximate, Physicochemical, Techno-Functional and Antioxidant Properties of Three Edible Insect (*Gonimbrasia belina, Hermetia illucens* and *Macrotermes subhylanus*) Flours

Nthabeleng Vanqa [1], **Vusi Vincent Mshayisa** [1,*] **and Moses Basitere** [2]

[1] Department of Food Science and Technology, Cape Peninsula University of Technology, Bellville 7535, South Africa; vanqanthabeleng@gmail.com

[2] Academic Support Program for Engineering (ASPECT) Cape Town, Centre of Higher Education Development University of Cape Town, Rondebosch, Cape Town 7701, South Africa; moses.basitere@uct.ac.za

* Correspondence: mshayisav@cput.ac.za

Abstract: In this study, edible insect flours from *Gonimbrasia belina* (Mashonzha), *Hermetia illucens* (black soldier fly larvae) and *Macrotermes subhylanus* (Madzhulu) were prepared and assessed in terms of proximal, physicochemical, techno-functional and antioxidant properties. The crude protein of the edible insect flours varied between 34.90–52.74%. The crude fat of the insect flours differed significantly ($p < 0.05$), with *H. illucens* (27.93%) having the highest crude fat. *G. belina* was lighter (L*) and yellower (+b*) compared to *H. illucens* and *M. subhylanus*, and there was no significant difference ($p > 0.05$) in the redness (+a*) of the edible insect flours. There were no significant differences ($p > 0.05$) in foam capacity and foam stability of all three edible insect flours. Moreover, the antioxidant activity against the DPPH radical was low for *H. illucens* (3.63%), with *M. subhylanus* (55.37%) exhibiting the highest DPPH radical. Principal component analysis (PCA) was applied to the techno-functional properties and antioxidant indices of the edible insect flours. PC1 accounted for 51.39% of the total variability, while component 2 accounted for 24.71%. In terms of PC1, the FS, OBC and FC were responsible for the major differences in the edible insect flours. The findings revealed that edible insect flours are a good source of antioxidants and can be used as an alternative protein source and a potential novel food additive due to their techno-functional qualities.

Keywords: edible insect flours; *G. belina*; *H. illucens*; *M. subhylanus*; nutritional properties; techno-functional properties; antioxidant activity; metal chelation; Mashonzha; Madzhulu; black soldier fly

Citation: Vanqa, N.; Mshayisa, V.V.; Basitere, M. Proximate, Physicochemical, Techno-Functional and Antioxidant Properties of Three Edible Insect (*Gonimbrasia belina, Hermetia illucens* and *Macrotermes subhylanus*) Flours. *Foods* **2022**, 11, 976. https://doi.org/10.3390/foods11070976

Academic Editors: Victor Benno Meyer-Rochow, Chuleui Jung and Theodoros Varzakas

Received: 8 February 2022
Accepted: 21 March 2022
Published: 28 March 2022

Publisher's Note: MDPI stays neutral with regard to jurisdictional claims in published maps and institutional affiliations.

1. Introduction

As vast as the challenge is to feed 9 billion people by 2050, increasing food availability is insufficient due to the increasingly limited resources, such as agriculturally cultivable land [1]. This, without a doubt, calls for innovative, alternative ways of ensuring that adequate, quality, safe and nutritious foods are available and accessible to all people at all times [2]. As early as 1975, Meyer-Rochow [3] argued and proposed that edible insects could play a role in alleviating food security and combating protein deficiency in some underdeveloped countries. Over the last two decades, there has been a renewed interest on edible insects for human consumption globally [4–6]. The FAO report titled *"Edible insects: Future prospects for food and feed"* [6] and other scientific literature seems to have re-invigorated the earlier call made by Meyer-Rochow in 1975. This is because, compared to conventional protein sources, edible insects have an excellent feed conversion ratio; a source of protein, fat and minerals, and this characteristic is particularly valuable given that future protein consumption is expected to increase with a declining food supply [7–10].

Entomophagy, the practice of consuming insects, has been practised worldwide for centuries, yet it has only recently gained momentum in Western cultures [11]. Insects are consumed prominently in Latin America, Asia, and Africa [10]. People throughout the world have been consuming insects as a regular part of their diets for millennia [12]. Considering the growing population worldwide and the increasing demand for additional sources of proteins, edible insects are seen as an economical alternative and as a sustainable source of nutrients and bioactive compounds [13]. *Hermetia illucens* (black soldier fly), *Gonimbrasia belina* (Mashonzha) and *Macrotermes subhylanus* (Madzhulu) are among edible insect species that have gained attention as alternative sources of protein; the latter two are indigenous to parts of South Africa and play a vital role in food security, rural livelihoods, and poverty eradication [4]. Black soldier fly larvae are commercially produced in South Africa by one of the largest industrial insect processing companies, AgriProtein. The European Food Safety Authority (EFSA) is currently considering black soldier fly as a novel ingredient to be used in food.

Gonimbrasia belina (*G. belina*) is an emperor moth species indigenous to Southern Africa's warmer areas. It is a giant edible caterpillar, known as the Mashonzha (in Venda), madora (in Shona) or mopane worm or amacimbi (Ndebele), which mainly feeds on mopane tree leaves but not exclusively. For millions in the region, Mashonzha are a significant source of protein. Emperor moth *G. belina* caterpillars are a significant natural resource for rural individuals residing in Botswana, Namibia, northern South Africa, and southern Zimbabwe's mopane forests [14].

Macrotermes subhylanus (*M. subhylanus*), known as Madzhulu in Venda and isusu in Nigeria are termites and are gregarious insects most common during the rainy season [4,15]. They are the second most eaten insects in South Africa and are harvested during the rainy season. At the same time, Mashonzha and Madzhulu are sold at informal markets predominantly in the Limpopo and KwaZulu Natal provinces, and in other parts of South Africa are considered a delicacy.

In addition to insects, algae and in vitro meat have also been considered as potential alternatives to conventional sources [16]. The inclusion of insects among these alternatives is highly recommended since they are widely incorporated in food cultures worldwide and have excellent nutritional qualities.

Nevertheless, it is essential to highlight that food neophobia is still directed to the consumption of edible insects, especially in western and urban societies. However, Schösler et al. [17] reported that edible insects, if incorporated in foods in a less obvious form, such as food ingredients (flours, powders, or pastes) in products that are indistinguishable from familiar food items, consumers would accept them. This indicates that insects could be used as food ingredients in the food supply chain, particularly in areas where traditional approaches are unlikely to be adopted owing to a lack of sensory appeal, and insect flour is one way to incorporate insects into food production systems

Therefore, it is crucial to note that the first step to large-scale industrial success is the exploration of the nutritional, techno-functional and antioxidant properties of proposed edible insect ingredients. Currently, available literature on the application of insect flour mainly focuses on *T. molitor (mealworm)* [13,18]. There has been little attention paid to the nutritional, techno-functional and antioxidant properties of Mashonzha, black soldier fly larvae and Madzhulu edible insect flours from South Africa.

Therefore, the aim of this study was to establish the proximate composition, physicochemical, techno-functional properties, and antioxidant activity of edible insect flours obtained from Mashonzha, black soldier fly larvae and Madzhulu with the view to find alternative protein sources for human consumption.

2. Materials and Methods

2.1. Source of Materials

The edible insects were sourced from different provinces of South Africa: Mashonzha (*G. belina*) and Madzhulu (*M. subhylanus*) were sourced in the Vhembe district, Limpopo, and the black soldier fly larvae (*H. illucens*) was sourced from Cape Town, Western Cape, South Africa. The chemical reagents, 2,2 diphenyl-1-picrylhydrazyl (DPPH), 2,2′ azobis (2-methyl, 2,2-Azinobis (3-ethylbenzothiazoline-6-sulphonic acid) diammonium salt (ABTS), Ferric (III) chloride, ethylenediaminetetraacetic acid (EDTA), tertiary butyl hydroquinone (TBHQ), ferrous (II) chloride and thiobarbituric acid (TBA) were obtained from Merk (Sigma-Aldrich, Kempton Park, South Africa). All the chemicals used in this study were of analytical grade, and chemical reagents were prepared according to standard analytical procedures. Prepared reagents were stored under conditions that prevented deterioration or contamination. The water used in the study was ultrapure water purified with a Milli-Q water purification system (Millipore, Microsep, Bellville, South Africa). The ethics committee of the faculty of applied sciences gave its approval to the study (215062965/05/2021).

2.2. Preparation of Insect Flours

Representative samples of sun-dried Mashonzha (hereinafter indicated as *G. belina*) and Madzhulu (hereinafter indicated as *M. subhylanus*) edible insects were purchased from street vendors from the Vhembe district (Limpopo province, South Africa). Black soldier fly larvae (hereinafter indicated as *H. illucens*) reared on clean larvae was purchased from AgriProtein (Cape Town, South Africa), and the flour was prepared following the method described by Zozo et al. [19] and freeze-dried (Wizard 2.0, SP Scientific, Johannesburg, South Africa). The dried edible insects were subjected to a grinding/milling process using a laboratory blender (Bamix, Checkers, Cape Town, South Africa). The flours were stored at room temperature under conditions that prevented deterioration.

2.3. Proximate Composition Analysis

Proximate composition, i.e., moisture (925.10), crude protein (920.87), crude fat (932.06), and ash content (923.03) of the insect flours were determined following standard methods recommended by the Association of Official Analytical Chemists (AOAC) [20]. The crude protein determination was performed using Dumas (TruSpec™ Leco Carbon/Hydrogen/ Nitrogen Series, Leco Africa) which was calibrated with EDTA according to Zozo et al. [19]. The crude protein was subsequently calculated by multiplying nitrogen content by a protein-to-nitrogen conversion factor of 5.60 as recommended by Janssen et al. [21]. Moisture percentage was calculated by drying the sample in a vacuum oven at 100 °C for two hours. The dried sample was placed into a desiccator, allowed to cool, and then re-weighed. The process was repeated until a constant weight was obtained. Crude fat was calculated by drying fats after extraction in a Soxhlet assembly using petroleum ether. The ash percentage was calculated by combusting the samples in a silica crucible placed in a muffle furnace at 550 °C. The percentage of carbohydrates on a dry basis was determined by subtracting all the components (moisture, crude protein, crude lipid, and ash) from 100. The energy was calculated using the formula [22]:

$$\text{Energy}\left(\frac{\text{Kcal}}{100\text{g}}\right) = 4\,(\%\text{ Carbohydrates} + \%\text{ Protein}) + (9 \times \%\text{fat})$$

2.4. Determination of Physicochemical Properties

2.4.1. Evaluation of Colour Properties of Edible Insect Flours

The colour of the edible insect flours was measured using spectrophotometry (Model CM-5, Konica Minolta Sensing, Tokyo, Japan) as described by [23], set at standard observer 10° and D65. The instrument was zero calibrated using a black tile ($L^* = 5.49$, $a^* = -7.08$, $b^* = 4.66$) and white calibration was performed using a white tile ($L^* = 93.41$, $a^* = -1.18$, $b^* = 0.75$). Edible insect flour samples were evenly placed in a petri-dish (30 mm diameter), and reflectance was measured for $L^*a^*b^*$ colour scales. The L^* coordinate is lightness, 100 represents white and closer to 0 represents black Measurements for each sample were performed in triplicate at three different positions in the samples, with the results recorded in L^* (lightness), a^* (chromaticity coordinate $+a^* =$ red and $-a^* =$ green), b^* (chromaticity coordinate $+b^* =$ yellow and $-b^* =$ blue).

2.4.2. Determination of Bulk Density

The procedure was described by Mintah et al. [24] with some modifications. First, 5 g of the sample was transferred into a weighed measuring cylinder (50 mL) (W_1) and then compressed by tapping until sample volume remained constant. The tube was again weighed (W_2) the new volume (V_1) was noted and the density (g/mL) was measured using the following formula:

$$\text{Bulk Density} = \frac{W_2 - W_1}{V1}$$

2.4.3. Determination of Water Activity

The water activity (A_w) of edible insect flours was measured using the method described by Benamara et al. [25] with minor modifications. Salt humidity standards of 53, 75 and 90% relative humidity were used to calibrate the measurement cell. A sample (5 g) of the insect flours was transferred into a sample dish and placed inside the (AW SPRINT TH500, Novasina analyser, Zurich, Switzerland), and the cell measuring protection filter was immediately closed. The reading was observed after a period of 60 to 80 s.

2.5. Determination of Techno-Functional Properties

2.5.1. Determination of Water Binding Capacity and Oil Binding Capacity

The water-binding capacity (WBC) of the edible insect flours was determined according to Mshayisa and van Wyk [26] with slight modifications. Briefly, a 0.5 g sample was mixed with 2.5 mL deionized water, vortexed for 60 s (Vortex-Genie 2, Scientific Industries, Bohemia, NY, USA), and centrifuged for 20 min at 3220 g at room temperature. The supernatant was removed by decantation and drainage of the residual non-bound water by placing the centrifugation tube upside-down on filter paper for 60 min. WBC was calculated as:

$$\text{WBC} = \frac{m_1 - m_0}{m_0}$$

where m_0 is the initial weight, m_1 is the final weight. The oil binding capacity (OBC) was analysed using sunflower oil instead of deionized water. Except for the vortexing step (120 s), the experimental procedure was performed in analogy to the WBC assay. OBC was similarly calculated.

2.5.2. Determination of Emulsion Capacity and Emulsion Stability

Emulsifying properties were determined according to the method of Mshayisa and van Wyk [26] The samples were dispersed in distilled water 1% (w/v), and 15 mL of the dispersion was homogenized with 15 mL of vegetable oil at a speed of 10,000 rpm for 3 min. Subsequently, the samples were centrifuged (Thermo Electron Corporation Jouan MR1812, Waltham, MA, USA) at 3220 g for 5 min and the volume of the individual layers were read. Emulsion stability was evaluated by heating the emulsion for 30 min at 80 °C. Then, the samples were centrifuged at 3200 g for 5 min. The emulsifying capacity (%) was expressed as a percentage of the volume of the emulsified layer (mL) against the volume of the whole layer (mL). Emulsion capacity and emulsion stability were calculated from the formula:

$$\% \text{ Emulsion capacity (EA)} = \frac{V_e}{V} \times 100$$

$$\% \text{ Emulsion stability (ES)} = \frac{V_{30}}{V_e} \times 100$$

where: V—total volume of tube contents, V_e—the volume of the emulsified layer, V_{30}—the volume of the emulsified layer after heating.

2.5.3. Determination of Foam Capacity and Foam Stability

Foaming capacity (FC) and foam stability (FS) were determined according to the method of Zielinska et al. [27]. First, 20 mL of a 1% sample was homogenized in a high shear homogenizer mixer (Polytron PT 2500E, United Scientific, Cape Town, South Africa) at a speed of 10,000 rpm for 4 min. The whipped sample was then immediately transferred into a graduated cylinder. The total volume was read at time zero and 30 min after homogenization. The foaming capacity and foam stability were calculated from the formula:

$$\% \text{ Foaming capacity (FC)} = \frac{V_0 - V}{V} \times 100$$

$$\% \text{ Foaming stability (FS)} = \frac{V_{30}}{V_0} \times 100$$

where: V—volume before whipping (mL), V_0—volume after whipping (mL), V_{30}—volume after standing (mL).

2.6. Determination of Antioxidant Activity

2.6.1. Preparation of Edible Insect Extract

Two grams of the edible insect flours was mixed with 40 mL Milli-Q water in a 50 mL centrifuge tube. The edible insect flour solution was centrifuged (Thermo Electron Corporation Jouan MR1812, Waltham, MA, USA) at room temperature for 15 min at 8000 rpm, and the supernatant was collected and stored at 4 °C until further analysis and the pellet was discarded.

2.6.2. Determination of DPPH Radical Scavenging Activity

The antioxidant activity of the extract was determined by the 1,1-diphenyl-2-picryl-hydrazyl radical scavenging (DPPH-RS) assay according to the method of Vhangani and van-Wyk [28]. The method uses a stable chromogen radical, DPPH in ethanol, which gives a deep purple colour. The reaction mixture was prepared by reacting 2 mL of edible insect extract with 4 mL of DPPH (0.12 mM) in 95% in ethanol. The reaction mixture was incubated for 30 min in the dark, and then the absorbance of the resulting solutions was measured at 517 nm using a spectrophotometer (Lambda 25, Perkin Elmer, Singapore). The control was prepared similarly, except that Milli-Q water was used, and TBHQ (0.1%) was used as a positive control. The percentage of inhibition was calculated using the formula:

$$\% \text{ DPPH} - \text{RS} = \frac{A_{0\ (517\text{nm})} - A_{1(517\text{nm})}}{A_{0(517\text{nm})}} \times 100$$

where: A_0 is the absorbance of the negative control (water) at 517 nm and A_1 is the absorbance of the edible insect extract at 517 nm test sample.

2.6.3. Determination of ABTS$^+$ Radical Scavenging Activity

The experiment was performed according to the method of Chatsuwan et al. [29] and Mshayisa and van Wyk [26]. The 2,2-Azinobis (3-ethylbenzothiazoline-6-sulphonic acid) diammonium salt (ABTS$^{\bullet+}$) radical was produced by reacting 7.4 mM ABTS stock solution with 2.45 mM potassium persulphate at a ratio of 1:1 (v/v). The mixture was allowed to react for 12–16 h at room temperature in the dark. This working solution of ABTS$^{\bullet+}$ solution was diluted with 95% ethanol at a ratio of 1:50 (v/v) in order to obtain an absorbance of 1.00 at 734 nm. A fresh ABTS$^{\bullet+}$ solution was prepared daily for each assay. The reaction mixture contained 0.15 mL of edible insect extract solution and 2.85 mL of ABTS$^{\bullet+}$ solution. The mixture was incubated at room temperature for 6 min in the dark. Then, the absorbance was measured at 734 nm in a spectrophotometer (Lambda 25, Perkin Elmer, Singapore). The control was prepared in the same manner, except that distilled water was used instead of the sample, and TBHQ (0.1%) was used as a positive control. The scavenging activity was determined according to the equation:

$$\% \text{ ABTS} - \text{RS} = \frac{A_{control(730mn)} - A_{sample(730nm)}}{A_{control(730nm)}} \times 100$$

where: $A_{control}$ is the absorbance of the control (water) at 730 nm and A_{sample} is the absorbance of edible insect extract at 730 nm.

2.6.4. Determination of Fe^{2+} Chelating Activity

The chelating effect on ferrous ions of the prepared extracts was estimated by the method of Sudan et al. [30] with slight modifications. Briefly, 1 mL of each edible insect extract was mixed with 1.85 mL of Milli-Q water and 0.05 mL of 2 mM $FeCl_2$. Next, the reaction was initiated by the addition of 0.1 mL of 5 mM ferrozine into the mixture, which was then left at room temperature for 10 min and the absorbance of the mixture was determined at 562 nm using a spectrophotometer (Lambda 25, Perkin Elmer, Singapore). The percentage of chelating activity was calculated as follows:

$$\% \text{ Chelating activity } \frac{A_0 - A_1}{A_0} \times 100$$

where: A_0 is the absorbance of the negative control (water) control and A_1 is the absorbance of the edible insect extract.

2.6.5. Determination of Reducing Power

The reducing power was determined according to the method of Athukorala et al. [31]. First, 1.0 mL aliquots of edible insect were mixed with 2.5 mL of phosphate buffer (0.2 mM, pH 6.6) and 2.5 mL of potassium ferricyanide. The reaction mixture was vortexed for 10 s and thereafter incubated at 50 °C in the water bath for 20 min. Thereafter, 2.5 mL of 10% trichloroacetic acid (TCA) was added to the reaction mixture, and then vortexed for 10 s, 2.5 mL of the solution was then pipetted out into beakers and mixed with 2.5 mL of distilled water and 0.5 mL of $FeCl_3$ was added and absorbance was measured at 700 nm in a spectrophotometer (Lambda 25, Perkin Elmer, Singapore).

2.7. Statistical Analysis

All assays were performed in triplicates, and the obtained data were presented as means ± standard deviation. Statistical analysis was performed by testing significant differences ($p < 0.05$) between treatments using multivariate analysis of variance (MANOVA), and Duncan's multiple range test was used to separate means where differences existed. Principal Component Analysis (PCA) was applied to extract the components that explained the variability in the edible insect flours antioxidant and functional properties. All quantitative data were analysed using SPSS 27.0 (2005) (SPSS Inc., Chicago, IL, USA).

3. Results and Discussion

3.1. Proximate Composition of Edible Insect Flours

The proximate composition of edible insect flours (*G. belina*, *H. illucens* and *M. subhylanus*) is depicted in Table 1. Protein is the dominant nutrient in all three edible insect flours, followed by crude fat. The protein content was significantly ($p < 0.05$) different between all the edible insect flours, and it ranged from 34.90–52.74%. This is superior to other protein sources, such as beef, eggs, milk, and soybeans, where protein constitutes approximately 30 and 45% of dry matter [32]. The protein content of *H. illucens* (34.90%) was significantly lower ($p < 0.05$) compared to *M. subhylanus* (52.74%). Our findings agreed with the results reported by Bußler et al. [11] on *H. illucens* (34.70%). In a literature review study conducted by Meyer-Rochow et al. [33] the protein content of the Macrotermes species ranged from 20.4–39.7%. Moreover, Kwiri et al. [34] reported the protein content of *G. belina* to be (55.41%). These values are higher than the result obtained in this study of the same insect flour. The differences in protein content can be attributed to differences in the edible insect flour, level of individual development, sex, feed type, climate, and geographical location. In this way, the edible insect flours are diversified nutritionally. The edible insects reported in this study may offer an affordable source of protein, especially for low-income communities and be used as ingredients in flour form to minimise the aversion towards consuming insects [35,36].

Table 1. Proximate composition of three edible insect flours.

Edible Insects	Crude Protein (%)	Ash (%)	Moisture (%)	Crude Fat (%)	Carbohydrates (%)	Energy (%)
G. belina	46.70 ± 0.82 [b]	11.38 ± 2.20 [b]	5.68 ± 0.25 [a]	14.04 ± 0.12 [b]	22.10 ± 1.45 [a]	399.38 ± 6.03 [a]
H. illucens	34.90 ± 0.47 [a]	7.50 ± 1.65 [a]	5.76 ± 0.01 [ab]	27.93 ± 6.13 [c]	23.66 ± 7.84 [a]	485.58 ± 26.69 [b]
M. subhylanus	52.74 ± 1.47 [c]	6.41 ± 0.07 [a]	6.40 ± 0.06 [b]	6.36 ± 0.05 [a]	27.27 ± 1.19 [a]	379.91 ± 1.06 [a]

Values are mean ± standard deviation. Means within a column followed by the same superscript are not significantly ($p > 0.05$) different.

The ash content of *G. belina* (11.38%), *H. illucens* (7.46%) and *M. subhylanus* (6.38%) was higher than the values reported for *M. nigeriensis* (3.24%) by Omotoso [37]. However, the values were comparable to those of *Macrotermes bellicosus* (*M. bellicosus*) (11.83%) reported by Adepoju and Omotayo [38]. Torruco-Uco et al. [39] also reported the ash of *Sphenarium purpurascens* (*S. purpurascens*) to be 2.31–3%, the values are much lower than the values reported in this study. Nyakeri et al. [40] reported *H. illucens* to contain 14.61% ash which is higher than the value of the similar species in this study which had 7.46% ash content. Considerable levels of ash indicate that the samples are a good source of minerals. The variation among the ash contents of samples may be driven by the difference in location, diet, and season in which the insects are reared and harvested [41]. Therefore, the addition of edible insect flour in processed food products has the potential to enhance the mineral content of food, especially where food fortification is essential. The considerable good ash content of the edible insect flours signifies good mineral composition that the edible insect flours might contain [42].

As shown in Table 1, the moisture of the three edible insect flours ranged from 5.77–6.59% and no significant differences ($p > 0.05$) were observed amongst all the edible insect flours. Siulapwa et al. [43] reported the moisture content of *G. belina* to be 9.1%, which is higher than the value reported in this study. Moreover, Anaduaka et al. [36] also reported high moisture values for *Zonocerus variegatus* (*Z. variegatus*) and *Oryctes rhinoceros* larva (*O. rhinoceros* larva) to be 11.85–26.17%, respectively. The low moisture values obtained in this study suggest that it likely results in low water activity and, therefore, can potentially extend the shelf-life of insect flours.

As illustrated in Table 1 the crude fat content in *G. belina*, *H. illucens*, and *M. subhylanus* was 13.91, 27.92 and 6.35%, respectively. *H. illucens* (27.92%) results were higher than those reported by Payne et al. [44] of the similar species (14%). Ganguly et al. [45] reported the fat of Oxya chinensis to be 2.2%, which is lower than the results obtained in this study. Moreover, Melo et al. [46] reported *S. purpurascens* to be 5.75%, which is comparable to *M. subhylanus*. However, Sogbesan and Ugwumba [47] reported the fat of *M. subhylanus* to be 10.6–22.2%, which is higher than the values of the similar species in this study. Fat is a major source of fuel in the body, and it is essential in the cell structures as well as in supplying some oil-soluble vitamins, such as vitamins A, D, E, K.

As the primary source of fibre and calories for humans, carbohydrates are essential components of proper nutrition [48]. The three edible insect flours (*G. belina*, *H. illucens* and *M. subhylanus*) showed no significant difference ($p > 0.05$) in their carbohydrate content and ranged from 22.33–28.10%, respectively. The observed carbohydrate content is low in comparison with those reported by Mishyna et al. [49] for *Schistocerca gregaria* (*S. gregaria*) and *Apis mellifera* (*A. mellifera*) flours which contained 47.2 and 54.10% carbohydrates, respectively.

Energy is primarily derived from carbohydrates, proteins, and fats in food, and because edible insects are high in these macromolecules, they have a high energy content [45]. As shown in Table 1 the energy values obtained for the edible insect flours ranged from 379.91–485.58 kJ. No significant differences ($p > 0.05$) were observed for *G. belina* and *M. subhylanus*. However, *H. illucens* was significantly different ($p < 0.05$) from the other two edible insect flours. The results reported in this study are similar to those reported by Montowska et al. [35] on edible insect flours of 486–524 kcal/100 g. Siulapwa et al. [43] reported *G. belina* energy values of (385 kcal/100 g), which is in the same range as *G. belina* energy value reported in this study.

3.2. Physicochemical Properties

3.2.1. Colour Properties of Edible Insect Flours

The colour attributes of edible insect flours measured were lightness (L*), greenness (−a*), redness (+a*), blueness (−b*), and yellowness (+b*). Lightness is the luminous intensity of colour measured on a scale of 0 to 100, with 0 indicating black and 100 indicating white [50]. Colour is a crucial factor influencing the acceptance of edible insects [18]. The descriptive colour determination of the three edible insect flours *G. belina*, *H. illucens* and *M. subhylanus* is shown in Table 2. There was a significant difference ($p < 0.05$) in the lightness of the edible insect flours, with *G. belina* (57.95) being the lighter in colour. No significant difference ($p > 0.05$) was observed in the redness of the three edible insect flours; however, *M. subhylanus* (5.72) was redder compared to *G. belina* (3.92) and *H. illucens* (4.46), respectively, as depicted in Figure 1.

Table 2. Physicochemical properties of three edible insect flours.

Edible Insects	L*	a*	b*	Bulk Density (g/mL)	pH
G. belina	57.95 ± 0.31 [c]	3.92 ± 1.49 [a]	20.02 ± 1.97 [b]	0.65 ± 0.01 [b]	6.12 ± 0.03 [a]
H. illucens	53.69 ± 0.54 [b]	4.46 ± 0.36 [a]	13.08 ± 2.68 [a]	0.51 ± 0.01 [a]	8.93 ± 0.05 [b]
M. subhylanus	43.52 ± 0.56 [a]	5.72 ± 3.90 [a]	12.00 ± 2.70 [a]	0.64 ± 0.00 [b]	6.14 ± 0.02 [a]

Values are mean ± standard deviation. Means within a column followed by the same superscript are not significantly ($p > 0.05$) different.

Figure 1. Ground edible insect flour of three different species. (**A**) *G. belina*; (**B**) *M. subhylanus*; and (**C**) *H. illucens*.

3.2.2. Bulk Density

Among other vital properties of powder products, bulk density (BD) has significant economic and functional importance, for example, in reducing packaging costs [51]. It is determined by particle density, internal porosity, and particle arrangement in the container [52]. Table 2 represents the bulk density of the three edible insect flours (*G. belina*, *H. illucens* and *M. subhylanus*). The bulk density of the edible insect flours varied from 0.51–0.64 g/mL and no significant difference ($p > 0.05$) was observed. Akpossan et al. [53] reported higher BD for *Imbrasia oyemensis* (*I. oyemensis*) to (1.1 g/mL), while in a study by [54] on *Imbrasia belina* (*I. belina*) the BD (0.65 g/mL) was comparable to that found in the present study. An apparent correlation exists between the bulk density and the protein content. Thus, the edible insect flours all had a low BD due to high protein content. Low BD of the flours is advantageous when storability and transportation are considered since the products could be easily transported and distributed [55]. Low BD flours also find application in the preparation of complementary foods and among the traditional techniques.

3.2.3. Water Activity and pH of Edible Insects

Water activity is a measure of how efficiently the water present can take part in a chemical (physical) reaction or the water available enough for microbial growth to occur in a food product [56,57]. Generally, food deterioration due to microbial growth (yeast and moulds to pathogens) occurs at a range of 0.6 to 1.0 [57]. The water activity of the three edible insect flours *G. belina*, *H. illucens*, and *M. subhylanus* is depicted in Figure 2. The A_w of the edible insect flours ranged from *M. subhylanus* (0.35 ± 0.26), *G. belina* (0.45 ± 0.01), to *H. illucens* (0.53 ± 0.01), and there were no significant differences ($p > 0.05$) within the different edible insect flours. This implies that the edible insect flours are not susceptible to microbial growth. However, some enzymatic reactions, such as browning, transpire at the range of 0.3 to 1.0 and increase rapidly at 0.6 to 0.8. In this study, *M. subhylanus* had the lowest A_w; therefore, it might be susceptible to enzymatic reactions rapidly compared to the other two edible insect flours.

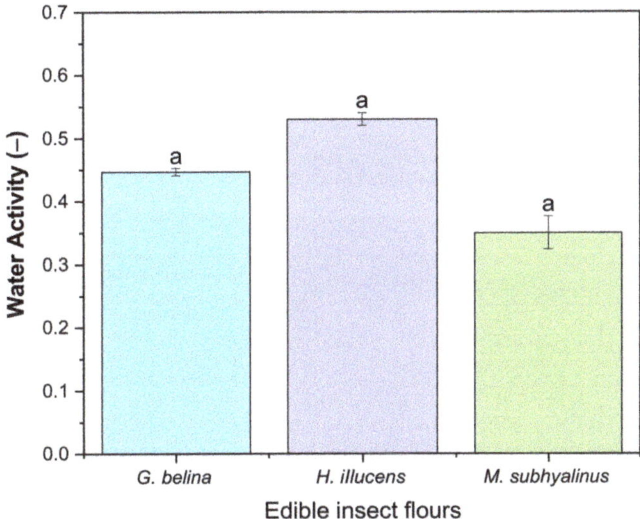

Figure 2. Water activity of three edible insect flours. Values are mean ± standard deviation, means with different superscripts are significantly different ($p < 0.05$).

In addition, pH in food contributes to reducing the growth of microorganisms, thereby ensuring food safety. The pH of *H. illucens* (8.93) had a significant difference ($p < 0.05$) between the pH of *G. belina* (6.12) and *M. subhylanus* (6.14), while there was no statistical difference ($p > 0.05$) between the pH of *G. belina* and *M. subhylanus* (Table 2). Lucas-González et al. [18] reported similar results for *Acheta domesticus* flour (6.31–6.48). The pH of these edible insect flours provides essential information since it determines which type of food matrix they can be added into without affecting their technological behaviour. Thus, potential food ingredients with pH values close to neutrality, such as those obtained in this study, will be better suited for application to neutral food matrices, such as meat replacers and baked products.

3.3. Techno-Functional Properties

3.3.1. Water Binding Capacity and Oil Binding Capacity

Water binding capacity (WBC) and oil binding capacity (OBC) are critical features of food ingredients in food processing and applications. They are related to the ability to take up and retain water and oil, respectively, which directly affect the texture and the flavour of the products, especially in meat and bakery [58]. There are several intrinsic factors affecting the water-binding properties of food flours with relatively high protein. These include amino acid composition, protein conformation, and surface polarity/hydrophobicity [53]. Table 3 depicts the water binding capacity of the edible insect flours. Higher WBC was notable for *M. subhylanus* (1.46 g/g); however, there was no significant difference ($p > 0.05$) between this edible insect flour and that of *G. belina* (1.30 g/g). While the lower WBC value was observed for *H. illucens* flour (1.11 g/g), Zielinska et al. [27] reported higher WBC of *Schistocerca gregaria* (*S. gregaria*) (2.18 g/g). Similarly, Lucas-González et al. [18] reported the WBC of *Acheta domesticus* flour to be (3.82 g/g). However, the WBC of *M. subhylanus* (1.46 g/g) was higher than that reported for *T. molitor* (0.4 g/g). The apparent difference in the WBC could be due to the higher protein content in the *M. subhylanus*, which contains more hydrophilic groups to bind to water molecules. The WBC of the edible insect flours is comparable to plant-based flours, such as wheat and rice, which were reported to have WBC from 1.4–1.9 g/g [59]. This information is crucial for the application of these flours in the food industry. The significant difference in water holding capacity between the edible insect flours might be an indication of the different applications they might have in food. This is the first study to report on the WBC of edible insects, such as *G. belina* and *M. subhylanus*, to our knowledge.

Table 3. Techno-functional properties of three edible insect flours.

Edible Insects	WBC (g/g)	OBC (g/g)	EC (%)	ES (%)	FC (%)	FS (%)
G. belina	1.30 ± 0.12 [ab]	0.89 ± 0.12 [a]	41.76 ± 2.84 [a]	33.75 ± 2.29 [a]	5.81 ± 3.69 [a]	95.32 ± 2.37 [a]
H. illucens	0.11 ± 0.02 [a]	1.35 ± 0.09 [b]	67.33 ± 8.49 [b]	42.45 ± 5.07 [b]	5.69 ± 1.41 [a]	97.38 ± 1.70 [a]
M. subhylanus	1.46 ± 0.06 [b]	1.48 ± 0.07 [b]	45.44± 4.28 [a]	32.80 ± 0.47 [a]	4.71 ± 2.46 [a]	97.51 ± 1.22 [a]

Values are mean ± standard deviation. Means within a column followed by the same superscript are not significantly ($p > 0.05$) different. WBC: water-binding capacity, OBC: oil biding capacity, EC: emulsion capacity, ES: emulsion stability, FS: foam stability, and FC: foam capacity.

The OBC is shown in Table 3. No significant difference was found ($p > 0.05$) between *H. illucens* (1.35 g/g) and *M. subhylanus* (1.48 g/g), and the lowest value was obtained for *G. belina* (0.89 g/g). These values are lower than those reported for *Gryllidae* sp. (2.02 g/g), *G. sigillatus* (2.82 g/g) and *A. domesticus* (3.37–3.52 g/g) [39]. Assielou et al. [60] reported the OBC of *O. owariensis* larvae flour to be 265.90% (2.65 g/g), which is higher than the OBC in this study. The OBC refers to the ability of the proteins in flour to physically bind to fat through capillary action, which is of great importance because fat is a flavour retainer and increases our ability to taste food. Akubor and Eze [61] illustrated that OBC has proven useful in the formulation of bakery products and sausages, and this shows that the studied flours (*M. subhylanus*, *H. illucens*, and *G. belina*), since they are low in OBC are, therefore, low flavour retainers and therefore may be useful in food systems that do not require high WBC/OBC values.

3.3.2. Emulsion Capacity and Emulsion Stability

Proteins are surface-active agents that can form and stabilise the emulsion by creating electrostatic repulsion on the oil droplet surface. Generally, the emulsifying activity of proteins is affected by their molecular weight, hydrophobicity, conformation stability, surface charge, and physicochemical properties, such as pH, ionic strength, and temperature [62]. The results obtained for emulsion capacity (EC) and emulsion stability (ES) of the edible insect flours are presented in Table 3. The emulsion capacity of *G. belina*, *M. subhylanus*, and *H. illucens* were 41.76, 45.44, and 67.33%, respectively. The results for EC in this study are higher than those reported by Mishyna et al. [49] for *S. gregaria* (39.5%) and *A. mellifera* (20.8%) insect flours. The protein emulsification properties are known to be influenced by their surface hydrophobicity, which affects the protein's ability to adsorb to the oil side of the interface. Higher emulsion capacities are usually associated with greater disintegration [11]. *M. subhylanus* had the highest EC (61.69%), which agrees with the macronutrient composition reported in Table 1.

In this study, the ES of *G. belina* (33.75%) and *M. subhylanus* (32.80%) were not significantly different ($p > 0.05$). The results are lower than those reported by Akpossan et al. [53] on *I. oyemensis* (84.76%). ES of *H. illucens* (42.45%) was comparable to that of the larva of *Cirina* (45.36%) reported by Omotoso. Adebowale et al. [63] reported adequate emulsification but poor stability in African cricket (*Gryllidae* sp.) flour. Food manufacturers have a growing demand for sustainable and secure protein sources. Currently, the most widely used emulsifiers are casein and whey [16]. Therefore, edible insect flours' high emulsion capacity and stability highlight the potential to effectively utilise them in food emulsions.

3.3.3. Foam Capacity and Foam Stability

Foams are colloidal systems that consist of a continuous aqueous phase and a dispersed gas phase [16]. Foam formation is governed by the transportation, penetration, and reorganisation of molecules at the air-water interface. To exhibit good foaming properties, a protein must be capable of migrating rapidly to the air-water interface, unfolding, and rearranging at the interface. Table 3 displays the FC and FS of the edible insect flours. The FC was higher for *G. belina* (5.81%); however, no significant differences ($p > 0.05$) were observed amongst all three edible insect flours. The FC values reported by Torruco-Uco et al. [39] for *Gryllidae* sp. (6%) were comparable to the reported values in this study. Zielinska et al. [27] reported FC of *G. sigillatus* (41%) while Assielou et al. [60] reported *Oryctes owariensis* (*O. owariensis*) larvae to have FC of (17.87%), which is also higher than the values reported in this study. This study shows that the low FC can be related to highly ordered globular proteins that resist surface denaturation [53,64].

There were no significant differences ($p > 0.05$) in FS of the edible insect flours in this study. However, the results obtained were higher than those reported by [54] on *I. belina* larvae flour (1.4–5.1%), whereas Omotoso [65] reported *Cirina forda* larva FS to be 3.00%, which is much lower than the FS reported in this study. There was a notable significant difference between the FC and FS values of the edible insect flours, and these results indicate that the proteins and other components of the edible insect flours have a greater ability to form a strong and cohesive film around air bubbles and greater resistance of air diffusion from the bubbles [66].

Presently, research is focused on finding alternatives to eggs, which are commonly used as a foaming agent in food products [16]. The data presented in this study showed that the three edible insect flours (*G. belina. H. illucens* and *M. subhylanus*) exhibited excellent foaming properties; hence, they can be a suitable foaming agent and has potential for such food applications.

3.4. Antioxidant Properties

3.4.1. DPPH-RS of Edible Insect Flours

The DPPH radical-scavenging (DPPH-RS) assay is a widely used method for evaluating the ability of food matrices to scavenge free radicals generated from the DPPH reagent, which undergo SET mechanism [67]. DPPH is a stable free radical that shows maximum absorbance at 517 nm in ethanol and changes from purple to yellow in the presence of antioxidants. When a DPPH radical encounters an electron-donating substrate, such as an antioxidant, the radical is scavenged [68]. As illustrated in Figure 3, the insect flours differed significantly ($p < 0.05$) from one another, with *M. subhylanus* (55.57%) exhibiting the highest radical scavenging activity followed by *G. belina* (37.44%) and *H. illucens* (3.63%), respectively. In a study reported by Navarro del Hierro et al. [69] *T. molitor* and *A. domesticus* extracts, the DPPH-RS was 57 and 72%, respectively, and the values for *T. molitor* are comparable to those of *M. subhylanus* from this study. Nabil et al. [70] also reported on *Moroccan cladode* flour, and the radical scavenging activity was between 7.18 and 72.37%, which is in line with the radical scavenging activity reported in this study.

The results, therefore, suggest that the edible insect flours could be scavenging agents and imply that they have the ability to react with free radicals. This study supports the observation of Mshayisa and van-Wyk [26], who proposed that edible insects can be used as novel functional components in food compositions.

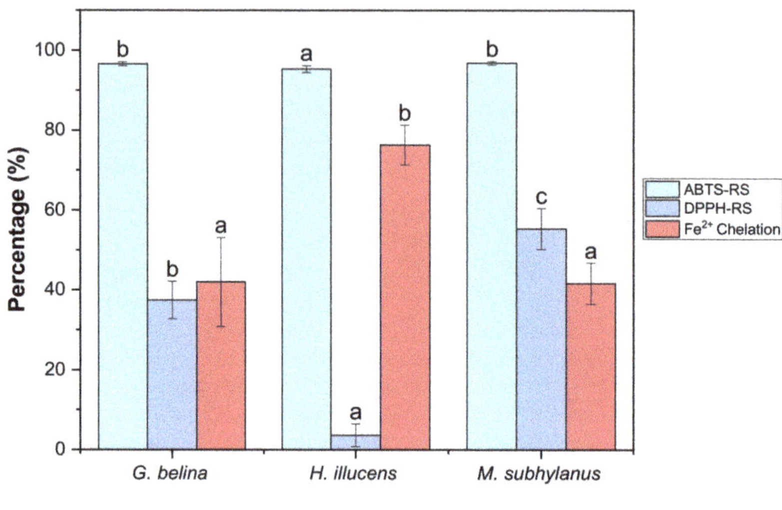

Figure 3. Scavenging effect of DPPH-RS, ABTS-RS and Fe^{2+} chelating activity of edible insect flours. Values are mean ± standard deviation; means with different superscripts are significantly different ($p < 0.05$).

3.4.2. ABTS-RS of Edible Insect Flours

The $ABTS^+$ radical scavenging activity was determined to assess the antioxidant potential of *H. illucens*, *G. belina* and *M. subhylanus*. As depicted in Figure 3, no significant differences ($p > 0.05$) were observed between *M. subhylanus* (96.81%) and *G. belina* (96.61%). However, a significant difference ($p < 0.05$) was observed between the two edible insect flours compared to *H. illucens* (95.32%). It was also observed that *H. illucens* showed lower DPPH-RS as compared to ABTS-RS. The difference in scavenging patterns of ABTS-RS and DPPH-RS could be responsible for these observations. ABTS is more accessible to hydrophilic peptides, while hydrophobic peptides can interact easily with peroxyl radicals, such as DPPH [71]. Most importantly, to our knowledge, this is the first study to establish the antioxidant indices of these three edible insect flours. This study's findings have implications for the utilization of edible insect flours as functional components in food.

3.4.3. Metal Chelation of Edible Insect Flours

The chelation of Fe^{2+} was used to determine the ability of edible insect flours in metal-chelating activity. Ferrozine quantitatively forms complexes with Fe^{2+} ions in the presence of chelating agents, the development of complexes is slowed in the presence of chelating substances disrupted, resulting in the decrease in colour formation [68]. As shown in Figure 3, all edible insects had a high ability to chelate Fe^{2+}. In this study, the highest chelating ability activity was observed in *H. illucens* (76.30%). Moreover, there were no significant differences ($p > 0.05$) in *G. belina* (42.00%) and *M. subhylanus* (41.61%). Ferrous ion (Fe^{2+}) is the most potent pro-oxidant among metal ions. This ion can interact with hydrogen peroxide in a Fenton reaction to produce the reactive oxygen species and hydroxyl free radical (OH), leading to the initiation and/or acceleration of lipid oxidation in food [72]. Therefore, the ability of these edible insect flours to chelate Fe^{2+} suggests they can reduce or avoid the free radical formation. To the best of our knowledge, this is the first study to empirically investigate the Fe^{2+} chelation of edible insect flours, such as *G. belina* and *M. subhylanus*. The results of this study are vital since they indicate that edible insect flours possess considerable meatal chelating activity, which is critical in antioxidant activity since it reduces the concentration of transition metals that catalyse lipid oxidation.

3.4.4. Reducing Power of Edible Insect Flours

Reducing power is a useful indicator of food component antioxidant activity. In this test, the ferric chloride/ferric cyanide complex is reduced to ferrous form (Fe^{2+}) in the presence of antioxidants, allowing the Fe^{2+} concentration to be measured spectrophotometrically by measuring the Prussian blue colour produced at 700 nm [73]. The reducing power assay is often used to evaluate the ability of antioxidants to donate an electron to the free radical [74]. In this study, the ability of edible insect flours to reduce Fe^{3+} to Fe^{2+} was investigated, and the results are depicted in Figure 4. A significant difference ($p < 0.05$) was observed between all the edible insect flours. *H. illucens* (0.61) had the highest RP, while *G. belina* (0.26) had the lowest RP. As articulated by Zielińska and Pankiewicz [75], due to their high protein nature, edible insects are, therefore, potential sources of bioactive proteins that could also possess antioxidant activity. In addition, due to the high reducing power, the obtained results suggest that *H. illucens* soluble proteins contain amino acids or peptides that act as electron donors and can react with free radicals to transform them into stable compounds.

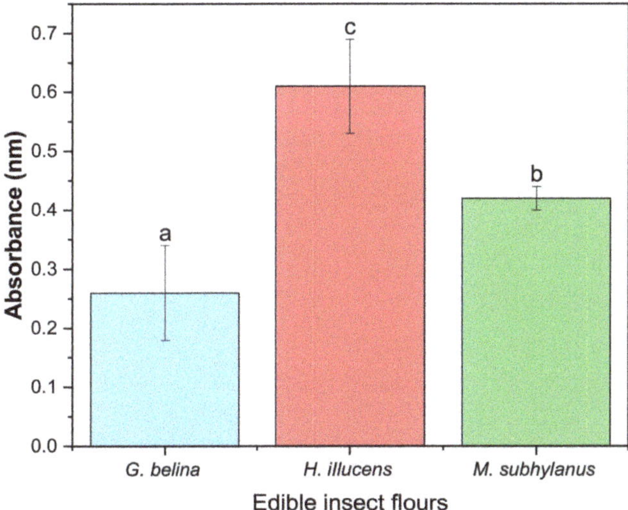

Figure 4. Reducing power activity of edible insect flours. Values are mean ± standard deviation; means with different superscripts are significantly different ($p < 0.05$).

3.5. Principal Component Analysis

Principal component analysis (PCA) was performed to understand the inter-relationships among the measured techno-functional properties and antioxidant activity indices and the similarities and differences among the edible insect samples. The suitability of data reduction by PCA was established by several factors, such as the high correlations between the variables (correlation matrix) and the significant ($p \leq 0.05$) Bartlett's test, as well as the Kaiser–Meyer–Olkin measure (0.68), which was significantly higher than the recommended minimum of 0.6. The PCA results were displayed using score and loading plots (Figure 5). To determine the relative contributions of the principal components in overall total variability, only the eigenvalues greater than one were considered. Thus, the first three principal components (PC1, PC2 and PC3) were found to be significant and explained 87.99% variability in the data set (Table S1). Component 1 accounted for 51.39% of the total variability, and represented EC (0.960), Fe Chelation (0.949), ES (0.897) and DPPH-RS (−0.897) while FS (0.837), FC (−0.080), ABTS-RS (−0.531) and WBC (0.515) contributed to PC2, with a total variability contribution of 24.71%. The PC3 accounted for 11.89% of the total variability due to OBC (0.745), FC (0.504), ABTS-RS (0.442) and RP (0.247), respectively, as shown in Table S1 (Supplementary Materials). The edible insects were clearly distributed into three clusters (Figure 5). It can be seen that *M. subhylanus* can be separated from *H. illucens* based on the DPPH-RS, WBC, and foam stability. In Figure 5, *H. illucens* were grouped in close proximity with values of component 1, whereas *M. subhylanus* and *G. belina* are diametrically opposed in PC2 (meaning they are on the negative and opposite sides). PCA showed that *M. subhylanus* and *G. belina* located on the opposite sides of PC2, the FS, OBC and FC were to be majorly responsible for the difference in the edible insect flours. This was due to the high FC and OBC exhibited by *M. subhylanus* samples, while *G. belina* exhibited the lowest OBC. Therefore, PCA could be helpful to provide valuable information on the classification and discrimination of edible insect flours and on relationships between antioxidant indices and techno-functional properties.

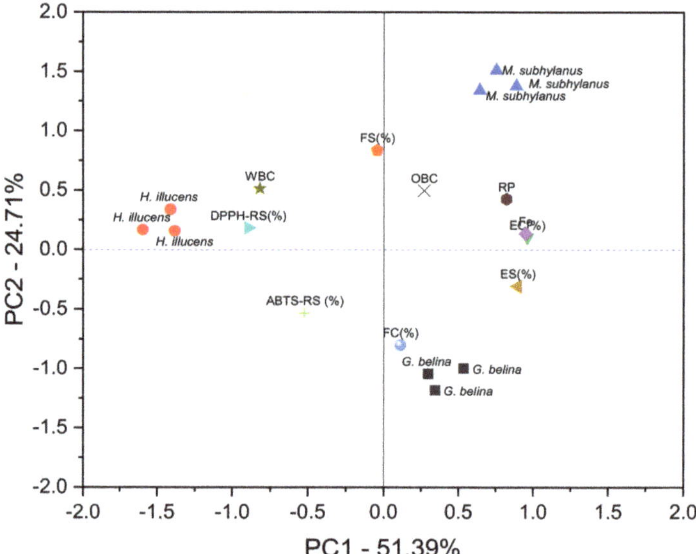

Figure 5. Principal components analysis plot for techno-functional properties and antioxidant indices of edible insect flours.

4. Conclusions

This study was undertaken to establish the potential for edible insect flours as a source of nutrients, as well as their techno-functional and antioxidants properties. The studied edible insect flour species were rich in protein and fat, which are essential nutrients required for the human diet. The results obtained for the physicochemical properties make the flours valuable to the food industry as potential fortifiers, such as *G. belina*, which was yellower and redder in colour since this characteristic is of importance in instances where a noticeable colour change to the product is not desired. *M. subhylanus* exhibited good water binding capacity, and the flour was generally found to have superior techno-functional properties among the studied species. This makes it useful for producing foods such as sausages and bakery products. The studied edible insects have unique techno-functional properties that can be exploited to provide functional ingredients. Future studies on the shelf life, rheological and structural properties of the edible insect flours are essential prior to incorporation in food product formulations.

Supplementary Materials: The following supporting information can be downloaded at: https://www.mdpi.com/article/10.3390/foods11070976/s1. Table S1: Principal components for illustrating the interpretation in Figure 5.

Author Contributions: Conceptualization, V.V.M., M.B. and N.V.; methodology, V.V.M. and N.V.; software, V.V.M. and M.B.; validation, V.V.M. and N.V.; formal analysis, N.V.; investigation, V.V.M. and N.V. resources, V.V.M. and M.B.; data curation, N.V.; writing—V.V.M., N.V.; writing—review and editing, V.V.M., N.V. and M.B. All authors have read and agreed to the published version of the manuscript.

Funding: This work was funded by the Cape Peninsula University of Technology (CPUT) research fund.

Institutional Review Board Statement: Ethical clearance obtained from the Faculty of Applied Sciences Ethics Committee, Cape Peninsula University of Technology, Bellville, South Africa.

Informed Consent Statement: Not applicable.

Data Availability Statement: The datasets used and/or analysed during the current study are available from the corresponding author on reasonable request.

Acknowledgments: As the authors of this paper, we would like to express our sincere gratitude to the Department of Food Science & Technology at the Cape Peninsula University of Technology, Cape Town, South Africa.

Conflicts of Interest: The authors declared that they have no conflict of interest.

References

1. Gahukar, R.T. Entomophagy and human food security. *Int. J. Trop. Insect Sci.* **2011**, *31*, 129–144. [CrossRef]
2. Imathiu, S. Benefits and food safety concerns associated with consumption of edible insects. *NFS J.* **2020**, *18*, 1–11. [CrossRef]
3. Meyer-Rochow, V.B. Meyer-Rochow (1975) Can insects help to ease the problem of world food shortage. *Search* **1975**, *6*, 261–262.
4. Hlongwane, Z.T.; Slotow, R.; Munyai, T.C. Indigenous knowledge about consumption of edible insects in South Africa. *Insects* **2021**, *12*, 22. [CrossRef] [PubMed]
5. Kinyuru, J.N.; Konyole, S.O.; Roos, N.; Onyango, C.A.; Owino, V.O.; Owuor, B.O.; Estambale, B.B.; Friis, H.; Aagaard-Hansen, J.; Kenji, G.M. Nutrient composition of four species of winged termites consumed in western kenya. *J. Food Compos. Anal.* **2013**, *30*, 120–124. [CrossRef]
6. Van-Huis, A.; Van-Itterbeeck, J.; Harmke, K.; Esther, M.; Afton, H.; Muir, M.; Paul, V. *Edible Insects: Future Prospects for Food and Feed Security*; Food and Agriculture Organization of the United Nations: Rome, Italy, 2013; Available online: http://edepot.wur.nl/258042 (accessed on 8 January 2021).
7. Kim, T.K.; Yong, H.I.; Kim, Y.B.; Kim, H.W.; Choi, Y.S. Edible insects as a protein source: A review of public perception, processing technology, and research trends. *Food Sci. Anim. Resour.* **2019**, *39*, 521–540. [CrossRef]
8. Toti, E.; Massaro, L.; Kais, A.; Aiello, P.; Palmery, M.; Peluso, I. Entomophagy: A Narrative Review on Nutritional Value, Safety, Cultural Acceptance and a Focus on the Role of Food Neophobia in Italy. *Eur. J. Investig. Health Psychol. Educ.* **2020**, *10*, 46. [CrossRef]
9. Orsi, L.; Voege, L.L.; Stranieri, S. Eating edible insects as sustainable food? Exploring the determinants of consumer acceptance in Germany. *Food Res. Int.* **2019**, *125*, 108573. [CrossRef]
10. Raheem, D.; Raposo, A.; Oluwole, O.B.; Nieuwland, M.; Saraiva, A.; Carrascosa, C. Entomophagy: Nutritional, ecological, safety and legislation aspects. *Food Res. Int.* **2019**, *126*, 108672. [CrossRef]
11. Bußler, S.; Rumpold, B.A.; Jander, E.; Rawel, H.M.; Schlüter, O.K. Recovery and techno-functionality of flours and proteins from two edible insect species: Meal worm (*Tenebrio molitor*) and black soldier fly (*Hermetia illucens*) larvae. *Heliyon* **2016**, *2*, e00218. [CrossRef]
12. Van-Huis, A.; van Itterbeeck, J.; Klunder, H.; Esther, M.; Halloran, A.; Muir, G.; Vantomme, P. *Edible Insects Future Prospects for Food and Feed Security*; Food and agriculture organization of the United Nations: Rome, Italy, 2013.
13. González, C.M.; Garzón, R.; Rosell, C.M. Insects as ingredients for bakery goods. A comparison study of H. illucens, A. domestica and T. molitor flours. *Innov. Food Sci. Emerg. Technol.* **2019**, *51*, 205–210. [CrossRef]
14. Hlongwane, Z.T.; Slotow, R.; Munyai, T.C. Nutritional composition of edible insects consumed in africa: A systematic review. *Nutrients* **2020**, *12*, 2786. [CrossRef]
15. Netshifhefhe, S.R.; Kunjeku, E.C.; Duncan, F.D. Human uses and indigenous knowledge of edible termites in Vhembe District, Limpopo Province, South Africa. *South Afr. J. Sci.* **2018**, *114*, 1–10. [CrossRef]
16. Hall, F.G.; Jones, O.G.; O'Haire, M.E.; Liceaga, A.M. Functional properties of tropical banded cricket (*Gryllodes sigillatus*) protein hydrolysates. *Food Chem.* **2017**, *224*, 414–422. [CrossRef]
17. Schösler, H.; de Boer, J.; Boersema, J.J. Can we cut out the meat of the dish? Constructing consumer-oriented pathways towards meat substitution. *Appetite* **2012**, *58*, 39–47. [CrossRef]

18. Lucas-González, R.; Fernández-López, J.; Pérez-Álvarez, J.A.; Viuda-Martos, M. Effect of drying processes in the chemical, physico-chemical, techno-functional and antioxidant properties of flours obtained from house cricket (*Acheta domesticus*). *Eur. Food Res. Technol.* **2019**, *245*, 1451–1458. [CrossRef]

19. Zozo, B.; Wicht, M.M.; Mshayisa, V.V.; Van-Wyk, J. Characterisation of black soldier fly larva protein before and after conjugation by the Maillard reaction. *J. Insects Food Feed* **2021**, *8*, 169–183. [CrossRef]

20. AOAC. *Official Methods of Analysis of AOAC*, 15th ed.; Association of Official Agricultural Chemists: Arlington, VA, USA, 2015. [CrossRef]

21. Janssen, R.H.; Vincken, J.P.; van den Broek, L.A.M.; Fogliano, V.; Lakemond, C.M.M. Nitrogen-to-Protein Conversion Factors for Three Edible Insects: *Tenebrio molitor*, *Alphitobius diaperinus*, and *Hermetia illucens*. *J. Agric. Food Chem.* **2017**, *65*, 2275–2278. [CrossRef]

22. Farzana, T.; Mohajan, S. Effect of incorporation of soy flour to wheat flour on nutritional and sensory quality of biscuits fortified with mushroom. *Food Sci. Nutr.* **2015**, *3*, 363–369. [CrossRef]

23. Larouche, J.; Deschamps, M.H.; Saucier, L.; Lebeuf, Y.; Doyen, A.; Vandenberg, G.W. Effects of killing methods on lipid oxidation, colour and microbial load of black soldier fly (*Hermetia illucens*) larvae. *Animals* **2019**, *9*, 182. [CrossRef]

24. Mintah, B.K.; He, R.; Agyekum, A.A.; Dabbour, M.; Golly, M.K.; Ma, H. Edible insect protein for food applications: Extraction, composition, and functional properties. *J. Food Process Eng.* **2020**, *43*, 1–12. [CrossRef]

25. Benamara, R.N.; Gemelas, L.; Ibri, K.; Moussa-Boudjemaa, B.; Demarigny, Y. Sensory, microbiological and physico-chemical characterization of Klila, a traditional cheese made in the south-west of Algeria. *Afr. J. Microbiol. Res.* **2016**, *10*, 1728–1738. [CrossRef]

26. Mshayisa, V.V.; Van-Wyk, J. Hermetia illucens Protein Conjugated with Glucose via Maillard Reaction: Antioxidant and Techno-Functional Properties. *Int. J. Food Sci.* **2021**, *2021*, 1–15. [CrossRef] [PubMed]

27. Zielińska, E.; Karaś, M.; Baraniak, B. Comparison of functional properties of edible insects and protein preparations thereof. *LWT-Food Sci. Technol.* **2018**, *91*, 168–174. [CrossRef]

28. Vhangani, L.N.; Van-Wyk, J. Antioxidant activity of Maillard reaction products (MRPs) derived from fructose-lysine and ribose-lysine model systems. *Food Chem.* **2013**, *137*, 92–98. [CrossRef]

29. Chatsuwan, N.; Nalinanon, S.; Puechkamut, Y.; Lamsal, B.P.; Pinsirodom, P. Characteristics, Functional Properties, and Antioxidant Activities of Water-Soluble Proteins Extracted from Grasshoppers, Patanga succincta and Chondracris roseapbrunner. *J. Chem.* **2018**, *2018*, 1–11. [CrossRef]

30. Sudan, R.; Bhagat, M.; Gupta, S.; Singh, J.; Koul, A. Iron (FeII) Chelation, Ferric Reducing Antioxidant Power, and Immune Modulating Potential of *Arisaema jacquemontii* (Himalayan Cobra Lily). *BioMed Res. Int.* **2014**, *2014*, 1–7. [CrossRef]

31. Athukorala, Y.; Kim, K.N.; Jeon, Y.J. Antiproliferative and antioxidant properties of an enzymatic hydrolysate from brown alga, Ecklonia cava. *Food Chem. Toxicol.* **2006**, *44*, 1065–1074. [CrossRef]

32. Churchward-Venne, T.A.; Pinckaers, P.J.M.; van Loon, J.J.A.; van Loon, L.J.C. Consideration of insects as a source of dietary protein for human consumption. *Nutr. Rev.* **2017**, *75*, 1035–1045. [CrossRef]

33. Meyer-Rochow, V.B.; Gahukar, R.T.; Ghosh, S.; Jung, C. Chemical composition, nutrient quality and acceptability of edible insects are affected by species, developmental stage, gender, diet, and processing method. *Foods* **2021**, *10*, 1036. [CrossRef]

34. Kwiri, R.; Winini, C.; Muredzi, P.; Tongonya, J.; Gwala, W.; Mujuru, F.; Gwala, S.T. Mopane Worm (*Gonimbrasia belina*) Utilisation, a Potential Source of Protein in Fortified Blended Foods in Zimbabwe: A Review. *Glob. J. Sci. Front. Res. D Agric. Vet.* **2014**, *14*, 55–66.

35. Montowska, M.; Kowalczewski, P.Ł.; Rybicka, I.; Fornal, E. Nutritional value, protein and peptide composition of edible cricket powders. *Food Chem.* **2019**, *289*, 130–138. [CrossRef]

36. Anaduaka, E.G.; Uchendu, N.O.; Osuji, D.O.; Ene, L.N.; Amoke, O.P. Nutritional compositions of two edible insects: Oryctes rhinoceros larva and Zonocerus variegatus. *Heliyon* **2021**, *7*, e06531. [CrossRef]

37. Omotoso, O. Nutrient Composition, Mineral Analysis and Anti-nutrient Factors of *Oryctes rhinoceros* L. (Scarabaeidae: Coleoptera) and Winged Termites, Macrotermes nigeriensis Sjostedt. (Termitidae: Isoptera). *Br. J. Appl. Sci. Technol.* **2015**, *8*, 97–106. [CrossRef]

38. Adepoju, O.; Omotayo, O. Nutrient Composition and Potential Contribution of Winged Termites (*Macrotermes bellicosus* Smeathman) to Micronutrient Intake of Consumers in Nigeria. *Br. J. Appl. Sci. Technol.* **2014**, *4*, 1149–1158. [CrossRef]

39. Torruco-Uco, J.G.; Hernández-Santos, B.; Herman-Lara, E.; Martínez-Sánchez, C.E.; Juárez-Barrientos, J.M.; Rodríguez-Miranda, J. Chemical, functional and thermal characterization, and fatty acid profile of the edible grasshopper (*Sphenarium purpurascens* Ch.). *Eur. Food Res. Technol.* **2018**, *245*, 285–292. [CrossRef]

40. Nyakeri, E.M.; Ogola, H.J.; Ayieko, M.A.; Amimo, F.A. An open system for farming black soldier fly larvae as a source of proteins for smallscale poultry and fish production. *J. Insects Food Feed.* **2017**, *3*, 51–56. [CrossRef]

41. Clarkson, C.; Mirosa, M.; Birch, J. Potential of extracted Locusta migratoria protein fractions as value-added ingredients. *Insects* **2018**, *9*, 20. [CrossRef]

42. Ghosh, S.; Jung, C.; Meyer-Rochow, V.B. Nutritional value and chemical composition of larvae, pupae, and adults of worker honey bee, Apis mellifera ligustica as a sustainable food source. *J. Asia-Pac. Entomol.* **2016**, *19*, 487–495. [CrossRef]

43. Siulapwa, N.; Mwambungu, A.; Lungu, E.; Sichilima, W. Nutritional Value of Four Common Edible Insects in Zambia. *Int. J. Sci. Res.* **2012**, *3*, 2319–7064.

44. Payne, C.L.R.; Scarborough, P.; Rayner, M.; Nonaka, K. A systematic review of nutrient composition data available for twelve commercially available edible insects, and comparison with reference values. *Trends Food Sci. Technol.* **2016**, *47*, 69–77. [CrossRef]
45. Ganguly, A.; Chakravorty, R.; Das, M.; Gupta, M.; Mandal, D.K.; Haldar, P.; Ramos-Elorduy, J.; Moreno, J.M.P. A preliminary study on the estimation of nutrients and anti-nutrients in *Oedaleus abruptus* (Thunberg) (Orthoptera-Acridida). *Int. J. Nutr. Metab.* **2013**, *5*, 50–65. [CrossRef]
46. Melo, V.; Garcia, M.; Sandoval, H.; Jiménez, H.D.; Calvo, C. Quality proteins from edible indigenous insect food of latin America and Asia. *Emir. J. Food Agric.* **2011**, *23*, 283–289.
47. Sogbesan, A.O.; Ugwumba, A.A.A. Nutritional Evaluation of Termite (*Macrotermes subhyalinus*) Meal as Animal Protein Supplements in the Diets of Heterobranchus longifilis (Valenciennes, 1840) Fingerlings. *Turk. J. Fish. Aquat. Sci.* **2008**, *8*, 149–157.
48. Rodríguez-Miranda, J.; Alcántar-Vázquez, J.P.; Zúñiga-Marroquín, T.; Juárez-Barrientos, J.M. Insects as an alternative source of protein: A review of the potential use of grasshopper (*Sphenarium purpurascens* Ch.) as a food ingredient. *Eur. Food Res. Technol.* **2019**, *245*, 2613–2620. [CrossRef]
49. Mishyna, M.; Martinez, J.J.I.; Chen, J.; Benjamin, O. Extraction, characterization and functional properties of soluble proteins from edible grasshopper (*Schistocerca gregaria*) and honey bee (*Apis mellifera*). *Food Res. Int.* **2019**, *116*, 697–706. [CrossRef]
50. Manera, F.J.; Legua, P.; Melgarejo, P.; Martínez, R.; Martínez, J.J.; Hernández, F. Effect of air temperature on rind colour development in pomegranates. *Sci. Hortic.* **2012**, *134*, 245–247. [CrossRef]
51. Piornos, J.A.; Burgos-Díaz, C.; Ogura, T.; Morales, E.; Rubilar, M.; Maureira-Butler, I.; Salvo-Garrido, H. Functional and physicochemical properties of a protein isolate from AluProt-CGNA: A novel protein-rich lupin variety (*Lupinus luteus*). *Food Res. Int.* **2015**, *76*, 719–724. [CrossRef]
52. Sharma, A.; Jana, A.H.; Chavan, R.S. Functionality of Milk Powders and Milk-Based Powders for End Use Applications-A Review. *Compr. Rev. Food Sci. Food Saf.* **2012**, *11*, 518–528. [CrossRef]
53. Akpossan, R.; Digbeu, Y.; Koffi, M.; Kouadio, J.; Dué, E.; Kouamé, P. Protein fractions and functional properties of dried Imbrasia oyemensis larvae full-fat and defatted flours. *Int. J. Biochem. Res. Rev.* **2015**, *5*, 116–126. [CrossRef]
54. Ekpo, K.E.E.; Ugbenyen, M.A.A.; Azeke, M.A.; Ugbenyen, A.M.; Ugbenyen, M.A.A.; Azeke, M.A.; Ugbenyen, A.M.; Ugbenyen, M.A.A.; Azeke, M.A.; Ugbenyen, A.M. Functional Properties of dried Imbrasia Belina larvae flour as affected by mesh size and pH. *Niger. Ann. Nat. Sci.* **2008**, *8*, 6–9.
55. Gupta, A.; Sharama, S.; Singh, B. Influence of Germination Conditions on the Techno-functional Properties of Amaranth flour. In Proceedings of the 2018 International Conference on Food Properties, Sharjah, United Arab Emirates, 22–24 January 2018.
56. Sandulachi, E.; Tatarov, P. Water Activity Concept and Its Role in Strawberries Food. *Chem. J. Mold.* **2012**, *7*, 103–115. [CrossRef]
57. Tapia, M.S.; Alzamora, S.M.; Chirife, J. Effects of Water Activity (aw) on Microbial Stability: As a Hurdle in Food Preservation. *Water Act. Foods Fundam. Appl.* **2008**, 239–271. [CrossRef]
58. Wang, J.; Jousse, M.; Jayakumar, J.; Fernández-Arteaga, A.; de Lamo-Castellví, S.; Ferrando, M.; Güell, C. Black soldier fly (*Hermetia illucens*) protein concentrates as a sustainable source to stabilize o/w emulsions produced by a low-energy high-throughput emulsification technology. *Foods* **2021**, *10*, 1048. [CrossRef]
59. Chandra, S. Assessment of functional properties of different flours. *Afr. J. Agric. Res.* **2013**, *8*, 4849–4852. [CrossRef]
60. Assielou, B.; Due, E.; Koffi, M.; Dabonne, S.; Kouame, P. Oryctes owariensis Larvae as Good Alternative Protein Source: Nutritional and Functional Properties. *Annu. Res. Rev. Biol.* **2015**, *8*, 1–9. [CrossRef]
61. Akubor, P.I.; Ike, E.Z.E.J. Quality evaluation and cake making potential of sun and oven dried carrot fruit. *Int. J. Biosci.* **2012**, *2*, 19–27.
62. Lucas, A.J.d.; de Oliveira, L.M.; da Rocha, M.; Prentice, C. Edible insects: An alternative of nutritional, functional and bioactive compounds. *Food Chem.* **2020**, *311*, 126022. [CrossRef]
63. Adebowale, Y.A.; Adeyemi, I.A.; Oshodi, A.A. Functional and physicochemical properties of flours of six Mucuna species. *Afr. J. Biotechnol.* **2005**, *4*, 1461–1468. [CrossRef]
64. Ndiritu, A.K.; Kinyuru, J.N.; Kenji, G.M.; Gichuhi, P.N. Extraction technique influences the physico-chemical characteristics and functional properties of edible crickets (*Acheta domesticus*) protein concentrate. *J. Food Meas. Charact.* **2017**, *11*, 2013–2021. [CrossRef]
65. Omotoso, O.T. Nutritional quality, functional properties and anti-nutrient compositions of the larva of *Cirina forda* (Westwood) (Lepidoptera: Saturniidae). *J. Zhejiang Univ. Sci. B* **2005**, *7*, 51–55. [CrossRef]
66. Gravel, A.; Doyen, A. The use of edible insect proteins in food: Challenges and issues related to their functional properties. *Innov. Food Sci. Emerg. Technol.* **2020**, *59*, 2–11. [CrossRef]
67. Liang, N.; Kitts, D.D. Antioxidant property of coffee components: Assessment of methods that define mechanism of action. *Molecules* **2014**, *19*, 19180–19208. [CrossRef]
68. Khaled, H.B.; Ktari, N.; Ghorbel-Bellaaj, O.; Jridi, M.; Lassoued, I.; Nasri, M. Composition, functional properties and in vitro antioxidant activity of protein hydrolysates prepared from sardinelle (*Sardinella aurita*) muscle. *J. Food Sci. Technol.* **2014**, *51*, 622–633. [CrossRef]
69. Del Hierro, J.N.; Gutiérrez-Docio, A.; Otero, P.; Reglero, G.; Martin, D. Characterization, antioxidant activity, and inhibitory effect on pancreatic lipase of extracts from the edible insects Acheta domesticus and Tenebrio molitor. *Food Chem.* **2020**, *309*, 125742. [CrossRef]

70. Nabil, B.; Ouaabou, R.; Ouhammou, M.; Essaadouni, L.; Mahrouz, M. Functional Properties, Antioxidant Activity, and Organoleptic Quality of Novel Biscuit Produced by Moroccan Cladode Flour *"Opuntia ficus-indica"*. *J. Food Qual.* **2020**, *2020*, 1–12. [CrossRef]
71. Chalamaiah, M.; Kumar, B.D.; Hemalatha, R.; Jyothirmayi, T. Fish protein hydrolysates: Proximate composition, amino acid composition, antioxidant activities and applications: A review. *Food Chem.* **2012**, *135*, 3020–3038. [CrossRef]
72. Khantaphant, S.; Benjakul, S.; Ghomi, M.R. The effects of pretreatments on antioxidative activities of protein hydrolysate from the muscle of brownstripe red snapper (*Lutjanus vitta*). *LWT-Food Sci. Technol.* **2011**, *44*, 1139–1148. [CrossRef]
73. Nooshkam, M.; Varidi, M.; Bashash, M. The Maillard reaction products as food-born antioxidant and antibrowning agents in model and real food systems. *Food Chem.* **2019**, *275*, 644–660. [CrossRef]
74. Khantaphant, S.; Benjakul, S. Comparative study on the proteases from fish pyloric caeca and the use for production of gelatin hydrolysate with antioxidative activity. *Comp. Biochem. Physiol. Part B Biochem. Mol. Biol.* **2008**, *151*, 410–419. [CrossRef]
75. Zielińska, E.; Pankiewicz, U. Characteristics of Shortcake Biscuits Enriched with *Tenebrio molitor* Flour. *Molecules* **2020**, *25*, 5629. [CrossRef]

Article

The Effects of Antioxidants and Packaging Methods on Inhibiting Lipid Oxidation in Deep Fried Crickets (*Gryllus bimaculatus*) during Storage

Jin Gan, Min Zhao, Zhao He, Long Sun, Xian Li and Ying Feng *

Key Laboratory of Breeding and Utilization of Resource Insects of National Forestry and Grassland Administration, Research Institute of Resource Insects, Chinese Academy of Forestry, Kunming 650233, China; ganjin638@126.com (J.G.); mzhao@caf.ac.cn (M.Z.); zhe999@hotmail.com (Z.H.); sunlong1818@126.com (L.S.); leexian99@163.com (X.L.)
* Correspondence: rirify@139.com

Abstract: This study aimed to investigate the effect of processing methods on inhibiting lipid oxidation of deep fried crickets (*Gryllus bimaculatus*) during storage. Four antioxidants and two packaging methods were used. The effects of different antioxidants and packaging methods on composition of fatty acids, contents of free fatty acids (FFA), peroxide value (PV), and thiobarbituric acid reactive substances (TBARSs) value of deep fried *Gryllus bimaculatus* were analyzed during 150 days of storage. The composition of fatty acids changed and the content of FFA, PV, and TBARs value also increased with the extension of storage time, indicating that the lipid oxidation dominated by oxidation of unsaturated fatty acids could occur in deep fried *Gryllus bimaculatus* during storage. In the same storage period, the total content of FFA, PV, and TBARs value of samples treated with antioxidants and vacuum-filling nitrogen packaging were lower than those of controls, suggesting that antioxidants and vacuum-filling nitrogen packaging have noticeable effects on inhibiting lipid oxidation and improving the quality of deep fried crickets, and dibutyl hydroxyl toluene (BHT) was found as the most effective antioxidant in this study. The results may provide a reliable reference for processing of deep fried edible insects.

Keywords: edible insects; *Gryllus bimaculatus*; processing and preservation; fatty acids; lipid oxidation

Citation: Gan, J.; Zhao, M.; He, Z.; Sun, L.; Li, X.; Feng, Y. The Effects of Antioxidants and Packaging Methods on Inhibiting Lipid Oxidation in Deep Fried Crickets (*Gryllus bimaculatus*) during Storage. *Foods* **2022**, *11*, 326. https://doi.org/10.3390/foods11030326

Academic Editors: Victor Benno Meyer-Rochow and Chuleui Jung

Received: 4 January 2022
Accepted: 21 January 2022
Published: 24 January 2022

Publisher's Note: MDPI stays neutral with regard to jurisdictional claims in published maps and institutional affiliations.

1. Introduction

Insects are a biological group with the largest number of species, huge biomass, fast reproduction capability, and high food conversion rate on earth. Insects are highly nutritious, containing a large amount of high-quality proteins, high levels of unsaturated fatty acids, and high levels of essential trace elements, such as iron and zinc [1]. Although the data of nutritional composition vary from species to species, most species of insects contain proteins, fat, vitamins, and minerals, corresponding to a human's nutritional needs [2]. In view of the increase of global population and the decrease of arable land, Meyer-Rochow [3] advocated the use of insects as human food and recommended that the FAO (Food and Agriculture Organization of the United Nations) regarded insects as a potential sustainable food source to deal with global food security problems, and encouraged more use of insects in daily diets [4]. More than 2000 kinds of insects are eaten in more than 110 countries/regions worldwide. The commonly edible insects include Coleoptera, Lepidoptera, Hymenoptera, Orthoptera, and Hemiptera. Among Orthoptera insects, crickets are the most widely consumed insect species [5].

Crickets belong to Gryllidae of Grylloidea in Orthopera, which are widely distributed universally. The commonly consumed varieties are *Brachytrupes membranaceus*, *Gryllus similis*, *Gryllus bimaculatus*, *Gryllotalpa orientalis*, and *Acheta domesticus* [6]. Crickets are also rich in nutrients, with a protein content of 55–70% [7], and the proportion of unsaturated fatty acids (UFAs) is more than 60% in lipids, which may even reach 80% in some varieties, such as *Gryllus testaceus* Walker and *Teleogryllus emma* [8,9]. Regarding minerals, calcium, iron, zinc, and copper contents in crickets are higher than those of conventional foods of animal origin, but not dangerously so [9]. Moreover, the low chitin content and hardness of crickets grown for six to seven weeks may lead to good taste and palatability [6]. Therefore, crickets promote the development of breeding industry in some countries in Asia, Europe, America, Australia, and Africa continents, especially in tropical regions (e.g., Southeast Asia). As reported previously, the total annual output of feeding crickets in Thailand is within 3000–7000 tons [10], and although the cricket farming scale in Cambodia and Laos is smaller than that in Thailand, it is promptly expanding and developing [7]. In Korea, cricket farming has a history of about 20 years. Farmed crickets are mainly used as feed for livestock and aquaculture, and the number of farmed crickets will also increase due to the use of cricket flour as a protein-rich additive to flour in the baking industry [11]. In recent years, cricket farming as the basis of food processing has markedly attracted scholars' attention as a result of the recognition of the nutritional value and food safety of edible crickets. Cambodian Center for Livestock and Agriculture Development evaluated the influences of different types of feed on the survival rate and growth of crickets [12]; Kenya aimed to introduce cool and high-altitude areas to expand the production of crickets [13]; United States Department of Agriculture—Agricultural Research Service National Biological Control Laboratory proposed an optimal harvesting age for reducing costs of cricket production by determining food conversion efficiency at different ages [14].

The development of cricket farming provides the possibility for exploiting and industrialization of cricket-based foods. Crickets are also consumed in making bread, biscuits, noodles, and meat sauce as additives [15–18]. However, deep frying is a common method of processing of crickets [19], and bagged fried cricket is a popular snack food in Thailand, Laos, and other countries. China is one of the first countries known to consume insects as food [20]. Although crickets have not been approved as a new resource of food in China, cricket farms and companies have recently appeared in some regions of China, and the scale is expanding. The breeding varieties are mainly *Gryllus bimaculatus* and *Acheta domesticus*, in which *Gryllus bimaculatus* is more popular due to its short life-cycle, being stronger, and having a cold resistance. In addition to being used as pet feed and serving at restaurants, farmed crickets are mainly processed with deep fried and packaging in small workshops, where they then attract consumers' attention because of their noticeable nutritional level and delicious taste. However, the lipid oxidation occurs easily during processing and storage as a result of abundance of UFAs in edible insects, influencing the quality and flavor of products in the period of storage. This is a prominent and urgent problem that needs to be eliminated in the production of edible insect, especially deep fried edible insects.

Therefore, regarding *Gryllus bimaculatus* as a food source, four antioxidants and two packaging methods were used in the present study, and fatty acid profile, free fatty acid content, and lipid oxidation indices, such as peroxide value (PV) and thiobarbituric acid reactive substances (TBARs), of deep fried crickets were analyzed during 150 days of storage. The effects of antioxidants and packaging methods on lipid oxidation were studied for improving processing technology of deep fried crickets.

2. Materials and Methods

2.1. Chemicals and Materials

A standard mixture of 37 fatty acid methyl esters, glyceryl triundecanoate standard, and undecanoic acid standard were purchased from Sigma-Aldrich Co., Ltd. (Shanghai branch, China); petroleum ether, n-heptane, boron trifluoride methanol solution, chloroform, and 2-isopropanol were purchased from Kunming Beijie Technology Co., Ltd. (Kunming, China); four antioxidants, including dibutyl hydroxyl toluene (BHT), rosemary extract (ET), and tert-butylhydroquinone (TBHQ), were purchased from Zhejiang Yinuo Biotechnology Co., Ltd. (Hangzhou, China), and *Phyllanthus emblica* polyphenol (PEP) was prepared by our laboratory; The packing bags made of PET/PE (polyethylene terephthalate/polyethylene, 0.16 mm of total thickness) and having excellent air tightness, oil resistance and fragrance retention, were purchased from Cangzhou Jingtian Plastic Industry Co., Ltd. (Cangzhou, China)

2.2. Preparation of Cricket Samples

The fresh *Gryllus bimaculatus* used in the experiment were provided by Yunnan Kuoyang Agricultural Science and Technology Co., Ltd. (Kunming, China), and identified by Research Institute of Resource Insects, Chinese Academy of Forestry (Kunming, China). The feeding life-cycle of crickets was 6–7 weeks. After fasting for 2 days, they were frozen at $-18\,^{\circ}\mathrm{C}$ for 1 h. The fresh *Gryllus bimaculatus* contain 71% water, 16.99% protein, and 9.98% lipid.

2.3. Preparation of Deep Fried Crickets

The dead crickets were washed with water and divided into 6 portions. One portion was 2 kg and that was deeply fried at 160 °C for 5 min in 20 kg of palm oil containing different antioxidants. The addition proportion of the four antioxidants was determined according to the maximum use limit of each antioxidant in the oil, and the amount of antioxidants used was based on the weight of palm oil. The lipid contents before and after deep frying of the crickets were 34.43% and 51.55% of dry weight, respectively. Two packaging methods were used for the deep fried crickets, including non-vacuum sealed packaging and vacuum-filling nitrogen packaging. Non vacuum packaging samples in bags made of PET/PE (polyethylene terephthalate/polyethylene, 0.16 mm of total thickness) were air-packed (50 g sample per bag) using a sealing machine (Guangzhou Feipu Packaging Machinery Co., Ltd., Guangzhou, China). Vacuum-filling nitrogen packaging samples in bags made of PET/PE were packed (50 g sample per bag) by vacuumizing and filling in with nitrogen using a vacuum sealing machine equipped a nitrogen cylinder (Zhejiang Baochun Packaging Machinery Co., Ltd., Wenzhou, China). The treatment methods of 6 samples were as follows: BHTV—0.2% BHT + vacuum-filling nitrogen packaging; ETV—0.5% ET + vacuum-filling nitrogen packaging; PEPV—0.5% PEP + vacuum-filling nitrogen packaging; TBHQV—0.2% TBHQ + vacuum-filling nitrogen packaging; control 1—no antioxidant + vacuum-filling nitrogen packaging; and control 2—no antioxidant + non vacuum sealed packaging. The samples were stored at room temperature in a dark place for the analysis of indices every 30 days.

2.4. Lipid Extraction

The lipids were extracted using Soxhlet extractor, in which 10 g of crushed sample and 300 mL of petroleum ether were transferred into a 500 mL flat bottom flask, and maintained in a 40 °C water bath for 8 h. The extract solution was concentrated using a rotary evaporator (Eyela, Tokyo, Japan) to obtain the lipid extract.

2.5. Analysis of Composition of Fatty Acids

As described by Chinese national standard GB 5009.168-2016 [21], briefly, glyceryl triundecanoate was used as an internal standard, and 60 mg of lipid extract and 2 mL of glyceryl triundecanoate solution (2.5 mg/mL) were mixed with 8 mL of 2% sodium hydroxide methanol solution. The mixture was incubated for 2 h on a water bath at 80 °C for saponification, and then 7 mL of 15% boron trifluoride methanol solution was added into the mixture and continually incubated for 6 min to achieve methyl esterification. After cooling, the esterification solution was mixed and shaken with 30 mL n-heptane and 30 mL saturated sodium chloride solution, kept for layering, and the upper solution was taken out for analysis.

Fatty acid methyl esters were analyzed by gas chromatography equipped with a polydicyanopropyl siloxane strong polar stationary phase capillary column (100 m × 250 µm ID × 0. 2 µm film). The temperature of the injection port and the detector was set to 270 °C and 280 °C, respectively. The carrier gas was nitrogen, and the flow rate was 1.5 mL/min.

The injection volume was 1 mL, and the split ratio was 10:1. The procedure was completed at the following thermal conditions: The initial temperature of 100 °C was kept for 13 min, and it increased to 180 °C at 10 °C/min and maintained for 6 min, and then the temperature was elevated to 200 °C at 1 °C/min and maintained for 20 min. Identification of fatty acid methyl esters was performed by comparing the retention time of the standard mixture of fatty acid methyl esters. The fatty acids were quantified by an internal standard, and the proportion of each fatty acid was calculated as the ratio of each fatty acid content to the total fatty acid content, and the result was expressed as percentage of total fatty acids. Each sample was analyzed in triplicate.

2.6. Determination of the Contents of Free Fatty Acids (FFAs)

Free fatty acids were separated by Wang et al.'s method [22], in which 240 mg of lipid extract was dissolved in 12 mL of chloroform, and 10 mL of chloroform solution was loaded onto an aminopropyl-silica minicolumn (500 mg/3 mL), which was previously activated with 10 mL of chloroform. The minicolumn was eluted with 10 mL of chloroform/2-isopropanol (2/1, *v/v*), and the eluent was discarded. The free fatty acids were eluted by 15 mL of ether solution, containing 2% acetic acid (*w/w*), the solvent was removed from eluent using a rotary evaporator, and the residue was free fatty acid.

The free fatty acids and 2 mL of undecanoic acid solution (50 μg/mL), as an internal standard, were mixed with 7 mL of 15% boron trifluoride methanol solution, and the mixture was incubated for 6 min on a water bath at 80 °C for methyl esterification. The cooled esterification solution was mixed and shaken with 20 mL n-heptane and 20 mL saturated sodium chloride solution, kept for layering, and the upper solution was taken out for analysis. Fatty acid methyl ester was analyzed according to the above-mentioned methods, and the result was expressed as mg/100 g sample. Each sample was analyzed in triplicate.

2.7. Determination of Peroxide Value (PV)

PV was determined according to Chinese national standard GB 5009.227-2016 [23]. Specifically, 10 g of crushed sample was mixed with 20 mL petroleum ether and was kept for 12 h. The filtrate was evaporated to remove petroleum ether under vacuum using a rotary evaporator in a water bath at 35 °C. The residue was dissolved in 30 mL of chloroform-glacial acetic acid mixture (2/3, *v/v*) and transferred to a 250 mL iodine-measuring bottle, and 1 mL of saturated potassium iodide solution was then added. After completion of reaction for 3 min in a dark place, 100 mL of deionized water and 1 mL of 1% starch solution, as an indicator, were added, and the reaction solution was titrated with sodium thiosulfate solution immediately until the intense blue color disappeared and solution became canary yellow. The blank test was performed according to the above-mentioned method. Each sample was analyzed in triplicate. The PV was calculated by Equation (1).

$$X = \frac{(V - V_0) \times c \times 0.1269 \times 1000}{m} \times 100 \tag{1}$$

where X represents peroxide value (mg/100 g); V and V_0 are the volume of sodium thiosulfate solution consumed by the sample and the blank, respectively (mL); c is the concentration of sodium thiosulfate solution (mol/L); 0.1269 denotes the mass of iodine equivalent to 1 mL of 1 mol/L sodium thiosulfate solution; and m indicates the sample weight (g).

2.8. Determination of Thiobarbituric Acid Reactive Substances (TBARs)

TBARs assay was performed according to Chinese national standard GB 5009.181-2016 [24]. First, 1,1,3,3-tetraethoxypropane was used as an internal standard. Then, 5 g of crushed sample and 50 mL of 7.5% trichloroacetic acid were placed into a 150 mL of conical flask, and shaken at 150 rpm for 30 min in an Ecotron shaking incubator (INFORS, Basel, Switzerland). The mixture was filtered with paper filters. Next, 5 mL of filtrate and standard solution (0.01, 0.05, 0.1, 0.15, 0.25 µg/L) was mixed with 5 mL of 0.02 mol/L thiobarbituric acid solution, respectively, and 5 mL of deionized water was used as blank. The mixture was incubated in a water bath at 90 °C for 30 min. The absorbance of mixture was measured using a microplate reader (Thermo Fisher Scientific, Waltham, MA, USA) at 532 nm after cooling. The standard curve was established with the standard solution concentration as the horizontal coordinate and the absorbance as the vertical coordinate. The TBARs value of the sample was calculated according to the standard curve and the dilution ratio of the sample. The result was expressed as milligrams malondialdehyde per kilogram of sample (mg MDA/kg). Each sample was analyzed in triplicate.

2.9. Statistical Analysis

All experiments were performed in triplicate. All data are expressed as mean ± standard deviations (SD). One-way analysis of variance (ANOVA) was used for comparing differences using the SPSS 13.0 software (IBM, Armonk, NY, USA). Bonferroni correction and Duncan's test were used for multiple comparisons.

3. Results

3.1. Fatty Acid Compositions of Raw Gryllus bimaculatus and Palm Oil

In order to analyze the changes of fatty acids in the samples after deep frying, the fatty acid compositions of raw *Gryllus bimaculatus* and palm oil were determined. The main fatty acids of different species of crickets are palmitic acid (C16:0), oleic acid (C18:1) and linoleic acid (C18:2), and the content of oleic acid in *Gryllus bimaculatus* is generally higher than that of other varieties [25]. As shown in Table 1, the fatty acid of *Gryllus bimaculatus* includes myristic acid (C14:0), palmitic acid (C16:0), palmitoleic acid (C16:1), stearic acid (C18:0), oleic acid (C18:1), linoleic acid (C18:2), and linolenic acid (C18:3). Among them, palmitic acid, oleic acid and linoleic acid are the main fatty acids of *Gryllus bimaculatus*. The results also showed that the proportion of UFAs was 65.79%, mainly consisting of oleic acid (22.75%) and linoleic acid (41.75%), the proportion of saturated fatty acids (SFAs) was 34.21%. Compared with the previously reported results on the analysis of fatty acids in *Gryllus bimaculatus* [26,27], the composition and content of main fatty acids were consistent, except for a trace of arachidic acid (C20:0) and arachidonic acid (C20:1), which were not detected in the present study.

Table 1. Composition of fatty acids in fresh *Gryllus bimaculatus* and palm oil (% of total fatty acids) [1].

	Content of Fatty Acids (% of Total Fatty Acids)	
	Gryllus bimaculatus	**Palm Oil**
Myristic acid (C14:0)	0.58 ± 0.01	0.97 ± 0.02
Palmitic acid (C16:0)	24.31 ± 0.14	40.67 ± 0.2
Palmitoleic acid (C16:1)	0.44 ± 0.00	-
Stearic acid (C18:0)	9.32 ± 0.07	4.24 ± 0.03
Oleic acid (C18:1)	22.75 ± 0.09	42.8 ± 0.04
Linoleic acid (C18:2)	41.75 ± 0.23	11.23 ± 0.03
Linolenic acid (C18:3)	0.85 ± 0.01	-
ΣUFA [2]	65.79 ± 0.19	54.03 ± 0.07
ΣSFA [3]	34.21 ± 0.2	45.89 ± 0.21
UFA/SFA	1.92 ± 0.02	1.18 ± 0.01

[1] "-" means that this fatty acid was not detected; [2] "UFA" indicates unsaturated fatty acids; [3] "SFA" indicates saturated fatty acids.

Palm oil contains five fatty acids—myristic acid, palmitic acid, stearic acid, oleic acid, and linoleic acid (Table 1)—and the main fatty acids are palmitic acid, oleic acid, and linoleic acid, which are the same as *Gryllus bimaculatus*. However, the contents of palmitic acid and oleic acid are remarkably higher than those of *Gryllus bimaculatus*, and the content of linoleic acid is much lower than that of *Gryllus bimaculatus*.

3.2. Changes in Fatty Acid Composition of Deep Fried Samples

The fatty acid composition of lipids is a critical indicator of nutritional value, and it is associated with the oxidative stability of lipids. The fatty acids of deep fried samples with different treatments during storage are summarized in Table 2. The results indicated that the fatty acid composition of deep fried crickets (0 days) significantly varied compared with the raw materials (Table 1). Palmitoleic acid and linolenic acid were not found, and the contents of palmitic acid and oleic acid increased significantly, while the content of linoleic acid markedly decreased, and the content of stearic acid was reduced as well. The proportion of UFAs decreased to 55% (Table 2) from 65.79% in raw samples (Table 1). Deep fried samples contain a large amount of palm oil after frying due to the penetration of palm oil into samples during frying. Therefore, the compositions and contents of fatty acids were similar to those of palm oil.

As shown in Table 2, during the storage period (150 days), there were significant differences between the contents of fatty acids in the same sample at different periods of storage, including palmitic acid, stearic acid, oleic acid, linoleic acid, ΣUFAs, ΣSFAs, and UFAs/SFAs in all samples ($p < 0.001$), as well as myristic acid in PEPV, TBHQV, control 1, and control 2 ($p < 0.001$). Meanwhile, the decline of ΣUFAs, the increase of ΣSFAs, and the decrease of UFAs/SFAs in all samples during storage were also observed, and the changes of ΣUFAs and ΣSFAs were the results of the decline of the content of oleic acid and the raise of the contents of myristic acid and palmitic acid. These results suggested that the oxidation of UFAs in the deep fried *Gryllus bimaculatus* occurred during storage, and UFAs were found more susceptible to oxidation than SFAs.

Regarding the differences in samples with different treatments in the same storage time, Table 2 shows that there were significant differences between the contents of stearic acid, oleic acid, and linoleic acid from different samples in the same storage time ($p < 0.001$). The content of myristic acid in different samples significantly changed after storing for 60 and 150 days ($p < 0.01$ and $p < 0.05$, respectively). The difference in the content of palmitic acid from different samples was statistically significant after storing for 60, 90, and 150 days ($p < 0.01$, $p < 0.05$, and $p < 0.001$, respectively). For ΣUFAs, ΣSFAs, and UFAs/SFAs, there was no significant difference among samples within 30–90 days after storage ($p > 0.05$), while the difference gradually increased with the extension of storage time. After 120 and 150 days of storage, the difference was significant ($p < 0.01$ and $p < 0.001$, respectively). The results indicated that different antioxidants and packaging treatments had certain effects on the changes of compositions of fatty acids during storage, and the effects gradually increased with the extension of storage time.

3.3. Contents of FFAs in Deep Fried Samples

FFA is attributed to the oxidation and hydrolysis of lipids under lipase and oxygen. The results from the analysis of FFAs in deep fried *Gryllus bimaculatus* (Table 3) indicated that palmitic acid, stearic acid, oleic acid, and linoleic acid were detected in samples. The contents of these FFAs from the same sample in different periods of storage were significantly different ($p < 0.05$, $p < 0.01$, and $p < 0.001$, respectively). The contents of FFAs noticeably increased with the prolongation of storage time, and the contents of ΣSFFAs were markedly higher than those of ΣUFFAs. The contents of a single FFA and total FFAs in BHTV reached the maximum values after 90 or 120 days of storage, and then decreased slightly. The contents of a single FFA and total FFAs in other samples basically reached the peak after 150 days of storage, due to the continuous accumulation of FFAs with the oxidation and hydrolysis of lipids.

Table 2. Composition of fatty acids in deep fried *Gryllus bimaculatus* during storage (% of total fatty acids) [1].

Fatty Acids	Treatments [2]	Storage Time/Days						Sign. [3]
		0	30	60	90	120	150	
Myristic acid (C14:0)	BHTV	0.88 ± 0.00	1.01 ± 0.02	0.73 ± 0.01 [B]	0.85 ± 0.00	1.34 ± 0.53	0.93 ± 0.02 [AB]	n.s.
	ETV	0.89 ± 0.00	1.02 ± 0.28	0.76 ± 0.00 [B]	0.86 ± 0.02	1.04 ± 0.05	0.94 ± 0.00 [AB]	n.s.
	PEPV	0.87 ± 0.00 [b]	0.93 ± 0.03 [c]	0.80 ± 0.01 [Ba]	0.86 ± 0.01 [b]	0.96 ± 0.02 [c]	0.94 ± 0.01 [ABc]	***
	TBHQV	0.89 ± 0.03 [b]	0.90 ± 0.02 [bc]	0.81 ± 0.01 [Ba]	0.89 ± 0.01 [b]	0.93 ± 0.01 [c]	0.93 ± 0.01 [ABc]	***
	control 1	0.87 ± 0.00 [b]	0.94 ± 0.06 [b]	0.68 ± 0.07 [Aa]	0.85 ± 0.01 [b]	0.95 ± 0.06 [b]	0.95 ± 0.01 [Ab]	***
	control 2	0.87 ± 0.00 [b]	090 ± 0.02 [c]	0.81 ± 0.01 [Ba]	0.87 ± 0.01 [b]	0.92 ± 0.02 [c]	0.91 ± 0.01 [Bc]	***
	Sign.	n.s.	n.s.	**	n.s.	n.s.	*	
Palmitic acid (C16:0)	BHTV	38.4 ± 0.04 [b]	37.69 ± 0.06 [a]	39.90 ± 0.10 [Cd]	39.26 ± 0.17 [ABc]	40.55 ± 0.37 [e]	40.66 ± 0.21 [BCe]	***
	ETV	38.37 ± 0.02 [b]	37.84 ± 0.08 [a]	39.32 ± 0.11 [BCd]	39.07 ± 0.20 [ABc]	40.16 ± 0.24 [e]	40.68 ± 0.02 [BCf]	***
	PEPV	37.84 ± 0.09 [a]	37.70 ± 0.26 [a]	39.09 ± 0.15 [ABCc]	38.57 ± 0.21 [Ab]	39.73 ± 0.10 [d]	40.75 ± 0.06 [BCe]	***
	TBHQV	37.94 ± 0.06 [ab]	37.80 ± 0.06 [a]	38.49 ± 0.26 [Ab]	39.39 ± 0.40 [ABc]	39.67 ± 0.47 [c]	40.38 ± 0.38 [Bd]	***
	control 1	38.3 ± 0.47 [b]	37.41 ± 0.07 [a]	39.51 ± 0.72 [BCc]	39.56 ± 0.19 [Bc]	40.45 ± 0.03 [d]	41.01 ± 0.06 [Cd]	***
	control 2	38.00 ± 0.01 [a]	37.77 ± 0.29 [a]	39.00 ± 0.22 [ABb]	39.04 ± 0.41 [ABb]	39.77 ± 0.45 [c]	39.87 ± 0.08 [Ac]	***
	Sign.	n.s.	n.s.	**	*	n.s.	***	
Stearic acid (C18:0)	BHTV	5.37 ± 0.00 [Dd]	5.17 ± 0.01 [Ac]	4.43 ± 0.06 [BCa]	5.07 ± 0.02 [CDb]	5.15 ± 0.01 [Bc]	5.25 ± 0.02 [Cd]	***
	ETV	5.21 ± 0.01 [Bc]	5.45 ± 0.02 [De]	4.29 ± 0.07 [Ba]	4.98 ± 0.01 [BCb]	5.05 ± 0.02 [Ab]	5.01 ± 0.02 [ABb]	***
	PEPV	5.43 ± 0.01 [Ee]	5.28 ± 0.05 [Bd]	4.76 ± 0.01 [Da]	5.17 ± 0.04 [Dec]	5.19 ± 0.02 [Bc]	5.08 ± 0.04 [Bb]	***
	TBHQV	5.19 ± 0.01 [Ad]	5.16 ± 0.02 [Ad]	4.62 ± 0.05 [CDa]	4.91 ± 0.03 [Bb]	5.06 ± 0.06 [Ac]	5.00 ± 0.07 [ABc]	***
	control 1	5.29 ± 0.02 [Cc]	5.16 ± 0.02 [A]	3.92 ± 0.29 [Aa]	4.70 ± 0.16 [Ab]	5.22 ± 0.06 [Bc]	4.98 ± 0.02 [Ac]	***
	control 2	5.44 ± 0.01 [Ed]	5.36 ± 0.04 [Cc]	5.18 ± 0.03 [Ea]	5.27 ± 0.04 [Eb]	5.20 ± 0.02 [Ba]	5.28 ± 0.01 [Cb]	***
	Sign.	***	***	***	***	***	***	
Oleic acid (C18:1)	BHTV	40.47 ± 0.02 [Ac]	41.40 ± 0.13 [BCd]	40.45 ± 0.03 [Bc]	40.09 ± 0.19 [Ab]	37.45 ± 0.09 [Aa]	37.39 ± 0.16 [Aa]	***
	ETV	41.07 ± 0.03 [Bd]	40.46 ± 0.41 [Ac]	41.66 ± 0.06 [Ce]	40.55 ± 0.22 [Bc]	38.25 ± 0.08 [Bb]	37.84 ± 0.03 [Ba]	***
	PEPV	40.36 ± 0.06 [Abc]	40.97 ± 0.37 [Abd]	40.47 ± 0.10 [Bc]	40.00 ± 0.22 [Ab]	37.88 ± 0.23 [Aa]	37.82 ± 0.13 [Ba]	***
	TBHQV	41.7 ± 0.07 [Cd]	41.71 ± 0.05 [Cd]	42.12 ± 0.15 [Cd]	40.71 ± 0.31 [Bc]	38.97 ± 0.32 [Cb]	38.31 ± 0.24 [Ca]	***
	control 1	40.62 ± 0.37 [Ab]	41.38 ± 0.02 [BCb]	40.95 ± 0.82 [Bb]	40.87 ± 0.23 [Bb]	37.80 ± 0.01 [Aa]	37.70 ± 0.08 [Ba]	***
	control 2	40.29 ± 0.01 [Ad]	40.88 ± 0.36 [Abe]	39.45 ± 0.13 [Ac]	39.62 ± 0.33 [Ac]	38.32 ± 0.27 [Bb]	37.43 ± 0.07 [Aa]	***
	Sign.	***	**	***	***	***	***	

Table 2. *Cont.*

Fatty Acids	Treatments [2]	Storage Time/Days						Sign. [3]
		0	30	60	90	120	150	
Linoleic acid (C18:2)	BHTV	14.88 ± 0.04 [Cd]	14.30 ± 0.04 [Aa]	14.48 ± 0.04 [Bb]	14.73 ± 0.07 [Cc]	15.51 ± 0.13 [Abe]	15.79 ± 0.07 [Bf]	***
	ETV	14.45 ± 0.04 [Bb]	15.22 ± 0.12 [Dc]	13.97 ± 0.05 [Aa]	14.54 ± 0.05 [Bb]	15.50 ± 0.18 [ABd]	15.53 ± 0.07 [Ad]	***
	PEPV	15.50 ± 0.03 [Ec]	15.12 ± 0.04 [CDb]	14.87 ± 0.06 [Ca]	15.40 ± 0.02 [Ec]	16.25 ± 0.16 [Cd]	15.42 ± 0.10 [Ac]	***
	TBHQV	14.28 ± 0.01 [Ac]	14.30 ± 0.05 [Ac]	13.95 ± 0.06 [Aa]	14.11 ± 0.14 [Aab]	15.38 ± 0.14 [Ad]	15.38 ± 0.09 [Ad]	***
	control 1	14.92 ± 0.09 [Cb]	14.70 ± 0.03 [Bb]	14.96 ± 0.27 [Cb]	14.01 ± 0.05 [Aa]	15.56 ± 0.05 [ABc]	15.36 ± 0.06 [Ac]	***
	control 2	15.39 ± 0.01 [Dbc]	15.04 ± 0.09 [Ca]	15.55 ± 0.11 [Dc]	15.20 ± 0.13 [Dab]	15.79 ± 0.19 [Bd]	16.50 ± 0.08 [Ce]	***
	Sign.	***	***	***	***	***	***	
ΣUFA [4]	BHTV	55.35 ± 0.04 [Ac]	55.95 ± 0.06 [d]	54.94 ± 0.05 [b]	54.82 ± 0.17 [b]	52.97 ± 0.20 [Aa]	53.17 ± 0.23 [Aa]	***
	ETV	55.53 ± 0.01 [ABd]	55.68 ± 0.36 [d]	55.63 ± 0.08 [d]	55.09 ± 0.22 [c]	53.75 ± 0.18 [BCb]	53.36 ± 0.03 [Aa]	***
	PEPV	55.86 ± 0.09 [Bd]	56.09 ± 0.34 [d]	55.34 ± 0.15 [c]	55.40 ± 0.22 [c]	54.13 ± 0.09 [Cb]	53.24 ± 0.08 [Aa]	***
	TBHQV	55.98 ± 0.08 [Bc]	56.01 ± 0.04 [c]	56.08 ± 0.21 [c]	54.82 ± 0.44 [b]	54.34 ± 0.45 [Cb]	53.68 ± 0.33 [Ba]	***
	control 1	55.54 ± 0.45 [ABbc]	56.31 ± 0.02 [c]	55.90 ± 1.09 [bc]	54.89 ± 0.27 [b]	53.37 ± 0.05 [Aba]	53.06 ± 0.06 [Aa]	***
	control 2	55.68 ± 0.01 [ABc]	55.91 ± 0.28 [c]	55.00 ± 0.24 [b]	54.82 ± 0.46 [b]	54.11 ± 0.46 [Ca]	53.94 ± 0.08 [Ba]	***
	Sign.	*	n.s.	n.s.	n.s.	**	***	
ΣSFA [5]	BHTV	44.65 ± 0.04 [Bb]	44.05 ± 0.06 [a]	45.06 ± 0.05 [c]	45.18 ± 0.17 [c]	47.03 ± 0.20 [Bd]	46.83 ± 0.23 [Bd]	***
	ETV	44.47 ± 0.01 [Aba]	44.32 ± 0.36 [a]	44.37 ± 0.08 [a]	44.91 ± 0.22 [b]	46.25 ± 0.18 [ABc]	46.64 ± 0.03 [Bd]	***
	PEPV	44.14 ± 0.09 [Aa]	43.91 ± 0.34 [a]	44.66 ± 0.15 [b]	44.60 ± 0.22 [b]	45.87 ± 0.09 [Ac]	46.76 ± 0.08 [Bd]	***
	TBHQV	44.02 ± 0.08 [Aa]	43.99 ± 0.04 [a]	43.92 ± 0.21 [a]	45.18 ± 0.44 [b]	45.66 ± 0.45 [Ab]	46.32 ± 0.33 [Ac]	***
	control 1	44.46 ± 0.45 [Aab]	43.69 ± 0.02 [a]	44.10 ± 1.09 [ab]	45.11 ± 0.27 [b]	46.63 ± 0.05 [BCc]	46.94 ± 0.06 [Bc]	***
	control 2	44.32 ± 0.01 [Aba]	44.09 ± 0.28 [a]	45.00 ± 0.24 [b]	45.18 ± 0.46 [b]	45.89 ± 0.46 [Ac]	46.06 ± 0.08 [Ac]	***
	Sign.	*	n.s.	n.s.	n.s.	**	***	
UFA/SFA	BHTV	1.24 ± 0.00 [Ac]	1.27 ± 0.00 [d]	1.22 ± 0.00 [b]	1.21 ± 0.01 [b]	1.13 ± 0.01 [Aa]	1.14 ± 0.01 [Aa]	***
	ETV	1.25 ± 0.00 [ABd]	1.26 ± 0.02 [d]	1.25 ± 0.00 [d]	1.23 ± 0.01 [c]	1.16 ± 0.01 [BCb]	1.14 ± 0.00 [Aa]	***
	PEPV	1.27 ± 0.00 [Bd]	1.28 ± 0.02 [d]	1.24 ± 0.01 [c]	1.24 ± 0.01 [c]	1.18 ± 0.00 [Cb]	1.14 ± 0.00 [Aa]	***
	TBHQV	1.27 ± 0.00 [Bc]	1.27 ± 0.00 [c]	1.28 ± 0.01 [c]	1.21 ± 0.02 [b]	1.19 ± 0.02 [Cb]	1.16 ± 0.02 [Ba]	***
	control 1	1.25 ± 0.02 [ABbc]	1.29 ± 0.00 [c]	1.27 ± 0.06 [bc]	1.22 ± 0.01 [b]	1.14 ± 0.00 [Aba]	1.13 ± 0.00 [Aa]	***
	control 2	1.26 ± 0.00 [ABc]	1.27 ± 0.01 [c]	1.22 ± 0.01 [b]	1.21 ± 0.02 [b]	1.18 ± 0.02 [Ca]	1.17 ± 0.00 [Ba]	***
	Sign.	*	n.s.	n.s.	n.s.	***	**	

[1] Different capital superscripts ([ABCDE]) after the mean in the same column indicate significant differences $p < 0.05$, different lowercase letter superscripts ([abcdef]) after the mean in the same row indicate significant differences $p < 0.05$. [2] BHTV = 0.2% BHT + vacuum-filling nitrogen packaging; ETV = 0.5% ET + vacuum-filling nitrogen packaging; PEPV = 0.5% PEP + vacuum-filling nitrogen packaging; TBHQV = 0.2% TBHQ + vacuum-filling nitrogen packaging; control 1 = no antioxidants + vacuum-filling nitrogen packaging; control 2 = no antioxidant + non vacuum sealed packaging. [3] Sign.: Significance; n.s.: Not significant; *, ** and *** indicate significance at $p < 0.05$, $p < 0.01$ and $p < 0.001$, respectively. [4] "UFA" indicates unsaturated fatty acids. [5] "SFA" indicates saturated fatty acids.

Table 3. The contents of free fatty acids (FFAs) in deep fired *Gryllus bimaculatus* during storage (mg/100 g) [1].

Free Fatty Acid	Treatments [2]	Storage Time/Day						Sign. [3]
		0	30	60	90	120	150	
Palmitic acid (C16:0)	BHTV	39.91 ± 2.66 ^{Ba}	46.98 ± 1.82 ^{Cab}	52.76 ± 5.47 ^{ABbc}	53.86 ± 1.15 ^{Bbc}	59.32 ± 7.50 ^{ABc}	55.43 ± 0.94 ^{ABbc}	**
	ETV	29.64 ± 1.00 ^{Aa}	42.22 ± 0.35 ^{ABb}	56.56 ± 5.71 ^{ABc}	58.07 ± 0.48 ^{Bc}	57.92 ± 3.48 ^{ABc}	63.90 ± 5.96 ^{Bc}	***
	PEPV	30.75 ± 3.31 ^{Aa}	44.10 ± 5.51 ^{ABb}	46.76 ± 1.00 ^{Ab}	61.92 ± 5.76 ^{Bc}	52.82 ± 6.57 ^{Abc}	62.91 ± 3.46 ^{Bc}	***
	TBHQV	33.15 ± 0.72 ^{Aa}	36.09 ± 1.46 ^{Aa}	46.38 ± 6.37 ^{Ab}	44.73 ± 5.96 ^{Ab}	53.83 ± 4.13 ^{Ab}	47.91 ± 4.28 ^{Ab}	**
	control 1	50.23 ± 3.82 ^{Ca}	47.05 ± 3.62 ^{Ca}	63.49 ± 6.62 ^{BCbc}	59.01 ± 0.1 ^{Bb}	68.72 ± 4.66 ^{BCc}	79.63 ± 3.88 ^{Cd}	***
	control 2	54.45 ± 1.39 ^{Ca}	64.25 ± 6.69 ^{Dab}	72.57 ± 3.42 ^{Cbc}	62.52 ± 5.37 ^{Bab}	76.41 ± 5.18 ^{Cc}	92.68 ± 6.58 ^{Dd}	***
	Sign. [2]	***	***	***	**	**	***	
Stearic acid (C18:0)	BHTV	14.83 ± 0.67 ^{Aa}	20.86 ± 2.08 ^{BCb}	27.27 ± 4.54 ^{Cc}	27.74 ± 0.69 ^{ABc}	19.52 ± 1.37 ^{ABb}	25.27 ± 1.96 ^{Bc}	**
	ETV	14.23 ± 0.66 ^{Aa}	17.61 ± 0.42 ^{ABa}	13.35 ± 1.55 ^{Aa}	23.78 ± 1.56 ^{Ab}	17.61 ± 3.41 ^{Aa}	37.82 ± 4.51 ^{Cc}	***
	PEPV	13.55 ± 4.20 ^{Aa}	17.47 ± 1.50 ^{ABa}	23.94 ± 0.47 ^{BCb}	29.05 ± 3.39 ^{Bb}	16.93 ± 1.75 ^{Cb}	15.20 ± 1.19 ^{Aa}	***
	TBHQV	16.73 ± 0.65 ^{ABb}	11.74 ± 1.70 ^{Aa}	19.88 ± 3.29 ^{Bb}	23.38 ± 1.48 ^{Ab}	23.12 ± 2.14 ^{BCb}	21.82 ± 4.43 ^{ABb}	**
	control 1	20.07 ± 1.76 ^{Ba}	21.76 ± 1.22 ^{BCa}	19.21 ± 0.87 ^{Ba}	35.82 ± 0.23 ^{Cc}	25.87 ± 1.13 ^{Cb}	19.12 ± 1.86 ^{ABa}	***
	control 2	24.61 ± 0.57 ^{Ca}	26.93 ± 6.48 ^{Ca}	28.30 ± 1.95 ^{Ca}	37.44 ± 3.54 ^{Cb}	31.56 ± 1.55 ^{Dab}	47.15 ± 3.60 ^{Dc}	***
	Sign.	***	**	***	***	***	***	
Oleic acid (C18:1)	BHTV	16.36 ± 2.33 ^{Aa}	23.64 ± 5.97 ^{ABab}	18.61 ± 0.70 ^{Aa}	27.53 ± 0.46 ^{ABb}	27.74 ± 4.63 ^{Ab}	24.58 ± 3.08 ^{Aab}	**
	ETV	14.86 ± 0.74 ^{Aa}	16.89 ± 0.11 ^{Aa}	35.83 ± 1.29 ^{BCb}	18.69 ± 4.12 ^{Aa}	30.11 ± 6.65 ^{Ab}	31.04 ± 4.98 ^{ABb}	***
	PEPV	14.71 ± 4.55 ^{Aa}	22.76 ± 7.40 ^{ABab}	27.75 ± 0.99 ^{Bab}	25.09 ± 7.66 ^{ABab}	25.48 ± 5.43 ^{Aab}	33.75 ± 3.89 ^{ABb}	*
	TBHQV	15.88 ± 0.51 ^{Aa}	15.35 ± 4.56 ^{Aa}	16.52 ± 2.24 ^{Aa}	21.72 ± 3.68 ^{ABa}	20.36 ± 2.48 ^{Aa}	47.38 ± 13.54 ^{Bb}	***
	control 1	20.54 ± 2.32 ^{Aa}	23.38 ± 2.95 ^{ABab}	28.26 ± 4.71 ^{Bbc}	32.72 ± 0.18 ^{Cc}	28.68 ± 2.06 ^{Abc}	39.45 ± 4.68 ^{ABd}	***
	control 2	24.61 ± 0.57 ^{Ca}	26.93 ± 6.48 ^{Ca}	28.30 ± 1.95 ^{Ca}	37.44 ± 3.54 ^{Cb}	31.56 ± 1.55 ^{Dab}	47.15 ± 3.60 ^{Dc}	***
	Sign.	***	*	***	**	***	*	
Linoleic acid (C18:2)	BHTV	11.39 ± 0.86 ^{Ba}	22.25 ± 3.19 ^{Bc}	16.55 ± 1.17 ^{ABb}	25.1 ± 0.32 ^{Bc}	29.79 ± 2.93 ^{Bd}	24.81 ± 2.17 ^{Bc}	***
	ETV	11.49 ± 0.64 ^{Ba}	13.64 ± 0.15 ^{ABa}	11.23 ± 2.99 ^{Aa}	16.34 ± 3.97 ^{Aa}	27.57 ± 4.08 ^{Bb}	35.08 ± 1.88 ^{Cc}	***
	PEPV	8.05 ± 0.20 ^{Aa}	13.61 ± 2.39 ^{ABb}	15.48 ± 0.14 ^{ABb}	16.42 ± 2.97 ^{Ab}	24.31 ± 1.90 ^{Bc}	28.23 ± 3.52 ^{BCc}	***
	TBHQV	11.19 ± 0.15 ^{Bb}	6.12 ± 0.30 ^{Aa}	12.83 ± 1.40 ^{Abc}	16.44 ± 1.52 ^{Abc}	14.10 ± 4.54 ^{Abc}	17.77 ± 3.29 ^{Ac}	**
	control 1	17.01 ± 1.03 ^{Ca}	20.52 ± 3.20 ^{Ba}	18.96 ± 1.78 ^{Ba}	18.32 ± 0.07 ^{Aa}	25.73 ± 7.61 ^{Bab}	29.02 ± 4.12 ^{BCb}	*
	control 2	41.48 ± 0.90 ^{Dab}	48.32 ± 8.70 ^{Cab}	39.79 ± 3.80 ^{Ca}	53.57 ± 3.62 ^{Cb}	45.80 ± 1.38 ^{Cab}	49.83 ± 5.18 ^{Dab}	*
	Sign.	***	***	***	***	***	***	

Table 3. *Cont.*

Free Fatty Acid	Treatments [2]	Storage Time/Day						Sign. [3]
		0	30	60	90	120	150	
ΣUFA [4]	BHTV	27.74 ± 2.82 Aa	45.89 ± 9.11 Bb	35.16 ± 1.85 ABa	52.64 ± 0.19 Bb	57.54 ± 7.56 Bb	49.40 ± 5.17 Ab	***
	ETV	26.36 ± 1.36 A	30.53 ± 0.21 ABa	47.06 ± 1.79 Cab	35.03 ± 8.08 Aa	57.68 ± 6.83 Bc	66.12 ± 6.84 Ac	***
	PEPV	22.76 ± 4.71 Aa	36.36 ± 8.04 ABb	43.24 ± 0.91 BCb	41.51 ± 10.55 ABb	49.79 ± 7.32 ABbc	61.98 ± 7.26 Ac	**
	TBHQV	27.07 ± 0.59 Aa	21.46 ± 4.35 Aa	29.35 ± 3.58 Aa	38.15 ± 5.20 ABa	34.46 ± 6.00 Aa	65.16 ± 14.53 Ab	***
	control 1	37.56 ± 3.32 Ba	43.90 ± 4.96 Bab	47.22 ± 6.48 Cab	51.04 ± 0.16 Bab	54.41 ± 9.20 Bb	68.46 ± 8.79 Ac	**
	control 2	73.24 ± 1.41 Ca	79.55 ± 11.58 Ca	78.77 ± 8.47 Da	85.88 ± 6.03 Ca	110.47 ± 13.87 Cb	87.69 ± 8.79 Ba	**
	Sign.	***	***	***	***	***	**	
ΣSFA [5]	BHTV	54.73 ± 2.34 Ba	67.84 ± 3.63 Bb	80.04 ± 9.90 Ab	81.6 ± 1.44 Bb	78.84 ± 8.85 Ab	80.70 ± 2.90 Ab	**
	ETV	43.87 ± 1.56 Aa	59.83 ± 0.75 ABb	69.92 ± 7.17 A	81.84 ± 1.69 Bd	75.54 ± 0.39 Acd	101.72 ± 10.45 Be	***
	PEPV	44.29 ± 5.23 Aa	61.57 ± 6.55 ABb	70.70 ± 1.44 Abc	90.97 ± 4.94 BCd	79.75 ± 8.32 Ac	78.10 ± 4.44 Ac	***
	TBHQV	19.89 ± 0.85 ABa	47.83 ± 2.87 Aa	66.26 ± 9.64 Ab	68.11 ± 7.15 Ab	76.95 ± 5.86 Ab	69.73 ± 6.79 Ab	***
	control 1	70.30 ± 5.15 Ca	68.82 ± 4.84 Ba	82.70 ± 7.49 Ab	94.83 ± 0.14 Cc	94.59 ± 5.48 Bc	98.76 ± 5.63 Bc	***
	control 2	79.07 ± 1.71 Da	91.19 ± 13.10 Cab	100.87 ± 4.46 Bb	99.96 ± 8.90 Cb	107.98 ± 5.50 Cb	139.84 ± 10.17 Cc	***
	Sign.	***	***	***	***	***	***	
Total FFAs	BHTV	82.48 ± 5.16 Ba	113.73 ± 12.01 Bb	115.20 ± 11.20 ABb	134.24 ± 1.52 BCb	136.37 ± 16.18 ABb	130.10 ± 7.84 Ab	***
	ETV	70.23 ± 2.91 ABa	90.36 ± 0.96 ABb	116.98 ± 8.88 ABc	116.87 ± 9.57 ABc	133.22 ± 7.23 ABc	167.84 ± 16.30 Ad	***
	PEPV	67.06 ± 9.82 Aa	97.93 ± 14.30 ABb	113.94 ± 2.35 ABbc	132.48 ± 9.66 BCc	129.54 ± 15.61 ABc	140.08 ± 10.33 Ac	***
	TBHQV	76.95 ± 0.72 ABab	69.29 ± 6.85 Aa	95.61 ± 11.58 Abc	106.26 ± 11.24 Ac	111.41 ± 9.00 Ac	134.89 ± 20.02 Ad	***
	control 1	107.86 ± 7.97 Ca	112.72 ± 9.78 Ba	129.92 ± 13.98 Bab	145.87 ± 0.26 Cbc	149.00 ± 14.61 Bbc	167.22 ± 13.46 Ac	***
	control 2	152.30 ± 3.11 Da	170.73 ± 24.56 Ca	179.63 ± 12.49 Ca	185.84 ± 14.85 Da	218.45 ± 17.46 Cb	227.53 ± 18.91 Bb	**
	Sign.	***	***	***	***	***	***	

[1] Different capital superscripts ([ABCDE]) after the mean in the same column indicate significant differences $p < 0.05$, different small letter superscripts ([abcdef]) after the mean in the same row indicate significant differences $p < 0.05$; [2] BHTV = 0.2% BHT + vacuum-filling nitrogen packaging; ETV = 0.5% ET + vacuum-filling nitrogen packaging; PEPV = 0.5% PEP + vacuum-filling nitrogen packaging; TBHQV = 0.2% TBHQ + vacuum-filling nitrogen packaging; control 1= no antioxidants + vacuum-filling nitrogen packaging; control 2 = no antioxidant + non vacuum sealed packaging; [3] Sign.: Significance; n.s.: not Significant; *, ** and *** indicate significance at $p < 0.05$, $p < 0.01$ and $p < 0.001$, respectively. [4] "UFA" indicates unsaturated fatty acids. [5] "SFA" indicates saturated fatty acids.

As presented in Table 3, the contents of a single FFA and total FFAs in different samples significantly varied ($p < 0.05$, $p < 0.01$, and $p < 0.001$, respectively) at the same storage time. The ΣUFFAs, ΣSFFAs, and total content of FFAs in control 2 were the highest among all the samples, especially the total content of FFAs (218.45 and 227.53 mg/100 g) was markedly higher than that of other samples after 120 and 150 days of storage. Meanwhile, the total content of FFAs in BHTV, TBHQV, and PEPV was lower than that of control 1 after 150 days of storage. These results demonstrated that antioxidants and vacuum-filling nitrogen packaging had inhibitory effects on the lipid oxidation in deep fried *Gryllus bimaculatus*. The total content of FFAs in BHTV was the lowest among samples treated with antioxidant, suggesting that the effect of BHT on lipid oxidation in deep fried *Gryllus bimaculatus* was more noticeable than that of other antioxidants.

3.4. PV of Deep Fried Samples

The PV was used to measure the primary lipid-oxidation products, especially hydroperoxides [28], which could be further decomposed into low-molecular-weight substances, such as aldehydes, ketones, and acids. The level of lipid oxidation can be judged from PV. Figure 1 shows that the PVs of all samples increased during storage, and the PVs of different samples in the same period were significantly different ($p < 0.001$). The PV of control 2 was the highest in each storage time point. The PVs of control 1 and control 2 were markedly higher than those of other samples after 150 days of storage (35.79 and 40.05 mg/100 g, respectively). Among samples that were treated with antioxidants, the PVs of BHTV and TBHQV were lower than those of other samples, which could be corresponded to the results of analysis of FFAs, indicating that the methods of using antioxidants and vacuum-filling nitrogen packaging could inhibit the oxidation of lipids, in which the effects of BHT and TBHQ were more noticeable.

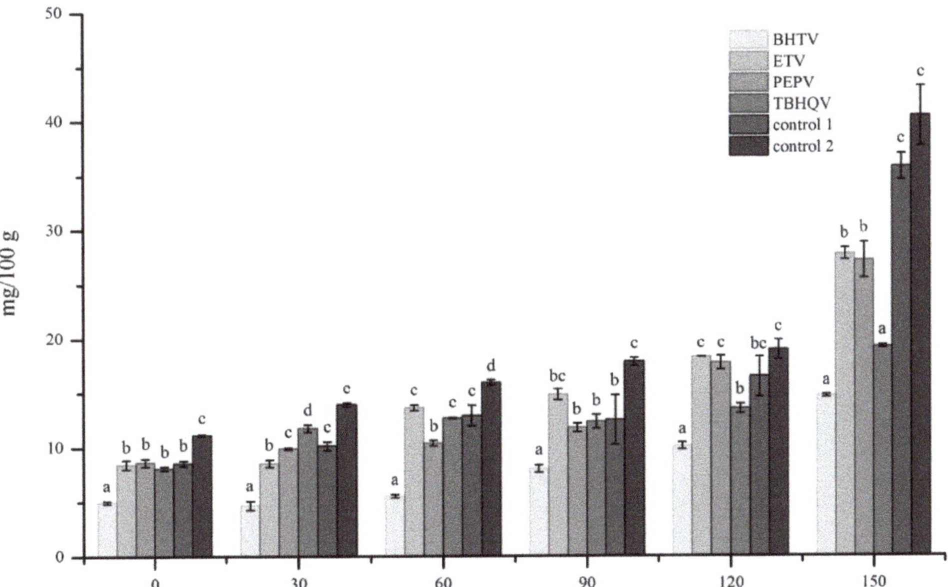

Figure 1. Changes in peroxide values (PVs) of deep fried *Gryllus bimaculatus*. BHTV = 0.2% BHT + vacuum-filling nitrogen packaging; ETV = 0.5% ET + vacuum-filling nitrogen packaging; PEPV = 0.5% PEP + vacuum-filling nitrogen packaging; TBHQV = 0.2% TBHQ + vacuum-filling nitrogen packaging; control 1 = no antioxidants + vacuum-filling nitrogen packaging; control 2 = no antioxidant + non vacuum sealed packaging; different lowercase letters on the column chart showed significant differences in the same storage time ($p < 0.001$).

3.5. TBARs of Deep Fried Samples

TBARs is an important indicator of oxidation of fatty acids during storage and processing, which could be characterized with the content of malondialdehyde (MDA) [29]. MDA is one of the main end-products of lipid oxidation. A continuous increase in TBARs values of all samples was observed during storage (Figure 2), and the differences in TBARs values of different samples at the same storage time were statistically significant ($p < 0.001$), in which the TBARs values of control 1 and control 2 were higher than those of other samples, especially the differences in TBARs values of control 2 and other samples increased gradually with the extension of storage time. After 150 days of storage, the TBARs value of control 2 was notably higher than that of other samples. In samples treated with antioxidants, the TBARs values of BHTV and TBHQV were low after 90 days of storage, and the TBARs values of BHTV, ETV, and PEPV were approximately lower than those of TBHQV after 150 days of storage. The results indicated that the methods of using antioxidants and vacuum-filling nitrogen packaging could be advantageous for controlling lipid oxidation in deep fried *Gryllus bimaculatus*. Among antioxidants used in the present study, BHT, ET, and PEP were more appropriate for reducing TBARs value.

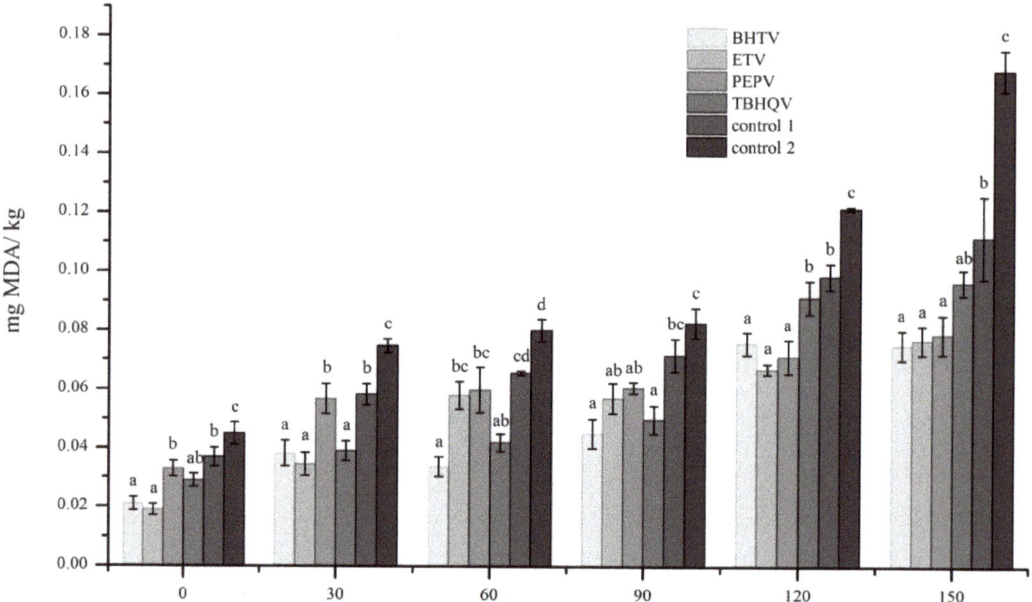

Figure 2. Changes in thiobarbituric acid reactive substances (TBARs) values of deep fried *Gryllus bimaculatus*. BHTV = 0.2% BHT + vacuum-filling nitrogen packaging; ETV = 0.5% ET + vacuum-filling nitrogen packaging; PEPV = 0.5% PEP + vacuum-filling nitrogen packaging; TBHQV = 0.2% TBHQ + vacuum-filling nitrogen packaging; control 1 = no antioxidants + vacuum-filling nitrogen packaging; control 2 = no antioxidant + non vacuum sealed packaging; different lowercase letters on the column chart showed significant differences in the same storage time ($p < 0.001$).

4. Discussion

Cooking and processing methods have a great influence on the composition of fatty acids in the products. In the frying process, the composition of fatty acids in the deep fried products may greatly change compared with the raw materials due to absorbing a large amount of frying oil. Weber et al. [29] and Garcia-Arias et al. [30] analyzed fatty acids of silver catfish and sardine fillets processed by different methods, and found that the composition of fatty acids in samples processed by frying and baking did not significantly vary, while the ratio of UFAs to SFAs in deep fried fish slices significantly increased as a result of the high contents of unsaturated fats in soybean oil used for frying. Palm oil is widely used in the fried food industry and the fried cricket enterprises due to its stability in the frying process, which is not easy to be hydrolyzed and oxidized, and can make the fried products have a good taste, as well as the price of palm oil is relatively cheap. As a result of its stability, palm oil has little impact on the quality of fried food, but the fatty acid composition of the fried food may change in different levels due to the adsorption of products to palm oil. In this study, fried crickets may adsorb about 17% of palm oil, therefor, the data from the analysis of composition of fatty acids in deep fried *Gryllus bimaculatus* were consistent with previously reported results, in which the composition of fatty acids in processed samples significantly differed from that of raw materials, and the contents of palmitic acid and oleic acid increased, while the contents of stearic acid and linoleic acid decreased, as the palm oil used for frying crickets has high contents of palmitic acid and oleic acid and low contents of stearic acid and linoleic acid.

During the processing and storage of animal products, the oxidation and hydrolysis of lipid occur easily under the action of lipase and oxygen. The stability of various fatty acids is different, and polyunsaturated fatty acids are more susceptible to oxidation, followed by monounsaturated fatty acids, while SFAs are the most stable type [31]. In the present study, the decline of ΣUAFs and the increase of ΣSFAs in deep fried cricks during storage were observed, and the analysis of FFAs showed that the contents of ΣSFFAs were higher than those of ΣUFFAs. These results also proved that UFAs in deep fried crickets were unstable and prone to oxidation.

As a result of lipid oxidation, UFAs containing double bonds with very unstable properties is hydrolyzed to hydroperoxides [32]. The peroxides accumulate gradually when the generation of peroxides is greater than their decomposition, resulting in the increase of PV. Therefore, TBARs value also increases gradually. The results of the current study showed that PV and TBARs values of deep fried crickets increased during 150 days of storage, which are consistent with the results of previous studies on beef [33], chicken [34], pork [35], and fish [28], indicating that the oxidation of fatty acids in deep fried crickets is similar to other meat, and is consistent with the variations of fatty acids and FFAs.

Compared with other meat products, such as pork [32], beef [33], and goose [22], edible insects have higher contents of UFAs, leading to lipid oxidation easily. Although there are relatively few studies on fatty acid oxidation of edible insects during processing and storage, a number of relevant researches have been conducted. For instance, Kinyuru [36] studied the composition of fatty acids in deep fried termites and long-horned grasshoppers, and it was revealed that the proportion of SFAs increased, whereas the proportion of UFAs significantly decreased during frying, indicating the occurrence of lipid oxidation and degradation. The temperature of processing has a significant effect on the composition of fatty acids of crickets, and the contents of polyunsaturated fatty acids in Acheta domesticus decreased from 45% to 30%, while the contents of monounsaturated fatty acids increased from 21% to 39% under freeze-drying and heat-drying respectively [37]; for black crickets, the content of palmitic acid in the sample dried at 120 °C was markedly higher than that of freeze-drying, while the contents of linoleic acid and linolenic acid were significantly lower than those of freeze-drying [38]. Kim et al. [39] and Ssepuuya et al. [40] analyzed the effects of storage conditions on lipid oxidation of edible insects, although there was no significant difference in the contents of main fatty acids of *Gryllus bimaculatus* powder after storage at 40 °C for six months, the indicator of lipid oxidation such as acid value varied significantly; vacuum packaging and refrigeration could inhibit the lipid oxidation of fried grasshoppers and prolong the shelf-life to 22 weeks, in which the PV of the product was <21.50 mEq O_2/kg, and the TBARs was <0.079 mg MDA/kg 40]. These studies demonstrated that different degrees of lipid oxidation and decomposition may occur in edible insects during processing and storage, which are similar to other animal foods. Lipid oxidation is an important factor, influencing the quality and shelf-life of animal foods.

Antioxidants inhibit the oxidation of lipids by scavenging the peroxide reaction matrix, complexing metal ions, reducing the concentration of active oxygen and blocking the dehydrogenation of fatty acids [41,42]. Synthetic antioxidants commonly used in meat processing include dibutyl hydroxyl toluene (BHT), tert-butylhydroquinone (TBHQ), and butyl hydroxyanisole (BHA), which have good antioxidant effects and can delay lipid oxidation by eliminating free radicals and chelating metal ions to increase the shelf life of meat products [43]. Natural antioxidants are increasingly used in meat products due to they are safer than chemically synthesized antioxidants, in which rosemary extract is more widely used [44,45]. The strong antioxidant activity of rosemary is related to substances such as carnosic acid, carnosol, diterpenes and rosemary diphenols [46], which slow down the oxidation process by combining with hydroxyl radicals and peroxides to form stable quinones [47,48] and chelating metal ions [49]. Plant polyphenols have been proven to have a good inhibitory effect on lipid oxidation in meat products [50], and *Phyllanthus emblica* polyphenols have activities such as scavenging free radicals and anti-lipid oxidation [51]. Therefore, two chemically synthesized antioxidants, including BHT and TBHQ, and two natural antioxidants, including oil-soluble rosemary extract (ET) and *Phyllanthus emblica* polyphenols (PEP), were used for the investigation in this study. The analysis results of free fatty acids content and PV value show that the antioxidant effects of BHT and TBHQ are better than that of ET and PEP, while the antioxidant effects of BHT, ET, and PEP are better than that of TBHQ in the analysis results of TBARs, in which the most effective antioxidant is BHT. Although some studies have reported that the antioxidant effects of some natural antioxidants in meat products are equivalent to BHT [52], and even better than BHT [53], the antioxidant effects of chemically synthesized antioxidants are significantly better than that of natural antioxidants in this study. The difference between the results of this study and the literatures may be related to different processing methods. The processing method in this study is high-temperature frying. At high temperatures, natural antioxidants are more unstable and easier to decompose. Therefore, chemical synthesis of antioxidants has greater advantages in the processing of fried food. In this study, only single antioxidants were carried out, and the combination of different antioxidants may have better effects, which can be further studied.

5. Conclusions

In order to study and inhibit lipid oxidation in deep fried *Gryllus bimaculatus*, four antioxidants were added to the palm oil used for deep frying, and the non-vacuum sealed packaging and vacuum-filling nitrogen packaging methods were used. Besides, the composition of fatty acids, contents of FFAs, PV, and TBARs value of samples that were treated with different methods during 150 days of storage were analyzed. The results showed that the contents of UFAs decreased, while the contents of SFAs increased. Meanwhile, the total content of FFAs, PV, and TBARs value increased during 150 days of storage, indicating that the lipid oxidation dominated by oxidation of UFAs could occur in deep fried *Gryllus bimaculatus* during storage. Additionally, at the same storage time, there were significant differences in the contents of UFAs and SFAs, as well as the total content of FFAs, PV, and TBARs value of samples treated with different methods. The total content of FFAs, PV, and TBARs value of samples treated with antioxidants and vacuum-filling nitrogen packaging were lower than those of the control, suggesting that antioxidants and vacuum-filling nitrogen packaging could significantly inhibit the lipid oxidation in deep fried *Gryllus bimaculatus*, which is an effective method to improve the quality and shelf-life of deep fried crickets, and BHT was found as the most effective antioxidant in the present study.

Author Contributions: Conceptualization, J.G. and Y.F.; Formal Analysis, J.G. and M.Z.; Funding Acquisition, Y.F.; Investigation, J.G., Z.H., L.S. and X.L.; Methodology, J.G. and Z.H.; Project Administration, Y.F.; Resources, M.Z.; Writing— Original Draft, J.G.; Writing— Review and Editing, Y.F. All authors have read and agreed to the published version of the manuscript.

Funding: This research was funded by the Cooperation Project of Zhejiang Province and CAF, grant number 2019SY02.

Institutional Review Board Statement: Ethical review and approval were waived for this study due to research procedures in this work did not violate any relevant laws and regulations, and no relevant ethics committee or Institution was applicable for the material (cricket) in this study.

Informed Consent Statement: Not applicable.

Data Availability Statement: All data and materials are available on request.

Acknowledgments: Authors would like to thank the reviewers for their time, effort, insightful comments and suggestions that helped improve this manuscript.

Conflicts of Interest: The authors declare that they have no conflict of interest.

References

1. Baiano, A. Edible insects: An overview on nutritional characteristics, safety, farming, production technologies, regulatory framework, and socio-economic and ethical implications. *Trends Food Sci. Technol.* **2020**, *100*, 35–50. [CrossRef]
2. Feng, Y.; Chen, X.M.; Zhao, M.; He, Z.; Sun, L.; Wang, C.Y.; Ding, W.F. Edible insects in China: Utilization and prospects. *Insect Sci.* **2017**, *25*, 184–198. [CrossRef]
3. Meyer-Rochow, V.B. Can insects help to ease the problem of world food shortage? *Search* **1975**, *6*, 261–262.
4. Vantomme, P. Way forward to bring insects in the human food chain. *J. Insects Food Feed.* **2015**, *1*, 121–129. [CrossRef]
5. Magara, H.J.O.; Tanga, C.M.; Ayieko, M.A.; Hugel, S.; Mohamed, S.A.; Khamis, F.M.; Salifu, D.; Niassy, S.; Sevgan, S.; Fiaboe, K.K.M.; et al. Performance of newly described native edible cricket *Scapsipedus icipe* (Orthoptera: Gryllidae) on various diets of relevance for farming. *J. Econ. Entomol.* **2019**, *112*, 653–664. [CrossRef]
6. Magara, H.J.O.; Niassy, S.; Ayieko, A.; Mukundamago, M.; Egonyu, J.P.; Tanga, C.M.; Kimathi, E.K.; Ongere, J.O.; Fiaboe, K.K.M.; Hugel, S.; et al. Edible crickets (Orthoptera) around the world: Distribution, nutritional value, and other benefits-a review. *Front. Nutr.* **2021**, *7*, 537915. [CrossRef]
7. Halloran, A.; Caparros-Megido, R.; Oloo, J.; Weigel, T.; Nsevolo, P.; Francis, T. Comparative aspects of cricket farming in Thailand, Cambodia, Lao People's Democratic Republic, Democratic Republic of the Congo and Kenya. *J. Insects Food Feed.* **2018**, *4*, 101–114. [CrossRef]
8. Wang, D.; Bai, Y.; Li, J.; Zhang, C. Nutritional value of the field cricket (*gryllus testaceus* walker). *Entomol. Sin.* **2004**, *11*, 275–283. [CrossRef]
9. Ghosha, S.; Leeb, S.; Jung, C.; Meyer-Rochow, V.B. Nutritional composition of five commercial edible insects in South Korea. *J. Asia Pac. Entomol.* **2017**, *20*, 686–694. [CrossRef]
10. Reverberi, M. Edible insects: Cricket farming and processing as an emerging market. *J. Insects Food Feed.* **2020**, *6*, 211–220. [CrossRef]
11. Meyer-Rochow, V.B.; Ghosh, S.; Jung, C. Farming of insects for food in South Korea: Tradition and innovation. *Berl. Und Muenchener Tieraerztliche Wochenschr.* **2019**, *131*, 236–244.
12. Miech, P.; Berggren, A.; Lindberg, J.E.; Chhay, T.; Khieu, B.; Jansson, A. Growth and survival of reared Cambodian field crickets (*Teleogryllu stestaceus*) fed weeds, agricultural and food industry by-products. *J. Insects Food Feed.* **2016**, *2*, 285–292. [CrossRef]
13. Kinyuru, J.N.; Kipkoech, C. Production and growth parameters of edible crickets: Experiences from a farm in a high altitude, cooler region of Kenya. *J. Insects Food Feed.* **2018**, *4*, 247–251. [CrossRef]
14. Morales-Ramos, J.A.; Rojas, M.G.; Dossey, A.T. Age-dependent food utilization of *Acheta domesticus* (Orthoptera: Gryllidae) in small groups at two temperatures. *J. Insects Food Feed.* **2018**, *4*, 51–60. [CrossRef]
15. Machado, C.D.R.; Thys, R.C.S. Cricket (*Gryllus assimilis*) as a new alternative protein source for gluten-free breads. *Innov. Food Sci. Emerg. Technol.* **2019**, *56*, 102180. [CrossRef]
16. Homann, A.M.; Ayieko, M.A.; Konyole, S.O.; Roos, N. Acceptability of biscuits containing 10% cricket (*Acheta domesticus*) compared to milk biscuits among 5–10-year-old Kenyan schoolchildren. *J. Insects Food Feed.* **2017**, *3*, 95–103. [CrossRef]
17. Duda, A.; Adamczak, J.; Chełminska, P.; Juszkiewicz, J.; Kowalczewski, P. Quality and nutritional/ textural properties of durum wheat pasta enriched with cricket powder. *Foods* **2017**, *8*, 46. [CrossRef]
18. Kim, H.; Setyabrata, D.; Lee, Y.; Jones, O.G.; Kim, Y.B. Effect of house cricket (*Acheta domesticus*) flour addition on physicochemical and textural properties of meat emulsion under various formulations. *J. Food Sci.* **2017**, *81*. [CrossRef]
19. Halloran, A.; Roos, N.; Flore, R.; Hanboonsong, Y. The development of the edible cricket industry in Thailand. *J. Insects Food Feed.* **2016**, *2*, 91–100. [CrossRef]
20. Feng, Y.; Zhao, M.; Ding, W.F.; Chen, X.M. Overview of edible insect resources and common species utilisation in China. *J. Insects Food Feed.* **2020**, *6*, 13–25. [CrossRef]
21. The National Health and Family Planning Commission of the People's Republic of China. *National Food Safety Standard: Determination of Fatty Acid in Food GB 5009.168–2016*; China Standards Press: Beijing, China, 2016; pp. 1–6.
22. Wang, Y.; Jiang, Y.; Cao, J.; Chen, Y.; Sun, Y.; Zeng, X.; Pan, D.; Ou, C.; Gan, N. Study on lipolysis-oxidation and volatile flavour compounds of dry cured goose with different curing salt content during production. *Food Chem.* **2016**, *190*, 33–40. [CrossRef]
23. The National Health and Family Planning Commission of the People's Republic of China. *National Food Safety Standard: Determination of Peroxide Value in Food GB 5009.227–2016*; China Standards Press: Beijing, China, 2016; pp. 1–4.

24. The National Health and Family Planning Commission of the People's Republic of China. *National Food Safety Standard: Determination of Malondialdehyde in Food GB 5009.181–2016*; China Standards Press: Beijing, China, 2016; pp. 3–5.
25. Tzompa-Sosa, D.A.; Dewettinck, K.; Provijn, P.; Brouwers, J.F.; Meulenaer, B.; Oonincx, D. Lipidome of cricket species used as food. *Food Chem.* **2001**, *349*, 129077. [CrossRef] [PubMed]
26. He, Z.; Sun, L.; Wang, C.; Feng, Y.; Zhao, M. Nutritional composition analysis and evaluation of the two-spotted cricket *Gryllus bimaculatus* (Orthoptera:Gryllidae). *Biot. Resour.* **2021**, *43*, 303–308. [CrossRef]
27. Kim, E.; Kim, D.; Lim, J.; Chang, Y.; Lee, Y.; Park, J.J.; Ahn, M. Physicochemical properties of edible cricket (*Gryllus bimaculatus*) in different districts. *Korean J. Food Preserv.* **2015**, *22*, 831–837. [CrossRef]
28. Li, J.; Wang, F.; Li, S.; Peng, Z. Effects of pepper (*Zanthoxylum bungeanum* Maxim.) leaf extract on the antioxidant enzyme activities of salted silver carp (*Hypophthalmichthys molitrix*) during processing. *J. Funct. Foods* **2015**, *18*, 1179–1190. [CrossRef]
29. Weber, J.; Bochi, V.; Ribeiro, C.P.; Victorio, A.M.; Emanuelli, T. Effect of different cooking methods on the oxidation, proximate and fatty acid composition of silver catfish (*Rhamdia quelen*) fillets. *Food Chem.* **2008**, *106*, 140–146. [CrossRef]
30. Garcia-Arias, M.T.; Pontes, E.A.; Garda-Linares, M.C.; Garcfa-Femandez, M.C.; Sanchez-Muniz, F.J. Cooking-freezing-reheating (CFR) of sardine (*Sardina pilchardus*) fillets. Effect of different cooking and reheating procedures on the proximate and fatty acid compositions. *Food Chem.* **2003**, *83*, 349–356. [CrossRef]
31. Timón, M.L.; Ventanas, J.; Carrapiso, A.I.; Jurado, A.; GarcõÂa, C. Subcutaneous and intermuscular fat characterisation of dry-cured Iberian hams. *Meat Sci.* **2001**, *58*, 85–91. [CrossRef]
32. Huang, L.; Xiong, Y.; Kong, B.; Huang, X.; Li, J. Influence of storage temperature and duration on lipid and protein oxidation and flavour changes in frozen pork dumpling filler. *Meat Sci.* **2013**, *95*, 295–301. [CrossRef]
33. Li, Y.; Wang, J.; Liu, C.; Guo, A.; Han, D. Effect of drying temperature on fatty acid composition in air-dried beef during chilled storage. *Food Sci.* **2019**, *40*, 14–21. [CrossRef]
34. Shin, D.; Yang, A.; Min, B.; Narciso-Gaytán, C.; Sánchez-Plata, M.; Ruiz-Feria, C. Evaluation of antioxidant effects of vitamins C and E alone and in combination with sorghum bran in a cooked and stored chicken sausage. *Korean J. Food Sci. Anim. Resour.* **2011**, *31*, 693–700. [CrossRef]
35. Cuong, T.V.; Chin, K.B. Effects of annatto (*Bixa orellana* L.) seeds powder on physicochemical properties, antioxidant and antimicrobial activities of pork patties during refrigerated storage. *Korean J. Food Sci. Anim. Resour.* **2016**, *36*, 476–486. [CrossRef] [PubMed]
36. Kinyuru, J.N. Oil characteristics and influence of heat processing on fatty acid profile of wild harvested termite (*Macrotermes subhylanus*) and long-horned grasshopper (Ruspolia differens). *Int. J. Trop. Insect Sci.* **2021**, *41*, 1427–1433. [CrossRef]
37. Lucas-González, R.; Fernández-López, J.; Pérez-Álvarez, J.; Viuda-Martos, M. Effect of drying processes in the chemical, physico-chemical, techno-functional and antioxidant properties of flours obtained from house cricket (*Acheta domesticus*). *Eur. Food Res. Technol.* **2019**, *245*, 1451–1458. [CrossRef]
38. Dobermann, D.; Field, L.M.; Michaelson, L.V. Impact of heat processing on the nutritional content of *Gryllus bimaculatus* (black cricket). *Nutr. Bull.* **2019**, *44*, 116–122. [CrossRef]
39. Kim, D.; Kim, E.; Chang, Y.; Ahn, M.; Lee, Y.; Park, J.J.; Lim, J. Determination of the shelf life of cricket powder and effects of storage on its quality characteristics. *Korean J. Food Preserv.* **2016**, *23*, 211–217. [CrossRef]
40. Ssepuuya, G.; Aringo, R.O.; Mukisa, I.M.; Nakimbugwe, D. Effect of processing, packaging and storage-temperature based hurdles on the shelf stability of sautéed ready-to-eat *Ruspolia nitidula*. *J. Insects Food Feed.* **2016**, *2*, 245–253. [CrossRef]
41. Huang, D.J.; Ou, B.X.; Prior, R.L. The chemistry behind antioxidant capacity assays. *J. Agric. Food Chem.* **2005**, *53*, 1841–1856. [CrossRef]
42. Leopoldini, M.; Russo, N.; Chiodo, S.; Toscano, M. Iron chelation by the powerful antioxidant flavonoid quercetin. *J. Agric. Food Chem.* **2006**, *54*, 6343–6351. [CrossRef]
43. Jayathilakan, K.; Sharma, G.K.; Radhakrishna, K.; Bawa, A.S. Antioxidant potential of synthetic and natural antioxidants and its effect on warmed over-flavour in different species of meat. *Food Chem.* **2007**, *105*, 908–916. [CrossRef]
44. Nissen, L.R.; Byrne, D.V.; Bertelsen, G.; Skibsted, L.H. The antioxidative activity of plant extracts in cooked pork patties as evaluated by descriptive sensory profiling and chemical analysis. *Meat Sci.* **2004**, *68*, 485–495. [CrossRef] [PubMed]
45. Pintado, T.; Herrero, A.M.; Jiménez-Colmenero, F.; Ruiz-Capillas, C. Strategies for incorporation of chia (*Salvia hispanica* L.) in frankfurters as a health-promoting ingredient. *Meat Sci.* **2016**, *114*, 75–84. [CrossRef]
46. Hernández-Hernández, E.; Ponce-Alquicira, E.; JaramilloFlores, M.E.; Guerrero Legarreta, I. Antioxidant effect rosemary (*Rosmarinus officinalis* L.) and oregano (*Origanum vulgare* L.) extracts on TBARS and colour of model raw pork batters. *Meat Sci.* **2009**, *81*, 410–417. [CrossRef]
47. Fujimoto, A.; Masuda, T. Antioxidation mechanism of rosmarinic acid, identification of an unstable quinone derivative by the addition of odourless thiol. *Food Chem.* **2012**, *132*, 901–906. [CrossRef]
48. Zhang, Y.; Yang, L.; Zu, Y.; Chen, X.; Wang, F.; Liu, F. Oxidative stability of sunflower oil supplemented with carnosic acid compared with synthetic antioxidants during accelerated storage. *Food Chem.* **2010**, *118*, 656–662. [CrossRef]
49. Sampaio, G.R.; Saldanha, T.; Soares, R.A.M.; Torres, E.A.F.S. Effect of natural antioxidant combinations on lipid oxidation in cooked chicken meat during refrigerated storage. *Food Chem.* **2012**, *135*, 1383–1390. [CrossRef]
50. Lu, M.; Yuan, B.; Zeng, M.; Chen, J. Antioxidant capacity and major phenolic compounds of spices commonly consumed in China. *Food Res. Int.* **2011**, *44*, 530–536. [CrossRef]

51. Gan, J.; He, Z.; Zhang, H.; Feng, Y. Optimization of polyphenols extraction from *Phyllanthus emblica* L. pomace and its antioxidant activity analysis. *Sci. Technol. Food Ind.* **2019**, *40*, 171–177. [CrossRef]
52. Armenteros, M.; Morcuende, D.; Ventanas, J.; Estévez, M. The application of natural antioxidants via brine injection protects Iberian cooked hams against lipid and protein oxidation. *Meat Sci.* **2016**, *116*, 253–259. [CrossRef]
53. Jia, Y.; Liu, X.; Hu, Y.; Zhou, J.; Liu, Y. Inhibitory activity of tea polyphenols against lipolysis and lipid oxidation in white fish. *Jiangsu J. Agric. Sci.* **2016**, *32*, 216–221. [CrossRef]

 *foods*

Article

Nutritional Properties of Larval Epidermis and Meat of the Edible Insect *Clanis bilineata tsingtauica* (Lepidoptera: Sphingidae)

Ying Su [1,†], Ming-Xing Lu [1,†], Li-Quan Jing [2], Lei Qian [3], Ming Zhao [1], Yu-Zhou Du [1,4,*] and Huai-Jian Liao [3,*]

[1] College of Horticulture and Plant Protection & Institute of Applied Entomology, Yangzhou University, Yangzhou 225009, China; wulisuying@163.com (Y.S.); lumx@yzu.edu.cn (M.-X.L.); zhaoming@yzu.edu.cn (M.Z.)

[2] Jiangsu Key Laboratory of Crop Genetics and Physiology/Co-Innovation Center for Modern Production Technology of Grain Crops, Yangzhou University, Yangzhou 225009, China; lqjing@yzu.edu.cn

[3] Institute of Leisure Agriculture, Jiangsu Academy of Agricultural Sciences, Nanjing 210014, China; qianlei14930@sina.cn

[4] Joint International Research Laboratory of Agriculture and Agri-Product Safety, Yangzhou University, Yangzhou 225009, China

* Correspondence: yzdu@yzu.edu.cn (Y.-Z.D.); lhj@jaas.ac.cn (H.-J.L.)

† Both authors contributed equally to this work.

Abstract: Insects represent a sustainable, protein-rich food source widely consumed in Asia, Africa, and South America. Eating *Clanis bilineata tsingtauica* Mell is common in the eastern part of China. A comparative characterization of nutrients in the meat and epidermis of *C. bilineata tsingtauica* was performed in this study. The results showed this insect to be high in nutrients, particularly in the epidermis where protein total was 71.82%. Sixteen different amino acids were quantified in *C. bilineata tsingtauica*, and the ratio of essential to nonessential amino acids in the epidermis and meat was 68.14% and 59.27%, respectively. The amino acid composition of *C. bilineata tsingtauica* is balanced, representing a high-quality protein source. Eight minerals were quantified in *C. bilineata tsingtauica*, including four macro and four trace elements. Fe in the epidermis and Zn in the meat were abundant at 163.82 and 299.31 µg/g DW, respectively. The presence of phytic acid impacted the absorption of mineral elements in food. We also detected phytic acid in *C. bilineata tsingtauica*. The molar ratio of phytic acid to zinc (PA/Zn) in *C. bilineata tsingtauica* was very low (3.28) compared to *Glycine max* and *Cryptotympana atrata*, which indicated that mineral utilization was high. In conclusion, this study confirms that *C. bilineata tsingtauica* is a highly nutritious food source for human consumption, and the results provide a basis for further consumption and industrialization of this edible insect.

Keywords: *Clanis bilineata tsingtauica* Mell; edible insects; nutritional composition; phytic acid

Citation: Su, Y.; Lu, M.-X.; Jing, L.-Q.; Qian, L.; Zhao, M.; Du, Y.-Z.; Liao, H.-J. Nutritional Properties of Larval Epidermis and Meat of the Edible Insect *Clanis bilineata tsingtauica* (Lepidoptera: Sphingidae). *Foods* **2021**, *10*, 2895. https://doi.org/10.3390/foods10122895

Academic Editors: Victor Benno Meyer-Rochow and Chuleui Jung

Received: 24 October 2021
Accepted: 17 November 2021
Published: 23 November 2021

Publisher's Note: MDPI stays neutral with regard to jurisdictional claims in published maps and institutional affiliations.

1. Introduction

As the world population continues to increase, food shortage has become a serious global problem [1]. The possibility of using edible insects as food and feed was suggested in 1975 as a route to easing global food shortages [2,3] and has also been addressed by the Food and Agricultural Organization (FAO) [4,5]. Although insects are the largest and most diverse group of organisms on Earth, their implementation as a food source has not been fully utilized [6,7]. Over one million species of named insects are estimated to exist worldwide, and approximately 1900–2000 species are edible [8–10]. Humans primarily consume arthropods from one of the following five orders: Isoptera, Orthoptera, Coleoptera, Hymenoptera, and Lepidoptera [11–13]. Numerous studies have shown that these insects are of high nutritional value and contain considerable amounts of protein (20% to 76% of dry matter), amino acids, vitamins, and minerals [10,14–20]. Nutritional values are highly variable, however, because of the wide range of edible species and the variation in developmental stages and feeding substrates [3,17,21–23].

Clanis bilineata tsingtauica Mell, 1922, (Insecta: Lepidoptera: Sphingidae) is an important pest of *Glycine max* (Linn.) Merr. The history of *C. bilineata tsingtauica* as a food source dates back to the Chinese Qing Dynasty about two hundred years ago [24–26]. The consumption of the fifth instar larvae of *C. bilineata tsingtauica* is common in many regions of China, especially in the Jiangsu, Shandong, and Henan provinces [27], and Lianyungang (Jiangsu province) is a major production center. The production of *C. bilineata tsingtauica* in *G. max* amounted to approximately 3×10^4 t/year, with an output value of nearly RMB 4.5 billion. Since the annual demand for larvae is about 10×10^4 t/year, production has failed to satisfy consumption needs [28]. Recently, farmers in Jiangsu, Shandong, and Henan have begun growing *G. max* as a *C. bilineata tsingtauica* food source to meet consumer demand.

Many consumers believe that *C. bilineata tsingtauica* is rich in nutrients, primarily because it feeds on nutrient-dense *G. max* leaves. Tian et al. reported that the crude protein content of whole larvae was 65.5%, essential amino acids accounted for 52.84% of total amino acids, and the content of unsaturated fatty acids was as high as 64.17% [29]. It is important to mention that different regions of China exhibit different ways of preparing and eating *C. bilineata tsingtauica*. For example, the processing method in Lianyungang involves removing the larval head, squeezing the body with a round wooden stick to retain the internal meat, discarding the external epidermis, and then processing the remaining body parts into a delicious soup. However, in Xuzhou (Jiangsu Province), the intact larvae with skin are chopped and fried for consumption, whereas the larvae are consumed after grilling in Shandong [27]. The variation in preparation methods indicates that a comprehensive analysis of *C. bilineata tsingtauica* is warranted.

In humans, the scarcity of mineral elements such as iron and zinc has been a difficult problem to solve [30,31]. The bioavailability of minerals in plants and plant products is inhibited by phytic acid [32–34], a derivative of inositol with six phosphate groups [35,36]. Phytic acid (PA) negatively impacts the absorption of zinc and iron in compound diets [37–40]; however, there are relatively few studies on the presence of PA in edible insects that feed on plants [41,42]. In this study, we focused on the preparation of *C. bilineata tsingtauica* using protocols typical in Lianyungang. Changes in protein and amino acid contents in the meat and epidermis of *C. bilineata tsingtauica* were determined, along with analysis of minerals and phytic acid. At the same time, compared with *G. max* and *Cryptotympana atrata*, (recorded in the Pharmacopoeia of the People's Republic of China and the Taiwan Herbal Pharmacopoeia to have medicinal value), the results of this study provide a scientific basis and reference for the nutritional value, processing methods, and further industrialization of *C. bilineata tsingtauica*.

2. Materials and Methods

2.1. Collection and Preparation of Materials

The fifth instar larvae of *C. bilineata tsingtauica* reared on *G. max* leaves were obtained from the Dongming Yellow River Beach Ecological Agriculture Co. Ltd. in Heze, Shandong Province, China ($35°14'$ N, $115°26'$ E). The internal meat of the larvae was removed with round wooden sticks to separate the inner material and organs from the epidermis (skin). The meat included all insect tissues except the epidermis. Larval parts were then frozen at $-70\,°C$ and transferred to a drying oven (Hangzhou Lantian Instrument Co., Ltd., Zhejiang, China) at 65 °C for 24 h. Body parts were then macerated with a stainless steel grinder (FW-100, Zhejiang, China) fitted with a 100-mesh sieve, and stored in brown, wide-mouth bottles until needed.

G. max leaves were collected from the same field as *C. bilineata tsingtauica*. A rice huller (OHYA-25, OHYA Corporation, Tokyo, Japan) was used to remove *G. max* hulls, and seeds were then pulverized into powder with a stainless steel grinder (FW-100, Zhejiang, China). For comparison, cicada, *Cryptotympana atrata*, was obtained from the Dongming Yellow River Beach Ecological Agriculture Co., Ltd. (Shandong, China), and the grinding and storage processes were identical to that of *C. bilineata tsingtauica*.

2.2. Determination of Protein Content

Crude protein was determined by the Kjeldahl method using an automated Kjeldahl System (Buchi, Flawil, Switzerland). The protein in each sample was disintegrated by catalytic heating, which released ammonia that subsequently reacted with sulfuric acid to produce ammonium sulfate [43]. Ammonia was freed, gasified by alkaline distillation, absorbed with boric acid, and further titrated with sulfuric acid [42]. Nitrogen content was calculated as described previously, and protein content was calculated by multiplying the result by a conversion factor of 6.25 [43,44].

The contents of albumin, globulin, glutelin, and prolamin were determined as described by Ju et al. [45]. In brief, the sample was first defatted with hexane and then subjected to protein extraction. The above protein was extracted with distilled water, 50 g/L NaCl, 0.02 mol/L NaOH, and 700 mL/L ethanol in sequence [46].

2.3. Determination of Amino Acid Content

AccQ·Tag high-performance liquid chromatography (HPLC) was used to measure amino acid content, and all HPLC analyses were performed on a Waters 2695 HPLC system (Waters Corporation, Milford, MA, USA). About 0.10 g of the powder sample was weighed into a 10 mL glass bottle; then 5.00 mL of 6 mol/L HCl was added and the mixture was well-shaken and sealed. After the bottle was placed in an oven at 110 °C for 24 h, the samples were run through filters for quantification and the volume was adjusted to 50 mL. The filtrate (2.00 mL) was transferred into a cuvette, and low-pressure evaporation was used to remove HCl in a vacuum freeze drier (Labconco Corporation, Kansas, MO, USA). The concentrate thus obtained was dissolved in 2.00 mL of pure water, and the solution was filtrated with a 0.45 μm membrane filter. Then, 10 μL of filtrate, 70 μL of AccQ·Fluor buffer, and 20 μL of derivatizing agent (Waters Corporation, Milford, MA, USA) were transferred successively into the derivatization tube, and the mixture was warmed at 55 °C for 10 min in an oven (Hangzhou Lantian Instrument Co., Ltd., Zhejiang, China). Finally, amino acid content was determined by high-performance liquid chromatography (Waters 2695 HPLC separation unit, 2487 UV detector, Waters Co., Milford, MA, USA, and Empower management system, X&Y Solutions, Inc., Boston, MA, USA). The AccQ·Tag reversed-phase analysis column was 3.9 mm × 150 mm; mobile phase A was sodium acetate (140 mmol/L)-ethylamine (17 mmol/L, pH 4.95, adjusted with phosphoric acid); mobile phase B was acetonitrile; and mobile phase C was pure water. Flow velocity was 1.00 mL/min; column temperature was 37 °C; UV detector wavelength was 248 nm; sample volume was 10 μL. The standards of 17 amino acid components were provided by Waters Co. Ltd. (Milford, MA, USA), except for tryptophan (Trp) [44].

A Waters 2695 HPLC detection system was used to determine the quantity of seven essential amino acids, including threonine (Thr), valine (Val), methionine (Met), isoleucine (Ile), leucine (Leu), phenylalanine (Phe), and lysine (Lys), but not tryptophan (Trp). The content of the following nine nonessential amino acids was also determined: aspartic acid (Asp), serine (Ser), glutamic acid (Glu), glycine (Gly), alanine (Ala), arginine (Arg), proline (Pro), histidine (His) tyrosine (Tyr), and cysteine (Cys) [44].

The following equations were used for evaluating amino acids:

$$\text{Amino acid score (AAS)} = \frac{mg\ of\ amino\ acid\ in\ test\ protein}{mg\ of\ amino\ acid\ in\ FAO\ model}$$

$$\text{Chemical score (CS)} = \frac{mg\ of\ amino\ acid\ in\ test\ protein}{mg\ of\ amino\ acid\ in\ eggs}$$

$$\text{Essential amino acid index (EAAI)} = \sqrt[n]{\frac{Lys^t}{Lys^s} \times 100 \times \frac{Trp^t}{Trp^s} \times 100 \times \cdots \times \frac{Thr^t}{Thr^s} \times 100}$$

where n is the total number of essential amino acids, t is the amino acid content of the sample, and s is the amino acid content of the egg.

2.4. Determination of Mineral Content

Inductively coupled plasma–atomic emission spectroscopy (ICP-AES, ICAP 6300, Thermo, Waltham, MA, USA) was used for rapid and precise determinations of macro and trace mineral content in the samples. Briefly, pulverized samples (0.50 g) were weighed and combined with ultrapure-grade HNO_3 (5.00 mL) and H_2O_2 (2.00 mL). A microwave digester (CEM MARS 5, Matthews, NC, USA) and the easy prep microwave digestion program were used to digest the samples [47]. After complete digestion, the mixed sample was cooled at room temperature and increased to a final volume of 20 mL with ultrapure water. Then, macro elements (Ca, K, Mg, and P) and trace elements (Cu, Fe, Mn, and Zn) were determined using ICP-AES. The optimal instrumental conditions were maintained at 15 L/min for the stable plasma gas flow rate. The auxiliary and the nebulizer gas flow rate were kept at 0.2 and 0.8 L/min, respectively. The sample flow rate was 1.5 mL/min, and the power was 1500 W.

2.5. Determination of Phytic Acid Content

Pulverized samples (0.25 g) were weighed and combined with a 0.7% HCl solution (5.00 mL). This mixture was incubated in a constant-temperature oscillator at 25 °C and 150 rpm for 1 h. After centrifugation at 4000 rpm for 15 min at 4 °C, supernatants were collected. Supernatants (0.60 mL) were combined with 2.40 mL of deionized water and 0.50 mL of $FeCl_3$ (0.1%, *w/v*); the mixture was shaken and then centrifuged at 3400 rpm for 10 min. The absorbance of supernatants was measured with a spectrophotometer (UV-1601 UV-VIS Spectrophotometer, Shimadzu Corporation, Tokyo, Japan) at 500 nm, and the amount of phytic acid was calculated using a phytic acid standard curve [44].

2.6. Statistical Analyses

Results are shown as means ± standard deviation (SD). Analysis of *C. bilineata tsingtauica* was compared with *G. max* and *C. atrata*. Data were stored in Microsoft Excel (2013) and analyzed with SPSS v. 16.0 statistical software.

3. Results

3.1. Classification and Determination of Total Protein

The total protein content of *C. bilineata tsingtauica* was higher than *G. max*, and more abundant in the larval epidermis (71.82%) than in the meat (64.24%) (Table 1). Four proteins, including albumin, globulin, glutelin, and prolamin, were identified in *C. bilineata tsingtauica*, and had different concentrations in the larval meat and epidermis. For example, glutelin was the predominant protein (meat, 34.35%; epidermis, 34.23%) and globulin was the least abundant (meat, 5.13%; epidermis, 3.70%). Prolamin content in the *C. bilineata tsingtauica* epidermis (11.67%) was double that found in the meat (5.27%). Interestingly, prolamin levels in the *C. bilineata tsingtauica* epidermis and *C. atrata* (11.02%) were eight-fold greater than levels in *G. max* (1.28%) (Table 1).

Table 1. The protein content (% dry weight) in *Clanis bilineata tsingtauica*, *Glycine max*, and *Cryptotympana atrata*.

	Clanis bilineata tsingtauica Meat	*Clanis bilineata tsingtauica* Epidermis	*Glycine max* (CK$_1$)	*Cryptotympana atrata* (CK$_2$)
Albumin	19.50 ± 2.40	22.25 ± 8.56	33.55 ± 2.05	12.75 ± 4.03
Globulin	5.27 ± 1.20	3.70 ± 0.57	4.56 ± 0.30	3.53 ± 0.39
Glutelin	34.35 ± 4.43	34.23 ± 3.49	8.17 ± 0.52	39.07 ± 3.20
Prolamin	5.13 ± 0.98	11.67 ± 1.32	1.28 ± 0.35	11.02 ± 2.01
Total protein	64.25 ± 0.21	71.85 ± 3.18	47.55 ± 1.63	66.35 ± 4.74

Note: Data are expressed as means ± SD (standard deviation).

3.2. Essential and Nonessential Amino Acid Content

Sixteen amino acids were analyzed in *C. bilineata tsingtauica*, including seven essential amino acids (EAA) and nine nonessential amino acids (NEAA). Trp was not measured (an alkaline hydrolysis method is needed), and Cys could not be determined due to analytical methods and not necessarily because it was absent. The total amino acid content (TAA) of the *C. bilineata tsingtauica* meat and epidermis was 455.62 and 710.07 mg/g DW, respectively. Both EAA and NEAA were higher in the larval epidermis than in the meat (Table 2). In the epidermis, the EAA and NEAA contents were 287.75 and 422.33 mg/g DW, respectively. Furthermore, the EAA/NEAA ratio (meat, 59.27%; epidermis, 68.14%) and EAA/TAA ratio (meat, 37.18%; epidermis, 40.51%) were higher in the epidermis than in the meat. Among the seven EAAs, Leu content was the highest in the *C. bilineata tsingtauica* meat (40.81 mg/g DW), Lys was highest in the epidermis (59.05 mg/g DW), and the Met content was the lowest of the four EAAs in both the meat and epidermis. Among the nine NEAAs, the Glu content was highest in the *C. bilineata tsingtauica* meat (78.40 mg/g DW) and epidermis (99.21 mg/g DW), and the Pro content was the lowest (meat, 3.60; epidermis, 3.69 mg/g DW). All 16 amino acids exhibited a higher content in the epidermis than in the *C. bilineata tsingtauica* meat. The TAA, EAA, and NEAA contents in the four food sources were ranked from high to low as follows: *G. max*, *C. bilineata tsingtauica* epidermis, *C. atrata*, and *C. bilineata tsingtauica* meat. The contents of Thr, Met, Ile, Leu, Lys, Glu, and Arg in the epidermis of *C. bilineata tsingtauica* were higher than in *G. max* and *C. atrata*; but the Phe content of *G. max* (107.26 mg/g DW) was over four-fold higher than that of the *C. bilineata tsingtauica* meat (24.81 mg/g DW), and more than twice that of the *C. bilineata tsingtauica* epidermis (40.64 mg/g DW) (Table 2).

The amino acid score (AAS), chemical score (CS), and essential amino acid index (EAAI) are important indicators to further evaluate the quality of amino acids. With respect to EAAs, our results showed the AAS of Met in *C. bilineata tsingtauica* was the lowest (meat, 0.27; epidermis, 0.54), which defined the first limiting amino acid (Table 3). The highest AAS score was that of the aromatic amino acids (Phe + Tyr) in the meat and epidermis of *C. bilineata tsingtauica*, where values were 0.72 (meat) and 1.33 (epidermis). The content of Ile, Lys, Phe + Tyr, and Thr in the epidermis of *C. bilineata tsingtauica* exceeded the amino acid content in the FAO standard, and the content of total essential amino acids was higher in the epidermis of *C. bilineata tsingtauica* compared to the FAO standard. Furthermore, the AAS and CS of essential amino acids in the epidermis of *C. bilineata tsingtauica* were higher than those in the meat (Table 3). *C. bilineata tsingtauica* possessed an excellent amino acid index of 87.67 in the meat and 94.91 in the epidermis (Figure 1). The overall EAAI was slightly lower for *C. bilineata tsingtauica* compared to *G. max*.

Table 2. Amino acid content in *C. bilineata tsingtauica*, *Glycine max*, and *Cryptotympana atrata* (mg/g DW).

	Clanis bilineata tsingtauica **Meat**	*Clanis bilineata tsingtauica* **Epidermis**	*Glycine max* **(CK$_1$)**	*Cryptotympana atrata* **(CK$_2$)**
Thr *	23.35 ± 4.34	44.22 ± 2.81	42.37 ± 5.59	29.95 ± 1.51
Val *	22.85 ± 1.00	33.17 ± 4.12	37.85 ± 4.91	25.05 ± 1.78
Met *	9.49 ± 1.22	18.81 ± 1.21	10.40 ± 1.09	6.73 ± 0.74
Ile *	21.02 ± 1.71	40.79 ± 2.08	36.74 ± 4.41	29.73 ± 2.76
Leu *	40.81 ± 2.55	50.89 ± 6.54	41.23 ± 6.50	39.67 ± 2.67
Phe *	24.81 ± 3.10	40.64 ± 3.21	107.26 ± 6.93	34.65 ± 1.76
Lys *	26.91 ± 1.82	59.05 ± 2.38	48.73 ± 14.81	42.03 ± 1.51
Trp *	-	-	-	-
Asp	45.49 ± 7.47	66.59 ± 2.47	69.51 ± 7.28	64.38 ± 2.86
Ser	24.53 ± 4.26	35.12 ± 3.02	40.15 ± 1.53	31.56 ± 2.89
Glu	78.40 ± 2.18	99.21 ± 2.04	93.21 ± 4.42	97.59 ± 1.44

Table 2. *Cont.*

	Clanis bilineata tsingtauica Meat	*Clanis bilineata tsingtauica* Epidermis	*Glycine max* (CK$_1$)	*Cryptotympana atrata* (CK$_2$)
Gly	28.43 ± 1.84	42.37 ± 3.56	48.97 ± 2.99	30.74 ± 2.65
Ala	35.52 ± 3.28	47.76 ± 2.98	52.52 ± 4.02	28.34 ± 2.58
Tyr	18.38 ± 0.78	39.11 ± 3.04	58.96 ± 2.26	24.50 ± 2.69
Arg	30.92 ± 3.08	54.82 ± 4.94	40.30 ± 1.54	48.68 ± 0.52
Pro	3.60 ± 0.48	3.69 ± 0.11	6.28 ± 0.39	3.79 ± 0.45
His	21.13 ± 1.25	33.68 ± 2.24	39.60 ± 4.16	21.29 ± 3.14
Cys	-	-	-	-
EAA	169.23 ± 1.97	287.75 ± 15.49	324.56 ± 3.43	207.72 ± 7.87
NEAA	286.39 ± 20.95	422.33 ± 1.22	449.48 ± 16.70	350.84 ± 7.29
TAA	455.62 ± 22.92	710.07 ± 14.27	774.04 ± 20.13	558.55 ± 15.16
EAA/NEAA (%)	59.23 ± 3.64	68.14 ± 3.86	72.24 ± 1.92	59.20 ± 1.01
EAA/TAA (%)	37.18 ± 1.44	40.51 ± 1.37	41.94 ± 0.65	37.19 ± 0.40

Abbreviations: * essential amino acids (EAAs); -, not determined; EAA/NEAA, ratio of EAA and nonessential amino acids; EAA/TAA, ratio of EAA and total amino acids (TAA). Data are expressed as means ± SD.

Table 3. Comparison of the amino acid score and chemical score in *Clanis bilineata tsingtauica* with other sources.

	Content (mg/g DW)				AAS		CS	
	FAO *	Eggs *	Meat	Epidermis	Meat	Epidermis	Meat	Epidermis
Ile	40	52.4	21.02	40.99	0.53	1.02	0.40	0.78
Leu	70	84.1	40.81	50.89	0.58	0.73	0.49	0.61
Lys	55	64.9	26.91	59.05	0.49	1.07	0.41	0.91
Met	35	62.7	9.49	18.81	0.27	0.54	0.15	0.30
Phe + Tyr	60	95.5	43.19	79.75	0.72	1.33	0.45	0.84
Thr	40	53.9	23.35	44.22	0.58	1.11	0.43	0.82
Trp	10	16.2	-	-	-	-	-	-
Val	50	57.6	22.85	33.17	0.46	0.66	0.40	0.58
TAA	360	487.3	217.94	362.60	-	-	-	-

Note: * amino acid content as reported by Qiao et al. [48]; -, not determined. Abbreviations: AAS, amino acid score; CS, chemical score; FAO, Food and Agricultural Organization; TAA, total amino acids.

3.3. Determination of Mineral Content

The contents of four macro elements (calcium, potassium, magnesium, and phosphorus) and four trace elements (copper, iron, manganese, and zinc) were determined in *C. bilineata tsingtauica* (Table 4). Among the macro elements, Ca content was the highest in the meat (0.57 mg/g DW) and epidermis (0.73 mg/g DW), whereas Mg was the lowest (meat, 13.92; epidermis, 11.24 mg/g DW). As for trace elements in the *C. bilineata tsingtauica* meat, Cu content was the lowest (6.79 μg/g DW) and Zn was the highest (299.31 μg/g DW), whereas Mn was the lowest (11.04 μg/g DW) and Fe was the highest (163.82 μg/g DW) in the epidermis. The K, Mg, P, and Zn concentrations were higher in the *C. bilineata tsingtauica* meat than in the epidermis; however, the Ca, Cu, Fe, and Mn concentrations were higher in the epidermis than in the meat. Notable differences in the mineral content of *C. bilineata tsingtauica* and *G. max* included Fe in the epidermis and Zn content in the meat, which were about two- and six-fold higher than *G. max*, respectively (Table 4).

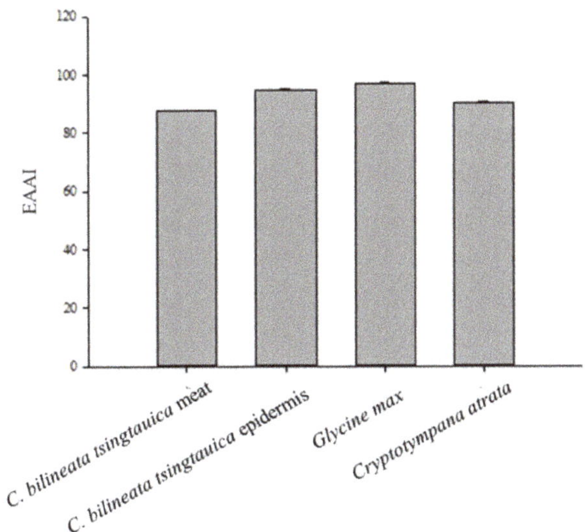

Figure 1. The essential amino acid index (EAAI) in *Clanis bilineata tsingtauica, Glycine max,* and *Cryptotympana atrata.*

Table 4. Mineral element content in *Clanis bilineata tsingtauica, Glycine max,* and *Cryptotympana atrata* (DW).

	Clanis bilineata tsingtauica Meat	Clanis bilineata tsingtauica Epidermis	Glycine max (CK$_1$)	Cryptotympana atrata (CK$_2$)
Ca (mg/g) *	0.57 ± 0.12	0.73 ± 0.01	0.98 ± 0.03	1.59 ± 0.15
K (mg/g) *	12.53 ± 2.08	9.52 ± 0.13	13.56 ± 1.02	5.45 ± 0.16
Mg (mg/g) *	13.92 ± 2.21	11.24 ± 0.48	21.75 ± 0.49	12.09 ± 1.45
P (mg/g) *	7.91 ± 0.44	3.06 ± 0.15	5.10 ± 0.50	5.30 ± 0.24
Cu (µg/g)	6.79 ± 0.62	11.48 ± 0.29	14.00 ± 0.91	33.15 ± 0.16
Fe (µg/g)	63.05 ± 7.94	163.82 ± 4.08	87.88 ± 4.19	532.80 ± 30.92
Mn (µg/g)	10.57 ± 2.39	11.04 ± 0.02	23.37 ± 1.44	391.98 ± 13.51
Zn (µg/g)	299.31 ± 39.69	94.80 ± 20.57	44.71 ± 3.11	296.06 ± 33.54

Note: * macro element. Data are expressed as means $\pm$ SD.

3.4. Phytic Acid (PA) Content and Mineral Bioavailability

Analysis showed that PA was present in both *C. bilineata tsingtauica* and *C. atrata*, but the concentration was lower than in *G. max* (Figure 2). The PA content of the *C. bilineata tsingtauica* epidermis (16.47 mg/g DW) was higher than that of the meat (Figure 2). The PA/Zn ratio in *C. bilineata tsingtauica* meat was 3.28, lower than the epidermis, *G. max*, and *C. atrata* (Figure 3). The PA/Fe ratio in *C. atrata* was 2.95, lower than *C. bilineata tsingtauica* and *G. max*. It is also important to note that the PA/Fe ratios in the *C. bilineata tsingtauica* meat and epidermis were lower than *G. max* at 13.13 and 8.52, respectively (Figure 3).

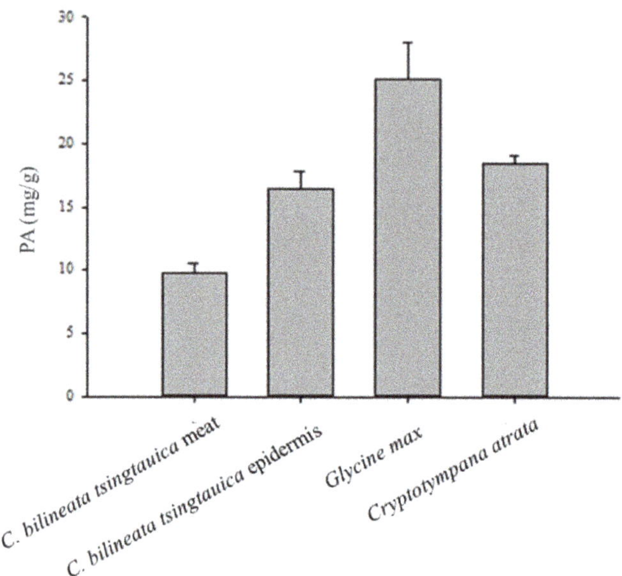

Figure 2. Phytic acid content in *Clanis bilineata tsingtauica*, *Glycine max*, and *Cryptotympana atrata* (DW).

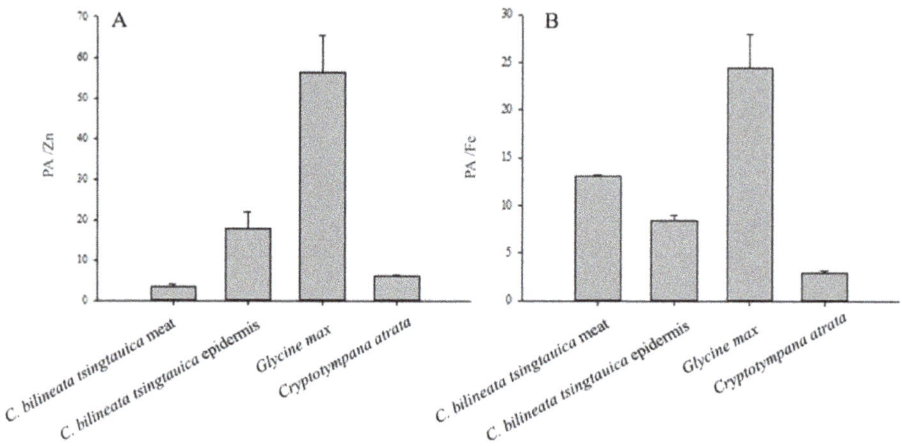

Figure 3. Bioavailability of Zn (**A**) and Fe (**B**) in *Clanis bilineata tsingtauica*, *Glycine max*, and *Cryptotympana atrata*.

4. Discussion

According to the FAO, insects are an environmentally friendly food source for the growing world population [49]. As an edible insect, *C. bilineata tsingtauica* has a long history in China and is loved by the Chinese people. In this study, we found that *C. bilineata tsingtauica* is rich in protein, and the glutelin content is much higher than in *G. max*. Edible insects generally have a rich protein content and supply energy for various physiological functions [50–54]. Although many researchers have tried to substitute edible insect protein for meat protein [55], it is important to consider the risks of food allergy following insect ingestion [56], and possible pathogens in insects, too.

The essential amino acid composition of *C. bilineata tsingtauica* is relatively balanced and comprehensive. According to FAO standards, the ratios of EAA/NEAA and EAA/TAA in the *C. bilineata tsingtauica* epidermis meet the FAO/WHO recommended values (60% and 40%, respectively), while ratios in its meat are slightly lower [57]. Previous reports demonstrated that other edible insects, namely *Tenebrio molitor*, *Acheta domesticus*, and *Locusta migratoria*, contain seven essential amino acids, but the balance of EAAs is not as good as in *C. bilineata tsingtauica*. For example, the EAA/NEAA ratios in *T. molitor*, *A. domesticus*, and *L. migratoria* are only 57.32%, 55.04%, and 56.09%, respectively [58].

Minerals are essential micronutrients for animals and humans [59]. In this study, we measured the concentrations of eight minerals in *C. bilineata tsingtauica* and found that Zn is the highest in *C. bilineata tsingtauica* meat and Fe is highest in the epidermis. The recommended human intake of Fe and Zn is 15 mg/day [60,61], and Zn deficiency is highly prevalent in children and women in developing countries [62,63]. Additionally, phytic acid acts as an anti-nutrient by binding to Fe and Zn; this prevents the absorption of minerals in the gastrointestinal tract and decreases their bioavailability [64,65]. Phytic acid is present in numerous edible insect species (see Table 7 in [41]), and the PA/Zn and PA/Fe molar ratios of cereal and legumes are important considerations [66]. This study showed that the PA/Zn and PA/Fe ratios in *C. bilineata tsingtauica* are lower than in *G. max*, indicating that Zn, Fe, and bioavailability are higher in *C. bilineata tsingtauica* than in *G. max*. Therefore, the consumption of *C. bilineata tsingtauica* can alleviate the Fe and Zn deficiency caused by cereal- and bean-based diets in some areas [64,67], and can be used as a zinc supplement to reduce childhood morbidity and mortality in developing countries [68,69].

In conclusion, this study evaluated the nutritional value of *C. bilineata tsingtauica*, a rich source of protein and minerals. PA is present in *C. bilineata tsingtauica* and *C. atrata*, but the ratios showed that the bioavailability of minerals in these insects is superior to *G. max*. It is important to note that nutrition is distributed throughout the insect body, including the fat body in the abdomen and beneath the epidermis [70,71]. Based on our results, we advocate that the traditional way of eating *C. bilineata tsingtauica* in Lianyungang should be changed, and recommend the consumption of the entire larvae as is common in Xuzhou and Shandong. Future research should address the determination of toxic and beneficial substances in *C. bilineata tsingtauica*, and breeding technologies need to be improved to meet consumer demands for *C. bilineata tsingtauica* products.

Author Contributions: Conceptualization, Y.S. and M.-X.L.; methodology, M.-X.L.; software, L.-Q.J.; validation, M.Z., L.Q. and L.-Q.J.; formal analysis, M.-X.L.; investigation, M.Z.; resources, H.-J.L.; data curation, Y.S.; writing—original draft preparation, Y.S. and M.-X.L.; writing—review and editing, Y.S., M.-X.L., H.-J.L. and Y.-Z.D.; visualization, M.Z.; supervision, Y.-Z.D.; project administration, M.-X.L.; funding acquisition, H.-J.L. All authors have read and agreed to the published version of the manuscript.

Funding: This research was funded by Jiangsu Agricultural Science and Technology Innovation Fund (CX (19) 2036).

Institutional Review Board Statement: Not applicable.

Informed Consent Statement: Not applicable.

Data Availability Statement: Not applicable.

Acknowledgments: We thank the Key Laboratory of Crop Genetics and Physiology of Jiangsu Province, College of Agriculture, Yangzhou University for their assistance with sample determination. We also sincerely thank Carol L. Bender and the anonymous reviewers for their English editing and helpful comments regarding the manuscript.

Conflicts of Interest: The authors declare no conflict of interest.

References

1. Singh, V.K.; Ellur, R.K.; Singh, A.K.; Nagarajan, M.; Singh, B.D.; Singh, N.K. Effect of *qGN4.1* QTL for grain number per panicle in genetic backgrounds of twelve different mega varieties of rice. *Rice* **2018**, *11*, 8. [CrossRef]

2. Meyer-Rochow, V.B. Can insects help to ease the problem of world food shortage? *Search* **1975**, *6*, 261–262.
3. Hong, J.; Han, T.; Kim, Y.Y. Mealworm (*Tenebrio molitor* Larvae) as an alternative protein source for monogastric animal: A review. *Animals* **2020**, *10*, 2068. [CrossRef]
4. Huis, V.A.; Itterbeeck, J.V.; Klunder, H.; Mertens, E.; Vantomme, P. *Edible Insects: Future Prospects for Food and Feed Security*; FAO: Rome, Italy, 2013; Volume 171, pp. 1–201.
5. Kim, S.Y.; Kwak, K.-W.; Park, E.-S.; Yoon, H.J.; Kim, Y.-S.; Park, K.; Kim, E.; Kim, S.-D. Evaluation of subchronic oral dose toxicity of freeze-dried skimmed powder of *Zophobas atratus* Larvae (frpfdZAL) in Rats. *Foods* **2020**, *9*, 995. [CrossRef] [PubMed]
6. Vilcinskas, A.; Schwabe, M.; Brinkrolf, K.; Plarre, R.; Wielsch, N.; Vogel, H. Larvae of the clothing moth *Tineola bisselliella* maintain gut bacteria that secrete enzyme cocktails to facilitate the digestion of keratin. *Microorganisms* **2020**, *8*, 1415. [CrossRef]
7. Nowak, V.; Persijn, D.; Rittenschober, D.; Charrondiere, U.R. Review of food composition data for edible insects. *Food Chem.* **2016**, *193*, 39–46. [CrossRef] [PubMed]
8. Nakane, W.; Nakamura, H.; Nakazato, T.; Kaminaga, N.; Nakano, M.; Sakamoto, T.; Nishiko, M.; Bono, H.; Ogiwara, I.; Kitano, Y.; et al. Construction of TUATinsecta database that integrated plant and insect database for screening phytophagous insect metabolic products with medicinal potential. *Sci. Rep.* **2020**, *10*, 1–11. [CrossRef]
9. Elhassan, M.; Wendin, K.; Olsson, V.; Langton, M. Quality aspects of insects as food—Nutritional, sensory, and related concepts. *Foods* **2019**, *8*, 95. [CrossRef]
10. Huis, A.V. Potential of insects as food and feed in assuring food security. *Annu. Rev. Entomol.* **2013**, *58*, 563–583. [CrossRef]
11. Lyke, M.M.; Di Fiore, A.; Fierer, N.; Madden, A.A.; Lambert, J.E. Metagenomic analyses reveal previously unrecognized variation in the diets of sympatric Old World monkey species. *PLoS ONE* **2019**, *14*, e0218245. [CrossRef]
12. McGrew, W.C. The 'other faunivory' revisited: Insectivory in human and non-human primates and the evolution of human diet. *J. Hum. Evol.* **2014**, *71*, 4–11. [CrossRef] [PubMed]
13. Rothman, J.M.; Raubenheimer, D.; Bryer, M.A.; Takahashi, M.; Gilbert, C. Nutritional contributions of insects to primate diets: Implications for primate evolution. *J. Hum. Evol.* **2014**, *71*, 59–69. [CrossRef] [PubMed]
14. Claudia, C.; Miranda, M.; John, B. Potential of extracted Locusta Migratoria protein fractions as value-added ingredients. *Insects* **2018**, *9*, 20.
15. Wendin, K.; Mårtensson, L.; Djerf, H.; Langton, M. Product quality during the storage of foods with insects as an ingredient: Impact of particle size, antioxidant, oil content and salt content. *Foods* **2020**, *9*, 791. [CrossRef]
16. Zielińska, E.; Baraniak, B.; Karaś, M.; Rybczyńska-Tkaczyk, K.; Jakubczyk, A. Selected species of edible insects as a source of nutrient composition. *Food Res. Int.* **2015**, *77*, 460–466. [CrossRef]
17. Zielińska, E.; Baraniak, B.; Karaś, M. Antioxidant and anti-inflammatory activities of hydrolysates and peptide fractions obtained by enzymatic hydrolysis of selected heat-treated edible insects. *Nutrients* **2017**, *9*, 970. [CrossRef]
18. Gasco, L.; Biasato, I.; Dabbou, S.; Schiavone, A.; Gai, F. Animals fed insect-based diets: State-of-the-art on digestibility, performance and product quality. *Animals* **2019**, *9*, 170. [CrossRef] [PubMed]
19. Ghosh, S.; Jung, C. Nutritional value of bee-collected pollens of hardy kiwi, *Actinidia arguta* (Actinidiaceae) and oak, *Quercus* sp. (Fagaceae). *J. Asia-Pac. Entomol.* **2017**, *20*, 245–251. [CrossRef]
20. Chakravorty, J.; Ghosh, S.; Jung, C.; Meyer-Rochow, V.B. Nutritional composition of *Chondacris rosea* and brachytrupes orientalis: Two common insects used as food by tribes of arunachal pradesh, india. *J. Asia-Pac. Entomol.* **2014**, *17*, 407–415. [CrossRef]
21. Terova, G.; Gini, E.; Gasco, L.; Moroni, F.; Antonini, M.; Rimoldi, S. Effects of full replacement of dietary fishmeal with insect meal from Tenebrio molitor on rainbow trout gut and skin microbiota. *J. Anim. Sci. Biotechnol.* **2021**, *12*, 1–14. [CrossRef]
22. Rumpold, B.A.; Schlüter, O.K. Nutritional composition and safety aspects of edible insects. *Mol. Nutr. Food Res.* **2013**, *57*, 802–823. [CrossRef]
23. Fontaneto, D.; Tommaseo-Ponzetta, M.; Galli, C.; Risé, P.; Glew, R.H.; Paoletti, M.G. Differences in fatty acid composition between aquatic and terrestrial insects used as food in human nutrition. *Ecol. Food Nutr.* **2011**, *50*, 351–367. [CrossRef]
24. Feng, Y.; Chen, X.-M.; Zhao, M.; He, Z.; Sun, L.; Wang, C.-Y.; Ding, W.-F. Edible insects in China: Utilization and prospects. *Insect Sci.* **2018**, *25*, 184–198. [CrossRef]
25. Feng, Y.Y.; Ma, G.C.; Jin, Q.A.; Lü, B.Q.; Peng, Z.Q.; Jin, T.; Wen, H.B. Effect of temperature on developmental duration and feeding amount of *Clanis bilineata tsingtauica*. *Chin. J. Trop. Crop.* **2014**, *35*, 2442–2444.
26. Guo, M.M.; Li, X.F.; Deng, P.; Li, D.W.; Li, J.L.; Fan, J.W.; Chen, F. Diapause termination and post-diapause of overwintering *Clanis bilineata tsingtauica* larvae. *Chin. J. Appl. Entomol.* **2021**, *58*, 966–972.
27. Tian, H.; Zhang, Y.M. Advances of integrated utilization on resource insect *Clanis bilineata tsingtauica*. *Guizhou Agric. Sci.* **2009**, *37*, 111–113.
28. Guo, M.M.; Liao, H.J.; Deng, P.; Li, D.W.; Li, J.L.; Zhang, J.Q.; Fan, J.W.; Chen, F. Effects of soybean varieties (lines) and planting density on survival and development of *Clanis bilineata tsingtauica* Mell larvae. *J. Environ. Entomol.* **2020**, *42*, 1401–1408.
29. Tian, H.; Zhang, Y.M. The nutritional components analysis and evaluation of *Clanis bilineata tsingtauica* Mell. *Acta Nutr. Sin.* **2012**, *34*, 289–291.
30. Doumani, N.; Severin, I.; Dahbi, L.; Bou-Maroun, E.; Tueni, M.; Sok, N.; Chagnon, M.-C.; Maalouly, J.; Cayot, P. Lemon juice, sesame paste, and autoclaving influence iron bioavailability of hummus: Assessment by an in vitro digestion/Caco-2 cell model. *Foods* **2020**, *9*, 474. [CrossRef] [PubMed]

31. Jobarteh, M.L.; McArdle, H.J.; Holtrop, G.; Sise, E.A.; Prentice, A.M.; Moore, S.E. mRNA levels of placental iron and zinc transporter genes are upregulated in Gambian women with low iron and zinc status. *J. Nutr.* **2017**, *147*, 1401–1409. [CrossRef]

32. Rebholz, C.M.; Lichtenstein, A.H.; Zheng, Z.; Appel, L.J.; Coresh, J. Serum untargeted metabolomic profile of the Dietary Approaches to Stop Hypertension (DASH) dietary pattern. *Am. J. Clin. Nutr.* **2018**, *108*, 243–255. [CrossRef]

33. Clements, R.S.; Darnell, B. Myo-inositol content of common foods: Development of a high-myo-inositol diet. *Am. J. Clin. Nutr.* **1980**, *33*, 1954–1967. [CrossRef] [PubMed]

34. Steadman, K.J.; Burgoon, M.S.; Schuster, R.L.; Lewis, B.A.; Edwardson, S.E.; Obendorf, R.L. Fagopyritols, D-chiro-inositol, and other soluble carbohydrates in buckwheat seed milling fractions. *J. Agric. Food Chem.* **2000**, *48*, 2843–2847. [CrossRef] [PubMed]

35. Silva, E.O.; Bracarense, A.P.F.; Bracarense, A.P. Phytic acid: From antinutritional to multiple protection factor of organic systems. *J. Food Sci.* **2016**, *81*, R1357–R1362. [CrossRef]

36. Joseph, L. D-chiro-inositol—Its functional role in insulin action and its deficit in insulin resistance. *Int. J. Exp. Diabetes Res.* **2002**, *3*, 47–60.

37. Gupta, S.; Habeych, E.; Scheers, N.; Merinat, S.; Rey, B.; Galaffu, N.; Sandberg, A.-S. The development of a novel ferric phytate compound for iron fortification of bouillons (part I). *Sci. Rep.* **2020**, *10*, 5340. [CrossRef]

38. Sandberg, A.-S.; Andersson, H. The effect of dietary phytase on the digestion of phytate in the stomach and small intestine of humans. *J. Nutr.* **1988**, *118*, 469–473. [CrossRef]

39. Hurrell, R.F. Phytic acid degradation as a means of improving iron absorption. *Int. J. Vitam. Nutr. Res.* **2004**, *74*, 445–452. [CrossRef] [PubMed]

40. Brune, M.; Rossander-Hultén, L.; Hallberg, L.; Gleerup, A.; Sandberg, A.-S. Iron absorption from bread in humans: Inhibiting effects of cereal fiber, phytate and inositol phosphates with different numbers of phosphate groups. *J. Nutr.* **1992**, *122*, 442–449. [CrossRef]

41. Meyer-Rochow, V.; Gahukar, R.; Ghosh, S.; Jung, C. Chemical composition, nutrient quality and acceptability of edible insects are affected by species, developmental stage, gender, diet, and processing method. *Foods* **2021**, *10*, 1036. [CrossRef]

42. Chakravorty, J.; Ghosh, S.; Megu, K.; Jung, C.; Meyer-Rochow, V.B. Nutritional and anti-nutritional composition of *Oecophylla smaragdina* (Hymenoptera: Formicidae) and *Odontotermes* sp. (Isoptera: Termitidae): Two preferred edible insects of Arunachal Pradesh, India. *J. Asia-Pac. Entomol.* **2016**, *19*, 711–720. [CrossRef]

43. Liu, X.; Chen, X.; Wang, H.; Yang, Q.; ur Rehman, K.; Li, W.; Cai, M.; Li, Q.; Mazza, L.; Zhang, J.; et al. Dynamic changes of nutrient composition throughout the entire life cycle of black soldier fly. *PLoS ONE* **2017**, *12*, e0182601. [CrossRef] [PubMed]

44. Zhou, X.; Zhou, J.; Wang, Y.; Peng, B.; Zhu, J.; Yang, L.; Wang, Y. Elevated tropospheric ozone increased grain protein and amino acid content of a hybrid rice without manipulation by planting density. *J. Sci. Food Agric.* **2014**, *95*, 72–78. [CrossRef] [PubMed]

45. Ju, Z.; Hettiarachchy, N.; Rath, N. Extraction, denaturation and hydrophobic properties of rice flour proteins. *J. Food Sci.* **2001**, *66*, 229–232. [CrossRef]

46. Toshio, S.; Kunisuke, T.; Zenzaburo, K. Improved extraction of rice prolamin. *Agric. Biol. Chem.* **1986**, *50*, 2409–2411.

47. Jing, L.; Chen, C.; Hu, S.; Dong, S.; Pan, Y.; Wang, Y.; Lai, S.; Wang, Y.; Yang, L. Effects of elevated atmosphere CO_2 and temperature on the morphology, structure and thermal properties of starch granules and their relationship to cooked rice quality. *Food Hydrocoll.* **2021**, *112*, 106360. [CrossRef]

48. Qiao, T.S.; Tang, H.C.; Liu, J.X.; Li, L. Analysis of nutritional components and evaluation of protein quality of *Oxya chinensis*. *Chin. Bull. Entomol.* **1992**, *29*, 113–117.

49. Huis, A.V. Prospects of insects as food and feed. *Org. Agric.* **2021**, *11*, 301–308. [CrossRef]

50. Khampakool, A.; Soisungwan, S.; You, S.G.; Park, S.H. Infrared assisted freeze-drying (IRAFD) to produce shelf-stable insect food from *Protaetia brevitarsis* (white-spotted flower chafer) larva. *Food Sci. Anim. Resour.* **2020**, *40*, 813–830. [CrossRef]

51. Mi, Y.C.; Gwon, E.Y.; Hwang, J.S.; Goo, T.W.; Yun, E.Y. Analysis of general composition and harmful material of *Protaetia brevitarsis*. *J. Life Sci.* **2013**, *23*, 664–668.

52. Melis, R.; Braca, A.; Mulas, G.; Sanna, R.; Spada, S.; Serra, G.; Leonarda, F.M.; Roggio, T.; Uzzau, S.; Anedda, R. Effect of freezing and drying processes on the molecular traits of edible yellow mealworm. *Innov. Food Sci. Emerg. Technol.* **2018**, *48*, 138–149. [CrossRef]

53. Kwon, Y.J.; Lee, H.S.; Park, J.Y.; Lee, J.W. Associating intake proportion of carbohydrate, fat, and protein with all-cause mortality in Korean adults. *Nutrients* **2020**, *12*, 3208. [CrossRef]

54. Carreiro, A.L.; Dhillon, J.; Gordon, S.; Higgins, K.A.; Jacobs, A.G.; McArthur, B.M.; Redan, B.W.; Rivera, R.L.; Schmidt, L.R.; Mattes, R.D. The macronutrients, appetite, and energy intake. *Annu. Rev. Nutr.* **2016**, *36*, 73–103. [CrossRef] [PubMed]

55. Kim, T.K.; Yong, H.I.; Chang, H.J.; Han, S.G.; Kim, Y.B.; Paik, H.D.; Choi, Y.S. Technical functional properties of water- and salt-soluble proteins extracted from edible insects. *Food Sci. Anim. Resour.* **2019**, *39*, 643–654. [CrossRef]

56. Pali-Scholl, I.; Meinlschmidt, P.; Larenas-Linnemann, D.; Purschke, B.; Hofstetter, G.; Rodriguez-Monroy, F.A.; Einhorn, L.; Mothes-Luksch, N.; Jensen-Jarolim, E.; Jager, H. Edible insects: Cross-recognition of IgE from crustacean and house dust mite allergic patients, and reduction of allergenicity by food processing. *World Allergy Organ. J.* **2019**, *12*, 100006. [CrossRef]

57. Li, Y.C.; Li, K.; Dai, R.H. Analysis and evaluation of nutrition of *Opisthoplata orientalis* Buron. *J. Environ. Entomol.* **2013**, *35*, 466–472.

58. Boulos, S.; Tnnler, A.; Nystrm, L. Nitrogen-to-protein conversion factors for edible insects on the Swiss market: *T. molitor*, *A. domesticus*, and *L. migratoria*. *Front. Nutr.* **2020**, *7*, 89. [CrossRef]

59. Liu, J.; Huang, L.; Li, T.X.; Liu, Y.X.; Yan, Z.H.; Tang, G.; Zheng, Y.L.; Liu, D.C.; Wu, B.H. Genome-wide association study for grain micronutrient concentrations in wheat advanced lines derived from wild emmer. *Front. Plant Sci.* **2021**, *12*, 651283. [CrossRef]
60. Calayugan, M.; Formantes, A.K.; Amparado, A.; Descalsota-Empleo, G.I.; Nha, C.T.; Inabangan-Asilo, M.A.; Swe, Z.M.; Hernandez, J.E.; Borromeo, T.H.; Lalusin, A.G.; et al. Genetic analysis of agronomic traits and grain iron and zinc concentrations in a doubled haploid population of rice (*Oryza sativa* L.). *Sci. Rep.* **2020**, *10*, 2283. [CrossRef]
61. Welch, R.M.; Graham, R.D. Breeding for micronutrients in staple food crops from a human nutrition perspective. *J. Exp. Bot.* **2004**, *55*, 353–364. [CrossRef]
62. Sandstead, H.H. Is zinc deficiency a public problem? *Nutrition* **1995**, *11*, 87–92.
63. Baqui, A.H.; Black, R.E.; Arifeen, S.E.; Yunus, M.; Chakraborty, J.; Ahmed, S.; Vaughan, J.P. Effect of zinc supplementation started during diarrhoea on morbidity and mortality in Bangladeshi children: Community randomised trial. *BMJ* **2002**, *325*, 1059–1062. [CrossRef]
64. Caproni, L.; Raggi, L.; Talsma, E.F.; Wenzl, P.; Negri, V. European landrace diversity for common bean biofortification: A genome-wide association study. *Sci. Rep.* **2020**, *10*, 19775. [CrossRef]
65. Luzak, A.; Heier, M.; Thorand, B.; Laxy, M.; Nowak, D.; Peters, A.; Schulz, H. Physical activity levels, duration pattern and adherence to WHO recommendations in german adults. *PLoS ONE* **2017**, *12*, e0172503. [CrossRef]
66. Lowe, N.M.; Khan, M.J.; Broadley, M.R.; Zia, M.H.; Mcardle, H.J.; Joy, E.J.; Ohly, H.; Shahzad, B.; Ullah, U.; Kabana, G.; et al. Examining the effectiveness of consuming flour made from agronomically biofortified wheat (Zincol-2016/NR-421) for improving Zn status in women in a low-resource setting in Pakistan: Study protocol for a randomised, double-blind, controlled cross-over trial (BiZiFED). *BMJ Open* **2018**, *8*, e021364.
67. Petry, N.; Boy, E.; Wirth, J.P.; Hurrell, R.F. Review: The potential of the common bean (*Phaseolus vulgaris*) as a vehicle for iron biofortification. *Nutrients* **2015**, *7*, 1144–1173. [CrossRef] [PubMed]
68. Walker, C.F.; Black, R.E. Zinc and the risk for infectious disease. *Annu. Rev. Nutr.* **2004**, *24*, 255–275. [CrossRef] [PubMed]
69. Cole, C.R.; Grant, F.K.; Dawn, S.; Smith, J.L.; Anne, J.; Northrop-Clewes, C.A.; Calawell, K.L.; Pfeiffer, C.M.; Ziegler, T.R. Zinc and iron deficiency and their interrelations in low-income African American and Hispanic children in Atlanta. *Am. J. Clin. Nutr.* **2010**, *91*, 1027–1034. [CrossRef] [PubMed]
70. Li, L.F.; Gao, X.; Lan, M.X.; Yuan, Y.; Guo, Z.J.; Tang, P.; Li, M.Y.; Liao, X.B.; Zhu, J.Y.; Li, Z.Y.; et al. De novo transcriptome analysis and identification of genes associated with immunity, detoxification and energy metabolism from the fat body of the tephritid gall fly, *Procecidochares utilis*. *PLoS ONE* **2019**, *14*, e0226039. [CrossRef] [PubMed]
71. Arrese, E.L.; Soulages, J.L. Insect fat body: Energy, metabolism, and regulation. *Annu. Rev. Entomol.* **2010**, *55*, 207–225. [CrossRef]

Article

Extending the Storage Time of *Clanis bilineata tsingtauica* (Lepidoptera; Sphingidae) Eggs through Variable-Temperature Cold Storage

Chenxu Zhu [1], Ming Zhao [1], Haibo Zhang [2], Fang Zhang [2], Yuzhou Du [1,3] and Mingxing Lu [1,*

[1] College of Horticulture and Plant Protection & Institute of Applied Entomology, Yangzhou University, Yangzhou 225009, China; syzzzgalaxy@gmail.com (C.Z.); zhaoming@yzu.edu.cn (M.Z.); yzdu@yzu.edu.cn (Y.D.)

[2] Plant Protection and Quarantine Station of Jiangsu Province, Nanjing 210009, China; zhanghbjs@126.com (H.Z.); cebaoke@126.com (F.Z.)

[3] Joint International Research Laboratory of Agriculture and Agri-Product Safety, Yangzhou University, Yangzhou 225009, China

[*] Correspondence: lumx@yzu.edu.cn

Abstract: *Clanis bilineata tsingtauica* Mell, 1922 (Lepidoptera, Sphingidae), also known as "Doudan" in China, is an important pest in legume crops. As an edible insect, it is most commonly consumed in Jiangsu, Shandong, and Henan Provinces. Mass rearing requires access to large numbers of eggs. This stage, however, is of short duration and supplies are frequently not sufficient for insect production. Therefore, we identified the cold storage conditions for *C. bilineata tsingtauica* that can effectively prolong the storage time of the eggs, to make supplies more readily available. We found that when stored at 4 °C, only 7.5% of the eggs hatched after 7 days, while at 10 °C the hatch rate was 78.3%. At 15 °C, the egg hatch rate remained at this same level (77.8% even after 14–20 days). Considering various combinations, we found that optimal egg hatch occurred if eggs were stored at 15 °C for 11 days, and then held at 15–20 °C under dark conditions. Stored as described above, the egg hatch rate was not significantly different from the control group (at 28 °C). These conditions allow for easier mass rearing of *C. bilineata tsingtauica* by providing a stable supply of eggs.

Keywords: *Clanis bilineata tsingtauica*; edible insects; cold storage; temperature; artificial breeding

Citation: Zhu, C.; Zhao, M.; Zhang, H.; Zhang, F.; Du, Y.; Lu, M. Extending the Storage Time of *Clanis bilineata tsingtauica* (Lepidoptera; Sphingidae) Eggs through Variable-Temperature Cold Storage. *Foods* **2021**, *10*, 2820. https://doi.org/10.3390/foods10112820

Academic Editor: Victor Benno Meyer-Rochow

Received: 5 October 2021
Accepted: 12 November 2021
Published: 16 November 2021

Publisher's Note: MDPI stays neutral with regard to jurisdictional claims in published maps and institutional affiliations.

1. Introduction

Due to the increase in the world's population, the need for proteins is increasing [1]. Due to limitations in the growth of the production of livestock, alternative meat products are of interest. In nutritional value, insects are comparable to traditional meat products and their production offers some advantages compared to increasing livestock production. Insects emit less greenhouse gas than vertebrate animals and the cost of insect rearing is lower than that of keeping vertebrates [1–3]. The interest in insects as food has increased ever since it was first suggested by Meyer-Rochow in 1975 that edible insects could ease the problem of global food shortages and that the FAO and WHO should support the use of insects as food and feed [3–5]. There are at least 2100 species of insects that are consumed in various parts of the world, especially species of beetles, crickets, caterpillars, bees, and wasps [3]. Insects may be fried, boiled, or processed as an ingredient added to flour, beer, candy, noodles, and protein bars [6–11].

Clanis bilineata tsingtauica Mell, 1922 (Lepidoptera, Sphingidae), in contrast to most edible insects, has a long history of consumption in China of 300 years plus [12]. *C. bilineata tsingtauica* larvae are rich in proteins, amino acids, fatty acids, vitamins, and trace elements, and they are a good source of nutrition [13–16]. *C. bilineata tsingtauica* is widely distributed in China. Breeding of *C. bilineata tsingtauica* has become an important source of income from agricultural activities for farmers in some areas of Jiangsu province. Currently, the market for *C. bilineata tsingtauica* products is gradually expanding, and has radiated to Shandong province, Nanjing city, Shanghai city, Guangzhou city, and other places in China. The annual output of *C. bilineata tsingtauica* through artificial breeding in the country is 3×10^4 t, with an output value of nearly 4.5 billion RMB (0.7 billion USD). However, the annual demand for *C. bilineata tsingtauica* is about 10×10^4 t, and the artificial breeding of *C. bilineata tsingtauica* is far from meeting people's consumption needs [15]. Methods of cooking *C. bilineata tsingtauica* larvae include deep-frying, adding larvae into soups, cooking larvae together with vegetables, and stir frying them with chili (Figure 1).

Figure 1. A common *Clanis bilineata tsingtauica* dish in a restaurant in Lianyungang City, Jiangsu Province.

Clanis bilineata tsingtauica, whose larvae are also known as "Doudan (豆丹)" in China, feed as larvae on legumes, for example soybeans, among other plants [17]. In some years, this species can be a pest of soybeans. In nature, *C. bilineata tsingtauica* has 1 or 2 generations per year in China. In Jiangsu Province, there is 1 generation per year. There are 5 instars, and the biomass increases dramatically in the 3rd and 4th instars. [18]. The weight of a 5th instar larva is around 8.15 g, and approximately 120 larvae weigh one kilogram. It is the 5th instar larvae that are usually sold in the market, and they are eaten in restaurants. To meet the demand for these larvae as human food, larvae are being reared in plastic greenhouses on soybean plants. The level of such production is expanding, and under ideal conditions three generations can be produced each year in Jiangsu Province.

For mass production of this insect, a crucial problem is the need for a reliable supply of high-quality eggs that can be synchronized with the production of soybean plants under greenhouse conditions. The duration of the egg stage of *C. bilineata tsingtauica* is between 3 and 5 days. Unless these eggs are held under suitable cold conditions, losses occur when trying to match the egg supply to development of the plant hosts. Meanwhile, due to the short duration of the egg stage, larvae may hatch during the long-distance transportation and cause losses. Thus, cold storage is an appropriate way of preserving *C. bilineata tsingtauica* eggs when there is a lack of host plants or transportation. Furthermore, considering the demand in the market, sometimes when supply exceeds demand, people can also cold store *C. bilineata tsingtauica* eggs and sell them when the market demand increases, which would lead to more profit.

For ectotherms, temperature is a major environmental factor that affects insect development and growth and almost all ecological and physiological processes [19–24]. Low temperature can harm insects being stored, causing reduced vitality or slow body metabolism [25–27]. Cold storage has been used to preserve and manipulate several stages of various insects, including eggs and pupae of *Chilo suppressalis* (Walker, 1863) (Lepidoptera, Crambidae) and eggs of *Corcyra cephalonica* (Stainton, 1866) (Lepidoptera, Pyralidae) and *Sitotroga cerealella* (Oliver, 1789) (Lepidoptera, Geechiidae) [27,28]. The desired temperature is one that will maintain the insects in a live and suitable stage, with adequate viability. For example, after cold-storing *Carposina sasakii* (Matsumura, 1900) (Lepidoptera, Carponisidae) eggs at 4 °C for 7 days, the hatch rate showed no significant difference from the control group [29]. The goal of this study was to determine the optimal temperature for cold storage of *Clanis bilineata tsingtauica* eggs to support mass-rearing facilities producing this species for human consumption. We developed and assessed a method of cold storage based on the use of variable storage temperatures, which we found to effectively prolong the storage time and retain egg viability.

2. Material and Methods

2.1. Source of Insect Materials

Clanis bilineata tsingtauica eggs were supplied from Suntiao Village, Yangji Town, Guanyun County, Jiangsu Province, China. *C. bilineata tsingtauica* larvae in our colony were reared on soybean leaves in glass containers.

2.2. Storage Container

The device for storing eggs consisted of a layer of moist filter paper on the bottom of a closed Petri dish to prevent the larvae hatched crawling away and affecting the statistics, with approximately 30 newly laid eggs scattered in the Petri dish. The Petri dish diameter is around 11.5 cm.

2.2.1. Storage at Fixed Temperatures

Viability after storage was assessed for 4, 10, and 15 °C, where 4 °C and 10 °C are common temperatures in the cold storage of insects. Furthermore, 4 °C is the temperature of household refrigerators. It will be convenient if eggs could be stored at 4 °C with a high egg hatch rate and long storage time. Eggs were held at 4 and 10 °C for 7, 14, and 21 days in the dark all the time. The control group was held at 28 °C, which is the optimal development temperature for *C. bilineata tsingtauica*. Each treatment was replicated 4 times, with a replicate being one dish of 30 eggs.

Eggs were also held at 15 °C, which is the threshold for development [30]. Eggs held at 15 °C were kept in the dark until the last larva hatched. The control group was held at 28 °C. Each treatment (15 °C) and control group (28 °C) was replicated 3 times.

After storage, eggs were observed daily at the same time and any hatched or dead eggs were noted. A blackened or shriveled or empty egg indicates the death of the egg. The egg hatch rate, mortality rate, and stage duration were then calculated. Eggs were held under moist conditions to maintain humidity by spraying the filter paper with distilled water. Relative humidity in the test petri dishes and the incubator were held at 60−70%. Larvae from hatched eggs were reared at 28 ± 1 °C, 60−70% relative humidity, and L:D = 16:8 photoperiod, and were fed on soybean leaves in order to further observe the feeding intake and growth status. Larvae under different treatments consumed soybean leaves and molted regularly, as in the control group.

2.2.2. Storage at Variable Temperatures

For this experiment, there were 5 treatments and 1 control, all with three replicates. The control group was held in the dark at 28 °C. This control group is the same as the one mentioned above when eggs were held at 15 °C. Before the experiment, eggs destined for the treatment groups were held in the dark at 20 °C for one day. Then, eggs were placed in the respective treatments in the dark, which were as follows:

Treatment 1: 15 °C for 5 days, 10 °C for 7 days, then put them under 28 °C
Treatment 2: 15 °C for 5 days, 10 °C for 7 days, then put them under 15 °C
Treatment 3: 15 °C for 10 days, 10 °C for 7 days, then put them under 28 °C
Treatment 4: 15 °C for 10 days, 10 °C for 7 days, then put them under 15 °C
Treatment 5: 15 °C for 11 days, then put them under 20 °C
Control: 28 °C all the time

Egg hatch and egg mortality were observed daily at the same time and, from these data, we calculated the overall egg hatch and mortality rates for each treatment and the control. Eggs were held under moist conditions to maintain humidity by spraying the filter paper with distilled water. Relative humidity in the test Petri dishes and the incubator were held at 60−70%. When the larvae hatched, they were also observed and reared at 28 ± 1 °C, 60−70% relative humidity, and L:D = 16:8 photoperiod, and were fed on soybean leaves. Larvae under different treatments consumed soybean leaves and molted regularly, as in the control group. The routine of the whole experiment is showed below (Figure 2).

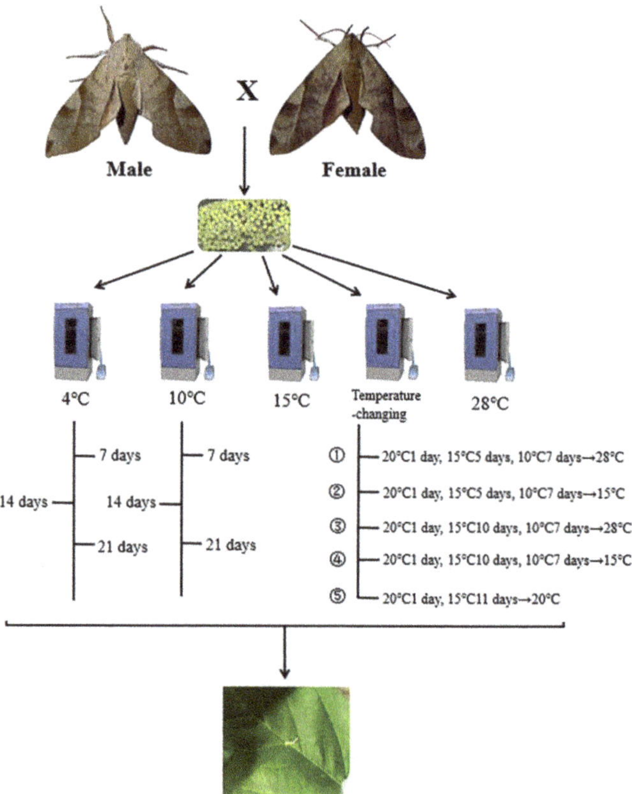

Figure 2. The routine of the whole experiment.

2.3. Statistical Analysis

The experimental data were analyzed by GraphPad Prism 8 and SPSS 16.0 (IBM Inc., New York, NY, USA). Significant differences were determined using the LSD (homogeneity of variances) or Dunnett T3 test (non-homogeneous). The hatch rates of eggs stored at 4 °C and 10 °C were analyzed using Dunnett T3. Student's t-test ($p < 0.05$) was used to test whether the hatch rate of eggs stored at 15 °C was significantly different from the control group. The hatch rates of eggs stored at variable temperatures were analyzed using LSD. Dunnett T3 was used to analyze the storage time of eggs stored at 15 °C and variable temperatures.

3. Results

3.1. Effects of Storage at Fixed Temperatures on Egg Hatch

When eggs were stored at 4 °C for 7 days and then placed at 28 °C after storage, only 7.5% (±3.4%) (mean ± SE) of the eggs hatched. When eggs were stored at 4 °C for 14 days or longer, there was no hatch after storage. The hatch rate for the control group was 79.6% (±3.3%) (mean ± SE) (Figure 3). Eggs stored at 10 °C for 7 days and then returned to 28 °C had a hatch rate of 78.3% (±3.2%) (mean ± SE), which was not significantly different from the control ($p = 1.000$). When eggs were stored at 10 °C for 14 or 21 days, no eggs hatched after storage. (Figure 4). At 15 °C, 77.8% (±5.6%) (mean ± SE) of the eggs hatched, which was not different from the control group ($p = 0.074$) (Figure 5). Meanwhile, the feeding and growth status of the larvae from 15 °C were similar to the larvae from the control group. Therefore, 15 °C can be used as a temperature for cold storage.

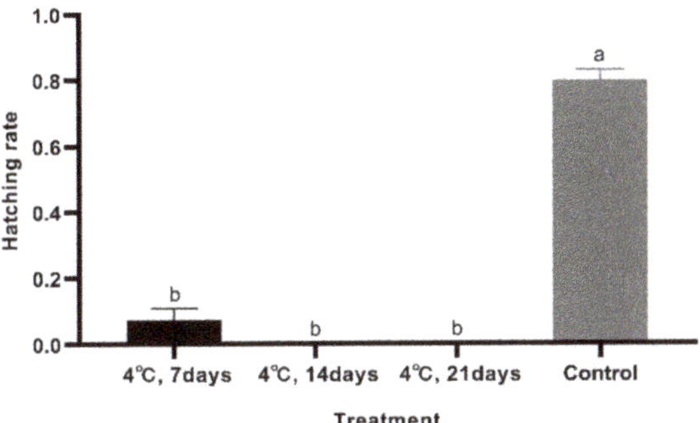

Figure 3. Difference analysis among the different treatments (4 °C). Bars represent the standard error. The hatch rates of eggs stored at 4 °C analyzed using Dunnett T3. Different letters on the bars indicate significant differences ($F_{3,12} = 264.61$; $p < 0.05$).

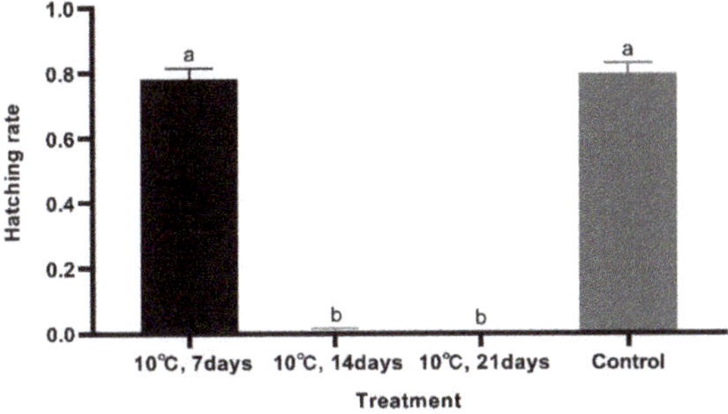

Figure 4. Difference analysis among the different treatments (10 °C). Bars represent the standard error. The hatch rates of eggs stored at 10 °C were analyzed using Dunnett T3. Different letters on the bars indicate significant differences ($F_{3,12} = 378.74$; $p < 0.05$).

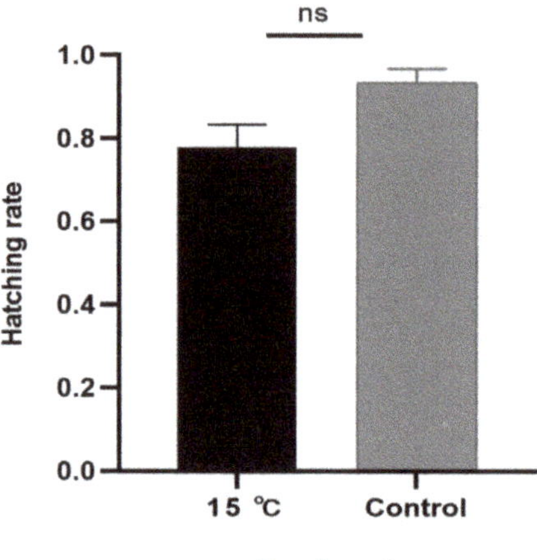

Figure 5. Difference analysis among the different treatments (15 °C). Bars represent the standard error. Student's t-test ($p < 0.05$) was used to test whether the hatch rate of eggs stored at 15 °C was significant different from the control group ($p = 0.074$).

3.2. Effects of Storage at Variable Temperatures on Egg Hatch

Except for the fifth treatment, none of the treatments with variable combinations of colder and warmer temperatures produced eggs with hatch rates similar to the control (Figure 6). Only treatment 5, in which eggs were held at 15 °C for 11 days (just at the developmental threshold) and then brought up to 20 °C, had a high hatch rate (84.5% ± 4.0%) (mean ± SE), which exhibited a remarkable difference compared to the control hatch rate of 93.3% (±3.3%) (mean ± SE) ($p = 0.216$) (Figure 6). The feeding and growth status of the larvae from treatment 5 were not significantly different from the larvae from the control group.

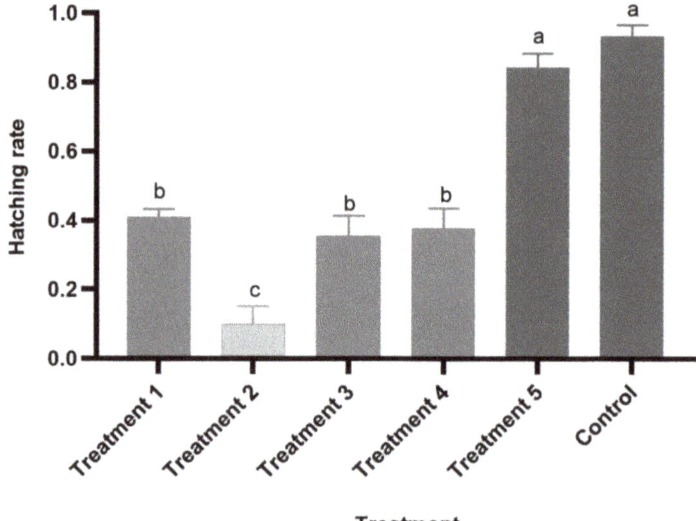

Figure 6. Effect of five variable temperature regimes on the hatching rate of *Clanis bilineata tsingtauica* eggs. Bars represent the standard error. The hatch rates of eggs stored at variable temperatures were analyzed using LSD. Different letters on the bars indicate significant differences ($F_{5,11} = 44.43$; $p < 0.05$).

3.3. Storage Times under Various Temperature Regimes

Storage time of the control group (at 28 °C) was only 5 days until hatching occurred. The storage times of treatments 1, 2, 3, and 4 were 15–18, 19–21, 17–19, and 18–21 days, respectively, but all of these treatments reduced the hatch rates to under 40% (Figure 7). The storage time of treatment 5 averaged 15.5 days but hatched similarly to the control. For eggs held continuously at 15 °C, the storage time averaged 17 days (Figure 8).

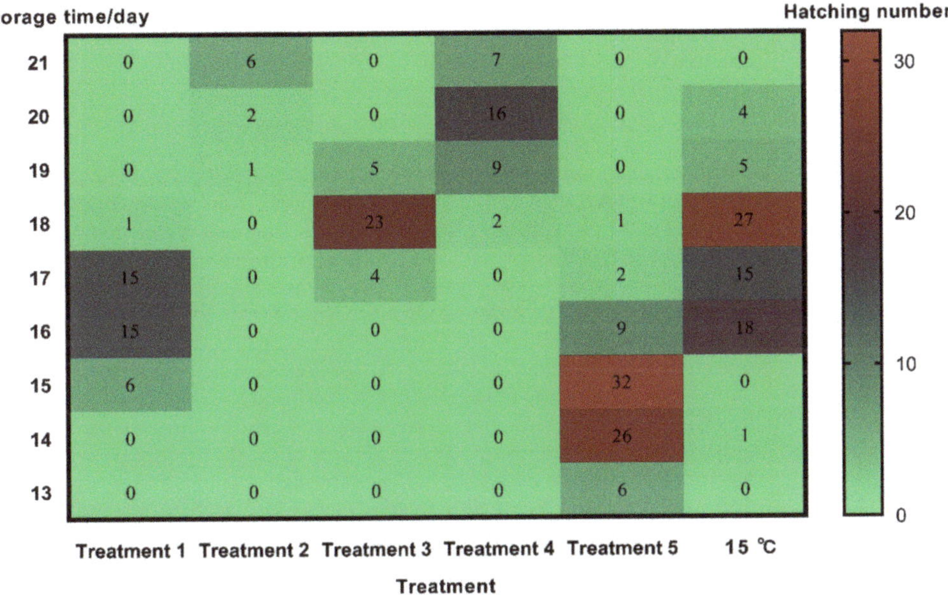

Figure 7. Heat map of the storage time and hatching number of each treatment.

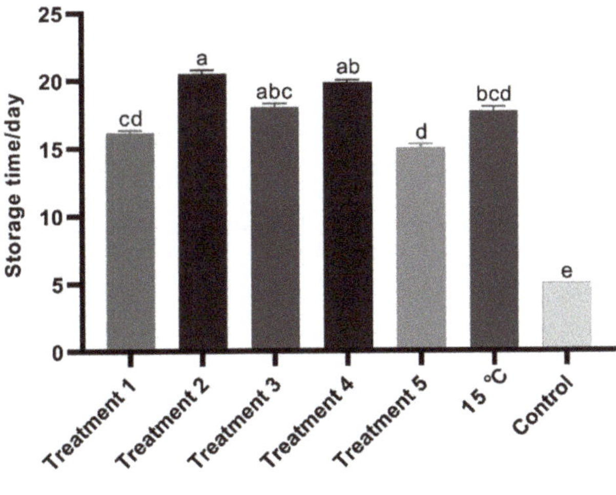

Figure 8. Storage times of *Clanis bilineata tsingtauica* eggs under different temperature regimes. Bars represent the standard error. Dunnett T3 was used to analyze the storage time of eggs stored at 15 °C and variable temperatures. Different letters on the bars indicate significant differences ($F_{6,14} = 366.58$, $p < 0.05$).

4. Discussion

Clanis bilineata tsingtauica is an edible insect with an expanding commercial market. As the demand for this product increases, the rearing scale will have to increase. Cold storage can be used to store *C. bilineata tsingtauica* eggs. We assessed various possible storage temperatures or combinations of temperature. Use of 4 °C for storage was examined because it is the temperature of household refrigerators and if suitable would be extremely convenient. However, from eggs stored at this temperature, almost none hatched, likely because a larva, not egg, is the overwintering stage in nature. The cold resistance of larvae is better than that of eggs and the developmental threshold temperature of larvae is lower than that of eggs [30]. At 10 °C, some larvae hatched when stored for 7 days; but, when stored longer at this temperature, the hatching rate declined. This phenomenon may be due to the fact that at low temperature eggs produce anti-cold biochemical substances to improve their cold tolerance [25,26,31], and with a longer storing time, the accumulation of cold-resistant substances reaches saturation. At this point they will consume a high fraction of metabolic energy to adapt to the external low-temperature environment. Even given sufficient and suitable growth conditions, they will not restore their original vitality. At 15 °C, which is the normal threshold temperature for egg developmental, many eggs hatched after a period of storage of 11 days. Combining this temperature with subsequent storage at 10 °C did not improve egg hatch. In general, storage at a temperature below the lower developmental threshold temperature lowered the hatch rate of eggs significantly. If eggs are held at a temperature above the developmental threshold temperature (15 °C), the ultimate hatching rate of the eggs is not harmed, and the storage time can be prolonged for over 2 weeks. In the same time, the feeding and growth status of the larvae were similar to those of the control group. This finding is of great importance to mass reproduction systems and will allow eggs to be held until needed over a period of several weeks between production and use. Being able to store eggs when market demand for them is low or during the transportation can help increase profitability and make possible shipping eggs between different regions. The storage time, however, is not long (14–20 days), and future work on selection of eggs best adapted to longer storage should be pursued.

Alternatively, markets could use larvae rather than eggs as the stage to initiate rearing, given that *C. bilineata tsingtauica* overwinters as diapausing larvae [30]. However, breaking of the larval diapause after having received the material would take time, which would be disadvantageous in mass-rearing operations; also, even after diapause termination, the growth and feeding of the resulting larvae may be affected [32], or larvae may remain in quiescence for several months [33]. In general, technology for effectively breaking diapause in insects is not well developed. However, if methods to break diapause effectively could be developed, it would solve a major problem limiting the mass reproduction of this insect and likely would be valuable for the cold storage of other insects as well.

Author Contributions: Conceptualization, M.L. and C.Z.; Methodology, M.L.; formal analysis, M.L., C.Z. and M.Z.; investigation, M.L., C.Z. and M.Z.; data curation, M.L. and C.Z.; writing—original draft preparation, C.Z.; writing—review and editing, C.Z., M.L. and Y.D.; visualization, H.Z. and F.Z.; supervision, M.L. and Y.D.; project administration, M.L.; funding acquisition, M.L. and Y.D. All authors have read and agreed to the published version of the manuscript.

Funding: This research was funded by Jiangsu Agricultural Science and Technology Innovation Fund, grant number CX (20) 3179 and Dongming huanghetan Ecological Agriculture Co., Ltd., grant number 204032897.

Acknowledgments: This research was founded by Jiangsu Agricultural Science and Technology Innovation Fund (CX (20) 3179), and Dongming huanghetan Ecological Agriculture Co., Ltd. (204032897).

Conflicts of Interest: The authors declare no conflict of interest.

References

1. Murefu, T.R.; Macheka, L.; Musundire, R.; Manditsera, F.A. Safety of wild harvested and reared edible insects: A review. *Food Control* **2019**, *101*, 209–224. [CrossRef]
2. Oonincx, D.G.; van Itterbeeck, J.; Heetkamp, M.J.; van den Brand, H.; van Loon, J.J.; van Huis, A. An exploration on greenhouse gas and ammonia production by insect species suitable for animal or human consumption. *PLoS ONE* **2010**, *5*, e14445. [CrossRef] [PubMed]
3. van Huis, A. Edible insects. In *Handbook of Eating and Drinking*; Springer: Cham, Switzerland, 2020; pp. 965–980.
4. Schouteten, J.J.; De Steur, H.; De Pelsmaeker, S.; Lagast, S.; Juvinal, J.G.; De Bourdeaudhuij, I.; Verbeke, W.; Gellynck, X. Emotional and sensory profiling of insect-, plant- and meat-based burgers under blind, expected and informed conditions. *Food Qual. Prefer.* **2016**, *52*, 27–31. [CrossRef]
5. Meyer-Rochow, V.B. Can insects help to ease the problem of world food shortage? *Search* **1975**, *6*, 261–262.
6. Han, S.R.; Lee, B.S.; Jung, K.J.; Yu, H.J.; Yun, E.Y.; Sam, J.H.; Wang, J.S.; Moon, K.S. Safety assessment of freeze-dried powdered *Tenebrio molitor* larvae (yellow mealworm) as novel food source: Evaluation of 90-day toxicity in Sprague-Dawley rats. *Regul. Toxicol. Pharm.* **2016**, *77*, 206–212. [CrossRef]
7. Sun-Waterhouse, D.; Waterhouse, G.I.N.; You, L.J.; Zhang, J.N.; Liu, Y.; Ma, L.; Gao, J.; Dong, Y. Transforming insect biomass into consumer wellness foods: A review. *Food Res. Int.* **2016**, *89*, 129–151. [CrossRef]
8. Grabowski, N.T.; Klein, G. Microbiology of cooked and dried edible Mediterranean field crickets (*Gryllus bimaculatus*) and superworms (*Zophobas atratus*) submitted to four different heating treatments. *Food Sci. Technol. Int.* **2017**, *23*, 17–23. [CrossRef]
9. Mutungi, C.; Irungu, F.G.; Nduko, J.; Mutua, F.; Affognon, H.; Nakimbugwe, D.; Ekesi, S.; Fiaboe, K.K.M. Postharvest processes of edible insects in Africa: A review of processing methods, and the implications for nutrition, safety, and new products development. *Crit. Rev. Food Sci.* **2017**, *59*, 276–298. [CrossRef]
10. Zhang, J.S.; Zhou, X.Y.; Cai, M.M.; Zheng, L.Y.; Yu, Z.N.; Zhang, J.B. Research and application of edible insects. *Biot. Resour.* **2018**, *40*, 232–239. (In Chinese)
11. Zhang, J.L.; Wang, J.; Ren, G.X.; Gao, M.X. Function of insect protein preparation method and development status of related foods. *Farm Prod. Process.* **2019**, *8*, 75–77. (In Chinese)
12. Zou, S.W. *History of Chinese Entomology*; Science Press: Beijing, China, 1981; pp. 1–186. (In Chinese)
13. Cantrill, R.C.; Huang, Y.S.; Ells, G.W.; Horrobin, D.F. Comparison of the metabolism of α-linolenic acid and its Δ6 desaturation product, stearidonic acid, in cultured NIH-3T3 cells. *Lipids* **1993**, *28*, 163–166. [CrossRef]
14. Tian, H.; Zhang, Y.M. The nutritional components analysis and evaluation of *Clanis bilineata tsingtauica* Mell. *Acta Nutr. Sin.* **2012**, *34*, 289–291. (In Chinese)
15. Guo, M.M.; Liao, H.J.; Deng, P.; Li, D.W.; Li, J.L.; Zhang, J.Q.; Fan, J.W.; Chen, F. Effects of soybean varieties (lines) and planting density on survival and development of *Clanis bilineata tsingtauica* Mell larvae. *J. Environ. Entomol.* **2020**, *42*, 1401–1408. (In Chinese)
16. Ghosh, S.; Lee, S.M.; Jung, C.; Meyer-Rochow, V.B. Nutritional composition of five commercial edible insects in South Korea. *J. Asia Pac. Entomol.* **2017**, *20*, 686–694. [CrossRef]
17. Zhu, H.F. *Chinese Zoology Insecta Volume 11 Lepidoptera Sphingidae*; Science Press: Beijing, China, 1997; pp. 234–236. (In Chinese)
18. Lü, F.; Liu, Y.S.; Wang, Z.P.; Zhang, X.B. Advances in production and comprehensive utilization of *Clanis bilineata tsingtauica*. *Entomol. J. East China* **2006**, *15*, 192–195. (In Chinese)
19. Hoffmann, A.A.; Sorensen, J.G.; Loeschcke, V. Adaptation of Drosophila to temperature extremes: Bringing together quantitative and molecular approaches. *J. Therm. Biol.* **2003**, *28*, 175–216. [CrossRef]
20. Nelson, W.A.; Bjornstad, O.N.; Yamanaka, T. Recurrent insect outbreaks caused by temperature-driven changes in system stability. *Science* **2013**, *341*, 796–799. [CrossRef]
21. Ouyang, F.; and Ge, F. Methodology of measuring and analyzing insects cold hardiness. *Chin. J. Appl. Entomol.* **2014**, *51*, 1646–1652. (In Chinese)
22. Kerr, J.T.; Pindar, A.; Galpern, P.; Packer, L.; Potts, S.G.; Roberts, S.M.; Rasmont, P.; Schweiger, O.; Colla, S.R.; Richardson, L.L.; et al. Climate change impacts on bumblebees converge across continents. *Science* **2015**, *349*, 177–180. [CrossRef]
23. Garcia-Robledo, C.; Kuprewicz, E.K.; Staines, C.L.; Erwin, T.L.; Kress, W.J. Limited tolerance by insects to high temperatures across tropical elevational gradients and the implications of global warming for extinction. *Proc. Natl. Acad. Sci. USA* **2016**, *113*, 680–685. [CrossRef]
24. Song, J.; Cao, S.S.; Lu, M.X.; Du, Y.Z. Mortality and HSP Genes Expression in the Endoparasitoid *Cotesia chilonis* (Hymenoptera: Braconidae) After Cold Acclimation at Different Temperatures. *Ann. Entomol. Soc. Am.* **2020**, *113*, 171–175. [CrossRef]
25. Guz, N.; Toprak, U.; Dageri, A.; Gurkan, M.O.; Denlinger, D.L. Identification of a putative antifreeze protein gene that is highly expressed during preparation for winter in the sunn pest, *Eurygaster maura*. *J. Insect Physiol.* **2014**, *68*, 30–50. [CrossRef] [PubMed]
26. Shi, C.H.; Hu, J.R.; Li, C.R.; Zhang, Y.J. Research progress in the cold tolerance mechanism of insects under environmental stress. *Plant Prot.* **2016**, *42*, 21–28. (In Chinese)
27. Dai, C.G.; Zhang, C.R.; Li, H.B.; Hu, Y. Impacts of low-temperature refrigeration on development of *Chilo suppressalis* (Walker) at egg and pupal stage. *China Plant Prot.* **2018**, *38*, 17–20. (In Chinese)
28. Mao, G.; Lu, X.; Zhou, S.X.; Li, L.J.; Zhang, G.H.; Sun, K.N.; Li, G.X.; Chang, X.; Ding, Y. Influence of cold storage on the quality of *Corcyra cephalonica* eggs for rearing at egg and pupal stage. *J. Environ. Entomol.* **2019**, *41*, 1118–1124. (In Chinese)

29. Meng, T.L. Effetcs of Egg Density and Low-Temperature Refrigeration on *Carposine sasakii* Growth and Development. Master's Thesis, Shanxi Agricultural University, Taigu, China, 2017.
30. Feng, Y.Y.; Ma, G.C.; Ji, Q.A.; Lü, B.Q.; Peng, Z.Q.; Jin, T.; Wei, H.B. Developmental duration and life table of the laboratory populations of *Clanis bilineata tsingtauica* (Lepidoptera: Sphingidae) at different temperatures. *Plant Prot.* **2014**, *40*, 80–83. (In Chinese)
31. Duman, J.G.; Bennett, T.; Sformo, T.; Hochstrasser, R.; Barnes, B.M. Antifreeze proteins in Alaskan insects and spiders. *J. Insect Physiol.* **2004**, *50*, 259–266. [CrossRef]
32. Liu, Y.; Denlinger, D.L.; Piermarini, P.M. The diapause program impacts renal excretion and molecular expression of aquaporins in the northern house mosquito, *Culex pipiens*. *J. Insect Physiol.* **2016**, *98*, 141–148.
33. Lu, M.X.; Cao, S.S.; Du, Y.Z.; Liu, Z.X.; Liu, P.Y.; Li, J.Y. Diapause, signal and molecular characteristics of overwintering *Chilo suppressalis* (Insecta: Lepidoptera: Pyralidae). *Sci. Rep.* **2013**, *3*, 3211. [CrossRef]

 *foods*

Article

Effects of Defatting Methods on the Physicochemical Properties of Proteins Extracted from *Hermetia illucens* Larvae

Tae-Kyung Kim [1], Jae-Hoon Lee [1], Hae In Yong [1], Min-Cheoul Kang [1], Ji Yoon Cha [1], Ji Yeon Chun [2] and Yun-Sang Choi [1,*]

[1] Research Group of Food Processing, Korea Food Research Institute, Wanju 55365, Korea; kimtaekyung@kfri.re.kr (T.-K.K.); leejaehoon@kfri.re.kr (J.-H.L.); awsm_y@kfri.re.kr (H.I.Y.); mckang@kfri.re.kr (M.-C.K.); chajiyoon@kfri.re.kr (J.Y.C.)

[2] Department of Food Bioengineering, Jeju National University, Jeju 63243, Korea; chunjiyeon@jejunu.ac.kr

* Correspondence: kcys0517@kfri.re.kr; Tel.: +82-63-219-9387

Abstract: In this study, we investigated the effects of various defatting methods, including organic solvent (aqueous, acetone, ethanol, and hexane) extraction and physical (cold pressure) extraction, on the nutritional, physicochemical, and functional properties of proteins extracted from *Hermetia illucens* larvae. The total essential amino acid contents were higher with cold pressure protein extraction than other treatments. The surface hydrophobicity with cold pressure treatment was the lowest, and there were no significant differences among the other treatments. The protein solubility after defatting with organic solvent was higher than for other treatments. The nonreduced protein band at 50 kDa of the defatted protein prepared using organic solvent was fainter than in the cold pressure treatment. The cold pressure-defatted protein showed the highest emulsifying capacity, and the water extracted protein showed the lowest emulsifying capacity. Although organic solvents may be efficient for defatting proteins extracted from insects, organic solvents have detrimental effects on the human body. In addition, the organic solvent extraction method requires a considerable amount of time for lipid extraction. Based on our results, using cold pressure protein extraction on edible insect proteins is ecofriendly and economical due to the reduced degreasing time and its potential industrial applications.

Keywords: insect protein; cold pressure; protein characteristics; functional properties

Citation: Kim, T.-K.; Lee, J.-H.; Yong, H.I.; Kang, M.-C.; Cha, J.Y.; Chun, J.Y.; Choi, Y.-S. Effects of Defatting Methods on the Physicochemical Properties of Proteins Extracted from *Hermetia illucens* Larvae. *Foods* **2022**, *11*, 1400. https://doi.org/10.3390/foods11101400

Academic Editors: Victor Benno Meyer-Rochow and Chuleui Jung

Received: 9 March 2022
Accepted: 9 May 2022
Published: 12 May 2022

Publisher's Note: MDPI stays neutral with regard to jurisdictional claims in published maps and institutional affiliations.

1. Introduction

Globally, the insect industry has recently been in the spotlight because of it has a very high potential for development [1]. Insects are already being used as a substitute for meat in many countries and insects can be used as a source of high-quality protein [2,3]. Insect-derived protein is easier to digest and absorb because it has a lower molecular weight than conventional meat [4]. However, the use of insects as food is limited because of the stigma of consuming them in Western society [2,5,6]. Therefore, converting edible insects into a powder or extract form is necessary to increase marketability and consumer acceptance [7]. Several studies have reported that powder or extract proteins derived from dried edible insects can be added to nutritionally enhance food [2,8–10]. However, lipid oxidation caused by 20–30 g lipid/100 g edible insects can affect the rancid odor of products [11]. Thus, a defatting process is required to remove edible insect lipids.

Organic solvents, such as ethanol, acetone, and hexane, have been used to remove fat components [11]. Mishyna et al. [12] reported that the protein extracted from edible insects used n-hexane as the defatting solvent to improve functional properties. Hexane is widely used in the food industry as an extraction solvent; however, residual hexane can be problematic. Zhao et al. [13] also defatted yellow mealworm protein using ethanol as a solvent. However, this use of organic solvents is costly and results in environmental or health problems, owing to residual solvents [14]. Despite these harmful effects, organic solvent-based defatting has only minimal effects on the functional properties of proteins [1,4]. Therefore, alternative defatting methods need to be developed.

Cold pressure extraction is an extraction method that does not require heat or chemical treatment to obtain natural and safe products [14]. Cold pressure extraction is used to extract or remove vegetable oil and can inhibit protein denaturation by extracting lipids at controlled temperatures to inhibit lipid oxidation [14,15]. The use of cold pressure instead of degreasing using organic solvents should suppress the denaturation of insect proteins and thereby improve their functional properties.

Therefore, the objective of this study was to investigate the physicochemical properties and rheological properties of defatted proteins from *Hermetia illucens* larvae. Furthermore, we attempted to establish an ecofriendly protein extraction process for insects by defatting using cold pressure extraction without an organic solvent.

2. Materials and Methods

2.1. Materials

Ten kilograms of frozen *H. illucens* at the second instar stage were obtained in triplicate from a Real Nature Farm (Jeju, Korea). The larvae were fed on formulas of feed mixtures obtained from bio-waste and grown in cement boxes, and starved before freezing at $-20\,^{\circ}\text{C}$. Obtained frozen larvae were stored at $-20\,^{\circ}\text{C}$ and defatting was carried out within a week after obtained frozen larvae. Acetone, ethanol, and hexane were obtained from Sigma-Aldrich Chemical Co. (St. Louis, MO, USA).

2.2. Defatting, Processing, and Protein Extraction

All extraction processes were conducted on frozen and ground larvae. Chemical defatting processes were performed as described by Kim et al. [16]. For defatting, ground larvae (200 g) and a 99% organic solvent (1000 mL acetone, ethanol, or n-hexane) were stirred for 1 h at 20 °C. After discarding the organic solvent that contained insect fat, the same process was repeated five times until a clear solvent was obtained. The residual solvent was volatilized in a fume hood for 12 h at 20 °C. Cold pressure (Cold press-30; National Engineering, Goyang, Korea) was used for defatting at a feed rate of 4.18879 rad/s with temperatures of 80, 80, and 70 °C for the upper, middle, and lower layers, respectively. The oil cake thickness of cold pressure was fixed at 0.5 mm. Extracted fat components were discarded, and the residual sample was obtained to extract protein (46.2 °C). For protein extraction, 100 g defatted sample and 200 mL distilled water were homogenized at 1047.197551 rad/s for 2 min and centrifuged at 15,000× *g* for 30 min. The supernatant was filtered using a 500 µm pore-sized sieve, frozen at $-20\,^{\circ}\text{C}$, then freeze-dried using a freeze-dryer (Ilshinbiobase Co., Dongducheon, Korea). A water extract group was used as the control, and protein extraction was performed similarly to the treatment groups but without defatting. All extracted protein powder was stored at $-20\,^{\circ}\text{C}$ before the experiments within a week.

2.3. Amino Acid and Essential Amino Acid Index

The amino acid composition of the protein powder extracted from larvae was determined according to Kim et al. [17]. An L-8800 amino acid analyzer (Hitachi, Tokyo, Japan; ion-exchange resin column (4.6 mm inner diameter × 60 mm)) was used to determine the amino acid composition. The essential amino acid index was measured by calculating the essential amino acid index (EAAI; FAO/WHO/UNU, 1985).

2.4. Surface Hydrophobicity

The surface hydrophobicity of the defatted insect proteins was estimated using the method of Chelh et al. [18]. Insect powder (0.5 g) was dissolved in 100 mL phosphate buffer (0.025 M, pH 7.4, 4 °C), 1 mL of the sample, and 200 µL of 0.1 mg/100 mL bromophenol blue (Bio-Rad Laboratories, Hercules, CA, USA). This was mixed and reacted at 20 °C for 10 min. A mixture of 100 mL phosphate buffer and 200 µL bromophenol blue was prepared as a blank. All reacted samples and blanks were centrifuged at 19,613.3 m/s^2 for 15 min. The absorbance of 10-fold diluted solution in phosphate buffer was estimated at 595 nm using an Optizen 2120 UV plus UV/VIS spectrophotometer (Mecasys Co. Ltd., Daejeon, Korea). Bromophenol blue-bound protein (µg), which can show surface hydrophobicity, was calculated using the following formula:

$$\text{Bromophenol blue bound (µg)} = \frac{\text{absorbance of control at 595 nm} - \text{absorbance of sample at 595 nm}}{\text{absorbance of control at 595 nm}} \times 200 \text{ µg} \quad (1)$$

2.5. Protein Solubility

The protein solubility of the proteins extracted from larvae was determined following the methods of Kim et al. (2019). Briefly, freeze-dried protein from larvae was diluted with distilled water (pH 6.86), and the concentration was adjusted to 10 mg/mL. Then, the protein solubility of the extract was measured using Bradford reagent (Sigma-Aldrich, St. Louis, MO, USA).

2.6. Sodium Dodecyl Sulfate–Polyacrylamide Gel Electrophoresis (SDS-PAGE)

The molecular weight distribution of insect proteins was estimated using SDS-PAGE [19]. After adjusting the protein concentration to 1 mg/mL in sample buffer, 20 µL of solution was poured into Mini-PROTEIN TGX Gels (Bio-Rad Laboratories, Hercules, CA, USA) after heating at 100 °C for 5 min. Precision Plus Protein dual-color standards (Bio-Rad Laboratories) were used as standard markers. After running at 80 mA, protein bands were stained with Coomassie Brilliant Blue R 250 (Bio-Rad Laboratories). Samples were reduced using 2-mercaptoethanol.

2.7. pH Measurements

The pH of protein extracted from defatted *H. illucens* larvae was measured using a model 340 pH meter (Mettler-Toledo GmbH, Schwerzenbach, Switzerland), and pH 4, 7, and 10 buffers (Mettler-Toledo GmbH) were used as pH standard buffers.

2.8. Color Measurements

The color values of insect protein solutions were estimated using a CR-410 colorimeter (Minolta, Tokyo, Japan) attached to a CR-A50 granular attachment model (Minolta, Tokyo, Japan) according to the manufacturer's instructions. Color values were presented using the International Commission on Illumination (CIE) L* a* b*color values. The illumination source was D65, and the observer degree was 2. The instrument was calibrated using a calibration plate (Y = 87.1, x = 0.3166, y = 0.3338). According to the CIE76 standard, the color difference (ΔE) was calculated, and the water extract was used as the standard.

2.9. Foam Capacity and Stability

After adjusting the protein concentration to 1 mg/mL in distilled water, foaming capacity and foam were determined according to the method of Mishyna et al. [12]. Ten milliliters of the sample were poured into a 50 mL conical tube and homogenized at 1256.637061 rad/s for 2 min. To estimate the foaming capacity, the foam volume was immediately compared with the initial volume of the solution. Changes in foam volume were recorded after 2, 5, 10, 20, 30, and 60 min and were calculated as mL/100 mL to estimate the foam stability.

2.10. Emulsion Capacity and Emulsion Stability

The emulsion capacity and emulsion stability were determined as previously described by Pearce and Kinsella [20] with modifications. For emulsion capacity, 1 mL olive oil was added to 10 mL protein that was extracted from defatted *H. illucens* larvae, and homogenized at 1884.955592 rad/s for 2 min. The homogenized sample was allowed to stand for 10 min, and the emulsified layer and initial volume were compared and expressed as a percentage.

To determine the emulsion stability of the protein extracted from defatted *H. illucens* larvae, a 50 µL emulsion was added to 10 mL SDS solution (0.3 mg/100 mL) and inverted several times. Changes in the absorption of each emulsion were detected at a wavelength of 500 nm. The emulsion stability was calculated by dividing the absorbance after the interval time by the initial absorbance and multiplying by 100. The interval times were 10, 20, 30, 40, 50, 60, 90, and 120 min.

2.11. Statistical Analysis

Statistical Package for Social Sciences 20 software (SPSS Inc., Chicago, IL, USA) was used to analyze the data statistically. A one-way analysis of variance with the Duncan's range test was performed ($p < 0.05$). Fixed effects were considered when the effects of defatting methods were compared. All experiments were performed in triplicate, and protein extraction from *H. illucens* larvae was evaluated for each replicate. Replicates were considered as random effects.

3. Results and Discussion

3.1. Amino Acid Composition and Essential Amino Acid Index

When calculating the crude protein content of insects, the conventional nitrogen conversion factor (6.25) should be modified because of nonprotein nitrogen compounds, such as chitin [21]. Protein extraction could enhance the factor value by decreasing chitin content, and different amino acid components have been observed in previous studies by using different defatting methods and insect species [1,17]. Therefore, the amino acid composition should be estimated to obtain the nutritional value in insects. The amino acid composition of the protein extracted from *H. illucens* larvae using various defatting methods is shown in Table 1. Among essential amino acids, the contents of His, Ile, Leu, Phe + Tyr, and Val were highest when the cold pressure protein extraction method was used ($p < 0.05$), whereas that of Lys was highest when the water extraction method was used ($p < 0.05$). In addition, the total essential amino acid contents were highest when cold pressure was used for the defatting proteins that were extracted from *H. illucens* larvae ($p < 0.05$).

The total essential amino acid was lowest in the hexane treatment ($p < 0.05$). Essential amino acids cannot be synthesized by the organism and must be obtained through the diet. Thus, essential amino acid intake is important for human nutrition, and their recommended intake has been investigated and widely used [22]. For nonessential amino acids, the contents of Als, Arg, Asp, and Gly were higher in the cold pressure treatment group than in the group treated with organic solvent ($p < 0.05$). However, the total nonessential amino acid contents were highest in the group treated with hexane ($p < 0.05$). These different amino acid compositions may be related to the chemical or physical denaturation of insect proteins. Protein denaturation could affect protein solubility during drying, defatting, and extraction. The organic solvent, pH condition, or heating denatured the insect proteins [1,23]. These changes in protein structure can affect protein characteristics, and changes in protein solubility can affect amino acid profile [1]. However, according to Queiroz et al. [24], the unfolding temperature of extracted protein from *H. illucens* ranged from 110 to 140 °C. Therefore, protein cold pressure treatment might not denature proteins, or result in a small amount of protein being denatured by cold pressure instrument.

Although the protein contents in the extracts were similar, their nutritional value could differ, and cold pressure treatment yielded the highest nutritional value among all treatments ($p < 0.05$). Protein hydrophobicity depends on the hydrophobic amino acid composition exposed at the surface. The balance of hydrophilic or hydrophobic amino acids can affect protein functionality, such as foaming and emulsifying properties [23]. In this study, the composition of hydrophilic and hydrophobic amino acids in all treatments were compared, and the most similar ratio was observed under cold pressure treatments (data not shown). Therefore, cold pressure may have good nutritional and functional qualities.

Table 1. Effects of different defatting methods on amino acid profiles and essential amino acid index for proteins extracted from *Hermetia illucens* larvae.

	Water Extract	Acetone	Ethanol	Hexane	Cold Pressure	Reference Value [1]
	Essential amino acids					
His	44.87 ± 1.16 [bc]	44.06 ± 0.01 [c]	45.78 ± 0.23 [b]	44.56 ± 0.47 [bc]	56.02 ± 0.44 [a]	15
Ile	27.17 ± 0.01 [b]	26.89 ± 0.10 [b]	26.63 ± 0.56 [bc]	25.86 ± 0.12 [c]	28.82 ± 0.64 [a]	30
Leu	38.77 ± 0.57 [b]	38.05 ± 0.45 [bc]	36.80 ± 0.79 [c]	36.84 ± 0.10 [c]	41.16 ± 0.53 [a]	59
Lys	95.09 ± 0.31 [a]	87.93 ± 0.02 [d]	88.65 ± 0.28 [c]	91.86 ± 0.26 [b]	86.48 ± 0.27 [e]	45
Met + Cys	11.69 ± 0.06 [c]	12.09 ± 0.02 [b]	12.10 ± 0.01 [b]	12.57 ± 0.03 [a]	10.21 ± 0.06 [d]	22
Phe + Tyr	99.50 ± 0.35 [b]	98.27 ± 0.40 [bc]	97.00 ± 0.40 [c]	93.68 ± 1.40 [d]	108.30 ± 0.13 [a]	38
Thr	35.31 ± 0.18 [c]	35.68 ± 0.17 [c]	38.13 ± 0.58 [a]	36.61 ± 0.03 [b]	36.61 ± 0.11 [b]	23
Val	37.02 ± 0.47 [bc]	37.57 ± 0.34 [b]	36.67 ± 0.20 [c]	36.85 ± 0.04 [bc]	42.08 ± 0.26 [a]	39
Sum of EAA	389.4 ± 0.34 [b]	380.51 ± 0.26 [cd]	381.74 ± 1.29 [c]	378.8 ± 0.69 [d]	409.64 ± 1.95 [a]	271
	Nonessential amino acids					
Ala	123.81 ± 0.23 [b]	123.98 ± 0.69 [b]	117.3 ± 0.17 [c]	123.17 ± 0.18 [b]	135.78 ± 0.47 [a]	
Arg	47.12 ± 0.09 [d]	47.67 ± 0.21 [cd]	49.15 ± 0.51 [b]	47.94 ± 0.17 [c]	52.58 ± 0.09 [a]	
Asp	85.55 ± 0.36 [c]	88.64 ± 0.71 [b]	92.76 ± 0.11 [a]	89.36 ± 0.07 [b]	91.95 ± 0.09 [a]	
Glu	185.79 ± 0.83 [c]	191.50 ± 1.20 [b]	196.44 ± 0.81 [a]	193.12 ± 0.10 [b]	155.88 ± 0.60 [d]	
Pro	73.94 ± 1.52 [a]	73.16 ± 3.38 [a]	64.25 ± 0.67 [b]	73.53 ± 0.60 [a]	62.94 ± 3.09 [b]	
Gly	48.86 ± 0.05 [b]	49.34 ± 0.42 [b]	50.47 ± 0.03 [a]	49.04 ± 0.14 [b]	52.11 ± 0.09 [a]	
Ser	45.56 ± 0.36 [b]	45.24 ± 0.84 [b]	47.94 ± 0.8 [a]	45.07 ± 0.37 [b]	39.16 ± 0.02 [c]	
Sum of nonessential amino acids	610.61 ± 0.34 [c]	619.5 ± 0.26 [ab]	618.27 ± 1.29 [b]	621.21 ± 0.69 [a]	590.37 ± 1.95 [d]	
Essential amino acid index	1.26 ± 0.01 [b]	1.25 ± 0.01 [bc]	1.25 ± 0.01 [bc]	1.24 ± 0.01 [c]	1.31 ± 0.01 [a]	1

All values are means ± standard deviations of three replicates (n = 3). The units for amino acids are mg amino acid/1 g protein. [a–d] Means within a row with different letters are significantly different ($p < 0.05$). Hydrophobic amino acids are presented in bold text. [1] Reference values of essential amino acid requirements for humans were from FAO/WHO/UNU (1985).

3.2. Surface Hydrophobicity, Protein Solubility, pH, and Color

Structural characteristics, protein solubility, and pH should be considered to improve protein techno-functionality [23]. The effects of the defatting method on protein characteristics, such as surface hydrophobicity, protein solubility, and pH, are shown in Table 2. The surface hydrophobicity was lowest for the samples treated with cold pressure ($p < 0.05$), and there were no significant differences among the other samples ($p > 0.05$). Buried hydrophobic residues can be exposed to protein denaturation, affecting the tension of the protein film [25]. Therefore, higher hydrophobicity can result in higher protein functionalities because of changes in the interfacial or surface tension [23]. However, if the surface hydrophobicity is greater than a certain degree, binding forces between proteins can be stronger, and unpredictable protein aggregation can occur [1]. Organic solvents can change the structural characteristics of the protein, affecting surface hydrophobicity and solubility [26]. Therefore, the protein solubility obtained after the defatting treatment with organic solvents was higher than the other treatments ($p < 0.05$). However, the water extract had similar surface hydrophobicity values when compared with the organic solvents treatments ($p > 0.05$). This may be because of insufficient defatting when compared with the other treatments. Kim et al. [1] also reported that the surface hydrophobicity of aqueous extracts was higher than the other methods because of the high-fat content in the extract.

Table 2. Effects of different defatting methods on the surface hydrophobicity, protein solubility, pH, and color characteristics of an extracted protein solution from *Hermetia illucens* larvae.

	Water Extract	Acetone	Ethanol	Hexane	Cold Pressure
Surface hydrophobicity (Bromophenol blue bound, μg)	35.69 ± 3.79 [a]	33.90 ± 6.84 [a]	30.58 ± 6.47 [a]	37.45 ± 2.80 [a]	11.65 ± 1.75 [b]
Protein solubility (mg/mL)	41.60 ± 0.17 [d]	52.42 ± 0.74 [a]	52.80 ± 0.78 [a]	50.40 ± 0.29 [b]	44.19 ± 0.02 [c]
pH	7.14 ± 0.01 [ab]	7.13 ± 0.01 [b]	7.14 ± 0.01 [ab]	7.14 ± 0.00 [a]	7.12 ± 0.01 [c]
CIE L*	16.67 ± 0.01 [c]	16.37 ± 0.01 [d]	16.83 ± 0.02 [c]	18.01 ± 0.01 [a]	17.24 ± 0.31 [b]
CIE a*	1.86 ± 0.06 [d]	1.82 ± 0.06 [d]	1.95 ± 0.08 [c]	2.22 ± 0.02 [a]	2.09 ± 0.09 [b]
CIE b*	4.07 ± 0.04 [b]	3.98 ± 0.03 [c]	3.85 ± 0.06 [d]	4.14 ± 0.03 [a]	3.87 ± 0.03 [d]
Color difference (ΔE)	- [1]	0.32 ± 0.03 [c]	0.31 ± 0.06 [c]	1.39 ± 0.01 [a]	0.65 ± 0.31 [b]

All values are means ± standard deviations of three replicates (n = 3). [a–d] Means within a row with different letters are significantly different ($p < 0.05$). [1] The color value of water extract was used as a standard to compare color differences between other treatments, and the color difference was calculated according to the CIE76 ΔE formula $(\Delta L^{*2} + \Delta a^{*2} + \Delta b^{*2})^{1/2}$.

Generally, protein functionality is increased away from the isoelectric point of a protein, which is typically close to 5.0 for insect proteins [12,23]. Therefore, changes in pH during extraction should be controlled, and differences in pH values could explain the different functions of the proteins [27]. Although there was a significant difference between cold pressure and the other methods ($p < 0.05$; Table 2), the difference was very small (0.02 points). Therefore, protein functionality differences among treatments caused by pH may be minimized.

Color pigments such as melanin can be affected by various conditions during the extraction of insect proteins [1]. In this study, CIE L*, CIE a*, and CIE b* values, as well as color differences of the hexane treatment, were higher than for the other treatments ($p < 0.05$). In a related study, a protein obtained from *Protaetia brevitarsis* was affected by organic solvents, and the highest color difference was also observed in the hexane treatment [1]. As shown in Table 2, the color difference for the hexane treatment only exceeded one, and that of the others was lower than one. The human eye can detect color differences when the color difference value between samples exceeds one [28]. Therefore, although some significant differences were detected in our statistical analysis of the treatments, consumers could not detect differences among treatments, except for the hexane treatment.

3.3. SDS-PAGE

The molecular weight distributions of insect proteins are presented in Figure 1. Because it was difficult to separate parts from tiny insects, the various proteins in insects, such as muscle tissues, organs, and hemolymph, were extracted without separation in this study [29]. Similar patterns of protein bands were observed but with varying intensities (Figure 1). When compared to the control, the protein band at 50 kDa was fainter for the nonreducing organic solvent and cold pressure treatment (Figure 1a). The interfacial films of proteins can be affected by protein molecular weight, and the formation of high-molecular-weight protein polymers could help achieve good protein functionalities [23]. Disulfide bonds can occur via protein denaturation, which can be observed through a reducing method using mercaptoethanol [19]. Figure 1b shows that all treatments had a stronger intensity at 37 kDa, and fainter bands were observed above 100 kDa after reduction. Protein resistance to denaturation may differ depending on the protein source and can be affected by chemical reagents, temperature, pH, and processing method [30]. Protein denaturation during extraction induces changes in protein solubility owing to structural changes [23]. According to the data shown in Table 1, the amino acid composition differed according to the defatting method used, and these different compositions may be correlated with variations in the molecular weight distributions of the proteins. Overall, high-molecular-weight proteins with high resistance to denaturation could be extracted after removing fat using cold pressure, and this could help enhance protein functionalities.

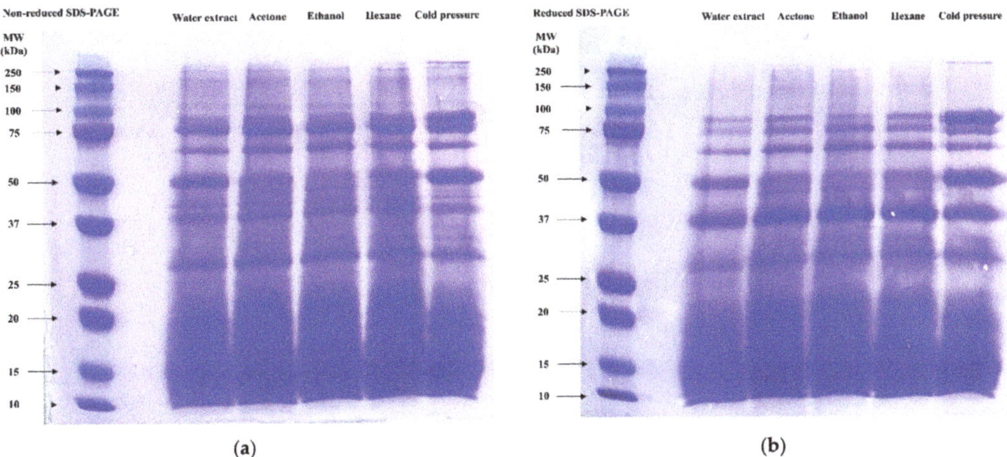

Figure 1. Effects of different defatting methods on sodium dodecyl sulfate-polyacrylamide gel electrophoresis (SDS-PAGE) of extracted nonreduced (**a**) and reduced (**b**) proteins from *Hermetia illucens* larvae.

3.4. Foaming Capacity and Foam Stability

Foaming properties were assessed to measure foaming capacity and foam stability. As shown in Figure 2a, there were no significant differences among treatments in foaming capacity ($p > 0.05$). Foaming capacity depends on protein solubility, surface flexibility, and hydrophobicity [12]. Although the protein solubility of the control and cold pressure groups was lower than that of the other groups, the large amount of high-molecular-weight proteins in these extracts could support the high foaming capacity, because thick films can physically hold their structure better than thin films [23]. However, the foams collapsed rapidly, and all treatments lost over 90 mL/100 mL of foam after 10 min (Figure 2b). Among treatments, the foam collapse speed of the water extract was the fastest (loss of >85 mL/100 mL of foam after 2 min; $p < 0.05$). Mishyna et al. [12] reported that the dominance of low-molecular-weight (5–30 kDa) components, such as proteins and polysaccharides, in insects could enhance foam stability. As shown in Figure 1, most protein bands were concentrated below 37 kDa. However, because the water extract was prepared without any defatting step, the remaining fat components may have inhibited the formation of a stable foam structure [1].

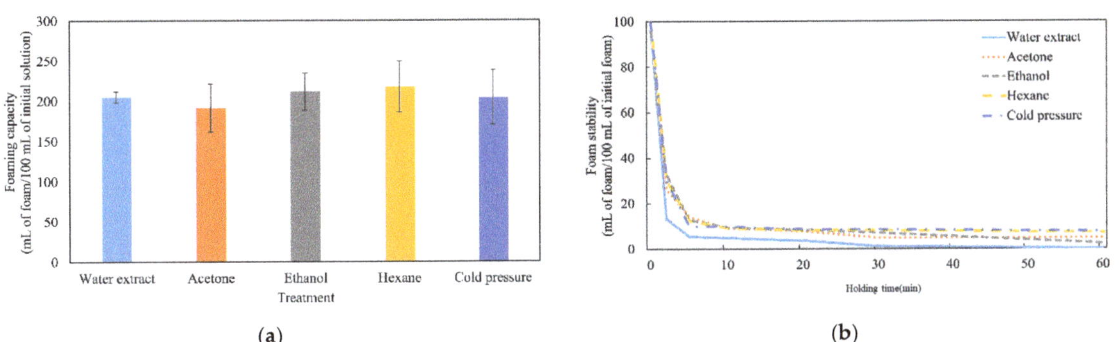

Figure 2. Effects of different defatting methods on foaming capacity (**a**) and foam stability (**b**) on the extracted protein solutions from *Hermetia illucens* larvae. Control and treatments (acetone, ethanol, hexane, cold pressure) are indicated using different line types and colors (——, ..., — — —, — — —, — ·· —).

3.5. Emulsifying Capacity and Emulsion Stability

Emulsifying capacity plays an important role in forming the frame and texture of food [17]. The emulsifying capacity and emulsion stability results are shown in Figure 3. Among the protein samples, cold pressure-defatted protein showed the highest emulsifying capacity ($p < 0.05$), and the water extract protein showed the lowest emulsifying capacity ($p < 0.05$). Mishyna et al. [12] reported that excessively hydrophobic proteins have strong binding forces between proteins and are unsuitable for emulsion formation. In addition, the cold pressure-defatted proteins had a lower surface hydrophobicity than the other defatted proteins and the water extract. Thus, the cold pressure-defatted proteins had the highest emulsifying capacity.

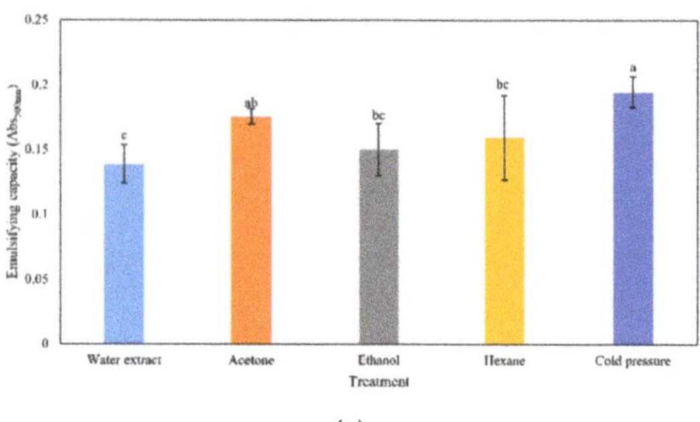

(a)

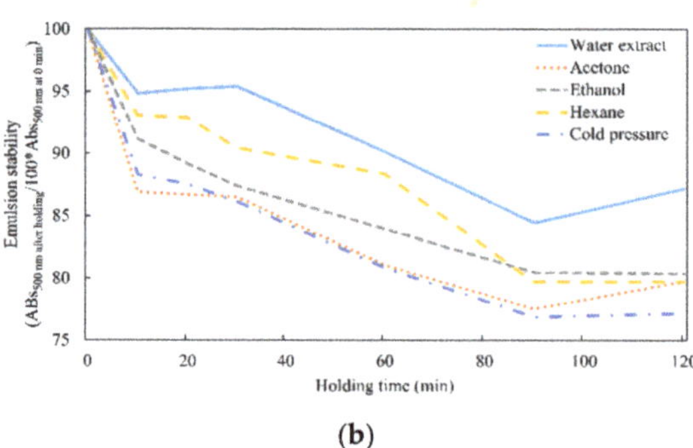

(b)

Figure 3. Effects of different defatting methods on the emulsifying capacity (**a**) and emulsion stability (**b**) of extracted proteins from Hermetia illucens larvae. Control and treatments (acetone, ethanol, hexane, cold pressure) are indicated using different line types and colors (——, ·········, -----, — — —, — · —). Different letters on top of columns indicated significant differences among treatments ($p < 0.05$).

Furthermore, the electrophoresis results (Figure 1) showed there were no significant changes in the protein composition when different defatting methods were used. Damodaran [31] reported that low-molecular-weight proteins are more effective at forming emulsions. In addition, other studies have reported that low-molecular-weight emulsifiers show better emulsion capacity [32,33]. In this study, however, the abundance of low-molecular-weight proteins was high, and this dispersion may have affected the emulsifying properties of the solution. The emulsion stability results are shown in Figure 3b. At 120 min, protein samples that were defatted using acetone, ethanol, and hexane extraction and cold pressure treatment showed a lower emulsion stability than the protein samples subjected to water extraction. During the defatting process, the interaction between proteins is weakened, and the interfacial tension is decreased, reducing emulsion stability [34,35].

4. Conclusions

The effects of various defatting methods, such as organic solvent extraction (aqueous, acetone, ethanol, and hexane) and physical extraction (cold pressure) on the amino acid profiles, physicochemical characteristics, and functional properties of proteins extracted from *H. illucens* larvae were investigated. The cold pressure extraction method with nonorganic solvent extraction showed excellent characteristics in terms of the nutritional properties of insect protein extracts. The physicochemical characteristics and functional properties of the insect protein treatment group that underwent extraction with an organic solvent were high. However, there were no significant differences when compared with the cold pressure extraction group. Organic solvents can have detrimental effects on the human body when ingested. Thus, cold pressure could be used at the industrial level when extracting edible insect protein as an ecofriendly extraction method. In addition, the organic solvent extraction method requires more time for lipid extraction; however, the cold pressure method has the advantage of shortening the extraction time. Therefore, the cold pressure method could be used as an appropriate and effective defatting method. Based on these findings, this approach may have industrial applications as a defatting process in the preparation of edible insect proteins.

Author Contributions: Conceptualization, T.-K.K. and Y.-S.C.; methodology, T.-K.K., J.-H.L., H.I.Y. and M.-C.K.; formal analysis, T.-K.K., J.-H.L., H.I.Y., J.Y.C. (Ji Yoon Cha) and Y.-S.C.; investigation, Y.-S.C.; data curation, J.Y.C. (Ji Yeon Chun) Y.-S.C.; writing-original draft preparation, T.-K.K. and Y.-S.C.; writing-review and editing, T.-K.K. and Y.-S.C.; supervision, J.Y.C. (Ji Yeon Chun) and Y.-S.C. All authors have read and agreed to the published version of the manuscript.

Funding: This research was supported by the Main Research Program (E21200-02) of the Korea Food Research Institute (KFRI), funded by the Ministry of Science and ICT (Korea). In addition, this research was also partially supported by the Livestock Industrialization Technology Development Program (321079-3) by the Ministry of Agriculture, Food, and Rural Affairs (Republic of Korea).

Institutional Review Board Statement: Not applicable.

Informed Consent Statement: Not applicable.

Data Availability Statement: Data are contained within the article.

Conflicts of Interest: The authors declare no conflict of interest.

References

1. Kim, T.K.; Yong, H.I.; Kim, Y.B.; Jung, S.; Kim, H.W.; Choi, Y.S. Effects of organic solvent on functional properties of defatted proteins extracted from Protaetia brevitarsis larvae. *Food Chem.* **2021**, *336*, 127679. [CrossRef]
2. Kim, T.K.; Yong, H.I.; Kim, Y.B.; Kim, H.W.; Choi, Y.S. Edible insects as a protein source: A review of public perception, processing technology, and research trends. *Food Sci. Anim. Resour.* **2019**, *39*, 521–540. [CrossRef] [PubMed]
3. Ham, Y.-K.; Kim, S.-W.; Song, D.-H.; Kim, H.-W.; Kim, I.-S. Nutritional composition of white-spotted flower chafer (*Protaetia brevitarsis*) larvae produced from commercial insect farms in Korea. *Food Sci. Anim. Resour.* **2021**, *41*, 416–427. [CrossRef] [PubMed]
4. Lee, J.H.; Kim, T.-K.; Jeong, C.H.; Yong, H.I.; Cha, J.Y.; Kim, B.-K.; Choi, Y.-S. Biological activity and processing technologies of edible insects: A review. *Food Sci. Biotechnol.* **2021**, *30*, 1003–1023. [CrossRef] [PubMed]
5. Kim, T.-K.; Cha, J.Y.; Yong, H.I.; Jang, H.W.; Jung, S.; Choi, Y.-S. Application of edible insects as novel protein sources and strategies for improving their processing. *Food Sci. Anim. Resour.* **2022**, *42*, 372–388. [CrossRef]
6. Ghosh, S.; Jung, C.; Meyer-Rochow, V.B. What Governs Selection and Acceptance of Edible Insect Species? In *Edible Insects in Sustainable Food Systems*; Springer: Cham, Switzerland, 2018; pp. 331–351.
7. Purschke, B.; Bruggen, H.; Scheibelberger, R.; Jager, H. Effect of pre-treatment and drying method on physico-chemical properties and dry fractionation behaviour of mealworm larvae (*Tenebrio molitor* L.). *Eur. Food Res. Technol.* **2018**, *244*, 269–280. [CrossRef]
8. da Rosa Machado, C.; Thys, R.C.S. Cricket powder (*Gryllus assimilis*) as a new alternative protein source for gluten-free breads. *Innov. Food Sci. Emerg. Technol.* **2019**, *56*, 102180. [CrossRef]
9. Barsics, F.; Megido, R.C.; Brostaux, Y.; Barsics, C.; Blecker, C.; Haubruge, E.; Francis, F. Could new information influence attitudes to foods supplemented with edible insects? *Br. Food J.* **2017**, *119*, 2027–2039. [CrossRef]
10. Castro, M.; Chambers, E. Willingness to eat an insect based product and impact on brand equity: A global perspective. *J. Sens. Stud.* **2019**, *34*, e12486. [CrossRef]
11. Jeong, M.-S.; Lee, S.-D.; Cho, S.-J. Effect of three defatting solvents on the techno-functional properties of an edible insect (*Gryllus bimaculatus*) protein concentrate. *Molecules* **2021**, *26*, 5307. [CrossRef]
12. Mishyna, M.; Martinez, J.I.; Chen, J.; Benjamin, O. Extraction, characterization and functional properties of soluble proteins from edible grasshopper (*Schistocerca gregaria*) and honey bee (*Apis mellifera*). *Food Res. Int.* **2019**, *116*, 697–706. [CrossRef] [PubMed]
13. Zhao, X.; Vázquez-Gutiérrez, J.L.; Johansson, D.P.; Landberg, R.; Langton, M. Yellow mealworm protein for food purposes-Extraction and functional properties. *PLoS ONE* **2016**, *11*, e0147791. [CrossRef] [PubMed]
14. Russin, T.A.; Boye, J.I.; Arcand, Y.; Rajamohamed, S.H. Alternative techniques for defatting soy: A practical review. *Food Bioprocess Technol.* **2011**, *4*, 200–223. [CrossRef]
15. Melo, D.; Álvarez-Ortí, M.; Nunes, M.A.; Costa, A.S.; Machado, S.; Alves, R.C.; Pardo, J.E.; Oliveira, M.B.P. Whole or defatted sesame seeds (*Sesamum indicum* L.)? The effect of cold pressing on oil and cake quality. *Foods* **2021**, *10*, 2108. [CrossRef] [PubMed]
16. Kim, T.-K.; Yong, H.I.; Kang, M.-C.; Jung, S.; Jang, H.W.; Choi, Y.-S. Effects of high hydrostatic pressure on technical functional properties of edible insect protein. *Food Sci. Anim. Resour.* **2021**, *41*, 185–195. [CrossRef] [PubMed]
17. Kim, T.K.; Yong, H.I.; Chun, H.H.; Lee, M.A.; Kim, Y.B.; Choi, Y.S. Changes of amino acid composition and protein technical functionality of edible insects by extracting steps. *J. Asia-Pac. Entomol.* **2020**, *23*, 298–305. [CrossRef]
18. Chelh, I.; Gatellier, P.; Sante-Lhoutellier, V. Technical note: A simplified procedure for myofibril hydrophobicity determination. *Meat Sci.* **2006**, *74*, 681–683. [CrossRef]

19. Laemmli, U. SDS-page Laemmli method. *Nature* **1970**, *227*, 680–685. [CrossRef]
20. Pearce, K.N.; Kinsella, J.E. Emulsifying properties of proteins: Evaluation of a turbidimetric technique. *J. Agric. Food Chem.* **1978**, *26*, 716–723. [CrossRef]
21. Janssen, R.H.; Vincken, J.P.; van den Broek, L.A.; Fogliano, V.; Lakemond, C.M. Nitrogen-to-protein conversion factors for three edible insects: *Tenebrio molitor, Alphitobius diaperinus*, and *Hermetia illucens*. *J. Agric. Food Chem.* **2017**, *65*, 2275–2278. [CrossRef]
22. FAO; WHO; UNU. *Energy and Protein Requirements: Report of a Joint FAO/WHO/UNU Expert Consultation*; Food and Agriculture Organization; World Health Organization; The United Nations University: Geneva, Switzerland, 1985; p. 206.
23. Zayas, J.F. *Functionality of Proteins in Food*; Springer Science & Business Media: Berlin/Heidelberg, Germany, 2012.
24. Queiroz, L.S.; Regnard, M.; Jessen, F.; Mohammadifar, M.A.; Sloth, J.J.; Petersen, H.O.; Ajalloueian, F.; Brouzes, C.M.C.; Fraihi, W.; Fallquist, H.; et al. Physico-chemical and colloidal properties of protein extracted from black soldier fly (*Hermetia illucens*) larvae. *Int. J. Biol. Macromol.* **2021**, *186*, 714–723. [CrossRef] [PubMed]
25. Jiang, L.; Wang, Z.; Li, Y.; Meng, X.; Sui, X.; Qi, B.; Zhou, L. Relationship between surface hydrophobicity and structure of soy protein isolate subjected to different ionic strength. *Int. J. Food Prop.* **2015**, *18*, 1059–1074. [CrossRef]
26. L'Hocine, L.; Boye, J.I.; Arcand, Y. Composition and functional properties of soy protein isolates prepared using alternative defatting and extraction procedures. *J. Food Sci.* **2006**, *71*, C137–C145. [CrossRef]
27. Salaün, F.; Mietton, B.; Gaucheron, F. Buffering capacity of dairy products. *Int. Dairy J.* **2005**, *15*, 95–109. [CrossRef]
28. Mokrzycki, W.; Tatol, M. Colour differenceΔ E-A survey. *Mach. Graph. Vis.* **2011**, *20*, 383–411.
29. Williams, J.; Williams, J.; Kirabo, A.; Chester, D.; Peterson, M. Nutrient Content and Health Benefits of Insects. In *Insects as Sustainable Food Ingredients*; Academic press: San Diego, CA, USA, 2016; pp. 61–84.
30. De Graaf, L.A. Denaturation of proteins from a non-food perspective. *J. Biotechnol.* **2000**, *79*, 299–306. [CrossRef]
31. Damodaran, S. Protein stabilization of emulsions and foams. *J. Food Sci.* **2005**, *70*, R54–R66. [CrossRef]
32. Amagliani, L.; O'Regan, J.; Kelly, A.L.; O'Mahony, J.A. Influence of low molecular weight surfactants on the stability of model infant formula emulsions based on hydrolyzed rice protein. *LWT-Food Sci. Technol.* **2022**, *154*, 112544. [CrossRef]
33. Jiang, J.; Jin, Y.; Liang, X.; Piatko, M.; Campbell, S.; Lo, S.K.; Liu, Y. Synergetic interfacial adsorption of protein and low-molecular-weight emulsifiers in aerated emulsions. *Food Hydrocoll.* **2018**, *81*, 15–22. [CrossRef]
34. O'sullivan, J.; Murray, B.; Flynn, C.; Norton, I. The effect of ultrasound treatment on the structural, physical and emulsifying properties of animal and vegetable proteins. *Food Hydrocoll.* **2016**, *53*, 141–154. [CrossRef]
35. Li, F.; Wang, B.; Kong, B.; Shi, S.; Xia, X. Decreased gelling properties of protein in mirror carp (*Cyprinus carpio*) are due to protein aggregation and structure deterioration when subjected to freeze-thaw cycles. *Food Hydrocoll.* **2019**, *97*, 105223. [CrossRef]

 *foods*

Article

Nutritional, Techno-Functional and Structural Properties of Black Soldier Fly (*Hermetia illucens*) Larvae Flours and Protein Concentrates

Vusi Vincent Mshayisa [1,*], Jessy Van Wyk [1] and Bongisiwe Zozo [2]

[1] Department of Food Science and Technology, Cape Peninsula University of Technology, Bellville 7535, South Africa; vanwykj@cput.ac.za
[2] Department of Chemistry, Cape Peninsula University of Technology, Bellville 7535, South Africa; 214050645@mycput.ac.za
* Correspondence: mshayisav@cput.ac.za

Abstract: Due to their protein content and balanced amino acid profile, edible insects have been described as an excellent alternative protein source to combat malnutrition. As the global population continues to grow, edible insects such as the black soldier fly larvae (BSFL) may contribute to food security. The effect of different protein extraction methods, i.e., alkaline solution and acid precipitation (BSFL-PC1) and extraction with an alkali (BSFL-PC2), on the nutritional, techno-functional, and structural properties of BSFL flours and protein concentrates were studied. The highest protein content (73.35%) was obtained under alkaline and acid precipitation extraction (BSFL-PC1). The sum of essential amino acids significantly increased ($p < 0.05$) from 24.98% to 38.20% due to the defatting process during extraction. Protein solubility was significantly higher in protein concentrates (85–97%) than flours (30–35%) at pH 2. The emulsion capacity (EC) was significantly higher ($p < 0.05$) in the protein concentrates (BSFL-PC1 and BSFL-PC2) compared to the freeze-dried and defatted BSFL flours, while the emulsion stability (ES) was significantly ($p < 0.05$) higher in BSFL-PC1 (100%) compared with BSFL-PC2 (49.8%). No significant differences ($p > 0.05$) were observed in foaming stability (FS) between freeze-dried and defatted BSFL flours. Fourier transform infrared spectroscopy (FT-IR) analysis revealed distinct structural differences between BSFL flours and protein concentrates. This was supported by surface morphology through scanning electron microscopy (SEM) images, which showed that the protein extraction method influenced the structural properties of the protein concentrates. Therefore, based on the nutritional and techno-functional properties, BSFL flour fractions and protein concentrates show promise as novel functional ingredients for use in food applications.

Keywords: functional properties; black soldier fly larvae; protein extraction; foaming; insect protein; novel food ingredients

Citation: Mshayisa, V.V.; Van Wyk, J.; Zozo, B. Nutritional, Techno-Functional and Structural Properties of Black Soldier Fly (*Hermetia illucens*) Larvae Flours and Protein Concentrates. *Foods* **2022**, *11*, 724. https://doi.org/10.3390/foods11050724

Academic Editors: Victor Benno Meyer-Rochow and Chuleui Jung

Received: 8 February 2022
Accepted: 25 February 2022
Published: 28 February 2022

Publisher's Note: MDPI stays neutral with regard to jurisdictional claims in published maps and institutional affiliations.

1. Introduction

The United Nations (UN) has predicted that the world population will increase from seven to nine billion people by the year 2030 [1], and with this increase, about 60% of people are expected to migrate and live in cities [2]. Animal protein is expensive and is becoming beyond the reach of many people, especially in developing countries. To feed this growing population, a paradigm shift towards producing sustainable and cost-effective food products is required now more than ever before. Entomophagy, or the consumption of insects, has been practised by humankind on every continent in the world throughout history and continues today. However, the advancement of the scientific impetus for large-scale insect rearing, production, and utilisation began in 1975 with the call by Meyer-Rochow [3]. Over 2000 species have been deemed edible since 2012 in 116 countries, and there has been an increase in industrial insect rearing companies in recent years [4,5]. Large industrial-scale insect farming companies include AgriProtein in South Africa, Ynsect in France, Enviroflight in Ohio, USA, and HaoCheng Mealworms Inc. in China [6]. Edible insects have recently been suggested as a food source that could address economic, environmental, and health concerns as the population expands. Due to the emerging demand in edible insect consumption, the industrial rearing of edible insects such as the black soldier fly larvae (*Hermetia illucens*, hereinafter referred to as BSFL) has been expanding due to its stable supply, cost-effectiveness, and hygienic production. Currently, edible insects are primarily marketed as whole insects, ground pastes or flours, protein powders, and oil fractions which can further be used as ingredients in food applications [2,7,8]. Edible insects are a good source of minerals and vitamins and fat, and, most importantly, protein. According to La Barbera et al. [9], consumers may be willing to consume insects when they are added as an ingredient in an unrecognisable form. The food industry's further processing of insects as an alternative protein source and acceptance depends mainly on their ability to fulfil tailored techno-functional properties required in food systems. The potential use of an insect ingredient as a techno-functional food ingredient is dependent on chemical, physical, techno-functional, and structural properties [10].

An ingredient's techno-functionality has been described as any food property, excluding its nutritional value, affecting its utilisation [11,12]. Emulsification, solubility, water and oil binding, foam capacity and stability, gelation, and viscosity are among the techno-functional properties vital in food processing. Yi et al. [13] studied five different acid-extracted insect proteins and reported that these exhibited poor foaming and gelling. Protein-enriched fractions of *Schistocerca gregaria* (*S. gregaria*) and *Apis mellifera* (*A. mellifera*) were reported to have a higher foam stability after alkaline and sonication extraction, respectively. Moreover, Akpossan et al. [14] demonstrated that *Imbrasia oyemensis* (*I. oyemensis*) protein fractions from full fat and defatted flours possessed poor solubility due to their isoelectric point. However, both studied flours exhibited good water absorption and emulsification characteristics. According to Omotoso [15], the dried powders of *Cirina forda* (*C. forda*) possessed good emulsion and solubility properties. Kim et al. [7] evaluated the effects of added defatted *Tenebrio molitor* (*T. molitor*) and *Bombyx mori* (*B. mori*) in sausages. Including edible insect flours increased cooking yield and firmness in the emulsion-based sausages. For the successful application of insect-derived ingredients in food formulations, it is essential to understand their nutritional and techno-functional properties and how processing affects them.

To date, an in-depth analysis of the nutritional value of commercially available *H. illucens* available in South Africa has not been reported. For the possible use of *H. illucens* flours and protein concentrates as foodstuffs, information on the nutritional value and techno-functional properties are extremely essential. Therefore, the aim of this study was to determine the effect of different extraction methods on the nutritional, techno-functional, and structural properties of BSFL flour and protein concentrates in order to identify new protein sources for human nutrition.

2. Materials and Methods

2.1. Chemicals

If not stated otherwise, all chemicals were purchased from either Merck (Modderfontein, South Africa) or Sigma-Aldrich (Aston Manor, South Africa). All the chemicals used in this study were of analytical grade, and chemical reagents were prepared according to standard analytical laboratory procedures. Ultrapure water purified with a Milli-Q water purification system was used throughout the experiments conducted (Millipore, Microsep, Cape Town, South Africa). The study was approved by the faculty of applied sciences ethics committee (Ref: 208176519/01/20220 date: 4 February 2020).

2.2. Edible Insects Flour Preparation

Previously fasted *Hermetia illucens* (black soldier fly—BSF) in the larval stage were obtained from AgriProtein, Cape Town, South Africa. They were immediately blanched in boiling water for two minutes to prevent browning and washed with cold water (Figure 1). To obtain a paste, the clean larvae were frozen at −75 °C in a blast freezer prior to grinding in a laboratory blender. This paste was then freeze-dried (Genesis, Virtis, New York, NY, USA) to obtain a stable powder. Some of the freeze-dried (BSFL-FD) powder was ground in a laboratory blender (Kenwood, Titanium, South Africa) and was passed through a 40-mesh US Standard sieve (Endecotts, London, UK) to separate the integument and stored at −20 °C until further analysis. The sieved BSFL flour was defatted using hexane and isopropanol (3:2) mixture. One part of the insect flour and five parts of defatting solvent were stirred in a magnetic stirrer for six hours. After the solids were sedimented, the solvent-fat mixture was decanted, and the procedure was repeated twice. Residual hexane was removed by evaporation overnight in a fume hood, and the defatted flour (BSFL-DF) was stored at −20 °C until further analysis.

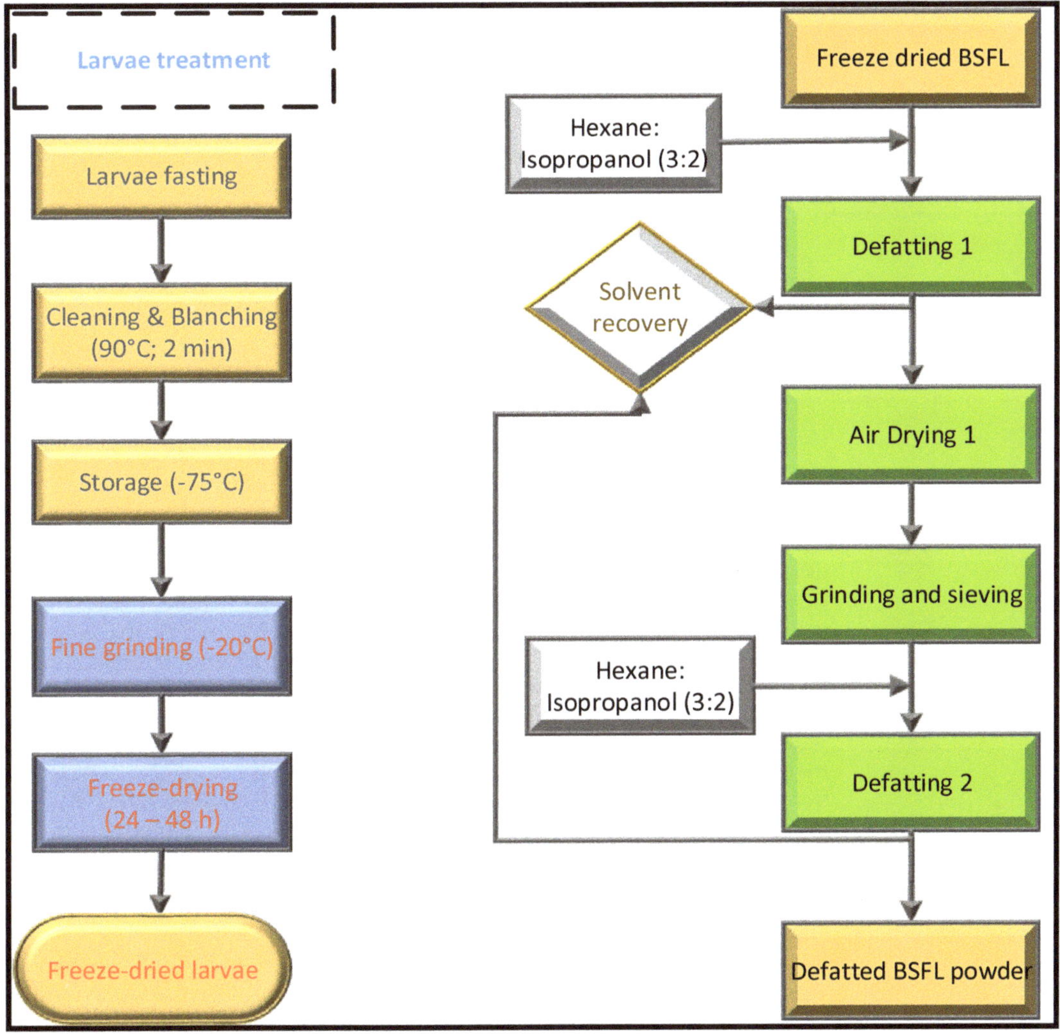

Figure 1. BSFL pre-treatment and defatting.

2.3. Preparation of Insect Proteins

This study used two chemical techniques, alkaline solution and isoelectric precipitation (IEP) and alkaline extraction, to extract protein concentrates from defatted BSFL flour (Figure 2).

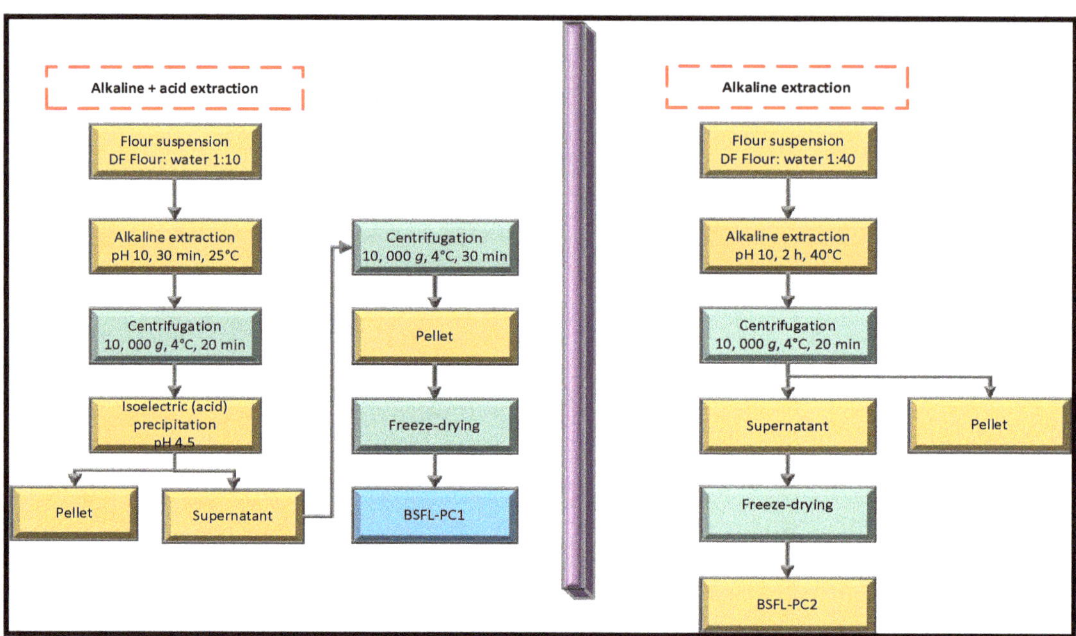

Figure 2. Alkaline and acid precipitation extraction (BSFL-PC1) and alkaline extraction (BSFL-PC2) of protein concentrates.

2.3.1. Alkaline and Isoelectric Precipitation

The BSFL protein concentrate (BSFL-PC1) was prepared as previously described [16]. Briefly, after defatting, BSFL flour (BSFL-DF) was mixed with Milli-Q water at a ratio of 1:10 (w/w), and the pH of the mixture was adjusted to pH 10 using 1 M NaOH solution (Figure 2). With the aid of a magnetic stirrer, the slurry was stirred for 30 min (25 °C $\pm$ 1.0) on a laboratory mixer at a rate intended to prevent the formation of a vortex. Subsequently, the slurry was centrifuged at 10,000$\times$ g for 20 min (4 °C). The pH of the supernatant was then adjusted to 4.5 with 1 M HCl, and the suspension was left at 4 °C overnight to allow protein precipitation. Centrifugation at 10,000$\times$ g for 30 min (4 °C) was used to recover the precipitated soluble proteins. The BSFL protein concentrate (BSFL-PC1) was then freeze-dried (0.094 mBar, −74.5 °C, 48 h, Genesis, Virtis, New York, NY, USA) and stored in vacuum bags at −20 °C.

2.3.2. Alkaline Extraction

Following the method of Azagoh et al. [10], with some modifications, BSFL-PC2 was extracted by mixing defatted BSFL flour with Milli-Q water at a ratio of 1:40 (w/w), and the pH of the mixture was adjusted initially to pH 10 using 1 M N NaOH solution (Figure 2). Subsequently, the slurry was stirred at a rate designed to prevent the formation of a vortex for 2 h at 40 °C. The pH was monitored intermittently and maintained at 10 throughout the stirring period. The mixture was centrifuged at 10,000$\times$ g for 30 min (4 °C). The BSFL protein concentrate (BSFL-PC2) supernatants were freeze-dried under similar experimental conditions with BSFL-PC1.

2.4. Proximate Composition Analysis

The proximate composition, i.e., the moisture (925.10), crude protein (920.87), crude fat (932.06), and ash content (923.03) of the insect flour and protein concentrates, was determined following standard methods recommended by the Association of Official Analytical Chemists [17]. Moisture percentage was calculated by drying the sample in an oven at 100 °C for two hours. The dried sample was placed into a desiccator, allowed to cool, and then re-weighed. The process was repeated until a constant weight was obtained. Crude protein content was analysed by the high-temperature combustion process according to the Dumas combustion method (TruSpec-N Leco, St. Joseph, MI, USA) using a protein-to-nitrogen conversion factor of 5.60 as recommended by Janssen et al. [18]. EDTA was used as a standard. Crude fat was determined by drying fats after extraction in a Soxhlet assembly using diethyl ether. The ash percentage was calculated by combusting the samples in a silica crucible placed in a muffle furnace at 550 °C. The percentage of carbohydrates on a dry basis was determined by subtracting all of the components (moisture, crude protein, crude lipid, and ash) from 100.

2.5. Amino Acid Analysis

Five hundred milligrams of each insect flour and protein concentrate were hydrolysed with 6 mL of 6 N HCl at 110 °C for 23 h. Then the internal standard (7.5 mL of 5 mM norleucine in water) was added. The hydrolysed samples were analysed by High-Performance Liquid Chromatography with a fluorescence detector (HPLC/FLD, Waters Alliance 2695, Agilent technologies, Chemetrix, Midrand, South Africa) after derivatisation with 6-aminoquinolyl-N-hydro-xysuccinimidyl carbamate (AQC). The amino acid profiles of insect flour and protein concentrates were compared with data in the literature on egg white and cow's milk proteins used for human consumption. The FAO/WHO method was used to calculate the amino acid score (AAS) as shown below:

$$AAS = \frac{\text{mg of AA in 1 g of test protein}}{\text{mg of AA in 1 g of the FAO/WHO reference pattern}} \times 100$$

2.6. Bulk Density

Bulk density was measured as a ratio of mass to volume. A graduated cylinder, previously tared, was gently filled up to the ten mL mark with BSFL flour or protein. The sample was then packed by gently tapping the cylinder on the bench-top from a height of five cm until there was no further diminution of the sample level, and the volume was noted. The weight of the filled cylinder was taken, and the bulk density was calculated as the weight of sample per unit volume (g/mL).

2.7. Colour

Colour was determined by the method as described by Diedericks and Jideani [19] and Bußler et al. [20], with minor modifications. The colour of the insect flours and protein concentrates was measured with a spectrophotometer (Konica Minolta Sensing Americas Inc., Tokyo, Japan) using the CIEL* a* b* colour space system. The instrument was calibrated by using the white calibration plate followed by zero calibration. Powdered samples were placed evenly in the provided cuvette (diameter 30 mm), covering the bottom of the dish, to allow for reflectance measurement. Measurements for each sample were performed in triplicate at 3 different positions in the samples (one reading = average of 3 readings per rotated position), with the results recorded in L* (lightness), a* (+a* = red and −a* = green), b* (+b* = yellow and −b* = blue). The colour change (ΔE) was calculated, whereas the indices 0 and s indicate measured values of unprocessed (larvae) and processed insects (flour fractions), respectively.

$$\Delta E = \sqrt{(L_0 - L_s)^2 + (a_0 - a_s)^2 + (b_0 - b_s)^2}$$

where: L—is L* (lightness), a—a* (difference in green and red), b—b* (difference in yellow and blue), o—indicates measured values of BSFL-FD, s—indicates measured values of processed flour or protein concentrates.

2.8. Determination of Techno-Functional Properties

2.8.1. Water and Oil Binding Capacity

The water-binding capacity (WBC) was determined according to Purschke et al. [21], with slight modifications. Briefly, 0.5 g insect flour or protein concentrate powder was mixed with 2.5 mL Milli-Q water at an ambient temperature (25 °C), vortexed for 1 min (Vortex-Genie 2, Scientific Industry Inc., Bohemia, NY, USA), and centrifuged for 2 min at $3330\times g$. The non-bound water was decanted, and the tube was placed upside down on a Whatman No 1 filter paper for one hour and re-weighed. WBC was calculated as:

$$\mathrm{WBC} = \frac{W_b - W_a}{W_{a,DM}}$$

where: W_a is the initial weight, W_b is the final weight, and $W_{a,DM}$ is the initial weight of the sample based on dry matter. The oil binding capacity (OBC) was analysed using commercial sunflower oil instead of Milli-Q water. Except for the vortexing step (2 min), the experimental procedure was identical to the WBC assay, and OBC was calculated similarly.

2.8.2. Emulsion Capacity and Stability

Emulsifying properties were determined based on the method of Coelho and Salas-Mellado [22] and Zielińska et al. [23] with minor revisions. Briefly, the edible insect sample was dispersed in distilled water (5% w/v) and centrifuged (Thermo Electron Corporation Jouan MR1812, Waltham, MA USA) at $9000\times g$ speed for 15 min. The supernatant was mixed with sunflower oil (1:1 v/v) and homogenised (Polytron PT 2500 E, Thermofisher, Cape Town, South Africa) at 18,000 rpm for 1 min. Aliquots (15 mL) of the emulsion samples were centrifuged at $3000\times g$ for five min, and the volume of the individual layers was read using a 50 mL scaled tube. The emulsifying activity was calculated as:

$$\text{Emulsion capacity (EC)} = \frac{V_{el}}{V} \times 100$$

The emulsion was heated in a water bath (80 °C for 30 min) to determine the stability values. Emulsion stability was calculated as:

$$\text{Emulsion stability (EC)} = \frac{V_{30}}{V_{el}} \times 100$$

where: V—total volume of tube contents, V_{el}—volume of the emulsified layer, V_{30}—volume of the emulsified layer after heating.

2.8.3. Solubility

The solubility of the insect flours and protein concentrates was determined using a modified version of the method by Hall et al. [24]. Each sample (400 mg) was dispersed in 20 mL phosphate buffers of pH 2–10, respectively. Each buffer mixture was stirred with a magnetic stirrer bar at room temperature for 30 min and centrifuged at $7500\times g$ for 20 min at 4 °C. The protein content of the supernatant and total protein in the sample was determined using the bicinchoninic acid protein assay (BCA) method with bovine serum albumin as a standard, following the manufacturer's protocol (Sigma, St. Louis, MO, USA). Protein solubility was expressed as a percentage and calculated as follows:

$$\text{Solubility (ES)} = \frac{\text{protein content in supernatant}}{\text{total protein content in sample}} \times 100$$

2.8.4. Foaming Capacity and Stability

The foaming properties of the edible insects were measured according to Mshayisa and Van Wyk [16] with slight modifications. The edible insect and protein concentrates were rehydrated in Milli-Q water (5% w/v) followed by centrifugation at 10,000 rpm for 15 min. Then 20 mL of the supernatant was homogenised in a high shear homogeniser mixer (Polytron PT 2500E) at a speed of 16,000 rpm for three minutes. The homogenised solution was poured into a 50 mL graduated cylinder. The total volume was read at time zero and 30 min after homogenisation. The foaming capacity and foam stability were calculated from the formula:

$$\text{Foaming Capacity (FC)} = \frac{V_a - V}{V} \times 100$$

$$\text{Foam Stability (FS)} = \frac{V_{30}}{V_a} \times 100$$

where: V—volume before whipping (mL), V_a—volume after whipping (mL), V_{30}—volume after standing (mL).

2.9. Surface Charge (Zeta Potential)

The ζ-Potential of the protein concentrates (BSFL-PC1 and BSFL-PC2) were determined using a Zetasizer Nano Series (Malvern Instruments, Malvern, Worcestershire, UK) as described by Ladjal-Ettoumi et al. [25] and Mshayisa et al. [26], respectively. The protein powders were diluted to 1% (w/v) with Milli-Q water. The Smoluchowski model was used to process data collected over least five of the experiments [27].

2.10. Scanning Electron Microscopy (SEM)

Scanning electron microscopy (TM-3000, Hitachi Corporation, Tokyo, Japan) was used to study the surface morphology of the BSFL flours and protein concentrates. The edible insect powders were placed onto double-sided carbon adhesive tape attached to the specimen stubs, respectively. The sample's surface structure was observed at 320× magnification and in the secondary electron mode at 15.0 kV following the procedure described by Mshayisa et al. [26].

2.11. Fourier Transform Infrared Spectroscopy (FT-IR)

All flour and protein concentrate samples were analysed using a Perkin Elmer Fourier transform infrared spectroscope (FT-IR) equipped with a universal attenuated total reflectance (UATR) polarisation accessory for spectra as described by Mshayisa et al. [26]. A background spectrum was collected prior to data collection of each sample, and then the sample powders obtained by grinding in a mortar were placed directly, covering the surface of the ATR crystal. All spectra were acquired by the co-addition of 32 scans at a resolution of 4 cm^{-1} in the range of 400–4000 cm^{-1}. Acetone was used to clean the UATR crystal to remove any residual contribution from previous samples.

2.12. Data Analysis

All statistical tests were performed using multivariate analysis of variance (MANOVA), the level of significance was $p < 0.05$, followed by Duncan's multiple comparisons using the statistical package for social sciences (SPSS 27, SPSS, Inc., Chicago, IL, USA).

3. Results and Discussion

3.1. Nutritional Properties

The proximate compositions of freeze-dried BSFL (BSFL-FD), defatted BSFL (BSFL-DF), alkali and isoelectric precipitation BSFL protein concentrate (BSFL-PC1), and alkaline extraction protein concentrate (BSFL-PC2) are depicted in Table 1. The ash content of all samples ranged between 2.08 and 10.81%, and BSFL-FD had the highest ($p < 0.05$) content. BSFL-PC1 had the highest protein (73.35%) content, and BSFL-FD (44.47%) had the lowest protein content ($p < 0.05$), while BSFL-PC1 had a significantly ($p < 0.05$) higher protein content compared to BSFL-PC2. This signifies that the protein extraction method affected the protein content. The protein content obtained for BSFL-FD was within a similar range to that reported by Huang et al. [28] for oven-dried BSFL (40–44%). The protein contents measured in this study were much higher than those reported by Bußler et al. [20] for fresh and freeze-fried BSFL, namely, 31.7% and 34.7%, respectively. The differences in protein content can be attributed to the different feeding regimes (diets), age, and size of the black soldier fly larvae prior to analysis. Moreover, in this study, the conversion factor of 5.60, as advised by Janssen et al. [18], was used instead of 6.25, which has been reported to overestimate the protein content since edible insects contain chitin which contributes to the nitrogen content. The protein and lipid composition of BSFL is highly impacted by what they consume [29]. Edible insects reported in this study were fed on clean, standardised feed and were fasted and blanched prior to further processing. The protein content of all the fractions investigated in this study was higher than that of common food products such as cow's milk (3.5%), eggs (13%), fish (18.3%), and chicken (22%) [13,30].

Table 1. Proximate composition of BSFL flour and protein concentrates.

Sample	Crude Protein (%)	Crude Fat (%)	Carbohydrates (%)	Moisture (%)	Ash (%)
BSFL-FD	44.47 ± 1.77 [a]	22.60 ± 1.39 [b]	21.13 ± 0.57 [a]	9.48 ± 0.34 [c]	10.81 ± 0.26 [d]
BSFL-DF	50.12 ± 0.66 [b]	0.83 ± 0.06 [a]	40.80 ± 0.39 [d]	6.07 ± 0.25 [b]	8.25 ± 0.33 [c]
BSFL-PC1	73.35 ± 0.88 [d]	0.37 ± 0.12 [a]	22.92 ± 1.02 [b]	1.48 ± 0.01 [a]	3.36 ± 0.28 [b]
BSFL-PC2	68.47 ± 0.93 [c]	0.27 ± 0.06 [a]	29.19 ± 0.92 [c]	1.57 ± 0.03 [a]	2.08 ± 0.01 [a]

Freeze-dried BSFL flour—BSFL-FD, defatted BSFL flour—BSFL-DF, alkaline and acid precipitation extraction BSFL protein concentrate—BSFL-PC1, alkaline extraction BSFL protein concentrate—BSFL-PC2. Results are reported as mean ± standard deviation. Different superscripts in the column indicate significant differences between treatments ($p \leq 0.05$).

As shown in Table 1, the defatting step significantly ($p < 0.05$) decreased the crude fat present in BSFL from 22.60 to 0.83%, while protein concentrates contained 0.27–0.37% fat. The fat content of BSFL-FD (22.60%) was comparable to the data reported by Bußler et al. [20] for *H. illucens* (21.1%) and *T. molitor* (20.0%). The current practice in the food industry is geared towards the extensive use of soy protein products as non-meat proteins in processed meat products. These are classified based on their protein concentration as soy flour (50–54%), concentrated soy protein (62–69%), and isolated protein (86–87%) [7]. The results of this study clearly demonstrate that BSFL flours and protein concentrates were comparable to that of soy flour and soy concentrated protein, both of which are widely used as food ingredients commercially.

3.2. Bulk Density and Colour

The bulk density of BSFL flour and protein concentrates is shown in Table 2. Bulk density can be described as the weight of powder per unit volume (expressed as g/mL). The bulk density of BSFL flours and protein concentrates ranged from 0.83 to1.04 g/mL. No significant differences were observed between the freeze-dried and defatted BSFL flours ($p > 0.05$). Akpossan et al. [14] reported similar bulk density values in full-fat and defatted ground *I. oyemensis* flours. In terms of the protein concentrates, a significant ($p < 0.05$) reduction in bulk density was observed compared to the flour fractions, signifying that the extraction process employed affected bulk density. In assessing packaging requirements, material handling, and applications in wet manufacturing, bulk density is a critical parameter for consideration by food processors [31].

Processing of the BSFL affected the visual appearance of the BSFL flour and protein concentrates produced. The colour changes are summarised in Table 2. Defatting with the hexane:isopropanol mixture led to a significantly higher ($p < 0.05$) lightness (L) in the BSF-DF than the freeze-dried BSFL flour. Similar results were reported by Mishyna et al. [32] in *A. mellifera* defatted with hexane. In this study, the defatting process significantly increased lightness ($p < 0.05$), redness ($p < 0.05$), and yellowness ($p < 0.05$). In terms of a* (red/green), only BSFL-PC2 fell within the greener quadrant (negative quadrant), while no significant ($p > 0.05$) differences were observed in the yellowness b* (positive quadrant) of all the flour and protein concentrate samples. In this study, the ΔE was in the range 11.95–23.87, indicating perceptible colour differences at a glance; in other words, the colour changes could be clearly perceived without closer inspection. The high ΔE value for BSFL-PC2 can partly be attributed to phenolic compounds situated in the insect cuticle or integument that can undergo oxidation, protein–polyphenol interaction, and enzymatic browning catalysed by phenoloxidase [7,33]. In a study conducted by Janssen et al. [33], *H. illucens* exhibited a distinct black appearance after grinding, while *T. molitor* exhibited a more deep brown colour. Therefore, the bright colour of the defatted BSFL flours could be due to the removal of compounds such as phenoloxidase, which catalyses the browning of insect flours. However, the lightness, as well as the redness, decreased significantly after further extraction. The reaction mechanism resulting in the extracted protein fractions' dark and brown colour is still not yet well understood. Nonetheless, edible insect flours can also be used in baked goods where colour or visual appeal may not be a critical problem.

Table 2. Bulk density and colour attributes of black soldier fly larvae flours and protein concentrates.

Sample	Bulk Density (g/mL)	L *	a *	b *	ΔE
BSFL-FD	1.01 ± 0.02 [b]	51.24 ± 0.34 [c]	0.84 ± 1.34 [ab]	17.99 ± 1.70 [a]	Control
BSFL-DF	1.04 ± 0.02 [b]	62.95 ± 1.01 [d]	1.67 ± 0.12 [b]	18.43 ± 2.67 [a]	11.95 ± 1.16 [a]
BSFL-PC1	0.84 ± 0.01 [a]	44.56 ± 0.63 [b]	0.85 ± 1.35 [ab]	17.99 ± 1.70 [a]	16.16 ± 1.27 [a]
BSFL-PC2	0.86 ± 0.01 [a]	35.20 ± 1.87 [a]	−2.14 ± 3.28 [a]	18.78 ± 6.56 [a]	23.87 ± 4.16 [b]

Freeze-dried BSFL flour—BSFL-FD, defatted BSFL flour—BSFL-DF, alkaline and acid precipitation extraction BSFL protein concentrate—BSFL-PC1, alkaline extraction BSFL protein concentrate—BSFL-PC2. Results are reported as mean ± standard deviation. Different superscripts in the column indicate significant differences between treatments ($p \leq 0.05$).

3.3. Amino Acid Composition

The amino acid composition of BSFL flours and protein concentrates compared to that of cow's milk, egg, and FAO protein intake recommendations for adults are shown in Table 3. The essential amino acids, leucine, and lysine contents were predominant in BSFL-DF compared with BSFL-FD. Lysine is considered a limiting amino acid in staple cereals such as maize, cassava, rice, and wheat [12,29]. Thus, incorporating BSFL-DF (6.76% lysine) and BSFL-PC1 (9.16% lysine) in food products can serve as a source of lysine, especially in developing countries. Histidine, an essential amino acid regarded as vital for infants and toddlers, was significantly higher ($p < 0.05$) in BSFL-DF (3.84%) and BSFL-PC1 (3.64%) than that in cow's milk and egg (2.70% and 2.44%, respectively). The sum of the essential amino acids increased due to defatting from 24.98% to 38.20%. Moreover, BSFL-PC1 had a significantly ($p < 0.05$) higher sum of amino acids than BSFL-PC2. These values align with those of Leni et al. [34] and Huang et al. [28]. Among the non-essential amino acids in Table 3, glutamic acid had the highest concentration in BSFL-DF (13.2%), BSFL-PC1 (12.4%), and BSFL-PC2 (12.13%), respectively. These results are in agreement with the work conducted by Köhler et al. [35], who reported high levels of glutamic acid in whole house cricket (*Acheta domesticus*) flour. In this study, tryptophan was not detectable, possibly due to the fact that this amino acid is destroyed during acid hydrolysis or it was not present in the samples under investigation. The content of essential amino acids for BSFL flours and protein concentrates was comparable or exceeded FAO's recommended amount [36] as a basic human dietary requirement. In the case of BSFL-PC1, the sum of the essential amino acids for BSFL was more than double (45.52%) the FAO requirements. The results of this study confirm the claims of other studies that, in general, edible insects are good sources of amino acids (Table 3), proteins, and lipids (Table 1) [13,35,37–40]. The high levels of essential amino acids are particularly pertinent as they cannot be synthesised by the human body in sufficient quantities and thus should be provided by the diet. Moreover, protein functionality and bioavailability are governed by the amino acid composition and the amino acid sequence. Table 4 shows the amino acid scores of BSFL Flours and protein concentrates. Leucine was the first limiting amino acid in all the samples with the exception of BSFL-FD. In terms of the BSFL protein concentrates, lysine and valine were the second limiting amino acids for BSFL-PC1 and BSFL-PC2, respectively. The essential amino acid scores for BSFL flours and protein met FAO/WHO [41] standards for older children, adolescents, and adults.

Table 3. Amino acid composition (g.100 g^{-1}) of BSFL flour fractions and protein concentrates compared to cow's milk, egg protein, and FAO requirements for human consumption.

Amino Acid	BSFL-FD	BSFL-DF	BSFL-PC1	BSFL-PC2	Cow's Milk	Egg Protein	FAO [36]
Essential							
Histidine	1.82 ± 0.01 [a]	3.84 ± 0.01 [d]	3.64 ± 0.02 [c]	2.48 ± 0.00 [b]	2.70	2.40	1.50
Isoleucine	2.49 ± 0.01 [a]	4.32 ± 0.44 [bc]	5.18 ± 0.78 [c]	3.99 ± 0.45 [b]	4.90	5.60	3.00
Leucine	4.01 ± 0.12 [a]	7.29 ± 0.15 [c]	7.99 ± 0.29 [d]	4.61 ± 0.17 [b]	9.10	8.30	5.90
Lysine	3.73 ± 0.01 [a]	6.79 ± 0.19 [c]	9.16 ± 0.34 [d]	5.37 ± 0.20 [b]	7.40	6.30	4.5
Methionine	2.53 ± 0.00 [a]	1.72 ± 0.45 [a]	2.53 ± 0.00 [a]	2.67 ± 0.08 [a]	2.60	3.20	1.60
Phenylalanine	4.14 ± 0.01 [a]	4.36 ± 0.20 [a]	7.18 ± 0.05 [c]	5.41 ± 0.21 [b]	4.90	5.10	Not supplied
Threonine	2.87 ± 0.53 [a]	4.09 ± 0.49 [b]	4.95 ± 0.12 [bc]	4.65 ± 0.00 [b]	4.40	5.10	2.30
Valine	3.39 ± 0.01 [a]	5.80 ± 0.51 [b]	5.61 ± 0.89 [c]	5.09 ± 0.51 [b]	6.60	7.60	3.90
Sum	**24.98 ± 0.80 [a]**	**38.20 ± 2.44 [b]**	**46.52 ± 2.48 [c]**	**34.27 ± 2.19 [b]**	**42.60**	**43.60**	**22.70**
Non-essential							
Alanine	3.58 ± 0.01 [a]	5.95 ± 0.48 [c]	4.66 ± 0.04 [b]	5.41 ± 0.48 [bc]	3.60	5.40	-
Arginine	3.38 ± 0.00 [a]	4.85 ± 0.36 [b]	4.57 ± 0.02 [b]	5.72 ± 0.02 [c]	3.60	6.10	-
Aspartic acid	5.89 ± 0.01 [a]	10.38 ± 0.39 [b]	12.56 ± 0.69 [c]	9.62 ± 0.40 [a]	7.70	10.70	-
Glycine	3.08 ± 0.06 [a]	4.80 ± 0.13 [c]	3.88 ± 0.06 [b]	4.97 ± 0.15 [c]	3.00	2.00	-
Glutamic acid	6.89 ± 0.05 [a]	13.62 ± 0.05 [d]	12.13 ± 0.09 [b]	12.40 ± 0.05 [c]	12.0	20.60	-
Proline	2.57 ± 0.01 [a]	5.71 ± 0.23 [d]	4.38 ± 0.01 [c]	3.38 ± 0.24 [b]	8.50	3.80	-
Serine	2.13 ± 0.04 [a]	4.42 ± 0.10 [d]	3.99 ± 0.28 [c]	3.13 ± 0.17 [b]	5.20	7.90	-
Tyrosine	6.89 ± 0.00 [c]	6.47 ± 0.07 [b]	6.28 ± 0.15 [a]	9.34 ± 0.09 [b]	4.10	7.60	-
Sum	**34.41 ± 0.21 [a]**	**56.21 ± 1.74 [b]**	**52.45 ± 1.85 [b]**	**53.98 ± 2.00 [b]**	**55.30**	**56.50**	**-**

Freeze-dried BSFL flour—BSFL-FD, defatted BSFL flour—BSFL-DF, alkaline and acid precipitation extraction BSFL protein concentrate—BSFL-PC1, alkaline extraction BSFL protein concentrate—BSFL-PC2. Results are reported as mean ± standard deviation of triplicate analysis. Different letters indicate significant ($p < 0.05$) differences between the means across the rows.

Table 4. Amino acid score of BSFL flours and protein concentrates and the FAO/WHO/UNU (2007) consultation.

Amino Acids	FAO/WHO/UNU 2007 [41] (mg/g Protein)	Chemical Score (%)			
		BSFL-FD	BSFL-DF	BSFL-PC1	BSFL-PC2
Histidine	15	191.84	254.03	175.72	230.52
Isoleucine	30	139.86	152.62	150.66	174.96
Leucine	59	110.73 [b]	126.51 [c]	85.67 [b]	132.72 [b]
Lysine	45	130.74 [c]	149.90	126.75 [c]	193.37
Methionine	16	265.73	113.68 [b]	188.91	160.22
Threonine	23	193.08	173.10	210.92	200.63
Valine	39	142.76	153.46	144.30	142.11 [c]
Phenylalanine + tyrosine	38	488.83	301.99	439.71	358.91
Total EAA	277	151.8	146.1	140.2	169.1

[b] First limiting amino acid, [c] Second limiting amino acid, Freeze-dried BSFL flour—BSFL-FD, defatted BSFL flour—BSFL-DF, alkaline and acid precipitation extraction BSFL protein concentrate—BSFL-PC1, alkaline extraction BSFL protein concentrate—BSFL-PC2.

3.4. Techno-Functional Properties

3.4.1. Water and Oil Binding Capacity

Owing to their effects on the flavour and the textural properties of food products, water and oil interactions with proteins play a significant role in food systems [12]. Protein ingredients' water-binding capacity (WBC) and oil-binding capacity (OBC) may be influenced by intrinsic factors such as the amino acid sequence, protein conformation, hydrophobicity, and polarity. The ability of protein ingredients to interact with water under restricted conditions is expressed by its WBC. The WBCs of BSFL flour and protein fractions are displayed in Figure 3. The removal of the fat did not have a significant effect ($p < 0.05$) on the WBC of BSFL flours, whereas the BSFL-PC1 showed a higher ($p > 0.05$) WBC compared to BSFL-PC2. The differences in WBC between the protein fractions can be attributed to the differences in amino acids reported in Table 2 as influenced by the extraction method. Few studies have been conducted on the WBCs of BSFL flours and protein fractions extracted using different chemical techniques. The results of this study were higher than the WBC values described for BSFL flour fractions (0.4–0.8 g/g-) by Bußler et al. [20]. The differences in the WBC values of BSFL flours and protein concentrates can be ascribed to the differences in the methods of extraction used by the authors and the insect origin and diet. Among the BSFL flour and protein fractions, the highest WBC was observed in BSFL-PC1 (5.6 g/g). This value was comparatively higher than the WBC of other insect protein fractions reported in the literature, such as *T. molitor* (1.87 g/g) [30] and *S. gregaria* (2.18 g/g) [23]. A high water-binding capacity value for a protein allows the moistness mouthfeel and freshness of baked goods to be maintained and is correlated with a decreased moisture loss in bakery products. Thus, information about the WBC of insect-derived ingredients is essential for future application in food systems.

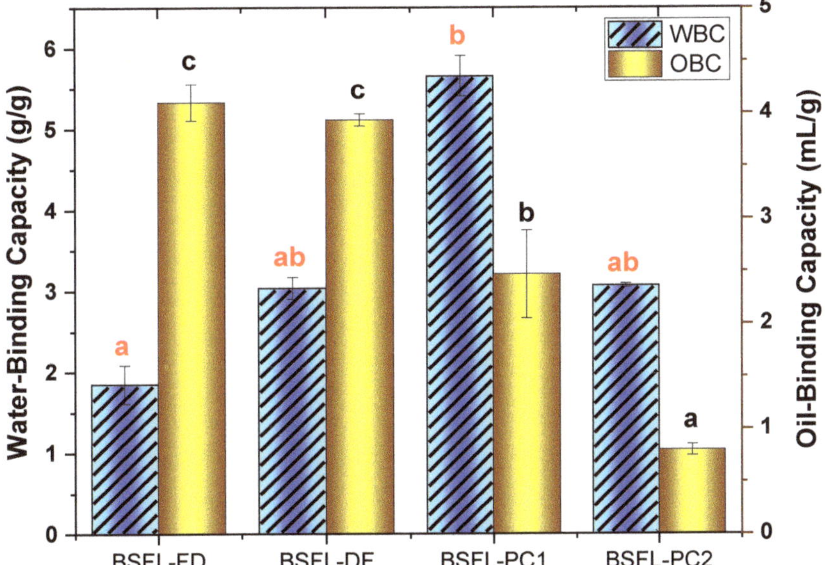

Figure 3. Water and oil binding capacity of flour and protein fractions. Freeze-dried BSFL flour—BSFL-FD, defatted BSFL flour—BSFL-DF, alkaline and acid precipitation extraction BSFL protein concentrate—BSFL-PC1, alkaline extraction BSFL protein concentrate—BSFL-PC2. Values are mean ± standard deviation; means with different superscript are significantly different ($p < 0.05$).

Another primary techno-functional attribute of food ingredients used in processed foods is the OBC. To enhance the palatability and flavour retention of foods, a high OBC is desirable [12]. Figure 3 depicts the result of the OBC. Similar to the WBC, the defatting step did not have a significant ($p > 0.05$) effect on the OBC of BSFL flours, but for the BSFL protein concentrates, it was significantly lower ($p < 0.05$). The OBC of BSFL-FD flour in this study was (4.1 mL/g). This was superior to the OBC value obtained for hydrolysed migratory locust protein flour (1.5 mL/g) [21] and *T. molitor* (1.71 mL/g) [23].

The capacity of a protein to bind oil or fat is critical in the formulation of meat substitutes and extenders, as well as cake batters, sausages, and emulsions [42]. In terms of the protein fractions, the highest ($p < 0.05$) OBC was observed in BSFL-PC1 compared to BSFL-PC2 (Figure 4). The differences in OBC are possibly due to the different surface hydrophobicity or hydrophilicity of these proteins, which may be influenced by the extraction method [43]. In this research, the WBC and OBC values showed that BSFL flour and protein concentrates could be useful for various food applications, such as enhancing the palatability and texture of formulated foods.

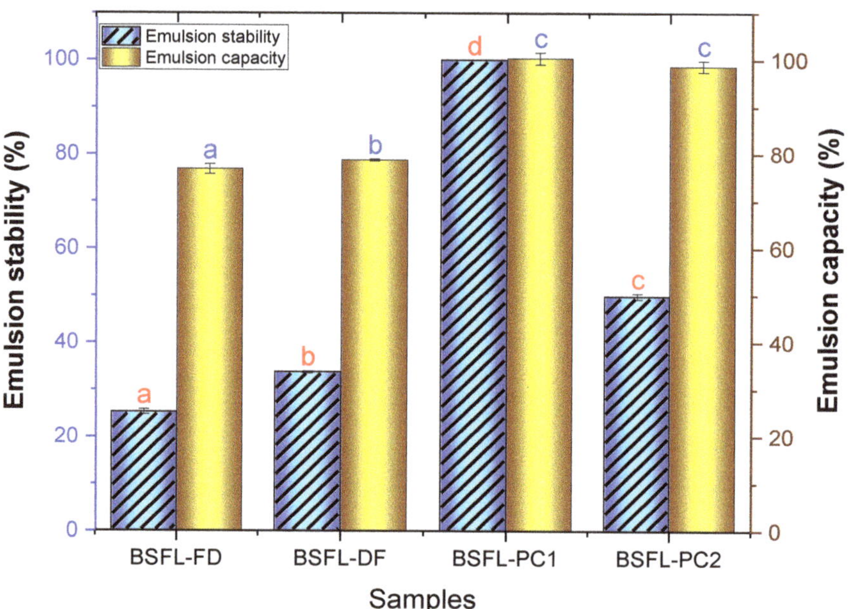

Figure 4. Emulsification capacity and stability of BSFL flour fractions and protein concentrates. Freeze-dried BSFL flour—BSFL-FD, defatted BSFL flour—BSFL-DF, alkaline and acid precipitation extraction BSFL protein concentrate—BSFL-PC1, alkaline extraction BSFL protein concentrate—BSFL-PC2. Values are mean ± standard deviation; means with different superscript are significantly different ($p < 0.05$).

3.4.2. Emulsifying Capacity (EC) and Stability (ES)

The amphiphilic nature of proteins allows them to form and stabilise food emulsions. In the development of formulated foods, emulsifying properties are among the most vital properties, and strong emulsifying properties are needed to produce meat analogues and milk-like drinks [44]. Figure 4 exhibits the emulsion capacity and stability of the BSFL flour and protein concentrates. The emulsion capacity of defatted BSFL flour (BSFL-DF, 78.73%) was significantly higher ($p < 0.05$) than the full-fat flour (BSFL-FD, 76.80%). Similar results were reported for the EC of I. oyemensis by Akpossan et al. [14], while Kim et al. [8] observed a lower EC of cricket flour (39.17–45%). In a study conducted by Mishyna et al. [32], no significant differences were observed in raw and defatted S. gregaria flours. The variations in the EC of edible insect flours could be attributed to protein content and molecular structure differences. In general, the EC of a food protein is based on the protein–oil and protein–water interactions. The highest EC was determined for the protein concentrates BSFL-PC1 (100%) and BSFL-PC2 (100%). These results show that the processes used to obtain protein concentrates did not have a significant ($p > 0.05$) impact on the ability of BSFL-PC1 and BSFL-PC2 to aid in the creation of emulsions.

In terms of ES, BSFL-FD (25.27%) formed emulsions with a significantly ($p < 0.05$) lower ES compared with BSFL-DF (33.73%), signifying that its protein did not effectively interact at the interface to form a strong interfacial membrane (Figure 4). The ES of the protein concentrates (49.83–100%) was significantly higher ($p < 0.05$) than the flour fractions (25.27–33.73%). The BSFL protein concentrate results reported in this study indicate that these novel proteins may be appropriate for use in the formulation of a wide range of processed food products. Currently, there is a paucity of studies investigating the emulsification properties of insect protein concentrates obtained using different extraction techniques (alkaline and acid precipitation and alkaline extraction), which makes the comparison of the results difficult. Proteins currently used by the food industry due to their emulsifying abilities are mainly derived from soybean, milk (whey or casein), and egg. These are widely used in various food formulations due to their commercial availability and good functional properties [45]. However, the major drawback is that they all have been identified as common food allergens. Therefore, further studies on the emulsifying properties of edible insect concentrates are required, especially using different extraction methods. Based on the findings of this study, it could be concluded that BSFL flours and protein concentrates possess emulsifying properties.

3.4.3. Protein Solubility

In order to provide knowledge on the successful use of insect-derived ingredients in different food applications, the protein solubility of edible insect flours and protein concentrates was investigated. Solubility at different pH values and the level of protein denaturation due to heat or chemical treatment serve as a measure of how well insect flours and protein concentrates can perform when integrated into food systems [46].

In general, BSFL-FD and BSFL-DF flours had a low protein solubility at pH 2–3 compared to the protein concentrates (Figure 5). The high solubility at low pH values for the protein concentrates (BSFL-PC1, 95% and BSFL-PC2, 85%) makes them ideal candidates for use in acidic beverages. The protein solubility of all BSFL samples ranged from a minimum at the isoelectric point (pI) to its maximum at pH 11 (Figure 5). For all the insect flours, the pI was in the region of pH 4.0–4.5. During the extraction process, protein solubility was highly dependent on the pH. The results of this study resemble the pI of common essential food proteins such as casein (4.6), soybean (4.5), and meat products (5.0) [12,13,46]. These results can be attributed to the reduced interaction between protein and water at pH 4.5–5.0, and this phenomenon enhances the protein–protein interaction in foods resulting in protein aggregation and precipitation [46]. The protein solubility profiles of BSFL flours and protein concentrates against pH were generally similar to each other and consistent with previously published data for insect species [14,24,47] and plant legumes such as peas, Kabuli chickpeas, and kidney beans [12,31]. Protein concentrates must be highly soluble in order to be used as functional ingredients in a wide variety of foods, including confections, salad dressings, coffee whiteners, whipped toppings, and beverages [12,32]. These findings further support the possible application of BSFL flour, and BSFL protein concentrates over a broad pH spectrum, in addition to acidic foods.

3.4.4. Foaming Capacity and Foam Stability

Food foams can be defined as air bubbles imprisoned in a liquid and stabilised by protein at the air–liquid interface. As foaming agents, proteins play a significant role in the distribution of fine air cells in the processed food structure. They are also responsible for imparting smoothness and lightness and allowing flavours to be volatilised to improve the palatability of food products [12]. The foam capacity and stability results for BSFL flour and protein concentrates are exhibited in Figure 6. The FC of BSFL-FD (40%) was not statistically different ($p > 0.05$) compared with that of BSFL-DF (55%). The foaming capacity of BSFL-PC2 (78.43%) after alkaline extraction was significantly higher ($p < 0.05$) than BSFL-PC1 (75.97%). Overall, the protein concentrates exhibited an improved foaming capacity. The high foamability may be attributed to increased protein content and the possibility of changes in protein characteristics after the extraction. The FC results of this study show that BSFL-FD and BSFL-DF flours were higher than those described by Adebowale et al. [37] for the large African cricket (*Gryllidae* sp.) that had an FC of only 6%. Previous studies on freeze-dried *A. mellifera* and *S. gregaria* conducted by Mishyna et al. [32] also exhibited low FC of 5.8% and 45%, respectively. These results are consistent with the observation of Akpossan et al. [14], who reported a poor FC of the edible full-fat insect *I. oyemensis* flour. The variations in FC may be due to the different conformation characteristics of edible insect proteins.

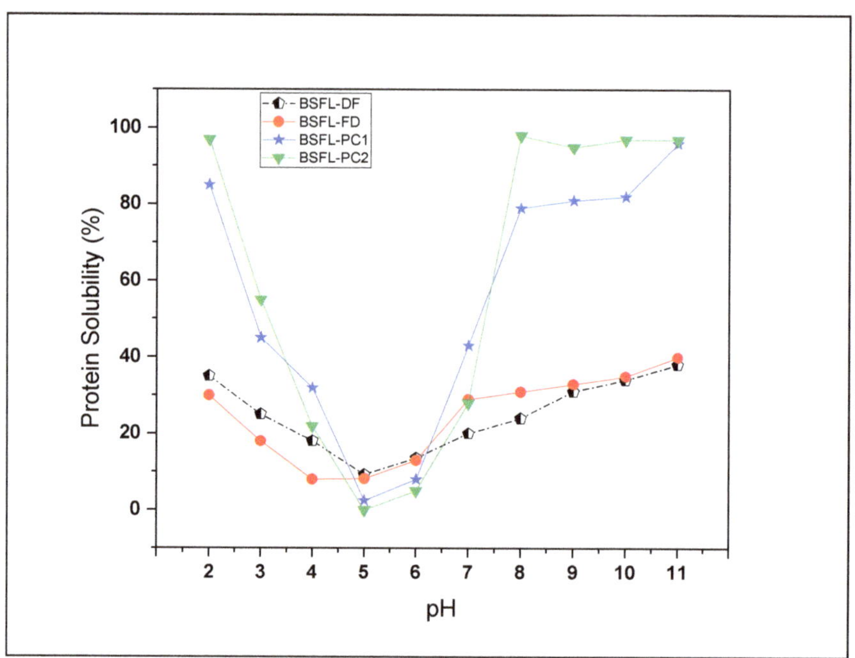

Figure 5. Solubility profile of BSFL flour fractions and protein concentrates as a function of pH. Freeze-dried BSFL flour (pentagon), defatted BSFL flour (circle), alkaline and acid precipitation extraction BSFL protein concentrate (star), and alkaline extraction BSFL protein concentrate (triangle).

The stabilisation of foams is primarily dependent on the formation of a thick cohesive viscoelastic film involving each gas bubble. The present study supported earlier findings that foam stability increases with the removal of fat. To our best knowledge, this is the first study to investigate the effect of protein extraction methods on the FC and FS of BSFL protein concentrates. This work will generate fresh insights into the effect of extraction methods on edible insect protein techno-functional properties. An increased FC and FS of protein fractions is also consistent with higher protein solubility (Figure 5) compared to BSFL flours, which likely contributes to the formation of a cohesive viscoelastic film at the interface via intermolecular interactions. The high FC and FS of BSFL-PC1 and BSFL-PC2 reported in this study suggest that they can be used in bakery and confectionery products. Currently, the food industry utilises wheat, soy, and dairy-based protein concentrates and isolates as ingredients. However, consumers and food processors are searching for new novel protein sources to alleviate the allergenicity challenges posed by the common eight priority allergens (wheat, peanut, soy, fish, dairy, tree nuts, crustaceans, and egg).

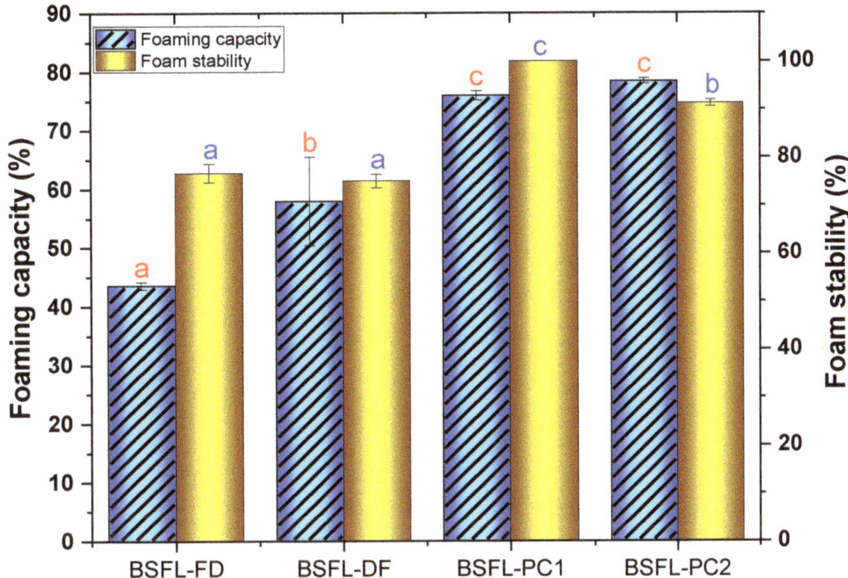

Figure 6. Foaming capacity and stability of BSFL flour and protein concentrates. Freeze-dried BSFL flour—BSFL-FD, defatted BSFL flour—BSFL-DF, alkaline and acid precipitation extraction BSFL protein concentrate—BSFL-PC1, alkaline extraction BSFL protein concentrate—BSFL-PC2. Values are mean $\pm$ standard deviation; means with different superscript are significantly different ($p < 0.05$).

3.4.5. Effect of pH on the ζ-Potential of BSFL Protein Concentrates

A significant feature of proteins which determines their functional properties is the surface charge. In most cases, protein molecules carry a charge, which plays an essential role in interacting with other food matrices components [48]. Protein Zeta potential (ζ) is a critical analysis tool that can be used to optimize food product formulations for new ingredients, predict interactions with surfaces, and predict long-term stability. Zeta (ζ) potential measurements of protein dispersions for different pH values provide information about the isoelectric point (pI). The ζ-potential of BSFL protein concentrates is displayed in Figure 7. The apparent isoelectric point (pI) of all proteins together was determined at a zero net charge. The pH-dependent droplet ζ-potential of BSFL proteins extracted with both methods was zero at pH 4.5, representing the pI of the isolates at this pH. Form the results, it is clear that the chemical extraction method does not affect the isoelectric point of the BSFL protein concentrates. Zeta potential values (positive or negative) at the curve's extremes typically indicate increased electrostatic repulsion, increasing protein solubility. Protein solubility around the apparent pI, on the other hand, was the lowest since there was slight repulsion. These results are in agreement with previously reported data for BSFL [49]. Typically, proteins (e.g., from milk, soy, egg) used as ingredients in the food industry have an isoelectric point in the pH range 4–6 [45]. Therefore, the respective droplet ζ-potentials converged to zero in this pH range to finally change from positive to negative values at pH values above the pI. The pI tended to coincide with the minimum solubility, as previously shown in Figure 3.

3.5. Scanning Electron Microscopy (SEM)

The study of the microstructure of BSFL flours and protein concentrates provided further information on the results of the physicochemical and functional parameters and allowed a more complete interpretation of the effects produced by the different treatments. The surface morphology and microstructure of BSFL flours and protein concentrates are shown in Figure 8. The BSFL freeze-dried and defatted flours showed particle morphology and distribution differences, with BSFL-DF exhibiting large particles. The microstructure of BSFL-FD flour was less dense and had a smooth surface, although some irregular, cracked, or shrunk particles could also be observed, while the BSFL-DF was largely irregularly shaped. The BSFL-PC1 microstructure exhibited a thin flaky plate-like surface morphology, whereas the surface morphology of BSFL-PC2 appeared large and blocky. There were distinct differences in the microstructure of the protein concentrates. These findings suggest that the protein extraction methods changed or modified the microstructure of the BSFL proteins and further explain the observed differences in the functional and physicochemical properties. Moreover, more work on higher resolution or magnification should be further conducted with the view to characterise the BSFL flours and protein concentrates fully.

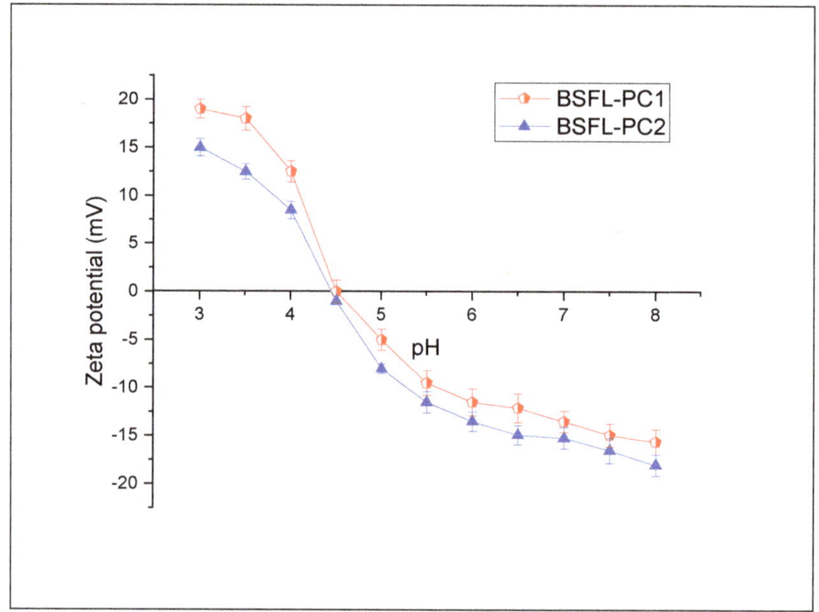

Figure 7. Effect of pH on the zeta potential of BSFL protein concentrates. The point where the line crosses the x-axis represents the apparent pI of the protein solution (prepared in MilliQ water).

3.6. Fourier Transform Infrared (FT-IR) Spectrometer Analysis

FT-IR, a precise, low-cost, and non-destructive analytical technique, was employed to examine the effect of the protein extraction technique on the protein secondary structure (functional groups). Figure 9 presents representative FT-IR absorption spectra for BSFL flours and protein concentrates in the 400–4000 cm^{-1} region. The major peaks for BSFL samples in this study were found at wavenumbers 3278, 2931, 2580, 1742, 1627, and 1534 cm^{-1} for amide A, amide B, amide I, amide II, and amide III, respectively. The absorption peaks of all samples at 2931 and 2850 cm^{-1} represent the functional groups O-H and C-H, respectively. After defatting and protein extraction, the intensity of these peaks decreased significantly. This can be attributed to the chemical treatment applied. Vital information on the protein secondary structure is provided by the Amide I band (1650–1800 cm^{-1}) resulting from the stretching of the C=O of amide in protein. Its intensity decreased as a function of protein extraction and thus indicated an alteration of the protein structure. The Amide I and II bands are the most important. The sensitivity of C=O peptide bonds to the different conformations of protein secondary structures is mainly due to these bands. The protein concentrate (BSFL-PC1 and BSFL-PC2) samples showed a significantly lower peak intensity at 1627 cm^{-1} according to the FT-IR spectrum presented in Figure 9, which corresponds to native intramolecular β-sheets and had marginally lower α-helices intensities (1652 cm^{-1}) compared to the BSFL-FD sample [50,51]. The IR spectra of BSFL-FD and BSFL-DF were composed of the distinct regions that correspond to the amide I band (1600–1700 cm^{-1}), which is mostly carbonyl stretching (C=O) vibrations and amide II which is essentially the combination of the N-H plane boundary and C-N stretching vibrations. It is widely acscepted that the functional and digestive properties of proteins are directly related to their molecular structure, wherein the configurations of the α-helix and beta-sheet are associated with their performance in food systems, particularly their absorption of water and digestion in vitro [50,51]. The BSFL flour and protein concentrate FT-IR spectra were largely comparable. However, variations were observed in a few characteristic peaks and intensities (Figure 9), suggesting minor differences in structure, amino acids, and protein function groups. In addition, the prolonged exposure of proteins during freeze-drying to low temperatures may have caused an increase in disordered structures. Thus, the extraction process influences the secondary structure of the proteins.

Figure 8. Scanning electron micrographs of (**a**) BSFL-FD flour, (**b**) BSFL-DF flour, (**c**) BSFL-PC1, and (**d**) BSFL-PC2.

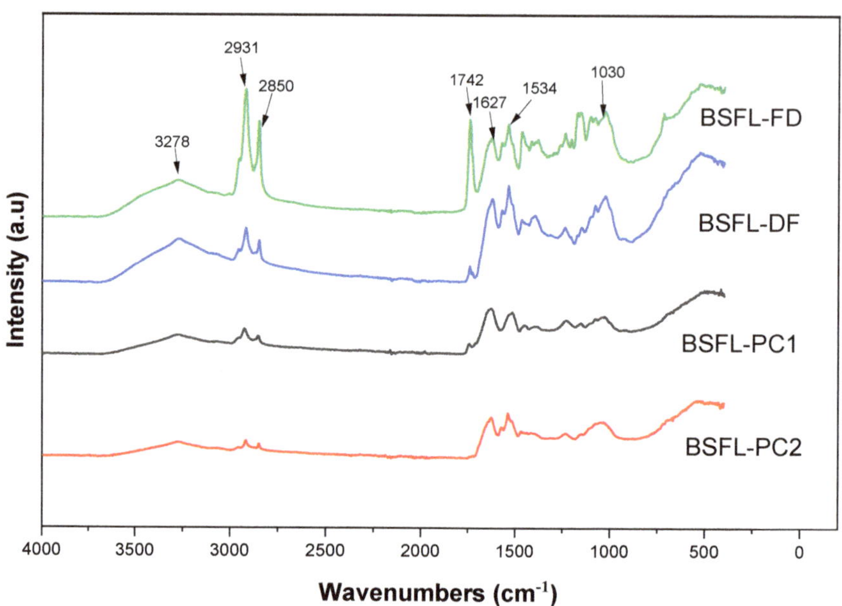

Figure 9. FT-IR Spectra of BSFL flours and protein concentrates. Freeze-dried BSFL flour—BSFL-FD, defatted BSFL flour—BSFL-DF, alkaline and acid precipitation extraction BSFL protein concentrate—BSFL-PC1, alkaline extraction BSFL protein concentrate—BSFL-PC2.

4. Conclusions

This study established the nutritional properties of BSFL flour fractions and protein concentrates. All fractions met the recommended FAO requirements for a well-balanced essential and non-essential amino acid content for human consumption. The overall results indicated that alkaline and acid precipitation extractions of BSFL protein concentrates resulted in enhanced nutritional and functional properties. The protein extraction method appeared to have altered the molecular structure and characteristics of proteins such as surface charge and functional groups and thus contributed to the improved functionality of protein fractions. As a general trend, an improvement in water binding capacity, solubility, and emulsifying capacity were observed. Protein concentrates (BSFL-PC1 and BSFL-PC2) extracted from BSFL exhibited a high emulsion capacity. This offers the possibility of the industrial processing of these edible insect protein concentrates and their use in suitable commercial food applications. More research on the interaction of edible insect ingredients with other food components and on the microbiological, rheological, and sensory properties of new insect-based food proteins is recommended for future studies.

Author Contributions: Conceptualization, V.V.M., J.V.W. and B.Z.; methodology, V.V.M. and B.Z.; software, V.V.M. and B.Z.; validation, V.V.M. and J.V.W. formal analysis, V.V.M. and B.Z.; investigation, V.V.M. and B.Z. resources, J.V.W. data curation, V.V.M.; writing—V.V.M.; writing—review and editing, V.V.M. and J.V.W.; V.V.M. All authors have read and agreed to the published version of the manuscript.

Funding: This research received no external funding.

Institutional Review Board Statement: Ethical clearance was obtained from the Faculty of Applied Sciences Ethics Committee, Cape Peninsula University of Technology, Bellville, South Africa.

Informed Consent Statement: Not applicable.

Data Availability Statement: The datasets used and/or analyzed during the current study are available from the corresponding author on reasonable request.

Acknowledgments: As the authors of this paper, we would like to express our sincere gratitude to the Department of Food Science and Technology at the Cape Peninsula University of Technology, Cape Town, South Africa.

Conflicts of Interest: The authors declared that they have no conflict of interest.

References

1. Van-Huis, A.; Van-Itterbeeck, J.; Harmke, K.; Esther, M.; Afton, H.; Muir, M.; Paul, V. *Edible Insects: Future Prospects for Food and Feed Security*; Food and Agriculture Organization of the United Nations: Rome, Italy, 2013; Available online: http://edepot.wur.nl/258042 (accessed on 1 February 2020).
2. Gmuer, A.; Guth, J.N.; Hartmann, C.; Siegrist, M. Effects of the degree of processing of insect ingredients in snacks on expected emotional experiences and willingness to eat. *Food Qual. Prefer.* **2016**, *54*, 117–127. [CrossRef]
3. Meyer-Rochow, V.B. Can insects help to ease the problem of world food shortage. *Search* **1975**, *6*, 261–262.
4. Rumpold, B.A.; Schlüter, O.K. Potential and challenges of insects as an innovative source for food and feed production. *Innov. Food Sci. Emerg. Technol.* **2013**, *17*, 1–11. [CrossRef]
5. Van Huis, A. Potential of insects as food and feed in assuring food security. *Annu. Rev. Èntomol.* **2013**, *58*, 563–583. [CrossRef] [PubMed]
6. Rumpold, B.A.; Schlüter, O. Insect-based protein sources and their potential for human consumption: Nutritional composition and processing. *Anim. Front.* **2015**, *5*, 20–24. [CrossRef]
7. Kim, H.-W.; Setyabrata, D.; Lee, Y.J.; Jones, O.G.; Kim, Y.H.B. Pre-treated mealworm larvae and silkworm pupae as a novel protein ingredient in emulsion sausages. *Innov. Food Sci. Emerg. Technol.* **2016**, *38*, 116–123. [CrossRef]
8. Kim, H.-W.; Setyabrata, D.; Lee, Y.; Jones, O.G.; Kim, Y.H.B. Effect of house cricket (*Acheta domesticus*) flour addition on physicochemical and textural properties of meat emulsion under various formulations. *J. Food Sci.* **2017**, *82*, 2787–2793. [CrossRef]
9. La Barbera, F.; Verneau, F.; Amato, M.; Grunert, K. Understanding Westerners' disgust for the eating of insects: The role of food neophobia and implicit associations. *Food Qual. Prefer.* **2018**, *64*, 120–125. [CrossRef]
10. Azagoh, C.; Ducept, F.; Garcia, R.; Rakotozafy, L.; Cuvelier, M.-E.; Keller, S.; Lewandowski, R.; Mezdour, S. Extraction and physicochemical characterization of *Tenebrio molitor* proteins. *Food Res. Int.* **2016**, *88*, 24–31. [CrossRef]
11. Kostić, A.Ž.; Barać, M.B.; Stanojević, S.P.; Milojković-Opsenica, D.M.; Tešić, Ž.L.; Šikoparija, B.; Radišić, P.; Prentović, M.; Pešić, M.B. Physicochemical composition and techno-functional properties of bee pollen collected in Serbia. *LWT Food Sci. Technol.* **2015**, *62*, 301–309. [CrossRef]
12. Boye, J.; Zare, F.; Pletch, A. Pulse proteins: Processing, characterization, functional properties and applications in food and feed. *Food Res. Int.* **2010**, *43*, 414–431. [CrossRef]
13. Yi, L.; Lakemond, C.M.; Sagis, L.M.; Eisner-Schadler, V.; van Huis, A.; van Boekel, M.A. Extraction and characterisation of protein fractions from five insect species. *Food Chem.* **2013**, *141*, 3341–3348. [CrossRef] [PubMed]
14. Akpossan, R.A.; Digbeu, Y.D.; Koffi, M.D.; Kouadio, J.P.E.N.; Dué, E.A.; Kouamé, P.L. Protein fractions and functional properties of Dried *Imbrasia oyemensis* Larvae full-fat and defatted flours. *Int. J. Biochem. Res. Rev.* **2015**, *5*, 116–126. [CrossRef]
15. Omotoso, O. Nutritional quality, functional properties and anti-nutrient compositions of the larva of *Cirina forda* (Westwood) (Lepidoptera: Saturniidae). *J. Zhejiang Univ. Sci. B* **2005**, *7*, 51–55. [CrossRef] [PubMed]
16. Mshayisa, V.V.; Van Wyk, J. *Hermetia illucens* protein conjugated with glucose via maillard reaction: Antioxidant and techno-functional properties. *Int. J. Food Sci.* **2021**, *2021*, 5572554. [CrossRef]
17. AOAC. *Official Methods of Analysis of AOAC*, 15th ed.; Association of Official Agricultural Chemists: Arlington, VA, USA, 2015. [CrossRef]
18. Janssen, R.H.; Vincken, J.-P.; Van Den Broek, L.A.M.; Fogliano, V.; Lakemond, C.M.M. Nitrogen-to-protein conversion factors for three edible insects: *Tenebrio molitor*, *Alphitobius diaperinus*, and *Hermetia illucens*. *J. Agric. Food Chem.* **2017**, *65*, 2275–2278. [CrossRef]
19. Diedericks, C.; Jideani, V.A. Physicochemical and functional properties of insoluble dietary fiber isolated from bambara groundnut (*Vigna subterranean* [L.] Verdc.). *J. Food Sci.* **2015**, *80*, C1933–C1944. [CrossRef] [PubMed]
20. Bußler, S.; Rumpold, B.A.; Jander, E.; Rawel, H.M.; Schlüter, O.K. Recovery and techno-functionality of flours and proteins from two edible insect species: Meal worm (*Tenebrio molitor*) and black soldier fly (*Hermetia illucens*) larvae. *Heliyon* **2016**, *2*, e00218. [CrossRef] [PubMed]
21. Purschke, B.; Meinlschmidt, P.; Horn, C.; Rieder, O.; Jäger, H. Improvement of techno-functional properties of edible insect protein from migratory locust by enzymatic hydrolysis. *Eur. Food Res. Technol.* **2018**, *244*, 999–1013. [CrossRef]
22. Coelho, M.S.; Salas-Mellado, M.D.L.M. How extraction method affects the physicochemical and functional properties of chia proteins. *LWT Food Sci. Technol.* **2018**, *96*, 26–33. [CrossRef]
23. Zielińska, E.; Karaś, M.; Baraniak, B. Comparison of functional properties of edible insects and protein preparations thereof. *LWT Food Sci. Technol.* **2018**, *91*, 168–174. [CrossRef]
24. Hall, F.G.; Jones, O.G.; O'Haire, M.E.; Liceaga, A.M. Functional properties of tropical banded cricket (*Gryllodes sigillatus*) protein hydrolysates. *Food Chem.* **2017**, *224*, 414–422. [CrossRef] [PubMed]

25. Ladjal-Ettoumi, Y.; Boudries, H.; Chibane, M.; Romero, A. Pea, Chickpea and lentil protein isolates: Physicochemical characterization and emulsifying properties. *Food Biophys.* **2016**, *11*, 43–51. [CrossRef]

26. Mshayisa, V.V.; Van Wyk, J.; Zozo, B.; Rodríguez, S.D. Structural properties of native and conjugated black soldier fly (*Hermetia illucens*) larvae protein via Maillard reaction and classification by SIMCA. *Heliyon* **2021**, *7*, e07242. [CrossRef]

27. Schwenzfeier, A.; Helbig, A.; Wierenga, P.A.; Gruppen, H. Emulsion properties of algae soluble protein isolate from *Tetraselmis* sp. *Food Hydrocoll.* **2013**, *30*, 258–263. [CrossRef]

28. Huang, C.; Feng, W.; Xiong, J.; Wang, T.; Wang, W.; Wang, C.; Yang, F. Impact of drying method on the nutritional value of the edible insect protein from black soldier fly (*Hermetia illucens* L.) larvae: Amino acid composition, nutritional value evaluation, in vitro digestibility, and thermal properties. *Eur. Food Res. Technol.* **2019**, *245*, 11–21. [CrossRef]

29. Wang, Y.-S.; Shelomi, M. Review of black soldier fly (*Hermetia illucens*) as animal feed and human food. *Foods* **2017**, *6*, 91. [CrossRef]

30. Zhao, X.; Vázquez-Gutiérrez, J.L.; Johansson, D.; Landberg, R.; Langton, M. Yellow mealworm protein for food purposes—Extraction and functional properties. *PLoS ONE* **2016**, *11*, e0147791. [CrossRef]

31. Wani, I.A.; Sogi, D.S.; Wani, A.A.; Gill, B.S. Physico-chemical and functional properties of flours from Indian kidney bean (*Phaseolus vulgaris* L.) cultivars. *LWT Food Sci. Technol.* **2013**, *53*, 278–284. [CrossRef]

32. Mishyna, M.; Martinez, J.-J.I.; Chen, J.; Benjamin, O. Extraction, characterization and functional properties of soluble proteins from edible grasshopper (*Schistocerca gregaria*) and honey bee (*Apis mellifera*). *Food Res. Int.* **2019**, *116*, 697–706. [CrossRef]

33. Janssen, R.H.; Canelli, G.; Sanders, M.G.; Bakx, E.J.; Lakemond, C.M.M.; Fogliano, V.; Vincken, J.-P. Iron-polyphenol complexes cause blackening upon grinding *Hermetia illucens* (black soldier fly) larvae. *Sci. Rep.* **2019**, *9*, 2967. [CrossRef] [PubMed]

34. Leni, G.; Caligiani, A.; Sforza, S. Killing method affects the browning and the quality of the protein fraction of Black Soldier Fly (*Hermetia illucens*) prepupae: A metabolomics and proteomic insight. *Food Res. Int.* **2019**, *115*, 116–125. [CrossRef] [PubMed]

35. Köhler, R.; Kariuki, L.; Lambert, C.; Biesalski, H. Protein, amino acid and mineral composition of some edible insects from Thailand. *J. Asia Pac. Entomol.* **2019**, *22*, 372–378. [CrossRef]

36. FAO. Dietary Protein Quality Evaluation in Human Nutrition. Report of an FAO Expert Consultation; FAO: Auckland, New Zealand, 2013; Available online: http://www.fao.org/ag/humannutrition/35978-02317b979a686a57aa4593304ffc17f06.pdf (accessed on 7 February 2022).

37. Adebowale, Y.A.; Adebowale, K.O.; Oguntokun, M.O. Evaluation of nutritive properties of the large African Cricket (*Gryllidae* sp.). *Pak. J. Sci. Ind. Res.* **2005**, *48*, 274–278.

38. Kouřimská, L.; Adámková, A. Nutritional and sensory quality of edible insects. *NFS J.* **2016**, *4*, 22–26. [CrossRef]

39. Meyer-Rochow, V.; Gahukar, R.; Ghosh, S.; Jung, C. Chemical composition, nutrient quality and acceptability of edible insects are affected by species, developmental stage, gender, diet, and processing method. *Foods* **2021**, *10*, 1036. [CrossRef] [PubMed]

40. Ghosh, S.; Jung, C.; Meyer-Rochow, V.B. Nutritional value and chemical composition of larvae, pupae, and adults of worker honey bee, *Apis mellifera ligustica* as a sustainable food source. *J. Asia-Pac. Èntomol.* **2016**, *19*, 487–495. [CrossRef]

41. FAO/WHO/UNU. Protein and Amino Acid Requirements in Human Nutrition. Report of an FAO Expert Consultation; FAO: Geneva, Switzerland, 2007. Available online: http://www.ncbi.nlm.nih.gov/pubmed/14258992 (accessed on 8 January 2019).

42. Gravel, A.; Doyen, A. The use of edible insect proteins in food: Challenges and issues related to their functional properties. *Innov. Food Sci. Emerg. Technol.* **2020**, *59*, 102272. [CrossRef]

43. Ndiritu, A.K.; Kinyuru, J.N.; Kenji, G.M.; Gichuhi, P.N. Extraction technique influences the physico-chemical characteristics and functional properties of edible crickets (*Acheta domesticus*) protein concentrate. *J. Food Meas. Charact.* **2017**, *11*, 2013–2021. [CrossRef]

44. Garcia-Vaquero, M.; Lopez-Alonso, M.; Hayes, M. Assessment of the functional properties of protein extracted from the brown seaweed *Himanthalia elongata* (Linnaeus) S. F. Gray. *Food Res. Int.* **2017**, *99*, 971–978. [CrossRef]

45. Burger, T.G.; Zhang, Y. Recent progress in the utilization of pea protein as an emulsifier for food applications. *Trends Food Sci. Technol.* **2019**, *86*, 25–33. [CrossRef]

46. Ma, M.; Ren, Y.; Xie, W.; Zhou, D.; Tang, S.; Kuang, M.; Wang, Y.; Du, S.-K. Physicochemical and functional properties of protein isolate obtained from cottonseed meal. *Food Chem.* **2018**, *240*, 856–862. [CrossRef] [PubMed]

47. Omotoso, O.T. An evaluation of the nutrients and some anti-nutrients in Silkworm, *Bombyxmori* L. (Bombycidae: Lepidoptera). *Jordan J. Biol. Sci.* **2015**, *8*, 45–50. [CrossRef]

48. Schwenzfeier, A.; Wierenga, P.A.; Eppink, M.; Gruppen, H. Effect of charged polysaccharides on the techno-functional properties of fractions obtained from algae soluble protein isolate. *Food Hydrocoll.* **2014**, *35*, 9–18. [CrossRef]

49. Janssen, R.H. Potential of Insect Proteins for Food and Feed: Effect of Endogenous Enzymes and Iron-Phenolic Complexation. Ph.D. Thesis, Wageningen University, Wageningen, The Netherlands, 2018; pp. 1–157. [CrossRef]

50. Hadnađev, M.; Dapčević-Hadnađev, T.; Lazaridou, A.; Moschakis, T.; Michaelidou, A.-M.; Popović, S.; Biliaderis, C. Hempseed meal protein isolates prepared by different isolation techniques. Part I. physicochemical properties. *Food Hydrocoll.* **2018**, *79*, 526–533. [CrossRef]

51. Wang, W.-D.; Li, C.; Bin, Z.; Huang, Q.; You, L.-J.; Chen, C.; Fu, X.; Liu, R.H. Physicochemical properties and bioactivity of whey protein isolate-inulin conjugates obtained by Maillard reaction. *Int. J. Biol. Macromol.* **2020**, *150*, 326–335. [CrossRef]

 *foods*

Article

Effect of the Rearing Substrate on Total Protein and Amino Acid Composition in Black Soldier Fly

Andrea Fuso [1], Silvia Barbi [2], Laura Ioana Macavei [3], Anna Valentina Luparelli [1], Lara Maistrello [3], Monia Montorsi [4], Stefano Sforza [1] and Augusta Caligiani [1,*]

1. Food and Drug Department, University of Parma, Via Parco Area delle Scienze 17/A, 43124 Parma, Italy; andrea.fuso@unipr.it (A.F.); annavalentina.luparelli@unipr.it (A.V.L.); stefano.sforza@unipr.it (S.S.)
2. Interdepartmental Research Center for Industrial Research and Technology Transfer in the Field of Integrated Technologies for Sustainable Research, Efficient Energy Conversion, Energy Efficiency of Buildings, Lighting and Home Automation—En&Tech, University of Modena and Reggio Emilia, Via Amendola 2, 42122 Reggio Emilia, Italy; s.barbi@unimore.it
3. Department of Life Sciences, University of Modena and Reggio Emilia, Via Giovanni Amendola 2, 42122 Reggio Emilia, Italy; lauraioana.macavei@unimore.it (L.I.M.); lara.maistrello@unimore.it (L.M.)
4. Department of Sciences and Methods for Engineering, University of Modena and Reggio Emilia, Via Giovanni Amendola 2, 42122 Reggio Emilia, Italy; monia.montorsi@unimore.it
* Correspondence: augusta.caligiani@unipr.it

Abstract: Insects are becoming increasingly relevant as protein sources in food and feed. The Black Soldier Fly (BSF) is one of the most utilized, thanks to its ability to live on many leftovers. Vegetable processing industries produce huge amounts of by-products, and it is important to efficiently rear BSF on different substrates to assure an economical advantage in bioconversion and to overcome the seasonality of some leftovers. This work evaluated how different substrates affect the protein and amino acid content of BSF. BSF prepupae reared on different substrates showed total protein content varying between 35% and 49% on dry matter. Significant lower protein contents were detected in BSF grown on fruit by-products, while higher contents were observed when autumnal leftovers were employed. BSF protein content was mainly correlated to fibre and protein content in the diet. Among amino acids, lysine, valine and leucine were most affected by the diet. Essential amino acids satisfied the Food and Agricultural Organization (FAO) requirements for human nutrition, except for lysine in few cases. BSF could be a flexible tool to bio-convert a wide range of vegetable by-products of different seasonality in a high-quality protein-rich biomass, even if significant differences in the protein fraction were observed according to the rearing substrate.

Keywords: *Hermetia illucens*; insect rearing; vegetable leftovers; protein fraction; amino acids composition; growth substrate

Citation: Fuso, A.; Barbi, S.; Macavei, L.I.; Luparelli, A.V.; Maistrello, L.; Montorsi, M.; Sforza, S.; Caligiani, A. Effect of the Rearing Substrate on Total Protein and Amino Acid Composition in Black Soldier Fly. *Foods* **2021**, *10*, 1773. https://doi.org/10.3390/foods10081773

Academic Editors: Theodoros Varzakas, Victor Benno Meyer-Rochow and Chuleui Jung

Received: 17 May 2021
Accepted: 27 July 2021
Published: 30 July 2021

Publisher's Note: MDPI stays neutral with regard to jurisdictional claims in published maps and institutional affiliations.

1. Introduction

By 2050, the world will host more than 9 billion people, and the availability of proteins is the main concern in feeding the increasing population, as already pointed out as long ago as 1975 by Meyer-Rochow [1]. Presently, in Western diets the proteins are predominantly introduced through products of animal origin, although the zootechnical production constitutes an important issue from an ecological point of view due to its impact on the environment [2]. A valuable and more sustainable source of protein is represented by insects. They have a high nutritional value [3–5] and, compared with traditionally farmed animals, insects have a much higher conversion efficiency and require much less water [2], and their rearing involves much less greenhouse gas emissions [6].

Nowadays, more than one-third of the edible parts of food produced gets lost or wasted; insects could bear an important role in managing and valorizing food waste, since they can be reared on a large variety of bio-waste substrates [7,8], thus contributing to their mass reduction and preventing unnecessary waste of resources and further emissions of greenhouse gas [9]. Therefore, insects fit perfectly in the perspective to valorise bio-waste to create a sustainable food and feed production system that embraces the concept of circular economy and increased sustainability [10,11]. The ability of insects to convert waste materials into high-quality nutrients has long been known [12], and the Black Soldier Fly (BSF, *Hermetia illucens* L., *Diptera, Stratiomiyidae*) is known as one of the most efficient bioconverters among insects, being able to reduce the weight of organic waste up to 75% and converting it into a biomass rich in proteins and lipids [13]. This makes the BSF suitable to be used as feed for farmed animals [14], for biodiesel production [15] and also for cosmetics or pharmaceuticals industries, thanks to its high chitin content [16]. All these applications could be simultaneously tackled through an appropriate fractionation method [17]. Last but not least, the residual larval frass can be employed as a quality soil improver [18,19].

Many studies have shown that BSF larvae can live on many substrates with different characteristics, ranging from mushrooms [20] and winery by-products [21], restaurant waste [8], municipal waste [22], animal manure [10,23] and human faeces [24]. However, according to European legislation (Regulation (EC) No. 1069/2009), when invertebrates are industrially reared, they are considered as "farmed animals"; therefore, the use of animal manure, catering waste or former foodstuffs containing meat and fish as feeding substrates is totally forbidden [25,26]. As a consequence, despite the lower content of nutrients, vegetable and fruit leftovers are increasingly used as rearing substrates for BSF, also due to their high availability in industrialized regions that are fully compliant with the legislation on feed for farmed animals [9]. European Food Safety Authority (EFSA) highlighted that no additional microbiological risks are expected for insect rearing on authorized substrates with respect to other animal farming [26]. In addition, a few studies demonstrated that the eventual pesticides and mycotoxins do not accumulate in BSF biomass after the breeding process [27–30].

An important point in the use of vegetal substrates as feed for insects is their seasonal availability. As a matter of fact, most of the vegetal leftovers are not constantly available throughout the year, thus it is important, both from the economic and technical point of view, to understand the feasibility of rearing BSF on a variety of substrates and a combination thereof, in order to have a constant BSF production and composition throughout the year.

Many authors have studied the influence of rearing substrates on BSF protein fractions in terms of total amount and of amino acid profile. The protein amount showed marked differences, varying from 32 to 58% on dry matter [13], while for the amino acid composition, studies in the literature are often inconsistent, and this is true for other insect species as well [31]. According to Newton et al., BSF larvae reared on animal manure were lacking some essential amino acids, such as cysteine, methionine and threonine [12], whereas according to Liland et al., BSF larvae had sufficient threonine amounts when reared on a conventional diet supplemented with increasing quantities of seaweed [32]. When fed on a vegetable mix, BSF showed high amounts of aspartic acid, glutamate and arginine [7], but also of leucine [8].

In this paper, a wide range of data was collected on the content and quality of BSF pre-pupae protein in relation to the rearing substrates used. A total of 49 different combinations of vegetable rearing substrates were tested in order to obtain the most complete picture of the effect of diet on BSF proteins and on the possibility of using diverse substrates for their rearing. All these vegetable by-products have been chosen focusing on their availability in a specific region (Regione Emilia-Romagna, Italy) and mostly on their seasonal availability. The diets were formulated according to a Design of Experiment approach (Mixture Design) based on the nutrient composition of each substrate and their seasonal availability, as reported in Barbi et. al, in the optic of a possible scale-up application of BSF rearing on vegetable by-products across the whole year [33]. To put this approach into practice and fulfil the requirements of a circular economy, it is of primary importance to verify the composition of BSF reared on the different substrates, and in particular their protein fraction that nowadays is considered as the most valuable component of insects.

2. Materials and Methods

2.1. Materials

The vegetable by-products were collected from different suppliers in the Emilia Romagna Region (northern Italy): Agribologna (Bologna, Italy), Conserve Italia (Bologna, Italy) and Cooperativa Agricola Brisighellese (Ravenna, Italy). The by-products were stored at −20 °C until use, and further grinded with IKA A10 laboratory grinder.

Chemicals: an AccQ-Fluor reagent kit was obtained from Waters (Milford, MA, USA). DL-norleucine, amino acid standard mixture, L-tryptophan, 5-Methyl-DL-tryptophan, L-Cysteic acid, DL-Dithiothreitol, Ethylenediaminetetraacetic acid (EDTA) and Tris-HCl were purchased from Sigma-Aldrich (St. Louis, MO, USA). Sodium dodecyl sulfate was purchased from Biorad (Hercules, CA, USA). All the other solvents, salts, acids and bases were of analytical grade and were purchased from Sigma-Aldrich or Carlo Erba (Milan, Italy).

2.2. Insect Rearing Conditions and Substrate Selection

BSF prepupae for the analyses were obtained from a series of experiments carried out in the Applied Entomology laboratory of the University of Modena and Reggio Emilia (Italy). Leftovers selection, rearing substrates design and larvae rearing experiments have been managed by the University of Modena and Reggio Emilia (Italy), as discussed in another study [33]. Essentially, BSF larvae were reared under controlled conditions (27 ± 0.5 °C, 60–70% Relative Humidity, RH) in various mixtures of fruit and vegetable by-products with a different seasonal availability (All-year, Summer and Autumn, Table S1), and the mixtures were designed through a Mixture Design approach. Details on the chemical composition of the substrates are reported in Table S1. The "Gainesville House Fly" diet (50% wheat bran, 30% alfalfa meal and 20% corn meal) [34] was used as the control diet (CTR). Forty-nine experiments were conducted, inoculating exactly 300 BSF larvae (of second and third stage) for each substrate. The rearing experiments was conducted for a maximum period of 65 days, checking regularly the prepupae developed. At each control, the prepupae observed were collected, killed by freezing and stored at −20 °C until further analysis.

2.3. Proximate Composition of Rearing Substrates and BSF Prepupae

Proximate composition of agri-food leftovers was determined using standard procedures [35]. Moisture was determined in an oven at 105 °C for 24 h. Crude fat content was determined using an automatized Soxhlet extractor (SER 148/3 VELP SCIENTIFICA, Usmate Velate, Italy) using diethyl ether. Total ash was determined after mineralization at 550 °C for 5 h. Total nitrogen was determined with a Kjeldahl system (DKL heating digestor and UDK 139 semiautomatic distillation unit, VELP SCIENTIFICA) using 6.25 as the mean nitrogen coefficient conversion for vegetable rearing substrates. Dietary fibres of vegetable by-products were determined through the official method AOAC 991.43, while total polyphenol content was determined by using the Folin–Ciocalteu method. Digestible carbohydrates were determined by difference.

Regarding BSF prepupae, the total protein content was determined by the Kjeldahl method, utilizing 4.67 as the nitrogen to protein conversion factor, in order to exclude chitin contributing to total nitrogen, as previously reported [36].

2.4. Amino Acid Profile of BSF Prepupae

2.4.1. Sample Preparation

The total amino acid profile was evaluated according to the protocol proposed by Caligiani et al. with some modifications [17]. An amount of 0.5 g of BSF prepupae was hydrolysed with 6 mL of HCl 6 N at 110 °C for 23 h, then the internal standard (7.5 mL of 5 mM Norleucine in HCl 0.1 M) was added. Cysteine was determined as cysteic acid after performic acid oxidation followed by acid hydrolysis. In this case, an amount of 0.5 g of BSF was added to performic acid freshly prepared (by mixing 9 volumes of formic acid with 1 volume of hydrogen peroxide) and samples were kept in an ice bath for 16 h at 0 °C.

Then, 0.3 mL of hydrobromidric acid was added and the bromine formed was removed under nitrogen flow. Then, acid hydrolysis was performed as described above.

2.4.2. UPLC/ESI-MS Analysis

The hydrolysed samples were analysed by ultra-performance liquid chromatography with electrospray ionization and mass spectrometry detector (UPLC/ESI-MS, WATERS ACQUITY) after derivatization with 6-aminoquinolyl-N-hydroxysuccinimidyl carbamate (AQC). In particular, UPLC/ESI-MS analysis was performed by using an ACQUITYU-PLC separation system with an Acquity BEH C18 column (1.7 µm, 2.1 × 150 mm). The mobile phase was composed of H_2O + 0.2% CH_3CN +0.1% HCOOH (eluent A) and CH_3CN + 0.1% HCOOH (eluent B). Gradient elution was performed: isocratic 100% A for 7 min, from 100% A to 75.6% A and 24.4% B by linear gradient from 8 to 28 min, isocratic 100% B from 29 to 32 min, isocratic 100% A from 33 to 45 min. The flow rate was set at 0.25 mL/min, injection volume 2 µL, column temperature 35 °C and sample temperature 18 °C. Detection was performed by using Waters SQ mass spectrometer: the ESI source was in positive ionization mode, capillary voltage 3.2 kV, cone voltage 30 V, source temperature 150 °C, desolvation temperature 300 °C, cone gasflow (N2) 100 L/h, desolvation gas flow (N2) 650 L/h, full scan acquisition(270–518 m/z) and scan duration 1 s. Calibration was performed with standard solutions prepared mixing norleucine, amino acids hydrolysate standard mixture, cysteic acid and deionized water.

2.4.3. Tryptophan Determination

Total tryptophan was determined following the protocol proposed by Delgado-Andrade et al. with some modifications [37]. An amount of 0.2 g of sample was weighed and dissolved in 3 mL of 4N NaOH. An amount of 150 µL of 5-methyl-tryptophan (16 mg/100 mL), used as internal standard, was added and mixed. Hydrolysis was then carried out for 18 h at 110 °C. The hydrolysates were aerated and cooled, then carefully acidified to a pH 6.5 with HCl 37%, diluted to 25 mL with sodium borate buffer (0.1 M, pH 9.0) and allowed to stand for 15 min. Samples were finally centrifuged at 4000 rpm for 5 min and supernatant filtered through a 0.45 µm nylon filter membrane into UPLC vials. UPLC/ESI-MS analysis was performed as for the other amino acids.

2.5. Statistical Analysis

All analyses on prepupae were carried out in duplicate. Data are expressed as the mean ± standard deviation. Protein and amino acid data were subjected to one-way analysis of variance (ANOVA) followed by a Tukey post hoc test using IBM SPSS software version 21.0 (SPSS Inc., Chicago, IL, USA) to determine differences between samples. Significant differences were compared at a level of $p < 0.05$.

The Pearson correlation coefficient was calculated to evaluate the linear correlations between the nutrients in rearing substrates and protein amounts in BSF prepupae biomass. Both values were taken in absolute terms, calculated according to the following equations:

$$\text{BSF total protein content} = \frac{\text{Protein\%*total grams of prepupae at the end of the breeding}}{100}$$

$$\textit{Leftovers total nutrient content} = \frac{\text{nutrient\% in diet} * \text{total grams of diet}}{100}$$

The second equation was applied separately for each nutrient present in the diet: proteins, lipids, fibres, carbohydrates, ashes and polyphenols.

Principal Component Analysis (PCA) was carried out on amino acid data using IBM SPSS software version 21.0 (SPSS Inc., Chicago, IL, USA). PCA was performed through a 52-point matrix (49 samples plus 3 replicates of the control substrate, one for each seasonal period) and 19 variables (18 amino acids and total protein content), and the principal components were derived with the correlation method. As an unsupervised learning approach, PCA allows us to describe the variation in the dataset. Data were visualised by plotting the score plot and the loading plot, the latter allowing the identification of the amino acids having influences on specific grouping of samples according to the rearing substrate.

The Design Expert 12.0 (Stat-Ease) code was used both to set up the experimental plan and to analyse the results. A mixture design was selected to obtain predictive reliability on the effect of the leftovers' composition on the amino acid content of BSF's prepupae. Six factors were considered: proteins, fibres, carbohydrates, polyphenols, lipids, and ashes, from which the experimental plan of Table S1 was derived. ANOVA was employed to estimate the influence of each factor over the responses observing the p-values (α-level of 0.05) and F-tests. The quality of the fit in terms of regression analysis and the predictive power of the model were evaluated by using the R2 and Pred-R2, respectively. R2 is the proportion of the dependent variable's variance predictable from the independent variables. In a similar way, Pred-R2 shows how well the model can predict the responses for new observations. Response contour plots were used as functional tools in explaining graphically the role of the main components on the final considered properties [38].

3. Results and Discussion

3.1. Rearing Substrate Composition and Related BSF Biomass

The composition of rearing substrates, both in terms of leftover combinations and specific nutrient composition is reported in Table S1, while in Table 1, the mean composition of the diets according to the seasonality (substrates available all year, substrates available in summer and substrates available in autumn) is reported and compared with the BSF standard diet (Gainsville diet, CTR). The three groups of tested diets differed between each other and in respect to the control diet. The differences in diet compositions are related to the vegetable by-products employed: (i) fruit peels and pulp (exotic fruits, apple, kiwi, pineapple, melon) for the All-Year diets, (ii) tomato peels/seed and peach pulp/peels for the Summer group and (iii) legume, corn and olive pomace for the Autumn group, although the exotic fruits and melon, as substrates that are available all year round, were introduced in some diets of the AUTUMN and SUMMER groups. Corn, wheat brans and alfalfa were instead the ingredients of the control diet (CTR diet), used as the BSF rearing substrate in the lab colony.

Table 1. Mean percentage composition of the different diets, expressed as g/100 g dry matter and corresponding amount of total prepupal biomasses obtained (g) starting from 300 Black Soldier Fly (BSF) larvae. Nd = not detected.

	All-Year (n = 21)	**Summer (n = 13)**	**Autumn (n = 15)**	**CTR**
Lipid	0.78 ± 0.35	3.28 ± 2.12	3.18 ± 1.46	3.37 ± 0.1
Ashes	6.66 ± 3.20	4.87 ± 0.71	4.02 ± 1.59	5.9 ± 0.5
Fibres	23.1 ± 17.51	56.86 ± 14.79	56.69 ± 6.53	41 ± 2.5
Polyphenols	0.03 ± 0.02	0.09 ± 0.02	0.04 ± 0.01	Nd
Protein	5.07 ± 1.20	9.97 ± 2.71	16.03 ± 6.57	17.28 ± 1
Available carbohydrates	64.36 ± 20.07	24.93 ± 15.61	17.05 ± 10.72	32.61 ± 1
Total prepupae biomass (g)	21.2 ± 2	39.8 ± 3	57.1 ± 5	52.0 ± 4

The different ingredients employed in the three groups are also related to a different proximate composition of the diets, as reported in Table 1. The All-Year diets are rich in available carbohydrates and poor in protein and lipids, while the Summer and Autumn groups contain larger amounts of fibres and protein, the latter especially represented in the Autumn group.

Despite differences in composition, all the diets tested allowed BSF to grow (more than 90% of the initial 300 BSF larvae reached the prepupal stage in all experiments), even if with consistent differences in total weight of final prepupae biomass. In Table 1, the mean amount of prepupae (g) obtained for each different rearing mixture is also reported, clearly outlining the prominent effect of seasonality on the BSF prepupae total biomass amount. The specific amount of prepupal biomasses in each rearing experiment is reported in Table S1 of the Supplementary Materials.

As a consequence, the BSF prepupae obtained can be roughly divided into three groups according to their biomass weight at the end of the growing process, which basically correspond to the three groups of vegetable diets administered, classified according to their seasonality. Indeed, the lower quantity of prepupal biomass were obtained by using the All-Year substrates, the intermediate weight with Summer substrates (both of them lower than when using the control diet), while the higher quantities were obtained by employing the Autumn group substrates, which seems to be the most suitable feeding substrate group to maximize the BSF biomass.

3.2. Effect of the Rearing Substrate Composition on BSF Prepupae Total Protein Content

Given that the composition of the diets influences the growth of the prepupae, the target of this work, as previously stated, was to verify how the different substrates, and in turn the different growth performance of prepupae (low, medium and high biomass-weights), influence the final BSF protein quality and its total amount. BSF samples were clustered on the basis of the prepupal biomass, corresponding to the three groups of BSF diets (All-Year, Summer and Autumn).

The total protein content of each BSF prepupa reared on the different substrates is reported in Figure 1 (see Table S1 of the supplementary material for the details about substrate typologies, composition and sample codes). Protein amount, calculated as g/100 g of BSF on dry matter (% DM), varied in the range of 35% to 49.5%. These values agree with the studies in the literature, reporting a range of 32–58% [13].

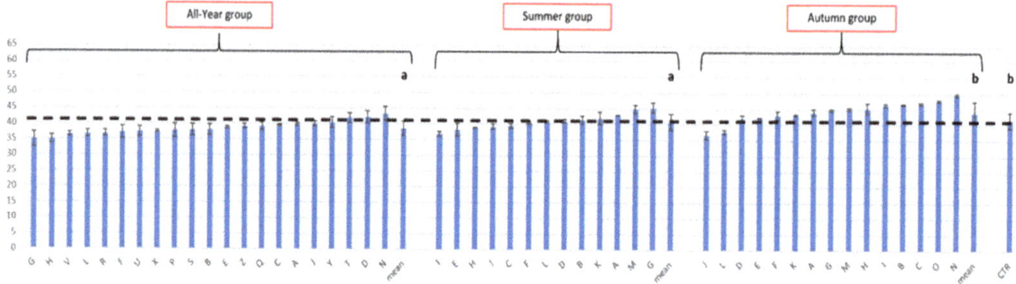

Figure 1. Total protein content (Nx4.67, g/100 g DM) for each BSF sample grown on the 49 diets considered and compared with the control samples (dotted line). Each datum is the mean of two replicates. Global mean of each group was compared with one-way ANOVA (*p*-level 0.05). Different letters (**a,b**) on the bars indicate significantly different values.

In general, most of the samples from the All-Year and Summer groups contain lower amounts of protein compared to the control group, although with a few exceptions (Sample N—All-Year, sample A, M, G—Summer). On the contrary, many of the Autumn group samples contain higher amounts of protein with respect to the control, indicating that a better growth performance, as expected, also corresponds to a higher protein content

Taken as a whole (mean), the Autumn group did not show a significant difference with respect to the control, outlining that the agrifood leftovers used in this group could fully replace the control diet, while in the All-Year and the Summer groups, a significantly lower amount of protein was recorded (one-way ANOVA, Tukey post hoc test, *p*-level = 0.05). This indicates that, while growing on these substrates is certainly possible, it would be less efficient and result in a smaller overall amount of protein. Thus, growing on these substrates should be performed by carefully balancing the advantage of re-using leftovers with the disadvantage of having a slightly lower amount of protein.

To better understand which relationship exists between the various diets administered and the production of protein in the prepupae biomass, we correlated the latter and the single nutrients of the diets through linear correlation analysis.

A positive, although moderate, correlation was observed between the lipid content of the diet and the protein content of prepupae (R = 0.54; *p* < 0.001), while non-structural carbohydrates (digestible carbohydrates) showed a moderate negative one (R = −0.56, $p < 0.001$). On the contrary, polyphenols (R = 0.39, $p < 0.001$) and ashes (R = 0.17, $p = 0.37$) in the diet did not affect the protein content of BSF prepupae.

A significant positive correlation was detected between the BSF prepupae protein content and the fibres content of the diet (R = 0.87, $p < 0.001$, Figure 2a). Fibres are typically difficult degrade for most insects, but this positive correlation could be the result of the increased availability of nutrients due to the action of microorganisms able to hydrolyse cellulose, both endogenous and exogenous (i.e., already present in the rearing substrates) [21,39–41]. Indeed, previous studies have identified cellulase genes in the gut microflora of BSF larvae [40,42]. Thus, a likely hypothesis for the correlation found might be that the ability of using more and better fibre biomass leads to better larval growth, which in turn also means a higher amount of protein produced.

A positive correlation was also identified between the BSF prepupae protein content and the diet's protein content in absolute terms (R = 0.84, $p < 0.001$, Figure 2b), confirming previous findings [21,23].

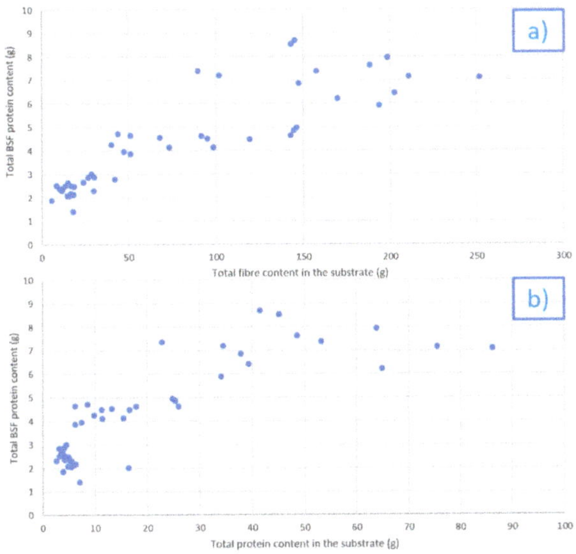

Figure 2. Correlation between total protein content (absolute amount) in the BSF prepupae biomass recorded in each experiment (starting from 300 BSF larvae) and (**a**) total fibre and (**b**) total protein contained in the rearing substrate (absolute amount).

Observing qualitatively the data in Figure 2a,b, a lack of a linear correlation (a plateau is reached) seems evident, especially when a high amount of fibre or protein was present in the diet, suggesting that there is a specific level of these nutrients in the substrates allowing them to reach the maximum amount of protein in insects.

More specifically, our data show that by increasing the amount of vegetable protein in the diet, BSF larvae convert progressively a smaller part of it into their own animal protein. In fact, whilst for low-protein diets they were able to convert more than 90%, for the most protein-rich diets, this percentage dropped to 10% (Figure 3). This suggests that for BSF larvae growth and protein content, the amount of protein in the rearing substrate is very important until a minimum threshold is reached, and then it becomes less relevant. These experimental data are in agreement with a recent work about digestive enzyme expression and production in BSF [43]. In this paper, the authors clearly demonstrate that the midgut of *H. illucens* larvae is able to adapt to diets with different nutrient content; an increase in proteolytic activity together with a decrease in α-amylase and lipase activity was observed as a consequence of nutritionally poor diet. Moreover, Barragán Fonseca (2018) demonstrated that larvae feeding on substrates rich in protein have a higher lipid content, and thereby a reduced protein content (in % dry mass) [44]. Consequently, in order to obtain prepupae with a high protein content, according to Figure 3, a good compromise to maintain an advantageous conversion rate would be to rear them on a substrate containing 7% by weight of protein. This value is also in accordance with a previous study [31].

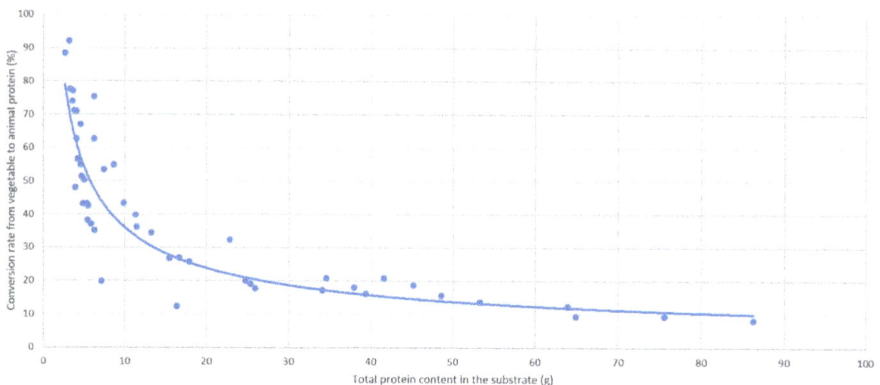

Figure 3. Conversion rate (%) of vegetable protein present in the substrate into BSF protein (absolute amounts, starting from 300 BSF larvae).

3.3. BSF Amino Acid Content

To better verify the influence of the rearing substrate composition on the BSF protein nutritional value, further insights into the complete amino acid profile were provided. As a matter of fact, information on the nutritional quality of proteins, and therefore on their amino acid composition, is of utmost importance to understand the possible uses of the BSF protein fraction. Results on the complete total amino acid profile of the BSF prepupae are reported in Table S2. An explorative Principal Component Analysis (PCA) was performed to assess if and how different rearing substrates would affect amino acid composition. Principal Components are new variables obtained by linear combinations of the original variables (amino acids and total protein), allowing us to describe the system variability using only a few components, thus reducing the complexity, enabling the visualization of the samples in a two-dimensional graph. The analysis showed that about 39% of the total variation is explained by the first component (PC1), 57% by the first two components and 69% by the first three components. The most important variables for each principal component are reported in Table S3 of the Supplementary Materials. Figure 4a shows the scatter plot of the scores of PC1 versus PC2. The loadings for the first two components are schematized in the component plot (Figure 4b). Glutamic acid/glutamine, lysine, phenylalanine, aspartic acid/asparagine, tyrosine, glycine, valine, serine and histidine turned out to be the most influencing variables for PC1, PC2 was predominantly characterized by cysteine, tryptophan, arginine and threonine, while leucine, isoleucine and methionine had the greatest effect for PC3. Figure 4a shows a partial separation of the BSF prepupae samples, based on the seasonality of the substrates. In particular, BSF prepupae that had been reared on substrates belonging to the Autumn group are found in correspondence with PC1 positive values, well separated from the others. In this group, the most represented essential amino acids are phenylalanine/tyrosine, valine and leucine.

On the other hand, the All-Year and Summer groups showed a less clear separation between each other and both were found at negative values of PC1. They differed from the Autumn group mainly in their higher amounts of lysine, aspartate and glutamate content. Finally, the control group was found in an intermediate position, closer to the All-Year and Summer groups and with greater differences compared to the Autumn one.

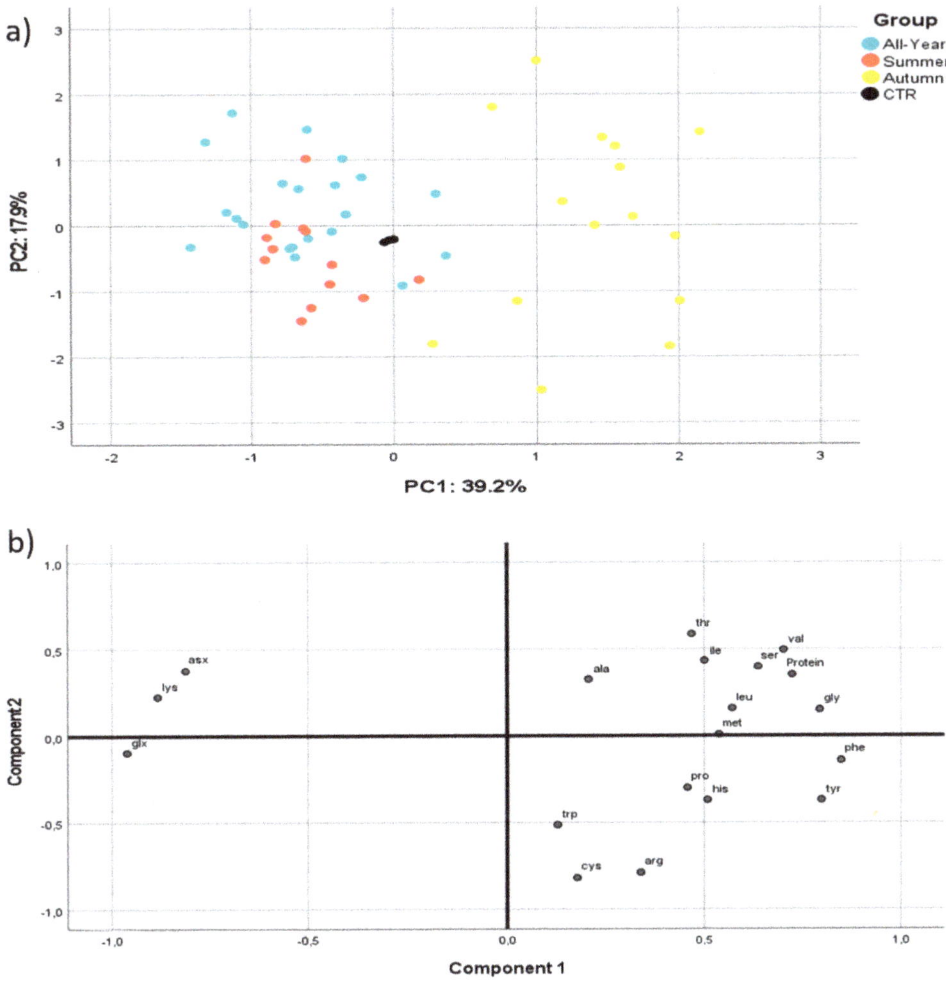

Figure 4. (**a**) Score plot of BSF samples on the first two principal components; (**b**) loadings values of the variables associated with the first two principal components.

In order to verify to what extent the nutritional properties of BSF prepupae proteins were affected by the composition of the rearing substrates, a one-way ANOVA was carried out on the essential amino acids (EAAs), dividing the samples according to the four groups of substrates (Table 2).

As a general consideration, ANOVA results confirm what is evidenced by PCA; in fact, the Autumn group is the one presenting the larger number of significant differences with respect to the other experimental diets (All-Year and Summer), and also with respect to the control diet. The BSF prepupae of the Autumn group contain the highest amounts of essential amino acids, except for lysine, which was detected in significant lower amounts compared to the BSF reared on the other substrates.

Isoleucine, methionine, cysteine and tryptophan did not differ significantly in any of the four diets.

On the other hand, specific differences in essential amino acid content of prepupae reared on the three diets were observed. Leucine turned out to be the EAA most affected from the diet, highlighting a sharp decline when samples of the All-Year group were considered. A similar trend was also observed for valine and histidine. Phenylalanine and tyrosine were lower in the All-Year and Summer groups when compared with the Autumn group, while threonine in the Summer group was present in a lower amount with respect to the Autumn group. Actually, several peculiar differences were also observed in the single samples (Table S2).

Table 2. Mean values (expressed as mg/g protein) for essential and semi-essential amino acids of BSF prepupae that had been reared on the different groups of substrates.

	All-Year (n = 21)	Summer (n = 13)	Autumn (n = 15)	CTR (n = 3)
His	33.50 a	37.86 ab	39.00 b	37.00 ab
Ile	42.89 a	43.11 a	44.10 a	45.39 a
Leu	71.24 a	77.66 bc	79.88 c	74.00 ab
Val	60.49 a	60.25 a	66.09 b	62.43 a
Lys	62.22 a	62.93 a	51.73 b	62.44 a
Cys	18.17 a	20.42 a	20.14 a	20.70 a
Met	18.37 a	17.54 a	19.16 a	18.40 a
Phe	41.69 a	40.03 a	46.47 b	42.53 ab
Tyr	61.45 a	63.16 ab	69.67 b	64.13 ab
Thr	39.50 ab	37.36 a	40.58 b	39.30 ab
Trp	14.80 a	15.76 a	15.84 a	17.60 a

Values followed by different letters within one row are significantly different (one-way ANOVA, Tukey post hoc test, $p < 0.05$).

Our results suggested a partial influence of the BSF diet on the total protein production and on their amino acid composition, and similar findings were also obtained by Spranghers et al. [8] and Soetemans et al. [45] on the tenebrionid *Alphitobius diaperinus*. However, it is not yet clear how BSF larvae convert the amino acids from the diet into amino acids useful for their metabolism and how they store and use differently the different essential amino acids. According to Liland et al. [31], BSF larvae were able to produce certain amino acids, such as tyrosine, which were almost absent in the seaweed-containing media.

Due to the complexity of the system and the possible interaction of diet components on specific essential amino acid content, a multivariate statistical analysis was used as further approach to better understand the influence of the diet composition (Table 1) on the essential amino acid profile. This approach was possible because the experimental diets were formulated according to a Design of Experiments (DoE) [32], allowing us to construct statistically reliable models describing the correlation among food leftovers' composition and amino acid content, and utilizing ANOVA to verify if the effects of the main factors and their interaction terms are statistically reliable. Results of ANOVA (Table S4) indicate the influence of many significant factors for each response, thus confirming the complex nature of these correlations. These relationships are in general better explained by the interaction between the factors rather than by an independent single factor, confirming the need for the multivariate approach. Furthermore, the fitting parameters show a fairly good fitting ($R^2 > 0.50$) only for some responses, in particular for leucine, valine and lysine (Table S4). The graphical representation of the results with acceptable fitting quality is reported in Figures 5 and 6. According to these results, for the leucine content in the BSF prepupae (Figure 5), a strong interaction emerges with the content of lipids and proteins in the rearing substrate. In particular, when the rearing diets were lacking carbohydrates, the highest content of leucine was detected when the content of proteins and lipids in the substrate was equal to or above 50 wt% (Figure 5a). In the presence of a higher content of carbohydrates (50 wt%), the highest content of leucine in BSF prepupae is obtained with slightly lower contents of both proteins and lipids in the rearing diet (Figure 5b). Overall, the increase in carbohydrate content leads to a reduction in the red area, thus indicating that a smaller combination of rearing substrates is suitable for an increase in the leucine content of the BSF prepupae. In conclusion, the highest leucine content in the BSF prepupae can be achieved by increasing the amounts of protein and lipids while reducing the carbohydrates content in the rearing diet. Similarly (data not shown), the valine content in BSF prepupae is enhanced by maximising the content of lipids and proteins and reducing that of carbohydrates in the rearing substrate. The correlation between the content of leucine and valine in the BSF prepupae with the protein content of the rearing substrates could indicate that these amino acids are essential for BSF development.

Finally, the contour plots related to the content of lysine in BSF prepupae (Figure 6) show that carbohydrate variation plays a crucial role in enhancing this amino acid content. In fact, the highest content of lysine in the insects can only be obtained when the rearing substrate has at least 90% DM of carbohydrates (Figure 6b) and a slightly higher amount of lipids compared to protein and fibre. A rearing substrate based on 75% DM of carbohydrates (Figure 6a) will result in prepupae with lower amounts of lysine. In any case, to obtain average values of lysine in the BSF prepupae, in both situations a well-balanced amount of protein and fibre in the rearing substrate is needed.

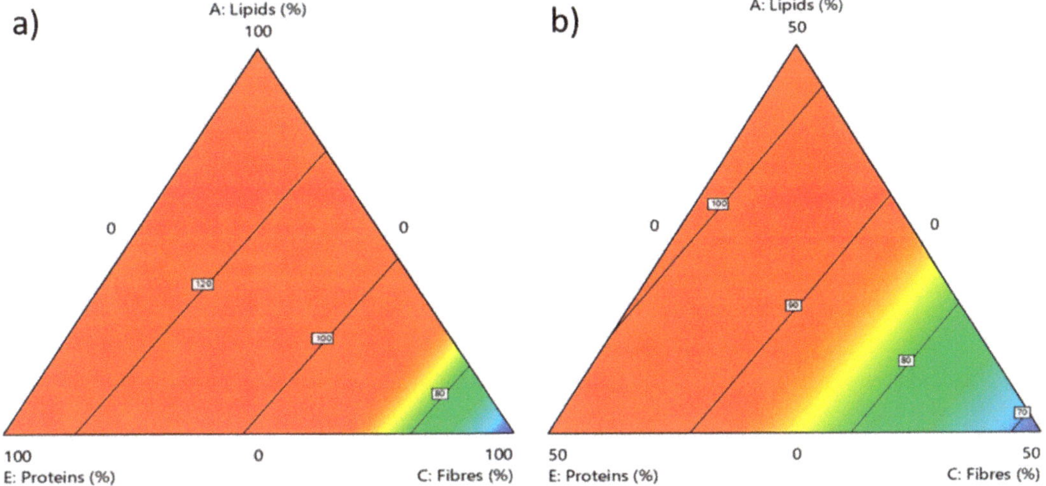

Figure 5. Graphical model variation of the amount of leucine in the proteins of BSF prepupae in relation to the composition of the rearing substrate in terms of lipids, proteins and fibres, considering two scenarios: (**a**) carbohydrates = 0 wt%; (**b**) carbohydrates = 50 wt%. The region representing the highest value of the response is shown in red colour whereas the lowest values are in blue. For each response, the most significant factor has been considered for the graphical model, expecting a higher variation in the response behaviour.

As a general consideration, it is important to highlight that none of the BSF prepupae reared on experimental diets contained significantly lower amounts of essential amino acids with respect to the control group, except in the above-mentioned case of lysine. Despite the specific differences, the majority of the proteins obtained from the BSF prepupae reared on the experimental substrates contain on average a sufficient quantity of each EAA required for human consumption, as shown in Table 3. Recent studies have shown that BSF prepupae contain optimal amounts of all the essential amino acids (EAA) to satisfy the human adult requirements, as established in the reference of FAO/WHO [7,17,46].

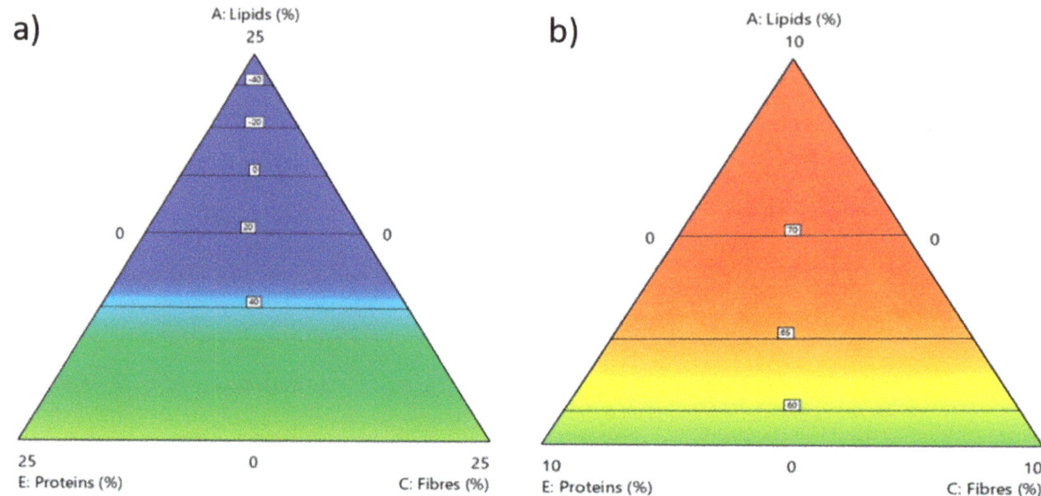

Figure 6. Graphical model variation of the amount of lysine in the proteins of BSF prepupae in relation to the composition of the rearing substrate in terms of lipids, protein and fibre, considering two scenarios: (**a**) carbohydrates = 75 wt%; (**b**) carbohydrates = 90 wt%. The region representing the highest value of the response is shown in red colour, whereas the lowest values are in blue. For each response, the most significant factor has been considered for the graphical model, expecting higher variation in the response behaviour.

Table 3. Highest, lowest and average essential amino acid content of BSF prepupae, compared with the FAO protein reference for human adults (2011).

	Reference Protein FAO 2011 (mg/g Protein)	Average Values Found in BSF Prepupae Protein (mg/g Protein)	Lowest Values Found in BSF Prepupae Protein (mg/g Protein)	Highest Values Found in BSF Prepupae Protein (mg/g Protein)
His	16	36	23	46
Ile	30	44	39	50
Leu	61	76	67	89
Lys	48	58	47	72
Cys + Met	23	38	32	45
Phe + Tyr	41	107	91	127
Thr	25	39	33	44
Trp	6.6	15	7	19
Val	40	62	56	72

According to these figures, all the analysed BSF prepupae can meet the FAO requirements, and, as in the case of histidine and tryptophan, the EAA content is even double the required amount. Moreover, the lowest values found in the samples turned out to be higher than the recommended amounts, except for lysine, which resulted in being at the lower limit in few cases (two of the BSF samples reared on the Autumn substrates). This finding is particularly relevant, considering that the analysed prepupae samples did not undergo any thermal treatment, which, on the other hand, is very likely to occur when BSF proteins are to be used in food and feed formulations, thus further lowering the lysine amount through the Maillard reaction. Moreover, one factor that could affect the lysine content is the killing method for BSF prepupae [47]. Killing by freezing, which was the method used in this experimental plan, leads to some alteration of the total amino acid fraction, with the notable decrease of lysine and cysteine, likely involved in the process of melanisation, reacting with quinones with their side chain.

4. Conclusions

Due to its physiological characteristics and excellent nutritional properties, the BSF is considered one of the most promising candidates in insect farming for feed and food purposes. Aiming at evaluating the effect of the nutrients of the rearing substrate on the protein content and composition of BSF prepupae, we examined 49 different substrates that consisted of variable proportions of different vegetable by-products and were divided into three groups according to their seasonal availability. The results showed that the total protein content of BSF prepupae ranged between 35% and 49% DM, with the highest values in the Autumn group substrates. It has also been observed that a higher protein content in the diet has resulted in a higher prepupae protein content up to a certain value; this vegetable-to-animal protein conversion lowered gradually as dietary proteins were increased, indicating the existence of a minimal critical amount of protein that has to be present in the diet, and that once reached makes any further protein addition to the rearing substrate much less relevant. Dietary fibres also seem to play a positive role in the achievement of BSF prepupae protein biomass. An important outcome of this work focuses on the essential amino acids. Higher amounts of essential amino acids in the BSF prepupae that had been reared with the substrates of the Autumn group were observed. Lysine, leucine and valine were found to be the most correlated with the presence of nutrients of the feeding diet. Leucine and valine were strongly dependent on the content of protein and lipid in the diet, while lysine is correlated to the amount of carbohydrates. Despite these important differences, the essential amino acid composition almost always fully satisfied the FAO nutritional requirements for humans. The only exception was lysine, and only in a very limited number of cases.

In conclusion, this study shows that by providing BSF larvae with substrates based on a very wide range of combinations of vegetable by-products, it is possible to obtain in the BSF prepupae a protein quality very similar to the one obtained with the control diet. However, when the employed leftovers have a very low-quality nutritional content, the development of BSF biomass is less efficient, and as a consequence a lower amount of protein is obtained. Thus, growing on these substrates should be performed by carefully balancing the advantage of re-using leftovers with the disadvantage of having a slightly lower amount of protein. These findings are very important in view of promoting BSF as a flexible tool able to bio-convert a wide range of vegetable by-products, which can vary according to seasonality or areas of production.

Supplementary Materials: The following are available online at https://www.mdpi.com/article/10.3390/foods10081773/s1, Table S1: Formulation and proximate composition of the rearing substrates, and total weight of corresponding prepupae biomass, Table S2: Complete amino acid composition (mg AA/g protein) of BSF prepupae grown on different diets, Table S3: PCA loadings, Table S4: Summary of ANOVA results.

Author Contributions: Conceptualization, A.C., S.S., L.M. and M.M.; methodology, A.F., S.B., L.I.M., L.M., M.M. and A.C.; software, S.B.; formal Analysis, A.F. and S.B.; investigation, A.F., L.I.M. and A.V.L.; writing—original draft preparation, A.F. and A.C.; writing—review and editing, L.M., M.M. and S.S.; supervision, A.C.; project administration, L.M., M.M. and A.C. All authors have read and agreed to the published version of the manuscript.

Funding: This work was supported by the Emilia-Romagna Region within the Rural Development Plan 2014–2020 Op. 16.1.01—GO EIP-Agri—FA 5C, Pr. "BIOECO-FLIES".

Data Availability Statement: The data presented in this study are available here and in the Supplementary Material.

Acknowledgments: The authors are grateful to Andrea Polacco for helping in laboratory analyses.

Conflicts of Interest: The authors declare that they have no known competing financial interest or personal relationship that could have appeared to influence the work reported in this paper.

References

1. Meyer-Rochow, V.B. Can insects help to ease the problem of world food shortage? *Search* **1975**, *6*, 261–262.
2. van Huis, A.; van Itterbeeck, J.; Klunder, H.; Mertens, E.; Halloran, A.; Muir, G.; Vantomme, P. *Edible Insects: Future Prospects for Food and Feed Security*; Food and agriculture Organization of the United Nation: Rome, Italy, 2013; ISBN 978-92-5-107596-8.
3. Rumpold, B.A.; Schlüter, O.K. Nutritional composition and safety aspects of edible insects. *Mol. Nutr. Food Res.* **2013**, *57*, 802–823. [CrossRef] [PubMed]
4. Chen, X.; Feng, Y.; Zhang, H. Review of the nutrition value of edible insects. In Proceedings of the Forest Insects as Food: Humans Bite Back, Chiang Mai, Thailand, 19–21 February 2008.
5. Bessa, L.W.; Pieterse, E.; Marais, J.; Hoffman, L.C. Why for feed and not for human consumption? The black soldier fly larvae. *CRFSFS* **2020**, *19*, 2747–2763. [CrossRef]
6. Oonincx, D.G.A.B.; van Itterbeeck, J.; Heetkamp, M.J.W.; van den Brand, H.; van Loon, J.J.A.; van Huis, A. An exploration on greenhouse gas and ammonia production by insect species suitable for animal or human consumption. *PLoS ONE* **2010**, *5*, e14445. [CrossRef] [PubMed]
7. Cappellozza, S.; Leonardi, M.G.; Savoldelli, S.; Carminati, D.; Rizzolo, A.; Cortellino, G.; Terova, G.; Moretto, E.; Badaile, A.; Concheri, G.; et al. A first attempt to produce proteins from insects by means of a circular economy. *Animals* **2019**, *9*, 278. [CrossRef] [PubMed]
8. Spranghers, T.; Ottoboni, M.; Klootwijk, C.; Ovyn, A.; Deboosere, S.; De Meulenaer, B.; Michiels, J.; Eeckhout, M.; De Clercq, P.; De Smet, S. Nutritional composition of black soldier fly (*Hermetia illucens*) prepupae reared on different organic waste substrates. *J. Sci. Food Agric.* **2017**, *97*, 2594–2600. [CrossRef] [PubMed]
9. Gustavsson, J.; Cederberg, C.; Sonesson, U.; van Otterdijk, R.; Meybeck, A. *Global Food Losses and Food Waste: Extent, Causes and Prevention*; Food and Agriculture Organization of the United Nations: Rome, Italy, 2011.
10. Bortolini, S.; Macavei, L.I.; Hadj Saadoun, J.; Foca, G.; Ulrici, A.; Bernini, F.; Malferrari, D.; Setti, L.; Ronga, D.; Maistrello, L. *Hermetia illucens* (L.) larvae as chicken manure management tool for circular economy. *J. Clean. Prod.* **2020**, *262*, 121289. [CrossRef]
11. Madau, F.A.; Arru, B.; Furesi, R.; Pulina, P. Insect farming for feed and food production from a circular business model perspective. *Sustainability* **2020**, *12*, 5418. [CrossRef]
12. Newton, G.L.; Booram, C.V.; Barker, R.W.; Hale, O.M. Dried *Hermetia illucens* Larvae Meal as a Supplement for Swine. *J. Anim. Sci.* **1977**, *44*, 395–400. [CrossRef]
13. Gold, M.; Tomberlin, J.K.; Diener, S.; Zurbrügg, C.; Mathys, A. Decomposition of biowaste macronutrients, microbes, and chemicals in black soldier fly larval treatment: A review. *Waste Manag.* **2018**, *82*, 302–318. [CrossRef]
14. Barragan-Fonseca, K.B.; Dicke, M.; van Loon, J.J.A. Nutritional value of the black soldier fly (*Hermetia illucens* L.) and its suitability as animal feed—A review. *J. Insects Food Feed* **2017**, *3*, 105–120. [CrossRef]
15. Li, W.; Li, Q.; Zheng, L.; Wang, Y.; Zhang, J.; Yu, Z.; Zhang, Y. Potential biodiesel and biogas production from corncob by anaerobic fermentation and black soldier fly. *Bioresour. Technol.* **2015**, *194*, 276–282. [CrossRef] [PubMed]
16. Gortari, M.C.; Hours, R.A. Biotechnological processes for chitin recovery out of crustacean waste: A mini-review. *Electron. J. Biotechnol.* **2013**, *16*, 14. [CrossRef]
17. Caligiani, A.; Marseglia, A.; Leni, G.; Baldassarre, S.; Maistrello, L.; Dossena, A.; Sforza, S. Composition of black soldier fly prepupae and systematic approaches for extraction and fractionation of proteins, lipids and chitin. *Food Res. Int.* **2018**, *105*, 812–820. [CrossRef]
18. Choi, Y.-C.; Choi, J.-Y.; Kim, J.-G.; Kim, M.-S.; Kim, W.-T.; Park, K.-H.; Bae, S.-W.; Jeong, G.-S. Potential Usage of Food Waste as a Natural Fertilizer after Digestion by *Hermetia illucens* (*Diptera: Stratiomyidae*). *Int. J. Ind. Entomol.* **2009**, *19*, 171–174.
19. Setti, L.; Francia, E.; Pulvirenti, A.; Gigliano, S.; Zaccardelli, M.; Pane, C.; Caradonia, F.; Bortolini, S.; Maistrello, L.; Ronga, D. Use of black soldier fly (*Hermetia illucens* (L.), *Diptera: Stratiomyidae*) larvae processing residue in peat-based growing media. *Waste Manag.* **2019**, *95*, 278–288. [CrossRef]
20. Cai, M.; Zhang, K.; Zhong, W.; Liu, N.; Wu, X.; Li, W.; Zheng, L.; Yu, Z.; Zhang, J. Bioconversion-Composting of Golden Needle Mushroom (*Flammulina velutipes*) Root Waste by Black Soldier Fly (*Hermetia illucens*, *Diptera: Stratiomyidae*) Larvae, to Obtain Added-Value Biomass and Fertilizer. *Waste Biomass Valorization* **2019**, *10*, 265–273. [CrossRef]
21. Meneguz, M.; Schiavone, A.; Gai, F.; Dama, A.; Lussiana, C.; Renna, M.; Gasco, L. Effect of rearing substrate on growth performance, waste reduction efficiency and chemical composition of black soldier fly (*Hermetia illucens*) larvae. *J. Sci. Food Agric.* **2018**, *98*, 5776–5784. [CrossRef]
22. Diener, S.; Zurbrugg, C.; Roa Gutiérrez, F.; Nguyen, H.D.; Morel, A.; Koottatep, T.; Tockner, K. Black soldier fly larvae for organic waste treatment-prospects and constraints. In Proceedings of the WasteSafe 2011 2nd International Conference on Solid Waste Management in the Developing Countries, Khulna, Bangladesh, 13–15 February 2011.
23. Oonincx, D.G.A.B.; van Huis, A.; van Loon, J.J.A. Nutrient utilisation by black soldier flies fed with chicken, pig, or cow manure. *J. Insects Food Feed* **2015**, *1*, 131–139. [CrossRef]
24. Banks, I.J.; Gibson, W.T.; Cameron, M.M. Growth rates of black soldier fly larvae fed on fresh human faeces and their implication for improving sanitation. *Trop. Med. Int. Health* **2014**, *19*, 14–22. [CrossRef]
25. European Commission Regulation. *Commission Regulation (EU) 2017/893-of 24 May 2017-Amending Annexes I and IV to Regulation (EC) No 999/2001 of the European Parliament and of the Council and Annexes X, XIV and XV to Commission Regulation (EU)*; EU: Brussels, Belgium, 2017.

26. EFSA Scientific Committee. Risk profile related to production and consumption of insects as food and feed. *EFSA J.* **2015**, *13*, 4257. [CrossRef]
27. Lalander, C.; Senecal, J.; Gros Calvo, M.; Ahrens, L.; Josefsson, S.; Wiberg, K.; Vinnerås, B. Fate of pharmaceuticals and pesticides in fly larvae composting. *Sci. Total Environ.* **2016**, *565*, 279–286. [CrossRef]
28. Purschke, B.; Scheibelberger, R.; Axmann, S.; Adler, A.; Jäger, H. Impact of substrate contamination with mycotoxins, heavy metals and pesticides on the growth performance and composition of black soldier fly larvae (*Hermetia illucens*) for use in the feed and food value chain. *Food Addit. Contam. Part A Chem. Anal. Control. Expo. Risk Assess.* **2017**, *34*, 1410–1420. [CrossRef] [PubMed]
29. Leni, G.; Cirlini, M.; Jacobs, J.; Depraetere, S.; Gianotten, N.; Sforza, S.; Dall'Asta, C. Impact of naturally contaminated substrates on alphitobius diaperinus and *Hermetia illucens*: Uptake and excretion of mycotoxins. *Toxins* **2019**, *11*, 476. [CrossRef] [PubMed]
30. Bosch, G.; Van Der Fels-Klerx, H.J.; De Rijk, T.C.; Oonincx, D.G.A.B. Aflatoxin B1 tolerance and accumulation in black soldier fly larvae (*Hermetia illucens*) and yellow mealworms (*Tenebrio molitor*). *Toxins* **2017**, *9*, 185. [CrossRef] [PubMed]
31. Gere, A.; Radványi, D.; Héberger, K. Which insect species can best be proposed for human consumption? *Innov. Food Sci. Emerg. Technol.* **2019**, *52*, 358–367. [CrossRef]
32. Liland, N.S.; Biancarosa, I.; Araujo, P.; Biemans, D.; Bruckner, C.G.; Waagbø, R.; Torstensen, B.E.; Lock, E.J. Modulation of nutrient composition of black soldier fly (*Hermetia illucens*) larvae by feeding seaweed-enriched media. *PLoS ONE* **2017**, *12*, e0183188. [CrossRef]
33. Barbi, S.; Macavei, L.I.; Fuso, A.; Luparelli, A.V.; Caligiani, A.; Ferrari, A.M.; Maistrello, L.; Montorsi, M. Valorization of seasonal agri-food leftovers through insects. *Sci. Total Environ.* **2020**, *709*, 136209. [CrossRef] [PubMed]
34. Sheppard, D.C.; Tomberlin, J.K.; Joyce, J.A.; Kiser, B.C.; Sumner, S.M. Rearing Methods for the Black Soldier Fly (*Diptera: Stratiomyidae*). *J. Med. Entomol.* **2002**, *39*, 695–698. [CrossRef]
35. AOAC. *Official Method of Analysis*, 16th ed.; Association of official analytical: Washington, DC, USA, 2002.
36. Janssen, R.H.; Vincken, J.P.; Van Den Broek, L.A.M.; Fogliano, V.; Lakemond, C.M.M. Nitrogen-to-Protein Conversion Factors for Three Edible Insects: *Tenebrio molitor*, Alphitobius diaperinus, and *Hermetia illucens*. *J. Agric. Food Chem.* **2017**, *65*, 2275–2278. [CrossRef]
37. Delgado-Andrade, C.; Rufián-Henares, J.A.; Jiménez-Pérez, S.; Morales, F.J. Tryptophan determination in milk-based ingredients and dried sport supplements by liquid chromatography with fluorescence detection. *Food Chem.* **2006**, *98*, 580–585. [CrossRef]
38. Montgomery, D.C. Design and Analysis of Experiments Eighth Edition. 2012. Available online: http://faculty.business.utsa.edu/manderso/STA4723/readings/Douglas-C.-Montgomery-Design-and-Analysis-of-Experiments-Wiley-2012.pdf (accessed on 20 December 2020).
39. Kim, E.; Park, J.; Lee, S.; Kim, Y. Identification and Physiological Characters of Intestinal Bacteria of the Black Soldier Fly, *Hermetia illucens*. *Korean J. Appl. Entomol.* **2014**, *53*, 15–26. [CrossRef]
40. Jeon, H.; Park, S.; Choi, J.; Jeong, G.; Lee, S.B.; Choi, Y.; Lee, S.J. The intestinal bacterial community in the food waste-reducing larvae of *Hermetia illucens*. *Curr. Microbiol.* **2011**, *62*, 1390–1399. [CrossRef] [PubMed]
41. Terra, W.R.; Ferreira, C. Biochemistry and Molecular Biology of Digestion. In *Insect Molecular Biology and Biochemistry*; Academic press: Cambridge, MA, USA, 2012; pp. 365–418. [CrossRef]
42. Lee, C.M.; Lee, Y.S.; Seo, S.H.; Yoon, S.H.; Kim, S.J.; Hahn, B.S.; Sim, J.S.; Koo, B.S. Screening and characterization of a novel cellulase gene from the gut microflora of *Hermetia illucens* using metagenomic library. *J. Microbiol. Biotechnol.* **2014**, *24*, 1196–1206. [CrossRef]
43. Bonelli, M.; Bruno, D.; Brilli, M.; Gianfranceschi, N.; Tian, L.; Tettamanti, G.; Caccia, S.; Casartelli, M. Black soldier fly larvae adapt to different food substrates through morphological and functional responses of the midgut. *Int. J. Mol. Sci.* **2020**, *21*, 4955. [CrossRef] [PubMed]
44. Barragán-Fonseca, K.B. Flies Are What They Eat. 2018. Available online: https://www.researchgate.net/publication/325956717_Flies_are_what_they_eat#fullTextFileContent (accessed on 15 January 2021).
45. Soetemans, L.; Gianotten, N.; Bastiaens, L. Agri-food side-stream inclusion in the diet of alphitobius diaperinus. Part 2: Impact on larvae composition. *Insects* **2020**, *11*, 190. [CrossRef] [PubMed]
46. De Marco, M.; Martínez, S.; Hernandez, F.; Madrid, J.; Gai, F.; Rotolo, L.; Belforti, M.; Bergero, D.; Katz, H.; Dabbou, S.; et al. Nutritional value of two insect larval meals (*Tenebrio molitor* and *Hermetia illucens*) for broiler chickens: Apparent nutrient digestibility, apparent ileal amino acid digestibility and apparent metabolizable energy. *Anim. Feed Sci. Technol.* **2015**, *209*, 211–218. [CrossRef]
47. Leni, G.; Caligiani, A.; Sforza, S. Killing method affects the browning and the quality of the protein fraction of Black Soldier Fly (*Hermetia illucens*) prepupae: A metabolomics and proteomic insight. *Food Res. Int.* **2018**, *115*, 116–125. [CrossRef] [PubMed]

 *foods*

Article

Yellow Mealworm and Black Soldier Fly Larvae for Feed and Food Production in Europe, with Emphasis on Iceland

Runa Thrastardottir [1], **Hildur Thora Olafsdottir** [2] **and Ragnheidur Inga Thorarinsdottir** [1,2,*]

[1] Faculty of Agricultural Sciences, The Agricultural University of Iceland, IS-311 Hvanneyri, Iceland; runa.th@simnet.is

[2] Faculty of Civil and Environmental Engineering, The University of Iceland, IS-107 Reykjavik, Iceland; htho4@hi.is

* Correspondence: ragnheidur@lbhi.is

Abstract: Insects are part of the diet of over 2 billion people worldwide; however, insects have not been popular in Europe, neither as food nor as a feed ingredient. This has been changing in recent years, due to increased knowledge regarding the nutritional benefits, the need for novel protein production and the low environmental impact of insects compared to conventional protein production. The purpose of this study is to give an overview of the most popular insects farmed in Europe, yellow mealworm, *Tenebrio molitor*, and black soldier fly (BSF), *Hermetia illucens*, together with the main obstacles and risks. A comprehensive literature study was carried out and 27 insect farming companies found listed in Europe were contacted directly. The results show that the insect farming industry is increasing in Europe, and the success of the frontrunners is based on large investments in technology, automation and economy of scale. The interest of venture capital firms is noticeable, covering 90% of the investment costs in some cases. It is concluded that insect farming in Europe is likely to expand rapidly in the coming years, offering new proteins and other valuable products, not only as a feed ingredient, but also for human consumption. European regulations have additionally been rapidly changing, with more freedom towards insects as food and feed. There is an increased knowledge regarding safety concerns of edible insects, and the results indicate that edible insects pose a smaller risk for zoonotic diseases than livestock. However, knowledge regarding risk posed by edible insects is still lacking, but food and feed safety is essential to put products on the European market.

Keywords: mealworm; black soldier fly larvae; insect farming; novel protein; Europe; food; feed; Iceland

Citation: Thrastardottir, R.; Olafsdottir, H.T.; Thorarinsdottir, R.I. Yellow Mealworm and Black Soldier Fly Larvae for Feed and Food Production in Europe, with Emphasis on Iceland. *Foods* **2021**, *10*, 2744. https://doi.org/10.3390/foods10112744

Academic Editors: Victor Benno Meyer-Rochow and Chuleui Jung

Received: 25 September 2021
Accepted: 3 November 2021
Published: 9 November 2021

Publisher's Note: MDPI stays neutral with regard to jurisdictional claims in published maps and institutional affiliations.

1. Introduction

The act of eating insects is called entomophagy and comes from the Greek terms "entomos", meaning "insects", and "phagein", meaning "to eat" [1]. Humans have eaten insects as a part of their diet for millennia all around the world [2] and today insects are part of the diet of over 2 billion people worldwide. About 2000 species of insects are eaten in the world today. Most of them are eaten in Central and South America (679 species), and 549 species of insects are consumed in Mexico alone. Entomophagy is also widely practiced in Africa (524 species), Asia (349 species) and Australia (152 species). However only 41 species of insects are eaten in Europe [3]. In 2019 it was estimated that 9 million Europeans consumed insects [4], which is about 1.2% of the European population in 2019 [5]. In comparison, a survey performed in Kinshasa in the republic of Congo in 2003 reported that 70% of the city's inhabitants consumed insects [6]. Additionally, about 25% of the world population consumes insects today [3,7]. However, the consumption of insects is declining in Asia but has also been reinvented in new forms and contexts, according to Andrew Müller [8]; this might be caused by several reasons as a form of modernity and globalization.

In spite of the high consumption of insects in the world, Europeans have abandoned entomophagy a long time ago and consider it to be a primitive behaviour [9]; however, this is slowly changing [4]. There are some speculations on why Europeans have abandoned this practice but one of the most likely reasons is the difference in the weather. Europe is in the temperate zone where insect species are smaller than in the tropics and insects are unavailable in the wintertime [10]. However, the ancient Greeks and Romans ate insects [11] and cockchafer (type of a beetle) soup was consumed in Central Europe even until the 20th Century [12]. There is an increased interest for edible insects in Europe today [10], both as a source of food and feed, which can be traced back to a 1975 publication by Meyer-Rochow [13], who urged Food and Agriculture Organization of the United Nations (FAO) and World Health Organisation WHO to take up the idea and support the use of edible insects as a food item for humans and animals. This interest increased rapidly when insects became regarded as a novel food in the European Union (EU) in 2015. The number of research papers on mealworms in Europe were 65 between 2012 and 2015 and 133 between 2016 and 2019 [3].

The global population is projected to increase to approximately 10 billion by 2050. Although the food produced could feed 10 billion people [14], 10% of the world still suffers from hunger every day [15]. Only two third of all food produced is consumed and the rest represents a huge waste of natural resources [14]. To counter this problem and to stop world hunger before 2030, new ways of producing and using food is required [15,16] along with alternative food and feed sources [17]. One way to reduce waste and world hunger is to grow insects on organic waste for the production of animal feed or food [6]. Studies have shown that both mealworms [18] and black soldier fly (BSF) larvae can be grown on waste [19], along with several other insect species [20].

Another thing to consider is the environmental effects of food production; the current food system is responsible for 80% of deforestation, 29% of all greenhouse gas emission, and agriculture uses 34% of all land on the planet, and withdraws 70% of freshwater and is responsible for 68% of animal extinction [14]. Growing insects requires less greenhouse gas production, water use and use of land per kg food produced than livestock, and thus can be produced in an environmentally sustainable manner [18].

Insects as feed are also considered to have less of an environmental impact compared with most-used protein sources today, soybean and fishmeal [21]. For example, oceans are overfished and 20% of all wild caught fish is used for aquaculture feed, fishmeal [14]. Most feed protein sources such as soybean and fishmeal are imported into the EU [17], with South America being the biggest producer of fishmeal [6] and South America and the United States being the biggest producers of soybean [22]. However, some fishmeal is produced in Iceland, Ireland, Denmark, Faroe Islands, Norway, United Kingdom, Estonia and Spain [23] and some soybean is produced in Ukraine, Russia, Italy, Serbia, France, Romania, Hungary and Austria [22]. The recent high demand has led to the high prices of these feed today [6], and the prices are expected to increase even more [17].

In 1975, edible insects were suggested to be able to counteract food shortage by increasing the use of them as food and feed [13]. Edible insects have been used as an alternative protein source for both humans and animals and research has shown that insects have a good nutrient value for humans, poultry, pork [18] and for aquaculture [24].

In this article, the production of mealworm and BSF larvae as food and feed in Europe is studied. Companies that have already started farming these species were contacted and asked key questions about their operations, investments and current status. The article also discusses the obstacles of using insects as feed or food in Europe and the food security related to insect farming. European regulations and how they are developing are presented as is how insect farming supports the Sustainable Development Goals (SDG) and European strategies. Moreover, the consumer acceptance of insects for feed and/or food is estimated. Finally, the article looks deeper into the situation in Iceland.

2. The Benefits of Breeding Insects

Edible insects have been shown to have nutritional, ecological and economic advantages, and increased insect farming is considered to promote increased food security across the world. Insect products are also considered to be beneficial for the health and welfare of livestock and they could lead to reduced antibiotic use in livestock production [25].

2.1. Health Benefits

2.1.1. Nutritional Value as Food

Edible insects are considered to be a valuable source of nutrients, with a high amount of energy, protein and fats. They are high in amino acids and monosatured fatty acids, which meets the requirements of humans. Besides being high in nutrients, edible insects are also rich in certain vitamins and minerals [25] and have a high content of fiber compared to livestock, as seen in Table 1 [26,27]. It has also been reported that insect protein has as many nutritional benefits as milk proteins [28] and that edible insects might decrease cholesterol levels in humans by 60% [29].

Insect species are highly variate in crude protein content, but on average, the crude protein content of edible insects ranges between 35–60% dry matter, which is higher than plant protein sources, including cereal, soybeans and lentils and the insects with a higher amount provide more protein than even meat and chicken eggs [30]. Live mealworm include about 20% protein, while dried mealworm includes about 53% protein [31]; the average protein content of BSF larvae is between 38–48% [27]. According to Liu et al. [32], the crude protein content of BSF varies between diverse lifecycle stages. In 1 day larvae, 14 days larvae, prepupae, pupae and adults it is 56.2%, 39.2%, 40.3%, 45% and 43.9% respectively. However, crude protein content is overestimated when using a nitrogen to protein conversion factor of 6.25, mainly due to the presence of chitin [33], which is not digestable, but insect protein digestibility is estimated to be between 77–98% [31].

Insect fat content varies between species, sex, reproduction stage, season, diet and habitat [31]. The average fat content of edible insects is between 2–50% dry matter [34]. The average fat content of mealworms is between 19.12–34.54% [18] and the average fat content of BSF larvae is between 15–35% [27]. Mealworms are reported to have a high level of polyunsaturated fatty acids [17]; however, studies have shown that edible insects are in general low in omega-3 fatty acids and have a high omega-6/omega-3 ratio. This can be changed by adding omega-3 fatty acids to insect diets [35]. Insects are reported to have a higher content of energy, sodium and saturated fat than conventional livestock. A high content of sodium and saturated fat in food can lead to over-nutrition=linked diseases such as heart diseases. However, insects tend to have a very high micronutrient content, especially in the micronutrients that are known to be deficient in many areas where food insecurity is high. This shows that meat products may be nutritionally preferable to certain insects in the context of overnutrition, and that severeal insects are potentially superior to meat in the context of undernutrition. However, nutritional composition of a product does not say everything about its effect on human health [36].

It has been reported that nutritional quality of edible insects varies greatly depending on the insect diet [18,27,31]. BSF larvae have been reported to be able to accumulate both lipid- and water-soluble nutrients from their diet, and BSF reared on brewery waste or a mixture of fruit and vegetables have been shown to have a higher protein content than BSF reared on fruit or winery by-products [27]. Futhermore, mealworms reared on plant waste have a higher protein content and a lower fat content than mealworms reared on a cereal-based diet [18] and, as mentioned earlier, adding omega-3 fatty acids into insect diets can decrease the omega-6/omega-3 ratio [35]. Additionally, it has been reported that processing methods can affect the nutritional quality of edible insects [27,37]. In a study by Nyangena et al. [27], it was reported that heat processing increases the protein content and decreases the fat content of BSF larvae.

Table 1. Nutritional level in raw products, mealworms and BSF larvae compared to ground pork, ground beef, ground chicken, and farmed Atlantic salmon.

	Mealworm	BSF Larvae	Pork	Beef Cattle	Poultry	Salmon	Daily Value [1]
General nutritional profile	[18,26,36,38]	[27,39–42]		[36,43]			[36,44]
Crude protein (g/100 g)	15.80–18.60	12.0–36.3	15.41–31.69	15.76–29.46	17.44–23.28	19.84–25.44	50
Fat (g/100 g)	10–26.6	12.25–29.8	4–33	3–30	8.1–13.9	6.34–13.42	65
Crude fiber (g/100 g)	0.68–1.29	7.9–8.1	0	0	0	0	
Energy (kcal/100 g)	152–268		121–393	121–332	143–198	142–208	2000
Crude ash (g/100 g)	1.13	3.9–15.8	0.79–1.49	0.7–1.71	1.17–1.57	1.13–3.26	
Fatty acids	[36,38,45]	[41,46]		[36,43]			[36,44]
Saturated fat (g/100 g)	2.58–8.97	6.14–23.50	1.42–11.31	1.48–11.75	0.8–4	0.98–3.05	20
Monounsaturated fatty acids (g/100 g)	3.79–14.29	1.49–8.60	1.89–15.33	1.13–14.17	3.61–4.88	2.10–4.18	
Polyunsaturated fatty acids (g/100 g)	1.10–3.17	2.07–6.39	0.66–4.32	0.22–0.70	1.51–2.08	2.54–4.55	
Vitamins	[26,36]	[41]		[36,43]			[36,44]
Vitamin A (µg/kg)	57–205		0–50	0–70	0	120–690	15,000
Niacin (mg/kg)	40.7–46.5		35.97–110.5	33.82–74.85	48.7–76.5	80.45–100.77	200
Pyridoxine (mg/kg)	6.9		1.67–7.17	2.78–4.35	5.12–5.38	6.36–9.44	
Riboflavin (mg/kg)	8.1–8.7		1.8–4.88	1.51–2.5	1.25–3.02	1.35–4.87	17
Folat (mg/kg)	1.55		0–0.06	0.040–0.21	0.01–0.02	0.25–0.34	
Biotin (mg/kg)	0.43						
Thiamin (mg/kg)	1.1–2.4		2.71–9.28	0.3–0.8	0.68–1.21	2.07–3.40	15
Vitamin B_{12} (µg/kg)	1.3		6.4–23	19.7–29	5.1–5.6	28–32.3	
Vitamin C (mg/kg)	99.0–120		0–23	0	0–20	0–39	600
Vitamin D (IU/kg)	<80		40–340	20–80		4410–5260	
Vitamin E (mg/kg)	33	53.3–248.8	2.6–4.7	0–4.3	2.7–3.9	11.4–35.5	

Table 1. *Cont.*

	Mealworm	BSF Larvae	Pork	Beef Cattle	Poultry	Salmon	Daily Value [1]
Minerals	[18,26,36,38,47]	[41,42,48]		[36,43]			[36,44]
Iron (mg/kg)	9.61–245	100–630	7–15.1	15.4–32.9	7–10	3.4–10.3	180
Zinc (mg/kg)	33.8–117.4	42–300	19.1–35.9	35.7–71.5	14.7–19.2	3.6–8.2	
Magnesium (mg/kg)	620–2027	2100–5610	160–270	140–290	210–280	270–370	
Calcium (mg/kg)	156–435	5360–61,620	60–200	50–410	60–120	90–150	10,000
Phosphorus (mg/kg)	2640–7061	6800–13,220	1610–2610	1320–2670	1780–2340	2000–2560	
Sodium (mg/kg)	225–3644	890–2500	555–940	525–960	600–895	440–610	24,000
Potassium (mg/kg)	3350–9480	10,200–18,790	2440–4280	2180–4700	5220–6770	3630–6280	
Copper (mg/kg)	8.3–20	7.5–34.25	0.33–1.31	0.5–1.08	0.62–0.65	0.45–3.21	
Manganese (mg/kg)	3.2	190–730	0.10–0.13	0.09–0.22	0.16	0.11–0.21	
Selenium (mg/kg)	0.12	0.1–1.2	0.25–0.49	0.13–0.23	0.10–0.14	0.24–0.47	

[1] Daily values from the US Food Labelling Guide [44]. All are daily reference values (DVRs) except for vitamins and minerals, which are recommended daily intake (RDI) value.

2.1.2. Nutritional Value as Feed

Insects are a natural part of the diet of some animals including pig, poultry and fish [49] and edible insects and can be used as a protein source in feed for these species along with other protein sources such as soybean meal and fishmeal [18,21,40,49,50]. Mealworms contain essential amino acid compositions sufficient to meet the dietary requirements of trout, where mealworms can replace fishmeal in the diet. However, mealworms lack sufficient methionine to meet the essential amino acid requirement of salmon, poultry and humans [21]. Studies have shown that edible insects can improve the growth rate and digestibility of poultry and pigs compared to other protein sources. Increasing the mealworm content to 15% in poultry diets increased the body weight and the daily feed intake of chicken in one study. In the same study, increasing mealworm content to 6% in pig diets increased the body weight, average daily gain, average daily feed intake and gain to feed ratio of weaning pigs [18]. Additionally, a study by Rawski et al. [40] showed that the replacement of fishmeal with BSF larvae can have positive effects on Siberian sturgeon growth performance.

2.1.3. Chitin

Besides protein and fat, insects contain fiber, mainly in the form of chitin. Chitin is a natural polysaccharide, which is probably one of the most abundant biopolymers in nature and is the second most abundant biomass in the world after cellulose. It plays a structural role in many organisms, including fungi, crustaceans, mollusks, coelomates, protozoa and green algae [51]. The composition and amount of chitin in insects can vary between species and developmental stages. Most of the chitin can be found in the exuviate (shed). Although chitin can play a role in the pathogenesis of asthma and allergies, it is considered to have potential positive effects on the immune system.

Chitosan, one of chitin's derivatives, is produced by deacetylation and can be a valuable by-product for biomedical use. Primex Iceland, an Icelandic company, has been producing chitosan from shrimp shells (*Pandalus borealis*) from the North Atlantic Ocean. They have been producing healing spray and gel for external use both for humans and animals, but also as a dietary supplement for weight management [52]. The composition and the amount of chitin varies with the species and developmental stages. The chitin contents of mealworms are considered to be between 4.92–13.0 g/100 g, with an average of 6.41 g/100 g [18]. The proportion of chitin in BSF larvae is similar, or is around 5.69–7.95 g/100 g [53].

2.1.4. Prebiotics and Probiotics

Prebiotics are defined as a fiber that stimulates the growth of preexisting good bacteria in the gut, but probiotics as a live microbial is a feed supplement that beneficially affects the host. Research has been carried out on whether chitin plays a role as a prebiotic of animal origin. It seems that chitin's derivatives, such as chitosan, chitin-glucan (GC), and chitin oligosaccharide (NACOS), show better results in the modulation of gut microbiota (GM) by enhancing the growth of beneficial bacteria and by inhibiting the growth of pathogenic bacteria along with anti-inflammatory effects. However, bacteria can have different roles in different species. The ratio between the types of bacteria seems to play an important role in human health. The derivatives showed more prebiotic activity when carried by low protein-containing food. Research has shown that fewer antibiotics are needed when insects are used as feed [51,54].

Probiotic bacteria in mealworm diets could be beneficial, because larvae are processed whole, so the residual microbiota is carried to the end consumer. It is estimated that the microbiota is up to 10% of the total insect biomass. The bacteria can also produce B vitamins and can decrease the need for antibiotics by stimulating the host immune system [17].

2.2. Environmental Benefits

2.2.1. Greenhouse Gas Emission

Edible insects have less of a negative effect on the environment than other livestock forms. Insect farming produces less greenhouse gas than livestock [25,55] and uses less land and water [25], as seen in Table 2. Moreover, insects are often considered to be environmentally friendly because their farming may have a low feed conversion ratio (FCR); however, this varies between insect species and the feed. For example, the FCR for mealworm ranges between 2.2 to 5.3, while the FCR for nymphal stage of *Acheta domesticus* ranges between 1.08 to 4.5 [56]. The FCR for edible insects can vary depending on the feed used [57]. When reared on an optimal diet, mealworms convert feed as efficient as poultry and the nitrogen use efficiency is higher than traditional livestock [17].

In a study by Oonincx et al. [55], greenhouse gasses (CO_2, CH_4 and N_2O) and ammonium (NH_3) emissions by five species of insects (including mealworms) were measured and compared to greenhouse gas and NH_3 emissions of cattle and pigs. The results showed that four of five insect species produced much less greenhouse gas than pigs and only 1% of greenhouse gas compared to ruminants. All of the insect species produced less NH_3 than cattle and pigs. Furthermore, mealworms do not produce methane (CH_4), contrary to pigs and cows. However, the true environmental effect of mass rearing of insects has not been determined [57] and a complete lifecycle analysis for edible insect species is lacking [55].

Breeding insects for feed is considered to be more environmentally friendly than the protein sources used for food and feed today, soybean and fishmeal [21]. A study by Smetana et al. [58] showed that the production of 1 kg of BSF larvae resulted in less land use, less CO_2 production and less water use than the production of both soybean meal and fishmeal. Fishmeal is produced through the overexploitation of fish in the oceans [14,25], as 20% of all wild caught fish are used for fishmeal [14]. Additionally, studies have shown that feeding chickens [59] and aquaculture [60] with mealworms can reduce these species' FCR.

Table 2. Environmental effect of mealworms and BSF compared to livestock.

Species	Total Greenhouse Gas Emission (g/kg Body Mass) [1] [55]	FCR Dry Matter [2]	Land Use m²/kg [61]	Water Footprint m³/Edible Ton [62]
Mealworm	0.45	1.6–2.1 [63]	3.56 [64]	4341
BSF	N/A	1.8 [65]	2 [66]	N/A
Pork	2.09–28.22	4.04–6.4 [67,68]	17.36	5988
Beef cattle	6.23–>7.53 [3]	18.9–25 [67,68]	326.21	15,415
Poultry	3.0–5.1 [69]	2.67–3.3 [67,68]	12.22	4325

[1] CO_2, CH_4, N_2O, NH_3; [2] FCR: Feed Conversion Ratio; [3] No data for N_2O.

2.2.2. Waste Management and Plastics

Almost 100 years ago, the idea of processing organic waste by using fly larvae was proposed. Studies have shown that several fly species are suited for the biodegradation of organic waste, e.g., BSF and house flies (*Musca domestica*) [70]. Furthermore, insects like mealworms can be used to biodegrade organic waste and plastic to proteins [18]. BSF larvae seems to be able to degrade a large variety of organic waste, ranging from food waste, agri-industry co-products, animal waste to meat-based products [71], aquaculture sludge [48], substrate containing up to 50% seaweed [41], and BSF larvae, and are also commonly found in rotten fruits and plant residues [70]. However, the composition of the substrate is important as it has a major effect on BSF development, survival, nutritional composition and the substrate bioconversion rate [71]. The composition of the substrate also affects food and feed safety, as it might contain metals and pathogens that can accumulate in the larvae [48].

A study by Tsochatzis et al. [72] showed that mealworms can be used to degrade plastics when reared on plastics, barley and water. The results indicated that plastic compounds do not bioaccumulate in mealworms and that a very low content is released into the frass. However, the consumption of plastics caused the mealworms more metabolic stress in comparison to their typical diet.

2.2.3. Environmental Risks

There are some concerns regarding the effect of a possible escape of edible insects into the environment and becoming locally invasive species to natural and production systems in non-native countries [57]. In Iceland, an environmental risk assessment for BSF was performed before receiving license from the Icelandic authorities for import and trials. The results showed that BSF poses no threat to the local insect environment, and it is highly unlikely that a wild population can form if an escape will happen. This is due to BSF being a tropical species that is not likely to survive in the cold climate of Iceland [73]. However, there is evidence that BSF could be established in Europe, especially with climate change making the establishment of many more non-native species more likely [57]. The northernmost region where wild BSF has been recorded is in the Czech Republic [74].

3. Food Security

For novel food to be placed on the European market, it must be safe, meaning it must not have any harmful effects on health or be unfit for human consumption according to the EU general food law [75]. This is also important regarding the novel products used for animal feed, as this is the most important factor to guarantee the sustainable production of safe and affordable animal proteins [76]. To prevent possible harmful effects, the risk regarding novel food and feed products must be known, and there must be techniques in place to prevent those risks [75]. The risk associated with edible insects can be allergies, toxins and pathogens [73].

3.1. Allergies

Food allergy is an adverse immunological response to a foreign substance. Further research is needed, but some studies report the potential allergy risk posed by mealworms or other insects. One study showed that mealworm proteins cross-reacted in vitro with IgE produced by patients who were allergic to house dust mites or crustaceans (crabs, lobsters, crayfish etc.) in response to tropomyosin (a structural protein found, e.g., in the cytoskeleton). Heat processing of the product reduces the allergic response, but it still exists. A double-blinded placebo study in humans showed that mealworm allergy is most likely in people allergic to shrimp, with a potentially severe outcome. A safety assessment of freeze-dried mealworm powder in rats showed no adverse effects, allergy or toxicity [17].

Another potential risk is that insects may carry mold that can cause allergic reactions. This can affect the workers in production as well as the consumers. Additionally, allergens from the feed (e.g., gluten) may end up in the insect that is consumed [77].

3.2. Toxicity

Use of insects as food and feed have raised questions about toxicity. Mealworms may contain defence substances, such as toxins produced by the exocrine and defensive glands. Focus has been placed on benzoquinones, which are secreted into the abdominal cavity in adult beetles and have toxic effects, but these findings refer to *T. molitor* beetles and not to larvae. Other hazards are contained in aflatoxin, mycotoxin, heavy metals, organic pollutants, plasticisers, flame retardants and others. It seems that different species show a different accumulation behaviour, e.g., BSF accumulates cadmium, but mealworms accumulate arsenic in the larval body and therefore it is important to keep a regular monitoring of contaminants in their feed. It is important also to keep track of every step of the production [78,79]. According to the EFSA Panel, the toxicity studies of mealworm from the literature did not raise any safety concerns, but noted that the larvae should be reared separately from the adult beetles [33,80]. Eleven applications for other species are pending for safety evaluation by EFSA [29].

3.3. Antinutrients

Anti-nutrients or antinutritional factors (ANFs) are, contrary to nutrients, compounds that reduce the absorption of nutrients. They are found in common foods such as whole grains, soybeans, spinach, broccoli, tea and coffee, even in chocolate. Glucosinolates, one anti-nutrient found in mustard and cabbage, can prevent the absorption of iodine and thus disturb thyroid function and cause goiter. They are therefore also known as goitrogens, and this is of special concern if there are pre-existing hypothyroidism. Lectins (hemagglutinins), found in the vast majority of organisms, can reduce the absorption of calcium, copper, iron, phosphorus and zinc. They are carbohydrate-binding proteins and can cause the agglutination of red blood cells. Raw legumes and whole grains contain higher levels of lectins, so it is important to take into account how the food is processed. Phytates (phytic acid) in whole grains, seeds, legumes and even nuts can decrease the absorption of iron, zinc, magnesium and calcium. Oxalates, found, for instance, in tea and chocolate, can prevent the uptake of calcium by forming calcium oxalate and tannin in tea, and coffee and legumes can decrease iron uptake. The effect of anti-nutrients differs between people's health and metabolism and which food they otherwise consume and when. Interestingly, anti-nutrients are also thought to have benefits for health, such as phytates, which have been found to lower cholesterol and to increase balance in blood sugar as well as having antioxidant effects. There are limitations however, because it is difficult to study the role of anti-nutrients in various diets and their levels differ in how the food is processed [81]. Edible insects are mostly herbivorous, as they feed on plants and plant parts. Plants synthesize different types of secondary metabolites for their self-preservation, and these secondary metabolites are known as allelochemicals and accumulate in the bodies of plant matter-ingesting insects. Insects contain a wide variety of antinutrients, which is likely caused by the different chemical compositions of plants on which the insects feed [82]. According to Turck et al. [33], the levels in whole dried mealworm larvae are comparable to the occurrence levels in other food substances. The development of rearing techniques of edible insects under controlled conditions can minimize, or even avoid the contamination of insects with antinutrients. Furthermore, it has been reported that processing methods can help to remove antinutrients and other unhealthy components [82].

3.4. Zoonosis

Microbiological hazards associated with insects as food and feed are either part of the insect's lifestyle and gut flora, or are introduced via human contact, through farming and processing. The insect's gut flora is essential for the metabolism and survival of the insects. The gut flora varies depending on the species and it includes bacteria, viruses, and fungi. Most of mealworms' and BSF larvae's gut flora is not pathogenic to humans and other animals; however, microbiota introduced during farming and processing possess a greater risk to humans and other animals [73].

3.4.1. Bacteria

There have been few studies into the microbial content of mealworms and BSF larvae, and its effect on food safety. These studies indicate that there is a high level of bacteria on the surface and in the gut of these insects [83–90].

Bacteria pathogenic to insects are considered harmless to humans and other vertebrates, since insects are so phylogenetically different. Therefore, bacterial hazards for humans and vertebrates will mainly originate from the insect microbiota, related to rearing conditions, handling, processing, and preservation [73]. Mealworm's microbiota consists mostly of Proteobacteria, Firmicutes and Actinobacteria with Propionibacterium being the most abundant taxa [83]. BSF larvae microbiota consists mainly of Proteobacteria and Firmicutes with *Providencia*, *Klebsiella* and *Bacillus* being the most abundant taxa, while the microbiota of prepupae consists mainly of Bacteroidetes, Proteobacteria, Firmicutes and Actinobacteria, with many taxa dominating the microbiota, e.g., *Providencia*, *Myroides*,

Proteus and *Morganella*, that can all act as opportunistic pathogens and may carry drug resistance [84].

Currently, no microbiological criteria exist specifically for insects sold as food; however, hygiene criteria for the processing of minced meat described in EU Regulation EC No. 1441/2007 can be used for insects. According to these criteria, the limit for the total aerobic count is 5.7 log cfu/g [83,85], the average total aerobic count in fresh and powdered mealworms and BSF larvae was higher than this limit (>8 log cfu/g on average) in most studies [83,85–87]. The current food hygiene criteria include *Salmonella enterica (S. enterica)*, *Listeria monocytogenes (L monocytogenes)*, *Escherichia coli (E. coli)*, *Staphylococcus aureus (S. aureus)* and *Bacillus cereus (B. cereus)* [88]. According to several studies, mealworms and BSF larvae do not act as a vector for *S. enterica* [73,86–89]; however, in a study done by Raimondi et al., 2020, [84] on BSF, *S. enterica* was detected in some samples of prepupae. It is believed that BSF larvae possess antimicrobial capacities that make them able to reduce pathogenic bacteria such as *S. enterica* and *E. coli* [87,90]. In most studies, *L. monocytogenes* and *E. coli* are not detected in the mealworms BSF larvae [86–89]; however, these bacteria can be detected in these species if they are reared on substrates contaminated with *L. monocytogenes* [91,92] and *E. coli*, but *E. coli* will be reduced in the larvae [92]. *Staphylococcus aureus* was detected in mealworms in a study done by Stastnik et al., 2021, [89] and coagulase-positive staphylococcus was detected in BSF prepupae in a study done by Raimondi et al., 2020 [84]. However, both studies had coagulase-positive staphylococcus under the contamination limit. All staphylococci detected in two other studies were coagulase-negative, meaning that no *S. aureus* was detected [87,88].

Bacillus cereus seems to be one of the biggest hazards regarding the use of edible insects as food [87,88]. Some strains of *B. cereus* produce toxins that cause emesis or diarrhoea [88]. To produce toxins, the density of *B. cereus* is believed to have to be around 4–5 log cfu/g. Some studies suggest that an even lower density is needed, but the density of *B. cereus* in BSF larvae in one study went up to 3.8 log cfu/g [87]; however, in another study where *B. cereus* was investigated on BSF pupae, one dried sample and one powdered sample exceeded the limit—one over 5 log cfu/g and another over 6 log cfu/g [88]. In a small study on mealworms in the Netherlands, 93% of all samples had less than 2 log cfu/g of *B. cereus* [73]. In another study, the median values of *B. cereus* in mealworm samples were in general under 4 log cfu/g, with two outliers between 4–5 log cfu/g [93]. Since neither do all *B. cereus* produce toxins nor are all non-cereus bacilli toxin free, toxic gene profiling may be a better diagnostic tool to estimate the true hazard [88].

B. cereus is also a spore-forming bacteria [87]. A high content of bacterial endospores has been found on mealworm and BSF larvae [73,86,87]. Endospores and toxins produced by *B. cereus* are heat- and processing-resistant [85,87,94]. Endospores can also germinate and produce toxins when food is not cooked, cooled and stored properly [94]. Bacterial endospores highly differentiate in number between different rearing batches from the same company [83,86,87,89]; in one study, the number of endospores in mealworms varied between 1.7 log cfu/g to 5.0 log cfu/g [86]; in another study focused on mealworms, the endospores detected were between <1 log cfu/g to 3.5 log cfu/g [84]. In a study on BSF larvae reared on different substrates, the endospores varied between 3.7 log cfu/g to 7.5 log cfu/g [87]. It is unclear why different samples reared within similar conditions have such a high variance in endospore content [83], but it is believed that the rearing substrate and insect species influence this [73,87]. *Bacillus cereus* is widely spread in soil water and in plants [87], and insects farmed on soil are believed to be more likely to include bacterial endospores [83]. According to these studies, BSF larvae seem to be more contaminated with *B. cereus* and endospores than mealworms. This could be caused by the different substrate used for these two species. While mealworms are normally fed with various flour types, and are often complemented with carrots, BSF larvae are often fed with soil and food waste that usually contains high levels of *B. cereus* and endospores. According to to Wynants et al., 2019 [87], one strategy to avoid food poisoning through BSF larvae is to only use substrates that do not carry *B. cereus*. However, this would reduce the economic positivity and sustainability of BSF rearing. Therefore, more research into *B. cereus* and endospores-contaminated insects and how to reduce the risk is needed.

Other pathogenic bacteria that have been detected in mealworms and BSF larvae are *Clostridium* spp. [83,84,87,89] and *Campylobacter* sp. [84,90,95] However, only low levels of *Clostridium perfringens* have been detected in these studies, with an average concentration under <1 log cfu/g [84,89]. High levels of *Campylobacter* sp. have been isolated from BSF prepupae [84,90] and the lesser mealworm (closely related to mealworm) has been shown to be able to infect poultry through ingestion; however, *Campylobacter* sp. is only active in the larva for 3 days after exposure [95].

Even though zoonotic pathogens found in the substrates used to grow insects could lead to insects acting as a vector for these bacteria, no active replication seems to occur in insects. However, zoonotic pathogens are widely known to be able to replicate in farmed animals, e.g., *Salmonella* [73].

3.4.2. Virus

Most viruses on insects are insect-specific and are not pathogenic for vertebrates [73]. Insects' viral pathogens are considered to be safe for humans and are approved in some cases as biocontrol agents in agriculture. The biggest problem these viruses cause is a financial burden to the insect farms, since viruses associated with insects are only pathogenic to the insects themselves [96] and may cause a loss in production. Vertebrate viruses taxonomically related to insect viruses are unable to replicate in insects, and are not actively transmitted by insects as vectors to vertebrates [73]; therefore, these viruses are not considered to lead to a health risk in humans and other vertebrates [96].

Today, there are no studies on the pathogenicity of insect-specific viruses in humans, but it is believed that the specificity of insect viruses is mainly limited to the species taxon and are unable to replicate in vertebrates. Due to the lack of comparable viruses between insects and vertebrates, the risk of recombination and reassortment of an insect-specific virus strain leading to a new mammalian specific virus strain, as was the case of Swine flu and COVID-19, is almost non-existent. Therefore, an increased consumption of insects is likely to reduce the risk of a new pandemic in the future [97].

However, viruses in insects that are called arthropod-borne viruses, or arboviruses, can cause human diseases and can replicate in both insect vectors and vertebrates. Known diseases caused by arboviruses are, e.g., West Nile disease, dengue, rift valley fever, haemorrhagic fever, and chikungunya [96]. There is no evidence that such viruses occur in insects used for food and feed [73]. Arboviruses are believed to originate from insect-specific viruses, which indicates that an evolutionary process might lead to novel insect origin pathogens in the future following the introduction of insects into the diet [97].

Another issue is that insects can also act as passive vectors of vertebrate viral diseases, where the virus does not replicate in insect vectors, but is rather carried by the vector to the host [73]. Adenovirus, norovirus, rotavirus, hepatitis E, and hepatitis A could possibly be introduced with a substrate in insect farms and could be transferred further through the production [73,96]. However, there is a lack of information relating to the likelihood of such transmission from feedstock through residual insect gut contents. Studies have shown that adenovirus, norovirus, and hepatitis A could survive in untreated manure and litter for at least 60 days at 20 °C and 4 °C, and other temperatures were not tested [73].

It has been concluded that the risk of edible insects acting as a passive vector of COVID-19 is extremely low, which demonstrates that edible insects should not be a reservoir for viral diseases with epizootic potential [97]. In the case of insects acting as passive vectors of vertebrate viruses, processing and cooking will reduce the risk of transmission in most cases [73,96].

3.4.3. Fungi and Yeast

Fungi, such as yeast and mold, are a part of edible insects' normal microbiota. These microorganisms produce spores and can easily spread to different environments and can contaminate food. Fungi causes the deterioration of food, nutritional losses, discolouration, and an off flavour and are the major organisms responsible for food spoilage. Some species of fungi are pathogenic to vertebrates and can produce toxins, e.g., mycotoxin [94]. Studies have shown that insect-specific pathogenic fungi pose a small risk to humans and other vertebrates. However, these fungi have occasionally caused diseases in immunosuppressive individuals [73,97]. Insects might also be carriers of fungi and yeast pathogenics to vertebrates and a considerable amount of fungi and yeast have been found in fresh, freeze-dried and frozen mealworms [73]. It has been reported that dried mealworms can be carriers of *Penicillium* spp., and *Mucor* spp., while fat from BSF larvae can carry *Aspergillus* spp., and *Cryptococcus neoformans*. *Aspergillus* spp., *Penicillium* spp., and *Cryptococcus neoformans* have been found in many insects and can cause opportunistic infections in humans; however, no direct infection after consuming insects has been recorded [89].

Good hygiene in the entire production chain will reduce the risk of fungi infection introduced during farming processing and storage [73]. However, if hygiene is not acceptable, studies have shown that a short-blanching of 10–40 s can considerably reduce fungi [85,94]. Incorrect storage conditions of feed intended for insects can lead to fungi formation in the feed and this type of fungi may form mycotoxins [89]. According to studies done on the accumulation of the mycotoxin in mealworms and BSF larvae, very low-levels of mycotoxins accumulate in these species. Mealworms and BSF larvae fed with feed spiked with high mycotoxin levels showed an accumulation well below the limit value [89,98–100] in food and feed, according to Commission Regulation (EC) No 1881/2006, Commission Recommendation 2006/576/EC, and Directive 2002/32/EC. The mycotoxins studied where aflatoxin B1 [89,99,100], deoxynivalenol [89,98,100], ochratoxin A [89,100] and zearalenone [100]. These results indicate that mycotoxins should not be a concern regarding the use of mealworms and BSF larvae as food or feed.

3.4.4. Parasites

Insects have been known to be able to infect humans with parasites through consumption for a long time. In 1871, it was discovered that a common parasitic disease in Russia was caused by the consumption of a raw beetle larvae that was an intermediate host for this parasitic disease [101]. Most studies on parasites in insects are related to non-European areas and insects harvested in the wild but the results from these studies suggest it to be a problem. However, the risk will be reduced in farmed insects with a strict control over the environment [73]. In a study done on edible insects as a vector for parasites, several parasitic species were detected in mealworms that can be pathogenic to humans and animals. *Cryptosporidium* was the most prevalent pathogenic parasite detected in mealworms and it was found in 16% of all analysed mealworm farms and in 12% of all samples. *Cryptosporidium* was found in the gastrointestinal tract and other parts of the mealworm's body. It is possible that mealworms can infect humans with *Cryptosporidium* aerogenically, and infection can occur on farms that are lacking in proper hygiene regarding contact with insects. Other pathogenic parasites detected in mealworms were *Isospora* spp., *Balantidium* spp. *Entamoeba* spp. *Cestoda*, *Pharyngodon* spp. larva, *Physaloptera* spp. larva, *Spiroidea* spp., and *Acanthocephala* spp. However, some of these parasites came with mealworms, which were delivered from outside of Europe and some of the farms were guilty of unethical practices that would not be accepted if the insects were farmed for food or livestock feed [102]. Another parasite that has been detected in mealworm larvae is *Toxoplasma gondii* [103].

The results from a study done on endoparasites within invertebrates used as a live feed for wild caged birds indicates a low risk for parasite transmission associated with mealworm consumption by birds [104].

Not as much research has been done on BSF larvae working as vectors for parasites like mealworm. However, there is evidence that BSF larvae can act as a vector for *Eimeria* and *Ascaris suum* [105].

According to EFSA Scientific Committee [73], insects reared in a properly managed closed farm environment would lack all the hosts necessary for the completion of parasite life cycles. Beside proper management before consumption, freezing and cooking, would further eliminate potential parasitical risk.

Canthariasis is the invasion of a living beetle larva on a living or dead organism, making them act as parasites themselves. Different species of beetle larvae lead to different pathological changes and clinical signs; the main categorization of canthariasis relies on the invasion location in the host. Mealworms rarely cause canthariasis, but there are some reported cases in the world. Mealworm larvae usually lead to gastric canthariasis [106], which can affect both humans [107] and animals through the ingestion of eggs or larvae. The clinical signs of gastric canthariasis can be nausea, vomiting, stomach-ache, abdominal bloating, loss of appetite, weight loss, and diarrhoea, resembling intestinal parasite infection. In extreme cases, the larvae penetrate the intestinal organs and invade other organs. Gastric canthariasis can lead to death if untreated [106]. Other organs mealworm larvae are known to invade are umbilicus and tonsils and there is a one known case of mealworm larva invading bladder and causing urinary canthariasis in humans [107]. Mealworms feeding live to animals and humans therefore contain a danger, but if the larvae are killed before consumption, the danger will be neglectable as long as eggs are filtered away from the larvae used for consumption.

3.4.5. Prion

Prion disease or transmissible spongiform encephalopathies are naturally occurring infectious protein-misfolding disorders that characterise the accumulation of misfolded protein aggregates in the brain. Prion diseases affects several mammals, and they are always fatal, an example of this diseases is Creutzfeldt-Jakob disease (vCJD) in humans, scrapie in sheep, bovine spongiform encephalopathy (BSE) in cattle and chronic wasting disease (CWD) in deer and elk. On rare occasions, prion diseases can be transmitted between species [108]; therefore, there exist concerns relating to the possibility of prion diseases being transmitted from insects through food or feed.

There is no evidence that there exists a special prion disease in insects, since no gene encoding prion or prion-related proteins have been reported in insects [109,110]. Therefore, mammalian prion cannot replicate in insects and insects are not considered to be possible biological vectors of mammalian prion diseases [73]. However, research has shown that insects can possibly act as a mechanical vector of prion disease. Mites from Icelandic sheep farms with a known scrapie infection were able to infect mice via intracerebral injection [111,112] and larvae of *Sarcophaga carnaria* (*S. carnaria*) fed with brain material from scrapie-infected hamsters were able to infect hamsters through an oral route at different stages and after death [109]. Additionally, studies have shown that *Drosophila melanogaster* (*D. melanogaster*) can act as a mechanical vector for prion diseases [113]. Since replication of prions are not considered possible in insects, the number of prions in the substrate used to feed the insect affects the total prion infectivity of insects and cannot be higher than in the substrate. The substrate strongly influences the possible risk of prion disease transmission and must therefore be controlled at insect farms to counter this problem. The substrate used to feed the insects should not have a ruminant nor a human origin, but according to EFSA regarding the risks related to prion-derived diseases, the risk in non-processed insects is expected to be equal or lower than the proteins of another animal origin, as long as the insects are fed on substrates that do not harbour material of a ruminant or human origin [73]. However, no research has been done into the transmission of prion disease through the consumption of mealworms or BSF larvae.

One study shows that insect haemolymphs might have an anti prion effect, haemolymph from the beetle, *Trypoxylus dichotomus septentrionalis* (*T. d. septentrionalis*) showed anti-prion activity on a special strain of prions after being heated at 70 °C for 3 h [114]. Mealworms are genetically more related to *T. d. septentrionalis* than to *S. carnaria*, *D. melanogaster* or mites. Figure 1 shows the relationship between the insect species.

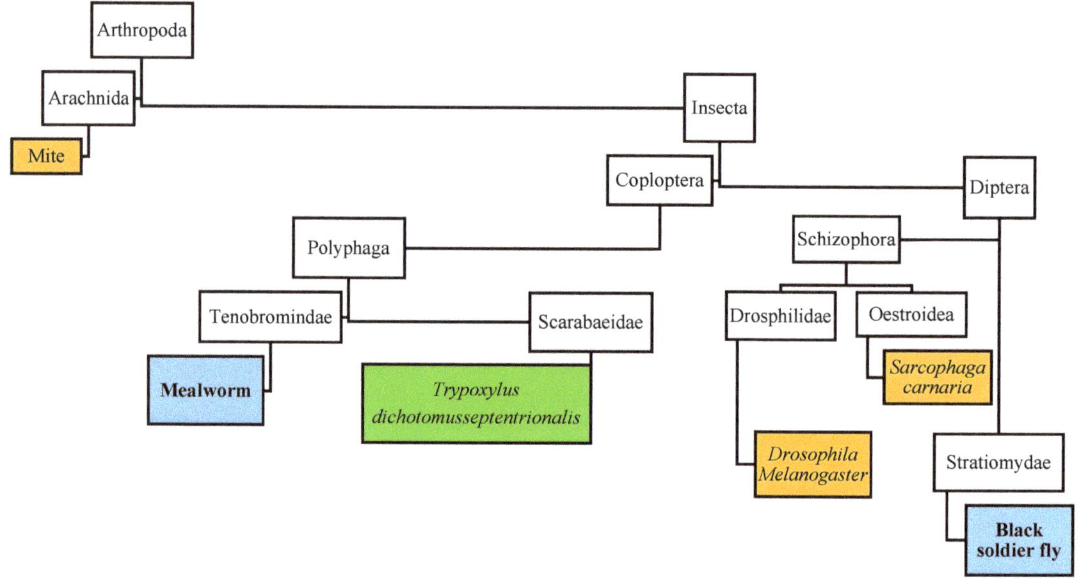

Figure 1. Relationship between mealworms and BSF (colored blue) to species that have been shown to be able to act as vector for prions (colored yellow) and to species that might possibly have a cure for prion diseases.

3.5. Food and Feed Safety Management

For food to be placed on the European market, it must be safe, meaning that it must not have harmful effects on health or be unfit for human consumption according to the EU general food law [75]. For this, there must be good hygiene practice in place (GHP) through the whole production chain and an ability to trace products. Production sites must be easy to clean and be constructed to eliminate pests and cross-contamination and must not contain hazardous chemicals [115]. Insects reared in a properly managed closed farm are less likely to act as a vector for parasites, since they would lack all host necessity to complete the lifecycle [73]. There is also a need for sufficient ventilation that can reduce air contamination and controlled temperature and humidity appropriate for the insect species. All equipment, vehicles, boxes, and tools used in the production site must be dedicated solely to insect-rearing activities and be cleaned thoroughly between batches [115].

Employees must be aware of hygiene requirements and be trained in GHP and other hygiene systems provided by the company. There has to be a separate area for staff to change to work clothes and staff are also required to use appropriate protective tools, e.g., people that have direct contact with products must wear gloves and people that work in the breeding chambers must use masks [115].

Insect producers in the EU must only use substrates that are accepted as feed for farmed animals within the EU. The substrate has to be traceable and of appropriate hygiene standards and must not contain any chemical contaminants. Insect producers should carry out regular checks of incoming substrate materials and substrates must be stored in dry, temperature appropriate and hygienic conditions [115]. Substrate control is an important part of safety management regarding insect breeding, because the substrates ingested can have a strong influence on insects' microbiota [90–92,116].

It is recommended to register rearing conditions and to test insects regularly for pathogens and chemicals [115]. 24 h before harvesting mealworms, it is recommended to remove them from the substrate for intestine cleaning [85,115,117,118]. This is performed because of the high microbial content in the insects gut [73,85]; however, there has not been documented benefits from this procedure [75]. When harvesting, foreign materials must be removed along with dead insects and frass. Chilling insects under controlled temperatures before harvesting has been reported to be beneficial for both mealworms and BSF larvae; it results in the maintenance of product properties and avoids microbial contamination [115].

Several killing methods have been researched regarding the food security of insects [85,119,120]. Farmed mealworms are often killed with blanching, boiling vapor, or freezing, while farmed BSF larvae are often killed with mincing and blanching. Blanching is performed by plunging insects into hot water, which will instantly kill the insects and destroy the microbial flora [115], and then they are often chilled by putting into clean water [115,119]. For mealworms, blanching is found to be the most successful heating method as it considerably reduces the bacterial content and fungi [85,94,117,119]; however, blanching is not sufficient to kill bacterial endospores [85,87,94,118]. As there is high endospore content in soil [87], which can be used as a substrate, and substrate influences the insects microbiota [90–92,116], management of the substrate is important, but the use of classical feed additives or fermentation has been shown to reduce spore forming [87]. Drying and acidifying techniques of insects are also promising to reduce endospores [118]. After blanching, mealworms can be stored in a refrigerator for 6 days without substantial microbial growth [119]. Not all time and temperature combinations will result in a sufficient reduction in microbial pathogens, and it is recommended to monitor the temperature used. An inadequate heat treatment can lead to bacterial proliferation. Another killing method, freezing, must be performed below 5 °C; however, most freezers operate at −20 °C and the appropriate freezing time to kill varies from species to species. Freezing has been shown to maintain the insects' nutritional value until they are further processed [115].

After killing, it has been shown that drying insects is important to reduce potential microbial, chemical, and allergenic hazards [121]. Freeze drying [115] and heat-based dehydration methods are used [115,119] and effective processing methods have been shown to further reduce the microbial load [73]. Sometimes, insects are processed through grinding for powder formation or fractioning, e.g., extracting chitin. These processing methods must be performed under GHP, and the grinding machine must be cleaned regularly. Water activity and storage temperature must also be appropriately monitored to reduce potential microbial contamination under processing and packaging [115].

Insects and insect-derived products must be stored in a close, clean and appropriate place and regard the product specification. There must be a prevention of accumulation of organic material and sampling plan for analysis of hazards for incoming raw materials and the outgoing product. If there is a transportation of food and feed products derived from insects, the same hygiene standards must be applied through the transportation as in other parts of the production chain [115].

One of the most important parts of the production chain is the packaging of insects, as it contributes to the condition that products will be in when they reach the consumers. Therefore, good hygiene, environment, security, and quality practices must be performed to ensure the safety of a product. The packaging must be clean and must not contain any chemical, physical or microbiological hazards. After the product is in the package, it must be closed immediately, and the operator must ensure that no external source of contamination is included. To prevent allergenic hazards, the product must be labelled with potential allergens in the product and a list of ingredients [115].

To simplify the control of potential hazards that can come up, there has been a developed system for the food industry that is called Hazard Analysis Critical Control Point (HACCP) [75]. However, no specific HACCP plan exists for the rearing of insects, but breeders have been working according to HACCP with company-specific approaches [73]. A properly designed HACCP can have control over all parts of food production that might pose a risk [122] and can prevent, eliminate and reduce to acceptable levels, microbial, chemical and physical hazards [115]. An HACCP system must be considered in the design, organization, and management of food production sites, along with the design of premises and equipment and a product-traceable system. With insects being considered as food or feed, it is necessary to guarantee their safety, but one of the main limitations of developing the insect farming industry involves guaranteeing the safety of the products. Therefore, manufactures must implement an HACCP plan to limit the risk for consumers' health [75].

When developing an HACCP, several things must be in place. There must be conducted hazard analysis for the production; Critical Control Points (CCP) must be determined; and there must be established critical limits and a system to monitor the CCP. Additionally, there must be established corrective actions when monitoring indicates that a particular CCP is not under control. There must also be established procedures of verification to confirm that the HACCP system is working effectively [115] and that should be documented [75].

Hazard analyses consist of hazard identification and an evaluation of the likelihood and severity of those hazards. It also consists of finding preventive measurements for these hazards. Hazards associated with insects as food and feed can be of a pathogen, chemical, allergenic and physical origin [115]. Whole insects, processed insect powder, and insects for food or feed can include different hazards [75].

Critical Control Points are defined as steps where control can be performed to prevent, eliminate or reduce a food and/or feed hazard to an acceptable level [75]. All CCPs require control measures, monitoring procedures, responsible staff, records and identified measurable critical limits to determine safe and unsafe conditions. In the insect industry, CCPs can be chilling, blanching, metal detectors in process lines [115], cooking after killing and hot drying [75] where the critical limits could be related to, e.g., temperature, pressure, time, water activity and pH [115]. When level outside of the critical limits are measured at one CCP, examples of corrective actions are, e.g., destroy the batch, readjust the temperature or time or restart the step [75].

4. Insect Farming in Europe

Many insect farming companies have emerged in the last few years. Some countries have had strict rules and regulations, but in other countries, a considerable experience is in marketing insects, i.e., for human consumption. This year, a step was taken by the European Commission by allowing yellow mealworm (*T. molitor*). Some countries, e.g., Belgium, Czech Republic, Denmark, Finland, the Netherlands and the UK, allow companies to keep selling whole insect-based products as long as applications for the species were made before 1st of January 2019 (transition period) [123].

4.1. Insect Farming in Europe

Insect farming is a growing industry in Europe [4] and it is a new business for Europeans [124]. Today, it is allowed in Europe to use insect-derived proteins, whole insects and insect-derived fats in pet food, feed for fur animals, and in aquaculture. Additionally, whole insects and insect-derived fats are allowed in feed for pigs and poultry [125–127]. Currently, in 2021, insect protein for feed is mostly produced as a pet food and for aquaculture [124]; however, this is believed to be about to change in the next few years in light of the recent authorisation of insect proteins for poultry and pigs on the 17 August 2021 [128]. By the end of the decade, new regulatory developments are expected to play a key role in increasing the production of insects and insect-derived products. Several tonnes of processed insect protein were produced in 2020 and the production of insects for feed is estimated to increase rapidly in the coming years. It is forecasted that the production of insect proteins for feed will reach 1 million tonnes of insect meal by 2030 [124]. At the same time, there has been a rapid change in the dietary habits of Europeans [4] and the willingness of consumers to try insect-based food is increasing [4,129–131]. This can be linked to an increased knowledge regarding the nutritional benefits and environmental effects of insects, alongside an increased willingness to consume environmentally friendly food [4,131–135]. This change in attitudes around food and growing demand for high protein food for sport nutrition, dietetic food or food supplements creates new opportunities for the production of insects as food. Currently, the use of insect-derived ingredients in food is low, but it is estimated to increase rapidly in the next few years [4] following mealworms being newly authorized for human consumption [33] and new insect products are expected to be authorized by the end of 2021 and by early 2022 [136–138]. In 2019, 500 tons of insect-based products for human consumption was produced in Europe, but the market for edible insect-based food products is estimated to produce 260,000 tonnes by 2030. Additionally, in 2019, 9 million Europeans consumed insects and insect-derived products, but by 2030 it is estimated that insects and insect-derived products will reach 390 million European consumers [4]. For insects to be a suitable alternative animal feed and for human consumption, insect farmers need to be able to produce large quantities of insects and insect-derived products and to have a steady production with sufficient quality. To be able to reach this level, insect farmers need to invest in capacity to offer satisfying quantity within costs that can compete with conventional animal feed used today, along with meat [139]. Increased availability of insect-derived products will lead to a decrease in prices [124]. The International Platform of Insects for Food and Feed (IPIFF) is an EU non-profit organisation that represents the interests of the insect production sector towards EU policy makers, European stakeholders and citizens. Within IPIFF that are 79 members [140], with 45 of them being insect companies in Europe today [141].

4.2. The Law in Europe

European law on insects in food and feed must strike the right balance between innovation and safety. The International Platform of Insects for Food and Feed (IPIFF) is an EU non-profit organization originally created in 2012 with its main mission to promote the wider use of insects and by advocating for EU legislative frameworks. The term 'Novel Food' is defined as food that has not been consumed to a significant degree by humans in the European Union before 15th of May 1997, when the first Regulation on novel food appeared. The main components are protein, fat and fiber (chitin). Since then, many new regulations have emerged (see Figure 2). Regulation no. 2283 from 2015 took over the regulations from 1997 and 2001 to update and develop guidance for applications for authorization of novel foods to the Commission, who may request a risk assessment from the European Food Safety Authority (EFSA) [142]. The marketing of dried mealworm, recently or on the 3rd of May 2021, got authorization to be placed on the market as a novel food. The European Food Safety Authority (EFSA) though, concluded that the consumption of the yellow mealworm may potentially lead to an allergic reaction, especially in individuals with pre-existing allergies to dust mites and crustaceans. Therefore, it is important to identify it on the food label. Toxicological and nutritional factors were also evaluated. The toxicity studies from the literature did not raise safety concerns and the consumption is not nutritionally disadvantageous [33].

Another EU Regulation, no. 2017/893, from 1st of July 2017, allowed a list of seven insect species to be included in the formulation of feeds for aquaculture. The species are BSF, *Musca domestica* (housefly), mealworm, *Alphitobius diaperinus* (lesser mealworm), *Acheta domesticus* (house cricket), *Gryllodes sigillatus* (tropical house cricket) and *Gryllus assimilis* (Jamaican field cricket sometimes referred to as a silent cricket) and even silkworm (*Bomby mori*) [142]. Previously, the addition of insects in feed for animals was not allowed due to potential prion-derived diseases. All other insect-based products were considered "Novel Food", and fell under EU regulation no. 2015/2283, where specific application to the European Commission is needed followed by EFSA scientific evaluation, before putting the product on the market as previously mentioned [143].

An interesting fact is that insects were already being sold as food in the EU, but there had been doubts among the Member States on whether whole insects were covered by previous Novel Food Regulation. The uncertainty was clarified by the European Court of Justice in October 2020, which concluded that the whole insects were not in the scope of previous regulation [78]. After contacting the companies in Europe, it was also clear that the legislation within each country in Europe can differ (see Figure 3) and that companies are obliged to follow their country's legislation [144–149] and some have not given their approval for mealworms for human consumption, e.g., in France [144]. However, insect protein was approved in September 2021 for poultry and pig feed [142] in Europe according to regulation no. 1372/2021 [142].

Insects as Novel Food in Europe

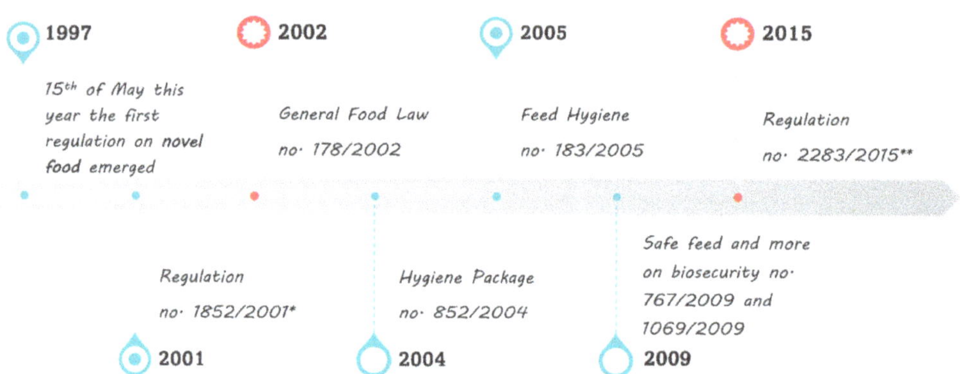

*Regulation no. 1852/2001 was supposed to lay down detailed rules for making certain information available to the public and for the protection of information submitted pursuant to Regulation (EC) No 258/97.
**Currently allowed feed materials must be used. Scientific opinion of October 2015 on the "Risk profile related to production and consumption of insects as food and feed", EFSA concluded that the substrate used as feed for the insects is one of the crucial elements with an impact on the occurrence and levels of biological and chemical contaminants and therefore needs to be assessed specifically.

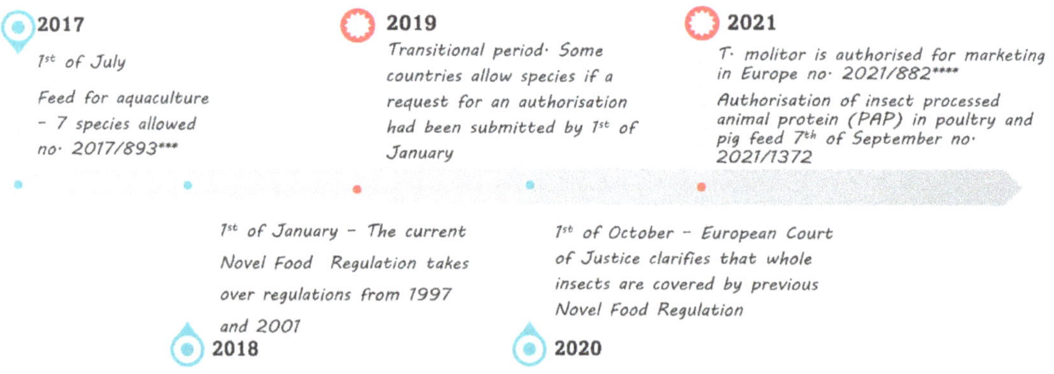

***The species are BSF, *Musca domestica* (housefly), yellow mealworm, *Alphitobius diaperinus* (lesser mealworm), *Acheta domesticus* (house cricket), *Gryllodes sigillatus* (tropical house cricket) and *Gryllus assimilis* (Jamaican field cricket sometimes referred to as silent cricket).
****Eleven applications were pending for safety evaluation by EFSA.

Figure 2. Insects as a Novel Food in Europe, timeline showing the main regulatory changes since 1997. Source: The European Commission [34,150–152].

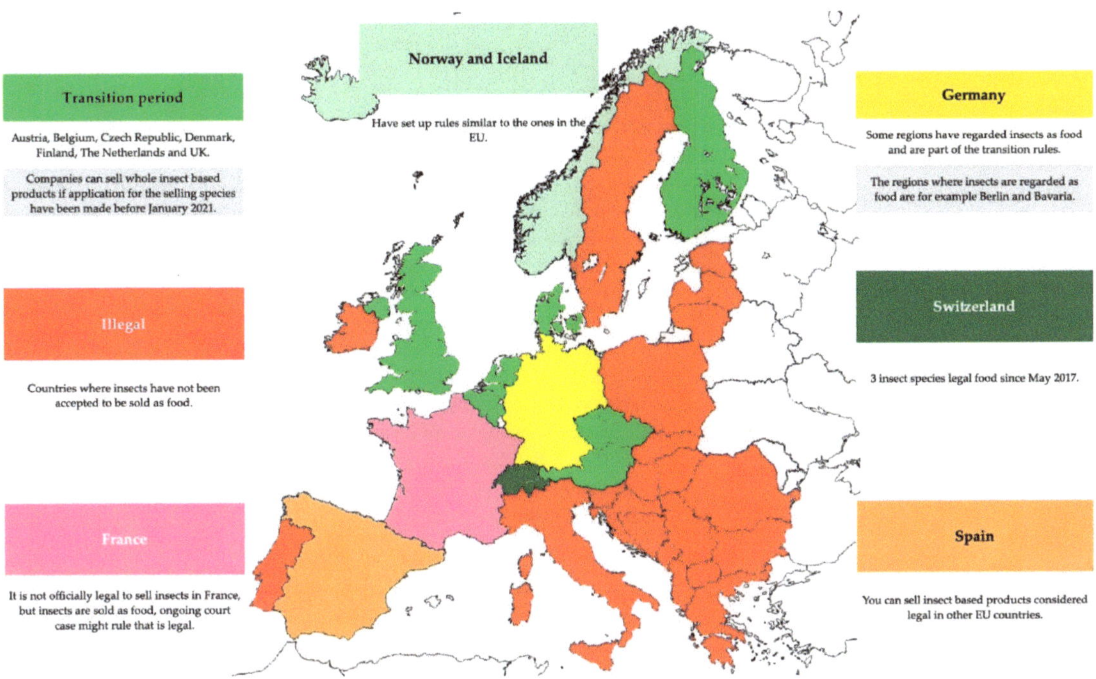

Figure 3. Insect Food Status in Europe. Adapted from Source: www.bugburger.se (accessed on 17 August 2021) [153].

Different rules apply in European countries, especially regarding the transition period. The main change from 2019 is that in 2021 Tenebrio molitor was allowed for human consumption.

4.3. Sustainable Development Goals (SDG) and European Strategies

SDG are goals that were first set in 2015 by the international community to pledge countries of the world to eradicate poverty, find sustainable and inclusive developmental solutions and ensure everyone's human right. There are 17 SDGs to be reached by 2030. These SDGs are: no poverty (SDG 1), zero hunger (SDG 2), good health and well-being (SDG 3), quality education (SDG 4), gender equality (SDG 5), clean water and sanitation (SDG 6), affordable and clean energy (SDG 7), decent work and economic growth (SDG 8), innovation, and infrastructure (SDG 9), reduced inequality (SDG 10), sustainable cities and communities (SDG 11), responsible consumption (SDG 12), climate action (SDG 13), life below water (SDG 14), life on land (SDG 15), peace, justice, and strong intuitions (SDG 16) and partnerships for the goals (SDG 17). The EU is committed to implement the SDGs in all of its policies and to encourage EU countries to do the same [154]. Insect rearing shows great potential to work towards the SDGs as it increases food security (SDG 2), improves waste management (SDG 12), and can have positive effects on human health and well-being (SDG 3). Furthermore, insect rearing with standardized techniques on an industrial scale is a novel economic sector able to improve the sustainability of the global food chain (SDG 9) [154].

The EU's strategies are developed and translated into policies and initiatives by the European Commission. The European Commission has set 6 priorities for 2019–2024; one of these priorities is A European Green Deal (EGD) [155]. The EGD aims to "transform the EU into a fair and prosperous society, with modern, resource-efficient and competitive economy where there are no net emissions of greenhouse gases in 2050 and where economic growth is decoupled from resource use". One of the strategies to reach the EGD goal is the Farm to Fork strategy, which was established to design a fair, healthy and environmentally friendly food system that has a global standard in sustainability and will contribute to achieving a circular economy [156]. The approval of mealworms as a novel food contributes to the objectives of the Green Deal and the Farm to Fork strategy [77].

4.4. Companies in Europe That Farm and Sell Insects

There exist several professional insect farming companies in Europe in various countries that farm mealworms and/or black soldier flies for either human or animal consumption, and some companies sell larva residues as fertilizer. In addition to mealworms and BSF, there are also some companies in Europe that farm other insect species, such as crickets [157]. Currently, the most used platform for the marketing of edible insect products are through companies' own websites, followed by fairs/events/conferences [4]. In March 2017, it was estimated that more than 200 start-up insect farming companies existed in Europe [158]. These start-up companies often consist of unique characteristics like the way they are organized, the growth plan or the financing structure. In spite of this, over 60% of start-up companies go bankrupt within 5 years [139]. Bug Burger lists 68 insect start-ups in Europe that have disappeared for various reasons, one of them being bankruptcy [157].

For this article, 27 insect-farming companies in Europe that farm mealworms and/or BSF larvae were contacted and 9 have answered [144–149,159–161]. However, 3 out of 9 answered companies did not provide proper answers, one was closed due to COVID-19 [159], one does not drive with insect rearing anymore [160] and one had no time to answer [161]. In Table 3, these 27 companies are compared; the companies that gave answers are yellow while the companies that did not provide answers are white. The data from the companies that did not answer or did not give proper answers was provided through the companies own websites along with newspaper articles about these companies and LinkedIn. When researching these companies, it can be estimated that there are equally as many companies in Europe that breed mealworms as there are that breed BSF larvae. It can also be estimated that most insect farming companies in Europe were founded between 2014 and 2018, as seen in Figure 4. Most insect farming companies investigated for this article were farming insects as feed mostly for either pets or aquaculture; this is in accordance with the IPIFF website. According to a survey performed in 2020 by IPIFF on the EU market in March 2020, most companies in Europe that sell insects for food are micro companies or 81%, which means that they have below 10 employees and only 3% are considered medium-sized companies with 50–250 employees [4]. However, according to the IPIFF survey on the market of insect as feed in 2021, over 40% of all insect farming companies that sell insects as feed were micro companies in 2020 and over 20% were medium-sized companies [124]. According to the companies that were investigated for this article, 42% were micro [146,149,162–167], 60% were small [145,148,168–174] and 11% were medium-sized companies [144,175]. Figure 5 shows the size of the companies researched for this article based on the type of industry. Ÿnsect in France is the biggest insect breeding company in Europe. It was founded in 2011 and has around 230 employees today. Currently, in 2021, Ÿnsect is building what will be the world's largest insect breeding facility and recently it also acquired an international branch in The Netherlands (Protifarm) [144].

The total investment into the majority of insect farming companies in Europe is below 500 K euros (€), around 30% of companies get between, 1 to 5 million €, with 6% getting over 10 million € [4]. As seen in Figure 6, insect farming companies in Europe investigated for this article were mostly been funded by Venture Capital. Most of these European insect farming companies are not economically sustainable today and are dependent on funding. However, both Ÿnsect in France [144] and Nasekomo in Bulgaria are estimated to become fully sustainable in 2022 [148].

Of the companies which answered, 50% were producing insects with automatic methods [144,145,148] and the other half with manual methods [146,147,149]. Automatic methods are considered beneficial on the grounds of increasing productivity, efficiency and consistency and of decreasing human labour [176].

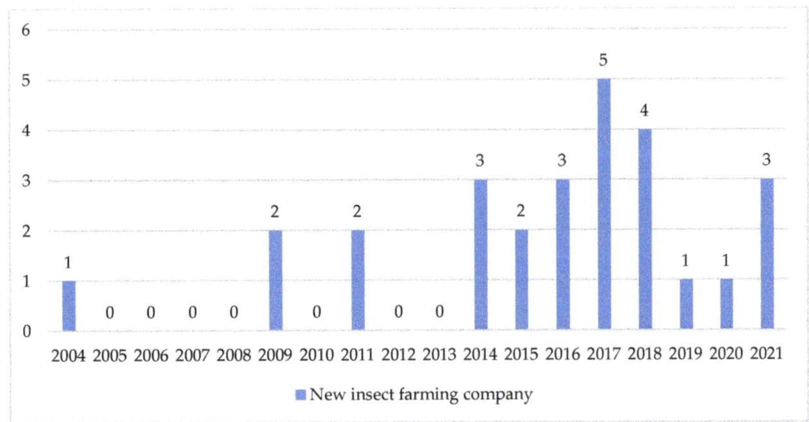

Figure 4. Foundation years of the 27 companies listed in Table 3.

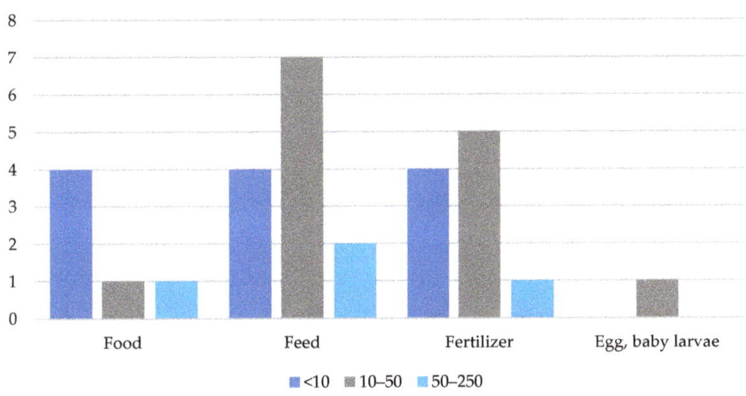

Figure 5. Number of employees based on the type of industry in the companies listed in Table 3.

Funding

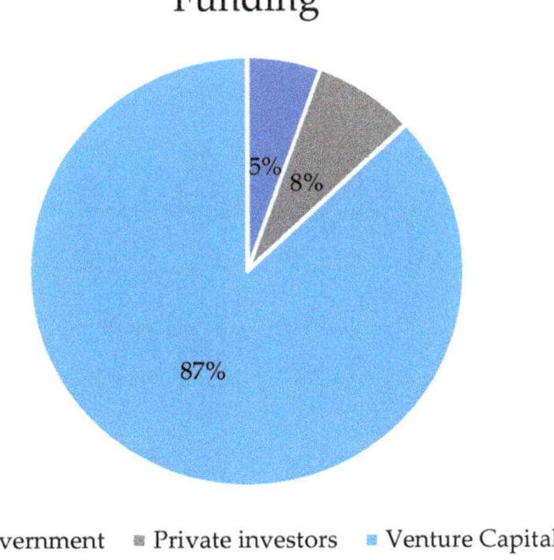

Figure 6. The prevalence of amount of investment in Ynsect, Nasekomo and Entomobio by the type of funding.

Table 3. 27 insect farming companies compared, yellow companies have given direct answers and the information from the white companies was provided through companies' website, articles, and LinkedIn.

Company	Species	Type of Farm	Location	Foundation	Annual Revenue €	Economic Sustainability	Use	Funding €	Number of Employees	Member of IPIFF
Ÿnsect [144]	Mealworm	Automated	France, The Netherlands, and USA	2011	89 M	Estimated to be profitable in 2022	Aquaculture, domestic animals, fertilizer, for human consume soon	360 M	230 will be close to 400 in 2022	Yes
Hexafly [145]	BSF	Automated,	Ireland	2016	200 tons meal production annually		Feed for aquaculture, pets, and animals and fertilizer	Equity funded	30	Yes
Verteco Farm [146]	Mealworm	Manual	Sweden	2020	Negative		Fertilizer	None	2	No
Syklus [147]	BSF	Manual	The Netherlands	2021	12 tons larva per year	Not yet	For ornamental fish	Yes		No
Nasekomo [148]	BSF	Automated	Bulgaria	2017	150 K	Not yet estimated to be sustainable in 2022	Aquaculture, fertilizer, pets	5 M	42	Yes
Entomobio [149]	Mealworm		Belgium	2018	7 K in 2020	Not yet	Human	60 K	1–3 depending on the period	No
Cricky [159,177]	Mealworm and crickets	Closed due to COVID-19	Croatia	2016		Closed due to COVID-19	Human			No
Urbanmat [160,178]	Mealworm, BSF and other species	Not rearing anymore, purchase and sale of insect and storage	Norway	2017			Human			No
Insectum [161,179]	BSF		Denmark	2018			Feed			No
Horizon Edible Insects [162,180]	Mealworm	manual	UK	2019			Human	Guided tours and cooking classes	1–10	No
Tebrito [163,181]	Mealworm	manual	Sweden	2016			Human, frass		1–10	Yes
Entobreed [182]	Mealworm		The Netherlands	April 2021, first egg arrived, have not started selling			Fertilizer, feed, human	130 K Crowdfunding		No
Marienlyst Ento [183]	Mealworm	Manual	Denmark	2017			Human mostly to companies but also for private use			No
Micronutris [172,184]	Mealworm		France	2011			Human		11–50	No
Entoinnov [164,185]	Mealworm		France	2021			Human, feed, fertilizer		1–10	No
Entocycle [173,186]	Black solder fly		UK	2014			Pets, frass	Governmental backing	11–50	Yes
Protix [175]	Mealworm, BSF, grasshopper, crickets	Automated	The Netherlands, active in 12 countries	2009			Pets, feed, food, and fertilizer	>45 M several private investors	50–250	Yes

Table 3. *Cont.*

Company	Species	Type of Farm	Location	Foundation	Annual Revenue €	Economic Sustainability	Use	Funding €	Number of Employees	Member of IPIFF
Enorm [165,187]	BSF	Automated	Denmark	2018	The goal is to produce 1.5 ton of living larva every day		Feed		1–10	Yes
Tebrio [174,188]	Mealworm	Automated	Spain	2014			Feed, pets, fertilizer, antibac, human in the future	Venture capital, government 40% interest	11–50	Yes
Bugimine [189]	Mealworm		Estonia	2017			Pets, aquaculture, poultry, fertilizer	Private capital		No
NextAlim [168]	BSF	Automated	France, plan to construct 3 additional product units in Europe and outside of Europe in 2022–2025	2014	2.4 tons of egg per year, estimated to be 12 tons of egg per year in 2022		Egg, neonates, and 5–7 days old larva ready to rear, to another companies	Yes	20	Yes
nextProtein [169]	BSF	Manual	France	2015			Pets, aquaculture, feed for other animals, fertilizer		15	Yes
Hermetia [166,190]	BSF	Manual	Germany	2009			Aquaculture		1–10	Yes
Illucens Gmbh [170]	BSF		Germany	2018			Pets and zoo animals		26	Yes
HiProMine [171]	BSF	Vertical	Poland	2015			Pets, aquaculture, feed for other animals in the future		15	Yes
EntoMass [167,191]	BSF	Manual	Denmark	2017			Pets, fertilizer		1–10	No
PAPEK s.r.o. [192]	Mealworm	Manual	The Czech Republic	2004			Export and zoos			

4.5. Consumer Acceptance

Consumer acceptance of entomophagy is important to start a large scale production of insects used for human food and it still remains the biggest challenge for the insect industry today [133]. In recent years, there has been a lot of research regarding the consumer's acceptance of entomophagy in Europe, and over 200 scientific papers about this topic have been written. This article focuses on 10 research articles [130–135,193–196] on consumers acceptance of entomophagy in Europe and the results from one review article [130] that focuses on other 38 research articles of the same topic. Most research regarding the consumer acceptance of entomophagy in Europe has focused on consumers from Italy, the Netherlands and Belgium. Few articles have focused on consumer acceptance in Germany, Switzerland, Finland, Denmark, Czech Republic, Poland, France, Hungary, Sweden, and Ireland [129]. These studies show an increased interest in commercializing insect-based foods [132], e.g., the Netherlands published the first article about consumers acceptance of insects in 2012 from a survey performed in 2010. The results showed a low acceptability towards entomophagy in Dutch consumers [130]. In 2020, another study on consumer acceptance among Dutch and German students was performed. The results showed a higher acceptability towards insects, as food by the Dutch students rather than in the study performed in 2012 [131]. However, the participants in the 2020 study were on average younger than the participants in the 2012 study [130,131], but studies suggest that younger people are generally more willing to consume insects than the older generation [129,133]. Despite this increased interest in entomophagy, the number of studies indicate that high proportions of Europeans still consider insects as a food to be taboo [132] and many do not know that insects are consumed in Europe [133]. Consumers in Northern Europe seem to be more accepting towards entomophagy than consumers in Central, Mediterranean and Western Europe [129,194]. According to several studies, men seem to be more accepting towards edible insects than women [133], while other studies reported no difference [132]. As these research articles seem to show increased consumer acceptance with time, it is highly likely that the consumer acceptance towards entomophagy will continue to increase in the near future, especially as younger participants seem to show more acceptance towards edible insects than older ones. In Bangkok, it seems that the young people are among the main drivers where significant revival of insect-eating is happening, along with increasing interest from tourists. The prices there are getting higher, but even so, people are buying them and the market for edible insects is growing [8]. Therefore, consumer acceptance will likely remain the biggest challenge for insect farming development into more financially viable businesses for the next five years, as many Europeans still regard insects as a taboo food.

Several explanations are considered regarding the negative attitude of Europeans towards entomophagy. One of these explanations is food neophobia, which is defined as the unwillingness to try new foods [132,133] and is related to human innate paradoxical behaviour towards unknown or unfamiliar food and consider it to be a potential threat to their organism [132]. Consumer's food neophobia tendencies have been shown to reduce the consumers' willingness to eat insects both as a whole and as an ingredient in food [131,133,193,194]. However, food neophobia seems to be an extremely complex attitude and can vary during the course of one's life [193]. It has also been stated that food neophobia is not as significant a barrier to insect consumption as it once it was, since edible insects are becoming more familiar to consumers [131].

Another explanation is disgust, but Europeans generally consider insects to be dirty [133], and view insects to be a pathogenic risk; therefore, food containing insects are considered disgusting [193] and repulsive [134]. Studies have shown that the feeling of disgust affects the willingness to consume insects negatively [133,134,193] and the feeling of disgust strongly influences perception, even before insect products are tasted [133]. The feeling of disgust is a complex phenomenon that could be associated with health risks posed by the consumption of a specific substance [134]. Disgust toward a specific food generally comes from culturally induced rejection [133,134]. It is conceptualized as an adaptive reaction and is closely connected to the information people have at the time. People can change their food preference through information, exchange, and experience [134]. According to several studies, food neophobia, along with disgust, have the most negative influence on acceptance of insect products [133].

A third explanation is insect phobia, but a study done by Moruzzo et al. [193] has shown that insect phobia has a more negative influence on the willingness towards tasting insects than food neophobia. In the future, when insects will be a more known food in Europe, insect phobia will have more of an effect than food neophobia on the intention to eat food containing insects and an increasing familiarity with insect food will not be enough for consumers to adopt insect-based food.

According to Meyer-Rochow et al. [195], Europeans attitudes towards edible insects might be influenced by idioms containing unfavourable references to insects. Idioms occur in all languages and can have an important influence on society and become integrated into feelings like irritation, contemptuous attitude, anger, and disgust. Idioms that exist in European languages convey predominantly negative attitudes, while the opposite is true in East Asia. Mirror neurons are believed to be activated when listening to idioms and could lead the listener of idioms towards a negative attitude towards insects and project it towards edible species. New idioms appear all the time and perhaps making more positive idioms towards insects might help change attitudes.

To increase consumer acceptance, persuasion tactics to reduce Europeans' anxiety towards entomophagy is important. These persuasion tactics can help to disguise insects in food, combining them with familiar ingredients or turning them into powder [132], but many studies have reported a higher consumer acceptance towards processed insect products rather than to whole insects [130,133–135,196]. Other tactics are to increase the familiarity with insects as a food by having them in grocery stores and talking to friends that have a positive experience with edible insects [132]. Increasing the knowledge about the positive environmental effects regarding edible insects and their health benefits has also been proven to improve consumer attitude towards entomophagy [131–135]. Even though insects are considered a taboo food today, this attitude might change. Sushi was once considered to be a taboo food, but it has increased in popularity in the recent years [132].

5. Iceland

Iceland has abundant resources, e.g., land, water and renewable energy (geothermal and hydro power). As an island in the North Atlantic Ocean, it is important to be sustainable in terms of food and feed. Today, the country imports fuel, fertilizer, feed raw materials, feed and food. Insects could be a valuable factor in supporting food security in Iceland and could serve an important role in food circulation.

5.1. Insect Trials in Recent Years

Limited research has been conducted regarding insect breeding and the use of insects as food and feed in Iceland. In 2014, black soldier flies were experimentally bred as a potential feed ingredient for fish farming in northwest Iceland, but the activity was discontinued. There were several reasons for this activity being discontinued. One of these reasons was that the EU laws regarding insect farming were under construction, and it seemed unlikely that the EU would allow the ideas on which the company based its business [197]. Moreover, according to the owners of the company, there was not enough underutilized food in Iceland to make it a stable feed for black soldier flies [198].

Another experiment in Iceland started in 2015 with the production of protein bars from crickets (Jungle Bar). The production went well, as well as the marketing, and the idea seemed to be well received by Icelanders [199]. However, only a few days after the product was launched in the stores the Icelandic government implemented European law prohibiting the sale of insect-based products for human consumption and the project had to be discontinued [200].

In 2018, a new start-up on breeding mealworms received funding. The plan was to produce mealworms for aquaculture and breeding the insects, at stable conditions with the use of geothermal heat [201]. However, the project was discontinued beyond 2019.

Additionally, BSF were experimentally bred in the governmentally owned, food and biotech R&D company in Iceland, Matís, from 2012 to 2014. One of the aims of the study was to examine the effect of different organic waste on the nutritional content of the larvae. Matís presented its findings in the international conference, Insect to feed the World, in 2014. The results showed that it is possible to have a great influence on the nutritional content of the larvae with different feeds [202]. In 2019, a new study relating to insect proteins was started in Matís. This project is estimated to take 4 years and it is about breeding crickets for the larvae to be used in bread [203]. Currently, crickets are not authorized for human consumption; however, in august 2021, The European Food and Safety Agency (EFSA) submitted an option on the safety of frozen and dried formulations from whole house crickets (*Acheta domesticus*) as a novel food pursuant to the European Commission, and the authorisation can be expected in early 2022 [136].

Iceland is also a part of the pan-European project NextGenProteins, which is optimizing the production of three alternative protein sources, with one of them deriving from black soldier flies and crickets [203].

Currently, mealworms and BSF are experimentally bred in the Agricultural University of Iceland. The mealworms are being fed with waste from Icelandic brewery production and carrots and the BSF larvae are being fed with kitchen waste [204].

As of today, no study of consumer acceptance of edible insects has been performed on Icelandic consumers. It is not unlikely that consumer acceptance in Iceland resembles the consumers acceptance in Europe mentioned in Section 4.3.

5.2. The Laws and Regulations

According to EU regulation no. 2015/2283, whole insects are in scope as well as parts of whole insects, powder and extracts. Insects cannot yet be fed with feed ingredients that are not authorized for farmed animals [205]. When the EU regulation 2015/2283 was implemented, the startup company Crowbar Protein in Iceland was working on the development of the previously mentioned Jungle Bar. The product, launched in January 2016, was requested to be pulled off the shelves, although the current regulation no. 2015/2283 was entered into force in Iceland 1st of January 2018 [206]. After only one week, the Jungle Bar was pulled off the shelves, requested by the Icelandic authorities referring to EU regulations. One of the owners of the startup company reported that they had submitted all documentation to the Directorate of Health and the Icelandic Food and Veterinary Authority to confirm that the products were safe for human consumption. The owners of Crowbar Protein subsequently signed a contract with a distributor in the USA [207]. Another company said at that time that their hands were tied by uncertain regulations [208].

Since then, new regulations have emerged in Europe but, as explained in Section 4.1, different countries have quite different rules. According to the Icelandic Food and Veterinary Authority, Iceland follows the regulations from the EU and evaluations from the EFSA. Therefore, it is now allowed to start marketing yellow mealworms in Iceland (Commission Implementing Regulation (EU) 2021/882 from 1st of June 2021) [206].

5.3. Importing Feed

Iceland imports feed, e.g., soybean meal and grain, especially for aquaculture, poultry and pigs, but based on work on food security, several opportunities for increased production in Iceland have been identified [209]. It is possible to make use of natural resources, become more sustainable, improve food security, e.g., if the import is uncertain as in pandemic times, use less currency, and create more jobs and use knowledge in Iceland. One goal could be establishing insect farming, but there is also exciting research going on with different grain cultivation in Iceland.

With insect farming, waste from agriculture as well as the essential food waste within the country and even preferentially from the next neighborhood could be used. By doing this, a sustainable cycle and food security is better maintained.

5.4. Food Waste in Iceland

According to the Food and Agriculture Organization of the United Nations (FAO), more than one third of food is wasted and at all stages of production. This waste contributes significantly to greenhouse gasses and climate change. At the same time, hunger exists in the world and the population has been growing 7-fold over the last 200 years and is still growing. According to the FAO (2018), a 60% increase will be needed to meet growing demands of the world population. Food waste is very complex and interdisciplinary research teams are starting to use model-driven integrative applied research approaches. The Recovery Food Hierarchy (EPA, 2019, appendix A) constitutes these main headlines: (1) Source Reduction; (2) Food for people; (3) Feed for animals; (4) Industrial Usage; (5) Composting; and at last (6), Landfill incineration [210].

Another primary idea is increased awareness, e.g., children who can participate in making the food they eat are less likely to throw it away. According to Icelandic research from 2016, it is estimated that the food waste is 23 kg of food that could have been used and 35 kg of food defined as non-useable (e.g., eggshell, coffee basket, bones and peel of vegetables and fruits) per individual in Iceland. The Icelandic information site *matarsoun.is* is a project led by The Environment Agency of Iceland (Umhverfisstofnun). There it is stated that the worth of the food wasted was 4.5 billion ISK in 2015 [211].

6. Conclusions

In recent years, the interest in insect farming has been increasing in Europe due to the need for new food sources with less of an environmental impact than conventional production. The United Nations identifies insects as a food product that could increase food security and human health and could reduce pollution, and the European Union supports innovation and research in these fields.

Studies have shown that the nutritional profile of both mealworms and BSF larvae is good for human consumption, as feed for animals, being high in protein, fat, fiber and several vitamins and minerals. Furthermore, studies have shown that eating insects might reduce cholesterol levels in the body. Insects are not only nutritious, but they also contain substances that could promote the immune system. Using insects as animal feed could lead to a reduction in antibiotic use in livestock. Insect farming is more environmentally friendly than the farming of traditional livestock and the production of soybean and fishmeal as insect production uses less arable land and water and results in lesser greenhouse gas emissions. Furthermore, studies have shown that insects, e.g., mealworms and BSF larvae, can be used to degrade several types of organic waste and that mealworms can be used to degrade plastics. However, the true environmental impact of large-scale insect farming is unknown, and more studies could focus on that.

For novel products such as food and feed to be placed on the European market, the product must be safe. Several studies have focused on the safety of edible insects and the safety risks that have been identified are allergies, toxins, and zoonotic pathogens. Studies have shown that insect protein has a cross reactivity with crustacea and mite allergies, concluding that people who suffer from these allergies should not consume or work around edible insects. Toxicity studies of mealworms from the literature do not raise safety concerns according to the EFSA and yellow mealworm has now been authorized for the market as a novel food product. More applications for other species are pending. It is interesting that different countries in Europe have different rules, but it is important to evaluate risk profiles. When hygiene and other safety needs are met, it seems that the risk profiles are like other products. To finish the feed and food circulation, it would be important to use other food waste. Research is also ongoing into how insects manage with plastics. As insects are so phylogenetically different from humans and other mammals, studies indicate that edible insects pose a smaller risk to humans than traditional livestock. However, some pathogenic bacteria, viruses and parasites have been identified in edible insect farms in Europe. Microbiota can be introduced during farming and processing and therefore good hygiene strategies should be on place on the farm. Additionally, the substrate used to feed the insects majorly affects the insect's microbiota and composition; therefore, it is important to choose the substrate well and to store it under proper conditions.

Interest in insect farming in Europe is growing and the biggest European company in this sector has approximately 230 employees. Regulations are developing, and investors and competition funds are supporting the development. Most importantly, European consumers are becoming more positive both toward insects as animal feed and as food for human consumption.

Author Contributions: Conceptualization, R.T., H.T.O. and R.I.T.; Formal analysis, R.T. and H.T.O.; Funding acquisition, R.I.T.; Investigation, R.T. and H.T.O.; Methodology, R.T., H.T.O. and R.I.T.; Project administration, R.I.T.; Resources, R.I.T.; Supervision, R.I.T.; Writing—original draft, R.T. and H.T.O.; Writing—review and editing, R.I.T. All authors have read and agreed to the published version of the manuscript.

Funding: This project has been subsidized through the ERANET Cofund GEOTHERMICA 2018-2021 (Project no. 731117), from the European Commission, Rannsoknamidstod Islands/Rannis, Ministerie van Economische Zaken/Rijksdienst voor Ondernemend Nederland/RVO and Ministrstvo za infrastrukturo, Direktorat za energijo/MzIDE and with support from Directorate of Labour in Iceland.

Informed Consent Statement: Not applicable.

Data Availability Statement: Data is contained within the article.

Acknowledgments: The Agricultural University of Iceland is acknowledged for administrative and technical support and for providing facilities for trials with mealworms and BSF larvae. Special thanks go to Stedji Brewery for providing barley for the tests, and to Libardo Lugo at Samraekt Ltd. for assisting with the setup and starting up the cultures of both species.

Conflicts of Interest: The authors declare no conflict of interest.

References

1. Evans, J.; Alemu, M.; Flore, R.; Frøst, M.; Halloran, A.; Jensen, A.; Maciel, G.; Meyer-Rochow, V.; Münke-Svendsen, C.; Olsen, S.; et al. 'Entomophagy': An evolving terminology in need of review. *J. Insects Food Feed* **2015**, *1*, 293–305. [CrossRef]
2. Raheem, D.; Carrascosa, C.; Oluwole, O.B.; Nieuwland, M.; Saraiva, A.; Millán, R.; Raposo, A. Traditional consumption of and rearing edible insects in Africa, Asia and Europe. *Crit. Rev. Food Sci. Nutr.* **2019**, *59*, 2169–2188. [CrossRef]
3. Bordiean, A.; Krzyżaniak, M.; Stolarski, M.J.; Czachorowski, S.; Peni, D. Will Yellow Mealworm Become a Source of Safe Proteins for Europe? *Agriculture* **2020**, *10*, 233. [CrossRef]
4. International Platform of Insects for Food and Feed. *Edible Insects on the European Market*; IPIFF: Brussels, Belgium, 2020.
5. Worldometer. Europe Population (LIVE). Available online: https://www.worldometers.info/world-population/europe-population/ (accessed on 30 August 2021).
6. Huis, A.; Van Itterbeeck, J.; Klunder, H.; Mertens, E.; Halloran, A.; Muir, G.; Vantomme, P. *Edible Insects Future Prospects fo Food and Feed Security*; Food and Agriculture Organization of the United Nations: Rome, Italy, 2013; Volume 171.
7. Worldometer. World Population. Available online: https://www.worldometers.info (accessed on 31 August 2021).
8. Müller, A. Insects as Food in Laos and Thailand. *Asian J. Soc. Sci.* **2019**, *47*, 204–223. [CrossRef]
9. De Carvalho, N.M.; Madureira, A.R.; Pintado, M.E. The potential of insects as food sources—A review. *Crit. Rev. Food Sci. Nutr.* **2020**, *60*, 3642–3652. [CrossRef]
10. Van Huis, A. Edible insects are the future? *Proc. Nutr. Soc.* **2016**, *75*, 294–305. [CrossRef] [PubMed]
11. Kiple, K.F. *A Movable Feast: Ten Millennia of Food Globalization*; Oxford University Press: Oxford, UK, 2007.
12. Massard, J.A. Maikäfer in Luxemburg: Historisches und Kurioses. *Lëtzebuerger J.* **2007**, *60*, 26–27.
13. Meyer-Rochow, V. Can insects help to ease the problem of world food shortage? *Search* **1975**, *6*, 261–262.
14. United Nations. Discussion Starter Action Track 3: Boost Nature-Positive Food Production at Scale. Available online: https://www.un.org/sites/un2.un.org/files/unfss-at3-discussion_starter-dec2020.pdf (accessed on 5 August 2021).
15. Food and Agriculture Organization; International Fund for Agricultural Development; United Nations Children's Fund; World Health Organization. *The State of Food Security and Nutrition in the World 2021: Transforming Food Systems for Food Security, Improved Nutrition and Affordable Healthy Diets for All*; Food & Agriculture Orgn: Rome, Italy, 2021; p. 240.

16. Hodson, E.; Niggli, U.; Kaoru, K.; Lal, R.; Sadoff, C. Boost Nature Positive Production at Sufficient Scale—A Paper on Action Track 3. In Proceedings of the United Nations Food Systems Summit 2021, New York, NY, USA, 23 September 2021.

17. Grau, T.; Vilcinskas, A.; Joop, G. Sustainable farming of the mealworm Tenebrio molitor for the production of food and feed. *Z. Für Nat. C* **2017**, *72*, 337–349. [CrossRef]

18. Hong, J.; Han, T.; Kim, Y.Y. Mealworm (*Tenebrio molitor* Larvae) as an Alternative Protein Source for Monogastric Animal: A Review. *Animals* **2020**, *10*, 2068. [CrossRef] [PubMed]

19. Dortmans, B.; Diener, S.; Verstappen, B.; Zurbrügg, C. *Black Soldier Fly Biowaste Processing—A Step-by-Step Guide*; Eawag: Dibendorf, Switzerland, 2017.

20. Sharma, R.; Kaur, R.; Rana, N.; Poonia, A.; Rana, D.C.; Attri, S. Termite's potential in solid waste management in Himachal Pradesh: A mini review. *Waste Manag. Res.* **2021**, *39*, 546–554. [CrossRef]

21. Thévenot, A.; Rivera, J.L.; Wilfart, A.; Maillard, F.; Hassouna, M.; Senga-Kiesse, T.; Le Féon, S.; Aubin, J. Mealworm meal for animal feed: Environmental assessment and sensitivity analysis to guide future prospects. *J. Clean. Prod.* **2018**, *170*, 1260–1267. [CrossRef]

22. The Donau Soja Association. Soya Bean History. Available online: https://www.donausoja.org/en/research/agriculture/soya-bean-history/ (accessed on 31 August 2021).

23. European Fishmeal and Fish Oil Producers. We Represent the European Fishmeal and Fish Oil Producers in Denmark, Faroe Islands, Iceland, Ireland, Norway, United Kingdom, Estonia and Spain. Available online: https://effop.org (accessed on 30 August 2021).

24. Belghit, I.; Liland, N.S.; Waagbø, R.; Biancarosa, I.; Pelusio, N.; Li, Y.; Krogdahl, Å.; Lock, E.-J. Potential of insect-based diets for Atlantic salmon (*Salmo salar*). *Aquaculture* **2018**, *491*, 72–81. [CrossRef]

25. Selaledi, L.; Hassan, Z.; Manyelo, T.G.; Mabelebele, M. Insects' Production, Consumption, Policy, and Sustainability: What Have We Learned from the Indigenous Knowledge Systems? *Insects* **2021**, *12*, 432. [CrossRef]

26. Finke, M.D. Complete nutrient content of four species of commercially available feeder insects fed enhanced diets during growth. *Zoo Biol.* **2015**, *34*, 554–564. [CrossRef] [PubMed]

27. Nyangena, D.; Mutungi, C.; Imathiu, S.; Kinyuru, J.; Affognon, H.; Ekesi, S.; Nakimbugwe, D.; Fiaboe, K. Effects of Traditional Processing Techniques on the Nutritional and Microbiological Quality of Four Edible Insect Species Used for Food and Feed in East Africa. *Foods* **2020**, *9*, 574. [CrossRef] [PubMed]

28. Hermans, W.J.H.; Senden, J.M.; Churchward-Venne, T.A.; Paulussen, K.J.M.; Fuchs, C.J.; Smeets, J.S.J.; van Loon, J.J.A.; Verdijk, L.B.; van Loon, L.J.C. Insects are a viable protein source for human consumption: From insect protein digestion to postprandial muscle protein synthesis in vivo in humans: A double-blind randomized trial. *Am. J. Clin. Nutr.* **2021**, *114*, 934–944. [CrossRef] [PubMed]

29. Gessner, D.K.; Schwarz, A.; Meyer, S.; Wen, G.; Most, E.; Zorn, H.; Ringseis, R.; Eder, K. Insect Meal as Alternative Protein Source Exerts Pronounced Lipid-Lowering Effects in Hyperlipidemic Obese Zucker Rats. *J. Nutr.* **2019**, *149*, 566–577. [CrossRef] [PubMed]

30. Kim, T.K.; Yong, H.I.; Kim, Y.B.; Kim, H.W.; Choi, Y.S. Edible Insects as a Protein Source: A Review of Public Perception, Processing Technology, and Research Trends. *Food Sci. Anim. Resour.* **2019**, *39*, 521–540. [CrossRef]

31. Mariod, A.A. Nutrient Composition of Mealworm (Tenebrio Molitor). In *African Edible Insects as Alternative Source of Food, Oil, Protein and Bioactive Components*; Adam Mariod, A., Ed.; Springer International Publishing: Cham, Switzerland, 2020; pp. 275–280.

32. Liu, X.; Chen, X.; Wang, H.; Yang, Q.; Ur Rehman, K.; Li, W.; Cai, M.; Li, Q.; Mazza, L.; Zhang, J.; et al. Dynamic changes of nutrient composition throughout the entire life cycle of black soldier fly. *PLoS ONE* **2017**, *12*, e0182601. [CrossRef]

33. EFSA Panel on Nutrition, N.F.; Allergens, F.; Turck, D.; Castenmiller, J.; De Henauw, S.; Hirsch-Ernst, K.I.; Kearney, J.; Maciuk, A.; Mangelsdorf, I.; McArdle, H.J.; et al. Safety of dried yellow mealworm (*Tenebrio molitor* larva) as a novel food pursuant to Regulation (EU) 2015/2283. *EFSA J.* **2021**, *19*, e06343. [CrossRef]

34. Kouřimská, L.; Adámková, A. Nutritional and sensory quality of edible insects. *NFS J.* **2016**, *4*, 22–26. [CrossRef]

35. Oonincx, D.G.A.B.; Laurent, S.; Veenenbos, M.E.; van Loon, J.J.A. Dietary enrichment of edible insects with omega 3 fatty acids. *Insect Sci.* **2020**, *27*, 500–509. [CrossRef]

36. Payne, C.L.; Scarborough, P.; Rayner, M.; Nonaka, K. Are edible insects more or less 'healthy' than commonly consumed meats? A comparison using two nutrient profiling models developed to combat over- and undernutrition. *Eur. J. Clin. Nutr.* **2016**, *70*, 285–291. [CrossRef]

37. Baek, M.; Kim, M.A.; Yun-Suk, K.; Hwang, J.S.; Goo, T.-W.; Jun, M.; Yun, E.-Y. Effects of processing methods on nutritional composition and antioxidant activity of mealworm (*Tenebrio molitor*) larvae. *Entomol. Res.* **2019**, *49*, 284–293. [CrossRef]

38. Wendin, K.; Mårtensson, L.; Djerf, H.; Langton, M. Product Quality during the Storage of Foods with Insects as an Ingredient: Impact of Particle Size, Antioxidant, Oil Content and Salt Content. *Foods* **2020**, *9*, 791. [CrossRef]

39. Mouithys-Mickalad, A.; Schmitt, E.; Dalim, M.; Franck, T.; Tome, N.M.; van Spankeren, M.; Serteyn, D.; Paul, A. Black Soldier Fly (*Hermetia illucens*) Larvae Protein Derivatives: Potential to Promote Animal Health. *Animals* **2020**, *10*, 941. [CrossRef] [PubMed]

40. Rawski, M.; Mazurkiewicz, J.; Kierończyk, B.; Józefiak, D. Black Soldier Fly Full-Fat Larvae Meal as an Alternative to Fish Meal and Fish Oil in Siberian Sturgeon Nutrition: The Effects on Physical Properties of the Feed, Animal Growth Performance, and Feed Acceptance and Utilization. *Animals* **2020**, *10*, 2119. [CrossRef] [PubMed]

41. Liland, N.S.; Biancarosa, I.; Araujo, P.; Biemans, D.; Bruckner, C.G.; Waagbø, R.; Torstensen, B.E.; Lock, E.J. Modulation of nutrient composition of black soldier fly (*Hermetia illucens*) larvae by feeding seaweed-enriched media. *PLoS ONE* **2017**, *12*, e0183188. [CrossRef]

42. Tschirner, M.; Simon, A. Influence of different growing substrates and processing on the nutrient composition of black soldier fly larvae destined for animal feed. *J. Insects Food Feed* **2015**, *1*, 249–259. [CrossRef]

43. USDA Database. Release 28. Available online: https://www.ars.usda.gov (accessed on 8 September 2021).

44. Office of Nutrition, Labeling and Dietary Supplements in the Center for Food Safety and Applied Nutrition at the US Food and Drug Administration. *A Food Labeling Guide: Guidance for Industry*; FDA: College Park, MD, USA, 2013; p. 132.

45. Alves, A.; Sanjinez-Argandoña, E.; Linzmeier, A.; Cardoso, C.; Macedo, L. Food Value of Mealworm Grown on Acrocomia aculeata Pulp Flour. *PLoS ONE* **2016**, *11*, e0151275. [CrossRef]

46. Kawasaki, K.; Hashimoto, Y.; Hori, A.; Kawasaki, T.; Hirayasu, H.; Iwase, S.I.; Hashizume, A.; Ido, A.; Miura, C.; Miura, T.; et al. Evaluation of Black Soldier Fly (*Hermetia illucens*) Larvae and Pre-Pupae Raised on Household Organic Waste, as Potential Ingredients for Poultry Feed. *Animals* **2019**, *9*, 98. [CrossRef]

47. Ravzanaadii, N.; Kim, S.; Choi, W.H.; Hong, S.-J.; Kim, N. Nutritional Value of Mealworm, Tenebrio molitor as Food Source. *Int. J. Ind. Entomol.* **2012**, *25*, 93–98. [CrossRef]

48. Schmitt, E.; Belghit, I.; Johansen, J.; Leushuis, R.; Lock, E.J.; Melsen, D.; Ramasamy Shanmugam, R.K.; Van Loon, J.; Paul, A. Growth and Safety Assessment of Feed Streams for Black Soldier Fly Larvae: A Case Study with Aquaculture Sludge. *Animals* **2019**, *9*, 189. [CrossRef]

49. Benzertiha, A.; Kierończyk, B.; Rawski, M.; Józefiak, A.; Kozłowski, K.; Jankowski, J.; Józefiak, D. Tenebrio molitor and Zophobas morio Full-Fat Meals in Broiler Chicken Diets: Effects on Nutrients Digestibility, Digestive Enzyme Activities, and Cecal Microbiome. *Animals* **2019**, *9*, 1128. [CrossRef] [PubMed]

50. Sogari, G.; Amato, M.; Biasato, I.; Chiesa, S.; Gasco, L. The Potential Role of Insects as Feed: A Multi-Perspective Review. *Animals* **2019**, *9*, 119. [CrossRef] [PubMed]

51. Islam, M.M.; Yang, C.J. Efficacy of mealworm and super mealworm larvae probiotics as an alternative to antibiotics challenged orally with Salmonella and E. coli infection in broiler chicks. *Poult. Sci.* **2017**, *96*, 27–34. [CrossRef]

52. Primex Iceland. Available online: http://www.primex.is (accessed on 1 August 2021).

53. Złotko, K.; Waśko, A.; Kamiński, D.M.; Budziak-Wieczorek, I.; Bulak, P.; Bieganowski, A. Isolation of Chitin from Black Soldier Fly (*Hermetia illucens*) and Its Usage to Metal Sorption. *Polymers* **2021**, *13*, 818. [CrossRef]

54. Lopez-Santamarina, A.; Mondragon, A.d.C.; Lamas, A.; Miranda, J.M.; Franco, C.M.; Cepeda, A. Animal-Origin Prebiotics Based on Chitin: An Alternative for the Future? A Critical Review. *Foods* **2020**, *9*, 782. [CrossRef]

55. Oonincx, D.G.; van Itterbeeck, J.; Heetkamp, M.J.; van den Brand, H.; van Loon, J.J.; van Huis, A. An exploration on greenhouse gas and ammonia production by insect species suitable for animal or human consumption. *PLoS ONE* **2010**, *5*, e14445. [CrossRef] [PubMed]

56. Halloran, A.; Roos, N.; Eilenberg, J.; Cerutti, A.; Bruun, S. Life cycle assessment of edible insects for food protein: A review. *Agron. Sustain. Dev.* **2016**, *36*, 57. [CrossRef] [PubMed]

57. Berggren, Å.; Jansson, A.; Low, M. Approaching Ecological Sustainability in the Emerging Insects-as-Food Industry. *Trends Ecol. Evol.* **2019**, *34*, 132–138. [CrossRef] [PubMed]

58. Smetana, S.; Schmitt, E.; Mathys, A. Sustainable use of Hermetia illucens insect biomass for feed and food: Attributional and consequential life cycle assessment. *Resour. Conserv. Recycl.* **2019**, *144*, 285–296. [CrossRef]

59. Khan, S.; Khan, R.U.; Alam, W.; Sultan, A. Evaluating the nutritive profile of three insect meals and their effects to replace soya bean in broiler diet. *J. Anim. Physiol. Anim. Nutr.* **2018**, *102*, e662–e668. [CrossRef]

60. Motte, C.; Rios, A.; Lefebvre, T.; Do, H.; Henry, M.; Jintasataporn, O. Replacing Fish Meal with Defatted Insect Meal (Yellow Mealworm *Tenebrio molitor*) Improves the Growth and Immunity of Pacific White Shrimp (*Litopenaeus vannamei*). *Animals* **2019**, *9*, 258. [CrossRef]

61. Poore, J.; Nemecek, T. Reducing food's environmental impacts through producers and consumers. *Science* **2018**, *360*, 987–992. [CrossRef] [PubMed]

62. Miglietta, P.P.; De Leo, F.; Ruberti, M.; Massari, S. Mealworms for Food: A Water Footprint Perspective. *Water* **2015**, *7*, 6190–6203. [CrossRef]

63. Alexander, P.; Brown, C.; Arneth, A.; Dias, C.; Finnigan, J.; Moran, D.; Rounsevell, M.D.A. Could consumption of insects, cultured meat or imitation meat reduce global agricultural land use? *Glob. Food Secur.* **2017**, *15*, 22–32. [CrossRef]

64. Oonincx, D.; Boer, I.J.M. Environmental Impact of the Production of Mealworms as a Protein Source for Humans—A Life Cycle Assessment. *PLoS ONE* **2012**, *7*, e51145. [CrossRef]

65. Gligorescu, A.; Fischer, C.H.; Larsen, P.F.; Nørgaard, J.V.; Heckman, L.-H.L. Production and Optimization of *Hermetia illucens* (L.) Larvae Reared on Food Waste and Utilized as Feed Ingredient. *Sustainability* **2020**, *12*, 9864. [CrossRef]

66. Bosch, G.; van Zanten, H.H.E.; Zamprogna, A.; Veenenbos, M.; Meijer, N.P.; van der Fels-Klerx, H.J.; van Loon, J.J.A. Conversion of organic resources by black soldier fly larvae: Legislation, efficiency and environmental impact. *J. Clean. Prod.* **2019**, *222*, 355–363. [CrossRef]

67. Mekonnen, M.M.; Neale, C.M.U.; Ray, C.; Erickson, G.E.; Hoekstra, A.Y. Water productivity in meat and milk production in the US from 1960 to 2016. *Environ. Int.* **2019**, *132*, 105084. [CrossRef] [PubMed]

68. Alexander, P.; Brown, C.; Arneth, A.; Finnigan, J.; Rounsevell, M.D.A. Human appropriation of land for food: The role of diet. *Glob. Environ. Chang.* **2016**, *41*, 88–98. [CrossRef]

69. Nielsen, I.; Jørgensen, M.; Bahrndorff, S. *Greenhouse Gas Emission from the Danish Broiler Production Estimated via LCA Methodology*; Knowledge Centre for Agriculture: Aarhus, Denmark, 2011; p. 30.

70. Čičková, H.; Newton, G.L.; Lacy, R.C.; Kozánek, M. The use of fly larvae for organic waste treatment. *Waste Manag.* **2015**, *35*, 68–80. [CrossRef] [PubMed]

71. Surendra, K.C.; Tomberlin, J.K.; van Huis, A.; Cammack, J.A.; Heckmann, L.-H.L.; Khanal, S.K. Rethinking organic wastes bioconversion: Evaluating the potential of the black soldier fly (*Hermetia illucens* (L.)) (Diptera: Stratiomyidae) (BSF). *Waste Manag.* **2020**, *117*, 58–80. [CrossRef] [PubMed]

72. Tsochatzis, E.D.; Berggreen, I.E.; Nørgaard, J.V.; Theodoridis, G.; Dalsgaard, T.K. Biodegradation of expanded polystyrene by mealworm larvae under different feeding strategies evaluated by metabolic profiling using GC-TOF-MS. *Chemosphere* **2021**, *281*, 130840. [CrossRef]

73. Hardy, A.; Benford, D.; Noteborn, H.; Halldorsson, T.; Schlatter, J.; Solecki, R.; Jeger, M.; Knutsen, H.; More, S.; Mortensen, A.; et al. Risk profile related to production and consumption of insects as food and feed. *EFSA J.* **2015**, *13*, 4257.

74. Roháček, J.; Hora, M. A northernmost European record of the alien black soldier fly *Hermetia illucens* (Linnaeus, 1758) (Diptera: Stratiomyidae). *Časopis Slez. Zemského Muz.—Ser. A* **2013**, *62*, 101–106. [CrossRef]

75. Kooh, P.; Jury, V.; Laurent, S.; Audiat-Perrin, F.; Sanaa, M.; Tesson, V.; Federighi, M.; Boué, G. Control of Biological Hazards in Insect Processing: Application of HACCP Method for Yellow Mealworm (*Tenebrio molitor*) Powders. *Foods* **2020**, *9*, 1528. [CrossRef] [PubMed]

76. Global Feed Safety Platform. Why Is Feed Safety Important. Available online: http://www.fao.org/feed-safety/background/why-feed-safety/en/ (accessed on 8 September 2021).

77. European Commission. Approval of First Insect as Novel Food. Available online: https://ec.europa.eu/food/safety/novel-food/authorisations/approval-first-insect-novel-food_en (accessed on 29 July 2021).

78. Poma, G.; Yin, S.; Tang, B.; Fujii, Y.; Cuykx, M.; Covaci, A. Occurrence of Selected Organic Contaminants in Edible Insects and Assessment of Their Chemical Safety. *Environ. Health Perspect.* **2019**, *127*, 127009. [CrossRef] [PubMed]

79. Schrögel, P.; Wätjen, W. Insects for Food and Feed-Safety Aspects Related to Mycotoxins and Metals. *Foods* **2019**, *8*, 288. [CrossRef] [PubMed]

80. Huis, A.V. Insects as food and feed, a new emerging agricultural sector: A review. *J. Insects Food Feed* **2020**, *6*, 27–44. [CrossRef]

81. Petroski, W.; Minich, D.M. Is There Such a Thing as "Anti-Nutrients"? A Narrative Review of Perceived Problematic Plant Compounds. *Nutrients* **2020**, *12*, 2929. [CrossRef] [PubMed]

82. Meyer-Rochow, V.B.; Gahukar, R.T.; Ghosh, S.; Jung, C. Chemical Composition, Nutrient Quality and Acceptability of Edible Insects Are Affected by Species, Developmental Stage, Gender, Diet, and Processing Method. *Foods* **2021**, *10*, 1036. [CrossRef] [PubMed]

83. Stoops, J.; Crauwels, S.; Waud, M.; Claes, J.; Lievens, B.; Van Campenhout, L. Microbial community assessment of mealworm larvae (*Tenebrio molitor*) and grasshoppers (*Locusta migratoria* migratorioides) sold for human consumption. *Food Microbiol.* **2016**, *53*, 122–127. [CrossRef]

84. Raimondi, S.; Spampinato, G.; Macavei, L.I.; Lugli, L.; Candeliere, F.; Rossi, M.; Maistrello, L.; Amaretti, A. Effect of Rearing Temperature on Growth and Microbiota Composition of Hermetia illucens. *Microorganisms* **2020**, *8*, 902. [CrossRef]

85. Caparros Megido, R.; Desmedt, S.; Blecker, C.; Béra, F.; Haubruge, É.; Alabi, T.; Francis, F. Microbiological Load of Edible Insects Found in Belgium. *Insects* **2017**, *8*, 12. [CrossRef]

86. Vandeweyer, D.; Crauwels, S.; Lievens, B.; Van Campenhout, L. Microbial counts of mealworm larvae (*Tenebrio molitor*) and crickets (*Acheta domesticus* and *Gryllodes sigillatus*) from different rearing companies and different production batches. *Int. J. Food Microbiol.* **2017**, *242*, 13–18. [CrossRef] [PubMed]

87. Wynants, E.; Frooninckx, L.; Crauwels, S.; Verreth, C.; De Smet, J.; Sandrock, C.; Wohlfahrt, J.; Van Schelt, J.; Depraetere, S.; Lievens, B.; et al. Assessing the Microbiota of Black Soldier Fly Larvae (*Hermetia illucens*) Reared on Organic Waste Streams on Four Different Locations at Laboratory and Large Scale. *Microb. Ecol.* **2019**, *77*, 913–930. [CrossRef] [PubMed]

88. Grabowski, N.T.; Klein, G. Microbiology of processed edible insect products—Results of a preliminary survey. *Int. J. Food Microbiol.* **2017**, *243*, 103–107. [CrossRef] [PubMed]

89. Stastnik, O.; Novotny, J.; Roztocilova, A.; Kouril, P.; Kumbar, V.; Cernik, J.; Kalhotka, L.; Pavlata, L.; Lacina, L.; Mrkvicova, E. Safety of Mealworm Meal in Layer Diets and their Influence on Gut Morphology. *Animals* **2021**, *11*, 1439. [CrossRef]

90. Tanga, C.M.; Waweru, J.W.; Tola, Y.H.; Onyoni, A.A.; Khamis, F.M.; Ekesi, S.; Paredes, J.C. Organic Waste Substrates Induce Important Shifts in Gut Microbiota of Black Soldier Fly (*Hermetia illucens* L.): Coexistence of Conserved, Variable, and Potential Pathogenic Microbes. *Front. Microbiol.* **2021**, *12*, 635881. [CrossRef]

91. Mancini, S.; Paci, G.; Ciardelli, V.; Turchi, B.; Pedonese, F.; Fratini, F. Listeria monocytogenes contamination of Tenebrio molitor larvae rearing substrate: Preliminary evaluations. *Food Microbiol.* **2019**, *83*, 104–108. [CrossRef]

92. Swinscoe, I.; Oliver, D.M.; Ørnsrud, R.; Quilliam, R.S. The microbial safety of seaweed as a feed component for black soldier fly (*Hermetia illucens*) larvae. *Food Microbiol.* **2020**, *91*, 103535. [CrossRef]

93. Fasolato, L.; Cardazzo, B.; Carraro, L.; Fontana, F.; Novelli, E.; Balzan, S. Edible processed insects from e-commerce: Food safety with a focus on the Bacillus cereus group. *Food Microbiol.* **2018**, *76*, 296–303. [CrossRef] [PubMed]

94. Garofalo, C.; Milanović, V.; Cardinali, F.; Aquilanti, L.; Clementi, F.; Osimani, A. Current knowledge on the microbiota of edible insects intended for human consumption: A state-of-the-art review. *Food Res. Int.* **2019**, *125*, 108527. [CrossRef] [PubMed]

95. Strother, K.O.; Steelman, C.D.; Gbur, E.E. Reservoir competence of lesser mealworm (Coleoptera: Tenebrionidae) for Campylobacter jejuni (Campylobacterales: Campylobacteraceae). *J. Med. Entomol.* **2005**, *42*, 42–47. [CrossRef] [PubMed]

96. Van der Fels-Klerx, H.J.; Camenzuli, L.; Belluco, S.; Meijer, N.; Ricci, A. Food Safety Issues Related to Uses of Insects for Feeds and Foods. *Compr. Rev. Food Sci. Food Saf.* **2018**, *17*, 1172–1183. [CrossRef]

97. Doi, H.; Gałęcki, R.; Mulia, R.N. The merits of entomophagy in the post COVID-19 world. *Trends Food Sci. Technol.* **2021**, *110*, 849–854. [CrossRef]

98. Ochoa Sanabria, C.; Hogan, N.; Madder, K.; Gillott, C.; Blakley, B.; Reaney, M.; Beattie, A.; Buchanan, F. Yellow Mealworm Larvae (*Tenebrio molitor*) Fed Mycotoxin-Contaminated Wheat-A Possible Safe, Sustainable Protein Source for Animal Feed? *Toxins* **2019**, *11*, 282. [CrossRef]

99. Bosch, G.; Fels-Klerx, H.J.V.; Rijk, T.C.; Oonincx, D. Aflatoxin B1 Tolerance and Accumulation in Black Soldier Fly Larvae (*Hermetia illucens*) and Yellow Mealworms (*Tenebrio molitor*). *Toxins* **2017**, *9*, 185. [CrossRef]

100. Camenzuli, L.; Van Dam, R.; De Rijk, T.; Andriessen, R.; Van Schelt, J.; Van der Fels-Klerx, H.J. Tolerance and Excretion of the Mycotoxins Aflatoxin B1, Zearalenone, Deoxynivalenol, and Ochratoxin A by Alphitobius diaperinus and Hermetia illucens from Contaminated Substrates. *Toxins* **2018**, *10*, 91. [CrossRef] [PubMed]

101. Schmidt, G.D. Acanthocephalan infections of man, with two new records. *J. Parasitol.* **1971**, *57*, 582–584. [CrossRef] [PubMed]

102. Gałęcki, R.; Sokół, R. A parasitological evaluation of edible insects and their role in the transmission of parasitic diseases to humans and animals. *PLoS ONE* **2019**, *14*, e0219303. [CrossRef] [PubMed]

103. Percipalle, M.; Salvaggio, A.; Pitari, G.M.; Giunta, R.P.; Aparo, A.; Alfonzetti, T.; Marino, A.M.F. Edible Insects and Toxoplasma gondii: Is It Something We Need To Be Concerned About? *J. Food Prot.* **2021**, *84*, 437–441. [CrossRef]

104. Da Cruz, C.E.F.; Marques, S.M.T.; Andretta, I. Endoparasites observed within invertebrates used as live food items for captive wild birds: Overview and potential risks. *Zoo Biol.* **2019**, *38*, 384–388. [CrossRef]

105. Müller, A.; Wiedmer, S.; Kurth, M. Risk Evaluation of Passive Transmission of Animal Parasites by Feeding of Black Soldier Fly (*Hermetia illucens*) Larvae and Prepupae. *J. Food Prot.* **2019**, *82*, 948–954. [CrossRef]

106. Gałęcki, R.; Michalski, M.M.; Wierzchosławski, K.; Bakuła, T. Gastric canthariasis caused by invasion of mealworm beetle larvae in weaned pigs in large-scale farming. *BMC Vet. Res.* **2020**, *16*, 439. [CrossRef]

107. Aelami, M.H.; Khoei, A.; Ghorbani, H.; Seilanian-Toosi, F.; Poustchi, E.; Hosseini-Farash, B.R.; Moghaddas, E. Urinary Canthariasis Due to Tenebrio molitor Larva in a Ten-Year-Old Boy. *J. Arthropod-Borne Dis.* **2019**, *13*, 416–419. [CrossRef]

108. Ma, J.; Wang, F. Prion disease and the 'protein-only hypothesis'. *Essays Biochem.* **2014**, *56*, 181–191. [CrossRef]

109. Post, K.; Riesner, D.; Walldorf, V.; Mehlhorn, H. Fly larvae and pupae as vectors for scrapie. *Lancet* **1999**, *354*, 1969–1970. [CrossRef]

110. Thackray, A.M.; Muhammad, F.; Zhang, C.; Denyer, M.; Spiropoulos, J.; Crowther, D.C.; Bujdoso, R. Prion-induced toxicity in PrP transgenic Drosophila. *Exp. Mol. Pathol.* **2012**, *92*, 194–201. [CrossRef] [PubMed]

111. Carp, R.I.; Meeker, H.C.; Rubenstein, R.; Sigurdarson, S.; Papini, M.; Kascsak, R.J.; Kozlowski, P.B.; Wisniewski, H.M. Characteristics of scrapie isolates derived from hay mites. *J. Neurovirol.* **2000**, *6*, 137–144. [CrossRef] [PubMed]

112. Wisniewski, H.M.; Sigurdarson, S.; Rubenstein, R.; Kascsak, R.J.; Carp, R.I. Mites as vectors for scrapie. *Lancet* **1996**, *347*, 1114. [CrossRef]

113. Raeber, A.J.; Muramoto, T.; Kornberg, T.B.; Prusiner, S.B. Expression and targeting of Syrian hamster prion protein induced by heat shock in transgenic Drosophila melanogaster. *Mech. Dev.* **1995**, *51*, 317–327. [CrossRef]

114. Hamanaka, T.; Nishizawa, K.; Sakasegawa, Y.; Kurahashi, H.; Oguma, A.; Teruya, K.; Doh-ura, K. Anti-prion activity found in beetle grub hemolymph of Trypoxylus dichotomus septentrionalis. *Biochem. Biophys. Rep.* **2015**, *3*, 32–37. [CrossRef]

115. International Platform of Insects for Food and Feed. *Guide on Good Hygiene Practices for European Union (EU) Producers of Insects as Food and Feed*; IPIFF: Brussels, Belgium, 2020; pp. 12–102.

116. Jensen, A.N.; Hansen, S.H.; Baggesen, D.L. Salmonella Typhimurium Level in Mealworms (Tenebrio molitor) After Exposure to Contaminated Substrate. *Front. Microbiol.* **2020**, *11*, 1613. [CrossRef]

117. Caparros Megido, R.; Poelaert, C.; Ernens, M.; Liotta, M.; Blecker, C.; Danthine, S.; Tyteca, E.; Haubruge, É.; Alabi, T.; Bindelle, J.; et al. Effect of household cooking techniques on the microbiological load and the nutritional quality of mealworms (*Tenebrio molitor* L. 1758). *Food Res. Int.* **2018**, *106*, 503–508. [CrossRef]

118. Huis, A.V. Potential of Insects as Food and Feed in Assuring Food Security. *Annu. Rev. Entomol.* **2013**, *58*, 563–583. [CrossRef]

119. Vandeweyer, D.; Lenaerts, S.; Callens, A.; Van Campenhout, L. Effect of blanching followed by refrigerated storage or industrial microwave drying on the microbial load of yellow mealworm larvae (*Tenebrio molitor*). *Food Control* **2017**, *71*, 311–314. [CrossRef]

120. Klunder, H.C.; Wolkers-Rooijackers, J.; Korpela, J.M.; Nout, M.J.R. Microbiological aspects of processing and storage of edible insects. *Food Control* **2012**, *26*, 628–631. [CrossRef]

121. Kröncke, N.; Grebenteuch, S.; Keil, C.; Demtröder, S.; Kroh, L.; Thünemann, A.F.; Benning, R.; Haase, H. Effect of Different Drying Methods on Nutrient Quality of the Yellow Mealworm (*Tenebrio molitor* L.). *Insects* **2019**, *10*, 84. [CrossRef]

122. Palsson, P.G.; Gissurarson, M. *HACCP Bókin—Fjölbreyttar og Gagnlegar Upplýsingar um HACCP og Framleiðslu Sjávarfangs (The HACCP Book—Diverse and Useful Information about HACCP and Seafood Production)*, 1st ed.; Matís: Reykjavik, Iceland, 2016; p. 49.

123. Engstrom, A. Är Det Lagligt Att Servera och sälja Insekter i Sverige? Här är Svaren! (Is It Legal to Serve and Sell Insects in Sweden? Here Are the Answers!). Available online: https://www.bugburger.se/politik/ar-det-lagligt-att-servera-insekter/?fbclid=IwAR1uixSogJqHPSp5nw68A5-S3eycvoFrkRiNHXUkQH13QACTy8qhjDOwqQo (accessed on 9 September 2021).

124. International Platform of Insects for Food and Feed. *An Overview of the European Market of Insects as Feed*; IPIFF: Brussels, Belgium, 2021.

125. European Commission. *No 999/2001*; European Commission: Brussels, Belgium, 2001.

126. European Commission. *Regulation (EU) No 189/2011*; European Commission: Brussels, Belgium, 2011.

127. European Commission. *Regulation (EU) No 2017/893*; European Commission: Brussels, Belgium, 2017.

128. European Commission. *Regulation (EU) No 2021/1372*; European Commission: Brussels, Belgium, 2021.

129. Mancini, S.; Moruzzo, R.; Riccioli, F.; Paci, G. European consumers' readiness to adopt insects as food. A review. *Food Res. Int.* **2019**, *122*, 661–678. [CrossRef] [PubMed]

130. Schösler, H.; Boer, J.D.; Boersema, J.J. Can we cut out the meat of the dish? Constructing consumer-oriented pathways towards meat substitution. *Appetite* **2012**, *58*, 39–47. [CrossRef] [PubMed]

131. Naranjo-Guevara, N.; Fanter, M.; Conconi, A.M.; Flotè Stammen, S. Consumer acceptance among Dutch and German students of insects in feed and food. *Food Sci. Nutr.* **2021**, *9*, 414–428. [CrossRef] [PubMed]

132. Sidali, K.L.; Pizzo, S.; Garrido-Pérez, E.I.; Schamel, G. Between food delicacies and food taboos: A structural equation model to assess Western students' acceptance of Amazonian insect food. *Food Res. Int.* **2019**, *115*, 83–89. [CrossRef]

133. Orsi, L.; Voege, L.L.; Stranieri, S. Eating edible insects as sustainable food? Exploring the determinants of consumer acceptance in Germany. *Food Res. Int.* **2019**, *125*, 108573. [CrossRef] [PubMed]

134. Lorini, C.; Ricotta, L.; Vettori, V.; Del Riccio, M.; Biamonte, M.A.; Bonaccorsi, G. Insights into the Predictors of Attitude toward Entomophagy: The Potential Role of Health Literacy: A Cross-Sectional Study Conducted in a Sample of Students of the University of Florence. *Int. J. Environ. Res. Public Health* **2021**, *18*, 5306. [CrossRef]

135. Tuccillo, F.; Marino, M.G.; Torri, L. Italian consumers' attitudes towards entomophagy: Influence of human factors and properties of insects and insect-based food. *Food Res. Int.* **2020**, *137*, 109619. [CrossRef] [PubMed]

136. International Platform of Insects for Food and Feed. The European Insect Sector Is very Pleased about the Third EFSA Opinion on Edible Insects. Available online: https://ipiff.org/the-european-insect-sector-is-very-pleased-about-the-third-efsa-opinion-on-edible-insects/ (accessed on 25 August 2021).

137. International Platform of Insects for Food and Feed. The European Insect Sector Welcomes the Latest EFSA Novel Food Opinion on Edible Insects! Available online: https://ipiff.org/2021/07/ (accessed on 25 August 2021).

138. International Platform of Insects for Food and Feed. IPIFF Welcomes the Fourth EFSA Opinion on Edible Insects. Available online: https://ipiff.org/ipiff-welcomes-the-fourth-efsa-opinion-on-edible-insects/ (accessed on 25 August 2021).

139. Kooistra, J. Financial Feasibility Analysis of Insect Farming in the Netherlands. Ph.D. Thesis, Wagening University & Research, Wageningen, The Netherlands, 2020.

140. International Platform of Insects for Food and Feed. Available online: https://ipiff.org (accessed on 25 August 2021).

141. International Platform of Insects for Food and Feed. IPIFF Members. Available online: https://ipiff.org/ipiff-members/ (accessed on 25 August 2021).

142. EFSA Panel on Dietetic Products; Nutrition and Allergies (NDA); Turck, D.; Bresson, J.-L.; Burlingame, B.; Dean, T.; Fairweather-Tait, S.; Heinonen, M.; Hirsch-Ernst, K.I.; Mangelsdorf, I.; et al. Guidance on the preparation and presentation of an application for authorisation of a novel food in the context of Regulation (EU) 2015/2283. *EFSA J.* **2016**, *14*, e04594. [CrossRef]

143. Garino, C.; Zagon, J.; Braeuning, A. Insects in food and feed—Allergenicity risk assessment and analytical detection. *EFSA J.* **2019**, *17*, e170907. [CrossRef]

144. Besnardeau, E.; (Ÿnsect, Paris, France). Personal Communication, 2021.

145. Hunt, A.; (Hexafly™, Meath, Ireland). Personal Communication, 2021.

146. Öhult, M.; (Verteco Farm, Piteå, Sweden). Personal Communication, 2021.

147. Kooistra, J.; (Syklus, Wageningen, The Netherlands). Personal Communication, 2021.

148. Marcenac, X.; (Nasekomo, Sofia, Bulgaria). Personal Communication, 2021.

149. Leseultre, X.; (Entomobio, Frasnes-les-Anvaing, Belgium). Personal Communication, 2021.

150. European Commision. Comission Implementing Regulation (EU) 2021/882 of 1 June 2021. *Off. J. Eur. Union* **2021**, *16*, 6.

151. Food Standards Australia New Zealand (Te Mana Kounga Kai—Ahitereiria me Aotearoa). Supporting Document 4—Overeview of International Regulatory Approaches—Proposal P1024 Revision of the Regulation of Nutirive Substances & Novel Foods. Available online: https://www.foodstandards.gov.au/code/proposals/Documents/P1024%20Nuts%20and%20novels%20SD4%20International%20regs.pdf (accessed on 12 September 2021).

152. IPIFF. EU Legislation. Available online: https://ipiff.org/insects-eu-legislation/ (accessed on 12 September 2021).

153. European Commision. Sustainable Developmental Goals. Available online: https://ec.europa.eu/international-partnerships/sustainable-development-goals_en (accessed on 9 September 2021).

154. Behre, E.; Heukels, B.; Mayayo, A.M.; Verschuur, X. *Insects as Livestock Feed*; UN Policy Analysis Branch, Division for Sustainable Development, Wageningen University & Research: Wageningen, The Netherlands, 2017.

155. European Commission. Strategy. Available online: https://ec.europa.eu/info/strategy_en (accessed on 9 September 2021).

156. European Commission. *Communication from the Commission to the European Parliament, the European Council, the Council, the European Economic and Social Committee and the Committee of the RegionS—The European Green Deal*; European Commission: Brussels, Belgium, 2019.

157. BugBurger. The Eatings Insects Startups: Here Is the List of Entopreneurs around. Available online: https://www.bugburger.se/foretag/the-eating-insects-startups-here-is-the-list-of-entopreneurs-around-the-world/ (accessed on 13 July 2021).

158. Van Huis, A.; Tomberlin, J.K. *Insects as Food and Feed: From Production to Consumption*; Wageningen Academic Press: Wageningen, The Netherlands, 2017.

159. Pikl, D.; (Cricky, Zagreb, Croatia). Personal Communication, 2021.

160. Rognstad, T.; (Urbanmat, Gjövik, Norway). Personal Communication, 2021.

161. Munk-Bogballe, D.; (Insectum ApS., Assens, Denmark). Personal Communication, 2021.

162. Horizon Insect LTD. About Us. Available online: https://www.linkedin.com/company/horizon-insects-ltd (accessed on 24 August 2021).

163. Tebrito. About Us. Available online: https://www.linkedin.com/company/tebrito/ (accessed on 24 August 2021).

164. Entoinnov. About Us. Available online: https://www.linkedin.com/company/entoinnov/ (accessed on 24 August 2021).

165. Enorm. About Us. Available online: https://www.linkedin.com/company/enorm/ (accessed on 24 August 2021).

166. Hermetia Baruth GmbH. About Us. Available online: https://be.linkedin.com/company/hermetia-baruth-gmbh?trk=similar-pages_result-card_full-click (accessed on 24 August 2021).

167. EntoMass. About Us. Available online: https://www.linkedin.com/company/entomass (accessed on 24 August 2021).

168. NextAlim. Welcome to NextAlim. Available online: https://www.nextalim.com (accessed on 13 July 2021).

169. NextProtein. NextProtein Feeding the Future. Available online: http://nextprotein.co (accessed on 13 July 2021).

170. Illucens Gmbh. Illucens. Available online: https://illucens.com/en/ (accessed on 13 July 2021).

171. HiProMine. Company. Available online: https://hipromine.com/company/?lang=en (accessed on 13 July 2021).

172. Micronutris. About Us. Available online: https://www.linkedin.com/company/micronutris/ (accessed on 24 August 2021).

173. Entocycle. About Us. Available online: https://www.linkedin.com/company/entocycle (accessed on 24 August 2021).

174. Tebrio. About Us. Available online: https://es.linkedin.com/company/tebri (accessed on 24 August 2021).

175. Protix. Food in Balance with Nature. Available online: https://protix.eu (accessed on 13 July 2021).

176. Ortiz, J.A.C.; Ruiz, A.T.; Morales-Ramos, J.A.; Thomas, M.; Rojas, M.G.; Tomberlin, J.K.; Yi, L.; Han, R.; Giroud, L.; Jullien, R.L. Chapter 6—Insect Mass Production Technologies. In *Insects as Sustainable Food Ingredients: Production, Processing and Food Applications*; Dossey, A.T., Morales-Ramos, J.A., Rojas, M.G., Eds.; Elsevier Science: San Diego, CA, USA; Academia Press: Cambridge, MA, USA, 2016; pp. 153–201.

177. Pikl, D. Beautiful Sunday Story about Cricky (Lijepa priča Nedjeljom) at Index.hr. Available online: https://cricky.eu/blog/beautiful-sunday-story-about-cricky-lijepa-prica-nedjeljom-at-index-hr/ (accessed on 13 July 2021).

178. Urbanmat. Om Oss (About Us). Available online: https://www.urbanmat.no/om-oss/ (accessed on 13 July 2021).

179. Insectum. About Us. Available online: https://insectum.farm/pages/about-insectum-aps (accessed on 13 July 2021).

180. Crowe, E. An Edible Insect Farm Has Launched in London. Available online: https://www.squaremeal.co.uk/restaurants/news/Edible-insect-farm-London-_9193 (accessed on 22 August 2021).

181. Jemthans, S. Moraföretaget Tebrito—Årets Nytänkare Producerar Insektsprotein för Framtiden (The Parent Company Tebrito—This Year's Innovator Produces Insect Protein for the Future). Available online: https://mora.se/aktuellt/lokalproducerad-mjolmask-kan-vara-framtidens-protein (accessed on 13 July 2021).

182. EntoBreed. EntoBreed—AgriTech Startup Voor Duurzame Meelwormenkweek (AgriTech Startup for Sustainable Mealworm Cultivation). Available online: https://entobreed.com (accessed on 13 July 2021).

183. Marienlyst Ento. Marienlyst Ento Organic Waste—Pure Food. Available online: https://www.marienlystento.dk (accessed on 13 July 2021).

184. Micronutris. Micronutris Créateur d'Alimentation Durable (Micronutris Creator of Sustainable Food). Available online: https://www.micronutris.com/en/home (accessed on 13 July 2021).

185. Entoinnov. Entoinnov We Are Changing the Paradigm of Insect Farming! Available online: https://www.entoinnov.com/en/homepage-entoinnov/ (accessed on 13 July 2021).

186. Entocycle. ENTOCYCLE. Available online: https://www.entocycle.com (accessed on 13 July 2021).

187. Enorm Biofactory. History & Vision. Available online: https://www.enormbiofactory.com/en/historie-og-vision (accessed on 13 July 2021).

188. Tebrio. Available online: https://tebrio.com/en/ (accessed on 13 July 2021).

189. Bugimine. Available online: https://bugimine.com (accessed on 13 July 2021).

190. Hermetia Futtermittel GbR. *Endbericht zum Forschungsvorhaben (Final Report on the Research Project)*; Hermetia Futtermittel GbR: Berlin, Germany, 2013; p. 27.

191. Entomass. About Us. Available online: https://www.entomass.dk/about-us (accessed on 13 July 2021).

192. PAPEK s.r.o. Available online: https://www.insect-papek.eu/en/ (accessed on 18 October 2021).

193. Moruzzo, R.; Mancini, S.; Boncinelli, F.; Riccioli, F. Exploring the Acceptance of Entomophagy: A Survey of Italian Consumers. *Insects* **2021**, *12*, 123. [CrossRef] [PubMed]

194. Piha, S.; Pohjanheimo, T.; Lähteenmäki-Uutela, A.; Křečková, Z.; Otterbring, T. The effects of consumer knowledge on the willingness to buy insect food: An exploratory cross-regional study in Northern and Central Europe. *Food Qual. Prefer.* **2018**, *70*, 1–10. [CrossRef]
195. Meyer-Rochow, V.B.; Kejonen, A. Could Western Attitudes towards Edible Insects Possibly be Influenced by Idioms Containing Unfavourable References to Insects, Spiders and other Invertebrates? *Foods* **2020**, *9*, 172. [CrossRef] [PubMed]
196. Roma, R.; Ottomano Palmisano, G.; De Boni, A. Insects as Novel Food: A Consumer Attitude Analysis through the Dominance-Based Rough Set Approach. *Foods* **2020**, *9*, 387. [CrossRef] [PubMed]
197. Víur Hafa Hætt Starfsemi. Available online: http://viur.gyl.fi (accessed on 19 August 2021).
198. Palsdottir, G.H. Framleiðsla Skordýra Reyndist ekki Arðbær (Insect Production Was Not Profitable). Available online: https://www.bbl.is/frettir/framleidsla-skordyra-reyndist-ekki-ardbaer (accessed on 19 August 2021).
199. Sverrisdottir, S.M. Innleiða Skordýraát í Vestræna Menningu: Búi og Stefán Segja Frá Sinni Gulleggshugmynd (Implementing Insect Eating in Western Culture: Bui and Stefan Talk about Their Golden Egg Idea). Available online: https://studentabladid.com/efni/innleida-skordyraat-i-vestraena-menningu (accessed on 19 August 2021).
200. Armannsson, B. Framleiðendur Matvæla úr Skordýrum Þurfa Leyfi Frá ESB (Insect Food Producers Need License from the EU). Available online: https://www.visir.is/g/2016160209242 (accessed on 19 August 2021).
201. Benediktsdottir, E.B. Rækta Mjölorma Heima Hjá Sér (Breed Mealworms in Their Home). Available online: https://www.ruv.is/frett/raekta-mjolorma-heima-hja-ser (accessed on 19 August 2021).
202. Munu Skordýr Fæða Heiminn? (Will Insects Feed the World?). Available online: https://matis.is/frettir/munu-skordyr-faeda-heiminn/ (accessed on 19 August 2021).
203. Sigurdardottir, K. Brauð úr Lirfum Krybba í Þróun Hjá MATÍS (Bread from Cricket's Larvae in Development at Matis). Available online: https://www.ruv.is/frett/braud-ur-lirfum-krybba-i-throun-hja-matis (accessed on 19 August 2021).
204. Hardarson, S.M. *Tilraunaræktun á Skordýrum til Manneldis eða Fóðurframleiðslu (Experimental Breeding of Insects for Human Consumption or Feed Production)*; Bændablaðið: Reykjavik, Iceland, 2021; pp. 26–27.
205. Salomone, R. Environmental impact of food waste bioconversion by insects: Application of Life Cycle Assessment to process using Hermetia illucens. *J. Clean. Prod.* **2017**, *140*, 890–905. [CrossRef]
206. Olafsson, G.; (MAST, Selfossi, Iceland). Personal Communication, 2021.
207. Hauksdottir, G. Cricket Snack to Go Jungle Bar on the Table in the USA. Available online: https://www.icenews.is/2016/01/28/cricket-snack-to-go-jungle-bar-on-the-table-in-the-usa/ (accessed on 12 September 2021).
208. The Reykavik Grapevine. Clearer Legislation Needed for Insect Based Food. Available online: https://grapevine.is/news/2016/01/20/clearer-legislation-needed-for-insect-based-food/ (accessed on 20 July 2021).
209. Sturludottir, E.; Thorvaldsson, G.; Helgadottir, G.; Gudnason, I.; Sveinbjornsson, J.; Sigurgeirsson, O.I.; Sveinsson, T. Fæðuöryggu á Íslandi. Skýrsla unnin fyrir atvinnuvega- og nýsköpunarráðuneytið (Food security in Iceland. Report prepared for the Ministry of Industry and Innovation). *Rit LbhÍ* **2021**, *139*, 56.
210. Saber, D.A.; Silka, L. Food Waste as a Classic Problem that Calls for Interdisciplinary Solutions: A Case Study Illustration. *J. Soc. Issues* **2020**, *76*, 114–122. [CrossRef]
211. Umhverfisstofnun. Matarsóun á Íslandi (Food Waste in Iceland). Available online: http://matarsoun.is/default.aspx?pageid=929ad605-0b03-11e6-a224-00505695691b (accessed on 10 August 2021).

Article

Meat Quality Parameters of Boschveld Indigenous Chickens as Influenced by Dietary Yellow Mealworm Meal

Letlhogonolo Selaledi [1], Josephine Baloyi [2], Christian Mbajiorgu [1], Amenda Nthabiseng Sebola [1], Henriette de Kock [2] and Monnye Mabelebele [1,*]

[1] Department of Agriculture and Animal Health, College of Agriculture and Environmental Science, University of South Africa, Florida Campus, 28 Pioneer Ave, Florida Park, Roodepoort 1709, South Africa; laselaledi@zoology.up.ac.za (L.S.); mbajica@unisa.ac.za (C.M.); nasebola@yahoo.com (A.N.S.)

[2] Department of Consumer and Food Sciences, Faculty of Natural and Agricultural Sciences, University of Pretoria, Pretoria 0002, South Africa; josephine.baloyi@up.ac.za (J.B.); riette.dekock@up.ac.za (H.d.K.)

* Correspondence: mabelebelem@gmail.com

Citation: Selaledi, L.; Baloyi, J.; Mbajiorgu, C.; Sebola, A.N.; Kock, H.d.; Mabelebele, M. Meat Quality Parameters of Boschveld Indigenous Chickens as Influenced by Dietary Yellow Mealworm Meal. *Foods* **2021**, *10*, 3094. https://doi.org/10.3390/foods10123094

Academic Editors: Victor Benno Meyer-Rochow and Chuleui Jung

Received: 5 October 2021
Accepted: 16 November 2021
Published: 14 December 2021

Publisher's Note: MDPI stays neutral with regard to jurisdictional claims in published maps and institutional affiliations.

Abstract: An experiment was conducted to examine the effects of yellow mealworm larvae (*Tenebrio molitor*) meal inclusion in diets of indigenous chickens. A total of 160 mixed-sex indigenous Boschveld chickens were randomly divided into four categories: control soybean meal (SBM) and yellow mealworm with percentage levels of 5, 10 and 15 (TM5, TM10 and TM15, respectively). Five replicate pens per treatment were used, with eight birds per pen/replicate. On day 60, two birds from each replicate were slaughtered and eviscerated. Meat quality parameters were measured out on raw carcass and cooked breast meat. The carcass weight, breast weight and gizzard weight of the control group was higher ($p < 0.05$) than the treatment group (TM15). The cooking loss was lower ($p < 0.05$) in the SBM control group but higher in the TM15 group. Colour characteristics of breast meat before cooking was lighter in the TM10 and TM15 group, ranged from 61.7 to 69.3 for L* and was significant ($p < 0.05$). The TM10 and TM15 groups showed a lighter colour than the SBM and TM5 groups. The breast meat pH taken after slaughter was different ($p < 0.05$) in TM5 and TM15, with the highest reading (pH 6.0) in the TM5 group. In conclusion, our experiment indicated that dietary *Tenebrio molitor* in growing Boschveld indigenous chickens' diets could be considered a promising protein source for Boschveld indigenous chickens.

Keywords: tenebrio molitor; local chicken; carcass characteristics; breast pH; edible insects

1. Introduction

Smallholder farmers in Southern Africa and many parts of the developing world generally rely on chicken meat to meet their dietary protein requirements [1,2]. The majority of these farmers breeds and keep indigenous chickens, which are mostly tolerant of local diseases and parasites; moreover, they provide huge economic benefits and income for rural communities [3,4]. The sudden increase in the demand for natural or organic meat could have influenced smallholder farmers to consider farming with indigenous chickens because these chickens need minimal use of additives and chemicals [1]. The indigenous chickens are known to be economically, socially and culturally important to the people of Africa and other developing countries, especially those from poor communities [5]. Although they are associated with poor productivity, most consumers prefer their flavoursome meat [5]. However, conventional protein sources, such as soybean and fishmeal, which are normally used in the poultry diets, are expensive, and thus, they lead to the search for alternative sources of protein [6].

Attempts to study alternative sources of protein have included insect meals. Insects are promising animal feed ingredients because they contain high levels of quality protein and are easy to produce [6]. Experimental results have shown that yellow mealworm used as a source of protein in the diets of fast-growing commercial broilers does not compromise growth performance [7,8]. However, there is evidence that indigenous chickens and commercial broiler have genetic differences that could affect growth rates [9]. The strain of the chicken affects the growth rate, feed conversion ratio, feed intake, and digestibility at different ages [9]. In Southern Africa, indigenous breeds, such as Boschveld chicken, which is a dual-purpose crossbreed of three indigenous breed, namely, Ovambo, Matabele and Venda, is normally produced for egg and meat [10]. It is not known how Boschveld indigenous chicken and their physico-chemical parameters could respond to the inclusion of yellow mealworms as a partial replacement of soybean in diets. Biasato et al. [11] studied the effect of yellow mealworm on female fast-growing broilers (Ross 708) and found that carcass weight increased quadratically with increasing levels of yellow mealworms. Furthermore, the abdominal fat weight showed linear responses to increasing levels of yellow mealworms; however, no significant effects were observed from other carcass traits. Cullere et al. [12] studied the effects of insect meal (Black soldier fly) larvae in fast-growing broilers' (Ross 708) finisher diet on the quality of meat and reported that breast meat displayed similar weight, thawing loss and pH.

To determine the acceptance of chicken meat, consumers consider several characteristics, such as its colour, chemical properties and sensory characteristics [1]. Chicken meat is one of the most consumed in the world due to a number of reasons, which are related to socio-economic conditions, digestion, nutritional value and ease of cooking [13]. Quantifiable properties of meat, such as cooking loss, pH, water holding capacity, protein solubility, fat binding capacity and drip loss, are the most important factors that influence final quality judgement by consumers [14]. Factors such as appearance, colour, tenderness, juiciness and flavour can lead the consumer to approve or not approve the meat product [14]. Therefore, the replacement of soybean meal with yellow mealworms (*Tenebrio molitor*) has the potential to have some effect on the cooking loss, shear force, meat colour and pH of fast-growing broiler chickens [7]. However, there is little information on the use of yellow mealworm meal in the diets of indigenous chickens which are genetically different from fast-growing broilers and their influence on the meat characteristics. Moreover, since the inclusion of yellow mealworms may interfere with the meat quality, studies that go beyond the examination of animal performance are needed. In most cases, the carcass traits of fast-growing broilers are not affected by the inclusion of yellow mealworm in poultry diet [15,16]. However, no studies are available in the literature on the effect of yellow mealworm on carcass traits and meat parameters of slow-growing Boschveld indigenous chickens fed yellow mealworms. Therefore, the aim of this research was to study the effect of yellow mealworm inclusion on the carcass traits and meat quality of Boschveld indigenous chickens.

2. Materials and Methods

The present study was conducted at the Proefplaas experimental farm of the University of Pretoria, South Africa. The chickens were reared in a climate-controlled poultry house. The general care and management of the chickens followed accepted guidelines as described by the South African Poultry Association. Furthermore, the experimental protocol was approved by the Human Research Ethics Committee of the College of Agriculture and Environmental Sciences at the University of South Africa with ethical clearance number: 2019/CAES_HREC and the University of Pretoria, Faculty of Veterinary Science, Animal Ethics committee, reference no.: NAS433/2019. The statistical report of the manuscript is added as Supplementary Material.

2.1. Birds and Management

A total of 160-day-old mixed-sex chicks (Boschveld Indigenous chickens) were divided into four equal groups of five replicates (eight birds per replicate). The birds had free access to water and feed and the brooding house was provided with fresh wood shavings as a bedding material. The total number of cages were 20; each cage was 2.5 m × 1 m. The chicks were brooded at 34 °C for a period of one week; thereafter, the temperature was gradually decreased to reach 23 °C. The infrared heat lamps (China Original manufacturer) were used to provide the supplementary heat and maintain stable housing temperature. The housing temperature and relative humidity was monitored continuously, and relative humidity was adjusted to 55. Chicks were vaccinated against Newcastle and Gumboro disease [17]. The chickens were kept in cages for a period of 60 days before slaughter.

2.2. Yellow Mealworm Larval Meal and Experimental Diets, Dietary Treatment

A preliminary proximate analysis of oven-dried yellow mealworm was conducted. The oven-dried yellow mealworm nutrient composition is presented in Table 1. Diets were formulated to meet growing indigenous chickens' requirements. A diet based on corn meal, wheat offal, full-fat soya, sunflower cake and fishmeal was formulated in a feed mill unit in crumble form and served as control. The dietary treatments were as follows: a control group containing 0% yellow mealworm (SBM), TM5, which had 5% inclusion of *tenebrio molitor* larvae, TM10, which had 10% inclusion, and TM15, which had 15% inclusion. The yellow mealworm larvae (20 kg) were collected from the breeding facility of the University of Pretoria, Department of Zoology and Entomology. The other 10 kg was purchased from the local supplier (Insectivore). The yellow mealworm was produced using wheat bran and carrots. Detailed information about the chemical composition of the mealworm is provided in the study by Selaledi and Mabelebele [18]. The yellow mealworms were kept in a −20 °C freezer and were subsequently oven-dried. Two EcoTherm ovens were used to dry the mealworms. A total of 1000 g of mealworm was oven-dried for an hour under a temperature of 120 °C. The experimental diets were isonitrogenous and isoenergetic and were formulated. The diets met or exceeded requirements and were adjusted according to NRC [19].

Table 1. Ingredients and the calculated analysis of yellow mealworm larvae meal and experimental diets [18].

Ingredient	SBM	TM5	TM10	TM15	Yellow Mealworm Larvae (g/100 g)
Maize	50.00	50.00	50.00	50.00	
Wheat offal	8.00	8.00	8.00	8.00	
Yellow Mealworm	0.00	5.00	10.00	15.00	
Full fat soya	23.73	20.00	13.00	6.00	
Sunflower cake	10.00	8.73	10.73	12.73	
Fish meal (72%)	3.00	3.00	3.00	3.00	
Limestone	1.20	1.20	1.20	1.20	
Monocalcium phosphate	1.50	1.50	1.50	1.50	
Salt	0.25	0.25	0.25	0.25	
Sodium Bicarbonate	1.50	1.50	1.50	1.50	
DL methionine	0.25	0.25	0.25	0.25	
L-Threonine	0.12	0.12	0.12	0.12	
Lysine HCL	0.25	0.25	0.25	0.25	
Tryptophan	0.05	0.05	0.05	0.05	
Vit TM Premix	0.05	0.05	0.05	0.05	
Min premix	0.05	0.05	0.05	0.05	
Coccidiostat	0.05	0.05	0.05	0.05	
	100.00	100.00	100.00	100.00	
Calculated analysis					
%CP	20.00	20.10	20.17	20.18	51.51
MEKcal/kg	3197.11	3144.25	3110.68	3177.11	24.63
%EE	16.31	16.24	16.72	16.20	25.73
% CF	18.69	18.06	18.29	18.52	6.11
%Ca	1.70	1.67	1.63	1.60	0.294
Avail P%	0.77	0.75	0.72	0.69	7.48
%Lysine	1.92	1.65	1.31	0.97	3.95
Met + Cys%	1.14	1.02	0.87	0.71	n/a

SMB: soybean meal; TM5, TM10 and TM15: *Tenebrio molitor* at 5, 10 and 15% of soybean protein substitution, respectively. CP = crude protein; EE = ether extract; CF = crude fibre; Ca = calcium; P = phosphorus.

2.3. Slaughter and Carcass Traits (Internal Parts)

On day 60, two birds were randomly selected from each pen (male and female) and transported to the University of Pretoria poultry abattoir, where they were humanely slaughtered. The total number of birds slaughtered was 40; birds were transported in live chicken crates with a lid. The dimensions of each crate were 74 × 53 × 31 cm, with five chickens per crate. Each treatment had two crates, with 5 males and 5 females per treatment. The distance from the poultry house to the abattoir was less than a one kilometre. Birds were transported in a safe reliable vehicle. The birds were stunned using a small electric stunner to induce immediate unconsciousness, which was then followed by severing the neck. After stunning, decapitation and bleeding, the carcasses were plucked, eviscerated and their feet were removed [20]. Carcass weight, eviscerated weight and breast, thigh and drumstick weight were recorded. Indexes of Slaughter rate were calculated as carcass weight/live weight × 100, Eviscerated weight/live weight × 100, Thigh muscle/live weight × 100, Breast muscle/live weight × 100 and wing muscle/live weight × 100 [21].

Immediately after slaughter, the carcasses were labelled and packed in transparent plastic bags and were then kept at 4 °C for 48 h. The carcass was immediately weighed USING THE Mettler PC 4400 SCALE 9Mettler-Toledo, Switzerland) to determine the carcass weight. The carcass was chilled for 48 h and the internal organs were collected and weighed separately. Breast and small intestine (Jejunum) pH was taken immediately after slaughter, followed by another reading after 24 and 48 h, respectively. The pH of the breast and small intestine was measured using a calibrated (standard buffers pH 4.0 and 7.0 at 25 °C) Portable HANNA HI 8424 to measure the pH. The pH meter was calibrated before the first measurement and every time we measured a new set of samples (for every treatment group). The electrode was placed in an incision that was made in the centre of the breast muscle, and for the small intestine, the electrode was placed in the jejunum.

2.4. Cooking Procedure

Frozen whole chickens (−20 °C) were thawed overnight at refrigeration temperatures (5 °C) before analysis. After thawing, chickens were placed in individual oven bags Glad® (Clorox Africa) that were tied at one end with a perforated strip. Holes were poked on the oven bag with chicken prior to placing it in the oven, to prevent the bag from bursting in the oven. Bagged chickens were placed in oven pans and cooked in a pre-heated oven at 180 °C for 24 min using Ambassade C505 CT Electric convection oven (Lacanche, France). After cooking and before cooling, the internal temperature of chicken was measured using the Testo 104-IR digital thermometer (Testo SE & Co. KGaA, Lenzkirch, Germany) by inserting the needle probe into the cranial side of both chicken breasts. After cooling, chickens were removed from their bags and weighed. Colour measurements were recorded before the chickens were wrapped in foil and stored in a refrigerator (5 °C) for at least 12 h before texture analysis.

To determine the cooking loss per sample, chickens were weighed before and after cooking. The following formula was used cook loss (%) = ([raw chicken weight–cooked chicken weight]/raw chicken weight) × 100. Cooking loss was calculated for eight chickens, as formula (1) [22]:

$$\text{Cooking Loss \%} = (\text{raw chicken weight} - \text{cooked chicken weight})/(\text{raw chicken weight}) \times 100 \tag{1}$$

2.5. Texture Properties of Chicken

Chickens were taken out of the refrigerator and left to equilibrate at room temperature (23.0 °C ± 2 °C) for one hour. For each whole chicken, two chicken breasts from the pectoralis major section were manually trimmed using a sharp knife. From each of the cooked chicken breast, one strip (section B), as presented by Lyon and Lyon [22], was cut parallel to the direction of the muscle fibres and used as a test sample. The length, width and thickness of the samples were measured by a Vernier calliper. A shear force test of chicken breasts was performed at room temperature using the texture analyser (Model EZ-L, Shimadzu Tokyo, Japan), with a 5000 N load cell and at a test speed of 50 mm/min.

A vertical force was applied on each sample using a Warner-Bratzler shear blade, at an angle perpendicular to the direction of the muscle fibre.

2.6. Instrumental Colour Analysis

The colour of raw and cooked chickens was measured using a CR 210 Minolta chromameter model CR-400 (Osaka, Japan), and an 8 mm aperture size, illuminant D65 and 10° standard observer angle was used. The chromameter was calibrated using a standard white plate (Y = 87.2, X = 0.3173; Y = 0.3348). The colour was measured on breast muscle with three measurements on each sample and an average of three readings per sample was taken. Readings were recorded as L*, a* and b* for four chickens for each treatment. L* represents lightness, a* represents redness (+) and greenness (−) and b* represents blueness (+) and yellowness (−), respectively. Colour measurements were determined for four chickens per treatment [23].

2.7. Statistical Analysis

Data were subjected to one-way analysis of variance procedures appropriate for a completely randomised design of IBM SPSS [24]. Means were compared using Tukey's test. The statement of statistical significance was based on $p < 0.05$. The following model was used for the analysis: $Y_{ij} = \mu + \alpha_j + \varepsilon_{ij}$, where Y_{ij} = Carcass and internal organ variables, μ = the population means, α_j = Treatments and ε_{ij} = the residual effect. Furthermore, the G-Power analysis was used to calculate the statistical power.

3. Results

The nutrient composition of the yellow mealworm used in this study and the ingredients used in the experimental diets is presented in Table 1. The carcass characteristics of the 60-day old mixed-sex Boschveld indigenous chickens are shown in Table 2. No significant difference between male and female birds was observed in statistical test. No significant interaction was detected for Eviscerated weight, bled weight, thigh, drumstick, neck and wing, but a significant effect was observed for the carcass weight (Control and TM15), back and breast weight.

Table 2. Effects of yellow mealworm meal on carcass characteristics of Boschveld indigenous chickens.

Parameter	SBM	TM5	TM10	TM15	SEM	*p*-Value
Bled weight (g)	807.8	721.6	708	676.4	19.421	0.094
Carcass weight	571.2 [a]	495.5 [ab]	530.8 [ab]	455.8 [b]	14.637	0.028
Eviscerated weight (g)	550.4	473.2	484.2	448.4	14.943	0.089
Thigh (g)	99	85.2	87	80	3.028	0.169
Drumstick (g)	89.8	74	76	69.4	15.067	0.185
Back (g)	135.2 [a]	118 [ab]	117 [ab]	107 [b]	3.546	0.045
Neck (g)	52.2	48.4	47.6	44.2	1.586	0.371
Breast (g)	145.2 [a]	120 [b]	122 [b]	115.8 [b]	3.670	0.015
Wing (g)	82.2	75.2	80	74	2.394	0.593

[a,b] Means in the same row not sharing a common superscript are different ($p < 0.05$). Diets: SBM = soybean meal, TM5% = a diet in which 5 g/kg of soybean was replaced with *Tenebrio molitor* meal, TM10% = a diet in which 10 g/kg of soybean was replaced with *Tenebrio molitor* meal. TM15% = a diet in which 15 g/kg of soybean was replaced with *Tenebrio molitor* meal. SEM: standard error of the mean.

The visceral weight, including the head and feet, of 60-day old mixed-sex Boschveld indigenous chickens is presented in Table 3. The dietary treatment ($p < 0.05$) affected the weight of the gizzard in favour of the control group; however, there were no differences between the treatment groups (TM5, 10 15).

Table 3. Effects of the percentage of yellow mealworm meal on organ weight characteristics of Boschveld indigenous chickens.

Parameter	SBM	TM5	TM10	TM15	SEM	*p*-Value
Head (g)	34.2 [a]	29.8 [ab]	31.2 [ab]	28.0 [b]	0.837	0.055
Feet (g)	37.6	31.2	35.6	31.6	1.403	0.304
Gizzard (g)	30.6 [a]	19.0 [b]	21.2 [b]	21.4 [b]	1.032	0.000
Lungs (g)	6.0	6.0	4.6	5.0	0.269	0.155
Liver (g)	18.4	14.8	16.4	16.2	0.626	0.245
Heart (g)	5.6	5.6	4.4	5.2	0.212	0.148
Proventriculus (g)	4.2	3.6	4.8	4.2	0.245	0.408
Small intestine (g)	20.4	18.4	17.8	17.2	0.570	0.220
Full intestine length (cm)	117.5	119.7	116.4	111.5	2.175	0.609

[a,b] Means in the same row not sharing a common superscript are different ($p < 0.05$). Diets: SBM = soybean meal, TM5% = a diet in which 5 g/kg of soybean was replaced with *Tenebrio molitor* meal, TM10% = a diet in which 10 g/kg of soybean was replaced with *Tenebrio molitor* meal. TM15% = a diet in which 15 g/kg of soybean was replaced with *Tenebrio molitor* meal. SEM: standard error of the mean.

The slaughter rate is one of many factors affecting the value of a slaughter animal. The indexes of slaughter rate are presented in Figure 1 and were calculated by dividing the weight of carcass by live weight and multiply by 100. The carcass rate was higher in the TM15 group followed by control group (SBM). There were slight variations in the percentages of other meat potions.

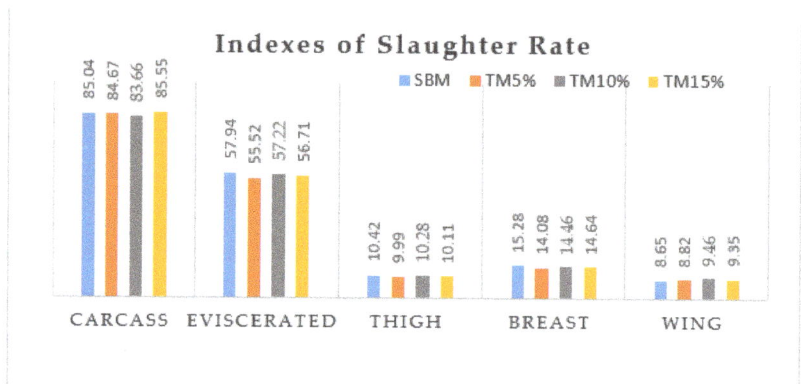

Figure 1. Indexes of slaughter rate of Boschveld chickens fed yellow mealworm.

Indexes of Slaughter rate were calculated as carcass weight/live weight × 100; Eviscerated weight/live weight × 100; Thigh muscle/live weight × 100; Breast muscle/live weight × 100 and wing muscle/live weight × 100.

Raw and cooked mixed-sex Boschveld indigenous chickens' meat quality parameters are presented in Table 4. The soybean fed group (SBM) were heavier ($p < 0.05$) than the experimental group (TM15). The cooking loss was lower in the SBM control group as compared to TM15 ($p < 0.05$). However, no differences existed between control, TM5 and TM10 group. The internal temperature was measured immediately after removing the breast meat from the oven. The breast (both left and right) internal temperature was higher in the TM15% group and lower in the SBM group. Colour characteristics was also measured before and after cooking. The values for lightness (before cooking) ranged from 61.7 to 69.3 and this was significant ($p < 0.05$). The TM10 and TM15 group showed lighter colours than the SBM and TM5 groups. No significant ($p > 0.05$) differences were observed for the a*, b* values; similarly, the colour of the breast meat after cooking did not differ in lightness, redness and yellowness. The breast meat from the TM5 group had a lower shear force value (33.7) than the other groups. The TM10 group had the highest shear force (55.6) and stress (0.0927).

Table 4. Physical properties of carcass without internal viscera of Boschveld indigenous chickens fed soybean meal (SBM) or *Tenebrio molitor* larvae meal (TML) at 60 day of age.

Parameters (Before Cooking)	SBM	TM5%	TM10%	TM15%	SEM	*p*-Value
Breast meat before cooking						
Lightness L* (before cooking)	61.8 [b]	61.7 [b]	69.3 [a]	68.8 [a]	1.201	0.008
Redness a* (before cooking)	1.9	1.7	3.0	2.7	0.354	0.549
Yellowness b* (before cooking)	2.9	6.3	5.0	5.0	0.651	0.352
Parameters (After cooking)						
Cooking loss %	8.82 [b]	12.23 [ab]	11.97 [ab]	13.70 [a]	0.650	0.049
Breast Internal temperature (left)	65.3 [b]	76.7 [a]	71.3 [ab]	80.7 [a]	1.702	0.004
Breast Internal temperature (right)	66.0 [b]	76.5 [a]	72.4 [ab]	80.8 [a]	1.570	0.003
Drained water (g)	29.0	33.1	34.3	36.4	1.516	0.388
Lightness L* (after cooking)	71.4	66.4	69.4	68.8	0.950	0.343
Redness a* (after cooking)	3.8	4.3	2.8	2.8	0.259	0.076
Yellowness b* (after cooking)	17.0	21.1	17.2	17.6	0.661	0.077
Texture (shear force)	40.9 [b]	33.7 [b]	55.6 [a]	39.8 [b]	2.166	0.001
Texture (stress)	0.0527 [b]	0.0574 [b]	0.0927 [a]	0.0682 [ab]	0.004	0.003

[a,b] Means in the same row not sharing a common superscript are different ($p < 0.05$). Diets: SBM = soybean meal, TM5% = a diet in which 5 g/kg of soybean was replaced with *Tenebrio molitor* meal, TM10% = a diet in which 10 g/kg of soybean was replaced with *Tenebrio molitor* meal. TM15% = a diet in which 15 g/kg of soybean was replaced with *Tenebrio molitor* meal. SEM: standard error of the mean.

Table 5 presents pH values taken 15 min after slaughter followed by another reading after 24 h and 48 h respectively. The breast meat pH taken after slaughter was different ($p < 0.05$) between the TM5 and TM15 groups, with the highest reading (pH 6.0) in the TM5 group. The small intestine pH taken from the jejunum after 24 h was the highest in the SBM group ($p < 0.05$).

Table 5. The pH values of breast and small intestine taken from 60-day-old mixed-sex Boschveld indigenous chickens.

Parameter	SBM	TM5%	TM10%	TM15%	SEM	*p*-Value
Breast pH (After slaughter)	5.83 [ab]	6.01 [a]	5.80 [ab]	5.74 [b]	0.037	0.055
Breast pH (After 24 h)	5.88 [a]	5.87 [ab]	5.80 [ab]	5.75 [b]	0.017	0.034
Breast pH (After 48 h)	5.86	6.02	5.86	5.98	0.028	0.093
Small intestine pH (After slaughter)	6.37	6.18	6.19	6.20	0.031	0.093
Small intestine pH (after 24 h)	6.45 [a]	6.23 [b]	6.29 [b]	6.30 [ab]	0.023	0.003
Small intestine pH (after 48 h)	6.28	6.17	6.20	6.27	0.020	0.174

[a,b] Means in the same row not sharing a common superscript are different ($p < 0.05$). Diets: SBM = soybean meal, TM5% = a diet in which 5 g/kg of soybean was replaced with *Tenebrio molitor* meal, TM10% = a diet in which 10 g/kg of soybean was replaced with *Tenebrio molitor* meal. TM15% = a diet in which 15 g/kg of soybean was replaced with *Tenebrio molitor* meal. SEM: standard error of the mean.

4. Discussion

Understanding factors that affect chicken meat quality and carcass characteristics are important in the poultry sector [21]. Unfortunately, knowledge on such factors is insufficient, particularly in the indigenous chicken production system. The success of the poultry production enterprise depends on the feed quality given to the chickens [25]. In this study, the inclusion of yellow mealworm meal did not significantly affect the carcass characteristics of the chicken, except on the back and breast meat including the head and gizzard. Similarly, Sedgh-Gooya et al. [15] reported that the carcass traits of broiler chickens fed yellow mealworm larvae powder as a dietary protein source did not significantly affect the carcass characteristics. The study reported that the carcass characteristics such as weight and length of different parts were not influenced by diets that had mealworm powder. Similar results were reported by Bovera et al. [16], which showed that the carcass traits of broilers fed yellow mealworm meal had no significant effect on the carcass traits.

However, Zadeh et al. [26] reported a higher carcass yield on the control group of Japanese quails at 35 days of age that were fed yellow mealworm larvae as an alternative protein source. The control group carcass yield for the study by Zadeh et al. [26] was 71.78 higher than TM7.5 (67.42), TM15 (64.47) and TM22.5 (62.24).

With respect to cooking loss, there were no differences among the breast meat between the control, TM5 and TM10 groups. However, the cooking loss in the control group was significantly lower as compared to the TM15 group. These results were consistent with those of Bovera et al. [16], who studied the use of yellow mealworm on broilers. Regarding Bovera's [16] findings, the cooking loss in the SBM group (21.4) was lower as compared to the TML group (23.6) at 62 d of age. Similarly, Mbhele et al. [27] reported a higher cooking loss in the treatment group (Black soldier fly), although the differences were not significant. Furthermore, Leiber et al. [28] also had a higher cooking loss in the insect fed group (*Hermetia* meal) as compared to the control. Cooking Loss is a significant indicator of meat quality, as it determines the technological yield of the cooking process [29]. Cooking loss is normally calculated as the per cent weight difference between fresh and cooked samples with respect to the weight of fresh meat samples [30]. Therefore, the difference in the cooking loss of the control group that was fed soybean basal diet and TM15 (*Tenebrio molitor*) could suggest that the inclusion of Yellow Mealworm on Boschveld indigenous chickens could have an effect on meat quality parameter (Cooking loss). This result suggests that TM5 and TM10 could be a better inclusion rate of yellow mealworm in Boschveld indigenous chickens' diet, as there were no significant differences between these treatment groups and the control.

Breast meat from Boschveld chickens that were fed 5% yellow mealworm had a slightly higher average ultimate pH ($p < 0.05$) than the TM15 group; however, the pH from the SBM group was lower than the TM5 group, but the differences were not significant. This reveals that feeding Boschveld indigenous chickens 5% or less of yellow mealworm could cause the breast meat to be less acidic, consumers could prefer meat with pH 5.96 [31]. The breast meat pH values obtained in this study are comparable to those reported by Nhlane et al. [32], who fed Boschveld chickens dietary treatment supplemented with seaweed. These results agree with earlier studies on broilers fed yellow mealworm [16], wherein the TML group (6.12) had the highest pH as compared to the SBM group (5.95). Shaviklo et al. reported that dietary inclusion of mealworm meal up to 3% may be appropriate; however, the meat quality of broilers may adversely be influenced in higher levels [33].

Meat colour can be influenced by environmental and genetic factors [34]. In this study, no difference existed ($p > 0.05$) after cooking among breast meat samples that were harvested from the control and treatment groups. A study conducted by Zadeh et al. [26] on 35-day-old Japanese quails also did not find any significant difference in the breast meat colour (L*), even though their values were lower (43.09–43.41) as compared to this study. The absence of differences in the meat colour after cooking is very important because colour can influence the consumer acceptance of meat (Bovera et al. [16]. The L*a*b* values for all treatments are comparable with the characteristics of normal indigenous chickens' breast meat; normal broiler meat is described as meat with L* values between 50 and 56, where dark meat will have an L* value <50 and pale meat will have an L* value >56 [35,36]. However, the meat samples from chickens that were fed 10–15% of yellow mealworm larvae were slightly lighter or paler before cooking (68 to 69) than the control group. Lighter breast meat colour is more preferred by consumers than dark meat [34]. Compounds such as myoglobin contribute to the colour of poultry meat [34]. Our L* a*b* findings are not similar to what Bovera et al. [16] reported; the lightness on the breast meat was (44–44.2), redness was (1.07–1.18), yellowness was (0.69–0.78) and chroma was (1.91–1.92). Furthermore, Schiavone et al. [37] did not find differences in the pH and lightness the breast meat between the control and the treatment groups that were fed Black soldier fly as a partial or total replacement of soybean oil.

Additionally, the findings of Bovera et al. [16] of the cooked breast meat shear force was 69.3 for the Soybean group and 73.2 *Tenebrio molitor* group. The shear force from our study was also higher (55.6) among the *Tenebrio molitor* group (TM 10%) but lower than the values reported by Bovera et al. [16]. Meat tenderness is considered a critical attribute in meat consumption. Therefore, it is important to meet the meat tenderness requirements that consumers demand, as this will result in customer satisfaction [9]. Moreover, Mbhele et al. [27] also reported higher shear force values (7.55) in the treatment group of Jumbo Quails that were fed Black soldier fly meal (BSFL50) as compared to the values recorded in the control group (6.20).

5. Conclusions

In conclusion, the present experiment indicated that increasing levels of dietary *Tenebrio molitor* in growing Boschveld indigenous chickens diets up to 10–15% could lighten the breast meat colour. It also could improve the pH of the breast meat (TM5) and meat texture (TM5). The yellow mealworms show potential to replace soymeal in slow growing Boschveld indigenous chickens; however, 15% or more could affect performance. Compared with the soybean control group (TM0), some cooking loss occurred in the TM15 group that was fed 15% yellow mealworm. Therefore, this could suggest that feeding Boschveld indigenous chickens 15% or more of yellow mealworm could be disadvantageous. The effect of insect meal on the cooking loss warrants further investigation, because this study and others have reported a higher cooking loss in the insect-fed group (TM15) as compared to the control group. The study confirms that insects can make a valuable contribution as chicken feed to global food security, as suggested already in 1975 by Meyer-Rochow [38].

Supplementary Materials: The following are available online at https://www.mdpi.com/article/10.3390/foods10123094/s1.

Author Contributions: Conceptualisation, L.S. and M.M.; methodology, L.S., M.M., H.d.K. and J.B.; software, L.S; formal analysis, L.S. and M.M.; writing—original draft preparation, L.S.; writing—review and editing, L.S, A.N.S.; visualisation, M.M. and L.S.; supervision, C.M.; project administration, M.M; funding acquisition, M.M. All authors have read and agreed to the published version of the manuscript.

Funding: This research received no external funding.

Institutional Review Board Statement: The experimental protocol was approved by the Human Research Ethics Committee of the College of Agriculture and Environmental Sciences at the University of South Africa with ethical clearance number: 2019/CAES_HREC and the University of Pretoria, Faculty of Veterinary Science, Animal Ethics committee, reference no.: NAS433/2019.

Informed Consent Statement: Informed consent was obtained from all subjects involved in the study.

Data Availability Statement: Data are contained within the article.

Acknowledgments: The authors would like to acknowledge the University of Pretoria, Department of Zoology and Entomology for providing the mealworm samples for this study. Our appreciation to Emmanuel Nekhudzhiga from the Department of Consumer and Food Science for assisting with the breast meat colour analysis.

Conflicts of Interest: The authors declare no conflict of interest.

References

1. Dyubele, N.L.; Muchenje, V.; Nkukwana, T.T.; Chimonyo, M. Consumer sensory characteristics of broiler and indigenous chicken meat: A South African example. *Food Qual. Prefer.* **2010**, *21*, 815–819. [CrossRef]
2. Mwale, M.; Masika, P.J. Ethno-veterinary control of parasites, management and role of village chickens in rural households of Centane district in the Eastern Cape, South Africa. *Trop. Anim. Health Prod.* **2009**, *41*, 1685–1693. [CrossRef]
3. Moges, F. *Indigenous Chicken Production and Marketing Systems in Ethiopia: Characteristics and Opportunities for Market-oriented Development*; ILRI (aka ILCA and ILRAD): Nairobi, Kenya, 2010; Volume 24.
4. Azimu, W.; Manatbay, B.; Li, Y.; Kaimaerdan, D.; Wang, H.E.; Reheman, A.; Muhatai, G. Genetic diversity and population structure analysis of eight local chicken breeds of Southern Xinjiang. *Br. Poult. Sci.* **2018**, *59*, 629–635. [CrossRef] [PubMed]
5. Manyelo, T.G.; Selaledi, L.; Hassan, Z.M.; Mabelebele, M. Local chicken breeds of Africa: Their description, uses and conservation methods. *Animals* **2020**, *10*, 2257. [CrossRef]

6. Ramos-Elorduy, J.; González, E.A.; Hernández, A.R.; Pino, J.M. Use of *Tenebrio molitor* (Coleoptera: Tenebrionidae) to recycle organic wastes and as feed for broiler chickens. *J. Econ. Entomol.* **2002**, *95*, 214–220. [CrossRef] [PubMed]

7. Elahi, U.; Wang, J.; Ma, Y.B.; Wu, S.G.; Wu, J.; Qi, G.H.; Zhang, H.J. Evaluation of yellow mealworm meal as a protein feedstuff in the diet of broiler chicks. *Animals* **2020**, *10*, 224. [CrossRef]

8. Biasato, I.; De Marco, M.; Rotolo, L.; Renna, M.; Lussiana, C.; Dabbou, S.; Capucchio, M.T.; Biasibetti, E.; Costa, P.; Gai, F.; et al. Effects of dietary *Tenebrio molitor* meal inclusion in free-range chickens. *J. Anim. Physiol. Anim. Nutr.* **2016**, *100*, 1104–1112. [CrossRef] [PubMed]

9. Mabelebele, M.; Ng'ambi, J.; Norris, D.; Ginindza, M. Comparison of gastrointestinal tract and pH values of digestive organs of Ross 308 broiler and indigenous Venda chickens fed the same diet. *Asian J. Anim. Vet. Adv.* **2014**, *9*, 71–76. [CrossRef]

10. Okoro, V.M.O.; Ravhuhali, K.E.; Mapholi, T.H.; Mbajiorgu, E.F.; Mbajiorgu, C.A. Effect of age on production characteristics of Boschveld indigenous chickens of South Africa reared intensively. *S. Afr. J. Anim. Sci.* **2017**, *47*, 157–167. [CrossRef]

11. Biasato, I.; Gasco, L.; De Marco, M.; Renna, M.; Rotolo, L.; Dabbou, S.; Capucchio, M.T.; Biasibetti, E.; Tarantola, M.; Bianchi, C.; et al. Effects of yellow mealworm larvae (*Tenebrio molitor*) inclusion in diets for female broiler chickens: Implications for animal health and gut histology. *Anim. Feed. Sci. Technol.* **2017**, *234*, 253–263. [CrossRef]

12. Cullere, M.; Schiavone, A.; Dabbou, S.; Gasco, L.; Dalle Zotte, A. Meat quality and sensory traits of finisher broiler chickens fed with black soldier fly (Hermetia illucens L.) larvae fat as alternative fat source. *Animals* **2019**, *9*, 140. [CrossRef] [PubMed]

13. Küçüközet, A.O.; Uslu, M.K. Cooking loss, tenderness, and sensory evaluation of chicken meat roasted after wrapping with edible films. *Food Sci. Technol. Int.* **2018**, *24*, 576–584. [CrossRef] [PubMed]

14. Mir, N.A.; Rafiq, A.; Kumar, F.; Singh, V.; Shukla, V. Determinants of broiler chicken meat quality and factors affecting them: A review. *J. Food Sci. Technol.* **2017**, *54*, 2997–3009. [CrossRef]

15. Sedgh-Gooya, S.; Torki, M.; Darbemamieh, M.; Khamisabadi, H.; Karimi Torshizi, M.A.; Abdolmohamadi, A. Yellow mealworm, *Tenebrio molitor* (Col: Tenebrionidae), larvae powder as dietary protein sources for broiler chickens: Effects on growth performance, carcass traits, selected intestinal microbiota and blood parameters. *J. Anim. Physiol. Anim. Nutr.* **2021**, *105*, 119–128. [CrossRef]

16. Bovera, F.; Loponte, R.; Marono, S.; Piccolo, G.; Parisi, G.; Iaconisi, V.; Gasco, L.; Nizza, A. Use of *Tenebrio molitor* larvae meal as protein source in broiler diet: Effect on growth performance, nutrient digestibility, and carcass and meat traits. *J. Anim. Sci.* **2016**, *94*, 639–647. [CrossRef] [PubMed]

17. Palya, V. *Manual for the Production of Marek's Disease, Gumboro Disease and Inactivated Newcastle Disease Vaccines (No. 89)*; Food & Agriculture Org: Rome, Italy, 1991.

18. Selaledi, L.; Mabelebele, M. The influence of drying methods on the chemical composition and body color of yellow mealworm (*Tenebrio molitor* L.). *Insects* **2021**, *12*, 333. [CrossRef]

19. NRC. *Nutrient Requirements of Poultry*, 9th ed.; National Academy of Sciences Press: Washington, DC, USA, 1994.

20. South Africa Poultry Association. Code of Practice. 2018. Available online: http://www.sapoultry.co.za/pdf-docs/code-of-practice-sapa.pdf (accessed on 5 November 2021).

21. Park, S.Y.; Byeon, D.S.; Kim, G.W.; Kim, H.Y. Carcass and retail meat cuts quality properties of broiler chicken meat based on the slaughter age. *J. Anim. Sci. Technol.* **2021**, *63*, 180. [CrossRef]

22. Lyon, B.G.; Lyon, C.E. Texture evaluations of cooked, diced broiler breast samples by sensory and mechanical methods. *Poult. Sci.* **1996**, *75*, 812–819. [CrossRef]

23. Rabeler, F.; Feyissa, A.H. Kinetic modeling of texture and color changes during thermal treatment of chicken breast meat. *Food Bioprocess Technol.* **2018**, *11*, 1495–1504. [CrossRef]

24. IBM Corp. *IBM SPSS Statistics for Windows, Version 27.0*; IBM Corp: Armonk, NY, USA, 2020; Available online: https://www.ibm.com/docs/en/SSLVMB_27.0.0/pdf/en/GPL_Reference_Guide_for_IBM_SPSS_Statistics.pdf (accessed on 12 September 2021).

25. Ravindran, V. Poultry Development Review. 2013. Available online: https://www.fao.org/3/i3531e/i3531e.pdf (accessed on 12 September 2021).

26. Zadeh, Z.S.; Kheiri, F.; Faghani, M. Use of yellow mealworm (*Tenebrio molitor*) as a protein source on growth performance, carcass traits, meat quality and intestinal morphology of Japanese quails (*Coturnix japonica*). *Vet. Anim. Sci.* **2019**, *8*, 100066. [CrossRef]

27. Mbhele, F.G.; Mnisi, C.M.; Mlambo, V. A nutritional evaluation of insect meal as a Sustainable protein source for Jumbo quails: Physiological and meat quality responses. *Sustainability* **2019**, *11*, 6592. [CrossRef]

28. Leiber, F.; Gelencsér, T.; Stamer, A.; Amsler, Z.; Wohlfahrt, J.; Früh, B.; Maurer, V. Insect and legume-based protein sources to replace soybean cake in an organic broiler diet: Effects on growth performance and physical meat quality. Renew. *Agric. Food Syst.* **2017**, *32*, 21–27. [CrossRef]

29. Kondjoyan, A.; Oillic, S.; Portanguen, S.; Gros, J.B. Combined heat transfer and kinetic models to predict cooking loss during heat treatment of beef meat. *Meat Sci.* **2013**, *95*, 336–344. [CrossRef]

30. Pathare, P.B.; Roskilly, A.P. Quality and energy evaluation in meat cooking. *Food Eng. Rev.* **2016**, *8*, 435–447. [CrossRef]

31. Droval, A.A.; Benassi, V.T.; Rossa, A.; Prudencio, S.H.; Paião, F.G.; Shimokomaki, M. Consumer attitudes and preferences regarding pale, soft, and exudative broiler breast meat. *J. Appl. Poult. Res.* **2012**, *21*, 502–507. [CrossRef]

32. Nhlane, L.T.; Mnisi, C.M.; Mlambo, V.; Madibana, M.J. Effect of seaweed-containing diets on visceral organ sizes, carcass characteristics, and meat quality and stability of Boschveld indigenous hens. *Poult. Sci.* **2021**, *100*, 949–956. [CrossRef] [PubMed]

33. Shaviklo, A.R.; Alizadeh-Ghamsari, A.H.; Hosseini, S.A. Sensory attributes and meat quality of broiler chickens fed with mealworm (*Tenebrio molitor*). *J. Food Sci. Technol.* **2021**, *58*, 4587–4597. [CrossRef] [PubMed]

34. Wideman, N.; O'bryan, C.A.; Crandall, P.G. Factors affecting poultry meat colour and consumer preferences-A review. *World's Poult. Sci. J.* **2016**, *72*, 353–366. [CrossRef]
35. Van Laack, R.L.J.M.; Liu, C.H.; Smith, M.O.; Loveday, H.D. Characteristics of pale, soft, exudative broiler breast meat. *Poult. Sci.* **2000**, *79*, 1057–1061. [CrossRef]
36. Petracci, M.; Betti, M.; Bianchi, M.; Cavani, C. Color variation and characterization of broiler breast meat during processing in Italy. *Poult. Sci.* **2004**, *83*, 2086–2092. [CrossRef]
37. Schiavone, A.; Cullere, M.; De Marco, M.; Meneguz, M.; Biasato, I.; Bergagna, S.; Dezzutto, D.; Gai, F.; Dabbou, S.; Gasco, L.; et al. Partial or total replacement of soybean oil by black soldier fly larvae (Hermetia illucens L.) fat in broiler diets: Effect on growth performances, feed-choice, blood traits, carcass characteristics and meat quality. *Ital. J. Anim. Sci.* **2017**, *16*, 93–100. [CrossRef]
38. Meyer-Rochow, V.B. Can insects help to ease the problem of world food shortage. *Search* **1975**, *6*, 261–262.

Article

Determination of Carbohydrate Composition in Mealworm (*Tenebrio molitor* L.) Larvae and Characterization of Mealworm Chitin and Chitosan

Yang-Ju Son [1,2], In-Kyeong Hwang [2], Chu Won Nho [1], Sang Min Kim [1] and Soo Hee Kim [3,*]

[1] Smart Farm Research Center, Korea Institute of Science and Technology, Gangneung Institute of Natural Products, Gangneung 25451, Korea; yangjuson@kist.re.kr (Y.-J.S.); cwnho@kist.re.kr (C.W.N.); kimsm@kist.re.kr (S.M.K.)

[2] Department of Food and Nutrition and Research Institute of Human Ecology, Seoul National University, Seoul 08826, Korea; ikhwang@snu.ac.kr

[3] Department of Culinary Arts, Kyungmin University, Uijeongbu 11618, Korea

* Correspondence: kshee@kyungmin.ac.kr; Tel.: +82-31-828-7271; Fax: +82-31-828-7951

Abstract: Mealworm (*Tenebrio molitor* L.) is a classic edible insect with high nutritional value for substituting meats from vertebrates. While interest in mealworms has increased, the determination of carbohydrate constituents of mealworms has been overlooked. Thus, the aim of the present study was to investigate the carbohydrate content and composition of mealworms. In addition, the characteristics of mealworm chitin were determined as these were the major components of mealworm carbohydrate. The crude carbohydrate content of mealworms was 11.5%, but the total soluble sugar content was only 30% of the total carbohydrate content, and fructose was identified as the most abundant free sugar in mealworms. Chitin derivatives were the key components of mealworm carbohydrate with a yield of 4.7%. In the scanning electron microscopy images, a lamellar structure with α-chitin configuration was observed, and mealworm chitosan showed multiple pores on its surface. The overall physical characteristics of mealworm chitin and chitosan were similar to those of the commercial products derived from crustaceans. However, mealworm chitin showed a significantly softer texture than crustacean chitin with superior anti-inflammatory effects. Hence, mealworm chitin and chitosan could be employed as novel resources with unique advantages in industries.

Keywords: edible insect; mealworm; sugar content; chitin; microscopy; degree of acetylation

Citation: Son, Y.-J.; Hwang, I.-K.; Nho, C.W.; Kim, S.M.; Kim, S.H. Determination of Carbohydrate Composition in Mealworm (*Tenebrio molitor* L.) Larvae and Characterization of Mealworm Chitin and Chitosan. *Foods* **2021**, *10*, 640. https://doi.org/10.3390/foods10030640

Academic Editor: Victor Benno Meyer-Rochow and Chuleui Jung

Received: 18 February 2021
Accepted: 15 March 2021
Published: 18 March 2021

1. Introduction

The global population is still increasing and, consequently, likely food shortcomings need to be addressed. Furthermore, the increasing consumption of meat in developing countries affects the sustainability issue because livestock-rearing requires excessive energy inputs [1]. Meat products are excellent protein sources for humans, and therefore, the search for substitutes has reached an impasse. Edible insects are attractive alternatives because of their high protein contents, and these show superior conversion efficiencies that are almost ten times higher than those of the ruminants [2]. Moreover, insects require less feedstuff and space, and omit fewer greenhouse gases than livestock [3]. Accordingly, the Food and Agriculture Organization (FAO), following a suggestion already made in 1975 by Meyer-Rochow [4], also identified edible insects as appropriate alternative protein sources to meats from vertebrates [5].

Among various types of edible insects, the mealworm (*Tenebrio molitor* L.) is one of the most popular species as a food and feed resource, and its extensive utilization has resulted in the advancement and establishment of commercial large-scale rearing systems [6]. Significant interest in mealworms has led to numerous studies on their growth and development [7], mass production systems [8], and nutrition value for use as feed [9,10]. In addition, the composition and nutritional value of mealworms have been verified for use in food products [11,12], and consumer perception [13] as well as industrial processing technologies have also been investigated [14,15]. However, most studies have focused on the composition of proteins or fats in mealworms, but the carbohydrate content has not been extensively studied. Although it is well known that mealworms have low carbohydrate contents, the constituents and properties of carbohydrates should be verified to further understand their characteristics. Moreover, chitin derivatives, the predicted major carbohydrates in mealworms, are important chemical compounds for functional nutrients or industrial materials.

The aim of the present study is to investigate the characteristics and composition of carbohydrate content in mealworms as it has not been extensively investigated in previous studies. To determine the carbohydrate content and its constituents, the contents of crude carbohydrate and total soluble sugar were examined, and the free sugar composition was identified. As chitin and its derivatives are the major constituents of the mealworm carbohydrate content, mealworm chitin and chitosan were extracted and prepared. Accordingly, their physical and structural properties were determined using diverse analytical methods. Thus, the analysis of carbohydrate compounds in mealworms was conducted, and the characteristics of chitin derivatives were determined to identify the possible applications of mealworm chitin.

2. Materials and Methods

All chemical compounds and solvents without provider information in the present study were purchased from Sigma-Aldrich Corporation (St. Louis, MO, USA).

2.1. Preparation of Mealworm Powder

Mealworm larvae were reared on a farm located at Gyeonggi-do, Republic of Korea, and mealworm powder was prepared according to the procedure described by Son et al. [16]. Briefly, mealworms were blanched using boiling water for 3 min (1:5, w/w), and the outer water was removed. The mealworms were dried in a hot-air dryer (LH.FC-PO-150; Lab house, Pocheon, Korea) at 60 °C for 12 h. The dried mealworms were milled using a HR-2860 blender (Philips, Amsterdam, The Netherlands). The pulverization process was repeated until all powders passed through a 535-μm diameter sieve.

2.2. Total Soluble Sugar and Reducing Sugar Contents

Mealworm powder (10 g) was mixed with distilled water (1:10, w/v) and extracted at 70 °C with shaking at 150 rpm for 2 h (BS-21; Jeiotech, Daejeon, Korea). The solution was centrifuged at $2000 \times g$ for 20 min (Combi-514R; Hanil Science Industrial, Incheon, Korea), and the supernatant was filtered using Whatman No. 1 filter paper (Whatman, Buckinghamshire, UK). The volume of the collected liquid was adjusted to 100 mL by adding distilled water.

The total soluble sugar content in the mealworms was analyzed using the phenol-sulfuric acid colorimetric method [17]. One milliliter of extract was mixed with 1 mL of 5% phenol solution and 5 mL of sulfuric acid, and it was placed at 20 °C for 30 min. The absorbance was detected at 470 nm using a spectrophotometer (Optizen 2120UV; Mecasys, Daejeon, Korea). The concentration of total soluble sugar was calculated using a standard curve constructed using D-(+)-glucose data (Sigma-Aldrich Corporation, St. Louis, MO, USA).

The content of reducing sugar in the mealworms was determined using the dinitrosalicylic acid reagent following the method described by Miller [18]. One milliliter of the extract and 1 mL of dinitrosalicylic acid agent were mixed and kept at 90 °C for 15 min. The solution was immediately cooled on ice, and its absorbance was detected at 570 nm (Optizen 2120UV; Mecasys, Korea). The concentration of reducing sugar was calculated by comparing the result with a standard curve of D-(+)-glucose.

2.3. Free Sugar Composition

The free sugar composition of the mealworms was determined using the modified method described by Kim et al. [19]. Mealworm powder (1 g) was mixed with 20 mL of 80% ethanol solution and extracted at 40 °C for 1 h, followed by centrifugation at $2000\times g$ for 20 min (Combi-514R; Hanil Science Industrial, Korea). The supernatant was filtered with a 0.2 μm syringe filter (Advantec MRF; Advantec, Tokyo, Japan). A column (3.9 × 300 mm, 10 μm; WAT084038; Waters, Milford, MA, USA) for carbohydrate analysis was used with a PU-2089 plus high-performance liquid chromatography (HPLC) system (Jasco, Tokyo, Japan). The mobile phase constituted a mixture of acetonitrile and water (87:13, *v/v*) with a flow rate of 1.2 mL/min, and the isocratic system was used for mobile phase. The injection volume was 20 μL, and a refractive index (RI) detector was used for detection. The concentration of each sugar was calculated using the standard curves of common monosaccharides (glucose, fructose, and galactose; Sigma-Aldrich, St. Louis, MO, USA) and disaccharides (sucrose, maltose, and lactose; Sigma-Aldrich, St. Louis, MO, USA). The R^2 of standard curves of mono- and di-saccharides were over 0.9999.

2.4. Total Glucosamine Content

The total glucosamine content of the mealworm powder was determined according to the methods described in previous studies [20,21]. Here, 30 mL of 2 M HCl was added to 5 g of mealworm powder and boiled at 95 °C in a reflux tube for 24 h. The solution was neutralized using 1 M NaOH solution, and its volume was adjusted to 100 mL by adding distilled water. One milliliter of the solution was mixed with acetylacetone solution (acetylacetone in 0.5 N sodium carbonate, 1:50, *w/w*) and reacted at 95 °C for 10 min. Then, 1 mL of dimethylaminobenzaldehyde solution (0.8 g of p-dimethylaminobenzaldehyde in 30 mL of ethanol and 30 mL of HCl solution) was added and the solution was maintained at 75 °C for 30 min. After cooling on ice, the volume of the solution was adjusted to 10 mL using ethanol. The absorbance of the final solution was measured at 530 nm using a spectrophotometer (Optizen 2120UV; Mecasys, Korea). The complete reaction procedure was similarly performed using a D-glucosamine standard (Sigma-Aldrich, St. Louis, MO, USA), and its standard curve was used to calculate the total glucosamine content in the mealworms.

2.5. Extraction of Chitin and Preparation of Chitosan from Mealworms

To remove oils from the mealworm powder, n-hexane was mixed with the powders (1:5, *w/v*) and the oils were extracted in a shaker at 170 rpm for 6 h (SI600R; Lab Companion, Daejeon, Korea). The liquid was separated by filtration using a filter paper (Whatman No. 1; Whatman, UK), and the same procedures were performed twice for the remaining powders. The defatted powders were dried in a hood at 20 °C for 24 h, and further dried using a speed vacuum evaporator (Maxivac alpha; Labogene ApS, Lynge, Denmark).

Chitin was extracted from defatted mealworm powders [22,23]. For protein degradation, 20 g of defatted powder was mixed with 400 mL of 1.25 M NaOH solution and maintained at 80 °C for 24 h. The solution was filtered using the ashless Whatman No. 5 filter paper (Whatman, UK) and rinsed with distilled water until neutralized, and the remaining powders were freeze-dried. The dried powders were mixed with 1.5 M HCl solution (1:10, *w/v*) and shaken at 20 °C in a shaker (120 rpm, 6 h) (SI600R; Lab Companion, Korea). The solution was filtered using a Whatman No. 5 filter paper (Whatman, UK) and rinsed with distilled water for neutralization. The lyophilized powder was used as the chitin sample extracted from the mealworms. The yields of mealworm chitin were determined by repeating the extraction procedure thrice.

The chitosan sample was prepared from chitin powder via deacetylation. Fifty milliliters of 50% NaOH solution and 5 g of mealworm chitin were mixed and reacted at 80 °C for 4 h. The liquids were removed by filtration using a Whatman No. 5 filter (Whatman, UK) and remained powders were rinsed with distilled water to neutralize them. The freeze-dried powder was used as the mealworm chitosan sample.

2.6. Moisture and Crude Protein Contents

The moisture and crude protein contents were analyzed using methods of the Association of Official Analytical Chemists (AOAC) [24]. The moisture content was determined using the oven drying method. One gram of chitin powder was placed in a drying oven at 105 °C for 12 h, and its weight was measured. The moisture content was determined by subtracting the powder weight before and after drying. The crude protein content was determined by the Kjeldahl method using an automatic Kjeldahl analyzer (KDN; NANBEI Instrument, Zhengzhou, China).

2.7. Color and Whiteness Index (WI)

The color values (Hunter L*, a*, and b*) of the chitin powders were measured using a colorimeter (CM-3500d; Minolta, Tokyo, Japan). The whiteness index (WI) of mealworm chitin was calculated using an equation described in a previous study [25]:
$WI = 100 - [(100 - L^*)^2 + (a^*)^2 + (b^*)^2]^{1/2}$.

2.8. Field-Emission Scanning Electron Microscopy (FESEM) Imaging and Energy Dispersive X-Ray Spectroscopy (EDS) Analysis

The surface images of chitin and chitosan extracted from the mealworms were obtained using FESEM (SUPRA 55VP; Carl-Zeiss-Strasse, Oberkochen, Germany), and EDS was simultaneously performed using the same instrument. α-Chitin from shrimp (Sigma-Aldrich, USA) was also analyzed as the reference sample. The chitin and chitosan samples were fixed on a stub and coated with platinum (Sputter coater BAL-TEC/SCD 005; BAL-TEC AG, Pfäffikon, Switzerland). The images of the samples were obtained at two magnifications ($\times 300$ and $\times 10,000$). The carbon and nitrogen contents of the samples were analyzed by EDS, and the ratio of carbon and nitrogen in the samples was used to calculate the degree of acetylation (DA) of chitin and the degree of deacetylation (DDA) of chitosan using the following equations [26,27]:

$$DA(\%) = ((C/N) - 5.14)/1.72 \times 100 \tag{1}$$

$$DDA(\%) = (6.857 - (C/N))/1.7143 \times 100 \tag{2}$$

where C and N are the carbon and nitrogen contents.

2.9. Nuclear Magnetic Resonance (NMR)

Proton (1H) NMR analysis was conducted using a 600 MHz AVANCE 600 NMR spectrometer (Bruker, Billerica, MA, USA). The chitin and chitosan samples were dissolved in 2% deuterated acetic acid in D2O at a concentration of 2 M, and 1H NMR spectra were obtained at 70 °C [23]. Additionally, 13C NMR spectra were obtained using a 500 MHz solid-state NMR spectrometer (AVANCE II; Bruker, Billerica, MA, USA) according to the method described by Van de Velde and Kiekens [28]. The obtained NMR peaks were used for calculating the DA of chitin and DDA of chitosan using the following formulas [29,30]:

$$DA(\%) = (100 \times I[CH_3])/((I[C1] + I[C2] + I[C3] + I[C4] + I[C5] + I[C6])/6) \times 100 \tag{3}$$

where I is the intensity of each peak and C represents the carbon atoms in the chitin monomer.

$$DDA(\%) = H1D/(H1D + HAc/3) \times 100 \tag{4}$$

where H1D is the integral of the peak of proton H1 of the deacetylated monomer and HAc corresponds to the peak of the three protons of the acetyl group.

2.10. Fourier Transform Infrared Spectroscopy (FT-IR)

FT-IR spectroscopy was performed using a Nicolet 6700 spectrometer (Thermo Scientific, Waltham, MA, USA) to determine the chemical structural characteristics of chitin and chitosan obtained from the mealworms. The samples were mixed with potassium bromide and placed in a ZnSe/diamond disc. The analyzed frequency range was 4000–650 cm^{-1} and the resolution was 8 cm^{-1} [23]. The DDA of the chitosan sample was calculated using the following equation [31]: DDA (%) = 100 − (A$_{1655}$/A$_{3450}$) × 115.

2.11. X-Ray Powder Diffraction (XRD)

XRD analysis was performed using a D8 ADVANCE instrument with DAVINCI (Bruker, Madison, WI, USA) according to the method described by Yen et al. [32]. The Cu radiation (40 kV, 40 mA) was used as the X-ray source with an LYNXEYE XE detector. The collected scanned data range of 2θ was 5–50°. The crystallinity index of the chitin samples was calculated using the following formula [33]:

$$\text{Crystallinity index (\%)} = [\text{Ic}/(\text{Ic} + \text{Ia})] \times 100 \tag{5}$$

where Ia is the intensity at 2θ = 16° and Ic is the intensity at 2θ = 20°.

2.12. Anti-Inflammatory Effect in Murine Macrophage Cell Line

The anti-inflammatory effect of mealworm chitosan was tested with the lipopolysaccharide (LPS)-induced murine macrophage cells (RAW 264.7). RAW 264.7 cells were obtained from American Type Culture Collection (ATCC, Rockville, MD, USA) and maintained in Dulbecco's modified Eagle's medium (DMEM) with 10% fetal bovine serum and 1% penicillin/streptomycin solution. The cells were incubated in an incubator with 5% CO$_2$ at 37 °C. The cells were seeded at a concentration of 5 × 10^4 cells per well in 96-well plates and maintained for 24 h. After removing the culture media and rinsing with phosphate buffer saline (PBS), 50 μL of mealworm chitosan-containing media was added to each well. Fifty microliters of LPS-containing media (2 μg mL^{-1}) was added after an hour, and incubated for 24 h in an incubator. Then, the supernatant of each well was mixed with the same volume of Griess reagent. After reaction for 20 min, the absorbance of the solution was measured at 550 nm using a microplate reader (SpectraMax 190; Molecular Devices, Sunnyvale, CA, USA).

2.13. Statistical Analysis

The data are presented as mean ± standard deviation (SD) based on triplicate measurements. After normality and homogeneity tests for data, the Student's *t*-test or one-way analysis of variance (ANOVA) was conducted using the IBM SPSS statistics v.25 software (IBM Inc., Armonk, NY, USA). Duncan's multiple-range test was also performed for the post-hoc analysis of ANOVA ($p < 0.05$).

3. Results and Discussion

3.1. Carbohydrate Content and Its Composition in Mealworms

While carbohydrates are the major components of most foods, animal foods have abundant proteins and lipids but low carbohydrate contents. Accordingly, the crude carbohydrate content in dried mealworms was 11.45 ± 0.38%, implying that the un-dried mealworms contained approximately 3.4% carbohydrate (Table 1). Among the total carbohydrates in the mealworms, the total soluble sugar content was 3.22 ± 0.10%. To further analyze the soluble sugar composition in the mealworms, the representative mono- or di-saccharide contents (glucose, fructose, galactose, maltose, sucrose, and lactose) were determined by HPLC analysis. Galactose, maltose, and lactose were not detected, and the contents of glucose, fructose, and sucrose were 31.02 ± 1.95, 77.36 ± 0.35, and 12.46 ± 0.94 mg/100 g dry basis, respectively. The glucose content of raw mealworms was approximately 12 mg/100 g, which was approximately a quarter of that determined in chicken breast [34,35] and 3–6% of that in pork [35,36]; in contrast, the glucose content of mealworms was similar to that found in fish (18–72 mg/100 g) [37]. Although free sugars are soporific chemical compounds in various ingredients and their content is associated with taste, particularly sweetness, the total free sugar content of the mealworms (120.84 ± 1.07 mg/100 g dry basis) was too low to affect the taste because the threshold of sugar is known to be approximately 0.125% [38]. Therefore, the savory taste of mealworms can be affected by the abundant free amino acids or peptides rather than the sugars [16].

Table 1. Carbohydrate content and its composition in dried mealworm.

	Dried Mealworm
Crude carbohydrate (g/100 g)	11.45 ± 0.38
Total soluble sugar (g/100 g)	3.22 ± 0.10
Reducing sugar (g/100 g)	0.19 ± 0.01
Total glucosamine (g/100 g)	4.84 ± 0.43
Free sugar (mg/100 g)	
Glucose	31.02 ± 1.95
Fructose	77.36 ± 0.35
Sucrose	12.46 ± 0.94
Total	120.84 ± 1.07

Data are expressed as mean ± SD.

Additionally, a considerable gap was observed between the contents of crude carbohydrates and total soluble sugars in the mealworms; therefore, mealworm bodies were predicted to consist of abundant insoluble carbohydrates. Chitin and its derivatives are typical insoluble carbohydrates that are found in the insect integument, and 4.84 ± 0.43% of glucosamine (the monomer of chitin) is present in dried mealworm powder. This allows the consumers to intake abundant chitins upon eating mealworms. Moreover, mealworm chitins can be applied for mass production in industries because these are easy to rear on a large scale. Therefore, chitins were extracted from the mealworms and their overall characteristics were examined to verify their availability.

3.2. Yield, Moisture and Crude Protein Contents, and Color Traits of Chitin Extracted from Mealworms

Common chitin products in industries constitute those originating from crustaceans such as shrimp and crab. Because of their tough textures and difficult digestion, their shells are dumped as waste; therefore, this waste is typically used for preparing chitins. Chitin and chitosan can be employed as therapeutic agents, pharmaceutical carriers, and packaging materials as natural biopolymers [39]. When chitin was extracted from mealworm powder on a laboratory scale, its yield was 4.72 ± 0.21 g/100 g dried mealworm (Table 2). The extraction may have performed properly because this amount was similar to the total glucosamine content of the mealworms (4.84 ± 0.43%); moreover, its yield was analogous to that reported by Song et al. [40] (4.91%), but considerably lower than that reported by Siregar and Suptijah [41] (10–13%). In a subsequent study, the authors reported a high yield for chitin preparation; however, we found that the color of their chitin samples was darker, and the DA (70–80%) was significantly lower than that observed in the present study. Moreover, the authors reported a high nitrogen content, which indicated an incomplete deproteinization process; therefore, it was expected that their yield could also be decreased provided sufficient purification was performed. The yield of chitin from the mealworms was significantly lower than that obtained from crustacean sources such as shrimp (approximately 20%) and crab (30–40%) [22,42], and slightly lower than that obtained from imago edible insects such as grasshoppers (4.71–11.84%) or crickets (approximately 8.7%) [43,44]. The differences between these edible insects are attributed to their growth stages because grasshoppers and crickets are adult insects that have well-developed cuticles. However, the chitin yield of mealworms was higher than that of the silkworm pupa (one of the less developed edible insects with a chitin yield of 2.5–4.2%) [45,46]. The color values of mealworm chitin were 82.21 ± 0.72, 1.81 ± 0.08, and 8.68 ± 0.37 for lightness, redness, and yellowness, respectively, whereas the color values of chitin from the crabs were 55.4–62.4, 0.3–1.1, and 14.7–16.8 for lightness, redness, and yellowness, respectively [32]. As mealworm chitin had a significantly higher lightness value but lower yellowness, it could exhibit white color compared to the chitin originating from crabs. Similarly, the WI value of mealworm chitin

(80.13 ± 0.81) was significantly higher than that of the chitin from crabs (43.9–59.6) [32], and its bright color can be potentially applicable in industries.

Table 2. Yield, moisture and protein contents and color characteristics of the extracted mealworm chitin.

	Chitin from Mealworm
Yield (g chitin/100 g dried mealworm powder)	4.72 ± 0.21
Moisture (g/100 g chitin)	2.38 ± 0.05
Crude protein (g/100 g chitin)	9.96 ± 0.36
Color	
L* (lightness)	82.21 ± 0.72
a* (redness)	1.81 ± 0.08
b* (yellowness)	8.68 ± 0.37
Whiteness index (WI)	80.13 ± 0.81

Data are expressed as mean ± SD.

3.3. Surface Microstructures of Chitin and Chitosan Derived from Mealworms

The surface microstructural images of mealworm chitin and chitosan samples were obtained using FESEM (Figure 1), and commercial α-chitin from shrimp (Sigma-Aldrich, St. Louis, MO, USA) was used as the reference. One of the prominent structures of the chitin matrix is the lamellar structure that develops by the arrangement and cross-linking of chitin polymers [47], and this structure was observed in both the mealworm chitin and shrimp chitin at ×300 magnification. At a higher magnification (×10,000), the surfaces of the chitins were slightly bumpy, but chitosan showed a smooth surface with multiple pores on it. The distinct polyporous structure of chitosan was also reported in a prior study [45]. The particle size of mealworm chitin was significantly larger than that of shrimp chitin, likely due to the fine milling of shrimp chitin during commercial processing. The DA of chitin and DDA of chitosan are important indicators that represent their physical characteristics, particularly their solubility, and a simple method for measuring these involves determining the ratio of carbon and nitrogen in various substances. Thus, the carbon and nitrogen ratios of the samples were determined using the EDS system of the FESEM instrument, and the results are included in Table 3. The DA of mealworm chitin was 95.02 ± 0.22% and no significant difference was observed compared to the DA of shrimp chitin (94.81 ± 0.05%), and its value was similar to that reported in a previous study (90–98%) [23]. The DDA of the mealworm chitosan sample was 94.69 ± 0.06% as calculated using the ratio of carbon and nitrogen (5.2237 ± 0.0010).

Table 3. C/N ratio determined based on energy dispersive X-ray spectroscopy (EDS) data and the predicted degree of acetylation for chitin or degree of deacetylation for chitosan.

	Chitin from Mealworm	**Chitin from Shrimp**	*t*-Value
C/N [†]	6.7743 ± 0.0038	6.7707 ± 0.0010	−1.310 [NS§]
DA [‡] (%)	95.02 ± 0.22	94.81 ± 0.05	

Data are expressed as mean ± SD. [†] C/N: ratio of carbon and nitrogen elements. [‡] DA: degree of acetylation. [§NS]: Not significant.

3.4. Determination of Precise DA and DDA of Mealworm Chitin and Chitosan Using NMR

NMR is a reliable method for verifying the DA and DDA of chitin derivatives, which can assist in the determination of their structural properties [28]. Therefore, solid-state 13C NMR was used for the analyses of chitin and chitosan samples, and 1H NMR was used to analyze mealworm chitosan (Figure 2). In a study by Heux et al. [48], the chemical shifts for the commercial chitins in the 13C solid-state NMR spectra were observed at 103.5–104.7 ppm (C1), 55.2–57.6 ppm (C2), 73.3–75.0 ppm (C3), 82.4–83.0 ppm (C4), 74.7–75.7 ppm (C5), 60.1–61.8 ppm (C6), and 22.8–23.2 ppm (CH3), and these data were consistent with the results of the present study. Interestingly, although the major chemical shift patterns were analogous for mealworm chitin and shrimp chitin, a distinctive chemical shift was observed at 30 ppm only in the data for the mealworm chitin. Zhang et al. [49] reported that this chemical shift was attributed to the catechol moiety, which is a hydroxyl-containing aromatic compound [50]. In the cuticle layer of the insects, catechol is a vital constituent that forms cross-linking networks with other structural polymers [51]. Thus, the presence of a peak at 30 ppm indicated that mealworms contained catechol molecules in their body, but this was not observed in the shrimp chitin. Meanwhile, the DA values calculated using NMR were 98.3% for mealworm chitin and 97.8% for shrimp chitin, and their values were slightly higher than the values calculated using the carbon and nitrogen ratio. This difference could be caused by the adulteration of proteins or other matter, but the prepared mealworm chitin showed a DA value similar to that of the commercially obtained shrimp chitin. Thus, it was expected that mealworm chitin could show physical characteristics similar to those of the shrimp chitin. The DDA of mealworm chitosan was calculated based on the 1H NMR data (89.4%), and the range for this value was reported in a previous study (83–93%) in which mealworm chitosan was prepared [32]. The DDA of mealworm chitosan was higher than that of the commercial chitosans derived from crustaceans (70–85%) [52].

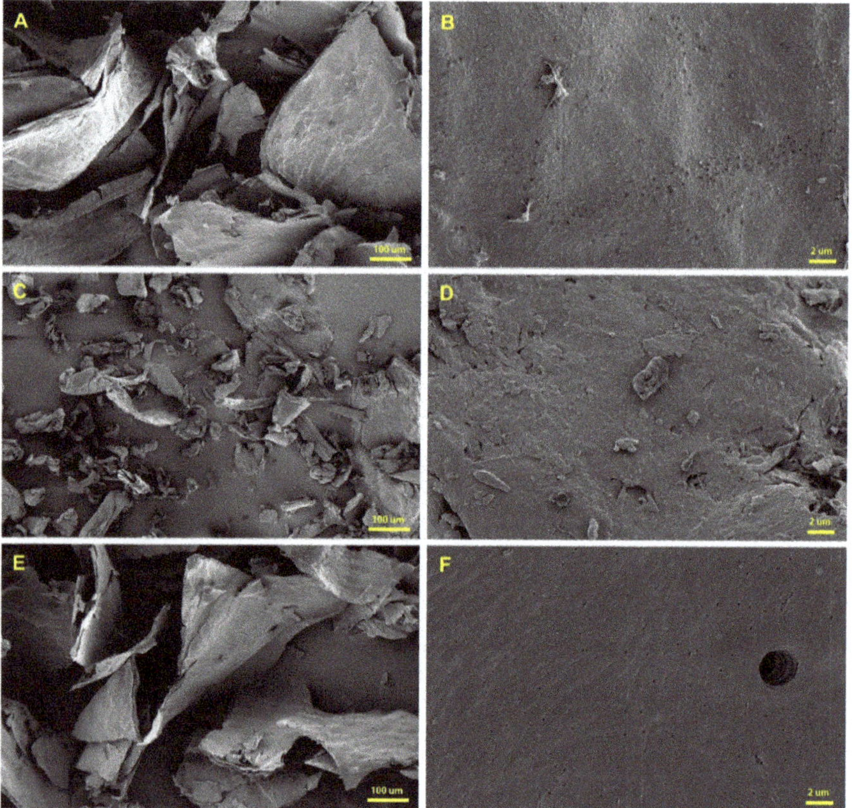

Figure 1. Field emission scanning electron microscopy (FESEM) data of chitin and chitosan samples. Images of mealworm chitin (**A**,**B**) (×300 and ×10,000), shrimp chitin (**C**,**D**) (×300 and ×10,000), and mealworm chitosan (**E**,**F**) (×300 and ×10,000).

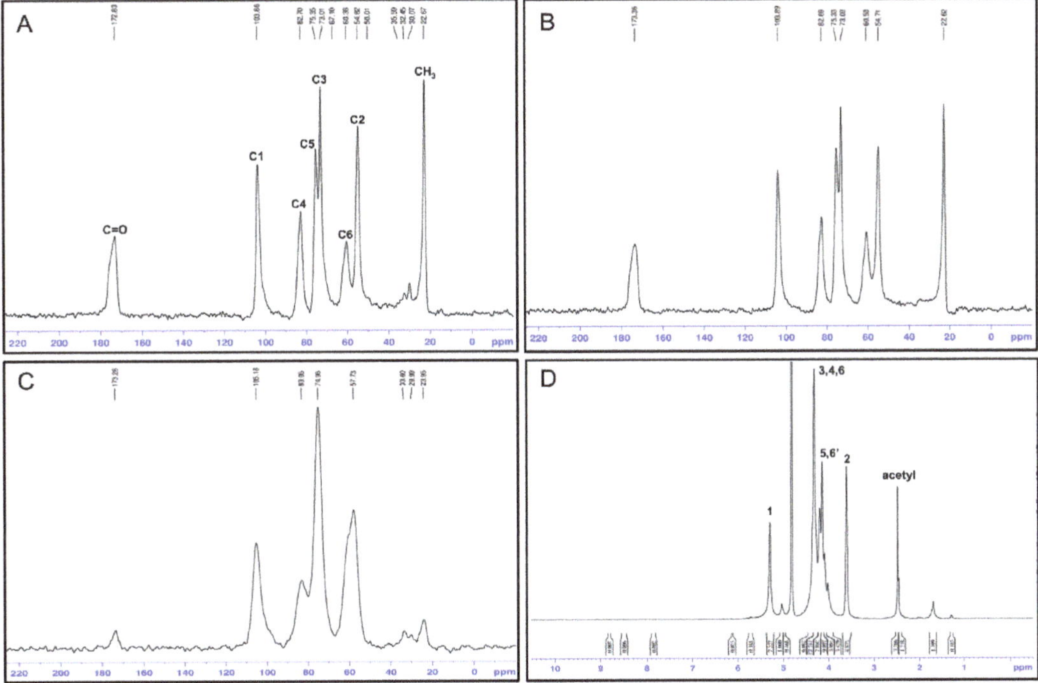

Figure 2. Nuclear magnetic resonance (NMR) data of chitin and chitosan. ^{13}C NMR data of mealworm chitin (**A**), shrimp chitin (**B**), and mealworm chitosan (**C**). ^{1}H NMR data of mealworm chitosan (**D**).

3.5. FT-IR Spectral Analyses of Mealworm Chitin and Chitosan

The FT-IR spectra of the samples are shown in Figure 3, and the relevant absorbance peaks have been reported in a prior study [28], including the signals for hydroxyl stretching (3450 cm^{-1}), C–H stretching (2878 cm^{-1}), C–H deformations (1420 cm^{-1}), C=O stretching in secondary amide (1550, 1625, 1655, and 1661 cm^{-1}), and C–N stretching in secondary amide (1320 cm^{-1}). Mealworm chitin shows a spectral pattern similar to that of shrimp α-chitin. The chitin fibers are commonly classified into three types (α-, β-, and γ-chitin) based on their crystal structures, and each type shows different physical properties. The chitin polymer is generally arranged in parallel, and α-chitin has an inverse configuration between the adjoined chitin chains, but β-chitin consists of equally aligned chitin fibers [53]. There are several distinguishing signals of β-chitin in comparison to those of α-chitin. First, separate peaks for the bands at 1660 cm^{-1} (corresponding to the hydrogen bond between C=O and NH groups) and 1625 cm^{-1} (corresponding to the hydrogen bond between C=O and HOCH2 groups) are not observed, and β-chitin shows a peak at 1430 cm^{-1} instead of 1416 cm^{-1} [54–56]. As mealworm chitin does not show spectral features of β-chitin but instead the standard FT-IR spectral patterns of α-chitin, the main chitin type in mealworms is identified as α-chitin. In contrast, mealworm chitin shows a relatively high absorbance at 2878 cm^{-1} (C–H stretching) and low absorbance at 3444 cm^{-1} (hydroxyl stretching) compared to the shrimp α-chitin. These distinct features are attributed to dibutyrylchitin (DBC), where chitin contains bonds of butyryl groups at the C3 and C6 positions [28,57]. DBC is typically obtained by the artificial conversion of chitins, and it has high solubility, suitability for fabricating films, and superior biodegradability [58]. While DBC is rarely found in natural sources, its distinct FT-IR spectral peaks (2878 and 3444 cm^{-1}) have been detected in the data for the chitin extracted from silkworm pupa, but it shows a significantly smaller peak than that observed in the present study [45]. Hence, DBC may be naturally contained in the chitin obtained from insects, and the mealworms show a notable signal; however, further confirmation is still required.

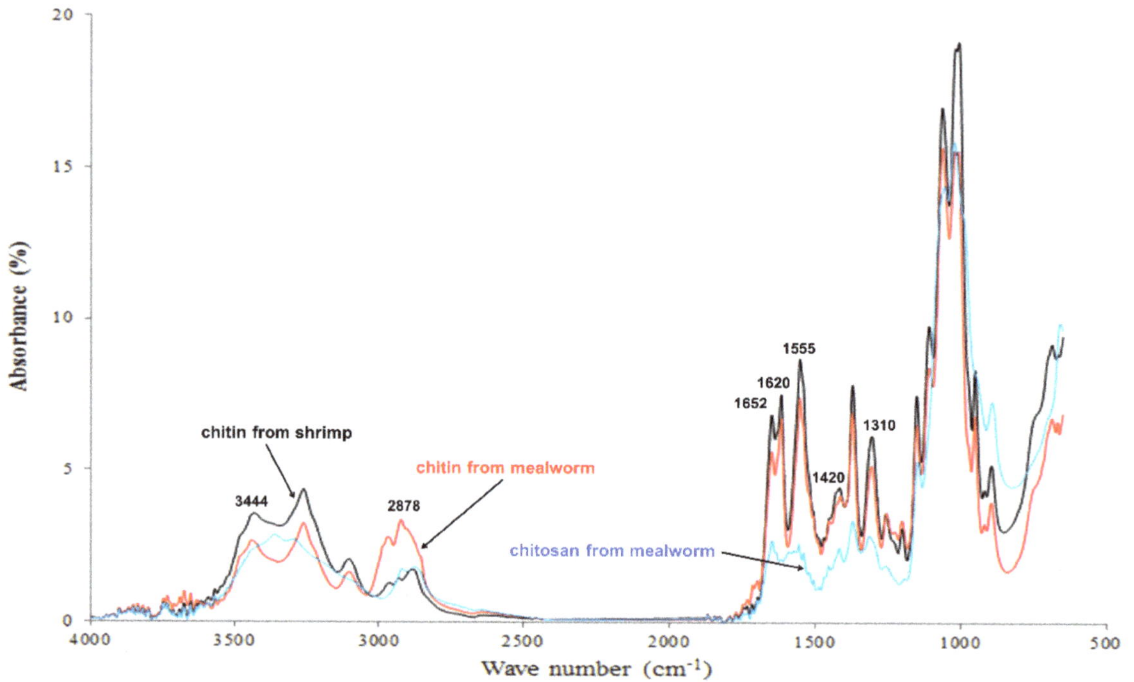

Figure 3. Fourier transform infrared spectroscopy (FT-IR) data of chitin and chitosan.

3.6. Determination of Crystalline Structures of Mealworm Chitin and Chitosan by XRD

To verify the crystalline structural characteristics of the chitin samples, XRD analysis was conducted (Figure 4). The α-chitin peaks were expected to be observed at 9.6°, 19.6°, 21.1°, 23.7°, and 36°, which were observed for chitins obtained from mealworm and shrimp [23]. In addition, other insect species (e.g., grasshopper, wasp, cockroach, and cicada) showed similar XRD patterns in the reported studies; therefore, α-chitin constitutes the prime structure of insect chitin [59,60]. It is well-known that the antiparallel arrangement of the chitin fibers of α-chitin forms a lamellar structure, which is observed in the SEM data obtained in this study. The intensities of the peaks at 26.3° and 12.7° in the data for mealworm chitin were lower than those of shrimp chitin, and this trend was also observed for the chitin obtained from silkworm pupa [49]. The XRD pattern of DBC was observed, as DBC showed clear peaks at approximately 9.6° and 21.1° but the peaks at 12.7° and 26.3° were not present [61]; therefore, the presence of DBC could decrease the corresponding peaks (12.7° and 26.3°). The crystallinity indices of chitin samples were 57.85 ± 0.17% and 62.04 ± 0.10% for mealworm and shrimp, respectively (Table 4). The crystallinities of silkworm pupa and beetle larvae were 54% and 58%, respectively [49], and their values were similar to that of mealworms (57.85%) observed in the present study. In contrast, the crystallinity indices of crustacean chitins had higher values (60–70%) than those of the mealworm chitin [62] as mealworms have a less firm shell. The tender structure is advantageous during the preparation of soft chitin materials because the crystallinity of chitin is associated with hardness [63]. Therefore, the use of mealworm chitin can have benefits in such specific applications.

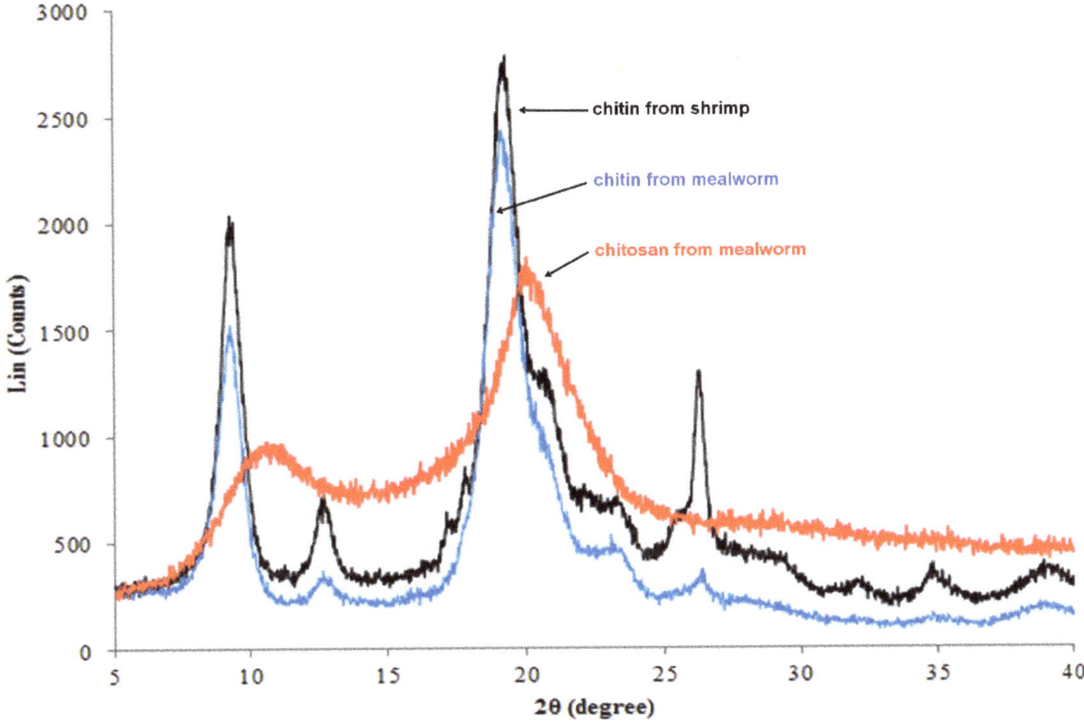

Figure 4. X-ray diffraction (XRD) data of chitin and chitosan.

Table 4. Degree of crystallinity of extracted chitin from mealworm or shrimp.

	Mealworm Chitin	Shrimp Chitin	*t*-Value
Degree of crystallinity (%)	57.85 ± 0.17	62.04 ± 0.10	−37.053 ***

Data are expressed as mean ± SD. *** Significant difference at $p < 0.001$.

3.7. NO Reduction Activity of Mealworm Chitosan

The anti-inflammatory and anti-degenerative arthritis activities of glucosamine and chitosan are well known, but the efficacy of mealworm chitosan has not been sufficiently proven [64,65]. To verify this, mealworm chitosan was treated at various concentrations in the LPS-induced RAW 264.7 cell line assay (Figure 5). Mealworm chitosan showed notable NO reduction effect on the macrophage cells, and its efficacy was relatively higher than that reported in other studies that used chitosan obtained from other animal sources [66,67]. In addition, the anti-inflammatory effect was attributed to not only chitosan, but also peptides, proteins, and unsaponifiable matter of the oils from mealworms in prior studies [16,66,67]. Hence, mealworm chitosan and mealworm-derived products can potentially be used for therapeutic applications in inflammatory disorders.

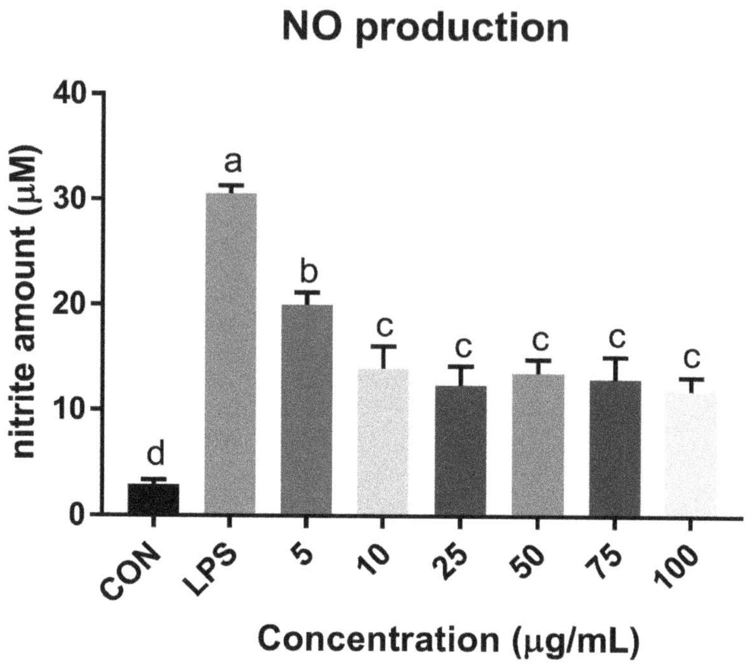

NO production

Figure 5. Nitric oxide reduction effects of mealworm chitosan on lipopolysaccharide (LPS)-stimulated murine macrophage cell line (RAW 264.7). Different superscripts (a–d) indicate significant differences at $p < 0.05$.

4. Conclusions

The aim of this study was to determine the carbohydrate content and composition of mealworms as well as examine the characteristics of mealworm chitin and chitosan because chitin derivatives are the major carbohydrate constituents in mealworms. The mealworms had a low carbohydrate content (11.5%), while the soluble sugar content was only 3.2%. Instead, chitins constituted almost half of the total carbohydrate content in the mealworms. The prepared mealworm chitin exhibited a white flake form, and its yield was 4.7 g per gram of dried mealworm. The prominent structural type of mealworm chitin was α-chitin, similar to that observed for other insects, and it developed lamellar structures in the matrices. The calculated DA of chitin and DDA of chitosan were 98.3% and 89.4%, respectively, based on NMR analysis, which were analogous to those of the commercial products. As the DA and DDA values are closely related to the physical properties of the chitin derivatives and those of mealworms showed similar values, mealworm chitin derivatives could replace the commercial chitins without considerable changes in the product quality. The crystallinity index of mealworm chitin was relatively lower than that of the crustacean-derived chitins, indicating that mealworm chitin could be applied to the products with soft textures. Considering the FT-IR and XRD patterns, it was assumed that DBC was compounded with α-chitin in the mealworms, and this could be highly beneficial because DBC showed superior properties such as high solubility, suitability for fabricating films, and superior biodegradability; however, it has been rarely found in natural sources. In addition, mealworm chitosan showed excellent anti-inflammatory effects in the LPS-induced murine macrophage cell line. Therefore, the physical characteristics of mealworm chitin and chitosan are expected to be suitable for use in industries, and afford additional advantages such as tender texture and potent anti-inflammatory effects.

Author Contributions: The study was designed by Y.-J.S., I.-K.H. and S.H.K. Y.-J.S. conducted the experiments and analyses of this study. Y.-J.S., C.W.N. and S.M.K. organized the figure and table data. The manuscript was written and revised by Y.-J.S., I.-K.H., C.W.N., S.M.K. and S.H.K. All authors have read and agreed to the published version of the manuscript.

Funding: This work was supported by the "Development of recipes and manufacturing technology for promoting the consumption of edible insects (Project No. 315060-3)" grant by the Ministry of Agriculture, Food and Rural Affairs, Republic of Korea. It was also supported by the Korea Institute of Science and Technology (Project No. 2Z05630).

Institutional Review Board Statement: Not applicable.

Informed Consent Statement: Not applicable.

Data Availability Statement: Not applicable.

Conflicts of Interest: The authors declare no conflict of interest.

References

1. de Boer, J.; Helms, M.; Aiking, H. Protein consumption and sustainability: Diet diversity in EU-15. *Ecol. Econ.* **2006**, *59*, 267–274. [CrossRef]
2. Van Huis, A.; Van Itterbeeck, J.; Klunder, H.; Mertens, E.; Halloran, A.; Muir, G.; Vantomme, P. *Edible Insects: Future Prospects for Food and Feed Security*; Food and Agriculture Organization of the United Nations: Rome, Italy, 2013.
3. van Huis, A.; Oonincx, D.G. The environmental sustainability of insects as food and feed. A review. *Agron. Sustain. Dev.* **2017**, *37*, 1–14. [CrossRef]
4. Meyer-Rochow, V.B. Can insects help to ease the problem of world food shortage. *Search* **1975**, *6*, 261–262.
5. Rumpold, B.A.; Fröhling, A.; Reineke, K.; Knorr, D.; Boguslawski, S.; Ehlbeck, J.; Schlüter, O. Comparison of volumetric and surface decontamination techniques for innovative processing of mealworm larvae (*Tenebrio molitor*). *Innov. Food Sci. Emerg. Technol.* **2014**, *26*, 232–241. [CrossRef]
6. Liu, C.; Masri, J.; Perez, V.; Maya, C.; Zhao, J. Growth performance and nutrient composition of mealworms (*Tenebrio molitor*) fed on fresh plant materials-supplemented diets. *Foods* **2020**, *9*, 151. [CrossRef]
7. Weaver, D.K.; McFarlane, J. The effect of larval density on growth and development of *Tenebrio molitor*. *J. Insect Physiol.* **1990**, *36*, 531–536. [CrossRef]
8. Rumbos, C.I.; Karapanagiotidis, I.T.; Mente, E.; Psofakis, P.; Athanassiou, C.G. Evaluation of various commodities for the development of the yellow mealworm, *Tenebrio molitor*. *Sci. Rep.* **2020**, *10*, 1–10.
9. Iaconisi, V.; Marono, S.; Parisi, G.; Gasco, L.; Genovese, L.; Maricchiolo, G.; Bovera, F.; Piccolo, G. Dietary inclusion of *Tenebrio molitor* larvae meal: Effects on growth performance and final quality treats of blackspot sea bream (Pagellus bogaraveo). *Aquaculture* **2017**, *476*, 49–58. [CrossRef]
10. Bovera, F.; Loponte, R.; Marono, S.; Piccolo, G.; Parisi, G.; Iaconisi, V.; Gasco, L.; Nizza, A. Use of *Tenebrio molitor* larvae meal as protein source in broiler diet: Effect on growth performance, nutrient digestibility, and carcass and meat traits. *J. Anim. Sci.* **2016**, *94*, 639–647. [CrossRef]
11. Wu, R.A.; Ding, Q.; Yin, L.; Chi, X.; Sun, N.; He, R.; Luo, L.; Ma, H.; Li, Z. Comparison of the nutritional value of mysore thorn borer (Anoplophora chinensis) and mealworm larva (*Tenebrio molitor*): Amino acid, fatty acid, and element profiles. *Food Chem.* **2020**, *323*, 126818. [CrossRef]
12. Ghosh, S.; Lee, S.-M.; Jung, C.; Meyer-Rochow, V.B. Nutritional composition of five commercial edible insects in South Korea. *J. Asia-Pac. Entomol.* **2017**, *20*, 686–694. [CrossRef]
13. Schäufele, I.; Albores, E.B.; Hamm, U. The role of species for the acceptance of edible insects: Evidence from a consumer survey. *Br. Food J.* **2019**, *121*. [CrossRef]
14. Kröncke, N.; Böschen, V.; Woyzichovski, J.; Demtröder, S.; Benning, R. Comparison of suitable drying processes for mealworms (*Tenebrio molitor*). *Innov. Food Sci. Emerg. Technol.* **2018**, *50*, 20–25. [CrossRef]
15. Son, Y.-J.; Lee, J.-C.; Hwang, I.-K.; Nho, C.W.; Kim, S.-H. Physicochemical properties of mealworm (*Tenebrio molitor*) powders manufactured by different industrial processes. *LWT* **2019**, *116*, 108514. [CrossRef]
16. Son, Y.-J.; Choi, S.Y.; Hwang, I.-K.; Nho, C.W.; Kim, S.H. Could defatted mealworm (*Tenebrio molitor*) and mealworm oil be used as food ingredients? *Foods* **2020**, *9*, 40. [CrossRef]
17. Dubois, M.; Gilles, K.A.; Hamilton, J.K.; Rebers, P.t.; Smith, F. Colorimetric method for determination of sugars and related substances. *Anal. Chem.* **1956**, *28*, 350–356. [CrossRef]
18. Miller, G.L. Use of dinitrosalicylic acid reagent for determination of reducing sugar. *Anal. Chem.* **1959**, *31*, 426–428. [CrossRef]
19. Kim, K.-S.; Park, J.-B.; Kim, S.-A. Quality characteristics of kochujang prepared with Korean single-harvested pepper (Capsicum annuum L.). *J. Korean Soc. Food Sci. Nutr.* **2007**, *36*, 759–765. [CrossRef]
20. Belcher, R.; Nutten, A.; Sambrook, C. The determination of glucosamine. *Analyst* **1954**, *79*, 201–208. [CrossRef]

21. Elson, L.A.; Morgan, W.T.J. A colorimetric method for the determination of glucosamine and chondrosamine. *Biochem. J.* **1933**, *27*, 1824. [CrossRef]

22. Manni, L.; Ghorbel-Bellaaj, O.; Jellouli, K.; Younes, I.; Nasri, M. Extraction and characterization of chitin, chitosan, and protein hydrolysates prepared from shrimp waste by treatment with crude protease from Bacillus cereus SV1. *Appl. Biochem. Biotechnol.* **2010**, *162*, 345–357. [CrossRef] [PubMed]

23. Al Sagheer, F.; Al-Sughayer, M.; Muslim, S.; Elsabee, M. Extraction and characterization of chitin and chitosan from marine sources in Arabian Gulf. *Carbohydr. Polym.* **2009**, *77*, 410–419. [CrossRef]

24. AOAC, H.W. *International A: Official Methods of Analysis of the AOAC International*; The Association: Arlington County, VA, USA, 2000.

25. Hsu, C.-L.; Chen, W.; Weng, Y.-M.; Tseng, C.-Y. Chemical composition, physical properties, and antioxidant activities of yam flours as affected by different drying methods. *Food Chem.* **2003**, *83*, 85–92. [CrossRef]

26. Xu, J.; McCarthy, S.P.; Gross, R.A.; Kaplan, D.L. Chitosan film acylation and effects on biodegradability. *Macromolecules* **1996**, *29*, 3436–3440. [CrossRef]

27. Kasaai, M.R.; Arul, J.; Charlet, G. Intrinsic viscosity–molecular weight relationship for chitosan. *J. Polym. Sci. Part B: Polym. Phys.* **2000**, *38*, 2591–2598. [CrossRef]

28. Van de Velde, K.; Kiekens, P. Structure analysis and degree of substitution of chitin, chitosan and dibutyrylchitin by FT-IR spectroscopy and solid state 13C NMR. *Carbohydr. Polym.* **2004**, *58*, 409–416. [CrossRef]

29. Ottøy, M.H.; Vårum, K.M.; Christensen, B.E.; Anthonsen, M.W.; Smidsrød, O. Preparative and analytical size-exclusion chromatography of chitosans. *Carbohydr. Polym.* **1996**, *31*, 253–261. [CrossRef]

30. Lavertu, M.; Xia, Z.; Serreqi, A.; Berrada, M.; Rodrigues, A.; Wang, D.; Buschmann, M.; Gupta, A. A validated 1H NMR method for the determination of the degree of deacetylation of chitosan. *J. Pharm. Biomed. Anal.* **2003**, *32*, 1149–1158. [CrossRef]

31. Baxter, A.; Dillon, M.; Taylor, K.A.; Roberts, G.A. Improved method for ir determination of the degree of N-acetylation of chitosan. *Int. J. Biol. Macromol.* **1992**, *14*, 166–169. [CrossRef]

32. Yen, M.-T.; Yang, J.-H.; Mau, J.-L. Physicochemical characterization of chitin and chitosan from crab shells. *Carbohydr. Polym.* **2009**, *75*, 15–21. [CrossRef]

33. Wan, Y.; Creber, K.A.M.; Peppley, B.; Bui, V.T. Ionic conductivity and related properties of crosslinked chitosan membranes. *J. Appl. Polym. Sci.* **2003**, *89*, 306–317. [CrossRef]

34. Liao, G.Z.; Wang, G.Y.; Xu, X.L.; Zhou, G.H. Effect of cooking methods on the formation of heterocyclic aromatic amines in chicken and duck breast. *Meat Sci.* **2010**, *85*, 149–154. [CrossRef] [PubMed]

35. Bruckner, S.; Albrecht, A.; Petersen, B.; Kreyenschmidt, J. Characterization and Comparison of Spoilage Processes in Fresh Pork and Poultry. *J. Food Qual.* **2012**, *35*, 372–382. [CrossRef]

36. EnfÄLt, A.-C.; Hullberg, A. Glycogen, glucose and glucose-6-phosphate content in fresh and cooked meat and meat exudate from carriers and noncarriers of the RN$^-$ allele. *J. Muscle Foods* **2005**, *16*, 330–341. [CrossRef]

37. Tarr, H. Postmortem degradation of glycogen and starch in fish muscle. *J. Fish. Board Can.* **1968**, *25*, 1539–1554. [CrossRef]

38. Jakinovich Jr, W.; Sugarman, D. Sugar taste reception in mammals. *Chem. Senses* **1988**, *13*, 13–31. [CrossRef]

39. Harish Prashanth, K.V.; Tharanathan, R.N. Chitin/chitosan: Modifications and their unlimited application potential—an overview. *Trends Food Sci. Technol.* **2007**, *18*, 117–131. [CrossRef]

40. Song, Y.-S.; Kim, M.-W.; Moon, C.; Seo, D.-J.; Han, Y.S.; Jo, Y.H.; Noh, M.Y.; Park, Y.-K.; Kim, S.-A.; Kim, Y.W.; et al. Extraction of chitin and chitosan from larval exuvium and whole body of edible mealworm, *Tenebrio molitor*. *Entomol. Res.* **2018**, *48*, 227–233. [CrossRef]

41. Siregar, H.C.; Suptijah, P. Mealworm (Tenebrio molitor) as Calcium, Phosphor, and Chitosan Source. Available online: https://repository.ipb.ac.id/handle/123456789/76877 (accessed on 11 January 2021).

42. Jung, W.; Jo, G.; Kuk, J.; Kim, Y.; Oh, K.; Park, R. Production of chitin from red crab shell waste by successive fermentation with Lactobacillus paracasei KCTC-3074 and Serratia marcescens FS-3. *Carbohydr. Polym.* **2007**, *68*, 746–750. [CrossRef]

43. Kaya, M.; Lelešius, E.; Nagrockaitė, R.; Sargin, I.; Arslan, G.; Mol, A.; Baran, T.; Can, E.; Bitim, B. Differentiations of chitin content and surface morphologies of chitins extracted from male and female grasshopper species. *PLoS ONE* **2015**, *10*, e0115531. [CrossRef]

44. Wang, D.; Bai, Y.y.; Li, J.h.; Zhang, C.x. Nutritional value of the field cricket (Gryllus testaceus Walker). *Insect Sci.* **2004**, *11*, 275–283. [CrossRef]

45. Paulino, A.T.; Simionato, J.I.; Garcia, J.C.; Nozaki, J. Characterization of chitosan and chitin produced from silkworm crysalides. *Carbohydr. Polym.* **2006**, *64*, 98–103. [CrossRef]

46. Hong, N.; Huaixin, C.; Yanyan, Y.; Liang, T. Studies on Extraction and Preparation Technique of Silkworm Chrysalis (*Bombyx mori* L.) Pupa Chitin and Chitosan. *J. Hubei Univ. (Nat. Sci. Ed.)* **1998**, *20*, 94–96.

47. Pesch, Y.-Y.; Riedel, D.; Behr, M. Drosophila Chitinase 2 is expressed in chitin producing organs for cuticle formation. *Arthropod Struct. Dev.* **2017**, *46*, 4–12. [CrossRef]

48. Heux, L.; Brugnerotto, J.; Desbrières, J.; Versali, M.F.; Rinaudo, M. Solid State NMR for Determination of Degree of Acetylation of Chitin and Chitosan. *Biomacromolecules* **2000**, *1*, 746–751. [CrossRef] [PubMed]

49. Zhang, M.; Haga, A.; Sekiguchi, H.; Hirano, S. Structure of insect chitin isolated from beetle larva cuticle and silkworm (Bombyx mori) pupa exuvia. *Int. J. Biol. Macromol.* **2000**, *27*, 99–105. [CrossRef]

50. Suresh, S.; Srivastava, V.C.; Mishra, I.M. Adsorption of catechol, resorcinol, hydroquinone, and their derivatives: A review. *Int. J. Energy Environ. Eng.* **2012**, *3*, 32. [CrossRef]

51. Kramer, K.J.; Kanost, M.R.; Hopkins, T.L.; Jiang, H.; Zhu, Y.C.; Xu, R.; Kerwin, J.L.; Turecek, F. Oxidative conjugation of catechols with proteins in insect skeletal systems. *Tetrahedron* **2001**, *57*, 385–392. [CrossRef]

52. Foster, L.J.R.; Ho, S.; Hook, J.; Basuki, M.; Marçal, H. Chitosan as a Biomaterial: Influence of Degree of Deacetylation on Its Physiochemical, Material and Biological Properties. *PLoS ONE* **2015**, *10*, e0135153. [CrossRef] [PubMed]

53. Moussian, B. Chitin: Structure, Chemistry and Biology. In *Targeting Chitin-containing Organisms*; Yang, Q., Fukamizo, T., Eds.; Springer: Singapore, 2019; pp. 5–18. [CrossRef]

54. Focher, B.; Naggi, A.; Torri, G.; Cosani, A.; Terbojevich, M. Structural differences between chitin polymorphs and their precipitates from solutions—evidence from CP-MAS 13C-NMR, FT-IR and FT-Raman spectroscopy. *Carbohydr. Polym.* **1992**, *17*, 97–102. [CrossRef]

55. Lavall, R.L.; Assis, O.B.; Campana-Filho, S.P. β-Chitin from the pens of Loligo sp.: Extraction and characterization. *Bioresour. Technol.* **2007**, *98*, 2465–2472. [CrossRef]

56. Rinaudo, M. Chitin and chitosan: Properties and applications. *Prog. Polym. Sci.* **2006**, *31*, 603–632. [CrossRef]

57. Blasinska, A.; Drobnik, J. Effects of nonwoven mats of Di-O-butyrylchitin and related polymers on the process of wound healing. *Biomacromolecules* **2008**, *9*, 776–782. [CrossRef]

58. Szosland, L.; Krucinska, I.; Cislo, R.; Paluch, D.; Staniszewska-Kus, J.; Solski, L.; Szymonowicz, M. Synthesis of dibutyrylchitin and preparation of new textiles made from dibutyrylchitin and chitin for medical applications. *Fibres Text. East. Eur.* **2001**, *9*, 54–57.

59. Badawy, R.M.; Mohamed, H.I. Chitin extration, composition of different six insect species and their comparable characteristics with that of the shrimp. *J. Am. Sci.* **2015**, *11*, 127–134.

60. Sajomsang, W.; Gonil, P. Preparation and characterization of α-chitin from cicada sloughs. *Mater. Sci. Eng. C* **2010**, *30*, 357–363. [CrossRef]

61. Muzzarelli, C.; Francescangeli, O.; Tosi, G.; Muzzarelli, R.A.A. Susceptibility of dibutyryl chitin and regenerated chitin fibres to deacylation and depolymerization by lipases. *Carbohydr. Polym.* **2004**, *56*, 137–146. [CrossRef]

62. Hajji, S.; Younes, I.; Ghorbel-Bellaaj, O.; Hajji, R.; Rinaudo, M.; Nasri, M.; Jellouli, K. Structural differences between chitin and chitosan extracted from three different marine sources. *Int. J. Biol. Macromol.* **2014**, *65*, 298–306. [CrossRef] [PubMed]

63. Di Nardo, T.; Moores, A. Mechanochemical amorphization of chitin: Impact of apparatus material on performance and contamination. *Beilstein J. Org. Chem.* **2019**, *15*, 1217–1225. [CrossRef]

64. Reginster, J.Y.; Deroisy, R.; Rovati, L.C.; Lee, R.L.; Lejeune, E.; Bruyere, O.; Giacovelli, G.; Henrotin, Y.; Dacre, J.E.; Gossett, C. Long-term effects of glucosamine sulphate on osteoarthritis progression: A randomised, placebo-controlled clinical trial. *Lancet* **2001**, *357*, 251–256. [CrossRef]

65. Huang, B.; Xiao, D.; Tan, B.; Xiao, H.; Wang, J.; Yin, J.; Duan, J.; Huang, R.; Yang, C.; Yin, Y. Chitosan oligosaccharide reduces intestinal inflammation that involves calcium-sensing receptor (CaSR) activation in lipopolysaccharide (LPS)-challenged piglets. *J. Agric. Food Chem.* **2016**, *64*, 245–252. [CrossRef] [PubMed]

66. Chang, S.-H.; Lin, Y.-Y.; Wu, G.-J.; Huang, C.-H.; Tsai, G.J. Effect of chitosan molecular weight on anti-inflammatory activity in the RAW 264.7 macrophage model. *Int. J. Biol. Macromol.* **2019**, *131*, 167–175. [CrossRef] [PubMed]

67. Chou, T.-C.; Fu, E.; Shen, E.C. Chitosan inhibits prostaglandin E2 formation and cyclooxygenase-2 induction in lipopolysaccharide-treated RAW 264.7 macrophages. *Biochem. Biophys. Res. Commun.* **2003**, *308*, 403–407. [CrossRef]

 *foods*

Article

In Vitro Study of Cricket Chitosan's Potential as a Prebiotic and a Promoter of Probiotic Microorganisms to Control Pathogenic Bacteria in the Human Gut

Carolyne Kipkoech [1,2], John N. Kinyuru [1], Samuel Imathiu [1], Victor Benno Meyer-Rochow [3,4,*] and Nanna Roos [5]

1 Department of Food Science and Technology, Jomo Kenyatta University of Agriculture and Technology, P.O. Box 62000-00200, Nairobi, Kenya; carolyne.kipkoech@bfr.bund.de (C.K.); jkinyuru@agri.jkuat.ac.ke (J.N.K.); samuel.imathiu@jkuat.ac.ke (S.I.)
2 Federal Institute of Risk assessment, D-12107 Berlin, Germany
3 Department of Plant Medicals (Agricultural Science and Technology), Andong National University, Andong 36729, Korea
4 Department of Genetics and Ecology, Oulu University, SF-90140 Oulu, Finland
5 Department of Nutrition, Exercise and Sports, University of Copenhagen, Rolighedsvej 26, 1958 Frederiksberg, Denmark; nro@nexs.du.dk
* Correspondence: meyrow@gmail.com

Abstract: In this study, cricket chitosan was used as a prebiotic. *Lactobacillus fermentum, Lactobacillus acidophilus,* and *Bifidobacterium adolescentis* were identified as probiotic bacteria. Cricket chitin was deacetylated to chitosan and added to either De Man Rogosa and Sharpe or *Salmonella/Shigella* bacterial growth media at the rates of 1%, 5%, 10%, or 20% to obtain chitosan-supplemented media. The growth of the probiotic bacteria was monitored on chitosan-supplemented media after 6, 12, 24, and 48 h upon incubation at 37 °C. Growth of *Salmonella typhi* in the presence of probiotic bacteria in chitosan-supplemented media was evaluated under similar conditions to those of the growth of probiotic bacteria by measuring growth inhibition zones (in mm) around the bacterial colonies. All chitosan concentrations significantly increased the populations of probiotic bacteria and decreased the populations of pathogenic bacteria. During growth, there was a significant pH change in the media with all probiotic bacteria. Inhibition zones from probiotic bacteria growth supernatant against *Salmonella typhi* were most apparent at 16 mm and statistically significant in connection with a 10% chitosan concentration. This study suggests cricket-derived chitosan can function as a prebiotic, with an ability to eliminate pathogenic bacteria in the presence of probiotic bacteria.

Keywords: human gut bacteria; growth inhibition; chitin; chitosan; pathogenic; diet; pre- and probiotics

Citation: Kipkoech, C.; Kinyuru, J.N.; Imathiu, S.; Meyer-Rochow, V.B.; Roos, N. In Vitro Study of Cricket Chitosan's Potential as a Prebiotic and a Promoter of Probiotic Microorganisms to Control Pathogenic Bacteria in the Human Gut. *Foods* **2021**, *10*, 2310. https://doi.org/10.3390/foods10102310

Academic Editor: Reza Ovissipour

Received: 24 August 2021
Accepted: 26 September 2021
Published: 29 September 2021

1. Introduction

Crickets are edible insects and farms to rear them commercially have been established in several countries around the world in recent years [1]. The main objectives for farming crickets have been their nutritive value and their therapeutic roles, for example lowering blood pressure and exerting anti-aging effects [2,3]. Since only a small fraction of humans, according to Paoletti et al. [4] possesses the enzyme chitinase, the chitin cuticle of these insects has generally been regarded as not very useful. Although chitosan, a polymer with applications in food [5], cosmetics [6], biomedical and pharmaceutical realms [7,8], can be obtained from crickets, it is commercially acquired mainly from crustaceans. In this paper, we focus on cricket chitin in connection with pre-and probiotics and show that this component of the cricket body when ingested can exert a beneficial influence on the gut flora of human consumers and need not be regarded as useless [4,9,10].

Prebiotics are fermentable fibres that can benefit the growth of beneficial bacteria in the host's colon [10]. Positive alteration of the composition and metabolic activity of the host colon is of great interest to human health promoters owing to the important role of the intestinal micro-flora to synthesize vitamins and stimulate the growth of bifidobacteria and lactobacilli. Gut beneficial bacteria, widely referred to as probiotics, have been defined as live microorganisms that positively affect the host's organism by improving the intestine's microbial balance [10]. Probiotic bacteria are known to suppress the growth of pathogenic bacteria by lowering the pH and by producing growth-suppressing metabolites. In this way they protect an organism against gastrointestinal illnesses [11].

Recent advances in understanding the important role of prebiotics include demonstrations of prebiotics to stimulate the growth of beneficial bacteria and to alleviate depression by reversing the pathophysiology of depression [12]. Consuming crickets has been shown to promote the growth of probiotic bacteria with reduced plasma TNF-α [13]. In a randomised control trial on the use of oligosaccharide as a prebiotic and *Bifidobacterium lactis* as a probiotic in infants, higher *Bifidobacterium* and *Lactobacillus* with lower clostridia counts occurred in the group that consumed prebiotics [14].

2. Materials and Methods

2.1. Preparation of Chitosan

Chitosan was obtained by deacetylation of chitin obtained from crickets of the species *Scapsipedus icipe*, farmed at Jomo Kenyatta University of Agriculture and Technology Farm (JIF); latitude 1.10325 S1°6′11.718″, longitude 37.0208 E 37°1′14.898″. A colony was initiated using 625 juveniles and 466 adult crickets (348 females and 118 males). Wild-caught juvenile and adult crickets were transferred separately into transparent Perspex cages (65 cm height × 50 cm width × 65 cm length) with vertically arranged cardboard egg trays to provide hiding sites for the crickets. Each cage had a rectangular opening (20 cm × 35 cm) made on the lid of the cage to which a net was fixed. Two additional openings (25 cm diameter) were also made on the front and backsides of the cage, screened with a net to allow for air circulation.

According to the protocol of Magara et al. [15], the crickets were fed a mixture of soybean flour, wheat bran, and maize diets daily. Besides, fresh plant leaves were also provided regularly. Wet cotton balls with approximately 60% moisture [confirmed using a moisture sensor with two 12-cm-long probes; HydroSenseTM CS620, Campbell Scientific, Inc., Logan, UT, USA] were introduced into the cages to provide water and to serve as oviposition sites for adult crickets. The cotton balls were replaced every 3 days. The colonies were maintained at 27 ± 1 °C, relative humidity (RH) of 65 ± 5%, and a photoperiod of 12:12 (L: D) h cycle.

This species of cricket typically undergoes 9–10 moults to maturity depending on the temperature. In the adult cages, the cotton balls were checked daily, and those containing eggs were carefully removed and transferred into transparent rectangular plastic containers (20 × 15 × 15 cm; Kenpoly Manufacturer Ltd., Nairobi, Kenya). Thereafter, the containers were placed in a climate-controlled chamber at 30 °C with an RH of 70% and a photoperiod of 12:12 (L: D) h light cycle. An opening (15 × 8 cm) was made on the lid of each container and covered with fine netting organza material capable of retaining emerging nymphs. The newly hatched nymphs were transferred into Perspex cages as described above and fed powdered soybean and maize diets ad libitum [15]. The rearing cultures were monitored daily to record and remove dead crickets. The colony was reared for 8–12 generations before the start of the experiment. Once every 6 months, wild-caught crickets were reared separately and the young neonates were transferred to cages, holding the newly hatched neonates of the stock culture to maintain the genetic vigour of the colonies and prevent inbreeding depression as well as disease transfer. In addition, cricket populations were kept at low densities to reduce the stressful crowding effect, which is very common in insect mass production [16].

Although not as well known as *Acheta domesticus* and *Gryllus bimaculatus*, general biology and life cycle of *Scapsipedus icipe* Hugel & Tanga have been described in detail by Otieno et al. [17] and are largely similar to the other commercially used cricket species.

Before the commencement of the experiment, the rearing room was maintained at 27±1°C using Xpelair heater: WH30, 3 KW Wall Fan Heater, and UK. The RH in the experimental room was maintained at 65 ± 5% using a diabatic atomizer humidifier Condair ABS3 and a photoperiod of 12:12 (L: D) h. The condition of the room was monitored daily using a digital thermohydrometer (Humidity/Temperature Traceable Dew Point Meter–4800 CC). From the adult cricket stock colony, eggs (~1 h old) were collected using Petri dishes (9 cm diameter × 1.2 cm height) filled with 70% moist autoclaved wood shavings (sawdust) screened with aluminum wire mesh netting (2 mm^2) to avoid cannibalism. The eggs were individually counted with the aid of entomological tweezers and a moist fine camel's hair brush under stereomicroscope (Leica MZ 125 Microscope; Leica Microsystems Switzerland Limited), fitted with Toshiba 3CCD camera using the Auto-Montage software (Syncroscopy, Synoptic Group, Cambridge, UK) at a magnification of 25× to avoid damage.

In total, 3000 eggs were subdivided into three groups (1000 each) and transferred into 4-L transparent rectangular plastic containers (21 × 14 × 15 cm; Kenpoly Manufacturer Ltd., Nairobi, Kenya) containing moist wood shavings (sawdust). The experimental setup was monitored at 6-h intervals daily until eclosion from the eggs commenced. An opening (14.5 × 8.3 cm) was made on the lid of each container and covered with fine netting organza material capable of retaining emerging nymphs. Crickets were sampled three times weekly from the JIF production site, from week 4 to week 13 of the cricket production period. The crickets were cleaned and oven-dried at 50 °C for 72 h and crushed to obtain cricket flour for chitosan extraction.

2.2. Extraction of Chitin and Chitosan

The extraction process of chitosan involved three major steps such as demineralisation, deproteinisation, and deacetylation.

2.2.1. Demineralisation

This step was performed using dilute hydrochloric acid (HCl) solution. The raw materials were ground to 20 mesh in a Wiley mill. The samples were demineralised with 2 N HCl at room temperature for 6 h and then were treated with 2 N HCl at 10 °C. In this case, the ratio of the raw material to 2 N HCl was 1 g/10 mL. Then, the samples were washed with distilled water and dried in an oven at 40 °C overnight to eliminate the calcium carbonate and calcium chloride, which constitute the main inorganic compounds of the crickets' exoskeleton. During the digestion reaction, the emission of carbon dioxide (CO_2) gas is an important indicator of the content of the mineral materials. The resulting materials were then filtered, washed to neutrality with distilled water, and dried in an oven overnight at 50 °C.

2.2.2. Deproteinisation

The samples were treated with 4% NaOH at 10°C for 24 h and then were washed with distilled water. In this case, the ratio of the sample to 4% NaOH was 1 g/10 mL. Deproteinisation was performed using alkaline treatment using dilute sodium hydroxide (NaOH) solution to remove proteins. The mixture was filtrated, washed several times with deionised water to remove the excess of NaOH, and then dried in an oven overnight. The product obtained was designated as purified chitin.

2.2.3. Deacetylation

This step was to convert chitin to chitosan by removal of the acetyl group. The preparation of chitosan was generally achieved by treatment with concentrated NaOH solution at elevated temperature. After the reaction, the material produced was washed several times with distilled water until neutrality and then dried in an oven overnight.

2.3. Structural Characterisation of Chitin

After being dried completely at 50 °C under a vacuum, the sample was used for analysis. FTIR spectra were obtained with a Shimadzu FTIR 8700 spectrophotometer (Tokyo, Japan) under dry air at room temperature with KBr pellets. The pellets were prepared via the thorough mixing of KBr (300 mg) and chitin (3 mg). Solid-state CP–MAS ^{13}C NMR spectra were obtained at a ^{13}C frequency of 500 MHz with a Bruker Avance-300 NMR spectrometer. Spectra were acquired with a contact time of 0.224 s. A repetition time of 10 s was used for all the samples. The spinning speed was 8567 Hz, and the number of scans was 876.

2.4. Antibacterial Activity and Mechanism of Chitosan

2.4.1. Microorganisms and Cultivation

All microbiological media and chemicals were obtained from Sigma Chemical Company Ltd. unless otherwise indicated. Commercial media for probiotic bacteria, de Man, Rogosa, and Sharpe Agar, and broth (MRS agar and broth) were used. For *Salmonella typhi*, *Salmonella/Shigella* (SS) agar was used. For the combined cultivation of probiotics and pathogenic bacteria, a commercial nutrient broth was used. Chitosan-amended media were prepared by substituting SS or MRS bacterial media with 1%, 5%, 10%, or 20% chitosan. The control medium consisted of media without chitosan, as either MRS, nutrient broth, or *Salmonella/Shigella* agar, which before use was sterilized at 121 °C for 15 min. Probiotic bacterial cultures: *Lactobacillus fermentum* ATCC 9338, *Lactobacillus acidophilus* ATCC 4356, and *Bifidobacterium adolescentis* ATCC 15703 were obtained from Chr. Hansen-Denmark through Promaco Ltd., Nairobi, Kenya. The pathogenic bacterium *Salmonella typhi* ATCC 6539 was obtained from Kenya Medical Research Institute, Nairobi, Kenya.

2.4.2. Antibacterial Assessment

Antibacterial activities of the series of chitosan and its derivatives against *Salmonella typhi* were evaluated. A representative colony was picked off with a wire loop and placed in a nutrient broth (peptone 10 g, beef extract 3 g, NaCl 3 g in distilled water 1000 mL; pH 7.0), which was then incubated at 37 °C overnight. Then, a culture where *S. typhi* grew in a logarithmic growth phase was prepared for an antibacterial test. 0.1 g of chitosan was dissolved in 4.9 g of nutrient medium containing 0.1 mol/L of acetic acid. After chitosan was completely dissolved, it was gradiently diluted by 5.0 g of nutrient medium to chitosan concentrations of 1%, 5%, 10%, and 20%. Its pH value was adjusted to 6.0 with dilute NaOH solution. These test tubes containing nutrient medium and chitosan were sterilised at 121 °C for 15 min. After cooling down, 50 µL of bacterial suspension was added to above each test tube and cultured at 37 °C for 24 h. The controlled test tube contained the nutrient medium (pH 6.0) with bacterial suspension but without chitosan. Then 0.1 mL of bacterial suspension was transferred to an agar plate (three plates for one sample) and cultured at 37 °C for 24 h.

A loopful of each culture was spread to obtain single colonies on the nutrient agar (agar 15 g, peptone 10 g, beef extract 3 g, NaCl 3 g in distilled water 1000 mL; pH 7.0) and incubated at 37 °C for 24 h. Colonies were then counted and colony-forming units calculated. Enumeration of bacteria was done after an incubation at 37 °C for 6 h, 12 h, 24 h, and 48 h, where 0.1 mL of 10^{-6} of each replicate was pour-plated and cultivated on MRS agar and *Salmonella/Shigella* (SS) agar for cell colony counts and calculation of cell colony-forming units. The number of colonies was read and the average value was obtained. The control was the bacterial suspension with the same pH value but without chitosan. The bactericidal rate (R) of every sample was calculated according to the following Formula: $R = B - A\,B \times 100$. A was the number of colonies of the tested plate (CFU/mL); B was the number of colonies of the controlled plate (CFU/mL).

To determine the growth of pathogenic and probiotic bacteria on different concentrations of chitosan, 0.1 mL samples of 16-h old bacterial cultures were incubated in nutrient broth with different concentrations of chitosan 1%, 5%, 10%, or 20%, incubated at 37 °C for 6 h, 12 h, 24 h, and 48 h. The cultures were serially diluted 10-fold and 0.1 mL of 10^{-6} dispensed and pour-plated on either SS agar or MRS agar and incubated at 37 °C for 48 h; the colonies were counted and colony-forming units (CFUs) calculated. Before each incubation, the pH was adjusted to neutral by the use of sodium hydroxide or hydrochloric acid and monitored during the experiment period.

2.4.3. Chitosan Inhibition of *S. typhi* Growth

The plate well diffusion method [18], was used to visualize the formation of a zone of inhibition in a solid culture medium, i.e., *Salmonella/Shigella* (SS) agar plates. The procedure carried out and used in this analysis follows the agar diffusion method, in which small circular cavities are punctured in the culture medium and filled with approximately 0.25 mL of chitosan for each concentration. Then 50 µL of bacterial suspension was spread and the plates were stored for 24 h at 37 °C to allow growth. Inhibition zones were measured in mm based on the average diameter of the clear area, directly on the dishes [19].

2.5. Data Analysis

Data collected were recorded in Microsoft excel® 2016 (16.0.5188.1000). Data normality was tested using the Kolmogovov—Smirnov test. Statistical analyses of the data were performed using the statistical methods of Motulsky [20]. Data are presented as means ± standard deviation (SD) and were analysed by one-way ANOVA followed by Tukey's multiple comparisons test across experimental groups. The difference between means was considered significant at $p \leq 0.05$. In the result and discussion section, the word 'significantly' is used to denote the statistically significant difference. For each species as well as for the sum of pathogenic and probiotic species, a linear fixed model with experimental run as a random factor was applied. In all cases, comparisons with the control were set up and corrected for simultaneous hypothesis testing according to Dunnett.

3. Results

The growths of *Salmonella typhi* and probiotic bacteria at different chitosan concentrations are shown in Figure 1. The bacterial growth proceeded as per the expected bacterial curve, except that when chitosan concentration in the media was increased, the growth of the pathogenic bacteria slowed down while the growth of the probiotic bacteria (especially that of *L. acidophilus*) expanded (Figure 1).

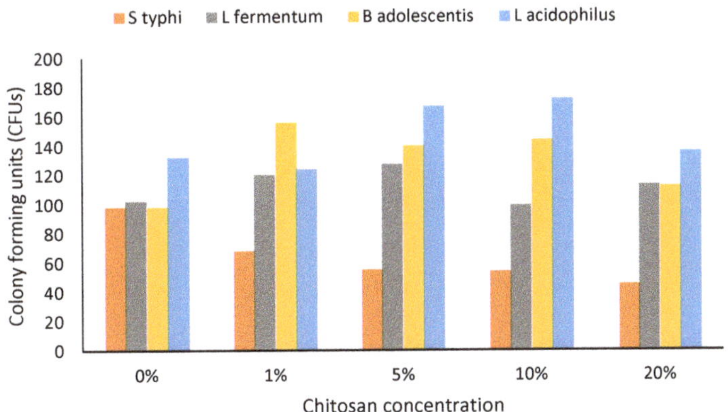

Figure 1. Growth of bacterial cells in chitosan-amended media (24 h). *n* = 3. Abbreviation: S typhi: *Salmonella typhi*, L fermentum: *Lactobacillus fermentum*, B adolescentis: *Bifidobacterium adolescentis*, L. acidophilus: *Lactobacillus acidophilus*. Bacterial growth in 0%, 1%, 5%, 10% and 20% chitosan-supplemented media was monitored at 24 h. Bacterial growth was measured in colony-forming unit per millilitre (CFU/mL).

Combined chitosan and probiotic bacteria effects on the growth of the pathogenic bacteria were assessed. An increase in growth was seen in the first 6 h before a drop in *Salmonella typhi* was noted (Figure 2). Even in media that did not contain chitosan but contained the probiotic bacteria, *Salmonella typhi* growth was highly and severely suppressed at 24 h; suppression time was shortened with increased chitosan concentration (Figure 2).

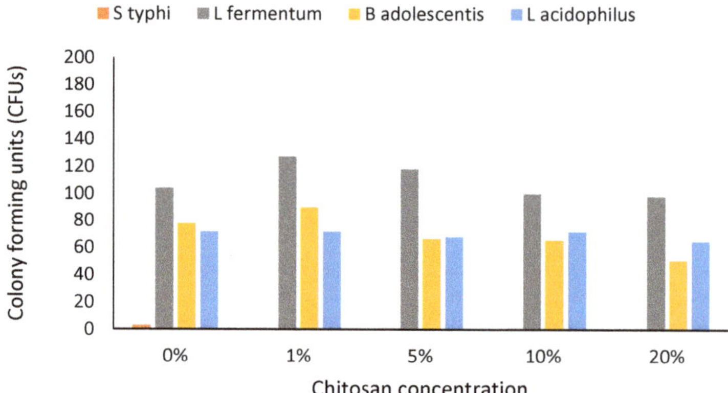

Figure 2. Effect of bacterial growth on chitosan-amended media in the presence of probiotic-prebiotic at a ratio of 1:1 (24HRS). *n* = 3. Abbreviation: S typhi: *Salmonella typhi*, L fermentum: *Lactobacillus fermentum*, B adolescentis: *Bifidobacterium adolescentis*, L. acidophilus: *Lactobacillus acidophilus.* Bacterial growth in 0%, 1%, 5%, 10% and 20% chitosan-supplemented media was monitored at 24 h. Bacterial growth was measured in colony-forming unit per millilitre (CFU/mL).

As the probiotic and pathogenic bacterial cells grew, the pH fell to around 4, being initially neutral. However, differences in connection with different chitosan concentrations were apparent (Figure 3).

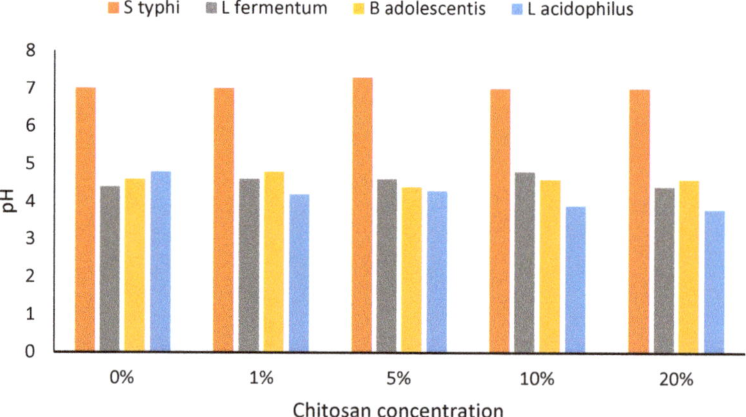

Figure 3. Change of pH in chitosan-amended media during bacterial cell growth (24 h). *n* = 3. Abbreviation: S typhi: *Salmonella typhi*, L fermentum: *Lactobacillus fermentum*, B adolescentis: *Bifidobacterium adolescentis*, L. acidophilus: *Lactobacillus acidophilus.*

When probiotic bacteria were cultivated on chitosan-supplemented media and the supernatant of the growth media was used to inhibit the growth of pathogenic bacteria, inhibition zones were seen even in the media that did not contain chitosan (although this inhibition did not reach statistical significance: $p = 0.05$; M = 10 mm, SD = 0.3). The highest and significantly most different inhibition was apparent in connection with the 16 mm zone seen in *L. fermentum* at a 10% chitosan concentration (Table 1). An increase in chitosan concentration to 20% also led to an increased diameter of the inhibition zone in *B. adolescentis*.

Table 1. Inhibition of pathogenic bacteria growth by probiotic bacteria supernatant derived from the growth of probiotic in chitosan-supplemented media.

Chitosan Concentration	Bacterial Species		
	L. fermentum	*B. adolescentis*	*L. acidophilus*
0%	10 ± 0.3 [a]	10 ± 0.2 [a]	10 ± 0.3 [a]
1%	11 ± 0.3 [a]	10 ± 0.3 [a]	11 ± 0.2 [a]
5%	11 ± 0.3 [a]	10 ± 0.1 [a]	12 ± 0.2 [b]
10%	16 ± 0.3 [b]	12 ± 0.2 [b]	12 ± 0.3 [b]
20%	13 ± 0.3 [c]	14 ± 0.3 [c]	10 ± 0.1 [a]

$n = 3$. Values displayed as means ± standard deviation Values within the same column under the same chitosan concentration with different superscripts are significantly different $p < 0.05$ Abbreviation: *L. Fermentum*: *Lactobacillus fermentum*, *B. adolescentis*: *Bifidobacterium adolescentis*, *L. acidophilus*: *Lactobacillus acidophilus*. Bacterial growth in 0%, 1%, 5%, 10% and 20% chitosan-supplemented media was monitored at 24 h. Bacterial inhibition zones were measured in millimeters' (mm).

4. Discussion

4.1. Chitosan and Chitin Characteristics

Chitosan is a natural antimicrobial agent found in the shells of crustaceans: majorly crabs, shrimp, and crayfish, and some studies have pointed to the possibility of chitosan production from squid and fungi [21]. In this study, chitosan was extracted from edible insects. There was a steady increase in bacterial growth between 0 and 12 h, which remained stable up to 48 h before dropping, following an expected normal bacterial growth curve. The lowest increase in growth was seen at 20% chitosan concentration especially in *Salmonella typhi*, this could be due to the depletion of available nutrients in the growth media and an inability of *Salmonella typhi* to ferment chitosan and utilize it for its growth. At 48 h in 20% chitosan concentration, all bacterial cells were suppressed with bacterial cells reduced to the initial number at 0 h. There was no significant change in *Salmonella typhi* growth in chitosan-supplemented media, and the normal bacterial growth curve was lacking, which may indicate that *Salmonella typhi* growth was suppressed by the presence of chitosan. In the absence of chitosan, *Salmonella typhi* growth was significant at 12 h. *Salmonella typhi* growth was limited by an increased concentration of chitosan, probably because *Salmonella typhi* was unable to break down chitosan. That is what could have affected the availability of favourable growth media and led to a limitation of nutrients amid increased bacterial concentration.

The exact mechanism of chitosan's antibacterial activity is yet to be fully understood. It is known, however, that chitosan's antimicrobial activity is influenced by several factors that mainly act in an orderly, yet independent way. Sea creatures have been the main sources of chitin and its derivative chitosan; insects are a new potential source [22]. Chitin yield, molecular weight and the degree of deacetylation determine its properties, [6,7,23,24] and insects have been shown to have a higher deacetylation percentage. The most likely antibacterial activity of chitosan is by binding to the negatively charged bacterial cell wall, disrupting the cell and, thus, altering the membrane permeability, followed by attachment to DNA causing inhibition of DNA replication and subsequently cell death. Another possible mechanism is that chitosan acts as a chelating agent that selectively binds to trace metal elements causing toxin production [21]. It could be postulated that chitosan disrupts the barrier properties of the cell wall structure of Gram-negative bacteria, as advanced by the measurement of a potassium release [25]. This mechanism could be explained by the presence of the free amino groups from the chitosan structure, causing variable mortality rates in different bacterial strains [26].

The probiotic bacteria were likely able to ferment chitosan in the media and then continue to use it for their normal growth. *Lactobacillus fermentum* and *Lactobacillus acidophilus* exhibited the highest growth although the growth of *Lactobacillus acidophilus* initially picked up slowly. This shows that probiotic bacteria can degrade chitosan only to a limited concentration and beyond this; higher chitosan concentration may not be beneficial in connection with the growth of probiotic bacteria. Past studies have shown prebiotics such as inulin and chitin can restrict the growth of pathogenic bacteria [14,27].

The gut microbiome plays an important role in the health of humans and animals. Beneficial microbes diversity can be modulated by diet [28]. Fermentable sources of fibre, and in particular insect chitin, often increase the abundance of beneficial microbes. This study demonstrates that chitosan can increase the growth of probiotic bacteria, but that its usefulness as a prebiotic to profoundly modify the gut microbial composition depends on an optimal concentration for beneficial bacteria to be able to ferment this carbohydrate. When food ingested contains chitin, which can pass through the digestive system and get to the colon almost unaltered, beneficial bacteria that reside in the colon can ferment and deacetylate [29] and utilise the products for their growth [30]. Chitin may be introduced through consumption of whole edible insect products, or in food as an additive to normally consumed food such as yogurt that may already contain probiotic bacteria.

4.2. Interactions

When *Salmonella typhi* and probiotic bacteria were cultivated in chitosan-supplemented media in the ratio of 1:1, the drop in bacterial cell growth paralleled chitosan concentration increases; only *Lactobacillus fermentum* seemed to thrive in high chitosan concentrations and the presence of the pathogen in the media (Figure 3). The numbers of colony-forming units became reduced in *Salmonella typhi* after 6 h of growth. *Bifidobacterium adolescentis* and *Lactobacillus acidophilus* growth in different concentrations did not change much over time. This was a different scenario from the initial growth of probiotic bacteria in chitosan-supplemented media where there was a significant increase in growth. This could have been due to the presence of *Salmonella typhi*, which was probably exerting a negative impact on probiotic bacterial growth as a survival strategy. Pathogenic bacteria have over time developed means of survival in the gut environment [31].

Probiotic bacteria grew well in chitosan-supplemented media, although the growth was slowed by the presence of *Salmonella typhi*, possibly due to a lack of sufficient media and unknown effects of *Salmonella typhi* exerted as a survival strategy, which may likely include toxins as a survival strategy [32]. On the other hand, in *Salmonella typhi*, populations were greatly reduced, which could have been caused by the combined effects of an inability to ferment chitosan and unknown factors involving products of the probiotics as part of a survival strategy. Recent studies have shown probiotic bacteria to be able to suppress pathogenic bacteria [33], and this would be one of the advantages of consuming chitin or its derivative chitosan together with probiotics to improve gut health. The probiotic bacteria already existing in the gut would benefit. There is currently more consciousness in functional food [34] and chitin from insects is likely to be a part of this. The use of insect chitin in the manufacture of an edible film should be encouraged owing to the importance of chitin and its role in modifying gut health.

Apart from nutrient depletion, there would be more metabolites produced by the growing probiotic bacteria, which would see a faster elimination of *Salmonella typhi*. Quite likely, this is why at 12 h the suppression of *Salmonella typhi* was high and with an increase in probiotic bacteria at 0%, 1%, and 5% chitosan concentrations. The presence of *Salmonella typhi* and high chitosan concentration beyond 5% did not favour the growth of *Lactobacillus acidophilus*. Despite the initial chitosan concentration, pathogenic bacteria were outgrown at the end, a likely result of the combined effects of the probiotic bacteria and low pH. Studies in fish fed with chitin showed increased survival, although there was no evidence of increased cellular immunity. The researchers pointed out that increased survival may have been the result of the suppression of pathogenic bacteria by chitin [35].The growth of probiotic bacteria may also have aided the suppression of pathogenic bacteria through the production of bacteriocin, exopolysaccharides, and biogenic amines [36]. These points to the importance of using chitin and chitosan as a prebiotic, and thus our findings encourage chitin or chitosan consumption to suppress pathogenic bacteria in the gut.

4.3. Prebiotics and Gut Health

Good prebiotic candidates would selectively support the growth of specific beneficial bacteria leading to a positive modulation of gut microbiota and according to Slomka et al. [37] in a study that was investigating the use of prebiotics in oral health, prebiotics greatly increased the proportion of beneficial but lowered that of pathogenic species. Pathogenic bacteria in the gut are a major cause of diarrhoea and the use of probiotic bacteria and chitosan that can suppress pathogenic bacteria by nourishing the probiotic species, which produce metabolites to kill off pathogenic species, has to be seen as an advance. Chitosan can also act directly as an antimicrobial in the human gut, suppressing the growth of pathogenic bacteria [33]. Many children in developing countries suffer greatly from childhood diarrhoea [38]. The addition of beneficial fibre such as chitin in the intervention that is aimed at helping children from the poor setup is likely to be more fruitful due to the additional benefits of chitin to gut health. Several antibacterial mechanisms of chitosan that have been proposed include: ionic surface interaction resulting in wall cell leakage; inhibition of the mRNA and protein synthesis via the penetration of chitosan into the nuclei of the microorganisms; formation of an external barrier, chelating metals and provoking the suppression of essential nutrients to microbial growth. It is likely that all events occur simultaneously but at different intensities [39].

The pH in chitosan containing *Salmonella typhi* growth media barely changed and was almost neutral at all chitosan concentrations ($p > 0.05$), but in *L. fermentum*, *B. adolescentis*, and *L. acidophilus* the pH decreased as the bacteria grew (Figure 3). In *L. acidophilus*, a species that seemed to thrive well at low pH, the highest drop in pH was noticed when the bacteria were growing in chitosan. This species was able to grow at a pH as low as pH 3.6, occurring in 20% chitosan-supplemented media at 24 h. All other probiotic bacteria growth media pH drops were seen after 12 h of growth, even in unmodified media. At 20%, chitosan concentration the expected pH drop was delayed up to 24 h (Figure 3). This might have been due to the slow growth of probiotic bacteria at high chitosan concentrations. A lower pH during growth is likely to be part of a survival strategy for probiotic bacteria and the lack to cope with the low pH in *Salmonella typhi* could be the reason why probiotic bacteria were able to suppress pathogenic bacteria. Fermented milk products have been consumed with perceived health benefits due to the presence of live bacteria responsible for milk fermentation, and recent years have seen an increased interest in fermented milk products due to their health benefit [40]. Combining fermented milk products with a prebiotic is likely to increase the benefits of its consumption.

Studies have shown that the antimicrobial activity of chitosan is pH-dependent since chitosan is soluble in an acidic environment. Yang et al., 2005 observed that the antibacterial activity of the N-alkylated chitosan derivatives against *E. coli* increased as the pH increased. These results verify that positive charge on the amino groups is not the sole factor resulting in antimicrobial activities because little is known about the antimicrobial activity of chitosan under alkaline conditions [41,42]

4.4. Altered Growth and Limitations

Nutrient limitations and a low pH have been indicated as the main inhibitors of the growth of pathogenic bacteria in the gut [2]. At the same time, they may have stimulated the growth of probiotic bacteria adapted to a low pH as such an adaptation is a well-documented survival strategy of many bacterial species [43]. Consumption of prebiotics would increase the colonization by probiotic bacteria, which in turn would suppress pathogenic bacteria and enhance their important role [44,45]. The gut microbiota would be highly regulated and therefore alleviate infections caused by microbial imbalances in the gut [46]. Intestinal microbiota depends on non-digestible fibre, [27] and cricket chitin is a possible prebiotic candidate. Fermentation of chitin leads to lower colonic pH due to the production of acetate, propionate, and butyrate by probiotic bacteria and these weak acids then influence the microbial composition, by suppressing pathogens, favouring the growth of probiotic bacteria [33,47,48]. The difference in the restriction distance was statistically significant at $p < 0.05$ for *L. fermentum* at 10% and 20% chitosan concentrations, for *B. adolescentis* at 10% and 20% chitosan concentrations, and for *L. acidophilus* at 5%, 10%, and 20% chitosan concentrations. In the media that did not contain chitosan, the difference was not significant $p > 0.05$. The presence of chitosan leads the bacteria to use the products of fermentation for their growth. During fermentation, weak acids and other metabolites restrict the growth of *Salmonella typhi*. There is a need to profile the metabolites in the media to ascertain the exact cause of pathogenic bacterial growth restriction, and the possibility of their further exploitation in the pharmaceutical industry.

Chitosan fermentation metabolites are known to inhibit bacterial growth [36,40,48,49], while chitosan itself exhibits antimicrobial activities that are useful in inhibiting gram-positive bacteria [50]. In a recent study, chitosan was shown to inhibit all of the bacterial strains tested [51]. Chitin hydrogels and other products have been developed to help in wound dressing and antimicrobial properties and to enhance healing [52]. In connection with the acceptance of insects as food and insect farming, suggested as early as 1975 [53], there is a need to fully utilise the chitin from these farmed insects in addition to the insects' high protein content.

Apart from the antimicrobial potential of chitosan, it has other applications in the food industry that have been widely discussed by a review on the application of chitosan for improvement of quality and shelf life of food [54]. The antimicrobial activity of chitosan against a wide range of food-borne filamentous fungi, yeast, and bacteria has made it a potential food preservative. Chitosan also possesses film-forming and barrier properties, thus making it a potential raw material for edible films or coatings. Inherent antibacterial/antifungal properties and the film-forming ability of chitosan make it ideal for use as a biodegradable antimicrobial packaging material that can be used to improve the storability of perishable food [54]. Numerous researches have demonstrated that chitosan can be used as an effective preservative or coating material for the improvement of quality and shelf life of various foods [55–59]. Recently, the research on chitosan has increased drastically because of its considerable potential owing to its antimicrobial activity, biodegradability, non-toxicity, biocompatibility, and versatile chemical and physical properties as well as abundance. Chitosan has a significant role in the health and food application area, given the growing concern regarding the negative environmental impact of materials, deemed toxic, currently in use such as vinyl chlorides in plastic [59]. Chitosan-based polymeric materials can be formed into films, fibres, gels, sponges, nanoparticles, or even beads [60].

The most important food applications of chitosan include the encapsulating material for probiotic stability in the production of functional food products and the formation of biodegradable films and enzyme binding [61]. Probiotics' mainly lactic acid bacteria (LAB) are widely used in the production of fermented dairy foods. They include yogurt and cheese and are the richest sources of probiotic foods [60,61]. For the probiotic microencapsulation technique used in connection with fruit juices, cereal-based products, chocolates, and cookies, chitosan is an ideal candidate as a coating material, since it does not affect the sensory properties of the encapsulated food [58–60,62].

5. Conclusions

Cricket chitin has a close similarity to commercial shrimp chitin. This study has demonstrated the ability of probiotic bacteria to break down chitosan, to lower the pH of growth media and thereby inhibit bacterial growth. Cricket-derived chitosan may be a functional prebiotic due to its ability to stimulate the growth of specific beneficial bacteria.

Cricket-derived chitosan can help solve gut health in children directly by acting as an antimicrobial substance, or as a prebiotic to nourish probiotic bacteria.

Author Contributions: Conceptualization, C.K., J.N.K., S.I., N.R.; Methodology, C.K., J.N.K., S.I., N.R.; Software, not applicable; Validation, C.K., J.N.K., S.I., V.B.M.-R., N.R.; Formal analysis, C.K., J.N.K., S.I., V.B.M.-R., N.R.; Investigation, C.K., J.N.K., S.I., N.R.; Resources, J.N.K., N.R.; Data curation, C.K., J.N.K., S.I., V.B.M.-R., N.R.; Writing—original draft preparation, C.K., J.N.K., S.I., V.B.M.-R., N.R.; Writing—review and editing, C.K., V.B.M.-R.; Visualization, C.K., J.N.K., S.I., V.B.M.-R., N.R.; Supervision, J.N.K., S.I., N.R.; Project administration, All authors have read and agreed to the published version of the manuscript.

Funding: Financial support for CK came from DANIDA funded by the Ministry of Foreign Affairs Denmark (2013–2018) under the funding no 13-06KU GREEiNSECT project. VBM-R was supported by a grant from the National Research Foundation of Korea (NRF-2018R1A6A1A03024862) to Prof. Chuleui Jung of Andong National University's Basic Science Research Program. Ethical approval for this study was issued by the Director of Mount Kenya University through Dr Francis W. Muregi of the university's Ethics Review Committee (ERC) on 09/01/2017 under the number AG422-4492/2015 to Principal Investigator Carolyne Kipkoech.

Informed Consent Statement: Not applicable.

Data Availability Statement: All data and materials are available on request.

Acknowledgments: The authors wish to acknowledge the support they have received for this research via funds from by DANIDA through the GREEiNSECT project. The authors would also like to express their thanks to Hansen-Denmark through Promaco Limited, Kenya for the provision of probiotic bacterial cultures.

Conflicts of Interest: The authors declare that there are no competing interest regarding this work.

References

1. Meyer-Rochow, V.B.; Sampat, G.; Jung, C. Farming of insects for food and feed in South Korea: Tradition and innovation. *Berl. und Münchener Tierärztliche Wochenschr.* **2019**, *132*, 236–244.
2. Di Gioia, D.; Biavati, B. *Probiotics and Prebiotics in Animal Health and Food Safety: Conclusive Remarks and Future Perspectives, in Probiotics and Prebiotics in Animal Health and Food Safety*; Springer: Berlin/Heidelberg, Germany, 2018; pp. 269–273.
3. Seabrooks, L.; Hu, L. Insects: An underrepresented resource for the discovery of biologically active natural products. *Acta Pharm. Sin. B* **2017**, *7*, 409–426. [CrossRef] [PubMed]
4. Paoletti, M.G.; Norberto, L.; Damini, R.; Musumeci, S. Human gastric juice contains chitinase that can degrade chitin. *Ann. Nutr. Metab.* **2007**, *51*, 244–251. [CrossRef]
5. Tian, B.; Liu, Y. Chitosan-based biomaterials: From discovery to food application. *Polym. Adv. Technol.* **2020**, *31*, 2408–2421. [CrossRef]
6. Aranaz, I.; Acosta, N.; Civera, C.; Elorza, B.; Mingo, J.; Castro, C.; Gandía, M.D.L.L.; Caballero, A.H. Cosmetics and cosmeceutical applications of chitin, chitosan and their derivatives. *Polymers* **2018**, *10*, 213. [CrossRef]
7. Wang, W.; Xue, C.; Mao, X. Chitosan: Structural modification, biological activity and application. *Int. J. Biol. Macromol.* **2020**, *164*, 4532–4546. [CrossRef] [PubMed]
8. Ibram, A.; Ionescu, A.-M.; Cadar, E. Comparison of extraction methods of chitin and chitosan from different sources. *Eur. J. Med. Nat. Sci.* **2021**, *5*, 44–56.
9. Mlcek, J.; Borkovcová, M.; Bednářová, M. Biologically active substances of edible insects and their use in agriculture, veterinary and human medicine. *J. Cent. Eur. Agric.* **2014**, *15*, 225–237. [CrossRef]
10. Gibson, G.R.; Hutkins, R.; Sanders, M.E.; Prescott, S.L.; Reimer, R.A.; Salminen, S.J.; Scott, K.; Stanton, C.; Swanson, K.S.; Cani, P.D.; et al. Expert consensus document: The International Scientific Association for Probiotics and Prebiotics (ISAPP) consensus statement on the definition and scope of prebiotics. *Nat. Rev. Gastroenterol. Hepatol.* **2017**, *14*, 491–502. [CrossRef]
11. Vogt, S.L.; Finlay, B.B. Gut microbiota-mediated protection against diarrheal infections. *J. Travel Med.* **2017**, *24*, S39–S43. [CrossRef]
12. Kazemi, A.; Noorbala, A.A.; Azam, K.; Eskandari, M.H.; Djafarian, K. Effect of probiotic and prebiotic vs placebo on psychological outcomes in patients with major depressive disorder: A randomized clinical trial. *Clin. Nutr.* **2019**, *38*, 522–528. [CrossRef]
13. Stull, V.J.; Finer, E.; Bergmans, R.S.; Febvre, H.P.; Longhurst, C.; Manter, D.K.; Patz, J.A.; Weir, T.L. Impact of edible cricket consumption on gut microbiota in healthy adults, a double-blind, randomized crossover trial. *Sci. Rep.* **2018**, *8*, 10762. [CrossRef]
14. Radke, M.; Picaud, J.-C.; Loui, A.; Cambonie, G.; Faas, D.; Lafeber, H.N.; De Groot, N.; Pecquet, S.S.; Steenhout, P.G.; Hascoet, J.-M. Starter formula enriched in prebiotics and probiotics ensures normal growth of infants and promotes gut health: A randomized clinical trial. *Pediatr. Res.* **2017**, *81*, 622–631. [CrossRef]

15. Magara, H.; Tanga, C.; Ayieko, M.; Hugel, S.; Mohamed, S.A.; Khamis, F.; Salifu, D.; Niassy, S.; Subramanian, S.; Fiaboe, K.K.M.; et al. Performance of newly described native edible cricket *Scapsipedus icipe* (Orthoptera: Gryllidae) on various diets of relevance for farming. *J. Econ. Entomol.* **2019**, *112*, 653–664. [CrossRef]

16. Sørensen, J.; Loeschcke, V. Larval crowding in Drosophila melanogaster induces Hsp70 expression, and leads to increased adult longevity and adult thermal stress resistance. *J. Insect Physiol.* **2001**, *47*, 1301–1307. [CrossRef]

17. Otieno, M.H.J.; Ayieko, M.A.; Niassy, S.; Salifu, D.; Abdelmutalab, A.G.A.; Fathiya, K.M.; Subramanian, S.; Fiaboe, K.K.M.; Roos, N.; Ekesi, S.; et al. Integrating temperature-dependent lifetable data into insect life cycle model for predicting the potential distribution of Scapsipedus icipe Hugel & Tanga. *PLoS ONE* **2019**, *14*, e0222941. [CrossRef]

18. Ryan, M.P.; Rea, M.C.; Hill, C.; Ross, R. An application in cheddar cheese manufacture for a strain of Lactococcus lactis producing a novel broad-spectrum bacteriocin, lacticin 3147. *Appl. Environ. Microbiol.* **1996**, *62*, 612–619. [CrossRef] [PubMed]

19. Goy, R.C.; Morais, S.T.; Assis, O.B. Evaluation of the antimicrobial activity of chitosan and its quaternized derivative on *E. coli* and *S. aureus* growth. *Rev. Bras. Farm.* **2016**, *26*, 122–127. [CrossRef]

20. Motulsky, H. *The GraphPad Guide to Comparing Dose-Response or Kinetic Curves*; GraphPad Software: San Diego, CA, USA, 1998.

21. Atay, H.Y. Antibacterial activity of chitosan-based systems. *Funct. Chitosan* **2019**, *March 6*, 457–489. [CrossRef]

22. Mărţău, G.A.; Mihai, M.; Vodnar, D.C. The use of chitosan, alginate, and pectin in the biomedical and food sector—biocompatibility, bioadhesiveness, and biodegradability. *Polymers* **2019**, *11*, 1837. [CrossRef] [PubMed]

23. Kumar, M.; Vivekanand, V.; Pareek, N. *Insect Chitin and Chitosan: Structure, Properties, Production, and Implementation Prospective, in Natural Materials and Products from Insects: Chemistry and Applications*; Springer: Cham, Switzerland, 2020; pp. 51–66.

24. Zainol Abidin, N.A.; Kormin, F.; Zainol Abidin, N.A.; Mohamed Anuar, N.A.F.; Abu Bakar, M.F. The potential of insects as alternative sources of chitin: An overview on the chemical method of extraction from various sources. *Int. J. Mol. Sci.* **2020**, *21*, 4978. [CrossRef]

25. Lee, H.; Zandkarimi, F.; Zhang, Y.; Meena, J.K.; Kim, J.; Zhuang, L.; Tyagi, S.; Ma, L.; Westbrook, T.F.; Steinberg, G.R.; et al. Energy-stress-mediated AMPK activation inhibits ferroptosis. *Nat. Cell Biol.* **2020**, *22*, 225–234. [CrossRef]

26. Anal, A.K.; Singh, H. Recent advances in microencapsulation of probiotics for industrial applications and targeted delivery. *Trends Food Sci. Technol.* **2007**, *18*, 240–251. [CrossRef]

27. Sawicki, C.M.; Livingston, K.A.; Obin, M.; Roberts, S.B.; Chung, M.; McKeown, N.M. Dietary fiber and the human gut microbiota: Application of evidence mapping methodology. *Nutrients* **2017**, *9*, 125. [CrossRef] [PubMed]

28. Jarett, J.K.; Carlson, A.; Serao, M.R.; Strickland, J.; Serfilippi, L.; Ganz, H.H. Diets with and without edible cricket support a similar level of diversity in the gut microbiome of dogs. *PeerJ* **2019**, *7*, e7661. [CrossRef] [PubMed]

29. Marzorati, M.; Maquet, V.; Possemiers, S. Fate of chitin-glucan in the human gastrointestinal tract as studied in a dynamic gut simulator (SHIME®). *J. Funct. Foods* **2017**, *30*, 313–320. [CrossRef]

30. Buruiana, C.-T.; Gómez, B.; Vizireanu, C.; Garrote, G. Manufacture and evaluation of xylo-oligosaccharides from corn stover as emerging prebiotic candidates for human health. *LWT* **2017**, *77*, 449–459. [CrossRef]

31. Álvarez-Ordóñez, A.; Begley, M.; Prieto, M.; Messens, W.; López, M.; Bernardo, A.; Hill, C. Salmonella spp. survival strategies within the host gastrointestinal tract. *Microbiology* **2011**, *157*, 3268–3281. [CrossRef]

32. Bäumler, A.J.; Sperandio, V. Interactions between the microbiota and pathogenic bacteria in the gut. *Nature* **2016**, *535*, 85–93. [CrossRef]

33. Satokari, R. Modulation of gut microbiota for health by current and next-generation probiotics. *Nutrients* **2019**, *11*, 1921. [CrossRef]

34. Díaz, L.D.; Fernández-Ruiz, V.; Cámara, M. An international regulatory review of food health-related claims in functional food products labeling. *J. Funct. Foods* **2020**, *68*, 103896. [CrossRef]

35. Dong, L.; Ariëns, R.M.C.; Tomassen, M.M.; Wichers, H.J.; Govers, C. In vitro studies toward the use of chitin as nutraceutical: Impact on the intestinal epithelium, macrophages, and microbiota. *Mol. Nutr. Food Res.* **2020**, *64*, e2000324. [CrossRef] [PubMed]

36. Prodeus, A.; Niborski, V.; Schrezenmeir, J.; Gorelov, A.; Shcherbina, A.; Rumyantsev, A. Fermented milk consumption and common infections in children attending day-care centers: A randomized trial. *J. Pediatr. Gastroenterol. Nutr.* **2016**, *63*, 534–543. [CrossRef] [PubMed]

37. Slomka, V.; Herrero, E.R.; Boon, N.; Bernaerts, K.; Trivedi, H.M.; Daep, C. Oral prebiotics and the influence of environmental conditions in vitro. *J. Peridontol.* **2018**, *89*, 708–717. [CrossRef] [PubMed]

38. Platts-Mills, J.A.; Liu, J.; Rogawski, E.T.; Kabir, F.; Lertsethtakarn, P.; Siguas, M.; Khan, S.S.; Praharaj, I.; Murei, A.; Nshama, R.; et al. Use of quantitative molecular diagnostic methods to assess the aetiology, burden, and clinical characteristics of diarrhoea in children in low-resource settings: A reanalysis of the MAL-ED cohort study. *Lancet Glob. Health* **2018**, *6*, e1309–e1318. [CrossRef]

39. Goy, R.C.; de Britto, D.; Assis, O.B.G. A review of the antimicrobial activity of chitosan. *Polímeros* **2009**, *19*, 241–247. [CrossRef]

40. Vasiljevic, T.; Shah, N.P. Fermented milk: Health benefits beyond probiotic effect. In *Handbook of Food Products Manufacturing*; John Wiley & Sons, Inc.: Hoboken, NJ, USA, 2007; Volume 2, pp. 99–115.

41. Su, Z.; Han, Q.; Zhang, F.; Meng, X.; Liu, B. Preparation, characterization and antibacterial properties of 6-deoxy-6-arginine modified chitosan. *Carbohydr. Polym.* **2020**, *230*, 115635. [CrossRef]

42. Yang, T.-C.; Chou, C.-C.; Li, C.-F. Antibacterial activity of N-alkylated disaccharide chitosan derivatives. *Int. J. Food Microbiol.* **2005**, *97*, 237–245. [CrossRef]

43. Singh, R.; Mishra, N.K.; Kumar, V.; Vinayak, V.; Joshi, K.B. Transition metal ion-mediated tyrosine-based short-peptide amphiphile nanostructures inhibit bacterial growth. *ChemBioChem* **2018**, *19*, 1630–1637. [CrossRef]

44. Hojsak, I. Probiotics in functional gastrointestinal disorders. *Adv. Exp. Med. Biol.* **2019**, *1125*, 121–137.

45. Vandenplas, Y.; Savino, F. Probiotics and prebiotics in pediatrics: What is new? *Nutrients* **2019**, *11*, 431. [CrossRef] [PubMed]

46. Gao, J.; Xu, K.; Liu, H.; Liu, G.; Bai, M.; Peng, C.; Li, T.; Yin, Y. Impact of the gut microbiota on intestinal immunity mediated by tryptophan metabolism. *Front. Cell. Infect. Microbiol.* **2018**, *8*, 13. [CrossRef] [PubMed]

47. Galdeano, C.M.; Cazorla, S.I.; Dumit, J.M.L.; Vélez, E.; Perdigón, G. Beneficial effects of probiotic consumption on the immune system. *Ann. Nutr. Metab.* **2019**, *74*, 115–124.

48. Jeong, C.-W.; Choi, H.-J.; Yoo, G.-Y.; Lee, S.-H.; Kim, Y.-C.; Okorie, O.E.; Lee, J.-H.; Jun, K.-D.; Choi, S.-M.; Kim, K.-W.; et al. Effects of dietary probiotics supplementation on juvenile Olive Flounder *Paralichthys olivaceus*. *Korean J. Fish. Aquat. Sci.* **2006**, *39*, 460–465.

49. Moreno-Vásquez, M.J.; Valenzuela-Buitimea, E.L.; Plascencia-Jatomea, M.; Encinas, J.C.; Rodríguez-Félix, F.; Sánchez-Valdes, S.; Rosas-Burgos, E.; Ocaño-Higuera, V.M.; Graciano-Verdugo, A.Z. Functionalization of chitosan by a free radical reaction: Characterization, antioxidant and antibacterial potential. *Carbohydr. Polym.* **2017**, *155*, 117–127. [CrossRef]

50. Cremar, L.; Gutierrez, J.; Martinez, J.; Materon, L.; Gilkerson, R.; Xu, F.; Lozano, K. Development of antimicrobial chitosan based nanofiber dressings for wound healing applications. *Nanomed. J.* **2018**, *5*, 6–14.

51. Abdelmalek, B.E.; Sila, A.; Haddar, A.; Bougatef, A.; Ayadi, M.A. β-Chitin and chitosan from squid gladius: Biological activities of chitosan and its application as clarifying agent for apple juice. *Int. J. Biol. Macromol.* **2017**, *104*, 953–962. [CrossRef]

52. Shao, W.; Wu, J.; Ye, S.; Wang, S.; Jiang, L.; Wei, S.; Jimin, W.; Shan, Y.; Shuxia, W.; Lei, J. A facile and green method to prepare chitin based composites with antibacterial activity. *J. Bionanosci.* **2017**, *11*, 75–79. [CrossRef]

53. Meyer-Rochow, V.B. Can insects help to ease the probem of world food shrtage? *Search* **1975**, *6*, 261–262.

54. Chien, P.-J.; Sheu, F.; Lin, H.-R. Coating citrus (Murcott tangor) fruit with low molecular weight chitosan increases postharvest quality and shelf life. *Food Chem.* **2007**, *100*, 1160–1164. [CrossRef]

55. Butnaru, E.; Stoleru, E.; Brebu, M.A.; Darie-Nita, R.N.; Bargan, A.; Vasile, C. Chitosan-Based bionanocomposite films prepared by emulsion technique for food preservation. *Materials* **2019**, *12*, 373. [CrossRef] [PubMed]

56. Noorbakhsh-Soltani, S.; Zerafat, M.; Sabbaghi, S. A comparative study of gelatin and starch-based nano-composite films modified by nano-cellulose and chitosan for food packaging applications. *Carbohydr. Polym.* **2018**, *189*, 48–55. [CrossRef] [PubMed]

57. Kumar, S.; Mukherjee, A.; Dutta, J. Chitosan based nanocomposite films and coatings: Emerging antimicrobial food packaging alternatives. *Trends Food Sci. Technol.* **2020**, *97*, 196–209. [CrossRef]

58. Freepons, D. Enhancing food production with chitosan seed-coating technology. In *Applications of Chitin and Chitosan*; CRC Press: Boca Raton, FL, USA, 2020; pp. 128–139.

59. da Costa, J.C.M.; Miki, K.S.L.; da Silva Ramos, A.; Teixeira-Costa, B.E. Development of biodegradable films based on purple yam starch/chitosan for food application. *Heliyon* **2020**, *6*, e03718. [CrossRef]

60. Piotrowska, B. *Toxic Components of Food Packaging Materials*; CRC Press: Boca Raton, FL, USA, 2005.

61. Kritchenkov, A.S.; Egorov, A.; Yagafarov, N.Z.; Volkova, O.V.; Zabodalova, L.A.; Suchkova, E.P.; Kurliuk, A.V.; Khrustalev, V. Efficient reinforcement of chitosan-based coatings for Ricotta cheese with non-toxic, active, and smart nanoparticles. *Prog. Org. Coat.* **2020**, *145*, 105707. [CrossRef]

62. Călinoiu, L.F.; Vodnar, D.C. Thermal processing for the release of phenolic compounds from wheat and oat bran. *Biomolecules* **2020**, *10*, 21. [CrossRef] [PubMed]

 *foods*

Article

Characterization of *Escherichia coli* from Edible Insect Species: Detection of Shiga Toxin-Producing Isolate

Anja Müller [1], Diana Seinige [2], Nils T. Grabowski [3], Birte Ahlfeld [3], Min Yue [4] and Corinna Kehrenberg [1,*]

1   Institute for Veterinary Food Science, Justus Liebig University Giessen, Frankfurter Str. 92, 35392 Giessen, Germany; anja.mueller@vetmed.uni-giessen.de
2   Lower Saxony State Office for Consumer Protection and Food Safety, 26203 Wardenburg, Germany; diana.seinige@gmx.de
3   Institute for Food Quality and Food Safety, University of Veterinary Medicine Hannover, Foundation, Bischofsholer Damm 15, 30173 Hannover, Germany; Nils.Grabowski@tiho-hannover.de (N.T.G.); birteahlfeld@web.de (B.A.)
4   Institute of Veterinary Sciences & Department of Veterinary Medicine, College of Animal Sciences, Zhejiang University, 866 Yuhangtang Road, Hangzhou 310058, China; myue@zju.edu.cn
*   Correspondence: corinna.kehrenberg@vetmed.uni-giessen.de; Tel.: +49-641-9938250

Abstract: Insects as novel foods are gaining popularity in Europe. Regulation (EU) 2015/2283 laid the framework for the application process to market food insects in member states, but potential hazards are still being evaluated. The aim of this study was to investigate samples of edible insect species for the presence of antimicrobial-resistant and Shiga toxin-producing *Escherichia coli* (STEC). Twenty-one *E. coli* isolates, recovered from samples of five different edible insect species, were subjected to antimicrobial susceptibility testing, PCR-based phylotyping, and macrorestriction analysis. The presence of genes associated with antimicrobial resistance or virulence, including *stx1*, *stx2*, and *eae*, was investigated by PCR. All isolates were subjected to genome sequencing, multilocus sequence typing, and serotype prediction. The isolates belonged either to phylogenetic group A, comprising mostly commensal *E. coli*, or group B1. One O178:H7 isolate, recovered from a *Zophobas atratus* sample, was identified as a STEC. A single isolate was resistant to tetracyclines and carried the *tet*(B) gene. Overall, this study shows that STEC can be present in edible insects, representing a potential health hazard. In contrast, the low resistance rate among the isolates indicates a low risk for the transmission of antimicrobial-resistant *E. coli* to consumers.

Keywords: Shiga toxin; *Escherichia coli*; antimicrobial resistance; edible insects

Citation: Müller, A.; Seinige, D.; Grabowski, N.T.; Ahlfeld, B.; Yue, M.; Kehrenberg, C. Characterization of *Escherichia coli* from Edible Insect Species: Detection of Shiga Toxin-Producing Isolate. *Foods* **2021**, *10*, 2552. https://doi.org/10.3390/foods10112552

Academic Editors: Victor Benno Meyer-Rochow and Chuleui Jung

Received: 29 September 2021
Accepted: 20 October 2021
Published: 22 October 2021

Publisher's Note: MDPI stays neutral with regard to jurisdictional claims in published maps and institutional affiliations.

1. Introduction

Edible insects are popular foods in many parts of the world such as Asia and Africa. In Europe, insects are still rarely consumed but are gaining in popularity. They are of particular interest to consumers due to their nutritional value as well as aspects of sustainability, including a lower need for feed, water, and space, when compared to traditional livestock [1,2].

In the European Union (EU), edible insects are included in the novel food regulation, Regulation (EU) 2015/2283, which lays the legal framework for placing novel foods on the European market. However, specific EU regulations regarding the production of insects intended for human consumption and controls thereof are still lacking, and the legal situation remains complicated. Edible insects can be novel food, as established in Regulation (EU) 2015/2283, if a corresponding application is answered favorably by the authorities. To date, several applications for edible insect products have been submitted. The European Food Safety Authority has already published a scientific opinion regarding the safety of dried yellow mealworm (*Tenebrio molitor*), which concluded that they are safe for human consumption [3], an important first step towards an approval for this species to be placed on the European market. More recently, the European commission has released the Commission Implementing Regulation (EU) 2021/882, authorizing the placing on the market of dried yellow mealworms as of 1 June 2021. This was the first time an insect species was officially approved to be marketed as food within the EU.

Beyond general regulations applicable to all foodstuffs, some countries have issued national guidelines concerning edible insects, while other countries do not currently have any specific national framework in place [4]. In 2019, the working group on meat and poultry hygiene and specific issues relating to food of animal origin of the German national working group on consumer health protection (Arbeitsgruppe Fleischhygiene und fachspezifische Fragen von Lebensmitteln tierischer Herkunft der Länderarbeitsgemeinschaft Verbraucherschutz, AFFL) issued corresponding recommendations [5]. To date, potential food safety concerns associated with insect-based products are still being evaluated. In 2015, the European Food Safety Authority (EFSA) published a scientific opinion stating which hazards were expected to be largely comparable to those posed by other foods of animal origin, while also highlighting remaining uncertainties and the need for further research [1]. This includes a recommendation to monitor bacteria and gather data regarding antimicrobial resistance [1]. This is particularly important as high microbial loads have previously been reported in insect-based foods [6,7]. Consequently, the presence of antimicrobial-resistant isolates could result in significant exposure for consumers. There is very little information regarding the use of antimicrobials in the rearing of edible insects; it has been recommended only as a temporary measure in case of emergency [8]. In addition, negative side-effects of antibiotic treatment, such as a lower number of eggs produced and poorer development in some insect species, can discourage the use of antimicrobials [9]. Consequently, this may result in a lower selection pressure and a lower frequency of antimicrobial-resistant bacteria in insects when compared to vertebrate livestock. However, to date, very few studies have been published regarding the characteristics of bacterial strains present on edible insect species, including their antimicrobial susceptibility and the presence of virulence genes [10,11].

Escherichia coli and Shiga toxin-producing *E. coli* (STEC), in particular, are among the most important and most frequently reported food-borne pathogens [12]. In humans, STEC can cause severe diarrhea, hemorrhagic colitis, and hemolytic uremic syndrome, which may ultimately be fatal [13]. Besides STEC, ESBL-producing and other antimicrobial resistant *E. coli* present on food of animal origin are of concern as they can facilitate the spread of antimicrobial resistance to the consumer [14,15].

However, to date, few studies have examined the presence of STEC in edible insects [11], and virtually no data are available regarding the antimicrobial resistance status of *E. coli* isolated from edible insects. Such data are crucial to elucidate their potential role in the transmission of antimicrobial resistance via the food chain and other health hazards associated with the consumption of insects. To the best of our knowledge, this is the first study specifically investigating the phenotypic and genotypic antimicrobial susceptibility, the presence of certain virulence genes, as well as the genetic relatedness of *E. coli* isolates obtained from different species of edible insects.

2. Materials and Methods

2.1. Insect Samples

A total of 36 samples of edible insect species were available to be included in this study. They were collected and tested for the presence of *E. coli* between 2014 and 2016 and consisted of two subsets. One subset was derived from the project "ZooGlow" that dealt with legally and illegally introduced and commercialized foodstuffs inside and outside Germany. It contained ten samples of frozen silkworms (Lepidoptera: Bombycidae: *Bombyx mori*) imported from Vietnam and purchased at an Asian supermarket in Germany. Two additional samples were powdered yellow mealworm larvae (Coleoptera: Tenebrionidae: *Tenebrio molitor*) bought online from a German retailer.

The other subset was insects bought at a local pet store, i.e., Mediterranean crickets (Orthoptera: Gryllidae: *Gryllus bimaculatus*; n = 1), migratory locusts (Orthoptera: Acrididae: *Locusta migratoria*; n = 4), *T. molitor* (n = 1), and superworms (Coleoptera: Tenebrionidae: *Zophobas atratus*, previously known as "*Z. morio*"; n = 3). More specimens and other species were also analyzed but did not yield *E. coli*, e.g., Jamaican field crickets (Orthoptera: Gryllidae: *Gryllus assimilis*), house crickets (Orthoptera: Gryllidae: *Acheta domesticus*), and desert locusts (Orthoptera: Acrididae: *Schistocerca gregaria*), with five samples (n = 5) each. The animals obtained from the pet store were explicitly not intended as foodstuff. However, many entomophagy aficionados in Western countries repurpose them as they point out the freshness of the product, the better taste, and a higher degree of culinary diversity, i.e., more different dishes can be made from hygienically processed pet feed insects than from freeze–dried (and sometimes spiced) entire or ground food insects (Nils Th. Grabowski, personal communication, 2017). For this reason, a microbiological evaluation of edible insects sold as pet feed is important for risk assessment. They usually let the animals fast for one day, kill them, and cook them thoroughly before processing them further. Samples for this analysis were raw and killed by freezing.

2.2. Isolate Collection and Species Identification

While doing a classical microbiological analysis of the insect samples, *E. coli* strains were detected and isolated according to standard methods described in DIN EN ISO 16649-2. In brief, samples were subjected to an enrichment step in NaCl-peptone water for 24 h at 37 ± 0.5 °C, and the broth was subsequently streaked on TBX agar (Oxoid, Wesel, Germany) for the selective detection of *E. coli*.

The species of the isolates was confirmed using a MALDI-TOF biotyper (Bruker Daltonics, Bremen, Germany) and by species-specific PCR targeting the *gadA* gene [16]. Only one isolate per sample was included in further investigations.

2.3. Antimicrobial Susceptibility Testing

Antimicrobial susceptibility testing was performed and evaluated in accordance with CLSI standards [17,18] using commercially available Sensititre microtiter plates (EUVSEC layout; Trek Diagnostic Systems Ltd., East Grinstead, UK) containing the following antimicrobials and concentrations: ampicillin (1–64 µg/mL), azithromycin (2–64 µg/mL), cefotaxime (0.25–4 µg/mL), ceftazidime (0.5–8 µg/mL), chloramphenicol (8–128 µg/mL), ciprofloxacin (0.06–8 µg/mL), colistin (2–16 µg/mL), gentamicin (0.5–32 µg/mL), meropenem (0.06–16 µg/mL), nalidixic acid (4–128 µg/mL), sulfamethoxazole (8–1024 µg/mL), tetracycline (2–64 µg/mL), tigecycline (0.25–8 µg/mL), and trimethoprim (0.25–32 µg/mL). *E. coli* ATCC 25922 was used for quality control purposes.

2.4. Molecular Analyses

E. coli isolates were assigned to the four major phylogenetic groups based on PCR assays detecting the genes *chuA*, *yjaA*, and the DNA fragment TSPEC4.C2 [16]. Clonal relatedness of the isolates was investigated by *Xba*I macorestriction analyses and subsequent pulsed-field gel-electrophoresis according to the PulseNet Protocol for *Salmonella*, *Shigella*, *E. coli* O157, and other Shiga toxin-producing *E. coli* [19]. The results were analyzed using BioNumerics V. 7.6 (Applied Maths, Sint-Martens-Latem, Belgium) applying the Dice coefficient with 0.5% optimization and 1% position tolerance.

All isolates were subjected to full genome sequencing (MicrobesNG, University of Birmingham, UK) for further typing, including multilocus sequence typing and serotype prediction. Sequencing was performed on an Illumina sequencing platform using 2 × 250 bp paired-end reads. The following pipelines were used for the bioinformatics analyses included in their standard sequencing service: Kraken, BWA mem, SPAdes for de novo assembly of reads, and Prokka for automated annotation (see Figure S1 for assembly statistics). Genome sequences were deposited in the GenBank database (https://www.ncbi.nlm.nih.gov/genbank/ (accessed on 14 September 2021)), and accession numbers for all isolates are listed in Figure 1. The presence of antimicrobial resistance- and virulence-associated genes was investigated using ResFinder 4.1 [20] and Virulence Finder 2.0 [21], respectively. Genome sequences were additionally uploaded to the Enterobase *Escherichia/Shigella* database (http://enterobase.warwick.ac.uk/ (accessed on 13 September 2021)) for multilocus sequence typing (MLST) and serotype prediction. Assembly of Illumina reads within the Enterobase database was performed using QAssembly, and subsequently the prokka pipeline was run for annotation. A minimum spanning tree was created using the GrapeTree program within Enterobase, using the cgMLST V1 + HierCC V1 scheme and the NinjaJ algorithm [22].

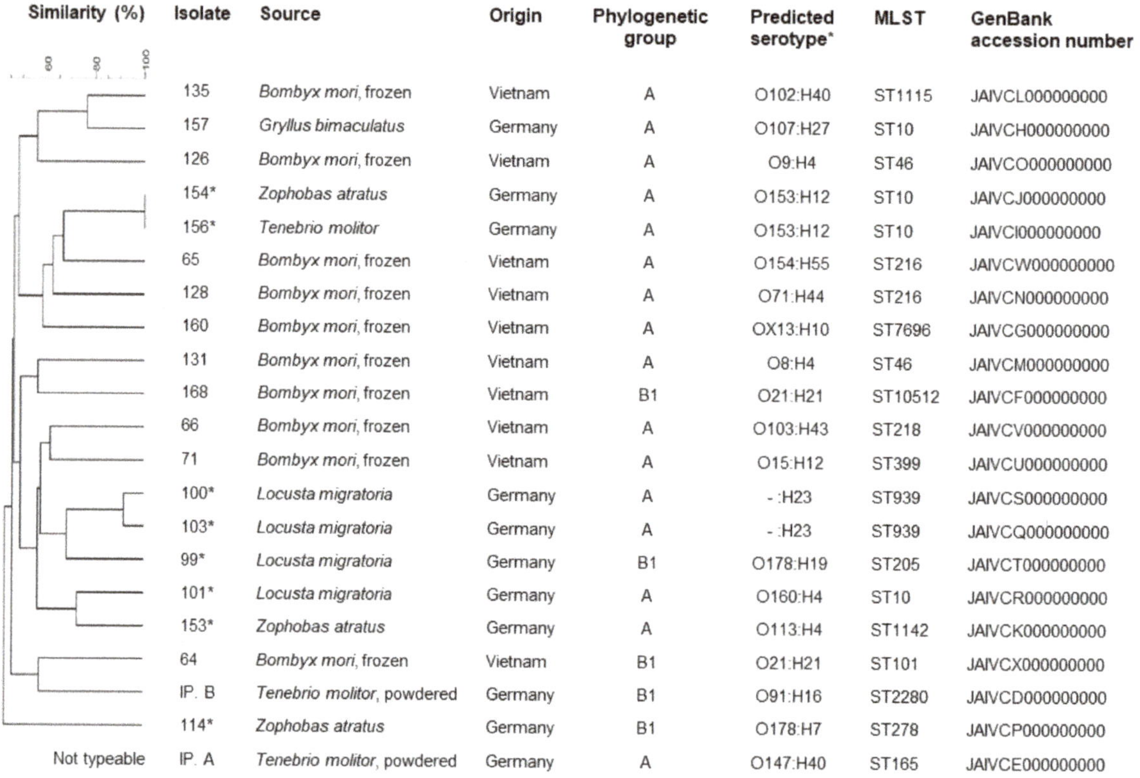

Similarity (%)	Isolate	Source	Origin	Phylogenetic group	Predicted serotype*	MLST	GenBank accession number
	135	*Bombyx mori*, frozen	Vietnam	A	O102:H40	ST1115	JAIVCL000000000
	157	*Gryllus bimaculatus*	Germany	A	O107:H27	ST10	JAIVCH000000000
	126	*Bombyx mori*, frozen	Vietnam	A	O9:H4	ST46	JAIVCO000000000
	154*	*Zophobas atratus*	Germany	A	O153:H12	ST10	JAIVCJ000000000
	156*	*Tenebrio molitor*	Germany	A	O153:H12	ST10	JAIVCI000000000
	65	*Bombyx mori*, frozen	Vietnam	A	O154:H55	ST216	JAIVCW000000000
	128	*Bombyx mori*, frozen	Vietnam	A	O71:H44	ST216	JAIVCN000000000
	160	*Bombyx mori*, frozen	Vietnam	A	OX13:H10	ST7696	JAIVCG000000000
	131	*Bombyx mori*, frozen	Vietnam	A	O8:H4	ST46	JAIVCM000000000
	168	*Bombyx mori*, frozen	Vietnam	B1	O21:H21	ST10512	JAIVCF000000000
	66	*Bombyx mori*, frozen	Vietnam	A	O103:H43	ST218	JAIVCV000000000
	71	*Bombyx mori*, frozen	Vietnam	A	O15:H12	ST399	JAIVCU000000000
	100*	*Locusta migratoria*	Germany	A	- :H23	ST939	JAIVCS000000000
	103*	*Locusta migratoria*	Germany	A	- :H23	ST939	JAIVCQ000000000
	99*	*Locusta migratoria*	Germany	B1	O178:H19	ST205	JAIVCT000000000
	101*	*Locusta migratoria*	Germany	A	O160:H4	ST10	JAIVCR000000000
	153*	*Zophobas atratus*	Germany	A	O113:H4	ST1142	JAIVCK000000000
	64	*Bombyx mori*, frozen	Vietnam	B1	O21:H21	ST101	JAIVCX000000000
	IP. B	*Tenebrio molitor*, powdered	Germany	B1	O91:H16	ST2280	JAIVCD000000000
	114*	*Zophobas atratus*	Germany	B1	O178:H7	ST278	JAIVCP000000000
Not typeable	IP. A	*Tenebrio molitor*, powdered	Germany	A	O147:H40	ST165	JAIVCE000000000

Figure 1. Origin and molecular typing results of the 21 *E. coli* isolates. Accession numbers refer to genome sequences deposited in the GenBank database. * Isolates recovered from samples purchased at pet feed store.

3. Results

A total of 21 *E. coli* isolates were recovered from the 36 samples. They were detected in five different insect species (Figure 1). No isolates were found in samples of Jamaican field crickets (*Gryllus assimilis*), house crickets (*Acheta domesticus*), or desert locusts (*Schistocerca gregaria*). The majority (n = 16) of the 21 isolates belonged to phylogenetic group A. The remaining isolates belonged to group B1 (Figure 1). PFGE analysis revealed identical band patterns of two isolates. These isolates were recovered from samples of superworms (*Zophobas atratus*) and mealworms (*Tenebrio molitor*), respectively, which were both purchased from a German pet feed store. In addition, a clustering of the four isolates recovered from migratory locusts (*Locusta migratoria*) was observed, with the two most similar isolates showing a similarity of band patterns of over 90% (Figure 1). These clusters were also apparent in the minimum spanning tree created based on cgMLST (Figure 2). In addition, isolates 126 and 131, both recovered from *B. mori* samples from Vietnam, showed a comparatively close relatedness, which was not apparent in their respective macrorestriction patterns. The remaining isolates showed very heterogenous PFGE band patterns, with less than 45% similarity between the most dissimilar isolates. One isolate (IP. A) did not produce bands after *Xba*I digestion and was thus classified as not typeable by this method.

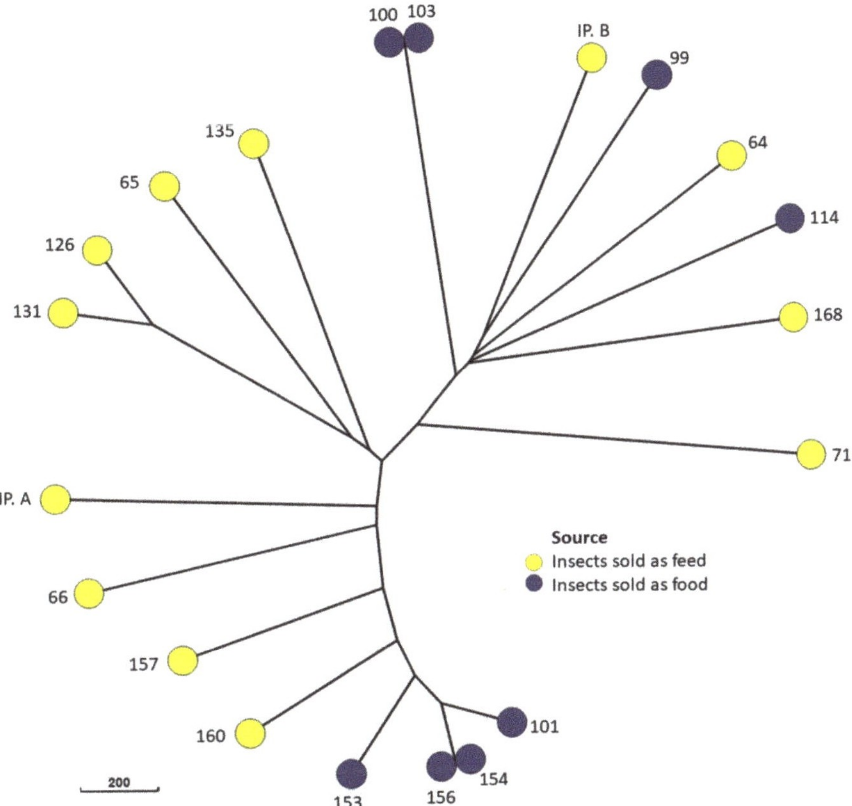

Figure 2. Minimum spanning tree of the 21 *E. coli* isolates based on core genome sequences.

The most commonly detected multilocus sequence type was ST10 (n = 4, Figure 1). In one isolate of phylogenetic group B1, a novel *fumC* allele was identified, and it was assigned to a novel sequence type, designated ST10512, in the Enterobase database. Serotype prediction based on genome sequences revealed a variety of serotypes. Three serotypes were identified in two isolates each. The two ST10 isolates sharing identical PFGE band patterns both belonged to O153:H12, and two ST939 isolates, which also shared a high similarity of PFGE band patterns, shared the same H-type and lack of detectable O-antigen (-:H12). Two other isolates from silkworms belonged to serotype O21:H21 but showed dissimilar band patterns (Figure 1). Other serotypes were only identified in single isolates. One O178:H7 isolate was classified as a STEC, carrying genes for Shiga toxin types 1 and 2 (*stx1c*, *stx2b*). Based on genome sequence analyses, the nucleotide sequence of *stx1* subunit A and B, as well as *stx2* subunit B, showed 100% identity to reference sequences deposited in the GenBank database under accession numbers Z36901 (*stx1c*) and AF043627 (*stx2b*) [23]. In contrast, there were seven nucleotide differences in the sequence of *stx2* subunit A compared to that of the reference strain (for the gene *stx2b*; GenBank acc. no. AF043627). Three of these resulted in amino acid exchanges in the deduced protein, which were identical to the variant of Stx2b subunit A deposited under Genbank accession number WP_001367518. The isolate lacked intimin encoding gene *eae* but carried several other virulence genes, including enterohaemolysin gene *ehxA*, subtilase cytotoxin encoding gene *subA*, and adherence-associated gene *iha*, among others (Table 1). The remaining isolates carried few virulence genes overall. Besides *gad* and *terC*, which were present in all isolates, the increased serum survival gene *iss* was present in several isolates, and two isolates carried *astA*, encoding a heat-stable enterotoxin (see Table 1 for full virulence profiles).

Table 1. Antimicrobial resistance phenotypes, antimicrobial resistance genes, and virulence-associated genes of 21 *E. coli* isolates.

Source	Isolate	Resistance Phenotype	Resistance Genotype	Virulence Genes
Insects sold as food	64	-	*mdf(A)*	*gad, lpfA, terC*
	65	-	*mdf(A)*	*gad, terC, traT*
	66	-	*mdf(A)*	*gad, kpsE, kpsMII, terC*
	71	-	*mdf(A)*	*gad, lpfA, terC*
	126	-	*qnrS2, mdf(A)*	*gad, celb, terC*
	128	-	*mdf(A)*	*gad, astA, terC*
	131	-	*mdf(A), sitABC*	*gad, sitA, terC, traT*
	135	-	*mdf(A)*	*gad, terC*
	157	-	*mdf(A)*	*gad, iss, terC*
	160	-	*mdf(A)*	*gad, astA, celb, fyuA, irp2, terC, traT*
	168	-	*mdf(A)*	*gad, lpfA, terC*
	IP. A	Tetracyclines	*tet(B), mdf(A)*	*gad, hra, terC*
	IP. B	-	*mdf(A)*	*gad, lpfA, terC*
Insects sold as feed	99	-	*mdf(A)*	*gad, hra, iss, lpfA, papC, terC*
	100	-	*mdf(A)*	*gad, hra, lpfA, terC*
	101	-	*mdf(A)*	*gad, iss, terC*
	103	-	*mdf(A)*	*gad, hra, lpfA, terC*
	114	-	*mdf(A)*	*gad, cba, celb, cia, cma, cvaC, ehxA, iha, ireA, iss, kpsE, lpfA, mchF, ompT, senB, stx1A, stx1B, stx2A, stx2B, subAB, terC, traT*
	153	-	*mdf(A)*	*gad, iss, ompT, terC*
	154	-	*mdf(A)*	*gad, ccl, iss, terC*
	156	-	*mdf(A)*	*gad, ccl, iss, terC*

Antimicrobial susceptibility testing revealed that only a single isolate showed resistance to tetracyclines. This isolate carried the *tet*(B) gene. The remaining isolates did not show resistance to any of the antimicrobial agents in our test panel. One isolate carried *qnrS2*, however, despite being in the susceptible range for both quinolones tested. The *mdf(A)* gene, encoding a multidrug efflux pump, was present in all isolates.

4. Discussion

The *E. coli* isolates included in this study were recovered from taxonomically diverse species and included holometabolous insects undergoing complete metamorphosis during development (Coleoptera, Lepidoptera), as well as hemimetabolous insects (Orthoptera). Previous research indicated that the microbiome of edible insects is usually dominated by Gram-positive bacteria, and in contrast to our current observations, *E. coli* were not typically detected [11,24]. Overall, *E. coli* does not appear to be a regular part of the microbiome of edible insects, and human pathogenic bacteria in general are likely obtained through feed, substrate, or during handling and processing [1]. While fecal bacteria in pet feed insects may be expected to some degree, the presence of *E. coli* in *B. mori* is more intriguing. They have been submitted to a heating process, and insect pupae should be sterile in the first place because larvae usually empty their intestinal tract content completely before pupation, and during it, the gut is dissolved and de novo synthesized. The larval gut turns into the so-called "yellow body" and is digested completely [25]. Thus, the presence of bacteria in silkworm pupa samples should be a result of secondary contamination during handling and processing. In fact, silkworm pupae from Vietnam were previously reported as positive for salmonellae via the European Rapid Alert System for Food and Feed [26].

4.1. Detection of STEC in Insects Sold as Feed

Notably, one isolate in our study was identified as a STEC. Shiga toxin-producing *E. coli* are most commonly transmitted via contaminated food, and infections can lead to diseases ranging from diarrhea to severe illness and death [27]. Insects living in farm environments have also been described to function as vectors for STEC, and houseflies were shown to harbor STEC for at least three days after experimental inoculation [28]. However, these insects (two muscid flies and one scarabeid beetle) feed on manure, something that does not happen in the case of the species tested here. *B. mori* exclusively consumes mulberry tree leaves, while locusts favor vegetables, and crickets and tenebrionid beetles are basically omnivorous. However, *Z. atratus* originates from Latin American caves inhabited by fruit-eating bats living in organic wastes and guano [29]. Yet, many beetle generations have passed since then. Still, to the best of our knowledge, no other reports about STEC recovered from edible insects have been published to date. Osimani and colleagues reported the absence of STEC among a variety of edible insect species marketed online [11].

The isolate in our study belonged to serotype O178:H7. O178 STEC have been associated primarily with food of bovine origin in Germany, as well as other countries [30]. Among them, O178:H19 STEC are most commonly reported, but O178:H7 have also been detected in meat and clinical samples from humans [30]. The prototype strain of O178:H7 also carried both *stx1* and *stx2* but lacked *eae* [31]. This strain was also reported to produce enterohaemolysin but, in contrast to our isolate, *ehxA* was not detected. More recently, Miko and colleagues described an O178:H7 variant associated with a *stx1c/stx2b/ehxA/subAB2/espI/[terE]/espP/iha* genotype isolated from deer meat and a diseased patient [30]. This is similar to the genotype of our isolate, sharing *stx1c/stx2b/ehxA/subAB/iha*. However, *espI*, *espP*, nor *terE* were detected in our study. It is also unclear if the other resistance determinants detected in our STEC isolate might be present in the isolates described by Miko et al., as these were not tested for.

In general, *stx2b* is a variant of *stx2* that is less frequently associated with human disease than the more potent variants *stx2a*, *stx2c*, and *stx2d* [32]. In addition, *eae* is present in the majority of isolates causing severe illness; however, it is not essential for pathogenicity, and *eae*-negative STEC can still cause severe disease, including hemolytic uremic syndrome [27,33]. Between 2012 and 2017, *stx1c/stx2b/eae*-negative STEC in particular has been associated with 234 human cases in the EU, including at least two cases of HUS, and at least 21 patients required hospitalization [27]. Therefore, a pathogenic potential of the detected STEC isolate cannot be excluded.

Our STEC isolate was obtained from *Z. atratus* sold in a German pet feed store. As these superworms were marketed as feed, they do not fall under the definition of "food" according to Regulation (EU) 178/2002 article 2 a. Besides, samples were raw. Considering the rising interest in edible insects and insect-based foods in the EU and the relative scarcity of currently available products on the market, live insects sold in pet feed stores can seem like an easy, versatile, and appealing alternative for consumers. From a legal point of view, although these products cannot be placed on the market as such, the habit of consuming pet feed insects by themselves seems not explicitly forbidden since Regulation (EU) 178/2002 does not apply to "the domestic preparation, handling or storage of food for private domestic consumption" (Art.1,3). However, the presence of STEC on the tested superworms shows that repurposing insects sold as pet feed to consumers is problematic. Following the recommendations issued for Germany, insects should be cooked for a minimum of 10 min [5]. Although this recommendation is not explicitly valid for pet feed insects, the method is effective to reduce the bacterial loads. Our findings also highlight the fact that appropriate hygiene should be observed by persons purchasing and handling live insects as feed.

4.2. Molecular Typing and Antimicrobial Resistance of E. coli from Edible Insect Species

Clustering among the isolates from the pet store according to their PFGE band patterns and cgMLST indicates a comparatively close genetic relationship between the respective isolates. This could be a result of closely related strains circulating in the production facility or within the store selling these animals. The STEC isolate, however, did not cluster with the other isolates.

Predicted serotypes and MLST results were diverse overall. Four isolates belonged to ST10, a very common ST that encompasses isolates from humans as well as animals and clinical as well as commensal strains [34,35]. Sequence types 46 and 101, to which three of our isolates from the Vietnamese silkworm samples belonged, have previously been reported to be very common in neighboring China [36].

In our study, only a single isolate from powdered mealworms was resistant to tetracyclines and carried the gene *tet*(B), which is a common mediator of tetracycline resistance in *E. coli* [37]. In addition, one isolate carried *qnrS2*. Plasmid-mediated quinolone resistance determinants such as *qnr* genes often do not confer clinical resistance to quinolones [38]. However, they can facilitate the development of chromosomal mutations in the quinolone resistance determining regions of the DNA gyrase and topoisomerase genes under treatment [38]. Previous studies have shown that insects such as flies in the farm environment can function as a reservoir and as vectors for the transmission of antimicrobial-resistant bacteria [39,40]. Regarding insects intended for consumption, the few published studies mostly focused on a screening of a certain subset of resistance genes and not on the characterization of individual isolates, including their resistance phenotypes. In three studies conducted in Italy, edible insects (*Tenebrio molitor*, *Locusta migratoria migratorioides*, and samples of a variety of species, respectively) were examined using a very similar method detecting a panel of antimicrobial resistance genes by PCR or nested PCR [41–43]. In contrast to our results, all three of these studies reported the presence of various resistance genes. However, the genes examined and detected in these studies are most commonly found in Gram-positive bacteria and are less frequently associated with Gram-negative species such as *E. coli* [37,44]. In a later study by Milanović and colleagues [45], the presence of five genes among the samples of grasshoppers and mealworms was investigated. The authors reported the presence of three different carbapenemase-encoding genes among samples from grasshoppers and mealworms, most notably *bla*$_{OXA-48}$ in over 50% of grasshopper samples. Regarding the clinical importance of carbapenems, these results are concerning. However, the results of these studies do not indicate the bacterial species of the isolates carrying the resistance genes. In contrast to their findings, the susceptibility testing of the *E. coli* isolates in our study showed that all were susceptible to all tested β-lactams.

Overall, it should be noted that several factors influence the microbiome composition and thus likely the resistome as well as the virulence of isolates present in edible insects. This includes the species of insect and the instar, the rearing practices and feed, as well as post-harvest processing, among others [1,46]. Thus, results obtained for samples from different manufacturers, and from different batches produced by the same manufacturer, can vary significantly.

5. Conclusions

The detection of a STEC shows that edible insect species, such as *Z. atratus*, can harbor foodborne zoonotic bacteria and highlights the need for hygienic rearing and processing practices to ensure the availability of safe insect-based foods for the interested consumer. On the other hand, the low resistance rate in our study indicates a favorable situation in edible insects in this regard and a low risk of a transmission of antimicrobial-resistant *E. coli*. Considering the high variability of the microbiome of insects, large-scale studies targeting more insect samples as well as different bacterial species are needed to further elucidate potential microbiological risks associated with these foods.

Supplementary Materials: The following are available online at https://www.mdpi.com/article/10.3390/foods10112552/s1, Figure S1: *Xba*I macrorestriction results including band patterns of individual isolates, Table S1: Assembly statistics and quality parameters of genome sequences.

Author Contributions: Conceptualization, A.M., N.T.G., B.A. and C.K.; methodology, A.M., D.S., N.T.G. and C.K.; validation, A.M., D.S., N.T.G., M.Y., C.K.; formal analysis, A.M. and D.S.; investigation, A.M., N.T.G. and B.A.; resources, C.K.; data curation, A.M.; writing—original draft preparation, A.M.; writing—review and editing, A.M., D.S., N.T.G., B.A., M.Y. and C.K.; visualization, A.M.; supervision, C.K.; project administration, A.M., C.K.; funding acquisition, N.T.G. All authors have read and agreed to the published version of the manuscript.

Funding: This work was in part funded by the German Federal Ministry of Education and Research within the research program for civil security, project ZooGloW (grant number 13N12697).

Institutional Review Board Statement: Not applicable.

Informed Consent Statement: Not applicable.

Data Availability Statement: The data presented in this study are available within the manuscript and the Supplementary Material (Figure S1 and Table S1).

Acknowledgments: We thank Claudia Walter, Cornelia Dürrschmidt, Inna Pahl, and Iris Oltrogge for expert technical assistance.

Conflicts of Interest: The authors declare no conflict of interest.

References

1. EFSA Scientific Committee. Risk profile related to production and consumption of insects as food and feed. *EFSA J.* **2015**, *13*, 4257. [CrossRef]
2. Zielińska, E.; Karaś, M.; Jakubczyk, A.; Zieliński, D.; Baraniak, B. Edible Insects as Source of Proteins. In *Bioactive Molecules in Food. Reference Series in Phytochemistry*; Mérillon, J.M., Ramawat, K., Eds.; Springer: Cham, Switzerland, 2018.
3. EFSA Panel on Nutrition, Novel Foods and Food Allergens (NDA); Turck, D.; Castenmiller, J.; De Henauw, S.; Hirsch-Ernst, K.I.; Kearney, J.; Maciuk, A.; Mangelsdorf, I.; McArdle, H.J.; Naska, A.; et al. Safety of dried yellow mealworm (*Tenebrio molitor* larva) as a novel food pursuant to Regulation (EU) 2015/2283. *EFSA J.* **2021**, *19*, e06343. [PubMed]
4. Grabowski, N.T.; Ahlfeld, B.; Lis, K.A.; Jansen, W.; Kehrenberg, C. The current legal status of edible insects in Europe. *Berl. Und Münchener Tierärztliche Wochenschr.* **2019**, *132*, 312–316.
5. *Arbeitspapier der AFFL Zum Inverkehrbringen von Insekten und Daraus Hergestellten Produkten als Lebensmittel [Working Paper of the AFFL regarding the Placing on the Market of Insects and Products thereof as Food]*, Internal Administrative Document; Länderarbeitsgemeinschaft Verbraucherschutz (AFFL): Wardenburg, Germany, 2019.
6. Stoops, J.; Crauwels, S.; Waud, M.; Claes, J.; Lievens, B.; Van Campenhout, L. Microbial community assessment of mealworm larvae (*Tenebrio molitor*) and grasshoppers (*Locusta migratoria migratorioides*) sold for human consumption. *Food Microbiol.* **2016**, *53*, 122–127. [CrossRef] [PubMed]
7. Van der Fels-Klerx, H.J.; Camenzuli, L.; Belluco, S.; Meijer, N.; Ricci, A. Food safety issues related to uses of insects for feeds and foods. *Compr. Rev. Food Sci. Food Saf.* **2018**, *17*, 1172–1183. [CrossRef] [PubMed]
8. Eilenberg, J.; Vlak, J.M.; Nielsen-LeRoux, C.; Cappellozza, S.; Jensen, A.B. Diseases in insects produced for food and feed. *J. Insects Food Feed.* **2015**, *1*, 87–102.
9. Grau, T.; Vilcinskas, A.; Joop, G. Sustainable farming of the mealworm *Tenebrio molitor* for the production of food and feed. *Z. Für Nat. C* **2017**, *72*, 337–349. [CrossRef]
10. Garofalo, C.; Milanovic, V.; Cardinali, F.; Aquilanti, L.; Clementi, F.; Osimani, A. Current knowledge on the microbiota of edible insects intended for human consumption: A state-of-the-art review. *Food Res. Int.* **2019**, *125*, 108527. [CrossRef]
11. Osimani, A.; Milanović, V.; Garofalo, C.; Cardinali, F.; Roncolini, A.; Sabbatini, R.; De Filippis, F.; Ercolini, D.; Gabucci, C.; Petruzzelli, A.; et al. Revealing the microbiota of marketed edible insects through PCR-DGGE, metagenomic sequencing and real-time PCR. *Int. J. Food Microbiol.* **2018**, *276*, 54–62. [CrossRef]
12. European Food Safety Authority (EFSA) and the European Centre for Disease Prevention and Control (ECDC). The European Union One Health 2018 Zoonoses Report. *EFSA J.* **2019**, *17*, e05926.
13. Alonso, M.Z.; Lucchesi, P.M.A.; Rodríguez, E.M.; Parma, A.E.; Padola, N.L. Enteropathogenic (EPEC) and Shigatoxigenic *Escherichia coli* (STEC) in broiler chickens and derived products at different retail stores. *Food Control* **2012**, *23*, 351–355.
14. European Food Safety Authority (EFSA) and the European Centre for Disease Prevention and Control (ECDC). The European Union summary report on antimicrobial resistance in zoonotic and indicator bacteria from humans, animals and food in 2016. *EFSA J.* **2018**, *16*, e05182.

15. Verraes, C.; Van Boxstael, S.; Van Meervenne, E.; Van Coillie, E.; Butaye, P.; Catry, B.; de Schaetzen, M.A.; Van Huffel, X.; Imberechts, H.; Dierick, K.; et al. Antimicrobial resistance in the food chain: A review. *Int. J. Environ. Res. Public Health* **2013**, *10*, 2643–2669. [CrossRef]

16. Doumith, M.; Day, M.J.; Hope, R.; Wain, J.; Woodford, N. Improved multiplex PCR strategy for rapid assignment of the four major Escherichia coli phylogenetic groups. *J. Clin. Microbiol.* **2012**, *50*, 3108–3110. [CrossRef] [PubMed]

17. Clinical and Laboratory Standards Institute (CLSI). *Methods for Dilution Antimicrobial Susceptibility Tests for Bacteria that Grow Aerobically: Approved Standard-Ninth Edition*; M07-A9; CLSI: Wayne, PA, USA, 2012.

18. Clinical and Laboratory Standards Institute (CLSI). *Performance Standards for Antimicrobial Susceptibility Testing*, 30th ed.; M100-ED30; CLSI: Wayne, PA, USA, 2020; Available online: http://em100.edaptivedocs.net/dashboard.aspx (accessed on 26 June 2020).

19. Ribot, E.M.; Fair, M.A.; Gautom, R.; Cameron, D.N.; Hunter, S.B.; Swaminathan, B.; Barrett, T.J. Standardization of pulsed-field gel electrophoresis protocols for the subtyping of *Escherichia coli* O157:H7, *Salmonella*, and *Shigella* for PulseNet. *Foodbourne Pathog. Dis.* **2006**, *3*, 59–67. [CrossRef] [PubMed]

20. Bortolaia, V.; Kaas, R.S.; Ruppe, E.; Roberts, M.C.; Schwarz, S.; Cattoir, V.; Philippon, A.; Allesoe, R.L.; Rebelo, A.R.; Florensa, A.F.; et al. ResFinder 4.0 for predictions of phenotypes from genotypes. *J. Antimicrob. Chemother.* **2020**, *75*, 3491–3500. [CrossRef]

21. Joensen, K.G.; Scheutz, F.; Lund, O.; Hasman, H.; Kaas, R.S.; Nielsen, E.M.; Aarestrup, F.M. Real-time whole-genome sequencing for routine typing, surveillance, and outbreak detection of verotoxigenic *Escherichia coli*. *J. Clin. Microbiol.* **2014**, *52*, 1501–1510. [CrossRef]

22. Zhou, Z.; Alikhan, N.F.; Sergeant, M.J.; Luhmann, N.; Vaz, C.; Francisco, A.P.; Carriço, J.A.; Achtman, M. GrapeTree: Visualization of core genomic relationships among 100,000 bacterial pathogens. *Genome Res.* **2018**, *28*, 1395–1404. [CrossRef] [PubMed]

23. Scheutz, F.; Teel, L.D.; Beutin, L.; Piérard, D.; Buvens, G.; Karch, H.; Mellmann, A.; Caprioli, A.; Tozzoli, R.; Morabito, S.; et al. Multicenter evaluation of a sequence-based protocol for subtyping Shiga toxins and standardizing Stx nomenclature. *J. Clin. Microbiol.* **2012**, *50*, 2951–2963. [CrossRef]

24. Grabowski, N.T.; Klein, G. Microbiology of processed edible insect products-Results of a preliminary survey. *Int. J. Food Microbiol.* **2017**, *243*, 103–107. [CrossRef] [PubMed]

25. Franzetti, E.; Huang, Z.J.; Shi, Y.X.; Xie, K.; Deng, X.J.; Li, J.P.; Li, Q.R.; Yang, W.Y.; Zeng, W.N.; Casartelli, M.; et al. Autophagy precedes apoptosis during the remodeling of silkworm larval midgut. *Apoptosis* **2012**, *17*, 305–324. [CrossRef] [PubMed]

26. European Rapid Alert System for Food and Feed (RASFF). 2013. Available online: https://webgate.ec.europa.eu/rasff-window/portal/?event=notificationDetail&NOTIF_REFERENCE=2013.1211 (accessed on 25 January 2021).

27. EFSA BIOHAZ Panel: Koutsoumanis, K.; Allende, A.; Alvarez-Ordóñez, A.; Bover-Cid, S.; Chemaly, M.; Davies, R.; De Cesare, A.; Herman, L.; Hilbert, F.; Lindqvist, R.; et al. Scientific Opinion on the pathogenicity assessment of Shiga toxin-producing *Escherichia coli* (STEC) and the public health risk posed by contamination of food with STEC. *EFSA J.* **2020**, *18*, e05967.

28. Persad, A.K.; LeJeune, J.T. Animal reservoirs of Shiga toxin-producing *Escherichia coli*. *Microbiol. Spectr.* **2014**, *2*, EHEC-0027-2014. [CrossRef]

29. Tschinkel, W.R. Larval dispersal and cannibalism in a natural population of *Zophobas atratus* (Coleoptera: Tenebrionidae). *Anim. Behav.* **1981**, *29*, 990–996. [CrossRef]

30. Miko, A.; Rivas, M.; Bentancor, A.; Delannoy, S.; Fach, P.; Beutin, L. Emerging types of Shiga toxin-producing *E. coli* (STEC) O178 present in cattle, deer, and humans from Argentina and Germany. *Front. Cell. Infect. Microbiol.* **2014**, *4*, 78. [CrossRef]

31. Scheutz, F.; Cheasty, T.; Woodward, D.; Smith, H.R. Designation of O174 and O175 to temporary O groups OX3 and OX7, and six new E. coli O groups that include Verocytotoxin-producing *E. coli* (VTEC): O176, O177, O178, O179, O180 and O181. *Apmis* **2004**, *112*, 569–584. [CrossRef] [PubMed]

32. Fuller, C.A.; Pellino, C.A.; Flagler, M.J.; Strasser, J.E.; Weiss, A.A. Shiga toxin subtypes display dramatic differences in potency. *Infect. Immun.* **2011**, *79*, 1329–1337. [CrossRef] [PubMed]

33. Newton, H.J.; Sloan, J.; Bulach, D.M.; Seemann, T.; Allison, C.C.; Tauschek, M.; Robins-Browne, R.M.; Paton, J.C.; Whittam, T.S.; Paton, A.W.; et al. Shiga toxin-producing *Escherichia coli* strains negative for locus of enterocyte effacement. *Emerg. Infect. Dis.* **2009**, *15*, 372–380. [CrossRef] [PubMed]

34. García-Meniño, I.; García, V.; Mora, A.; Díaz-Jiménez, D.; Flament-Simon, S.C.; Alonso, M.P.; Blanco, J.E.; Blanco, M.; Blanco, J. Swine enteric colibacillosis in Spain: Pathogenic potential of *mcr-1* ST10 and ST131 *E. coli* isolates. *Front. Microbiol.* **2018**, *9*, 2659. [CrossRef] [PubMed]

35. Yamaji, R.; Rubin, J.; Thys, E.; Friedman, C.R.; Riley, L.W. Persistent Pandemic Lineages of Uropathogenic Escherichia coli in a College Community from 1999 to 2017. *J. Clin. Microbiol.* **2018**, *56*, e01834-17. [PubMed]

36. Qiu, J.; Jiang, Z.; Ju, Z.; Zhao, X.; Yang, J.; Guo, H.; Sun, S. Molecular and Phenotypic characteristics of *Escherichia coli* isolates from farmed minks in Zhucheng, China. *BioMed Res. Int.* **2019**, *2019*, 3917841. [CrossRef]

37. Poirel, L.; Madec, J.Y.; Lupo, A.; Schink, A.K.; Kieffer, N.; Nordmann, P.; Schwarz, S. Antimicrobial resistance in *Escherichia coli*. *Microbiol. Spectr.* **2018**, *6*. [CrossRef] [PubMed]

38. Jacoby, G.A.; Strahilevitz, J.; Hooper, D.C. Plasmid-mediated quinolone resistance. *Microbiol. Spectr.* **2014**, *2*, 475–503. [CrossRef] [PubMed]

39. Fukuda, A.; Usui, M.; Okamura, M.; Dong-Liang, H.; Tamura, Y. Role of Flies in the maintenance of antimicrobial resistance in farm environments. *Microb. Drug Resist.* **2019**, *25*, 127–132. [CrossRef]

40. Zurek, L.; Ghosh, A. Insects represent a link between food animal farms and the urban environment for antibiotic resistance traits. *Appl. Environ. Microbiol.* **2014**, *80*, 3562–3567. [CrossRef] [PubMed]
41. Milanović, V.; Osimani, A.; Pasquini, M.; Aquilanti, L.; Garofalo, C.; Taccari, M.; Cardinali, F.; Riolo, P.; Clementi, F. Getting insight into the prevalence of antibiotic resistance genes in specimens of marketed edible insects. *Int. J. Food Microbiol.* **2016**, *227*, 22–28. [CrossRef]
42. Osimani, A.; Cardinali, F.; Aquilanti, L.; Garofalo, C.; Roncolini, A.; Milanović, V.; Pasquini, M.; Tavoletti, S.; Clementi, F. Occurrence of transferable antibiotic resistances in commercialized ready-to-eat mealworms (*Tenebrio molitor* L.). *Int. J. Food Microbiol.* **2017**, *263*, 38–46. [CrossRef] [PubMed]
43. Osimani, A.; Garofalo, C.; Aquilanti, L.; Milanović, V.; Cardinali, F.; Taccari, M.; Pasquini, M.; Tavoletti, S.; Clementi, F. Transferable antibiotic resistances in marketed edible grasshoppers (*Locusta migratoria migratorioides*). *J. Food Sci.* **2017**, *82*, 1184–1192. [CrossRef]
44. Chancey, S.T.; Zahner, D.; Stephens, D.S. Acquired inducible antimicrobial resistance in Gram-positive bacteria. *Future Microbiol.* **2012**, *7*, 959–978. [CrossRef]
45. Milanović, V.; Osimani, A.; Roncolini, A.; Garofalo, C.; Aquilanti, L.; Pasquini, M.; Tavoletti, S.; Vignaroli, C.; Canonico, L.; Ciani, M.; et al. Investigation of the dominant microbiota in ready-to-eat grasshoppers and mealworms and quantification of carbapenem resistance genes by qPCR. *Front. Microbiol.* **2018**, *9*, 3036. [CrossRef]
46. Ferri, M.; Di Federico, F.; Damato, S.; Proscia, F.; Grabowski, N.T. Insects as feed and human food and the public health risk-a review. *Berl. Münchener Tierärztliche Wochenschr.* **2019**, *132*, 191–218.

 *foods*

Article

Chemical Composition and Nutritional Value of Different Species of *Vespa* Hornets

Sampat Ghosh [1], Saeed Mahamadzade Namin [1,2], Victor Benno Meyer-Rochow [1,3] and Chuleui Jung [1,4,*]

1   Agriculture Science and Technology Research Institution, Andong National University, Andong 36729, Korea; sampatghosh.bee@gmail.com (S.G.); saeedmn2005@gmail.com (S.M.N.); meyrow@gmail.com (V.B.M.-R.)
2   Department of Plant Protection, College of Agriculture, Varamin-Pishva Branch, Islamic Azad University, Varamin 3381774895, Iran
3   Department of Ecology and Genetics, Oulu University, 90140 Oulu, Finland
4   Department of Plant Medicals, Andong National University, Andong 36729, Korea
*   Correspondence: cjung@andong.ac.kr; Tel.: +82-54-820-6229

Abstract: We genetically identified three different species of hornets and analyzed the nutrient compositions of their edible brood. Samples were collected from a commercial production unit in Shizong province of China and from forests near Andong City in Korea. The species were identified as *Vespa velutina*, *V. mandarinia*, and *V. basalis* from China and *V. velutina* from Korea. Farmed *V. velutina* and *V. mandarinia* were found to have similar protein contents, i.e., total amino acids, whereas *V. basalis* contained less protein. The *V. velutina* brood collected from the forest contained the highest amount of amino acids. Altogether 17 proteinogenic amino acids were detected and quantified with similar patterns of distribution in all three species: leucine followed by tyrosine and lysine being predominant among the essential and glutamic acid among the non-essential amino acids. A different pattern was found for fatty acids: The polyunsaturated fatty acid proportion was highest in *V. mandarinia* and *V. basalis*, but saturated fatty acids dominated in the case of *V. velutina* from two different sources. The high amounts of unsaturated fatty acids in the lipids of the hornets could be expected to exhibit nutritional benefits, including reducing cardiovascular disorders and inflammations. High minerals contents, especially micro minerals such as iron, zinc, and a high K/Na ratio in hornets could help mitigate mineral deficiencies among those of the population with inadequate nutrition.

Keywords: *Vespa velutina*; *Vespa mandarinia*; *Vespa basalis*; entomophagy; amino acids; fatty acids; minerals

Citation: Ghosh, S.; Namin, S.M.; Meyer-Rochow, V.B.; Jung, C. Chemical Composition and Nutritional Value of Different Species of *Vespa* Hornets. *Foods* **2021**, *10*, 418. https://doi.org/10.3390/foods10020418

Academic Editor: Anthony Fardet

Received: 27 January 2021
Accepted: 11 February 2021
Published: 14 February 2021

Publisher's Note: MDPI stays neutral with regard to jurisdictional claims in published maps and institutional affiliations.

1. Introduction

Numerous communities in the world traditionally include the broods of wasps and hornets (Hymenoptera) in their diets (Figure 1). Since ancient times it has been customary for Chinese people, especially those from Yunnan, to consume wasps, as is documented in a book from the Tang dynasty (618–907 C.E.) [1]. In Korea, *Vespa* hornet nests and larvae are harvested for medicinal and occasional edible use, especially *V. mandarinia* [2]. In Laos insects including wasps and hornets are consumed by 95% of the population, with the exception of two provinces for which a value of 85–90% has been reported [3]. The hornets consumed are of a different species, but *V. affinis* and *V. tropica* are particularly popular in Laos [4], whereas *V. cincta* (synonym to *V. tropica*, based on a report by [5]) is favored in Northern Thailand [6]. Wasp pupae as well as the adults belonging to the genus *Vespula*, locally known as *hebo*, are popular in the mountainous region of Honshu, Japan [7,8]. People commonly harvest the wasp nest from the forest, but in some cases the wasps are semi-domesticated [4,8] (Figure 1B).

Figure 1. (**A**) bottled hornet larvae sold in China; (**B**) *hebo* (*Vespula* sp.) contest in which judges evaluate wasp nests by weight in search for the largest nest of domestically raised wasps. [Photo credit: Soleil Ho; source: https://www.splendidtable.org/story/the-japanese-tradition-of-raising-and-eating-wasps, accessed on 31 August 2019]; (**C**) larvae of *Vespa velutina* collected from the forest near Andong (Korea); (**D**) pupae of *Vespa velutina* collected from the forest near Andong (Korea); (**E**) range expansion of the invasive hornet *Vespa velutina* in Korea [Source: 19]; (**F**) canned *hachinoko* (wasp brood) in Japan; (**G**) practice of eating wasps in Yunnan province [Source: http://teaurchin.blogspot.com/2011/10/weird-things-ive-eaten-in-yunnan.html, accessed on 31 August 2019].

The preparation of the wasp and hornet brood for consumption varies. The most common method of preparation includes deep-frying or frying with chicken eggs as in Yunnan, but the Dai people of southern Yunnan prefer to steam larvae and pupae together and mix them with vinegar and other seasonings [1]. Korean people also prefer frying and roasting [2], whereas Japanese people cook or preserve the wasps using soy sauce and mirin (sweet cooking rice wine) [8]. Although in Korea hornets are not considered a common food, they did, however, feature in the armamentarium of Korean traditional medicine. *Polistes* and *Vespa* larvae and adults are used as therapeutics for different ailments [9,10].

Vespa spp. have received attention not only as edible insects, but also as a pest affecting honeybee populations. In this context *V. velutina*, an invasive species, is regarded as the most obnoxious in Korea [11,12] and in Europe as well [13]. Although the Asian honeybee, *Apis cerana,* has evolved a "heat balling defense" and warning behavior when hornets are patrolling near their nests, the European honeybee, *A. mellifera*, does not possess this behavior and is therefore more susceptible to an attack by *V. velutina* [14–17]. A study of risk prediction in connection with the distribution of *V. velutina* [18,19] shows that the yearly dispersal of the species is 9.4 km northward in Korea and that this could have a serious impact on beekeeping as well as biodiversity and the ecosystem in general.

Nine native and one invasive hornet species occur in South Korea: *V. analis, V. binghami, V. mandarinia, V. simillima simillima, V. simillima xantothorax, V. crabro crobroformis, V. crabro flavofasciata, V. ducalis, V. dybowskii,* and *V. velutina nigrothorax* [20,21]. The lifecycle of the Vespa species includes different phases. Following overwintering, the emergence foundress, i.e., mated queen, searches for a suitable nest site and then constructs the primary nest for egg laying. About a month later, the eggs will have developed into female, non-reproductive adult workers and predation begins. During that time the workers build a secondary nest that is bigger in size than the primary nest. After that, large numbers of future queens are produced, which mate with drones that stem from unfertilized eggs. As the winter approaches, drones and workers die, while queens seek out overwintering places and the cycle continues the following spring [13].

Harvesting these wasps and hornets (*Vespa* spp.) can be a mode of biocontrol, but it can also lead to the use of these species as human food or animal feed. The present study was undertaken to assess the nutrient composition of three different species of *Vespa*. It was hoped that an investigation such as this would give us an opportunity to understand why there are differences in amino acid as well as fatty acid compositions between semi-domesticated hornets and specimens collected from the wild, e.g., the forest environment.

2. Materials and Methods

2.1. Sample Collection and Preparation

Three bottles of *Vespa* broods, representing three species, were obtained in frozen and dried form from a hornet-producing farm in Shizong County in China. The hornets were semi-domesticated primarily to be sold as food. All of the *Vespa* samples were packed into a freezing box and brought to the laboratory (Andong National University, Andong, South Korea). Samples were stored at $-20\ ^{\circ}$C until further processing. A few individuals ($n = 10$) from each bottle were taken for our DNA barcoding experiment in order to identify the species. Specimens used in the chemical analyses were not separated according to developmental stage; they included late instar larvae and pupae together, almost 50:50 (as this is the way they are sold and used by the consumer), but excluded adults.

A *V. velutina* nest was harvested in early morning from the forest behind the university (Andong City, Korea) in the month of July and brought to the laboratory. The broods were collected, similar to the Chinese commercial late instar larvae and pupae in a 50:50 proportion, separately from the nest and kept in a refrigerator ($-20\ ^{\circ}$C) until further processing ($n = 100$), which involved freeze drying and grinding up into a homogenous powder form.

2.2. Identification of the Species

The collected specimens i.e., broods of *Vespa*, were labelled VEUN20, VENU21, and VENU22 and identified based on DNA barcoding. The total DNA of each sample was extracted from the head and thorax using a DNeasy Blood & Tissue kit (QIAGEN, Inc., Dusseldorf, Germany) following the manufacturer's protocol. Two primers, LCO-1490 (5′-GGT CAA CAA ATC ATA AAG ATA TTG G-3′) and HCO-2198 (5′-TAA ACT TCA GGG TGA CCA AAA AAT CA-3′) targeting mitochondrial the *Cytochrome Oxidase I* (COI) gene [22] were used. The polymerase chain reaction (PCR) was conducted using AccuPower PCR PreMix (Bioneer, Daejeon, Korea) in order to amplify the COI gene corresponding to "DNA Barcode" region [23]. Sequencing was performed commercially by Macrogen (Seoul, South Korea). All three sequences were generated in both directions and assembled using Bioedit v7.0.5.2 [24] to annotate the species level identification using the BLAST (Basic Local Alignment Search Tool) database of the National Center for Biotechnology Information (NCBI) (http://www.ncbi.nlm.nih.gov, accessed on 31 August 2019). Based on the similarity (in %) the specimens were identified.

2.3. Nutritional Composition Analyses

2.3.1. Amino Acid Analysis

The amino acid composition was estimated using a Sykam Amino Acid analyzer S433 (Sykam GmbH, Eresing, Germany) following a standard method of AOAC (Association of Official Analytical Chemists) [25]. The *Vespa* samples in powder form were hydrolyzed in 6 N HCl for 24 h at 110 $^{\circ}$C under a nitrogen atmosphere followed by concentrating in a rotary evaporator. The concentrated samples were reconstituted with sample dilution buffer provided by the manufacturer (0.12N citrate buffer, pH 2.20). The hydrolyzed samples were analyzed for amino acid composition. The amino acid score was calculated considering the total estimated amino acid as protein, based on the WHO/FAO/UNU (World Health Organization/Food and Agriculture Organization/United Nations University) [26] report of a joint WHO/FAO/UNU Expert Consultation on protein and amino acid requirements in human nutrition following the formula [27]:

$$\text{Amino acid score} = \frac{(\text{mg of amino acid in 1g of test protein}) \times 100}{\text{mg of amino acid in reference pattern}}$$

2.3.2. Fatty Acid Composition Analysis

Fatty acid compositions of studied *Vespa* were determined and quantified using gas chromatography–flame ionization detection (GC-14B, Shimadzu, Tokyo, Japan) equipped with an SP-2560 column, following the recommended method of the Korean Food Standard Codex [28]. Briefly, the samples were derivatized into fatty acid methyl esters (FAMEs), which were then identified and quantified by comparing the retention time and peak areas of standards from Sigma (Yongin, Korea) and analyzed under the same conditions.

2.3.3. Mineral Analysis

Minerals of nutritional importance were analyzed following standard procedures of the Korean Food Standard Codex [28]. Dried *Vespa* powder samples were digested with nitric and hydrochloric acid (1:3) at 200 °C for 30 min in a high pressure microwave digestive system. The mineral contents, upon filtration with 0.45 micron filter paper, were analyzed using an inductively coupled plasma-optical emission spectrophotometer (ICP-OES 720 series; Agilent; Santa Clara, CA, USA). Recommended dietary allowance (RDA), population reference intake (PRI), and adequate intake (AI) values for the respective minerals were obtained from organizations such as the Linus Pauling Institute of Oregon State University and the European Food Safety Authority (EFSA).

2.3.4. Statistical Analysis

Composite sampling methods were followed including 100 samples for each group. In order to increase reliability the chemical analysis was carried out in at least duplicate and represented as mean ± standard deviation. To test the differences for individual nutrients of different *Vespa*, we carried out one-way ANOVA (analysis of variance) followed by a post hoc test (Tukey's Honestly Significant Difference (HSD)) using SPSS 16.0 (SPSS Inc., Chicago, IL, USA). If the *p*-value was found to be ≤0.05 (CI = 95%), the null hypothesis was rejected.

3. Results

3.1. Identification of the Species

With the help of the DNA barcoding method, based on the similarity between the sequences obtained in this study and sequences existing in the NCBI database, we identified the three species, VEUN20 as *V. mandarinia*, VENU21 as *V. basalis*, and VENU22 as *V. velutina*. The obtained sequences for the COI gene, corresponding to the "DNA Barcode" region, are available in GenBank under accession number MN477949- MN477951 (Appendix A).

3.2. Nutritional Composition of Vespa

3.2.1. Amino Acids

Amino acid compositions of the *Vespa* species studied are represented in Table 1. There were 17 amino acids in all *Vespa* samples. There was a significant difference in the total amino acid content of broods of different species of *Vespa* (df = 3,4, F = 19.135, *p* = 0.008). Almost all the amino acids were found to be higher in the *V. velutina* brood collected from the wild in Korea. However, the differences were not significant in all cases (Table 1).

Considering the totality of the amino acids as protein, the results show that *V. velutina* and *V. mandarinia* collected from the commercial production unit in Shizong province (China) had similar protein contents whereas *V. basalis* contained less protein. However, *V. velutina* broods collected from the forest near Andong (Korea) contained the highest amount of amino acids. Leucine was the predominating essential amino acid followed by lysine and valine. Tryptophan was not assessed and the amounts of cysteine and methionine were not measured in their entirety presumably because of the acid hydrolysis process [29]. Among the non-essential amino acids, glutamic acid was the most abundant one. The amino acid scoring pattern suggested that among all the estimated indispensable amino acids, methionine was found to be limiting; others satisfied (having a score of >100) the ideal protein pattern recommended by the WHO/FAO/UNU [26].

3.2.2. Fatty Acids

Table 2 represents the fatty acid compositions of the *Vespa* broods studied. Significant differences were found in the total fatty acid content of broods of different *Vespa* species (df = 3,4, F = 12.255, *p* = 0.017). Palmitic acid was the predominating saturated fatty acid; it was followed by stearic acid. Oleic acid was the most abundant monounsaturated fatty acid. However, there was apparently no consistency in the polyunsaturated group. Except for *V. velutina* from China, linoleic acid was always found in abundance among the polyunsaturated fatty acids, although the quantities varied widely (0.55 to 9.49 mg/100 g). Overall, the *V. mandarinia* and *V. basalis* broods were found to have the highest proportions of polyunsaturated fatty acids. However, no such difference was found between the saturated and monounsaturated fatty acid contents in these two species. By contrast, the *V. velutina* brood contained a higher amount of saturated fatty acids followed by monounsaturated and polyunsaturated fatty acids.

3.2.3. Mineral Content

Table 3 represents the comparative account of mineral contents of the species studied as well as the RDA values of respective elements. Significant differences in the mineral content, except manganese, of broods of different *Vespa* species were found (calcium: df = 3,4, F = 19.994, *p* = 0.007; magnesium: df = 3,4, F = 478.504, *p* = 0.000; sodium: df = 3,4, F = 161.653, *p* = 0.000; potassium: df = 3,4, F = 56.353, *p* = 0.001; phosphorus: df = 3,4, F = 42.778, *p* = 0.002; iron: df = 3,4, F= 38.232, *p* = 0.002; zinc: df = 3,4, F = 19.654, *p* = 0.007; manganese: df = 3,4, F = 4.086, *p* = 0.104; copper: df =3,4, F = 603.414, *p* = 0.000). The *V. velutina* brood collected from the wild in Korea was found to have a higher mineral content except for zinc and copper, however, the differences were not always significant. For magnesium, potassium, phosphorus, iron, and manganese, no significant differences were found between *V. velutina* broods from China and Korea. Potassium was the most abundant element. It was followed by phosphorus. Among the microminerals, iron and zinc were found to be dominating. The farmed *V. velutina* brood contained much less sodium than that detected in the wild-collected *V. velutina* brood. The potassium-to-sodium ratio (K/Na) in the wild *V. velutina* brood was as high as 11.7, but even higher ratios were noted in the farmed broods of *V. velutina*, *V. mandarinia*, and *V. basalis*, namely, 72.3, 13.7, and 45.5, respectively. It is noteworthy that the wild-collected *V. velutina* brood contained higher amounts of most of the minerals than the farmed hornets.

Table 1. Amino acid composition (g/100 g dry matter, mean ± SD, duplicate analysis with composite sampling process) of different *Vespa* species broods, requirement of indispensable amino acid and amino acid scoring pattern as per the WHO/FAO/UNU [26] and amino acid scoring pattern (%).

	China			Korea	Indispensable Amino Acid Requirements [1]		Amino Acid Scoring Pattern [1]			
							Amino Acid Score			
	Vespa velutina	*Vespa mandarinia*	*Vespa basalis*	*Vespa velutina*	mg/kg Per Day	mg/g Protein	*Vespa velutina*	*Vespa mandarinia*	*Vespa basalis*	*Vespa velutina*
Leucine *	3.3 ± 0.18 [b]	3.2 ± 0.33 [b]	2.4 ± 0.16 [c]	4.3 ± 0.32 [a]	39	59	145.8	149.3	145.0	143.7
Valine *	2.3 ± 0.15 [b]	2.3 ± 0.24 [b]	1.6 ± 0.11 [c]	3.2 ± 0.25 [a]	26	39	152.2	158.3	149.9	161.5
Isoleucine *	2.1 ± 0.17 [ab]	2.1 ± 0.19 [b]	1.5 ± 0.07 [c]	2.7 ± 0.28 [a]	20	30	188.2	185.8	177.1	174.9
Methionine *	0.6 ± 0.04 [a]	0.3 ± 0.33 [a]	0.3 ± 0.07 [a]	ND	10	16	92.3	57.8	73.5	ND
Cysteine	0.3 ± 0.28 [a]	0.7 ± 0.68 [a]	0.1 ± 0.01 [a]	2.0 ± 1.13 [a]	4	6	123.1	312.7	77.2	643.6
Lysine *	2.3 ± 0.03 [ab]	2.3 ± 0.22 [ab]	1.9 ± 0.23 [b]	2.7 ± 0.07 [a]	30	45	134.2	137.2	148.9	116.6
Threonine *	1.6 ± 0.05 [b]	1.6 ± 0.12 [b]	1.2 ± 0.13 [b]	2.3 ± 0.14 [a]	15	23	188.1	183.2	182.9	193.7
Histidine *	1.2 ± 0.09 [b]	1.2 ± 0.21 [b]	0.9 ± 0.06 [b]	1.7 ± 0.14 [a]	10	15	212.8	208.5	211.5	217.8
Phenylalanine *	1.6 ± 0.01 [ab]	1.6 ± 0.39 [ab]	1.2 ± 0.16 [b]	2.1 ± 0.07 [a]	25	38	286.7	311.2	296.5	283.0
Tyrosine **	2.5 ± 0.10 [ab]	2.7 ± 0.38 [ab]	2.0 ± 0.13 [b]	3.3 ± 0.46 [a]						
Arginine ***	1.7 ± 0.05 [ab]	0.8 ± 0.86 [b]	1.2 ± 0.13 [ab]	2.2 ± 0.18 [a]						
Aspartic acid	2.4 ± 0.06 [b]	2.4 ± 0.45 [b]	1.8 ± 0.28 [b]	3.7 ± 0.04 [a]						
Glutamic acid	7.6 ± 0.19 [a]	7.8 ± 1.6 [a]	6.2 ± 1.04 [a]	9.0 ± 0.96 [a]						
Serine	1.7 ± 0.12 [b]	1.6 ± 0.09 [bc]	1.2 ± 0.13 [c]	2.4 ± 0.28 [a]						
Proline	2.3 ± 0.23 [a]	2.1 ± 0.17 [a]	1.6 ± 0.05 [b]	2.4 ± 0.11 [a]						
Glycine	2.4 ± 0.60 [ab]	2.3 ± 0.34 [ab]	1.6 ± 0.13 [b]	3.4 ± 1.03 [a]						
Alanine	2.1 ± 0.69 [a]	2.0 ± 0.51 [a]	1.4 ± 0.16 [a]	3.4 ± 1.27 [a]						
Total	**37.9 ± 1.65** [b]	**36.8 ± 2.37** [b]	**28.1 ± 2.22** [c]	**50.5 ± 4.74** [a]						

* Essential amino acid. ** Conditional essential amino acid. *** Essential amino acid for children. [1] All the values were obtained for the adult population from the WHO/FAO/UNU 2007 report of a joint WHO/FAO/UNU Expert Consultation on protein and amino acid requirements in human nutrition. ND = Not detected. Superscripts indicate significant difference ($p \leq 0.05$).

Table 2. Fatty acid composition (g/100 g dry matter, mean ± SD, duplicate analysis with composite sampling process) of different *Vespa* species broods. Superscripts indicate a significant difference ($p \leq 0.05$) for selected predominating fatty acids.

	China			Korea
	Vespa velutina	*Vespa mandarinia*	*Vespa basalis*	*Vespa velutina*
Saturated Fatty Acids				
Capric acid	ND	ND	ND	<0.01
Lauric acid	0.2 ± 0.04	0.2 ± 0.04	0.1 ± 0.02	0.3 ± 0.02
Tridecanoic acid	ND	ND	ND	0.01 ± 0.00
Myristic acid	0.7 ± 0.14 [a]	0.5 ± 0.17 [ab]	0.3 ± 0.06 [b]	0.8 ± 0.11 [a]
Palmitic acid	3.7 ± 0.49 [a]	4.3 ± 0.24 [a]	3.5 ± 0.01 [a]	3.5 ± 0.28 [a]
Heptadecanoic acid	0.02 ± 0.00	0.02 ± 0.00	0.04 ± 0.00	0.02 ± 0.00
Stearic acid	0.9 ± 0.06	1.0 ± 0.17	1.2 ± 0.01	0.7 ± 0.00
Arachidic acid	0.1 ± 0.01	0.2 ± 0.03	0.2 ± 0.02	0.1 ± 0.00
Behenic acid	ND	0.1 ± 0.02	0.1 ± 0.01	0.1 ± 0.00
Lignoceric acid	ND	0.01 ± 0.01	0.03 ± 0.00	0.1 ± 0.01
Subtotal	**5.6 ± 0.71 [a]**	**6.2 ± 0.21 [a]**	**5.4 ± 0.05 [a]**	**5.4 ± 0.42 [a]**
Monounsaturated Fatty Acids				
Myristoleic acid	0.02 ± 0.01	ND	ND	0.03 ± 0.01
Palmitoleic acid	0.4 ± 0.10	0.2 ± 0.05	0.1 ± 0.00	0.4 ± 0.06
cis-10-Heptadecenoic acid	0.01 ± 0.02	ND	ND	0.01 ± 0.00
Oleic acid	4.1 ± 0.44 [b]	5.6 ± 0.55 [a]	5.3 ± 0.29 [a]	4.1 ± 0.28 [b]
cis-11-Eocosenic acid	ND	0.1 ± 0.06	0.14 ± 0.027	0.5 ± 0.01
Subtotal	**4.6 ± 0.56 [a]**	**5.9 ± 0.56 [a]**	**5.6 ± 0.31 [a]**	**5.1 ± 0.37 [a]**
Polyunsaturated Fatty Acids				
Linoleic acid	0.6 ± 0.11 [b]	6.8 ± 3.09 [a]	9.5 ± 1.96 [a]	0.6 ± 0.02 [b]
Linolenic acid	0.8 ± 0.11	1.2 ± 0.30	1.8 ± 0.01	ND
Arachidonic acid	0.04 ± 0.03	ND	ND	0.02 ± 0.00
cis-5,8,11,14,17-Eicosapentaenoic acid	0.03 ± 0.02	ND	ND	ND
Subtotal	**1.4 ± 0.05 [b]**	**8.1 ± 3.39 [a]**	**11.2 ± 1.98 [a]**	**0.6 ± 0.04 [b]**
Total	**11.5 ± 1.31 [b]**	**20.1 ± 3.75 [a]**	**22.2 ± 2.25 [a]**	**11.1 ± 0.82 [b]**

ND = Not detected. Superscripts indicate significant difference ($p \leq 0.05$).

Table 3. Mineral contents (mg/100 g dry matter, mean ± SD, duplicate analysis with composite sampling process) of different *Vespa* species broods and recommended dietary allowance (RDA) or population reference intakes (PRI) and satisfying the requirement in %

| | China | | | Korea | RDA [1] | | PRI/AI [2] | | Satisfying the Requirement as per PRI/AI[2] by 100 g of Consumption of Respective *Vespa* Brood (in %) [3] | | | | | | | |
| | | | | | | | | | *Vespa velutina* | | *Vespa mandarinia* | | *Vespa basalis* | | *Vespa velutina* | |
	Vespa velutina	*Vespa mandarinia*	*Vespa basalis*	*Vespa velutina*	M	F	M	F	M	F	M	F	M	F	M	F
Ca	38.8 ± 0.04 b	27.4 ± 0.20 c	31.8 ± 0.34 bc	46.3 ± 5.22 a	1000		950		4.1		2.9		3.3		4.9	
Mg	63.9 ± 0.01 a	33.0 ± 0.44 c	38.2 ± 0.24 b	66.3 ± 2.16 a	400	310	350 †	300 †	18.3	21.3	9.4	11.0	10.9	12.7	18.9	22.1
Na	10.4 ± 0.02 c	30.8 ± 0.40 b	8.9 ± 0.09 c	61.5 ± 5.44 a	1500 *	3800 *	–	–	0.7	0.3	2.1	0.8	0.6	0.2	4.1	1.6
K	751.6 ± 0.87 a	422.7 ± 6.58 b	404.4 ± 0.01 b	718.6 ± 69.87 a	3400 *	2600 *	3500 †		21.5		12.1		11.6		20.5	
P	561.2 ± 1.18 a	322.5 ± 2.93 b	318.4 ± 4.90 b	641.9 ± 71.37 a	700		550 †		102.0		58.6		57.9		116.7	
Fe	10.0 ± 0.12 a	7.2 ± 0.41 b	5.0 ± 0.18 c	9.1 ± 0.89 a	8	18	11	16	90.9	62.5	65.5	45.0	45.5	31.3	82.7	56.9
Zn	7.2 ± 0.02 a	4.7 ± 0.01 c	5.1 ± 0.04 c	6.1 ± 0.71 b	11	8	9.4	7.5	76.6	96.0	50.0	62.7	54.3	68.0	64.9	81.3
Mn	0.6 ± 0.02 ab	0.1 ± 0.01 b	1.2 ± 0.68 ab	2.8 ± 1.49 a	2.3 *	1.8 *	3 †		20.0		3.3		40.0		93.3	
Cu	2.2 ± 0.04 a	0.9 ± 0.01 d	1.1 ± 0.04 c	1.3 ± 0.04 b	900		1.6 †	1.3 †	137.5	169.2	56.3	69.2	68.8	84.6	81.3	100.0

[1] Recommended dietary allowance (RDA) values were obtained from the Micronutrient Information Center, Linus Pauling Institute, Oregon State University [www.lpi.oregonstate.edu/mic/minerals, accessed 9 February 2021]. All the values are for adult (>19–50) populations and provided in mg/day except for copper, which is in µg/day. M = male; F = female. * indicates the adequate intake (AI) values. [2] Population reference intakes (PRIs) and adequate intakes (AIs) for minerals were obtained from the European Food Safety Authority (EFSA) [www.efsa.europa.eu/sites/deafult/files/assets/DRV_Summary_tables_jan_17.pdf, accessed 9 February 2021]. All values are for adult (>25 years) populations and provided in mg/day. † indicates the adequate intake (AI) values. [3] Sodium (Na) was calculated based on the value provided by the Linus Pauling Institute, Oregon State University; all others were based on values provided by the European Food Safety Authority (EFSA). Superscripts indicate significant difference ($p \leq 0.05$).

4. Discussion

4.1. Amino Acid Composition

The amino acid distributions seen in our study are in general agreement with a previously published report, although the contents of some of the amino acids in the current study were a little less than what was reported earlier [1]. The total amino acid content as protein content of *Vespa* broods is comparable with other reports on edible insects, including wasps (*V. velutina nigrithorax*: 48.64 [30], *V. mandarinia*: 59.7 [2], *V. basalis*: 43.91, *V. mandarinia*: 52.20, *V. velutina auraria*: 49.03, *V. tropica ducalis*: 42.44 [1]). In the earlier study, as with ours, overall glutamic acid was the most abundant amino acid. Glutamic acid is the precursor of GABA (gamma aminobutyric acid), which is a neurotransmitter of inhibitory neurons [31] and thus might be responsible for docile behavior. However, the saliva of *Vespa* spp. larvae is not rich in glutamic acid [32], which could be a reason for the wasps' aggressiveness in defense and times of hunting prey. This aggressive behavior lessens in the nest where trophallaxis is performed, an essential behavioral trait for a social insect. The high proline content, amongst other effects, influences the flight of the insect, because it is metabolized to produce energy for the wing movements during flight [33]. It has been reported that hornet, with higher proline content in their saliva, generally build their nests higher up and also have a wider hunting range than hornets with less proline content in saliva and those live underground or nest in caves or tree holes and hunt over a smaller area [32].

Although *V. mandarinia* and *V. velutina* inhabit subterranean and open-air nests, respectively [34,35], they did not differ with respect to proline content, at least in the brood stage. Nonetheless, differences in body composition can be expected to exist between larvae, pupae, and adults as well as drones, workers, and queens with regard to developmental as well as physiological states, as has been demonstrated for the honeybee (*A. mellifera*) and the bumblebee (*Bombus terrestris*) [36–39]. Histidine, decarboxylated to histamine, is a major component of the venom and found in similar amounts in *V. velutina* and *V. mandarinia*, but less in the case of *V. basalis*. Among the indispensable amino acids, leucine was found to predominate and to be present in higher amounts in *V. velutina* and *V. mandarinia* than in *V. basalis*. Leucine and isoleucine are metabolized in the musculature.

From a nutritional standpoint, lysine deserves consideration as it is a limiting amino acid in cereals such as rice, wheat, and maize. Catabolism of lysine, an entirely ketogenic amino acid, includes the saccharopine pathway, which results in the formation of glutamate and α-aminoadipate. In addition, lysine is also a precursor for the biosynthesis of carnitine, which plays an important role in β-oxidation [40]. The aromatic amino acid tyrosine functions as precursor of catecholamines such as dopamine, norepinephrine, and epinephrine and is involved in melanogenesis. Tyrosine is a conditionally essential amino acid with phenylalanine playing a crucial role in tyrosine synthesis. Therefore, because of the presence of almost all proteinogenic amino acids and estimated indispensable ones except for methionine, satisfying the recommended protein pattern (Table 1), *Vespa* broods would supplement the nutritional requirements in people.

4.2. Fatty Acid Composition

The higher polyunsaturated acid contents of *V. mandarinia* and *V. basalis* are likely due to their diet, which consists of crickets and grasshoppers, which are often rich in polyunsaturated fatty acids (cf., grasshopper, *Chondacris rosea*: [41], and crickets, *Gryllus* sp. and *Teleogryllus* sp.: [42]). Compared to earlier analyses, the proportions of monounsaturated fatty acids were higher in the case of other hymenopteran species such as, to mention but a few, *B. ignitus* [37], *B. terrestris* [38], *Carebara vidua* [43], *Polyrhachis vicina* [44,45], and *Oecophylla smaragdina* [46]. However, diet manipulations can result in changes in the fatty acid composition in farmed versus wild insects [47,48].

Oleic acid was the dominant monounsaturated and palmitic acid, with stearic and myristic acids being the abundant saturated fatty acids. Linoleic acid was the most abundant polyunsaturated fatty acid followed by linolenic acid in *V. mandarinia* and *V. basalis*.

Primarily saturated fatty acids such as myristic, palmitic, and lauric acids increased the level of low density lipoprotein, the so-called bad cholesterol. However, stearic acid does not raise serum cholesterol [49]. Oleic acid inhibits the store-operated Ca^{+2} entry process (SOCE) that controls the Ca^{+2} influx pathway and is involved in several cellular and physiological processes, including cell proliferations that are often diagnostic for colorectal cancer [50]. Oleic acid also shows significant in vitro inhibition of prolyl-endopeptidase (PEP), an enzyme playing a crucial role in the formation of amyloid in the brain and implicated in disorders such as dementia and Alzheimer's disease [50].

Unsaturated fatty acids are being given special attention as they are seen to be beneficial to human health. Earlier studies, e.g., by Grundy [51], demonstrated the capacity of monounsaturated fatty acids to lower lipoprotein density and thus total cholesterol. Polyunsaturated fatty acids play a crucial role in the biosynthesis of cellular hormones such as eicosanoids and other signaling compounds that modulate human health [50]. Polyunsaturated fatty acids are of two types, i.e., $n-6$ and $n-3$, based on the position of the first unsaturated site in the chain. A higher ratio of $n-3$ to $n-6$, i.e., more $n-3$ and less $n-6$ in the diet, is preferable in connection with human health as it helps to reduce weight through the removal of intra-abdominal fat, a drop in adipocyte cell size, and the normalization of the heartbeat. Diets with high $n-3$ polyunsaturated fatty acids enhance the body's ability to reduce damaging inflammatory conditions, as they are usually converted into anti-inflammatory eicosanoids. Besides, epidemiological as well as clinical studies further suggest that $n-3$ polyunsaturated fatty acids, including $n-3$ from marine food resources, help lower cardiovascular mortality [52,53] through mechanisms that include the modulation of cellular metabolic functions and gene expression and exert beneficial effects on lipid profiles and blood pressure [54]. Thus, the presence of linolenic acid in *V. mandarinia* and *V. basalis* could have nutritional benefits for sufferers of cardiovascular complexities.

4.3. Mineral Content

Minerals are essential micronutrients that play critical roles in human health. Among the minerals of nutritional importance, potassium was found in abundance followed by phosphorus. The values were within the range of reports on Hymenoptera as well as edible insects belonging to other orders [42,55]. A substantial amount of evidence shows that potassium intake lowers blood pressure. An increased intake of potassium also plays a critical role in the management of hypercalciuria and is likely to decrease the risk of osteoporosis [56]. Low serum potassium is strongly related to glucose intolerance and increases the risk of lethal ventricular arrhythmias in patients suffering from ischemic heart disease. On the other hand, a high intake of dietary sodium is associated with a prevalence of hypertension [57].

From the human nutritional point of view, a high potassium-to-sodium ratio can be regarded as beneficial as it reduces cardiovascular risk and improves blood pressure [58]. All the hornet broods contained higher potassium and comparatively less sodium, and thus can be regarded as beneficial for human health. The calcium content of the *Vespa* brood was found to be within the range of other insects [55], but less than what had been reported for *A. mellifera* larvae and pupae [36]. Over 99% of total body calcium by virtue of its phosphate salt is present in the teeth and bones of mammals and the remainder is present in the blood, extracellular fluid, muscle, and other tissues. Calcium plays critical physiological roles, including mediating vascular contraction and vasodilation, muscle contraction, nerve transmission, and glandular secretion, to mention a few [59]. Chronic calcium deficiency often results in a reduction in bone mass and osteoporosis [60], the development of hypertension, and even colon cancer [61].

Among the minerals of nutritional importance, iron receives the most attention. Iron deficiency and anemia still exist and are major public health concerns worldwide, especially in many developing and low-income countries [62,63]. The most vulnerable sections of the population in developing countries regarding iron deficiency are women of childbearing age and children under five [64,65]. Although the iron content of the studied species was a little less than that was reported for *A. mellifera* workers [36], it could still supplement the iron requirement, especially for those in the population who cannot afford iron-rich food such as red meat, liver, fish, etc. The element zinc plays a role in several cellular processes, including catalytic, structural, and regulatory roles in many enzymes, gene transcription, signal transduction pathways, etc. [66]. Although severe zinc deficiency is considered rare [67], mild to moderate zinc deficiencies persist worldwide [68]. An inadequate dietary zinc intake is particularly common in Sub-Saharan Africa and South Asia [69]. Based on the RDA values provided by agencies, consumption of 100 g of *Vespa* brood could satisfy a significant proportion of daily dietary requirements for minerals, especially iron, zinc, and copper (Table 3). Even if it cannot meet the total mineral need, assuming good bioavailability, the consumption of *Vespa* brood could at least supplement the nutritional requirements of minerals and could help mitigate the problems of mineral deficiencies in those sections of the population most at risk for them.

5. Conclusions

Since it was first suggested in 1975 that insects could help ease the problem of global food shortages [70], the last two decades in particular have seen a remarkable increase in attention to edible insects as a nutrient-rich and healthy food resource. Numerous countries have formulated or are in the process of formulating legislation to regulate mass production and trade of edible insects. The global market value of edible insects is expected to exceed USD 522 million by 2023 [71]. Since 2012, the South Korean edible-insect market focusing on human consumption alone (not including insects as animal feed) has experienced major advances, with governmental support and successful research endeavors [42,72]. The country's tradition of using certain insects such as processed silkworm pupae, commonly known as *beondaegi*, as food [73] has further helped. Noting the competent nutrient composition of all three *Vespa* species examined by us and the feasibility of rearing these insects, we propose that these hornet broods can be a sustainable, high-quality nutritional source. However, the rearing process is yet to be established before any large-scale controlled production can commence.

Author Contributions: Conceptualization, C.J. and S.G.; methodology, S.G. and S.M.N.; software, S.G.; validation, S.G. and S.M.N.; formal analysis, S.G., S.M.N., and C.J.; investigation, S.G., C.J., and S.M.N.; data curation, C.J.; writing—original draft preparation, S.G.; writing—review and editing, C.J., V.B.M.-R., S.G., and S.M.N.; visualization, C.J.; supervision, C.J.; project administration, C.J.; funding acquisition, C.J. All authors have read and agreed to the published version of the manuscript.

Funding: This work was funded by the BSRP through the National Research Foundation of Korea (NRF), Ministry of Education (NRF-2018R1A6A1A03024862).

Institutional Review Board Statement: Not applicable.

Informed Consent Statement: Not applicable.

Data Availability Statement: Not applicable.

Acknowledgments: We appreciate Zhao Quanghui, owner of the *Vespa* farm in Shizong County and Professor Ken Tan, Chinese Academy of Science, Yunnan province, China, for the valuable assistance on this work. This study was partly supported by the BSRP through the National Research Foundation of Korea (NRF), Ministry of Education (NRF-2018R1A6A1A03024862).

Conflicts of Interest: The authors declare no conflict of interest.

Appendix A

Table A1. DNA sequences for the *COI* gene corresponding to the "DNA Barcode" region of VEUN20 as *Vespa mandarinia*, VENU21 as *Vespa basalis*, and VENU22 as *Vespa velutina*.

ID	Accession No.	Sequence
VEUN20	MN477949	TATATTATATTTTATTTTTGCCTTATGATCAGGAACTATCGGAGCATCCATAAGATTAA TTATTCGAATAGAGCTTAGATCTCCTGGCAATCTAATTAATAATGACCAAATTTACAA TTCTATTATTACTGCTCACGCATTTATTATAATTTTTTTTATAGTTATACCCTTTATAATT GGAGGGTTTGGAAATTGATTAATTCCATTAATACTAGGTATTCCAGATATGGCATTTC CTCGAATAAATAATATAAGATTTTGACTTCTACCTCCTTCATTATTCCTTCTAATTATAA GAAACTTTATTGGAGGGGGTGTTGGTACAGGATGAACCCTTTATCCCCCCCTATCATC CATTATTGGCCATAATTCTCCTTCAGTAGATCTAAGAATTTTCTCTCTCCATATTGCAGG AATTTCTTCAATTATAGGAGCAATTAATTTTATTGTAACAATTCTAAATATACATGTCAAA ACCCATTCATTAAATTTTTTACCATTATTCTCTTGGTCTGTCCTAATTACAGCATTCTTATT ACTTTTATCTTTACCTGTTTTAGCTGGCGCAATCACCATACTTTTAACAGATCGAAATTTT AATACATCCTTTTTCGATCCAACTGGAGGCGGAGACCCCATCTTATACCAACATTTATTC
VEUN21	MN477950	AATACTTTATTTTATTTTTGCTTTATGATCAGGATCTTTAGGAGCCTCTATAAGTTTAA TTATTCGTATAGAACTTAGATCCCCAGGAAGATTAATTAACAACGATCAAATCTATAAT TCTATTATTACCGCTCATGCTTTTATTATAATTTTTTTTTATAGTTATACCTTTTATAATTG GAGGATTTGGAAATTGATTAATTCCTTTAATATTAGGAATTCCAGATATAGCTTTTCCT CGAATAAATAATATAAGATTCTGACTATTACCTCCCTCTTTATTTCTATTAATTATAAGAA ATTTTATTGGAGGAGGAGTAGGAACTGGATGAACTCTTTACCCACCTCTATCATCAATT ACTGGTCATAATTCTCCAGCTGTTGATCTTAGAATCTTTTCATTACATATTGCAGGAATT TCATCAATTATAGGAGCCATTAATTTCATTGTTACAATTTTAAACATACACATTAAAACT CACTCACTAAGATTCTTACCTTTATTTTCATGATCAGTTTTAATTACAGCATTTTTACTAT TATTATCCTTACCTGTTCTAGCAGGAGCAATTACAATACTTCTTACCGATCGAAACTTTA ATACATCATTTTTTGATCCCACAGGAGGAGGAGACCCTATTTTATATCAACATTTATTT
VEUN22	MN477951	AATATTATACTTTATTTTTGCATTATGATCTGGAACATTGGGAGCATCAATAAGATTAA TTATTCGTATAGAATTAAGATCTCCCGGAAATTTAATTAATAATGATCAAATTTATAATT CAATTATCACTGCTCATGCTTTTATTATAATTTTTTTTATAGTTATACCTTTTATAATCGG AGGATTCGGTAACTGAATAATTCCTCTAATACTCGGAATTCCTGATATAGCTTTCCCTC GAATAAATAATATAAGATTCTGACTACTCCCTCCATCATTATTTATATTAATTATAAGAA ACTTTATTGGTGGAGGTGTAGGAACAGGATGAACTTTATATCCTCCTTTATCATCAATT ACTGGTCATAACTCACCATCAGTTGATTTAAGAATTTTCTCTTTACATATTGCAGGAAT TTCATCAATTATAGGTGCAATTAATTTTATTGTAACAATTCTGAATATACATGTAAAAAC ACACTCATTAAATTTTTTACCATTATTCTCATGATCAGTCTTAATTACTGCTTTTTTACTT TTATTATCACTCCCTGTATTAGCAGGAGCTATTACTATACTTTTAACAGATCGAAATTTTA ATACATCATTCTTCGATCCAACCGGAGGAGGAGACCCAATTCTATATCAACACTTATTT

References

1. Ying, F.; Xiaoming, C.; Long, S.; Zhiyong, C. Common edible wasps in Yunnan Province, China and their nutritional value. In *Forest Insects as Food: Human Bites Back*; Durst, P.B., Johnson, D.V., Leslie, R.N., Shono, K., Eds.; Food and Agriculture Organi-zation of the United Nations; Regional Office for Asia and the Pacific: Bangkok, Thailand, 2010; pp. 93–98.
2. Kim, H.S.; Jung, C. Nutritional characteristics of edible insects as potential food materials. *Korean J. Apic.* **2013**, *28*, 1–8.
3. Barennes, H.; Phimmasane, M.; Rajaonarivo, C. Insect Consumption to Address Undernutrition, a National Survey on the Prevalence of Insect Consumption among Adults and Vendors in Laos. *PLoS ONE* **2015**, *10*, e0136458. [CrossRef]
4. Hanboonsong, Y.; Durst, P.B. *Edible Insects in Lao PDR: Building on Tradition to Enhance Food Security*; Food and Agriculture Organization of the United Nations—Regional Office for Asia and the Pacific: Bangkok, Thailand, 2014.
5. Carpenter, J.M.; Kojima, J. Checklist of the species in the subfamily Vespinae (Insects: Hymenoptera: Vespinae). *Nat. Hist. Bull. Ibaraki Univ.* **1997**, *1*, 51–92.
6. Leksawasdi, P. Compendium of research on selected edible insects in northern Thailand. In *Forest Insects as Food: Human Bites Back*; Durst, P.B., Johnson, D.V., Leslie, R.N., Shono, K., Eds.; Food and Agriculture Organization of the United Nations—Regional Office for Asia and the Pacific: Bangkok, Thailand, 2010; pp. 183–188.
7. Nonaka, K. Feasting on insects. *Èntomol. Res.* **2009**, *39*, 304–312. [CrossRef]
8. Payne, C.L.R.; Evans, J.D. Nested Houses: Domestication dynamics of human-wasp relations in contemporary rural Japan. *J. Ethnobiol. Ethnomedicine* **2017**, *13*, 13. [CrossRef]
9. Pemberton, R.W. Insects and other arthropods used as drugs in Korean traditional medicine. *J. Ethnopharmacol.* **1999**, *65*, 207–216. [CrossRef]

10. Meyer-Rochow, V.B. Therapeutic arthropods and other largely terrestrial folk-medicinally important invertebrates: A comparative survey and review. *J. Ethnobiol. Ethnomedicine* **2017**, *13*, 9. [CrossRef] [PubMed]
11. Jung, C.; Kim, D.W.; Lee, H.S.; Baek, H. Some Biological Characteristics of a New Honeybee Pest, *Vespa velutina nigrithorax* Buysson, 1905 (Hymenoptera: Vespidae). *Korean J. Apic.* **2009**, *24*, 61–65.
12. Jung, C. Initial state risk assessment of an invasive hornet, *Vespa velutina nigrithorax* Buysson (Hymenoptera: Vespidae) in Korea. *Korean J. Apic.* **2012**, *27*, 95–104.
13. Monceau, K.; Bonnard, O.; Thiéry, D. *Vespa velutina*: A new invasive predator of honeybees in Europe. *J. Pest Sci.* **2014**, *87*, 1–16. [CrossRef]
14. Shah, F.A.; Shah, T.A. *Vespa Velutina*, a Serious Pest of Honey Bees in Kashmir. *Bee World* **1991**, *72*, 161–164. [CrossRef]
15. Abrol, D.P. Ecology, behaviour and management of social wasp, *Vespa velutina* Smith (Hymenoptera: Vespidae), attacking honeybee colonies. *Korean J. Apic.* **1994**, *9*, 5–10.
16. Tan, K.; Radloff, S.E.; Li, J.J.; Hepburn, H.R.; Yang, M.X.; Zhang, L.J.; Neumann, P. Bee-hawking by the wasp, *Vespa velutina*, on the honeybees *Apis cerana* and *A. mellifera*. *Naturwissenschaften* **2007**, *94*, 469–472. [CrossRef] [PubMed]
17. Ono, M.; Okada, I.; Sasaki, M. Heat production by balling in the Japanese honeybee, *Apis cerana* japonica as a defensive behavior against the hornet *Vespa simillima xanthoptera* (Hymenoptera: Vespidae). *Cell. Mol. Life Sci.* **1987**, *43*, 1031–1034. [CrossRef]
18. Jung, C. Spatial expansion of an invasive hornet, *Vespa velutina nigrithorax* Buysson (Hymenoptera: Vespidae) in Korea. *Korean J. Apic.* **2012**, *27*, 87–93.
19. Park, J.-J.; Jung, C. Risk Prediction of the Distribution of Invasive Hornet, *Vespa velutina nigrothorax* in Korea using CLIMEX Model. *J. Apic.* **2016**, *31*, 293–303. [CrossRef]
20. Jung, C.; Kang, M.S.; Kim, D.W.; Lee, H.S. Vespid Wasps (Hymenoptera) Occurring Around Apiaries in Andong, Korea-I. Taxonomy and life history. *Korean J. Apic.* **2007**, *22*, 53–62.
21. Choi, M.B.; Kim, J.K.; Lee, J.W. Checklist and Distribution of Korean Vespidae Revisited. *Korean J. Appl. Èntomol.* **2013**, *52*, 85–91. [CrossRef]
22. Folmer, O.; Black, M.; Hoeh, W.; Lutz, R.; Vrijenhoek, R. DNA primers for amplification of mitochondrial cytochrome c oxidase subunit I from diverse metazoan invertebrates. *Mol. Mar. Biol. Biotechnol.* **1994**, *3*, 294–299.
23. Hebert, P.D.; Ratnasingham, S.; de Waard, J.R. Barcoding animal life: Cytochrome c oxidase subunit 1 divergences among closely related species. *Proc. R. Soc. B Biol. Sci.* **2003**, *270*, S96–S99. [CrossRef]
24. Hall, T.A. BioEdit: A user-friendly biological sequence alignment editor and analysis program for Windows 95/98/NT. *Nucleic Acids Symp.* **1999**, *41*, 95–98.
25. AOAC. *Official Methods of Analysis of the AOAC*, 15th ed.; Association of official analytical chemists: Washington, DC, USA, 1990.
26. FAO/WHO/UNU. *Protein and Amino Acid Requirements in Human Nutrition, Report of a FAO/WHO/UNU Expert Consultation*; WHO Technical Report Series no. 935; World Health Organisation: Geneva, Switzerland, 2007.
27. Millward, D.J. Amino acid scoring patterns for protein quality assessment. *Br. J. Nutr.* **2012**, *108*, S31–S43. [CrossRef]
28. Ministry of Food and Drug Safety. *Korean Food Standard Codex*; Ministry of Food and Drug Safety: Cheongju, Korea, 2010.
29. Pickering, M.V.; Newton, P. Amino acid hydrolysis: Old problems, new solutions. *Lc Gc* **1990**, *8*, 778–781.
30. Jeong, H.; Kim, J.M.; Kim, B.; Nam, J.-O.; Hahn, D.; Choi, M.B. Nutritional Value of the Larvae of the Alien Invasive Wasp *Vespa velutina nigrithorax* and Amino Acid Composition of the Larval Saliva. *Foods* **2020**, *9*, 885. [CrossRef] [PubMed]
31. Curtis, C.R.; Watkins, J.C. The pharmacology of amino acids related to gamma-aminobutyric acid. *Pharmacol. Rev.* **1965**, *17*, 347–391.
32. Abe, T.; Tanaka, Y.; Miyazaki, H.; Kawasaki, Y.Y. Comparative study of the composition of hornet larval saliva, its effect on bahviour and role of trophallaxis. *Comp. Biochem. Physiol. Part C Comp. Pharmacol.* **1991**, *99C*, 79–84.
33. Beenakkers, A.M.T.; Horst, D.J.V.; Marrewijk, J.A.V. Biochemical process directed to flight muscle metabolism. In *Comprehensive Insect Physiology, Biochemistry and Pharmacology*; Kerkut, G.A., Gilbert, L.I., Eds.; Pergamon Press: Oxford, UK, 1985; Volume 10, pp. 451–486.
34. Matsuura, M.; Sakagami, S.F. A bionomic sketch of the giant hornet, *Vespa mandarinia*, a serious pest for Japanese apiculture. *J. Fac. Sci. Hokkaido Univ. Ser. VI Zoo.* **1973**, *19*, 125–162.
35. Sánchez, X.F.; Charles, R.J. Notes on the nest-architecture and colony composition in winter of the Yellow-legged Asian hornet, *Vespa velutina* Lepeletier 1836 (Hym.: Vespidae), in its introduced habitat in Galicia (NW Spain). *Insects* **2019**, *10*, 237. [CrossRef] [PubMed]
36. Ghosh, S.; Jung, C.; Meyer-Rochow, V.B. Nutritional value and chemical composition of larvae, pupae, and adults of worker honey bee, *Apis mellifera ligustica* as a sustainable food source. *J. Asia Pac. Èntomol.* **2016**, *19*, 487–495. [CrossRef]
37. Ghosh, S.; Jung, C.; Choi, K.; Choi, H.Y.; Kim, H.W.; Kim, S. Body fat and amino acid composition of a native bumblebee, *Bombus ignitus* relative to *B. terrestris* of foreign origin in Korea. *J. Apic.* **2017**, *32*, 111–117. [CrossRef]
38. Ghosh, S.; Choi, K.; Kim, S.; Jung, C. Body Compositional Changes of Fatty Acid and Amino Acid from the Queen of Bumblebee, *Bombus terrestris* during Overwintering. *J. Apic.* **2017**, *32*, 11–18. [CrossRef]
39. Ghosh, S.; Sohn, H.-Y.; Pyo, S.-J.; Jensen, A.B.; Meyer-Rochow, V.B.; Jung, C. Nutritional Composition of *Apis mellifera* Drones from Korea and Denmark as a Potential Sustainable Alternative Food Source: Comparison Between Developmental Stages. *Foods* **2020**, *9*, 389. [CrossRef] [PubMed]
40. Tomé, D.; Bos, C. Lysine Requirement through the Human Life Cycle. *J. Nutr.* **2007**, *137*, 1642S–1645S. [CrossRef]

41. Chakravorty, J.; Ghosh, S.; Jung, C.; Meyer-Rochow, V. Nutritional composition of *Chondacris rosea* and *Brachytrupes orientalis*: Two common insects used as food by tribes of Arunachal Pradesh, India. *J. Asia Pac. Èntomol.* **2014**, *17*, 407–415. [CrossRef]
42. Ghosh, S.; Lee, S.-M.; Jung, C.; Meyer-Rochow, V. Nutritional composition of five commercial edible insects in South Korea. *J. Asia Pac. Èntomol.* **2017**, *20*, 686–694. [CrossRef]
43. Ayieko, M.A.; Kinyuru, J.N.; Ndonǵa, M.F.; Kenji, G.M. Nutritional value and consumption of black ants (*Carebara vidua* Smith) from the Lake Victoria region in Kenya. *Adv. J. Food Sci. Technol.* **2012**, *4*, 39–45.
44. Sihamala, O.; Bhulaidok, S.; Shen, L.R.; Li, D. Lipids and fatty acid composition of dried edible red and black ants. *Agric. Sci. China* **2010**, *9*, 1072–1077.
45. Bhulaidok, S.; Sihamala, O.; Shen, L.; Li, D. Nutritional and fatty acid profiles of sun dried edible black ants (*Polyrhachis vicina* Roger). *Majeo Int. J. Sci. Technol.* **2010**, *4*, 101–112.
46. Chakravorty, J.; Ghosh, S.; Megu, K.; Jung, C.; Meyer-Rochow, V. Nutritional and anti-nutritional composition of *Oecophylla smaragdina* (Hymenoptera: Formicidae) and *Odontotermes* sp. (Isoptera: Termitidae): Two preferred edible insects of Arunachal Pradesh, India. *J. Asia Pac. Èntomol.* **2016**, *19*, 711–720. [CrossRef]
47. Lehtovaara, V.; Valtonen, A.; Sorjonen, J.; Hiltunen, M.; Rutaro, K.; Malinga, G.; Nyeko, P.; Roininen, H. The fatty acid contents of the edible grasshopper *Ruspolia differens* can be manipulated using artificial diets. *J. Insects Food Feed.* **2017**, *3*, 253–262. [CrossRef]
48. Rutaro, K.; Malinga, G.M.; Lehtovaara, V.J.; Opoke, R.; Nyeko, P.; Roininen, H.; Valtonen, A. Fatty acid content and composition in edible *Ruspolia differens* feeding on mixtures of natural food plants. *BMC Res. Notes* **2018**, *11*, 687. [CrossRef]
49. Mensink, R.P. Effects of the individual saturated fatty acids on serum lipids and lipoprotein concentrations. *Am. J. Clin. Nutr.* **1993**, *57*, 711S–714S. [CrossRef]
50. Aluko, R.E. *Functional Foods and Nutraceuticals*; Springer: New York, NY, USA, 2012.
51. Grundy, S.M. Comparison of Monounsaturated Fatty Acids and Carbohydrates for Lowering Plasma Cholesterol. *N. Engl. J. Med.* **1986**, *314*, 745–748. [CrossRef] [PubMed]
52. Abeywardena, M.Y.; Patten, G.S. Role of ω3 long-chain polyunsaturated fatty acids in reducing cardio-metabolic risk factors. *Endocr. Metab. Immune Disord. Drug Targets* **2011**, *11*, 232–246. [CrossRef] [PubMed]
53. Bagge, C.N.; Strandhave, C.; Skov, C.M.; Svensson, M.; Schmidt, E.B.; Christensen, J.H. Marine n-3 polyunsaturated fatty acids affect the blood pressure control in patients with newly diagnosed hypertension—A 1-year follow-up study. *Nutr. Res.* **2017**, *38*, 71–78. [CrossRef] [PubMed]
54. Cabo, J.; Alonso, R.; Mata, P. Omega-3 fatty acids and blood pressure. *Br. J. Nutr.* **2012**, *107*, S195–S200. [CrossRef]
55. Rumpold, B.A.; Schlüter, O.K. Nutritional composition and safety aspects of edible insects. *Mol. Nutr. Food Res.* **2013**, *57*, 802–823. [CrossRef]
56. He, F.J.; MacGregor, G.A. Beneficial effects of potassium on human health. *Physiol. Plant.* **2008**, *133*, 725–735. [CrossRef]
57. Doyle, M.E.; Glass, K.A. Sodium Reduction and Its Effect on Food Safety, Food Quality, and Human Health. *Compr. Rev. Food Sci. Food Saf.* **2010**, *9*, 44–56. [CrossRef]
58. Iwahori, T.; Miura, K.; Ueshima, H. Time to Consider Use of the Sodium-to-Potassium Ratio for Practical Sodium Reduction and Potassium Increase. *Nutrients* **2017**, *9*, 700. [CrossRef]
59. Institute of Medicine (US) Standing Committee on the Scientific Evaluation of Dietary Reference Intakes. *Dietary Reference Intakes for Calcium, Phosphorus, Magnesium, Vitamin D, and Fluoride*; National Academies Press: Washington, DC, USA, 1997.
60. Gennari, C.; Lezama, I.; Lockwood, K.; Bergmann, T. Calcium and vitamin D nutrition and bone disease of the elderly. *Public Health Nutr.* **2001**, *4*, 547–559. [CrossRef] [PubMed]
61. Power, M.L.; Heaney, R.P.; Kalkwarf, H.J.; Pitkin, R.M.; Repke, J.T.; Tsang, R.C.; Schulkin, J. The role of calcium in health and disease. *Am. J. Obstet. Gynecol.* **1999**, *181*, 1560–1569. [CrossRef]
62. Ramakrishnan, U.; Imhoff-Kunsch, B. Anemia and Iron Deficiency in Developing Countries. In *Handbook of Nutrition and Pregnancy*; Springer International Publishing: New York, NY, USA, 2008; pp. 337–354.
63. Lutter, C.K. Iron Deficiency in Young Children in Low-Income Countries and New Approaches for Its Prevention. *J. Nutr.* **2008**, *138*, 2523–2528. [CrossRef]
64. The Problem: About Iron Deficiency. Available online: http://www.unicef.org/nutrition/23964_iron.html (accessed on 27 August 2019).
65. Gupta, A.; Gadipudi, A. Iron deficiency anaemia in pregnancy: Developed versus developing countries. *EMJ Hematol.* **2018**, *6*, 101–109.
66. Aggett, P.J.; Comerford, J.G. Zinc and human health. *Nutr. Rev.* **1995**, *53*, S16–S22. [PubMed]
67. Caulfield, L.E.; Black, R.E. *Zinc deficiency*, In *Comparative Quantification of Health Risks: Global and Regional Burden of Disease Attributable to Selected Major Risk Factors*; Ezzati, M., Lopez, A.D., Rodgers, A., Murray, C.J.L., Eds.; World Health Organization: Geneva, Switzerland, 2004; Volume 1, pp. 257–279.
68. Maxfield, L.; Crane, J.S. Zinc Deficiency. Available online: https://www.ncbi.nlm.nih.gov/books/NBK493231/ (accessed on 28 August 2019).
69. Wessells, K.R.; Brown, K.H. Estimating the Global Prevalence of Zinc Deficiency: Results Based on Zinc Availability in National Food Supplies and the Prevalence of Stunting. *PLoS ONE* **2012**, *7*, e50568. [CrossRef]
70. Meyer-Rochow, V.B. Can insects help to ease the problem of world food shortage? *Search* **1975**, *6*, 261–262.
71. Han, R.; Shin, J.T.; Kim, J.; Choi, Y.S.; Kim, Y.W. An overview of the South Korean edible insect food industry: Challenges and future pricing/promotion strategies. *Èntomol. Res.* **2017**, *47*, 141–151. [CrossRef]

72. Kim, T.-K.; Yong, H.I.; Kim, Y.-B.; Kim, H.-W.; Choi, Y.-S. Edible Insects as a Protein Source: A Review of Public Perception, Processing Technology, and Research Trends. *Food Sci. Anim. Resour.* **2019**, *39*, 521–540. [CrossRef]
73. Meyer-Rochow, V.B.; Ghosh, S.; Jung, C. Farming of insects for food and feed in South Korea: Tradition and innovation. *Berl. Münchener Tierärztliche Wochenschr.* **2019**, *131*, 236–244. [CrossRef]

 *foods*

Review

Chemical Composition, Nutrient Quality and Acceptability of Edible Insects Are Affected by Species, Developmental Stage, Gender, Diet, and Processing Method

Victor Benno Meyer-Rochow [1,2,*], Ruparao T. Gahukar [3], Sampat Ghosh [2] and Chuleui Jung [2,4,*]

1   Department of Genetics and Ecology, Oulu University, SF-90140 Oulu, Finland
2   Agriculture Science and Technology Research Institute, Andong National University, Andong 36729, Gyeongsangbuk, Korea; sampatghosh.bee@gmail.com
3   103, Niyogi Bhavan, Abhyankar Marg, Dhantoli, Nagpur 4400012, Maharashtra, India; rtgahukar@gmail.com
4   Department of Plant Medicals, Andong National University, Andong 36729, Gyeongsangbuk, Korea
*   Correspondence: meyrow@gmail.com (V.B.M.-R.); cjung@andong.ac.kr (C.J.)

Abstract: Edible insects have been considered as either nutritious food itemsper se, or as wholesome ingredients to various dishes and components of traditional subsistence. Protein, fat, mineral and vitamin contents in insects generally satisfy the requirements of healthy food, although there is considerable variation associated with insect species, collection site, processing method, insect life stage, rearing technology and insect feed. A comparison of available data(based on dry weight) showed that processing can improve the nutrient content, taste, flavour, appearance and palatability of insects, but that there are additional factors, which can impact the content and composition of insect species that have been recommended for consumption by humans. This review focuses on factors that have received little attention in connection with the task to improve acceptability or choice of edible insects and suggests ways to guarantee food security in countries where deficiencies in protein and minerals are an acute and perpetual problem. This review is meant to assist the food industry to select the most suitable species as well as processing methods for insect-based food products.

Keywords: entomophagy; insect edibility; insect farming; insect diversity; acceptability; nutrients; food security; diet

Citation: Meyer-Rochow, V.B.; Gahukar, R.T.; Ghosh, S.; Jung, C. Chemical Composition, Nutrient Quality and Acceptability of Edible Insects Are Affected by Species, Developmental Stage, Gender, Diet, and Processing Method. *Foods* **2021**, *10*, 1036. https://doi.org/10.3390/foods10051036

Academic Editor: Christopher John Smith

Received: 15 April 2021
Accepted: 7 May 2021
Published: 10 May 2021

1. Introduction

Entomophagy (the habit of eating insects) has been practiced since time immemorial by humans [1–3] and their primate relatives [4,5]. Although entomophagy was not new to science, it was a paper by Meyer-Rochow [6], which for the first time suggested that edible insect species ought not to be neglected in the quest to safeguard future global food security. At present edible insects are still recognized as a sustainable food item by many residents of sub-Saharan Africa, South and Central America (including Mexico), South-East Asia and the Australia Papua New Guinea region. The consumption of insect species depends upon availability/access, suitability, preference, nutritional value, religious beliefs and social customs [7–12].

In North-East India, some highly appreciated species of edible insects are available (mostly seasonally) for sale at the local markets, but their cost is often higher than that of conventional animal meats or food of vertebrate origin [13,14]. This holds true also for Laos [15], Cameroon and many other African countries [16,17]. Nonetheless, the local people prefer the insects because of their taste and for traditional aspects [13,14,18]. However, insect consumption is declining, with one of the reasons being a shortage of the product due to a lack of facilities to efficiently and systematically rear suitable species and another reason in developing countries being an increasing "westernization" in terms of food choices [19]. As a result, sellers experience disruptions and delays in obtaining supplies and potential buyers are frustrated by the fluctuations of the product's condition and availability.

Insects contain easily digestible quality protein with all the essential amino acids readily identifiable (except for methionine and tryptophan, which are present in low levels). The absence of tryptophan and fractional recovery of methionine and cysteine are attributed to methods of analysis and not necessarily because they are actually absent. For example, based on the data of 5 insects, viz. yellow mealworm *Tenebrio molitor* L., house cricket *Acheta domesticus* L., superworm *Zophobas morio* Fabricius, lesser mealworm *Alphitobius diaperinus* Panzer and the roach *Blaptica dubia* Serville, Yi et al. [20] observed that the amount of essential amino acids (EAA) was high and that the content of protein was similar to that of conventional meat products. In China, the pupal powder of the silkworm *Antheraea pernyi* Guerin-Meneville is appreciated, because of its substantial amount of protein (71.9%), EAA, fat (20.0%) and ash (4.0%) [21]. Information on the composition and content of nutrients in edible insects is readily accessible through journals, special reports and dissertations and has been summarized repeatedly [22–26].

With established techniques such as HPLC (high-performance liquid chromatography) for extraction and quantification of nutrients and bioprospecting of new species of edible insects, studies on the nutritional value of insects are being intensified with an aim to search economic and efficient ways to supply processed insects [27]. Currently, nutritional contents are not yet known for the majority of the surveyed/collected insect species of the various geographical locations and eco-zones that they occur in. Furthermore, knowledge and perception of factors that are encountered during the rearing of domesticated insects or those collected in the wild is limited and available only for a certain number of species. More studies on the chemical compositions of edible insects in relation to factors like geography, climate, processing and preparation methods would facilitate the identification of species most suitable for mass rearing and a potential to ameliorate the state of health in humans in certain parts of the globe [28]. This is especially important in view of the fact that most insects used as food today may not be much better nutritionally than traditional meats and that their 'value' is actually more related to environmental issues rather than their nutritional content [29,30].

This review examines and summarises a variety of factors that either are known (or have the potential) to influence an insect's chemical composition and its nutritive value, such as a species' taxonomic position or ecotype, thereby enhancing or reducing its acceptability as a food item for humans. The review illuminates in particular the roles that developmental stages, castes, an insect's habitat and diet (whether natural or laboratory based) play in relation toaninsect's amino acid and fatty acid content profile and to what extent the amounts of fibre, soluble carbohydrates and minerals depend on environmental factors. We highlight the importance of different processing methods, the risks of contamination and allergies and relate such factors to consumer choice as well as general acceptability of edible insects and insect-containing products as an alternative to conventional food items.

2. Nutrient Contents

2.1. Biological Factors: Insect Species, Developmental Stage, Sex and Caste, Organ and Ecotype or Biological Variants

2.1.1. Insect Species

Edible insects generally belong to eight orders namely Blattodea (cockroaches, termites), Coleoptera (beetles), Diptera (flies), Hemiptera (true bugs), Hymenoptera (ants, bees and wasps), Lepidoptera (butterflies and moths), Odonata (dragonflies, damselflies) and Orthoptera (grasshoppers, crickets and locusts). Table 1 represents proximate nutrient composition of selected representative edible insect species and is not a compilation of all edible insect chemical analyses published to date. The results are based on dried insect samples, with the exception of the beetles *Oryctes boas* and *Oryctes rhinoceros*, which were not fully dry when analysed. The wide variation in the nutrient content among insect species generally depends on a variety of factors, of which geographic and climatic conditions as well as the insects' food intake seem to be the most important factors. Although not to a very great extent, chemical content can indeed vary among different species belonging to the same genus (Table 2). The feeding regime, physiology and even ecological factors are more important determinants of the nutrient content than the species' taxonomic proximity.

Table 3 contains comparative data on the amino acid composition of edible insect species. The results reveal that species belonging to the same genus may possess only somewhat different amino acid contents. To cite an example: various species of *Vespa* were found to differ with regard to the quantities of their amino acids [31,32]. Similar observations were made in connection with *Apis* spp. [33–35]. Moreover, the differences were attributed to the rearing system, including the insects' feed and ecological condition. Palm weevils were found to have slightly different protein and amino acid content, depending on where they had been collected from [22]. However, irrespective of the amounts, the relative distribution of the amino acids was found to follow an almost identical trend in all of the individuals.

Overall, glutamic acid was always found to be most abundant, but among the essential amino acids leucine predominated followed by lysine. Although the scope to discuss the nutritional benefits of individual nutrient is limited in the present manuscript, it is worth mentioning that lysine content of edible insects is advantageous as it is often limiting in the cereal-based diet of humans. Species-specific fat and fatty acid content was apparent in edible insects (Table 4). In general, palmitic acid followed by stearic acid was the predominating fatty acids among the saturated kinds (SFA), while oleic acid was the most abundant among the monounsaturated fatty acids (MUFA). Species-specific patterns were noticed for mineral content (Table 5). However, the differences in the mineral contents are primarily attributable to geographic and ecological factors as the minerals are not synthesized in the animal body but are obtained from the dietary sources.

Table 1. Proximate nutrient composition (g/100 g dry matter basis) of edible insects.

Insect	Developmental Stage	Protein	Fat	Fibre	NFE *	Ash	Reference
Blattodea (including infra order Isoptera)							
Edible cockroaches and termites		46.3	31.3	5.2	13.7	4.4	[22]
Macrotermes bellicosus	A	40.7	44.8	5.3	2.2	5.0	[36]
Macrotermes nigeriensis	A	37.5	48.0	5.0	2.1	3.2	[37]
Odototermes sp.	A	33.7	50.9	6.3	6.1	3.0	[38]
Syntermes sp. soldier	A	64.7	3.1	23.0	2.5	4.2	[36]
Coleoptera							
Edible beetles		40.7	33.4	10.7	13.2	5.1	[22]
Allomyrina dichotoma	L	54.2	20.2	4.0	17.7	3.9	[39]
Oryctes rhinoceros	L	52.0	10.8	17.9	2.0	11.8	[37]
Protaetia brevitarsis	L	44.2	15.4	11.1	22.5	6.9	[39]
Tenebrio molitor	L	53.2	34.5	6.3	1.9	4.0	
Tenebrio molitor	P	51.0	32.0	12.0	–	–	
Tenebrio molitor	L	52.0	31.0	13.0	–	–	[40]
Zophobas morio	L	46.0	35.0	6.0	–	–	
Diptera							
Edible flies		49.5	22.8	13.6	6.0	10.3	[22]
Caliphora vomitoria	A	64.9	0.7	16.6	12.2	5.6	
Hermetia illucens	Pre P	44.3	31.9	5.1	3.4	8.7	[41]
Hermetia illuscens	L	39.0	32.6	12.4	–	14.6	[42]
Hemiptera							
Edible bugs		48.3	30.3	12.4	6.1	5.0	[22]
Aspongopus nepalensis	A	10.6	38.4	33.5	15.3	2.2	[18]

Table 1. *Cont.*

Insect	Developmental Stage	Protein	Fat	Fibre	NFE *	Ash	Reference
Hymenoptera							
Edible ants, bees, wasps		46.5	25.1	5.7	20.3	3.5	[22]
Oecophylla smaragdina	A	55.3	15.0	19.8	7.3	2.6	[38]
Lepidoptera							
Edible moth		45.4	27.7	6.6	18.8	4.5	[22]
Cirina butyrospermi	L	62.7	14.5	5.0	12.6	5.1	[43]
Odonata							
Edible dragonfly, damselfly		55.2	19.8	11.8	4.6	8.5	[22]
Orthoptera							
Edible grasshoppers, crickets, locusts		61.3	13.4	9.6	13.0	3.9	[22]
Acheta domesticus	A	62.6	12.2	8.0	12.3	5.0	[41]
Brachytrupes sp.	A	65.4	11.8	13.3	2.5	4.9	[36]
Brachytrupes orientalis	A	65.7	6.3	8.8	15.2	4.3	[44]
Chondacris rosea	A	68.9	7.9	12.4	6.7	4.2	
Gryllus assimilis	A	56.0	32.0	7.0	–	–	[40]
Gryllus bimaculatus	A	58.3	11.9	9.5	10.6	9.7	[39]
Ruspolia nitidula	A	40.8	46.3	5.9	3.7	3.3	[41]
Schistocerca piceifrons piceifrons	A	80.3	6.2	12.6	–	3.4	[45]
Teleogryllus emma	A	55.7	25.1	10.4	0.7	8.2	[39]

L = Larva, P = Pupa, N = Nymph, A = Adult, B = Brood, NFE * = Nitrogen-free extract (indicative of soluble carbohydrates).

Table 2. Comparative account of proximate nutrient content (g/100g dry matter basis) of different species belonging to same genus.

Genus	Species	Developmental Stage	Protein	Fat	Fibre	NFE *	Ash	Reference
Blattodea								
Macrotermes	*bellicosus*	A	20.4	28.2	2.7	43.3	2.9	[46]
	notalensis		22.1	22.5	2.2	42.8	1.9	
	subhylanus		39.3	44.8	6.4	1.9	7.6	[47]
	bellicosus		39.7	47.0	6.2	2.4	4.7	
Periplaneta	*americana*	L,A	65.6	28.2	3.0	0.8	2.5	[48]
	australasiae		62.4	27.3	4.5	2.7	3.0	
Pseudacanthotermes	*militaris*	A	33.5	46.6	6.6	8.7	4.6	[47]
	spiniger		37.5	47.3	7.2	0.7	7.2	
Coleoptera								
Oryctes	*boas*	L	26.0	1.5	3.4	38.5	1.5	[46]
	rhinoceros		42.3	0.6	–	27.7	12.7	[49]
Hemiptera								
Edessa	*conspersa*	N,A	36.8	45.8	10.0	4.2	3.2	[50] (*cf.* [22])
	montezumae		37.5	45.9	10.9	2.1	3.7	
	petersii		37.0	42.0	18.0	1.0	2.0	[51]
	sp.		33.0	54.0	11.0	–	1.0	
Hymenoptera								
Atta	*mexicana*	A	46.0	39.0	11.0	0.0	4.0	
	cephalotes		43.0	31.0	10.0	14.0	2.0	
Brachygastra	*azteca*	B	63.0	22.0	3.0	9.0	3.0	[51]
	mellifica		53.0	30.0	3.0	11.0	3.0	
Polybia	*parvulina*	B	61.0	21.0	6.0	8.0	4.0	
	occidentalis nigratella		61.0	28.0	2.0	11.0	3.0	
	occidentalis bohemani		62.0	19.0	4.0	13.0	3.0	

Table 2. *Cont.*

Genus	Species	Developmental Stage	Protein	Fat	Fibre	NFE *	Ash	Reference
Lepidoptera								
Anaphe	*infracta*	L	20.0	15.2	2.4	66.1	1.6	[46]
	recticulata		23.0	10.2	3.1	64.6	2.5	
	venata		25.7	23.2	2.3	55.6	3.2	
	sp.		18.9	18.6	1.7	46.8	4.1	
Orthoptera								
Sphenarium	*purpurascens*	A	65.2	10.8	9.4	11.6	3.0	[48]
	mexicanum		62.1	10.8	4.1	22.6	0.3	
	purpurascens		56.0	11.0	9.0	21.0	3.0	[51]
	histrio		77.0	4.0	12.0	4.0	2.0	
	sp.		68.0	12.0	11.0	5.0	5.0	

L = Larva, P = Pupa, N = Nymph, A = Adult, B = Brood, NFE * = Nitrogen-free extract (indicative of soluble carbohydrates).

Table 3. Amino acid composition of different species belonging to the same genus.

Genus	Species	Amino Acid Composition (% of Total Amino Acids or Protein)																		Total Amino Acids or Protein (g/100 g Dry Matter)	Reference
		Val	Ile	Leu	Lys	Tyr	Thr	Phe	Trp	His	Met+Cys	Total EAA [††]	Arg	Asp	Ser	Glu	Gly	Ala	Pro		
Apis * (P)	mellifera	5.9	5.6	7.8	7.3	4.9	4.6	0.5	ND	2.7	1.0	40.3	5.6	8.6	4.9	20.5	6.1	7.1	ND	40.9	[33]
	cerana	6.1	4.7	8.6	5.9	3.7	4.3	4.1	ND	2.5	4.7	44.6	4.9	12.3	4.7	10.4	7.2	9.6	6.6	51.2	[34]
	dorsata	5.7	4.4	8.5	5.7	3.3	4.4	3.9	ND	2.6	4.9	43.4	4.9	13.4	4.9	11.1	7.5	8.5	6.9	38.9	
	florea	5.9	4.8	9.3	6.5	4.5	4.8	4.8	ND	2.8	4.8	48.2	5.3	10.4	5.1	14.0	6.2	8.1	7.6	35.6	[35]
Bombus * (A)	ignitus	7.0	5.7	9.3	6.1	3.0	2.3	2.7	ND	3.0	6.1	45.2	4.0	3.8	4.9	11.4	9.1	11.2	10.1	47.3	[52]
	terrestris	6.3	5.0	8.1	7.8	3.1	2.3	3.1	ND	2.6	6.3	44.6	5.0	3.9	6.3	12.5	8.1	10.2	9.9	38.3	
Brachygastra (B)	azteca	6.4	5.1	8.5	6.1	6.5	4.4	4.1	0.7	2.8	3.0	47.6	4.4	8.4	4.5	16.4	6.7	5.8	6.4	63.0	
	mellifica	5.4	4.4	7.8	3.6	7.5	4.4	4.0	0.7	3.6	3.8	45.2	5.7	8.6	4.2	16.0	6.7	6.1	7.1	53.0	[51]
Polybia (B)	occidentalis nigratella	5.9	4.5	7.8	7.4	5.6	4.0	3.3	0.7	3.0	5.0	47.2	5.7	8.4	4.5	12.9	7.1	6.5	6.3	61.0	
	parvulina	6.1	4.7	7.8	7.3	5.9	4.1	3.4	0.7	3.4	5.3	48.7	5.7	7.8	4.4	13.3	7.2	6.4	6.5	61.0	
Polistes *	sagittarius	6.6	5.5	7.8	4.4	5.0	4.2	5.0	ND	3.0	1.4	42.9	4.4	8.3	4.4	17.2	6.9	7.2	8.9	36.1	[31]
	sulcatus	6.7	6.2	8.0	4.2	4.9	4.2	4.4	ND	2.4	2.0	43.0	4.0	7.3	4.4	15.3	8.9	8.9	8.0	45.0	
Vespa * (B)	velutina	6.1	5.5	8.7	6.1	6.6	4.2	4.2	ND	3.2	2.4	47.0	4.5	6.3	4.5	20.1	6.3	5.5	6.1	37.9	[32]
	mandarinia	6.3	5.7	8.7	6.3	7.3	4.3	4.3	ND	3.3	2.7	48.9	2.2	6.5	4.3	21.2	6.3	5.4	5.7	36.8	
	basalis	5.7	5.3	8.5	6.8	7.1	4.3	4.3	ND	3.2	1.4	46.6	4.3	6.4	4.3	22.1	5.7	5.0	5.7	28.1	
Vespa * (L)	basalis	5.9	5.9	8.0	4.3	5.7	4.1	4.3	ND	2.5	2.1	42.8	3.9	7.7	4.3	17.1	8.2	7.7	8.4	43.9	[31]
	mandarinia mandarinia	5.0	4.6	6.1	16.5	4.0	3.3	10.5	ND	2.1	0.8	52.9	3.3	6.3	3.4	13.2	6.3	6.5	7.9	52.2	
	velutina auraria	6.9	5.9	7.6	2.9	7.6	4.3	4.1	ND	3.1	2.9	45.3	6.3	9.2	6.5	12.0	8.0	7.1	5.9	49.0	
	tropica duealis	7.5	5.4	8.3	3.3	5.4	4.5	4.2	ND	1.4	1.2	41.2	7.1	10.1	5.0	13.4	8.7	7.8	6.6	42.4	
Sphenarium	histrio	5.1	5.3	8.7	5.7	7.3	4.0	11.7	0.6	1.9	3.3	53.6	6.6	9.3	5.1	5.3	5.3	7.6	7.2	77.0	[51]
	purpurascens	5.7	4.2	8.9	5.7	6.3	3.1	10.3	0.7	2.2	4.3	51.4	6.0	8.7	4.8	10.7	6.8	6.4	6.2	56.0	

L = Larva, P = Pupa, A = Adult, B = Brood; ND = Not determined or not estimated; * Amino acid content (g/100 g dry matter) was obtained from the respective paper and recalculated as g/100 g of total amino acids or protein; [††] EAA: Essential amino acids, we include essential amino acids (Val, Ile, Leu, Lys, Thr, Trp, Phe, His, Met) and two conditional essential amino acids (Tyr, Cys).

Table 4. Fatty acid composition of selected edible insects.

Genus	Species	Developmental Stage	Fatty Acid Composition (% of Total Fatty Acids)								Total Fatty Acids or Fat (g/100 g Dry Matter)	Reference
			C14:0	C16:0	C18:0	SFA	C18:1	MUFA	C18:2	PUFA		
Apis [†]	*cerana*	L	3.9	38.2	8.1	50.7	46.9	48.7	0.5	0.7	6.1	[34]
		P	3.0	31.4	10.6	46.2	49.8	52.7	0.9	1.1	6.3	
		A	1.9	18.2	12.1	33.8	57.7	63.4	2.6	2.8	4.2	
	dorsata	P	3.2	33.3	11.8	49.4	47.7	49.8	0.8	0.8	6.2	
		A	1.0	14.4	14.4	31.3	61.0	66.5	2.2	2.2	3.1	
	mellifera	L	2.4	37.3	11.8	51.8	47.5	48.2	0.0	0.0	4.9	[33]
		P	2.9	35.1	12.6	51.1	47.6	48.9	0.0	0.0	5.5	
		A	0.6	14.4	9.3	25.2	45.2	67.0	7.8	7.8	1.7	
	florea	P	1.8	35.3	8.8	46.6	47.6	52.3	1.0	1.1	7.2	[35]
		A	1.5	30.7	9.7	43.2	49.7	55.7	1.1	1.1	5.4	
Aspongopus	*viduatus*	A	0.3	31.3	3.5	37.9	45.5	56.8	4.9	5.4	54.2	[53]
	nepalensis	A	0.4	32.3	4.8	37.5	46.4	56.1	6.1	6.1	35.9	[18]
Bombus [*,†]	*ignitus*	A	2.6	16.1	1.7	22.1	49.1	75.4	2.5	2.5	9.5	[52]
	terrestris	A	3.8	15.2	1.7	21.5	51.1	76.2	2.2	2.2	8.4	
Imbrasia	*belina*	L	1.2	31.9	4.7	37.9	34.2	36.0	6.0	26.1	23.4	[54]
	epimethea	L	0.6	23.2	22.1	46.1	8.4	9.0	7.0	42.5	13.3	[22]
	truncata	L	0.2	24.6	21.7	46.5	7.6	7.6	7.6	44.4	16.4	
	ertli	L	1.0	22.0	0.4	61.4	2.0	24.0	20.0	31.0	11.1	[55,56]
	oyemensis	L	0.5	46.0	7.2	54.2	34.6	34.6	11.2	11.2	25.4	[22]

Table 4. *Cont.*

Genus	Species	Developmental Stage	Fatty Acid Composition (% of Total Fatty Acids)								Total Fatty Acids or Fat (g/100 g Dry Matter)	Reference
			C14:0	C16:0	C18:0	SFA	C18:1	MUFA	C18:2	PUFA		
Macrotermes	*Bellicosus* **	A	2.2	42.5	2.9	49.0	15.8	17.9	24.2	33.1	36.1	[54]
	bellicosus	A	0.2	46.5	–	46.7	12.8	14.9	34.4	38.3	46.1	[56,57]
	nigeriensis	A	0.6	31.4	7.1	39.4	52.5	53.1	7.6	7.6	34.2	[58]
	subhylanus	A	1.1	27.7	6.3	35.1	48.6	52.8	10.8	12.2	44.8	
	bellicosus	A	1.2	38.4	9.5	49.5	41.7	44.6	5.0	5.9	47.0	[47]
Pseudacanthotermes	*militaris*	A		26.0	5.9	32.2	50.3	56.1	11.5	11.7	46.6	
	spiniger	A	0.8	28.0	6.1	35.8	49.3	52.9	10.5	11.3	47.3	
Oryctes	*owariensis*	L	2.5	0.2	0.2	3.1	5.2	43.6	45.5	50.9	53.8	[59]
	rhinoceros	L	3.5	28.7	2.1	34.4	41.5	45.9	14.1	19.7	38.1	[54]
Vespa [†]	*velutina*	B	6.0	31.9	7.8	48.3	35.3	39.7	5.2	12.1	11.6	
	mandarinia	B	2.5	21.3	5.0	30.7	27.7	29.2	33.7	40.1	20.2	[32]
	basalis	B	1.4	15.8	5.4	24.3	23.9	25.2	42.8	50.5	22.2	

L = Larva, P = Pupa, A = Adult; [†] Fatty acid content (mg/100g dry matter) was obtained from the respective paper and recalculated as % of total fatty acids; * Mated queen; ** Oil. SFA = Saturated fatty acids, MUFA = Monounsaturated fatty acids, PUFA = Polyunsaturated fatty acids

Table 5. Minerals content (mg/100g) of selected edible insects.

Genus	Species	Developmental Stage	Ca	Mg	Na	K	P	Fe	Zn	Cu	Mn	Reference
Anaphe	*infracta*	L	8.6	1.0			111.3	1.8				[46]
	reticulate	L	10.5	2.6			102.4	2.2				
	venata	L	8.6	1.6			100.5	2.0				
	sp.	L	7.6	1.0			122.2	1.6				
	venata	L	40.0	50.0	30.0	1150.0	730.0	10.0	10.0	1.0	40.0	[60]
Apis	*cerana*	L	63.1	86.6	37.2	823.1	715.6	5.9	7.3	1.0	1.1	[34]
		P	62.9	104.3	44.4	1153.2	931.5	7.1	7.7	1.2	0.2	
		A	91.1	148.8	77.1	1538.8	1283.9	11.1	12.9	1.9	0.2	
	dorsata	P	68.9	103.4	48.6	1136.6	905.0	5.8	6.4	1.1	0.1	
		A	78.5	113.3	53.9	1254.3	972.3	7.6	7.4	1.2	0.1	
Brachytupes	*orientalis*	A	76.3	87.2	112.0	412.3		18.7	8.5	1.5	5.0	[44]
	sp.	A	9.2	0.1			126.9	0.7				[22]
Imbrasia	*epimethea*	L	224.7	402.2	75.3	1258.1	666.7	13.0	11.1	1.2	5.8	[22]
	ertli	L	55.0	254.0	2418.0	1204.0	600.0	2.1		1.5	3.4	
	oyemensis	L	73.0		730.0	680.0						
Macrotermes	*subhylanus*	A	58.7					53.3	8.1			[47]
	bellicosus	A	63.6					116.0	10.8			
Pseudacanthotermes	*militaris*	A	48.3					60.3	12.9			
	spiniger	A	42.9					64.8	7.1			

L = Larva, P = Pupa, A = Adult.

2.1.2. Developmental Stage

Table 6 contains comparative data on the proximate nutrient content of different developmental stages of selected edible insect species. As already briefly touched upon, the contents of an insect can vary between adults, larvae, pupae and nymphs with regard to carbohydrates, protein, fat, fiber, ash, and minerals and there are complex reasons for this. In general, the protein content was found to be higher along with the more mature developmental stages. The opposite held true for fat content. Larvae and adults may feed on different foods and pupae usually do not consume any food at all, which would explain the differences in amino and fatty acid content as well as minerals seen in the different developmental stages of, for example, the honey bee [33]. To cite an example, in male bees (known as drones), the compositions of amino acids, protein and minerals all increase with development stage [61]. Saturated fatty acids dominated over monounsaturated fatty acids in the pupae but the reverse was reported for adults [61]. Variations in the nutrient composition of different post-embryonic developmental stages of nymphs and adults of the grasshopper *Zonoceros variegatus* were studied by Ademolu et al. [62]. Increments in protein but reductions in the amounts of fat from nymphs to adults were observed. Similar results, i.e., higher protein and lower fat content in parallel with the developmental stages hold true for three Blattodea species, namely *Blaptica dubia*, *Blaberus discoidalis* and *Blatta lateralis* [63].

The need to build up muscle tissue can change during developmental stages and is influenced by events of an insect's life cycle. For example, termite worker adults may have less fat than the more sedentary nymphs and that is presumably because of their more active lifestyles and the greater need to burn fat to meet the energy demands of their leg muscles. The sedentary termite queen, as the only egg producer of the colony, would require much less muscle tissue, but considerably more fat. When reared for 13 weeks, cricket (*A. domesticus*) nymphs contained 36–60% crude protein and 12–25% fat with maximum amounts of palmitic, oleic, linoleic, linolenic and a small amount of arachidic, EPA (eicosapentaenoic acid) and DHA (docosahexaenoic acid) fatty acids [64]. The concentrations of Mg, Ca and Zn reached their optimum after 9 weeks when they were 1.30–11.30 mg, 1.40–19.70 mg and 0.20–16.60 mg/100 mg, respectively. On that basis, Kipkoech et al. [64] suggested cricket harvesting to occur preferentially between 9–11 weeks, because only at that age the larvae are in their nutritionally best condition for consumption [65].

In this context another issue is related to differences between developmental stages, specifically in relation to their life cycle physiologies: during events like droughts and overwintering periods the conditions of an insect may change. Ghosh et al. [52], for example, demonstrated changes in the bodycomposition of bumblebee (*Bombus terrestris*) queens during overwintering and summer periods.

Table 6. Comparative account of proximate nutrient content (g/100g dry matter basis) of different developmental stages of edible insects.

Insect	Developmental Stage	Protein	Fat	Fibre	NFE *	Ash	Reference
Coleoptera							
Tebebrio molitor	L	47.7	37.7	5.0	7.1	3.0	[66]
	P	53.1	36.7	5.1	1.9	3.2	
	A	60.2	20.8	16.3	0.01	2.7	
Rhynchophorus phoenicis	Early L	9.1	61.5	22.1	4.9	2.4	[67]
	Late L	10.5	62.1	17.2	7.8	2.3	
	A	8.4	52.4	21.8	16.0	1.4	
Rhynchophorus phoenicis	L	23.4	54.2	3.4	5.0	5.2	[68]
	Immature P	33.1	42.7	3.1	6.7	7.4	
	Mature P	34.9	47.1	2.4	5.6	3	
	A	34.1	44.7	7.2	4.0	5.8	
Rhynchophorus phoenicis	Early L	9.1	24.2	5.8	13.0	2.4	[69]
	Late L	10.5	25.4	6.0	12.0	2.3	
Oryctes rhinoceros	L	70.8	7.5	5.4	7.0	8.3	[70]
	P	65.3	20.2	2.2	4.3	3.2	
	A	74.2	9.6	3.7	2.8	5.3	
Hymenoptera							
Apis mellifera	L	42.0	19.0	1.0	35.0	3.0	[51]
	P	49.0	20.0	3.0	24.0	4.0	
Apis mellifera ligustica	L	35.3	14.5		45.1	4.1	[33]
	P	45.9	16.0		34.3	3.8	
	A	51.0	6.9		30.5	11.5	

Table 6. *Cont.*

Insect	Developmental Stage	Protein	Fat	Fibre	NFE *	Ash	Reference
Orthoptera							
Acheta domesticus (as is basis)	N	15.4	3.3	5.8	0.9	1.1	[71]
	A	20.5	6.8		10.0	1.1	
Zonoceros variegatus	N1	18.3	4.3	0.9	0.4	1.9	[62]
	N2	14.4	4.8	0.9	0.4	1.0	
	N3	16.8	2.9	1.5	0.9	0.9	
	N4	15.5	0.7	0.9	9.7	1.6	
	N5	14.6	1.1	0.9	9.8	1.6	
	N6	16.1	0.9	1.0	8.8	1.5	
	A	21.4	0.9	1.2	10.0	1.4	

L = Larva, P = Pupa, N = Nymph, A = Adult; NFE * (nitrogen-free extract) indicates carbohydrate.

2.1.3. Sex and Caste

The powder of male silkworm (*B. mori*) pupae contained less protein than that of female pupae (reviewed by Mahesh et al. [72]), but there was no difference in the kinds of amino acid present between the two sexes [73,74]. Research by Cai et al. [75] and Kiuchi et al. [76] on male and female silkworm pupae has confirmed the presence of sex-related differences, adding information to the earlier reported differences in the amounts and compositions of the lipids in male and female pupae [77]. For example, more fatty acids were present in male than female pupae, but total lipid content of *B. mori* male pupae, on a fresh weight basis, was less (4.8%) than that of female pupae (9.0%) [78]. The content of unsaturated fatty acids was nearly the same in both sexes, but unsaturated acids were proportionately higher in female pupae [77]. In the case of *A. domesticus*, females have been shown to possess more lipids on a dry weight basis (18.3–21.7 g/100 g) than males (12.9–16.1 g/100 g), but less protein (63.1–65.7 g/100 g versus 69.9–71.9 g/100 g for female and male respectively). There was, however, no difference between the sexes with regard to the presence of essential amino acids (EAA) (72.3–77.1%), thrombogenicity (1.22–1.45%) and atherogenicity indices (0.53–0.58) [79].

In the subterranean termite, *Reticulitermes* sp., the reproductive caste had higher contents of the following nutrients than workers e.g., carbohydrates 2.7% versus 1.3%, protein 87.3% versus 81.7%, and amino acids 6.7% versus 4.7% [80]. Ntukuyoh et al. [81] evaluated the nutrient contents of the soldiers, workers and queens of the termite *Macrotermes bellicosus* in the Niger Delta region of Nigeria. Considering the average of the values provided, soldiers had the highest amount of protein (55.6%), lipid (2.7%) and fibre. Workers were especially rich in carbohydrates (65.1%), vitamin C (1.1 mg/kg), Fe (54.3 mg/kg), Mn (22.4 mg/kg) and Ca (58.3 mg/kg) whereas the queen contained higher amounts of vitamin A (7.0 mg/kg), Na (69.1 mg/kg), Mg (47.8 mg/kg) and Zn (25.2 mg/kg). Workers and queen had nearly the same amount of Cu (18.8 and 18.3 mg/kg respectively). In general agreement with this study by Ntukuyoh et al. [81], the same species collected from southwestern Nigeria by Idowu et al. [82] differed somewhat because of its higher content of ash, crude fibre, crude protein and carbohydrates in soldiers and workers rather than the reproductive caste which, on the contrary, had a higher fat content.

The weaver ant (*Oecophylla smaragdina*) exhibited a higher content of total lipid (average of annual values for larvae: 168.5, pupae: 140.7, and adults: 140.6 mg/g) in the queen while worker castes contained slightly more than half the amount (larvae: 112.0, pupae: 111.6, and adults: 100.8 mg/g) [83]. Lower contents of protein (37.5%) and ash (3.0%), but higher lipid content (36.9%) in the queen than that reported for other castes of weaver ants (presumably worker caste), were reported [84] for weaver ants in Thailand. In honey bees, considerable differences with regard to amino and fatty acids were documented not only for different developmental stages like larvae, pupae and adults, but also for the different sexes of the bees [33,61].

2.1.4. Organs

Dué et al. [85] analyzed oil extracted from the integument and the digestive tract fat content (DFC) of the larvae of the South American palm weevil *Rhynchophorus palmarum* (L.). Fat content obtained from the integument (="skin" in that paper) was lower (35.2%) than the DFC oil (49.1%). Oleic acid was highest in both oils (45.6–46.7%) followed by palmitic acid (39.9–40.4%). Saturated fatty acids were 45.1% and 45.0% for skin oil and DFC oil, respectively. Vitamin-A was found only in DFC oil. Regarding quality properties, the oil obtained from the integument was considered superior to that of the DFC oil, judged by their respective indices of iodine of 51.2 versus 48.4, peroxide of 6.9 versus 0.0 and oleic acidity (7.8 versus 0.6). However, data on the nutrient compositionsbased on specific insect organs such as fat body, ovaries, compound eyes, glands, etc.are extremely limited.

2.1.5. Ecotype or Biological Variations

When *Rhynchophorus phoenicis* larvae were obtained from plantations of the raffia palm (*Raphia* sp.), yellow larvae contained more fat (27.7%) than white wild (22.2%) or white breeding (17.4%) larvae, more protein (8.8% for yellow wild versus 7.8 and 8.7% for white wild and breeding kinds, respectively), more carotenoids (805.0 versus 391.0 and 276.0 µg/100 g for yellow wild, white wild and breeding respectively), but less polyunsaturated fatty acids (PUFA) (0.5 versus 0.8%) and tocopherol (2.3 versus 4.8 and 4.1 mg/100 g for yellow wild, white wild and breeding kinds, respectively) [86]. In Uganda, no significant differences were recorded for the dry matter and moisture contents between the brown ecotype and the green one of the cone-headed grasshopper, *Ruspolia nitidula* (Scopoli). High potassium content (5.55 mg/kg) was recorded [87] in brown grasshoppers collected in Kampala during the March–May season but not in the November–December season. These findings provide some information on the likely correlation between colour change (as a result of the ecotype) and chemical composition in grasshoppers influenced by climatic conditions and geographic location.

2.2. Ecosystem and Insect Habitat

Sustainability in insect diversity and year-round (or at least seasonal) availability is important for family livelihood of local communities [17]. Terrestrial insects are more abundant and easier to collect than aquatic species and may therefore be recommended for consumption in preference to the latter [88]. On the other hand, Williams and Williams [89] demonstrated the potential of aquatic insects to contribute to human diet. In a study of nutrient contents in both aquatic and terrestrial insects by using linear models, Fontaneto et al. [88] showed that terrestrial insects contained a significantly lower amount of monounsaturated fatty acids (22.5%) than that reported for aquatic edible insects (33.8%). In contrast, a higher amount (44.2%) of PUFA was found to be generally present in terrestrial insects rather than the aquatic species (27.9), although statistically the difference did not reach significance.

In Uganda, the effects of two swarming seasons (March–May and November–December) on the nutrient contents of *R. nitidula* were studied. No significant differences were found in the comparative contents of protein (39.7–40.4% in the March–May season and 37.0–39.1% in the November–December season), fat (41.9–42.4% in the March–May season and 41.2–43.0% in the November–December season) and carbohydrates, but carotenoids (2084.8–2273.1 µg/100 g in the March–May season versus 913.7–1389.4 µg/100g in the November-December season) and fibre (11.3–12.2% in the March–May season versus 13.1–14.3% in the November–December season) differed significantly with the seasons [87]. In an additional study, Ssepuuya et al. [90] reported that the geographical area was highly influential with regard to the insect's mineral content within a season, whereas the season alone affected significantly the variation in the contents of protein (34.2–45.8%), fat (42.2–54.3%), ash and minerals in *R. nitidula*. Geographical area or season, however, were not seen to affect the compositions of amino and fatty acids.

Maximum contents of fat in *O. smaragdina* ant queen larvae (249.2 mg/g), queen pupae (228.9 mg/g), worker larvae (129.9 mg/g), worker pupae (133.1 mg/g) and worker adults (123.2 mg/g) from Assam (India) were recorded in March, whereas maximum fat (207.3 mg/g) in queen adults was found in April and minimal amounts occurred during the November–February months [83]. These data, which were valid for various sites, can be used by collectors to decide the most suitable period to collect preferred insect species and their life stages based on nutrient contents.

In Botswana, Madibela et al. [91] studied contents of *Imbrasia belina* larvae sampled at three eco-sites (Mauntlala, Moreomabele, Sefophe). At Moreomabele, a high content of acid detergent fibre (ADF) (230.9 g/kg dry matter) and acid detergent insoluble nitrogen (ADIN) (18.0 g/kg DM) was noted, whereas these contents were least (155.5 g/kg for ADF and 11.8 g/kg for ADIN) at Sefophe. At Mauntlala, contents of ADF and ADIN were 175.1 and 12.2 g/kg respectively. In *M. bellicosus* termites from Nigeria, Idowu et al. [82] reported highest contents of vitamins (A, the B-complex, and C) in reproductive castes collected from farmland, whereas those from an industrial estate had the highest amount of Cu (0.076 mg/L). Lead was detected only in the soldiers. The highest value of Cr in workers was found in termites collected from farmland (0.226 mg/L) and a waste dumping site (0.223 mg/L). The chemical composition between hibernating queen bumblebees, *Bombus terrestris* (L.), during the winter season differed significantly from summer queens and featured an increased ratio from 3.6 to 4.9 between unsaturated and saturated fatty acids [52]. Thus, differences may be observed in species even if collected from a single site, but at different times of the year.

2.3. Insect Feed

Insects are reared on a synthetic diet (containing only chemical ingredients), semi-synthetic or meridic diet (containing synthetic and plant material). Both categories of food are classified as laboratory diets. The natural diet comes solely from natural sources, i.e., host plants or plant products. More details about these diets can be found in [92].

2.3.1. Plant Material

The nutritional status of the host plants that insects feed on affects the nutrient contents of the edible insects. When *vermiwash* (water washings of earthworm cocoons) at 10%, 25% or 50% was sprayed on mulberry leaves and fed to fifth instar *B. mori* larvae, significant increases on a dose-dependent basis were observed with regard to carbohydrates, protein and fat [93]. However, when Ebenebe et al. [94] reared *R. phoenicis* larvae on four organic substrates, e.g., sugarcane tops, split watermelon, split pineapple and raw papaya, their larvae had (on dry weight basis) normal carbohydrate, protein and fibre, contents that did not vary much and statistical significance was lacking. They concluded that split pineapple can be selected for feeding larvae as a potential source of protein and fluid.

Among 10 African host grasses of the longhorn grasshopper *Ruspolia differens* Serville, Malinga et al. [95] noted maximum survival of 65% on *Chloris guyana*, *Pennisetum purpureum*, *Setaaia sphacelata*, *Brachiaria ruziziensis* and *Sporoborus pyramicloris*. Fresh weight was highest (0.383 g/adult) on *P. purpurium* and *B. ruziziensis*. On a mixed diet basis, significantly shorter development time (16 days) from nymph to adult and higher survival (>65%) occurred in diversified diets compared to the use of single grass species. Contents of PUFA and fatty acid composition did not differ significantly among the diets. Therefore, a mixed feed can be recommended for mass rearing of grasshoppers.

Quaye et al. [96] evaluated four diets based on the oil palm yolk (*Elaeis guineensis* Jacq.) alone or mixed with banana and pineapple waste or millet waste. The highest protein (32.0%) and fibre (8.4%) content was found in *R. phoenicis* larvae fed on oil palm yolk. Practically, year-round non-availabilities of some plant materials and the costs involved to procure them make vegetative substrates often uneconomic for mass production by farmers and tribal communities [94].

Whenever an insect's natural food is altered, nutrient contents may be affected. For example, adding wheat bran to grass (natural food) to feed the locusts (*Locusta migratoria*) influence the nutrient composition, the protein content in adults varies from 555 g to 649 g/kg (dry matter) and the fat content varies from 186 g to 296 g/kg [97]. While rearing the Asian palm weevil, *Rhynchophorus ferrugineus* (Olivier) on three substrates, Cito et al. [98] recorded total fat of the order of 57.6%, 58.4% and 60.0% for apple juice, pineapple palm (*Phoenix canariensis* Chaubaud) and cocoa palm [*Syagrus romanzoffiana* (Cham.) Glassman] respectively. For the three substrates, the content of monounsaturated fatty acids was 43.9%, 41.6% and 44.7% while the unsaturated fatty acids reached 56.1%, 57.2% and 52.8% of the total fatty acids, respectively [98].

Larvae of *R. ferrugineus* fed on raffia palm [*Raphia farinifera* (Gaertn.)] were heavier (159 g) than those reared on oil palm, which weighed only 52 g [99]. These findings demonstrated how different substrates can be used and can be practical in relation to weevil rearing. Feeding neonate nymphs of the grasshopper (*R. differens*) till the imago stage on inflorescences of local grasses (8 species) did not strongly modify fatty acid content or composition or even adult body weight [100]. The ratio of n-6:n-3 fatty acids was generally low, a point of note in connection with the need for a healthy human diet. Significant differences in the composition of rare fatty acids (n-6/n-3, arachidonic acid, alpha-linolenic acid), however, were present in the grass species. In another study, Rutaro et al. [101] used inflorescences of four plants and found a high content (21%) of essential fatty acids in field-collected sixth instar nymphs compared to the low content (12–13%) in connection with less diversified diets. However, total lipid content and weight of grasshoppers did not differ among diets, but it showed that the fatty acid composition can be influenced by an insect's food uptake.

2.3.2. Laboratory Diets

Laboratory or artificial diets have certain advantages over natural plant material for rearing silkworms, because such diets are semi-synthetic or synthetic and can be used for several insect species. They help to rear insects that vary little between individuals which can then be made available whenever needed for bioassays or other purposes [92,102].

Rutaro et al. [103] formulated artificial diet for *R. differens* containing rice seed head, finger millet seed head, wheat bran, chicken egg buster, sorghum seed head, germinated finger millet, simsim cake, crushed dog biscuit pellet and shea butter. More diverse diets resulted in increased content of PUFA and linoleic acid. Fatty acid composition differed significantly among the diets. The workers concluded that essential fatty acid content can be increased by feeding grasshoppers on highly diversified diets, particularly when mass rearing is the objective.

Ghaly [104] prepared a diet of dry ingredients (corn flour + whole wheat flour + wheat bran + dried yeast powder mixed in a ratio of 3:3:3:1, by weight) and liquid ingredient (glycerine + honey mixed in 1:1 ratio, by weight). The two ingredients were then mixed in a 1:1 ratio. This diet proved superior to plant material in rearing *Gonimbrasia belina* and *Anthoaera zambezina* in Zambia. In Nigeria, the *G. belina* larvae reared on a semi-synthetic diet (corn starch + vegetable oil + glucose + cellulose + mineral mix + vitamin mix + protein) contained 7.1% carbohydrates, 35.2% protein, 15.2% fat and 7.4% ash [105]. These contents were comparatively low in larvae fed only natural plant biomass [105]. These diets should be tried in connection with other lepidopteran larvae consumed in Africa. Stull et al. [106] successfully reared *T. molitor* on a diet containing 40% stover (corn by product) by weight. Analyses after 32 days into rearing showed that the larvae contained all the essential amino acids. In another experimental series, the insects completed metamorphosis and all larvae survived on a 100% stover diet for multiple generations. Therefore, this diet can be recommended for the rearing of *Tenebrio* and possibly some other beetle species in the laboratory.

Indoor rearing can be further improved by fortifying the laboratory diet. For example, De Wit [107] prepared a vitamin D-enriched diet to rear *B. mori* larvae. There were significant changes in the content of the macro nutrients compared with diets that had not been fortified, e.g., increases in protein (61.2% versus 58.8%) and reduction in fat (37.3% versus 39.5%). The addition of the commercial protein supplement Nutrilite® increased the content of sericin and fibrous protein by 68% and 56%, respectively, with addition of 10% supplement compared to no addition for the late larval instars of *B. mori* [108]. This finding implies that laboratory diets should be preferred for edible insects if the objective is to obtain a greater amount of nutrients from the insect biomass.

2.3.3. Plant Based by-Products

While assessing 18 diets based on industry by-products (such as, potato protein, barley mash, leftover of turnip-rape and broad beans) for rearing of crickets such as *A. domesticus* and *G. bimaculatus*, Sorjonen et al. [109] reported yields of *A. domesticus* as high as 4.10 g on barley mash and 5.12 g of *G. bimaculatus* on turnip rape. The average weights of female and male *A. domesticus* were 0.459 g and 0.342 g, respectively, whereas in the case of *G. bimaculatus* the corresponding weights were 0.912 and 0.626 g. Thus, these protein-rich products can replace currently-used soybean in mass rearing. Further, Sorjonen et al. [110] added two more diets (mix of broad bean and pea, mix of potato, carrot and apple) for rearing of *R. differens*. Increasing protein level in the diet up to 17% enhanced growth, development time, and survival. Fatty acid content and composition differed as per diet. For example, high PUFA content was noted in connection with barley mash, barley feed and broad bean diets. Fresh weight was highest (0.507 g/adult) in the Suomalainen diet with vitamins and minerals as supplements [110]. However, the study suggested that the best options were barley feed, barley mash or potato protein.

Lehtovaara et al. [111] recorded nearly 10-fold increased contents of linoleic, alpha-linolenic, eicosapentaenoic and docosahexaenoicfatty acids in *R. differens* adults when these acids were present in the artificial diet. Development performance was also improved with n-6/n-3 acid ratio. Lack of protein and fat in the diet prolonged development and resulted in low final weight. Therefore, it is necessary to design nutritional content in artificial diets to obtain heavy, nutritious grasshoppers through mass rearing.

2.4. Insect Processing and Product Quality

Traditionally, insects are consumed raw or processed (dried, crushed, pulverized, ground, pickled, cooked, boiled, fried, roasted/grilled, toasted, smoked or extruded [112]. Besides these techniques, Kewuyemi et al. [113] suggested fermentation to enrich the inherent composition of insect-based products and to induce anti-microbial, nutritional and therapeutic properties. Similarly, defatted *T. molitor* larvae and oil could be used as food ingredients [114]. Defatted mealworm powder contains sufficient protein, minor amounts of minerals and bioactive compounds and has a savory taste due to plentiful amino acids. Oil is abundant in γ-tocopherol and possesses good shelf life [114].

Before processing, insects are often kept without food for fasting and large speci-mens are degutted or defatted, because the gut may contain undigested plant material, excreta, microbes etc.; moreover, degutted insects have higher contents of crude fibre protein [91,112]. This practice, being efficient and practical, has been routinely adopted by tribal communities particularly for large lepidopteran larvae. Processed insects can be preserved by freeze-drying or sun-drying and in canned form. Processing methods may differ as per consumer preference, availability and suitability of insect species, social custom, religious rituals, tribal ethics and family tradition [17]. The effects of four different drying temperatures (80, 100, 120, 140 °C) on antioxidant properties of silkworm powder was studied by Anuduang et al. [115] and showed that the lowest drying temperature preserved phenolic compounds and antioxidants best.

In selecting a food item on the basis of "post-ingestive fitness", the processing method can help to remove anti-nutrients and other unhealthy components as well as increasing the shelf life. Thus, processing is important to maintain the level of nutrient content, to extend the shelf life and to obtain functional and fortified foods [112]. In food processing units, products are enriched with insect chitosan (a polysaccharide derivative of chitin), which is more soluble and therefore preferred over raw chitin in food processing [116]. Local communities frequently know methods to improve insect-based foods with traditional wisdom built on generations of experience [117].

Methods can of course change and be replaced by others, because each method has certain advantages or disadvantages suiting regional needs.For example, roasting, cooking and frying are largely employed in North-East India, because of the better taste of insects compared to boiling and baking [13]. Generally, vitamins are susceptible to heating and heat processing decreases the level of these vital compounds entirely or partially. Storage conditions are important to prevent the deterioration of insects. However, tocopherol con-tent in *T. molitor* and *Zophobas mario* L. was not altered under different environments [118], but antioxidant properties in silkworm powder were [115]. Nyangena et al. [119] examined the effects of traditional processing techniques, i.e., boiling, toasting, solar-drying, oven-drying, boiling + oven drying, boiling + solar-drying, toasting + oven-drying, toasting + solar-drying on the proximate composition and microbiological quality of *Acheta domesticus*,

Ruspolia differens, *Hermetia illucens* and *Spodoptera litoralis*. They found that traditional processing improved microbial safety but altered the nutritional value. Moreover, species- and treatment patterns clearly existed. A few examples below may demonstrate the different processing techniques used in selected insect species.

2.4.1. Lepidoptera

Maceration of *Bombyx mori* pupae, being simple, practical and not at all costly, is the most common method employed by indigenous people. For example, Winitchai et al. [120] reported 72–79% of unsaturated fatty acids and 32–44% of alpha-linolenic acid in Soxhlet extractions of fresh pupae whereas respective contents after maceration were 75–80% and 40–46%. Defatting of pupae and turning them into a powder is practiced to retain nutrients as witnessed in comparisons with the powder of non-defatted pupae [121], e.g., crude protein 57.4% versus 48.3%, digestible protein 48.3% versus 40.1%, soluble protein 29.0% versus 16.4% and carbohydrates 15.8% versus 10.2% [74].

In India, fresh pupae of a bivoltine breed of *B. mori* contained 17.1% protein and 9.2% fat compared with 56.9–75.2% protein and 24.9% fat of dried pupae [122]. Oil extracted from pupae is an important source of unsaturated fatty acids (75%), essential linoleic acid (33%) and alpha-linoleic acid (35%) [121].

Once the silk threads are separated, silkworm pupae (often referred to as chrysalis) become waste matter, but the powder of the empty pupae contains important nutrients [72] as also reported by Rao [123] from India, e.g., 48.7% protein, 30.1% fat, 8.6% ash, and significant amount of minerals and vitamins; however, defatted spent pupae contained higher protein, e.g., 75.2%. In Brazil, Pereira et al. [124] recorded that chrysalis (pupae) toast contained protein (51.1%), fat (34.4%), linolenic (24.4% of total fatty acids), palmitic (24.6% of total fatty acids), stearic (7.6% of total fatty acids), oleic (34.8% of total fatty acids), and linoleic acids (7.0% of total fatty acids), Zn (244.0 µg/g) and K (4.8 mg/g). Because of the high content of major nutrients and essential fatty acids, chrysalis toast was recommended as an alternative dietary supplement in Brazil [124]. These results suggest that as far as possible, larvae should be defatted and de-oiled before processing. The powder is less perishable and can be stored in cool and dry places in a house/hut or refrigerator for a long time and may be consumed whenever needed. Currently, this processing technique remains invalidated although valuable for nutritional security. Furthermore, steamed and freeze-dried mature silkworm powder exhibits different pharmacological effects [125] as calorie restriction mimetics [126], including enhanced mitochondrial functions in the brain [127].

In Africa, when *Gonimbrasia* (=*Imbrasia*) *belina* larvae were degutted, washed and dried, the shelf life could be extended up to one year without any deterioration in quality [128]. Lautenschläger et al. [129] compared three traditional techniques of preparation before consumption, e.g., evisceration, cooking and drying of silk moth, *Imbrasia epimethea*,(Drury) larvae. No significant differences for protein, fat, amino acids and fatty acid composition were observed among the three treatments. Gut removal reduced carbohydrates originating from the leaves of the host plant and showed negative effects on the nutritional value. No changes in the content of nutrients (except reduction in MUFA) were observed when the larvae were exposed to thermal processing. The study by Medigo et al. [130] showed that processing can also involve adding foods containing high amounts of sugar and saturated fats, although such additions diminished the overall nutrient content of the insect product [130].

In Botswana, Madibela et al. [91] compared nutrient contents in *Gonimbrasia* (=*Imbrasia*) *belina* larvae without any processing and those that were degutted salted or degutted + salted. Degutted samples had a higher content of crude protein (567.5–579.3 g/kg DM) than non-degutted larvae (505.3–537.5 g/kg DM) but had a lower concentration of ash (41.2–39.0 g/kg in degutted versus 58.0–59.2 g/kg in non-degutted larvae). A similar trend was present with regard to acid detergent fibre (ADF) (148.5–173.2 g/kg in non-degutted versus 158.8–268.1 g/kg DM in degutted larvae). The addition of salt increased ADF, whereas the degutted + salted samples had a higher ADF than degutted or salted larvae or those without processing. Unprocessed insects diluted the concentration of protein but increased the fiber and tannin contents.

It is possible that the heating results in the formation of new components, and that solvents used for the extraction of protein at industrial level can affect the safety of novel protein products [131]. In two edible insects namely *Imbrasia truncate* Aurivillius and *I. epimethea*, appreciated by consumers in the Congo river basin, Fogang Mba et al. [132] reported respective contents (fresh weight) of protein as high as 19.1 and 20.1 g and of fat reaching 6.8 and 6.7 g/100 g in the larvae. Unsaturated fatty acids in *I. truncata* and *I.epimethea* were 2.6 and 2.2 g/100 g, of which alpha-linolenic acid amounted to 1.9 and 2.2 g/100 g respectively. Processing procedures are held to be responsible for the ranges and, therefore, need to be well planned in connection with food products that contain insects.

2.4.2. Coleoptera

There was a considerable increase in the content of carbohydrates, protein and fat of sun-dried larvae of *O. rhinoceros* and *R. phoenicis* over non-processed ones. Pith of coconut palm for *O. rhinoceros* [133] and raffia palm pith for *R. phoenicis* [134] served as natural feed source whereas a laboratory diet comprising of corn starch, vegetable oil, sucrose, glucose, cellulose, mineral mix and vitamin mix for *O. rhinoceros*, *Gonimbrasia (=Imbrasia) belina*, *M. bellicosus* and *R. phoenicis* [105] was successfully used. A greater increase in dry weight was seen in laboratory-diet fed larvae than those fed only natural plant material.

Wheat flour dough enriched with ground *T. molitor* larvae at 10% resulted in a softer dough with increased size and weight when baked at 200 °C for 22 min [135]. High moisture extruded meat substitutes can contain the biomass of another species of beetle, e.g., *Alphitobias diaperinus* (Panzer) 40% mixed with soy dry matter up to 60% [136]. Water content in the product was important for improving its physical properties, i.e., the sensation when biting or chewing it.

2.4.3. Orthoptera

In laboratory experiments, *R. differens* adults were frozen at about −50 °C for 96 h [137]. Another lot was oven-dried at 60 °C for 24 h. No significant differences in the two methods were present for the content of average crude protein (46.4 for freeze dried and 47.7% for oven fried samples) and fat content (35.6% and 35.5% for freeze and oven dried samples, respectively). The content of chitin, by contrast, varied from 11.3 to 13.4% for freeze and oven dried samples. The mineral contents (in 100 g DM) in oven-fried and freeze-dried grasshoppers were as follows. Na = 54.0 versus 69.1 mg, K = 779.2 versus 816.4 mg, Ca = 895.7 versus 1034.7 mg, Mg = 145.8 versus 161.0 mg, Zn = 14.6 versus 14.2 mg, Fe = 216.6 versus 220.1 mg, P = 652.3 versus 685.9 mg, Cu = 1.7 versus 1.7 mg and Mn = 7.4 versus 8.3 mg [137]. On the contrary, toasting + drying significantly reduced protein digestibility in *R. differens* (76.4% versus 82.3% in fresh specimens of the grasshopper: [138]). Boiling of grasshoppers resulted in significant increases of protein and elements such as Fe, Zn, Cu, Mn and Ca contents, but decreases in the fat content on dry matter basis. Amino and fatty acid profiles were minimally affected but a significant reduction in ash content was noted. In case of roasting, there was an increase in Ca and trace mineral elements. The colour was uniformly intensified in green and brown polymorphs when roasted together. The aroma of heat-processed grasshoppers was influenced by lipid oxidation [139].

Fresh dried grasshoppers had a maximum fat content of 43.1% compared with only 16.3% in fresh insects. Also, there was a reduction in niacin content when the grasshoppers were toasted, toasted + dried or fresh dried (3.06–3.28 mg/100 g versus 3.61 mg in fresh insects [138]. High protein and fat contents, which contributed to >75% of the dry mass, justify the high reputation of this nutritionally valuable species for human consumption.

Hassan et al. [140] compared two processing methods for the tree locust, *Anacridium melanorhodon* (Walker) and reported that frying of adults resulted in slightly increased fat absorption (1.3 mL/100 g by frying versus 1.0 mL/100 g by boiling).Boiling, however, resulted in a reduction in tannin content (9 mg/100 g by frying versus 5.8 mg/100 g by boiling) and high protein digestibility (41.1% by frying versus 49.9% by boiling). Better digestibility was associated with water absorption (2.5 mL/100 g in fried versus 2.9 mL/100 g in boiled insects). Farina [141] compared the broth prepared by cooking *A. domesticus* adults after freezing them with those adults that were alive when cooked. There was a significant difference in the pH, overall acceptance and perception of saltiness and umami/savory flavour. These qualities were associated with the breakdown of glycogen and the formation of lactic acid during the killing of the insect. Therefore, a proper processing method needs to be selected in connection with insect-based protein in the presence of sodium chloride.

2.4.4. Blattodea

Kinyuru et al. [138] reported no significant change in protein digestibility (range of 90.1–90.5%) in the winged termite *Macrotermes subhyalinus* Rambur, but a significant increase in fat content was present due to fresh drying (42.3 g versus 19.8 g/100 g in fresh collection). A significant reduction in retinol content (0.98–1.6 µg/g versus 2.2 µg/g in fresh insect) and riboflavin content (2.8 mg/100 g in toasted termites versus 4.2 mg in fresh stock) was recorded [138].

When the flour of the soldier caste of the termite *Syntermes* sp. was mixed with honey spread at 8, 16 and 24% and then processed by pan-frying at 80, 90 and 100 °C, the 24% mixture exhibited a significant increase in the content of protein from 5.6 to 15.9 g/100 g, Fe from 3.8 to 8.8 mg/100 g and Zn from 1.8 to 4.5 mg/100 g. It also led to improved sensory qualities, especially flavour and taste [142].

3. Insect Quality

3.1. Content of Anti-Nutrients

Table 7 represents anti-nutrient contents of selected edible insects based on currently available information. In comparison to the data on nutrient content of edible insects, data on anti-nutrients are limited and even controversial. By definition anti-nutrients hinder or inhibit the absorption of nutrients, especially minerals, but some may also provide anti-oxidants like polyphenols including tannins. Due to insufficient data, saponins, alkaloids, etc. have not been included in our list.

The available literature shows high variations in the amounts of individual anti-nutrient compounds. Edible insects are mostly herbivorous, feeding on plants and their parts. For self-preservation plants synthesize different types of secondary metabolites and these secondary metabolites are known as allelochemicals and accumulate in the bodies of plant matter-ingesting insects. Their primary action is to inhibit the absorption of necessary nutrients and they are therefore termed anti-nutrients. The wide variation of the insects' anti-nutrient content is likely to be due to the different chemical compositions of plants on which the insects feed. Primarily, it depends on the environment and the site that a plant is growing. However, a systematic protocol has to be developed in order to quantify the anti-nutrient contents. In this context it is also worth mentioning that the development of rearing techniques of edible insects under controlled conditions can minimize or even avoid the contamination of insects with these allelochemicals.

Table 7. Anti-nutrient content (mg/100 g) of selected edible insects.

	Phytate	Tannin	Oxalate	Trypsin Inhibitor	Lectin	Hydrocyanide	Reference
Ant [†]	2030.8	400.0					
Termite [†]	2482.1	948.3					
Winged termite [†]	1128.2	250.0					
Cricket [†]	3159.0	900.0					[143]
Meal bug	2256.4	1150.0					
Grasshopper [†]	1100.1	1050.0					
Anaphe venata [†]	1918.0	753.3					
Tree hopper		1000.0					
Rhynchophorus pheonicis * L	1.4	1.0	0.1	0.9	0.6		[144]
Gymnogryllus lucens [†] A	0.03	0.03	1.3			0.2	
Heteroligus meles [†]	0.03	0.04	2.8			0.3	[145]
Rhynchophorus [†] L	0.03	0.04	1.8			0.2	
Zonocerus variegatus [†] A	0.03	0.04	2.6			0.3	
Oedaleus abruptus [†] A		2450.0	600.0				[146]
Lethocerus indicus * N,A		372.3					
Laccotrephes maculatus * N,A		350.4					
Hydrophilus olivaceous * A		528.7					[147]
Cybister tripunctatus * A		301.7					
Crocothemes servillia * N		465.3					
Macrotermes nigeriensis [†] A	15.2	0.6	103.0				[37]
Oryctes rhinoceros [†] L	16.1	0.6	109.0				
Oecophylla smaragdina [†] A	171.0	496.7					[38]
Odontotermes sp. [†] A	141.2	615.0					
Oxya hyla hyla [†] A		2316.0	474.0				[148]
Oryctes rhinoceros [†] L	37.0	5.6	1.3				
Oryctes rhinoceros [†] P	39.4	6.8	1.3				[70]
Oryctes rhinoceros [†] A	41.1	4.2	1.2				

L = Larva, P = Pupa, N = Nymph, A = Adult; * Anti-nutrient content was estimated on the basis of wet weight; [†] Anti-nutrient content was estimated on the basis of dry weight.

3.2. Contamination with Chemical Pesticides, Inorganic Products and Infestations with Insect

Chemical pesticides sprayed on host plants of edible insects are often stored in the form of residue in the insect body. That chemical contamination causes a deterioration of, for example, the quality of edible insects such as the locust *Locusta* sp. in Kuwait (and the water bug, *Lethocerus indicus* in India was shown [149,150]. Poma et al. [151] reported low concentrations of heavy metals, DDT (Dichlorodiphenyltrichloroethane) and dioxins in edible insects compared with chicken egg, fish, and animal meat. In China, heavy metals have been found in *B. mori* larvae fed on mulberry leaves harvested from plants cultivated in soil treated with municipal solid waste compost [152] or grown in soil-polluted fields [153]. The hide beetle, *Dermestes maculatus* DeGeer, which feeds on dry animal matter, also attacks dry edible insects. In the laboratory, Fasunwon et al. [154] experimented with artificial inoculations by this beetle with larvae of the rhinoceros beetles *Oryctes boas* (Fabricius) and *R. phoenicis*. There were significant differences in the nutrient contents when containers were provided with a mixture of salt and pepper (10 g/container). Protected larvae of *O. boas* contained 56.1–60.6% protein versus 39.3% of the control and *R. phoenicis* had 34.8–37.3% protein versus 22.7% in the control. A similar trend was present for the fat content with 4.3–7.8% versus 4.5% in the control of *O. boas* larvae, and 20.1–30.7%

versus 13.5% in the control of *R. phoenicis* larvae. Therefore, storage of well dried edible insects mixed with salt and pepper has been recommended to maintain the nutritional quality [154].

3.3. Microbial Contamination

Contamination with microorganism is a major factor in the deterioration of the quality of insect-based food items. Numerous bacterial species are known to affect insects including *Bacillus cereus* Frankland andFrankland, *Staphylococcus aureus* Rosenbach, *Escherichia coli*, *Rickettsiella* spp. Some insects also act as carriers of human pathogens of the genera *Salmonella, Campylobacter, Shigella* [155]. Additionally, grasshoppers serve as intermediate hosts to several avian parasites, horsehair worms and tapeworms [156].

In Ghana, larvae of *R. phoenicis* fed on raffia palm or oil palm contained bacteria at 1.3×10^7–6.5×10^6 colony-forming units (CFU)/g body mass, which was higher than the acceptable level of 5.0×10^6 CFU/g [99].In Nigeria, Braide andNwaoguike [157] assessed the quality of processed larvae of this widely consumed species and reported a load of bacterial and fungal counts of the order of 1.68×10^5 CFU/g and 9.2×10^2 CFU/g, respectively. The major bacteria were *Lactobacillus plantarum* Bergey, *S. aureus* Rosenbacch, *Bacillus subtilis* (Ehrenberg) Cohn, *Pseudomonas aeruginosa* (Schroter) Migula and *Proteus vulgaris* Hauser.

Major fungal species were *Cladosporium* sp., *Penicillium verrucosum* Dierckx, *Aspergillus flavus* Link and *Fusarium poae* (Peck) Wollenweber. In Zimbabwe, stink bugs such as *Encosternum deregorguei* Spinola, stored in dung-smeared wooden baskets, were found contaminated with aflatoxin (a carcinogenic mycotoxin: [158]. Consequences of the contamination were, however, not studied. Braide et al. [159] recorded contamination of bacteria (4.49×10^7 CFU/g) and fungi (9.5×10^6 CFU/g) from collections in the wild of caterpillars of the emperor moth, *Bunaea alcinoe* (Stoll) in South Africa; caterpillars harbouring bacteria such as, *P. aeruginosa* and *Proteus mirabilis* produced undesirable flavours in food products, and *S. aureus, B. cereus* and *E. coli* produced toxins [159]. Furthermore, *S. aureus* was easily introduced during the handling of insects [160].

Processing (by traditional and innovative methods) can eliminate microbes or at least considerably reduce their load [112]. In lactic acid fermentation of a mixture of sorghum flour and *T. molitor* larvae, the level of spore forming bacteria (*B. subtilis, B. megaterium, B. licheniformis*) remained stable suggesting the bacteria were unable to germinate; their quantity remained at the acceptable level of $<10^3$ CFU/g [161]. Some of the effective and practical safety measures discussed below can be implemented. For example, thorough washing and heating can reduce microbial contamination to some extent [155]. Also, insect boiling before roasting proved effective to keep spore forming bacteria under check and insect dehydration can also reduce microbial contamination because at the lower humidity bacteria grow less.

Modified packaging systems are needed to prevent further contamination and to enhance shelf life of stored edible insects [159]. Regular monitoring and evaluation for bacterial contamination should be undertaken during storage as environmental changes can affect the insect quality. Therefore, refrigeration is employed to prevent contamination compared to outside storage at ambient temperature [162]. Packaging material is another critical aspect for safety. For example, boiled, solar dried and milled house crickets when stored for 2 months at ambient temperature in polypropylene (PP), plastic or polyethylene packages, the PP-packaged insects lasted only 45 days compared to 2 months with the other packaging materials. In all packages, iodine values, contents of SFA, MUFA and PUFA significantly decreased, peroxide, p-anisidine and saponification values increased and incidences of yeast and mould (*Aspergillus, Alternaria, Penicillium*) were high [162]. Although plastic packages with lids outperformed bags, adding a layer of polypropylene on the inner side can minimize permeability and exposure to both air and water vapour and thus can prolong the shelf life [162].

In food industries in Europe, a recent study has revealed antibiotic resistant genes in *A. domesticus*. As a preventive measure, Roncolini et al. [163] suggested standardization of the production processes and a prudent use of antimicrobials during the rearing of edible insects. Even though *Spiroplasma* sp. and *Erwinia* sp. in *T. molitor* and *Parabacteroides* sp. in the tropical house cricket, *Gryllodes sigillatus* (F. Walker) were the major pathogens during rearing of insects in the laboratory, Van der Weyer et al. [164] recommended that food safety should include also general bacteria like *Cronobacter* spp. or spoilage bacteria (*Pseudomonas* spp.) to be considered as potential human pathogens.

3.4. Allergenic Proteins

Allergenic reactions to the ingestion of insects and cross-reactivity with homologous proteins and co-sensitization between insects have been reported [165]. Allergens of edible insects identified as muscle proteins such as myosin, sarcoplasmic-Ca-binding protein, the major being tropomyosin and arginine-kinase (also known as pan-allergens).Persons who are allergic to dust mites and crustaceans could have an allergic reaction to foods containing *T. molitor* proteins [166]. It is possible that human susceptibility is due to immunoglobulin E-binding cross-reactions. For example, persons allergic to shrimps react with protein extract of *T. molitor* that shows Ig-E-binding cross-reacting allergens with other phylogenetically related groups of arthropods. Therefore, consumers allergic to shellfish should invariably be notified about the risk of developing an allergy by labelling the insect products accordingly [167].

It was shown that thermal processing and digestion did not eliminate insect protein allergenicity [168]. But the recent technique of high hygrostatic pressure coupled with enzymatic (pepsin) hydrolysis improvedin vitrodigestion of allergenic proteins of *T. molitor*. This technique can be an alternative strategy to conventional hydrolysis to generate a large quantity of peptide originating from allergenic *T. molitor* proteins [169].

3.5. Food Fortification

Insects as supplements for the predominant staples like corn, cassava, sorghum, pearl millet, beans and rice are commonly employed by indigenous folk in many developing countries. Insects also form a sustainable ingredient to produce new food items because fortification increases richness in nutrients. Corn being a major staple food crop in sub-Saharan Africa, members of local communities consume this tryptophan- and lysine-deficient product in considerable quantities. But this diet can be supplemented with termites and lysine-rich silkworms to overcome the deficiencies in these amino acids [170]. Similarly, natives of Papua New Guinea consume crop tubers with a low content of lysine and leucine. Vitamin deficiency can be remedied by supplementing the diet with *Rhynchophorus* spp. larvae, containing high amounts of vitamins and lysine. Tubers enriched with tryptophan and aromatic amino acids can render a diet more balanced and nutritious [170]. Ayensu et al. [171] mixed flour (70%) of *R. phoenicis* with orange-fleshed sweet potato and wheat flour to prepare biscuits. These biscuits had their energy, fat and protein content increased by fortification with palm weevil larvae powder compared with biscuits containing 100% wheat flour. Contents of Ca, Fe, Zn were also increased. The biscuits were highly appreciated by pregnant women in eastern Africa.

Winged termites such as *M. subhyalinus* added at 5.0% to cereal-based recipes in the Lake Victoria region of Kenya not only improved the food quality (protein, retinol, riboflavin, iron, zinc content) but made the food also more attractive [172]. In Mexico, corn bread ('tortilla') is sometimes supplemented with *T. molitor* larvae to improve consumer acceptance as well as nutritional content especially essential amino acids [173]. Likewise, Kwiri et al. [128] suggested that edible insect *G. belina* could be an alternative substitute of the current local plant-based supplements (beans, peanut, cowpea) in Zimbabwe.

Kim et al. [174] recommended up to 10% replacement of lean meat/fat portion with flour of the cricket *A. domesticus*. When compared with meat to which no cricket flour had been added, this level of mixing increased the content of protein from 14.0% to 20.7%, fat from 9.8% to 10.4%, potassium from 261.0 to 355.2 mg/100 g, phosphorus from 242.9 to 338.0 mg/100 g, magnesium from 20.4 to 33.5 mg/100 g, zinc from 1.7 to 3.8 mg/100 g and sodium from 967.0 to 1053 mg/100 g, but reduced Ca. The improved contents fulfilled the requirement of protein and micronutrients in meat emulsion [174]. In conclusion, lean meat can conveniently be replaced with cricket flour. This innovative step may encourage food industries to follow this mixture in food recipes not only to improve current entomophagy practices adopted by tribal communities but also to popularize mixing in commercial products sold by food companies.

4. Impact of Insect Quality on Consumers' Preference and Acceptability

For accepting insects as food, nutrient content (protein being a major component), quality of insects (particularly, taste, flavour, appearance, palatability) and external factors (availability, convenient pricing, conducive social environment) are important [12,175]. Forest-dwelling communities not only in developing countries [17] but also in rural places like, for instance, in Japan [176] have easy access to wild areas and prefer wild insects because of their taste. Currently, little is known about the consumers' reactions to wild insects and their food products, preference, acceptance and consumption of insect-based foods [29]. There are, however, anecdotal reports of preferences and greater acceptance of selected wild species of edible insects in Africa and India over reared species like silkworms or crickets.

In Kenya, Alemu et al. [177] found no significant difference in consumption of whole or powdered termites. At the local market, consumers checked the insect stock for freshness, presence of legs, cleanliness, species type and oil content before purchasing termites, for example, *Macrotermes falciger* (Ruelle). The majority of the buyers (77.6%) preferred fried adults [178]. Termite soldiers with long bodies were in great demand and highly preferred over alate forms [178]. The grasshopper (*R. differens*) is a traditional delicacy, a source of nutrients and tasty multipurpose insect in Tanzania, Kenya and Uganda [179]. Consumers preferred grasshoppers which were salted, boiled and smoked or deep fried in cotton seed oil over any other single processing methods (smoking, deep frying, sun-drying, toasting, boiling) [179]. In another survey, *R. differens* adults boiled with salt, onion and tomato and then dried, were preferred over those only deep-dried with salt and onion. The acceptability ranking was 7.2 and 5.2 for these products respectively (ranking scale of 0–9, with 9 being the maximum acceptance: [137]). In the case of *R. nitidula*, people in Uganda preferred the boiled and dried grasshoppers with salt, onion and tomatoes over those which were simply boiled and dried without tomatoes. In India whole insects are preferred, with the exception of grasshoppers, whose legs are sometimes removed; larvae, pupae and adult termites are often mixed and sold together by indigenous vendors [13].

Overall, although a greater acceptance was noted for insects without much attention to the species, fear of trying an unknown product, lack of taste experience and a belief of low social acceptance were considered as major constraints in popularizing edible insects [180] even though taste alone in more than 50% of probands tested did not enable them to distinguish insects from cheese or bread [181]. Correct labelling is also an important factor of acceptability. Recently, while assessing the accuracy of insect products in the UK market, Siozios et al. [182] found, by using DNA barcoding, frequent disparities between identity in packages containing mopane caterpillars, winged termites and grasshoppers. This may distract consumers from accepting and consuming insects or insect products.

In selecting an edible species information on entomophagy, prior experience and familiarity with edible insects, appearance, flavour and overall likability of a species are major factors [12]. Therefore, information and knowledge can influence attitudes towards insects as food and food supplemented with edible insects [183]. In fact, Van Thielen et al. [184] reported an increasing positive response in Belgium regarding acceptance, as revealed by a survey undertaken two years after the introduction of edible insects in that country. In a similar survey of Danish consumers revealed the fact that 23% of them were willing to eat insects [185].

Before experiencing the taste of cricket powder, Canadian consumers thought that consumption was undesirable, but their attitude changed after having consumed the powder and they were then willing to buy it [186]. In the USA, Mexico and Spain, replacing wheat flour with 15% and 30% of cricket (*A. domesticus*) powder in chocolate chip cookies was evaluated [187]. No difference between 15% and 30% was noted by USA consumers whilst Mexican and Spanish consumers liked the 15% sample significantly more than the control and 30% sample. From this survey, it was concluded that 15% powder did not negatively impact acceptability but improved liking and protein content in cookies. In Brazil, female consumers were more reluctant towards entomophagy than male subjects, but in Benin no such gender effect existed [188]. Preference was given to whole insects, although insect flour was liked by 40% of all consumers. Generally, insects were considered safe for consumption by educated consumers familiar with entomophagy [189]. Perception of entomophagy by residents of Korea and Ethiopia assessed through structured questionnaires revealed a positive note [190].

According to Medigo et al. [130], Belgians also have a positive attitude towards consuming insects and they have, moreover, developed a taste for certain species. Therefore, Sun-Waterhouse et al. [28] opined that practical approaches for transforming insect biomass into consumer food products are vital. Based on a survey and experiment in Australia and the Netherlands, Lensvelt and Steenbekkers [8] concluded that providing information on entomophagy and giving people the opportunity to try insects were two important aspects to influence consumers' attitude towards edible insects.

As another approach, Collins et al. [191] suggested educating school children, extending promotion for acceptance of the insect product to facilitate the adoption of insect food as a mainstream item and, thereby, as an available product through market chains. For example, the number of "burgers" with meal worms consumed by western people depended upon appearance, taste and flavour [192]. These criteria are important especially when mealworm products are considered inferior to the carrier products [193]. Medigo et al. [130] found that the processed mealworms with chocolate were the most popular insects whereas whole and crushed mealworms or boiled or baked crickets were least consumed. Similarly, despite the good nutritional qualities of mealworm larvae, it is uncertain that they would become a safe source of protein for Europeans because of their difficult to control, highly and variable microbial load [194].

Edible insects are not only something for developing countries, but equally important for developed countries (because of the problems of the latter with an increasingly obese populace) where they ought to find perhaps even greater acceptance than in the developing countries. Information about presentation, conservation, preservation of food products and local marketing is to some extent now available [17]. Studies are lacking, however, on likely changes in flavour, taste and texture of mealworm and other insect-containing products during storage; for the industrial production, moreover, information on suitable packaging and presentation of insect products is equally important. Effective advertising of edible insects could undoubtedly do with improvements and using catchy slogans like "Forget about the pork and put a cricket on your fork" or "Mealworms and spaghetti is food that makes you happy" could be expected to help as well [30].

5. Conclusions and Suggestions for Future Research

Insects are being considered as an alternative to conventional protein sources for both developing as well as developed countries. Edible insects have received attention from researchers in recent years because, firstly, the consumption of insects has spread to urban areas and the current concept is oriented towards health-related as well as ecological issues, and secondly, because conventional livestock rearing and certain systems of crop cultivation have proved environmentally disastrous [27,195–197]. Bioprospecting is currently limited to a few eco-zones in countries where insects have commonly been consumed. Intensive surveys may yet reveal more species that could be considered for consumption and farming. The shelf life of processed insects could possibly be improved if more research were devoted to this aspect. Factors responsible for nutrient content and quality of edible insects have not been explored sufficiently and to know how the chemical composition, handling and storage methods, contamination with micro-organisms, the insects' diet, feeding schedules, host plants and the plant's own nutrient content as well as the seasons affect food insect marketability would be of considerable benefit in selecting the most suitable species [198].

There is a need to develop rearing facility designs to be made available to small mass production units, which can help create a socially acceptable climate for the expansion of entomophagy [199]. Mass collections of aquatic insects by using nets woven by fishermen may be a remunerative venture, but encouraging locals and creating marketing channels as well as obtaining permits to fish for aquatic insects could be major hurdles. Recently, Oppert et al. [200] suggested sequencinggene transcripts from embryos, one-day hatchlings, nymphs and male and female adults of *A. domesticus* to use genetically modified crickets for improved insect production.

Regulations and legislation along with proper farming procedures, storage and hygiene would benefit consumers by way of healthier insects. Frameworks shared by different countries exist in Europe [201] but are lacking for most developing countries [202]. Proper processing and decontamination methods against micro-organisms during collection and storage, and preference of species should be included in surveys to ensure food safety [22]. A compilation of all information may be used to select a few insect species for mass rearing or augmenting their survival rate in natural habitats and a linkage through regional or international networks among countries/regions where entomophagy is practiced would be a first step. The network could facilitate exchanges and dissemination of information on insects and insect related recipes.

It is essential to conserve wild edible insect populations and to improve the survival of the most popular species [17]. This can be achieved by studying the population dynamics of these insects, identifying their host plants, and controlling their enemies. Other actions can include restrictions on over-harvesting, revoking the decreasing diversity of host plants, boosting the insects' resilience to adverse weather phenomena and seasonal effects and monitoring insect diseases.

A regulatory legal framework is required to guarantee that manufacturing practices, quality management, hazard analysis and other issues related to content and quality of edible insects are meeting acceptable standards [202]. Furthermore, proper labelling and documentation of the insect product would help to boost the consumers' knowledge and interest in entomophagy as would some cheeky and witty slogans to promote insect-containing food items [30]. The scientific guidelines explained by the European Food Security Authority [201] are worth studying to prepare a manual for insects consumed in developing countries, either on a regional or national basis to assure food and nutritional security.

In certain regions the people's diet may lack zinc, but in others there may be a shortage of magnesium, or iron, or calcium. To improve situations such as these, some species of insects could be promoted that are particularly rich in the minerals that are needed. Likewise, there may be reasons to boost certain fatty acids in the diet, acids that could be supplied by specific species of insects. To be able to select the appropriate species, it is of course essential to know precisely the chemical composition of the insect species, which demonstrates how important it is to have a detailed catalogue of the contents of as many species of insects as possible. If one extends this to animal feed, fish culturists may desire in particular protein-rich species, but pigs should perhaps be fed fatty insects and poultry farmers may wish to obtain insects with a high calcium content.

To promote insect-based functional foods as a platform for certain health-related properties is a promising option and is to some extent already taking place, e.g., larvae of the pallid emperor moth *Cirina forda* for protein solubility, oil absorption capacity and foaming stability [203], *T. molitor* larvae for their oil, foaming and emulsion capacity [204], black soldier fly for peptides with antimicrobial activity against the stomach ulcer bacterium *Helicobacter pylori* [205] and male silkworm pupal extract with its Viagra-like effect for erectile dysfunction [206].In fact, Meyer-Rochow [207] reviews hundreds of species that can be used therapeutically, but in many cases also serve as food for humans. This is an aspect certainly worth exploring further.

Author Contributions: Conceptualization, R.T.G., S.G. and V.B.M.-R.; Methodology, not applicable; Software, not applicable; Validation, R.T.G., S.G., C.J. and V.B.M.-R.; Formal analysis, R.T.G., S.G. and V.B.M.-R.; Investigation, R.T.G., S.G., C.J. and V.B.M.-R.; Resources, RTH, S.G., C.J. and V.B.M.-R.; Data curation, R.T.G., S.G., C.J. and V.B.M.-R.; Writing—original draft preparation, R.T.G. and V.B.M.-R.; Writing—review and editing, V.B.M.-R., S.G. and C.J.; Visualization, S.G. and V.B.M.-R.; Supervision, R.T.G. and C.J.; Project administration.All authors have read and agreed to the published version of the manuscript.

Funding: To complete this study Sampat Ghosh and Victor Benno Meyer-Rochow received support from Chuleui Jung of Andong National University's Insect Industry R&D Center via the Basic Science Research Program of the National Research Foundation of Korea (NRF) funded by the Ministry of Education (NRF-2018R1A6A1A03024862).

Institutional Review Board Statement: Not applicable

Informed Consent Statement: Not applicable.

Data Availability Statement: Not applicable.

Acknowledgments: The authors are thankful to two anonymous reviewers for their critical suggestions on an earlier draft of the manuscript.

Conflicts of Interest: The authors declare no conflict of interest.

References

1. Bequaert, J. Insects as food: How they have augmented the food supply of mankind in early and recent years. *Nat. Hist. J.* **1921**, *21*, 191–200.
2. Bergier, E. *Peuples Entomophages et Insectes Comestibles: Étude Sur les Moeurs de L'Homme et de L'Insecte*; Imprimérie Rullière Frères: Avignon, France, 1941.
3. Bodenheimer, F.S. Insects as human food. In *Insects as Human Food: A Chapter of the Ecology of Man*; Bodenheimer, F.S., Ed.; W. Junk: Hague, The Netherlands, 1951; pp. 7–38.
4. Bogart, S.L.; Pruetz, J.D. Insectivory of savanna chimpanzees (*Pan troglodytes verus*) at Fongoli, Senegal. *Am. J. Phys. Anthropol.* **2011**, *145*, 11–20. [CrossRef]
5. McGrew, W.C. The 'other faunivory' revisited: Insectivory in human and non-human primates and the evolution of human diet. *J. Human Evol.* **2014**, *71*, 4–11. [CrossRef] [PubMed]
6. Meyer-Rochow, V.B. Can insects help to ease the problem of world food shortage? *Search* **1975**, *6*, 261–262.
7. Gahukar, R.T. Entomophagy and human food security. *Int. J. Trop. Insect Sci.* **2011**, *31*, 129–144. [CrossRef]
8. Lensvelt, E.J.S.; Steenbekkers, L.P.A. Exploring consumer acceptance of entomophagy: A survey and experiment in Australia and the Netherlands. *Ecol. Food Nutr.* **2014**, *53*, 543–561. [CrossRef]
9. Shouteten, J.J.; De Steur, H.; De Pelsmaeker, S.; Lagast, S.; Juvinal, J.G.; De Bourdeaudhuij, L.; Verbeke, W.; Gellynck, X. Functional and sensory profiling of insect-, plant- and meat-based burgers under blind, expected and informed conditions. *Food Qual. Prefer.* **2016**, *52*, 27–31. [CrossRef]
10. Menozzi, D.; Sogari, G.; Veneziani, M.; Simoni, E.; Mora, C. Eating novel foods: An application of the theory of planned behaviour to predict the consumption of an insect-based product. *Food Qual. Prefer.* **2017**, *59*, 27–34. [CrossRef]
11. Tan, H.S.G.; House, J. Consumer acceptance of insects as food: Integrating psychological and socio-cultural perspectives. In *Edible Insects in Sustainable Food Systems*; Halloran, A., Flore, R., Vantomme, P., Roos, N., Eds.; Springer: Berlin, Germany, 2018; pp. 375–386. [CrossRef]
12. Ghosh, S.; Jung, C.; Meyer-Rochow, V.B. What governs selection and acceptance of edible insect species? In *Edible Insects in Sustainable Food Systems*; Halloran, A., Flore, R., Vantomme, P., Roos, N., Eds.; Springer: Berlin, Germany, 2018; pp. 331–351. [CrossRef]
13. Chakravorty, J.; Ghosh, S.; Meyer-Rochow, V.B. Practices of entomophagy and entomotherapy by members of the *Nyishi* and *Galo* tribes, two ethnic groups of the state of Arunachal Pradesh (North East India). *J. Ethnobiol. Ethnomed.* **2011**, *7*, 5. [CrossRef] [PubMed]
14. Chakravorty, J.; Ghosh, S.; Meyer-Rochow, V.B. Comparative survey of entomophagy and entomotherapeutic practices in six tribes of Eastern Arunachal Pradesh (India). *J. Ethnobiol. Ethnomed.* **2013**, *9*, 50. [CrossRef]
15. Meyer-Rochow, V.B.; Nonaka, K.; Boulidam, S. More feared than revered: Insects and their impact on human societies with some specific data on the importance of entomophagy in a Laotian setting. *Entomol. Heute* **2008**, *20*, 25.

16. Muafor, F.J.; Genetegha, A.A.; le Gall, P.; Levang, P. *Exploitation, Trade and Farming of Palm Weevil Grubs in Cameroon*; Center for International Forestry Research Content: Bogor, Indonesia, 2015.

17. Gahukar, R.T. Edible insects collected from forests for family livelihood and wellness of rural communities: A review. *Glob. Food Secur.* **2020**, *25*, 100348. [CrossRef]

18. Chakravorty, J.; Ghosh, S.; Meyer-Rochow, V.B. Chemical composition of *Aspongopus nepalensis* Westwood 1837 (Hemiptera; Pentatomidae), a common food insect of tribal people in Arunachal Pradesh (India). *Int. J. Vitam Nutr. Res.* **2011**, *81*, 49–56. [CrossRef] [PubMed]

19. Mueller, A. Insects as food in Laos and Thailand—A case of "Westernization"? *Asian J. Soc. Sci.* **2019**, *47*, 204–223. [CrossRef]

20. Yi, L.; Lakemond, C.M.M.; Sagis, L.M.G.; Eisner-Schadler, V.; van Huis, A.; van Boekel, M.J.S. Extraction and characterization of protein fractions from five insect species. *Food Chem.* **2013**, *141*, 3341–3348. [CrossRef]

21. Zhou, J.; Han, D. Proximate, amino acid and mineral composition of pupae of the silkworm, *Antheraea pernyi* in China. *J. Food Compos. Anal.* **2006**, *19*, 850–853. [CrossRef]

22. Rumpold, B.A.; Schlüter, O.K. Nutritional composition and safety aspects of edible insects. *Mol. Nutr. Food Res.* **2013**, *57*, 802–823. [CrossRef]

23. Van Huis, A.; van Itterbeeck, J.; Klunder, H.; Mertens, E.; Halloran, A.; Muir, G.; Vantomme, P. *Edible Insects: Future Prospects for Food and Feed Security*; FAO: Rome, Italy, 2013.

24. Nowak, V.; Persijn, D.; Rittenschober, D.; Charrondiere, U.R. Review of food composition data for edible insects. *Food Chem.* **2016**, *193*, 39–46. [CrossRef]

25. Paul, A.; Frederich, M.; Uyttenbroeck, R.; Hatt, S.; Malik, P.; Lebecque, S.; Hamaidia, M.; Miazek, K.; Goffin, D.; Willems, L.; et al. Grasshopper s a food resource? A review. *Biotechnol. Agron. Soc. Environ.* **2016**, *20*, 337–352.

26. Fogang Mba, A.R.; Kansci, G.; Viau, M.; Hafnaoui, N.; Meynier, A.; Demmano, G.; Genot, C. Lipid and amino acid profiles support the potential of *Rhynchophorus phoenicis* larvae for human food. *J. Food Compos. Anal.* **2017**, *60*, 64–73. [CrossRef]

27. Gahukar, R.T. Edible insects farming: Efficiency and impact on family livelihood, food security and environment compared to livestock and crops. In *Insects as Sustainable Food Ingredients: Production, Processing and Food Application*; Dossey, A.T., Morales-Ramos, J.A., Rojas, M.G., Eds.; Elsevier Inc.: New York, NY, USA, 2016; pp. 85–111.

28. Sun-Waterhouse, D.; Waterhouse, G.I.N.; You, L.; Zhang, J.; Liu, J.; Liu, Y.; Ma, L.; Gao, J.; Dong, Y. Transforming insect biomass into consumer wellness foods: A review. *Food Res. Int.* **2016**, *89*, 129–151. [CrossRef]

29. Payne, C.L.R.; Scarborough, P.; Rayner, P.; Nonaka, K. Are edible insects more or less 'healthy' than commonly consumed insects? A comparison using two nutrient profiling models developed to combat over- and under-nutrition. *Eur. J. Clin. Nutr.* **2016**, *70*, 285–291. [CrossRef]

30. Meyer-Rochow, V.B.; Jung, C. Insects used as food and feed: isn't that what we all need? *Foods* **2020**, *9*, 1003. [CrossRef] [PubMed]

31. Ying, F.; Xiaoming, C.; Long, S.; Zhiyong, C. Common edible wasps in Yunnan Province, China and their nutritional value. In *Forest Insects as Food: Human Bite Back*; Durst, P.B., Johnson, D.V., Leslie, R.N., Shono, K., Eds.; Food and Agriculture Organization of the United Nations Regional Office for Asia and the Pacific: Bangkok, Thailand, 2010; pp. 93–98.

32. Ghosh, S.; Namin, S.M.; Meyer-Rochow, V.B.; Jung, C. Chemical composition and nutritional value of different species of *Vespa* hornets. *Foods* **2021**, *10*, 418. [CrossRef]

33. Ghosh, S.; Jung, C.; Meyer-Rochow, V.B. Nutritional value and chemical composition of larvae, pupae and adults of worker honey bee, *Apis mellifera ligustica* as a sustainable food source. *J. Asia Pac. Entomol.* **2016**, *19*, 487–495. [CrossRef]

34. Ghosh, S.; Chuttong, B.; Burgett, M.; Meyer-Rochow, V.B.; Jung, C. Nutritional value of brood and adult workers of the Asia honeybee species *Apis cerana* and *Apis dorsata*. In *African Edible Insects as Alternative Source of Food, Oil, Protein and Bioactive Components*; Mariod, A.A., Ed.; Springer: Cham, Switzerland, 2020; pp. 265–273. [CrossRef]

35. Ghosh, S.; Jung, C.; Chuttong, B.; Burgett, M. Nutritional aspects of the dwarf honeybee (*Apis florea* F.) for human consumption. In *The Future Role of Dwarf Honeybees in Natural and Agricultural Systems*; Abrol, D.P., Ed.; CRC Press: Boca Raton, FL, USA, 2020; pp. 137–145.

36. Akullo, J.; Agea, J.G.; Obaa, B.B.; Okwee-Acai, J.; Nokimbugwe, D. Nutritional composition of commonly consumed edible insects in the Lango sub-region of northern Uganda. *Int. Food Res. J.* **2018**, *25*, 159–166.

37. Omotoso, O.T. Nutrition composition, mineral analysis and antinutrient factors of *Oryctes rhinoceros* L. (Scarabaeidae: Coleoptea) and winged termites, *Macrotermes nigeriensis* Sjostedt (Termitidae: Isoptera). *Br. J. Appl. Sci. Technol.* **2015**, *8*, 97–106. [CrossRef]

38. Chakravorty, J.; Ghosh, S.; Megu, K.; Jung, C.; Meyer-Rochow, V.B. Nutritional and anti-nutritional composition of *Oecophylla smaragdina* (Hymenoptera: Formicidae) and *Odontotermes* sp. (Isoptera: Termitidae): Two preferred edible insects of Arunachal Pradesh, India. *J. Asia Pac. Entomol.* **2016**, *19*, 711–720. [CrossRef]

39. Ghosh, S.; Lee, S.M.; Jung, C.; Meyer-Rochow, V.B. Nutritional composition of five commercial edible insects in South Korea. *J. Asia Pac. Entomol.* **2017**, *20*, 686–694. [CrossRef]

40. Adámková, A.; Mlček, J.; Kouřimska, L.; Borkovcová, M.; Bušina, T.; Adámek, M.; Bednářová, M.; Krajsa, J. Nutritional potential of selected insect species reared on the Island of Sumatra. *Int. J. Environ. Res. Public Health* **2017**, *14*, 521. [CrossRef] [PubMed]

41. Bbosa, T.; Ndagire, C.T.; Mukisa, I.M.; Fiaboe, K.K.M.; Nakimbugwe, D. Nutritional characteristics of selected insects in Uganda for use as alternative protein sources in food and feed. *J. Insect Sci.* **2019**, *19*, 23. [CrossRef] [PubMed]

42. Nyakeri, E.M.; Ogola, H.J.; Ayieko, M.A.; Amimo, F.A. An open system for farming black soldier fly larvae as a source of proteins for small scale poultry and fish production. *J. Insects Food Feed* **2017**, *3*, 51–56. [CrossRef]

43. Anvo, M.P.M.; Toguyeni, A.; Otchoumou, K.; Zoungrana-Kabore, C.Y.; Kouamelan, E.P. Nutritional qualities of edible caterpillars, *Cirina butyrospermi* in southeastern of Burkina Faso. *Int. J. Innov. Appl. Stud.* **2016**, *18*, 639–645.

44. Chakravorty, J.; Ghosh, S.; Jung, C.; Meyer-Rochow, V.B. Nutritional composition of *Chondacris rosea* and *Brachytrupes orientalis*: Two common insects used as food by tribes of Arunachal Pradesh, India. *J. Asia Pac. Entomol.* **2014**, *17*, 407–415. [CrossRef]

45. Pérez-Ramírez, R.; Torres-Castillo, J.A.; Barrientos-Lazano, L.; Almaguee-Sierra, P.; Torres-Acosta, R.I. *Schistocerca piceifrons piceifrons* as a source of compounds of biotechnological and nutritional interest. *J. Insect Sci.* **2019**, *19*, 10. [CrossRef]

46. Banjo, A.D.; Lawal, O.A.; Songonuga, E.A. The nutritional value of fourteen species of edible insects in southwestern Nigeria. *Afr. J. Biotechnol.* **2006**, *5*, 298–301.

47. Kinyuru, J.N.; Konyole, S.O.; Roos, N.; Onyango, C.A.; Owino, V.O.; Owuor, B.O.; Estambale, B.B.; Friis, H.; Aagaard-Hansen, J.; Kenji, G.M. Nutrient composition of four species of winged termites consumed in western Kenya. *J. Food Compos. Anal.* **2013**, *30*, 120–124. [CrossRef]

48. Ramos-Elorduy Blásquez, J.; Moreno, J.M.P.; Camacho, V.H.M. Could grasshoppers be a nutritive meal? *Food Nutr. Sci.* **2012**, *3*, 164–175. [CrossRef]

49. Onyeike, E.N.; Ayalogu, E.O.; Okaraonye, C.C. Nutritive value of the larvae of raphia palm beetle (*Orytes rhinoceros*) and weevil (*Rhynchophorus pheonicis*). *J. Sci. Food Agric.* **2005**, *85*, 1822–1828. [CrossRef]

50. Ramos-Elorduy, J.; Moreno, J.M.P.; Correa, S.C. Edible insects of the state of Mexico and determination of their nutritive values. *An. Inst. Biol. Univ. Nac. Auton. Mex. Ser. Zool.* **1998**, *69*, 65–104.

51. Ramos-Elorduy, J.; Moreno, J.M.P.; Prado, E.E.; Perez, M.A.; Otero, J.L.; de Guevara, O.L. Nutritional value of edible insects from the State of Oaxaca, Mexico. *J. Food Compos. Anal.* **1997**, *10*, 142–157. [CrossRef]

52. Ghosh, S.; Choi, K.C.; Kim, S.; Jung, C. Body compositional changes of fatty acid and amino acid from the queen bumblebee, *Bombus terrestris* during overwintering. *J. Apic.* **2017**, *32*, 11–18. [CrossRef]

53. Mariod, A.A.; Abdel-Wahab, S.I.; Ain, N.M. Proximate amino acid, fatty acid and mineral composition of two Sudanese edible pentatomid insects. *Int. J. Trop. Insect Sci.* **2011**, *31*, 145–153. [CrossRef]

54. Ekpo, K.E.; Onigbinde, A.O.; Asia, I.O. Pharmaceutical potentials of the oils of some popular insects consumed in southeast Nigeria. *Afr. J. Pharm. Pharmacol.* **2009**, *3*, 51–57.

55. Santos Oliveira, J.F.; Passos de Carvalho, J.; Bruno de Sousa, R.F.X.; Madalena, S.M. The nutritional value of four species of insects consumed in Angola. *Ecol. Food Nutr.* **1976**, *5*, 91–97. [CrossRef]

56. Bukkens, S.G.F. The nutritional value of edible insects. *Ecol. Food Nutr.* **1997**, *36*, 287–319. [CrossRef]

57. Ukhun, M.E.; Osasona, M.A. Aspects of the nutritional chemistry of *Macrotermes bellicosus*. *Nutr. Rep. Int.* **1985**, *32*, 1121–1130.

58. Igwe, C.U.; Ujowundu, C.O.; Nwaogu, L.A.; Okwu, G.N. Chemical analysis of an edible African termite, *Macrotermes nigeriensis*, a potential antidote to food security problem. *Biochem. Anal. Biochem.* **2011**, *1*, 1000105. [CrossRef]

59. Womeni, H.M.; Linder, M.; Tiencheu, B.; Mbiapo, F.T.; Villeneuve, P.; Fanni, J.; Parmentier, M. Oils of *Oryctes owariensis* and *Homorocoryphus nitidulus* consumed in Cameroon: Sources of linoleic acid. *J. Food Technol.* **2009**, *7*, 54–58.

60. Ashiru, M.O. The food value of the larvae of *Anaphe venata* Butler (Lepidoptera: Notodontidae). *Ecol. Food Nutr.* **1989**, *22*, 313–320. [CrossRef]

61. Ghosh, S.; Sohn, H.-Y.; Pyo, S.-J.; Jensen, A.B.; Meyer-Rochow, V.B.; Jung, C. Nutritional composition of *Apis mellifera* drones from Korea and Denmark as a potential sustainable alternative food source: Comparison between developmental stages. *Foods* **2020**, *9*, 389. [CrossRef] [PubMed]

62. Ademolu, K.O.; Idowu, A.B.; Olatunde, G.O. Nutritional value assessment of variegated grasshopper, *Zonocerus variegatus* (L.) (Acridoidea: Pyrgomorphidae) during post- embryonic development. *Afr. Entomol.* **2010**, *18*, 360–364. [CrossRef]

63. Kulma, M.; Plachý, V.; Kouřimská, L.; Vrabec, V.; Bubová, Y.; Adámková, A.; Hučko, B. Nutritional value of three Blattodea species used as feed for animals. *J. Anim. Feed Sci.* **2016**, *25*, 354–360. [CrossRef]

64. Kipkoech, C.; Kinyuru, J.N.; Imathiu, S.; Roos, N. Use of house cricket to address food security in Kenya; nutritional and chitin composition of farmed crickets as influenced by age. *Afr. J. Agric. Res.* **2017**, *12*, 3189–3197. [CrossRef]

65. Ombeni, B.J.; Munyuli, T.; Fideline, N.; Espoir, I.; Betu, M. Profile in amino acids and fatty acids of *Bunaeopsis aurantiaca* caterpillars eaten in South Kivu Province, eastern of the Democratic Republic of Congo. *Ann. Food Sci. Technol.* **2018**, *19*, 566–576.

66. Ramos-Elorduy, J.; Gonazález, E.A.; Hernández, A.R.; Pino, J.M. Use of *Tenebrio molitor* (Coleoptera: Tenebrionidae) to recycle organic wastes and as feed for broiler chickens. *J. Econ. Entomol.* **2002**, *95*, 214–220. [CrossRef] [PubMed]

67. Omotoso, O.T.; Adedire, C.O. Nutrient composition, mineral content and the solubility of the proteins of palm weevil, *Rhynchophorus phoenicis*. (Coleoptera: Curculionidae). *J. Zhejiang Univ. Sci. B.* **2007**, *8*, 318–322. [CrossRef] [PubMed]

68. Opara, M.N.; Sanyigha, F.T.; Ogbuewu, I.P.; Okoli, I.C. Studies on the production trend and quality characteristics of palm grubs in the tropical rainforest zone of Nigeria. *Int. J. Agric. Technol.* **2012**, *8*, 851–860.

69. Chinweuba, A.J.; Otuokere, I.E.; Opara, M.O.; Okafor, G.U. Nutritional potentials of *Rhynchophorus phoenicis* (Rahia palm weevil): Implications for food security. *Asian J. Res. Chem.* **2011**, *4*, 452–454.

70. Omotoso, O.T. The nutrient profile of the development stages of palm beetle, *Oryctes rhinoceros*. *Br. J. Environ. Sci.* **2018**, *6*, 1–11.

71. Finke, M.D. Complete nutrient composition of commercially raised invertebrates used as food for insectivores. *Zoo Biol.* **2002**, *21*, 269–285. [CrossRef]

72. Mahesh, D.S.; Vidharthi, B.S.; Narayanaswamy, T.K.; Subbarayappa, C.T.; Muthuraju, R.; Shruthi, P. Bionutritional science of silkworm pupal residue to mine: New ways for utilization. *Int. J. Adv. Res. Biol. Sci.* **2015**, *2*, 135140.

73. Trivedy, K.; Nirmal Kumar, S.; Mondal, M.; Bhat, A.K. Protein binding pattern and major amino acid component in de-oiled pupal powder of silkworm, *Bombyx mori* Linn. *J. Entomol.* **2008**, *5*, 10–16. [CrossRef]

74. Trivedy, K.; Nirmal kumar, S.; Quadri, S.M.H. Comparative study of major nutritional component of defatted and normal pupal powder of silkworm, *Bombyx mori*. *Indian J. Seric.* **2011**, *50*, 190–199.

75. Cai, J.R.; Yuan, L.M.; Liu, B.; Sun, L. Non-destructive gender identification of silkworm cocoon using x-ray imaging technology coupled with multivariate data analysis. *Anal. Methods* **2014**, *6*, 7224–7233. [CrossRef]

76. Kiuchi, T.; Koga, H.; Kawamoto, M.; Shoji, K.; Sakai, H.; Arai, Y.; Tshihara, G.; Kawaoka, S.; Sugano, S.; Shimada, T.; et al. A single female-specific piRNA is the primary determiner of sex in the silkworm. *Nature* **2014**, *509*, 633666. [CrossRef]

77. Nakasone, S.; Ito, T. Fatty acid composition of the silkworm, *Bombyx mori* L. *J. Insect Physiol.* **1967**, *13*, 1237–1246. [CrossRef]

78. Kotake-Nara, E.; Yamamoto, K.; Nozawa, M.; Miyashita, K.; Murakami, T. Lipid profiles and oxidative stability of silkworm pupal oil. *J. Oleo Sci.* **2002**, *51*, 681–690. [CrossRef]

79. Kulma, M.; Kouřimská, L.; Plachý, V.; Božik, M.; Adámková, A.; Vrabec, V. Effect of sex on the nutritional value of house cricket, *Acheta domestica* L. *Food Chem.* **2019**, *272*, 267–272. [CrossRef]

80. Paul, D.; Dey, S. Nutrient content of sexual and worker forms of the subterranean termite, *Reticulitermes* sp. *Indian J. Tradit. Knowl.* **2011**, *10*, 505–507.

81. Ntukuyoh, A.I.; Udiong, D.S.; Ikpe, E.; Akpakpan, A.E. Evaluation of nutritional value of termites (*Macrotermes bellicosus*), soldiers, workers and queen in the Niger Delta region of Nigeria. *Int. J. Food Nutr. Saf.* **2012**, *1*, 60–65.

82. Idowu, A.B.; Ademolu, K.O.; Bamidele, J.A. Nutrition and heavy metal levels in the mound termite, *Macrotermes bellicosus* (Smeathman) (Isoptera: Termitidae), at three sites under varying land use in Abeokuta, southwestern Nigeria. *Afr. Entomol.* **2014**, *22*, 156–162. [CrossRef]

83. Borgohain, M.; Borkotoki, A.; Mahanta, R. Total lipid, triglyceride, and cholesterol contents in *Oecophylla smaragdina* Fabricius consumed in upper Assam of northeast India. *Int. J. Sci. Res. Publ.* **2014**, *4*, 1–5.

84. Raksakantong, P.; Meeso, N.; Kubola, J.; Sirimornpun, S. Fatty acids and proximate composition of eight Thai edible terricolous insects. *Food Res. Int.* **2010**, *43*, 350–355. [CrossRef]

85. Dué, E.A.; Zabri, H.C.B.L.; Koudio, J.P.E.N.; Kouamé, L.P. Fatty acid compositionand properties of skin and digestive fat content oils from *Rhynchophorus palmarum* L. larva. *Afr. J. Biochem. Res.* **2009**, *3*, 89–94.

86. Mba, A.R.F.; Kansci, G.; Viau, M.; Ribourg, L.; Genot, C. Growing conditions and morphotypes of African palm weevil (*Rhynchophorus phoenicis*) larvae influence their lipophilic nutrient but not their amino acid composition. *J. Food Compos. Anal.* **2018**, *69*, 87–97. [CrossRef]

87. Ssepuuya, G.R.; Mukisa, J.M.; Nakimbugwe, D. Nutritional composition, quality and shelf stability of processed *Ruspolia nitidula* (edible grasshoppers). *Food Sci. Nutr.* **2017**, *5*, 103–112. [CrossRef] [PubMed]

88. Fontaneto, D.; Tommaseo-Ponzetta, M.; Galli, C.; Rise, P.; Glew, R.H.; Paoletti, M.G. Difference in fatty acid composition between aquatic and terrestrial insects used as food in human nutrition. *Ecol. Food Nutr.* **2011**, *50*, 351–367. [CrossRef]

89. Williams, D.D.; Williams, S.S. Aquatic insects and their potential to contribute to the diet of the globally expanding human population. *Insects* **2017**, *8*, 72. [CrossRef]

90. Ssepuuya, G.R.; Smets, R.; Nakimbugwe, D.; van der Borght, M.; Claes, J. Nutritional composition of long-horned grasshopper, *Ruspolia differens* Serville: Effect of swarming season and sourcing geographical area. *Food Chem.* **2019**, *301*, 125305. [CrossRef]

91. Madibela, O.R.; Mokwena, K.K.; Nsoso, S.J.; Thema, T.F. Chemical composition of mopane worm sampled at three different sites in Botswana subjected to different processing. *Trop. Anim. Health Prod.* **2009**, *41*, 935–942. [CrossRef]

92. Singh, P. *Artificial Diets for Insects, Mites and Spiders*; IFI/Plenum Data Company, Springer: New York, NY, USA, 1977; 594p.

93. Purushothaman, S.; Muthuvelu, S.; Balasubramanian, U.; Murugesan, P. Biochemical analysis of mulberry leaves (*Morus alba* L.) and silkworm, *Bombyx mori* enriched with vermiwash. *J. Entomol.* **2012**, *9*, 289–292. [CrossRef]

94. Ebenebe, C.I.; Okpoko, V.O.; Ufele, A.N.; Amobi, M.I. Survivability, growth performance and nutrient composition of the African palm weevil (*Rhynchophous phoenicis* Fabricius) reared on four different substrates. *J. Biosci. Biotechnol. Discov.* **2017**, *2*, 1–9. [CrossRef]

95. Malinga, G.M.; Valtonen, A.; Hiltunen, M.; Lehtovaara, V.J.; Nyeko, P.; Roininen, H. Performance of the African edible bush-cricket, *Ruspolia differens* on single and mixed diets containing inflorescences of their host plant species. *Entomol. Exp. Appl.* **2020**, *168*, 12932. [CrossRef]

96. Quaye, B.; Atuahene, C.C.; Donkoh, A.; Adjei, B.M.; Opoku, O.; Amankrah, M.A. Nutritional potential and microbial status of African palm weevil (*Rhynchophorus phoenicis*) larvae raised on alternative feed resources. *Am. Sci. Res. J. Eng. Technol. Sci.* **2018**, *48*, 45–52.

97. Oonincx, D.G.A.B.; van der Poel, A.F.B. Effects of diet on the chemical composition of migratory locusts (*Locusta migratoria*). *Zoo Biol.* **2011**, *30*, 9–16. [CrossRef]

98. Cito, A.; Longo, S.; Mazza, G.; Dreassi, E.; Francardi, V. Chemical evaluation of the *Rhynchophorus ferrugineus* larvae fed on different substrates as human food source. *Food Sci. Technol. Int.* **2017**, *23*, 529–539. [CrossRef]

99. Atuahene, C.C.; Adjei, M.B.; Adu, M.A.; Quaye, B.; Opare, M.B.; Benney, R. Evaluating potential of edible insects (palm weevil, *Rhynchophorus phoenicis* larvae) as an alternative protein source to humans. *Anim. Sci. Adv.* **2017**, *7*, 1897–1900.

100. Rutaro, K.; Malinga, G.M.; Lehtovaara, V.J.; Opoke, R.; Nyeko, P.; Roininen, H.; Valtonen, A. Fatty acid content and composition in edible *Ruspolia differens* feeding on mixtures of natural food plants. *BMC Res. Notes* **2018**, *11*, 687. [CrossRef] [PubMed]

101. Rutaro, K.; Malinga, G.M.; Lehtovaara, V.J.; Opoke, R.; Valtonen, A.; Kwetegyeka, J.; Nyeko, P.; Roininen, H. The fatty acid composition of edible grasshopper, *Ruspolia differens* (Serville) (Orthoptera; Tettigoniidae) feeding on diversifying diets of host plants. *Entomol. Res.* **2018**, *48*, 490–498. [CrossRef]

102. Meyer-Rochow, V.B.; Ghosh, S.; Jung, C. Farming of insects for food and feed in South Korea: Tradition and innovation. *Berl. Muenchener Tieraerztliche Wochenschr.* **2019**, *132*, 236–244. [CrossRef]

103. Rutaro, K.; Malinga, G.M.; Opoke, R.; Lehtovaara, V.J.; Omujal, F.; Nyeko, P.; Valtonen, A. Artificial diets determine fatty acid composition in edible *Ruspolia differens* (Orthoptera; Tettigoniidae). *J. Asia Pac. Entomol.* **2018**, *21*, 1342–1349. [CrossRef]

104. Ghaly, A.E. The use of insects as human food in Zambia. *Online J. Biol. Sci.* **2009**, *9*, 93–104. [CrossRef]

105. Ekpo, K.E. Nutritional and biochemical evaluation of the protein quality of four popular insects consumed in southern Nigeria. *Arch. Appl. Sci. Res.* **2011**, *3*, 24–40.

106. Stull, V.J.; Kersten, M.; Bergmans, R.S.; Patz, J.A.; Paskewitz, S. Crude protein, amino acid, and iron content of *Tribolium molitor* (Coleoptera: Tenebrionidae) reared on an agricultural byproduct from production: An exploratory study. *Ann. Entomol. Soc. Am.* **2019**, *112*, 533–543. [CrossRef]

107. De Wit, L. Rearing of *Bombyx mori* with Vitamin D-Enriched Diet. Ph.D. Thesis, University of Applied Sciences, Almere, The Netherlands, 2017.

108. Rani, A.G.; Premlatha, C.; Raj, R.S.; Ranjit Singh, A.J. Impact of supplementation of Amway protein on the economic characters and energy budget of silkworm, *Bombyx mori* L. *Asian J. Anim. Sci.* **2011**, *10*, 1–4. [CrossRef]

109. Sorjonen, J.M.; Valtonen, A.; Hirvisalo, E.; Karhapaa, M.; Lindgren, J.; Nilarmila, P.; Mooney, P.; Maki, M.; Sirjander-Rasi, H.; Tapio, M.; et al. The plant-based by-product diets for the mass rearing of *Acheta domesticus* and *Gryllus bimaculatus*. *PLoS ONE* **2019**, *14*, e0218830. [CrossRef] [PubMed]

110. Sorjonen, J.M.; Lehtovaara, V.J.; Immmonen, J.; Karhapää, M.; Valtonen, A.; Roininn, H. Growth performance and food conversion of *Ruspolia differens* on plant-based by product diets. *Entomol. Exp. Appl.* **2020**, *168*, 12915. [CrossRef]

111. Lehtovaara, V.J.; Valtonen, A.; Sorjonen, J.M.; Hiltunen, M.; Rutaro, K.; Malinga, G.M.; Roininen, H. The fatty acid contents of the edible grasshopper, *Ruspolia differens* can be manipulated using artificial diets. *J. Insects Food Feed* **2017**, *3*, 253–262. [CrossRef]

112. Melghar-Lalanne, G.; Hernandez-Alvarez, A.J.; Salinas-Castro, A. Edible insects processing: Traditional and innovative technologies. *Compr. Rev. Food Sci. Food Saf.* **2019**, *18*. [CrossRef] [PubMed]

113. Kewuyemi, Y.; Kesa, H.; Chinna, C.E.; Adebo, O.A. Fermented edible insects for promoting food security in Africa. *Insects* **2020**, *11*, 283. [CrossRef]

114. Son, Y.J.; Choi, S.Y.; Hwang, I.K.; Nho, C.W.; Kim, S.H. Could defatted mealworm (*Tenebrio molitor*) and mealworm oil be used as food ingredients? *Foods* **2020**, *9*, 40. [CrossRef] [PubMed]

115. Anuduang, A.; Loo, Y.Y.; Jomduang, S.; Lim, S.J.; Mustapha, W.A.W. Effect of thermal processing on physico-chemical and antioxidant propertie in mulberry silkworm (*Bombyx mori* L.) powder. *Foods* **2020**, *9*, 871. [CrossRef] [PubMed]

116. Williams, J.P.; Williams, J.R.; Kirabo, A.; Chester, D.; Peterson, M. Nutrient content and health benefits of insects. In *Insects as Sustainable Food Ingredients: Production, Processing and Food Applications*; Dossey, A.T., Morales-Ramos, J.A., Rojas, M.G., Eds.; Elsevier Inc.: New York, NY, USA, 2016; pp. 61–84.

117. Meyer-Rochow, V.B.; Chakravorty, J. Notes on entomophagy and entomotherapy generally and information on the situation in India in particular. *Appl. Entomol. Zool.* **2013**, *48*, 105–112. [CrossRef]

118. Sabolová, M.; Adámková, A.; Kouřimská, L.; Chrpová, D.; Pánek, J. Minor lipophilic compounds in edible insects. *Potravinarsmt* **2016**, *10*, 400–406. [CrossRef]

119. Nyangena, D.N.; Mutungi, C.; Imathiu, S.; Kinyuru, J.; Affognon, H.; Ekesi, S.; Nakimbugwe, D.; Fiaboe, K.K.M. Effects of traditional processing techniques on the nutritional and microbiological quality of four edible insect species used for food and feed in East Africa. *Foods* **2020**, *9*, 574. [CrossRef]

120. Winitchai, S.; Manoishori, T.; Abe, M.; Boonpisuttinant, K.; Manosroi, A. Free radical scavenging activity, tyrosinase inhibition activity and fatty acids composition of oils from pupae of native Thai silkworm (*Bombyx mori* L.). *Kasetsart J. Nat. Sci.* **2011**, *45*, 404–412.

121. Sangavi, M.; Sarath, S. Byproducts of seri-industry and their applications. *Kisan World* **2017**, *44*, 21–23.

122. Chavan, S.; Chinnaswamy, K.P.; Changalerayappa. Influence of mulberry varieties and silkworm breeds on biochemical constituent of oiled and de-oiled pupal powder. In Proceedings of the National Seminar on Tropical Sericulture, Bangalore, India, 28–30 December 1999; p. 57.

123. Rao, P.U. Chemical composition and nutritional evaluation of spent silkworm pupae. *J. Agric. Food Chem.* **1994**, *42*, 2201–2203. [CrossRef]

124. Pereira, N.R.; Ferrarese-Filho, O.; Matsushita, M.; de Souza, N.E. Proximate composition and fatty acid profile of *Bombyx mori* L. chrysalis toast. *J. Food Compos. Anal.* **2003**, *16*, 451–457. [CrossRef]

125. Ji, S.-D.; Nguyen, P.; Yoon, S.-M.; Kim, K.-Y.; Son, J.G.; Kweon, H.-Y.; Koh, Y.H. Comparison of nutrient composition and pharmacological effects of steamed and freeze-dried mature silkworm powders generated by four silkworm varieties. *J. Asia Pac. Entomol.* **2017**, *20*, 1410–1418. [CrossRef]

126. Kim, K.-Y.; Osabutey, A.F.; Nguyen, P.; Kim, S.B.; Jo, Y.-Y.; Kweon, H.Y.; Lee, H.-T.; Ji, S.-D.; Koh, Y.H. The experimental evidences of steamed amd freeze dried mature silkworm powder as the calorie restriction mimetics. *Int. J. Indust. Entomol.* **2019**, *39*, 1–8. [CrossRef]

127. Nguyen, P.; Kim, K.-Y.; Kim, A.-Y.; Choi, B.-H.; Osabutey, A.F.; Park, Y.H.; Lee, H.-T.; Ji, S.D.; Koh, Y.H. Mature silkworm powders ameliorated scopolamine-induced amnesia by enhancing mitochondrial functions in the brains of mice. *J. Func. Foods* **2020**, *67*, 103886. [CrossRef]

128. Kwiri, R.; Winini, C.; Muredzi, P.; Tongonya, J.; Gwala, W.; Mujuru, F.; Gwala, S.T. Mopane worm (*Goniobrasia belina*) utilization, a potential source of protein in fortified blended foods in Zimbabwe: A review. *Glob. J. Sci. Front. Res. D Agric. Vet.* **2014**, *14*, 55–67.

129. Lautenschläger, T.; Neinhuis, C.; Kikongo, E.; Henle, T.; Förster, A. Impact of different preparations on the nutritional value of the edible caterpillar, *Imbrasia epimethea* from northern Angola. *Eur. Food Res. Technol.* **2016**, *243*, 769–778. [CrossRef]

130. Megido, R.C.; Sablon, L.; Geuens, M.; Brostaux, Y.; Alabi, T.; Blecker, C.; Drugmand, D.; Haubruge, E.; Francis, F. Edible insects' acceptance by Belgian consumers: Promising attitude for entomophagy development. *J. Sens. Stud.* **2014**, *29*, 14–20. [CrossRef]

131. Van der Spieger, M.; Noordam, M.Y.; van der Fels-Klenx, H.J. Safety of novel protein sources (insects, microalgae, seaweed, duckweed and rapeseed) and legislative aspects for their application in food and feed production. *Compr. Rev. Food Sci. Food Saf.* **2013**, *12*, 662–678. [CrossRef]

132. Fogang Mba, A.R.; Kansci, G.; Viau, M.; Rougerie, R.; Genot, C. Edible caterpillars of *Imbrasia truncata* and *Imbrasia epimethea* contain lipids and proteins of high potential of nutrition. *J. Food Compos. Anal.* **2019**, *79*, 70–79. [CrossRef]

133. Okaraonyre, C.C.; Ikewuchi, J.C. Nutritional potential of *Oryctes rhinoceros* larva. *Pak. J. Nutr.* **2009**, *8*, 35–38. [CrossRef]

134. Womeni, H.M.; Tiencheu, B.; Linder, B.; Nabayo, E.M.C.; Tenyang, N.; Mbiupo, F.T.; Villeneuve, F.J.; Parmentier, M. Nutritional value and effect of cooking, drying and storage process on some functional properties of *Rhynchophorus phoenicis*. *Int. J. Life Sci. Pharm. Res.* **2012**, *2*, 203–219.

135. Severelini, C.; Azzolini, D.; Albenzio, M.; Deroesi, A. On printability, quality and nutritional properties of 3D printed cereal-based snacks enriched with edible insects. *Food Res. Int.* **2018**, *106*, 666–676. [CrossRef]

136. Smetana, S.; Larki, N.A.; Pernutz, C.; Franke, K.; Bindrich, I.J.; Toepfl, S.; Heinz, V. Structure design of insect-based meat analogs with high-moisture extrusion. *J. Food Eng.* **2018**, *229*, 83–85. [CrossRef]

137. Fombong, F.T.; van der Borght, M.; Broeck, J.V. Influence of freeze-drying and oven-drying post blanching on the nutrient composition of the edible insect, *Ruspolia differens*. *Insects* **2017**, *8*, 102. [CrossRef] [PubMed]

138. Kinyuru, M.; Kenji, G.M.; Njoroge, S.M.; Ayieko, M. Effect of processing methods on the *in vitro* protein digestibility and vitamin content of edible winged termite (*Macrotermes subhylanus*) and grasshopper (*Ruspolia differens*). *Food Bioprocess Technol.* **2010**, *3*, 778–782. [CrossRef]

139. Ssepuuya, G.R.; Nakimbugwe, D.; de Winne, A.; Smets, R.; Claes, J.; van dr Borght, M. Effect of heat processing on the nutrient composition, colour, and volatile odour compounds of the long-horned grasshopper, *Ruspolia differens* Serville. *Food Res. Int.* **2020**, *129*, 108831. [CrossRef] [PubMed]

140. Haassan, N.M.E.; Hamed, S.Y.; Hassan, A.B.; Mohamed, M.E.; Babiker, E.E. Nutritional evaluation and physiological properties of boiled and fried tree locust. *Pak. J. Nutr.* **2008**, *7*, 325–329. [CrossRef]

141. Farina, M.F. How method of killing crickets impact the sensory qualities and physiochemical properties when prepared in a broth. *Int. J. Gastron. Food Sci.* **2017**, *8*, 19–23. [CrossRef]

142. Akullo, J.; Agea, J.G.; Obaa, B.B.; Okwee-Acai, J.; Nokimbugwe, D. Process development, sensory and nutritional evaluation of honey spread enriched with edible insects' flour. *Afr. J. Food Sci.* **2017**, *11*, 30–39. [CrossRef]

143. Adeduntan, S.A. Nutritional and antinutritional characteristics of some insects foraging in Akure forest reserve Ondo state, Nigeria. *J. Food Technol.* **2005**, *3*, 563–567.

144. Ekpo, K.E. Nutrient composition, functional properties and anti-nutrient content of *Rhynchophorus phoenicis* (F.) larva. *Ann. Biol. Res.* **2010**, *1*, 178–190.

145. Ekop, E.A.; Udoh, A.I.; Akpan, P.E. Proximate and anti-nutrient composition of four edible insects in Akwa Ibom state, Nigeria. *World J. Appl. Sci. Technol.* **2010**, *2*, 224–231.

146. Ganguly, A.; Chakravorty, R.; Das, M.; Gupta, M.; Mandal, D.K.; Haldar, P.; Ramos-Elorduy, J.; Moreno, J.M.P. A preliminary study on the nutrients and antinutrients in *Oedaleus abruptus* (Thunberg) (Orthoptera: Acrididae). *Int. J. Nutr. Metab.* **2013**, *5*, 60–65. [CrossRef]

147. Shantibala, T.; Lokeshwari, R.K.; Debaraj, H. Nutritional and anti-nutritional composition of the five species of aquatic edible insects consumed in Manipur, India. *J. Insect Sci.* **2014**, *14*, 14. [CrossRef] [PubMed]

148. Ghosh, S.; Haldar, P.; Mandal, D.K. Evaluation of nutrient quality of a short-horned grasshopper, *Oxya hyla hyla* Serville (Orthoptera: Acrididae), in search of new protein source. *J. Entomol. Zool. Stud.* **2016**, *4*, 193–197.

149. Saeed, T.; Dagga, F.A.; Saraf, M. Analysis of residual pesticides present in edible insects captured in Kuwait. *Arab Gulf J. Sci. Res.* **1993**, *11*, 1–5.

150. Samom, S. Edible aquatic insects vanishing from Loktak. *The Assam Tribune*, 19 May 2016.

151. Poma, G.; Cuykx, M.; Amato, E.; Calaprice, C.; Focant, J.F.; Covaci, A. Evaluation of hazardous chemicals in edible insects and insect-based food intended for human consumption. *Food Chem. Toxicol.* **2017**, *100*, 70–79. [CrossRef] [PubMed]

152. Zhao, S.; Shang, X.; Duo, L. Accumulation and spatial distribution of Cd, Cr and Pb in mulberry from municipal solid waste compost following application of EDTA and $(NH_4)_2SO_4$. *Environ. Sci. Pollut. Res. Int.* **2013**, *20*, 967–975. [CrossRef]

153. Zhou, Z.; Zhao, Y.; Wang, S.; Han, S.; Liu, J. Lead in the soil—mulberry (*Morus alba* L.); silkworm (*Bombyx mori*) food chain: Translocation and detoxification. *Chemosphere* **2015**, *128*, 171–177. [CrossRef]

154. Fasunwon, B.T.; Banjo, A.D.; Jemine, T.A. Effect of *Dermetes maculatus* on the nutritional qualities of two edible insects (*Oryctes boas* and *Rhynchophorus phoenicis*). *Afr. J. Food Agric. Nutr. Dev.* **2011**, *11*, 5600–5613.
155. Grabowski, N.T.; Klein, G. Microbiology of processed edible insect products- results of a preliminary survey. *Int. J. Food Microbiol.* **2016**. [CrossRef]
156. Fink, M. An experimental infection model for *Tetrameres americana* (Cram, 1927). *Parasitol. Res.* **2005**, *95*, 179–185. [CrossRef]
157. Braide, W.; Nwaoguikpe, R.N. Assessment of microbiological quality and nutritional values of a processed edible weevil caterpillar (*Rhynchophorus phoenicis*) in Port Harcourt, southern Nigeria. *Int. J. Biol. Chem. Sci.* **2011**, *5*, 410–418. [CrossRef]
158. Musundire, R.; Osuga, I.M.; Cheseto, M.; Irungu, J.; Torto, B. Aflatoxin contamination detected in nutrient and anti-oxidant rich edible stink bug stored in recycled grain containers. *PLoS ONE* **2016**, *11*, e014914. [CrossRef]
159. Braide, W.; Oranusi, S.; Udegbunam, L.I.; Akobondu, C.; Nwaoguikpe, R.N. Microbiological quality of an edible caterpillar of emperor moth, *Bunaea alcinoe*. *J. Ecol. Nat. Environ.* **2011**, *3*, 176–180.
160. Mutungi, C.; Irungu, F.G.; Nduko, J.; Mutua, F.; Affoghon, H.; Nakjmbugwe, D.; Ekesi, S.; Fiabor, K.K.M. Post-harvest processes of edible insects in Africa. A review of processing methods and the implications for nutrition, safety and new products development. *Crit. Rev. Food Sci. Nutr.* **2019**, *59*, 276–298. [CrossRef]
161. Klunder, H.C.; Wolkers-Roojackers, J.; Korpela, J.M.; Nout, M.J.R. Microbiological aspects of processing and storage of edible insects. *Food Control* **2012**, *26*, 628–631. [CrossRef]
162. Kamau, E.; Mutungi, C.; Kinyuru, J.; Imathiu, S.; Tanga, C.; Affognon, H.; Ekesi, S.; Nakimbugwe, D.; Flaboe, K.K.M. Effect of packaging material, storage temperature and duration on the quality of semi-processed adult house cricket meal. *J. Food Res.* **2018**, *7*, 21–23. [CrossRef]
163. Roncolini, A.; Cardinali, F.; Aquilanti, I.; Millanović, V.; Garofalo, C.; Sabbatini, R.; Abaker, M.S.S.; Pandolfi, M.; Pasquini, M.; Tavoletti, S.; et al. Investigating antibiotic resistance genes in marketed ready-to-eat small crickets (*Acheta domesticus*). *J. Food Sci.* **2019**, *84*, 3222–3232. [CrossRef]
164. Vandeweyer, D.; Crauwels, S.; Lievens, B.; Van Campenhout, L. Metagenetic analysis of the bacterial communities of edible insects from diverse production cycles at industrial rearing companies. *Int. J. Food Microbiol.* **2017**, *261*, 11–18. [CrossRef]
165. Ribeiro, J.C.; Cunha, L.; Sousa-Pinto, B.; Fonseca, J. Allergic risks of consuming edible insects. A systematic review. *Mol. Nutr. Food Res.* **2018**, *62*. [CrossRef]
166. Barre, A.; Velazquez, E.; Delpianque, A.; Caze-Subra, S.; Bienvenu, F.; Bienvenu, J.; Maudouit, A.; Simplicien, A.; Gamier, L.; Benoist, H.; et al. Cross-reacting allergens of edible insects. *Rev. Fr. d'Allergologie* **2016**, *56*, 522–532. [CrossRef]
167. Barre, A.; Pichereau, C.; Velazquez, E.; Maudouit, A.; Simplicien, A.; Gamier, L.; Bienvenu, F.; Bienvenu, J.; Buriet-Schultz, O.; Auriol, C.; et al. Insights into the allergenic potential of the edible yellow mealworm (*Tenebrio molitor*). *Foods* **2019**, *8*, 515. [CrossRef]
168. De Gier, S.; Verhoeckx, K. Insect food allergy and allergens. *Mol. Immunol.* **2018**, *100*, 82–106. [CrossRef]
169. Boukil, A.; Perreault, B.A.; Chamberland, J.; Mezdour, S.; Poulilot, Y.; Doyen, A. High hygrostatic pressure-assisted enzymatic hydrolysis affects mealworm allergenic proteins. *Molecules* **2020**, *25*, 2685. [CrossRef]
170. Bukkens, S.G.F. Insects in the human diet: Nutritional aspects. In *Ecological Implications of Minilivestock: Potential of Insects, Rodents, Frogs and Snails*; Paoletti, M.G., Ed.; Science Publishers: Enfield, NH, USA, 2005; pp. 545–577.
171. Ayensu, J.; Lutterodt, H.; Annan, R.A.; Adusei, A.; Loh, S.P. Nutritional composition and acceptability of biscuits fortified with palm weevil (*Rhynchophorus phoenicis* Fabricius) and orange-fleshed sweet potato among pregnant women. *Food Sci. Nutr.* **2019**, *7*, 1807–1815. [CrossRef] [PubMed]
172. Kinyuru, M.; Kenji, G.M.; Njoroge, S.M. Process development, nutrition and sensory qualities of wheat buns enriched with edible termites (*Macrotermes subhyalinus*) from Lake Victoria region, Kenya. *Afr. J. Food Agric. Nutr. Dev.* **2009**, *9*, 1739–1750.
173. Aguilar-Miranda, E.D.; Lopez, M.G.; Escamilla-Santana, C.; De La Rosa, P.B. Characteristics of maize flour Tortilla supplemented with ground *Tenebrio molitor* larvae. *J. Agric. Food Chem.* **2002**, *50*, 192–195. [CrossRef]
174. Kim, H.W.; Setyabrata, D.; Lee, Y.J.; Jones, O.G.; Kim, Y.H.B. Effect of house cricket (*Acheta domesticus*) flour addition on physicochemical and textural properties of meal emulsion under various formulations. *J. Food Sci.* **2017**, *82*, 2787–2793. [CrossRef] [PubMed]
175. Van Huis, A. Edible insects: Marketing the impossible? *J. Insects Food Feed* **2017**, *3*, 67–68. [CrossRef]
176. Payne, C.L.R.; Evans, J.D. Nested houses: Domestication dynamics of human-wasp relations in contemporary rural Japan. *J. Ethnobiol. Ethnomed.* **2017**, *13*, 13. [CrossRef] [PubMed]
177. Alemu, M.H.; Olsen, S.B.; Vedel, S.E.; Pambo, K.O.; Owino, V.O. Combining product attributes with recommendation and shopping location attributes to assess consumer preferences for insect-based food products. *Food Qual. Prefer.* **2017**, *55*, 45–57. [CrossRef]
178. Netshifhethe, S.R.; Kunjeku, E.C.; Duncan, F.D. Human uses and indigenous knowledge of edible termites in Chombe district, Limpopo Province, South Africa. *S. Afr. J. Sci.* **2018**, *11*. [CrossRef]
179. Mmari, M.W.; Kinyuru, J.N.; Laswai, O.K.; Okoth, J.K. Traditions, beliefs and Indigenous technologies in connection with the edible longhorn grasshopper, *Ruspolia differens* (Serville) in Tanzania. *J. Ethnobiol. Ethnomed.* **2017**, *13*, 60. [CrossRef]
180. Schäufele, I.; Albores, E.B.; Hamm, U. The role of species for the acceptance of edible insects: Evidence from a consumer survey. *Br. Food J.* **2019**, *121*, 2190–2204. [CrossRef]

181. Meyer-Rochow, V.B.; Hakko, H. Can edible grasshoppers and silkworm pupae be tasted by humans when prevented to see and smell these insects? *J. Asia Pac. Entomol.* **2018**, *21*, 616–619. [CrossRef]

182. Siozios, S.; Massa, A.; Parr, C.l.; Verspar, R.L.; Hurst, G.D.D. DNA barcoding reveals incoorect labelling of insects sold as food in the UK. *PeerJ* **2020**, *8*, e8496. [CrossRef]

183. Barsics, F.; Megido, R.C.; Brostaux, Y.; Barsics, C.; Blecker, C.; Haubruge, E.; Francis, F. Could new information influence attitude to food supplemented with edible insects? *Br. Food J.* **2017**, *119*, 2027–2039. [CrossRef]

184. Van Thielen, L.; Vermuyten, S.; Storms, B.; Rumpold, B.; van Campenhout, L. Consumer acceptance of foods containing edible insects in Belgium two years after their introduction to the market. *J. Insects Food Feed* **2019**, *5*, 35–44. [CrossRef]

185. Videback, P.N.; Grunert, K.G. Disgusting or delicious? Examining attitudinal ambivalence towards entomophagy among Danish consumers. *Food Qual. Prefer.* **2020**, *83*, 103913. [CrossRef]

186. Barton, A.; Richardson, C.D.; McSweeney, M.B. Consumer attitudes toward entomophagy before and after evaluating cricket (*Acheta domesticus*)-based protein powders. *J. Food Sci.* **2020**, *35*, 781–788. [CrossRef]

187. Delgado, M.C.; Chambers, E.; Carbonell-Barrachina, A.; Artiaga, L.N.; Quintanar, R.V.; Hernadez, A.B. Consumer acceptability in the USA, Mexico, and Spain of chocolate chip cookies made with partial insect powder replacement. *J. Food Sci.* **2020**, *85*, 1621–1628. [CrossRef]

188. Ghosh, S.; Tchibozo, S.; Lammantchion, E.; Meyer-Rochow, V.B.; Jung, C. Observations on how people in two locations of the Plateau Département of Southeast Benin perceive entomophagy: A case study from West Africa. *Front. Nutr.* **2021**, *8*, 637385. [CrossRef] [PubMed]

189. Schardong, I.S.; Freiberg, J.A.; Santana, N.A.; dos Santos Richads, N.S.P. Brazilian consumers' perception of edible insects. *Ciência Rural* **2019**, *49*. [CrossRef]

190. Ghosh, S.; Jung, C.; Meyer-Rochow, V.B.; Dekebo, A. Perception of entomophagy by residents of Korea and Ethiopia revealed through structured questionnaire. *J. Insects Food Feed* **2020**, *6*, 59–64. [CrossRef]

191. Collins, C.M.; Vaskou, P.; Kountouris, Y. Insect food products in the western world: Assessing the potential of a new "green market". *Ann. Entomol. Soc. Am.* **2019**, *112*, 518–528. [CrossRef] [PubMed]

192. Megido, R.C.; Gierts, C.; Blecker, C.; Brostaux, Y.; Francis, F. Consumer acceptance of insect-based alternative meat products in western countries. *Food Qual. Prefer.* **2016**, *52*, 237–243. [CrossRef]

193. Tan, H.S.G.; van den Berg, E.; Stieger, M. The influence of product preparation, familiarity and individual traits on the consumer acceptance of insects as food. *Food Qual. Prefer.* **2016**, *52*, 222–231. [CrossRef]

194. Bordiean, A.; Krzyzaniak, M.; Stolarski, M.J.; Czachorowski, S.; Peni, D. Will yellow mealworm become a source of safe proteins for Europe? *Agriculture* **2020**, *10*, 233. [CrossRef]

195. Hedenus, F.; Wirsenius, S.; Johansson, D.J.A. The importance of reduced meat and dairy consumption for meeting stringent climate change targets. *Clim. Chang.* **2014**, *124*, 79–91. [CrossRef]

196. Gahukar, R.T. Insects as human food: Are they really tasty and nutritious? *J. Agric. Food Inf.* **2013**, *14*, 264–267. [CrossRef]

197. Rojas-Downing, M.M.; Nejadhashemi, A.P.; Harrigan, T.; Woznicki, S.A. Climate change and livestock: Impacts, adaptation and mitigation. *Clim. Risk Manag.* **2017**, *16*, 145–163. [CrossRef]

198. Belluco, S.; Losasso, C.; Maggioletti, M.; Alonzi, C.C.; Paoletti, M.G.; Ricci, A. Edible insects in a food safety and nutritional perspective: A critical review. *Compr. Rev. Food Sci. Food Saf.* **2013**, *12*, 296–313. [CrossRef]

199. Berggren, A.; Jansson, A.; Low, M. Using current systems to inform rearing facility design in the insects as food industry. *J. Insects Food Feed* **2018**, *4*, 167–170. [CrossRef]

200. Oppert, B.; Perkin, L.C.; Lorenzen, M.; Dossey, A.T. Transcriptome analysis of life stages of the house cricket, *Acheta domesticus*, to improve insect crop production. *Sci. Rep.* **2020**, *10*, 3471. [CrossRef]

201. EFSA (European Food Safety Authority). Risk profile related to production and consumption of insects as food and feed. *EFSA J.* **2015**, *13*, 4257. [CrossRef]

202. Grabowski, N.T.; Tchibozo, S.; Abdulmawjood, A.; Acheuk, F.; Guerfali, M.M.; Sayed, W.A.A.; Plötz, M. Edible insects in Africa in terms of food, wildlife resource, and pest management legislation. *Foods* **2020**, *9*, 502. [CrossRef]

203. Omotoso, O.T. Nutritional quality, functional properties and antinutrient composition of the larva of *Cirina forda* (Westwod) (Lepidoptera: Saturniidae). *J. Zhejiang Univ. Sci. B* **2006**, *7*, 51–55. [CrossRef]

204. Zielińska, E.; Karaś, M.; Baranaik, R. Comparison of functional properties of edible insects and preparation thereof. *LWT Food Sci. Technol.* **2018**, *91*, 168–174. [CrossRef]

205. Alvarez, D.; Wilkinson, K.A.; Treilhou, M.T.; Tene, M.; Castillo, D.; Sauvain, M. Prospecting peptide isolated from black soldier fly (Diptera: Stratiomycidae) with antimicrobial activity against *Helicobacter pylori* (Campylobacteriales: Helibactericeae). *J. Insect Sci.* **2019**, *19*, 17. [CrossRef]

206. Oh, H.G.; Lee, H.Y.; Kim, J.H.; Kang, Y.R.; Moon, D.I.; Seo, M.Y.; Back, H.I.; Kim, S.Y.; Oh, M.R.; Park, S.H.; et al. Effects of male silkworm pupa powder on the erectile dysfunction by chronic ethanol consumption in rats. *Lab. Anim. Res.* **2012**, *28*, 83–90. [CrossRef]

207. Meyer-Rochow, V.B. Therapeutic arthropods and other, largely terrestrial, folk medicinally important invertebrates: A comparative survey and review. *J. Ethnobiol. Ethnomed.* **2017**, *13*, 9. [CrossRef] [PubMed]

 *foods*

Article

Partial Substitution of Meat with Insect (*Alphitobius diaperinus*) in a Carnivore Diet Changes the Gut Microbiome and Metabolome of Healthy Rats

Sofie Kaas Lanng [1,2], Yichang Zhang [3], Kristine Rothaus Christensen [4], Axel Kornerup Hansen [4], Dennis Sandris Nielsen [3], Witold Kot [5] and Hanne Christine Bertram [1,2,*]

1   Department of Food Science, Aarhus University, Agro Food Park 48, 8200 Aarhus, Denmark; sofiekaaslanng@food.au.dk
2   CiFOOD, Centre for Innovative Food Research, Aarhus University, 8200 Aarhus, Denmark
3   Department of Food Science, University of Copenhagen, Rolighedsvej 30, 1958 Frederiksberg, Denmark; yichang@food.ku.dk (Y.Z.); dn@food.ku.dk (D.S.N.)
4   Department of Veterinary and Animal Sciences, University of Copenhagen, Grønnegårdsvej 15, 1958 Frederiksberg, Denmark; krs@sund.ku.dk (K.R.C.); akh@sund.ku.dk (A.K.H.)
5   Department of Plant and Environmental Sciences, University of Copenhagen, Thorvaldsensvej 40, 1871 Frederiksberg, Denmark; wk@plen.ku.dk
*   Correspondence: hannec.bertram@food.au.dk; Tel.: +45-616-87-389

Citation: Lanng, S.K.; Zhang, Y.; Christensen, K.R.; Hansen, A.K.; Nielsen, D.S.; Kot, W.; Bertram, H.C. Partial Substitution of Meat with Insect (*Alphitobius diaperinus*) in a Carnivore Diet Changes the Gut Microbiome and Metabolome of Healthy Rats. *Foods* **2021**, *10*, 1814. https://doi.org/10.3390/foods10081814

Academic Editors: Victor Benno Meyer-Rochow and Chuleui Jung

Received: 7 July 2021
Accepted: 3 August 2021
Published: 5 August 2021

Abstract: Insects are suggested as a sustainable protein source of high nutritional quality, but the effects of insect ingestion on processes in the gastrointestinal tract and gut microbiota (GM) remain to be established. We examined the effects of partial substitution of meat with insect protein (*Alphitobius diaperinus*) in a four-week dietary intervention in a healthy rat model (n = 30). GM composition was characterized using' 16S rRNA gene amplicon profiling while the metabolomes of stomach, small intestine, and colon content, feces and blood were investigated by ^{1}H-NMR spectroscopy. Metabolomics analyses revealed a larger escape of protein residues into the colon and a different microbial metabolization pattern of aromatic amino acids when partly substituting pork with insect. Both for rats fed a pork diet and rats fed a diet with partial replacement of pork with insect, the GM was dominated by *Lactobacillus*, *Clostridium cluster XI* and *Akkermansia*. However, Bray-Curtis dissimilarity metrics were different when insects were included in the diet. Introduction of insects in a common Western omnivore diet alters the gut microbiome diversity with consequences for endogenous metabolism. This finding highlights the importance of assessing gastrointestinal tract effects when evaluating new protein sources as meat replacements.

Keywords: NMR-based metabolomics; microbiota; insect protein; protein digestion; alternative proteins

1. Introduction

The population of the Earth is estimated to rise to 9.6 billion people by 2050 [1] with a resultant increased demand for food. The growing competition for land and water resources affects our ability to produce food and increases our need to minimize the impact of food production on our resources and environment [2]. Especially the production of protein sources leaves a heavy impact on the environment, where meat production has a high impact regarding water consumption and emission of greenhouse gases [3]. Therefore, there is a massive interest in exploring how to limit the effects on the environment when producing protein and concomitantly ensure protein with a high nutritional value. Compared to meat, plant proteins have a lower environmental impact but also a lower protein nutritional quality as some plant-derived proteins do not contain all essential amino acids and might also contain anti-nutritional compounds [4]. Insects have been reco, gnized as a new and alternative animal-based protein source [5]. The production of insect protein has a high feed conversion ratio [6], emits less greenhouse gases [7], uses less water, and requires less land compared to conventional livestock [6,8].

Insects are comparable in protein content to the conventional animal- and plant-based protein-dense foods, e.g., beef, eggs, milk and soy [9]. Moreover, the content of all essential amino acids in insects meets the requirements of WHO [10]. The total protein content as well as the amino acid composition can moreover be modulated by the feeding substrate [11] and also processing [12]. However, in vivo studies investigating the nutritional value of insects and insect protein are sparse. Few studies have investigated the use of insect protein on muscle protein synthesis [13,14] and muscle mass [15], showing insect protein suboptimal compared to whey both in amino acid availability and stimulation of muscle protein synthesis.

A study in pigs revealed lower ileal digestibility of most amino acids for diets including insect meal compared with a control diet and small changes in the metabolome when 10% of the conventional protein was substituted with insect protein [16]. In humans, a randomized controlled trial investigated the postprandial absorption of insect protein in young men and demonstrated that insect protein provides all essential amino acids, but insect protein was found to have lower bioaccessibility of amino acids than whey but similar to soy protein [17]. A two-week human intervention study investigated the effect of daily ingestion of 25 g cricket powder on the fecal microbiome and the metabolome. The insect supplementation was associated with a decrease in the fecal content of short-chain fatty acids, and minor changes in microbiota composition were observed at the species-level [18]. To our knowledge, these two studies are currently the only reported intervention studies focusing on the nutritional value of insects in a Western-type diet.

In the current study, insect protein isolate isolated from *Alphitobius diaperinus* (lesser mealworm) were used as this protein isolate has at high amount of protein (~82%). Moreover, *Alphitobius diaperinus* has been shown to have a high digestibility in rats (91–94%) [19].

In the present study, we investigated the use of insect protein (*Alphitobius diaperinus*) for a partial replacement (~13%) of meat in a conventional meat product. For this purpose, a four-week intervention study in healthy rats was conducted, where effects of partial insect substitution on growth, food consumption, the gut microbiome, as well as the metabolome were examined.

2. Materials and Methods

2.1. Animals and Diet

Thirty four-week old Sprague–Dawley (NTac:SD) rats (Taconic, Ll. Skensved, Denmark) were individually earmarked and weighed, before being randomly allocated with three animals into each of ten U1400 cages (Tecniplast, Buguggiate, Italy) with Tapvei® aspen bedding and enrichments such as chewing blocks, tunnels, and nesting material (Brogaarden, Lynge, Denmark) at temperatures of 22 ± 2 °C, humidity of $55 \pm 10\%$, air changing 15–20 times/hour and a 12-h light cycle. Health monitoring at the breeder and the experimental facility revealed no reportable infections in the rats [20]. For an adaption period of four days, all rats were fed ad libitum standard chow diet (Altromin 1324, Brogaarden, Denmark) and had free access to water. After the adaption period, the rats were weighted and fecal samples from each rat were obtained. Subsequently, the rats were cage-wise randomly allocated to one of three diets for a four-week intervention period: (1) standard chow diet (Chow) ($n = 6$), (2) insect-substituted pork sausage diet (Insect) ($n = 12$), or (3) pork sausage diet (Pork) ($n = 12$). The sausages were formulated from pork meat, pork fat and sunflower oil with added NaCl, phosphor, $NaNO_2$, AIN76 vitamin mix (CA40077) and AIN76 mineral mix (CA79055) (Table S1). To the insect-substituted pork sausages 13.16 (*w/w* %) insect protein isolate from Alphitobius Diaperinus (~82% protein, Protifarm, Ermelo, Netherlands) was added. The energy and macronutrient content areprovided in Table 1.

Table 1. Nutrient content of experimental diets.

	Pork Sausage Diet	Insect-Substituted Pork Sausage Diet	Chow Diet
Energy [Kcal/100 g]	288	286	319
Fat [g/100 g]	24.4	24.5	4.1
Carbohydrates [g/100 g]	4.2	1.5	40.8
Protein [g/100 g]	12.9	14.8	19.2

The experimental diets and water were offered ad libitum and replenished twice a week during the whole study period. During the intervention period, food and water intake was monitored for each cage, and the bodyweight of each rat was measured at day 7, 14, 21, and 28.

2.2. Sample Collection

Fecal samples were collected from the individual rats both at baseline and at the end of the intervention. After the four weeks of intervention, the rats were anesthetized with fentanyl/fluanison (Hypnorm®, Skanderborg Pharmacy, Skanderborg, Denmark) and midazolam (5 mg/mL, Accord Health Care, Solna, Sweden) (each diluted 1:1 with sterile water prior to mixing; 0.2 mL/g body weight), heart blood was collected in heparin tubes and the rats were finally euthanized by heart blood bleeding under supplementary fentanyl/fluanison/midazolam. Plasma was obtained by centrifugation at $10,000\times g$ for 5 min. Samples of stomach content, small intestinal content and colon content were carefully collected from the rats. All samples were snap-frozen in liquid nitrogen and stored at $-80\,^{\circ}$C until further analysis.

2.3. Sample Preparation for ^{1}H-NMR Spectroscopy

Samples of stomach content, small intestinal content, colon content and feces were weighted, thawed and diluted 1:9 with ddH$_2$O. The samples were whirl-mixed until they were dissolved, then centrifuged at $14,000\times g$ for 10 min at 4 $^{\circ}$C and the supernatant transferred to a new tube. Prior to this, 0.5 mL 10 K Amicon Ultra centrifugal filter units (Merck Millipore Ltd., Cork, Ireland) were prewashed four times with ddH$_2$O. Then the supernatant was centrifuged at $14,000\times g$ for 2 h at 4 $^{\circ}$C before the supernatant was transferred to the filters. The samples were centrifuged at $14,000\times g$ for 1 h at 4 $^{\circ}$C. A volume of 400 µL of the resulting filtrate was transferred to a 5 mm NMR tube, together with 200 µL D$_2$O and 50 µL phosphate buffer (8.66 % *w/v* K$_2$HPO$_4$, 1.812 % *w/v* NaH$_2$PO$_4$, H$_2$O) containing sodium trimethylsilylpropanesulfonate (DSS) (0.23 % *w/v* DSS).

Plasma samples were thawed and 500 µL transferred to prewashed 10 K Amicon Ultra centrifugal filters followed by centrifugation at $14,000\times g$ for 1 h at 4 $^{\circ}$C. A volume of 500 µL filtrate was transferred to 5 mm NMR tubes with 25 µL D$_2$O with 0.05 % Trimethylsilyl-propanoic acid (TSP), 75 µL D$_2$O and 100 µL phosphate buffer (300 mM Na$_2$HPO$_4$).

2.4. H-NMR Spectroscopy

^{1}H-NMR spectroscopy was conducted using a Bruker Avance III 600 MHz spectrometer operating at a ^{1}H frequency of 600.13 MHz with a 5 mm ^{1}H TXI probe (Bruker BioSpin, Rheinstetten, Germany). ^{1}H-NMR spectra were obtained at 298 K using a one-dimensional (1D) nuclear overhauser enhancement spectroscopy (NOESY) preset pulse sequence (noesypr1d) to ensure water suppression. The acquisition parameters used were: 128 scans (NS), spectral width (SW) = 7289 Hz (12.15 ppm), acquisition time (AQ) = 2.25 s, 32 768 data points (TD), and relaxation delay (D1) = 5 s. Prior to Fourier transformation, the free induction decays (FIDs) were multiplied by a line-broadening function of 0.3 Hz. The spectra obtained were subjected to baseline correction and phase correction in TopSpin 3.0 (Bruker BioSpin, Billerica, MA, USA).

2.5. Metabolome Analysis

The ^{1}H-NMR spectra were referenced to either TSP or DSS (0.00 ppm) and corrected for chemical shift by the interval correlation-shifting algorithm, icoshift [21] using MATLAB R2018b (Mathworks Inc., Natick, MA, USA). The spectral regions at the higher and lower chemical shift ranges without resonances and the spectral region containing the residual water signal were removed. Thereafter the spectral data were normalized to total area of the spectrum and binned into regions of 0.01 ppm. Multivariate data analysis was performed using SIMCA 16.0 (Sartorius Stedim Data Analytics AB, Umeå, Sweden), while Chenomx NMR Suite 8.13 (Chenomx Inc., Edmonton, Canada) was used for metabolite assignment and quantification using TSP or DSS as internal quantification standards. For plasma, the quantified concentrations were normalized to the sum of total quantified metabolites and are therefore relative values. Data were Pareto-scaled before principal component analysis (PCA) was conducted to examine variations in the data and possible differences between the diets. In addition, orthogonal projection to latent structures discriminant analysis (OPLS-DA) was performed to investigate the differences in the metabolome between the different diet groups. For each data group comparison, cross validation with seven cross validation groups was conducted and the Q_2-value describing the predictive ability of the model was calculated. Finally, the metabolic differences discriminating the experimental diet groups were elucidated by S-line plot.

2.6. Gut Microbiota Analysis

Total DNA was extracted from the feces using Bead-Beat Micro AX Gravity Kit (A&A Biotechnology, Gdynia, Poland) according to the instructions of the manufacturer (bead beating 2 times 30 s, at 6.5 M/s in a FastPrep-24 (MP Biomedicals, Irvine, California, USA). The prokaryotic microbial community was characterized using 16S rRNA gene amplicon profiling of the V3-region (Illumina NextSeq-based), as described previously [22]. The zOTU table was constructed using the UNOISE3 pipeline [23], and SINTAX [24] was used to predict taxonomy. Downstream data analysis in microbial communities was performed using packages Phyloseq [25] and Vegan [26] in R (version 4.0.2). For α-diversity analysis, Shannon index was calculated. For β-diversity analysis, the Bray-Curtis method was used to generate the dissimilarity matrix, which was subsequently used in principal coordinate analysis (PCoA).

2.7. Ethical Statement

The rat intervention study was carried out in accordance with Directive 2010/63/EU of the European Parliament and of the Council of 22 September 2010 on the protection of animals used for scientific purposes, as well as the Danish Animal Experimentation Act (LBK 474 15/05/2014). Specific approval was granted by the Animal Experiments Inspectorate under the Ministry of Environment and Food in Denmark (License No 2017-15-0201-01262).

2.8. Statistical Analysis

To identify significant differences in the metabolites among the diet groups, one-way analysis of variance (ANOVA) was applied. To correct for multiple testing, false discovery rate (FDR) by Benjamin and Hochberg was applied, and q-values < 0.05 were considered significant. If a significant difference was found, the experimental groups were compared by Tukey's honestly significant difference test, here a p-value < 0.05 were considered significant. Wilcoxon test was used to compare α-diversity differences. The component differences of different groups in PCoA were calculated by PERMANOVA tests. ANCOM [27] was used to detect differentially abundant taxa. The correlations between metabolites and differentially abundant bacteria genera were calculated by Pearson correlation, only metabolites and bacteria genera with at least one correlation coefficient larger than 0.7 were kept for downstream analysis.

3. Results

3.1. Feed Intake and Growth Is Not Changed by Insect Replacement

In the present study rats were fed ad libitum exclusively with one of the three diets: chow diet (Chow), insect-substituted pork sausage diet (Insect) or pork sausage diet (Pork) for a period of 4 weeks. Water consumption and feed intake were registered each week (Figure 1). No significant difference in water consumption was observed between the groups. In Weeks 1 and 2, the feed intake was significantly higher in the two sausage groups compared to the chow diet, but this was not reflected in the body weight of the rats where no significant differences among diet groups were observed. Similarly, no significant difference was found in the body weight after the four weeks of intervention.

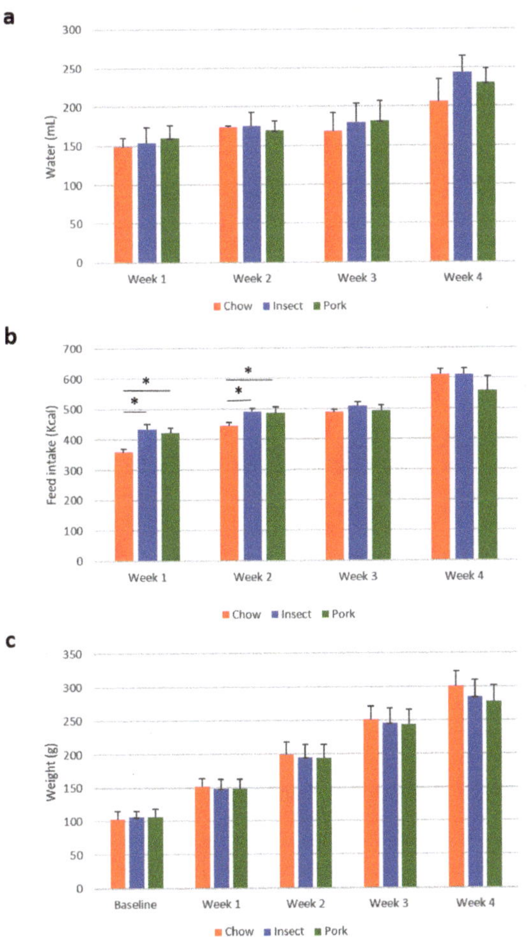

Figure 1. Water consumption, feed intake and body weight of the rats during the intervention period. (**a**) Water consumption (**b**) feed intake per week, and (**c**) the body weight of the rats was measured each week doing the intervention period. Data were analyzed by one-way ANOVA, values within the same cluster not sharing a common letter are significant different ($p < 0.05$). Chow diet (Chow, $n = 6$, red), insect-substituted pork sausage diet (Insect, $n = 12$, blue or pork sausage diet (Pork, $n = 12$, green). The groups found to be significantly different by ANOVA (p-value > 0.05) are marked with a star (*).

3.2. Gut Microbiome (GM) Composition and Diversity

For alpha diversity analysis, the Shannon diversity values were calculated both before and after diet intervention and between the different diets after intervention. The diet intervention resulted in a significantly lower Shannon index after the pork diet (p = 0.023), and chow diet (p = 0.0087), while no significant difference was observed after the insect diet intervention (Figure S1A). The Shannon diversity showed no significant difference between the three diet groups after the intervention (Figure S1B).

Bray-Curtis dissimilarity metrics (Figure 2A) showed that all interventions were significantly different form baseline (p = 0.001). In addition, the three different diets resulted in significantly different GM communities with the chow intervention resembling the baseline samples more as compared to the two sausage intervention groups (Figure 2B).

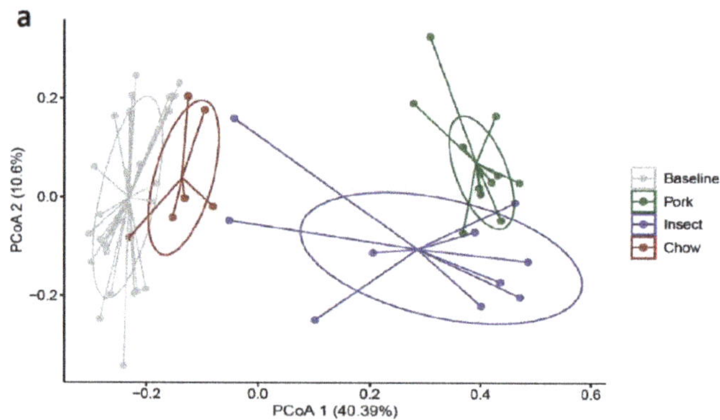

b

Bray-Curtis dissimilarity metrics		
	R^2	P-value
Baseline - intervention pork	0.435	0.001
Baseline - intervention insect	0.281	0.001
Baseline - intervention chow	0.086	0.001
Intervention pork - intervention insect	0.130	0.001
Intervention pork - intervention chow	0.486	0.001
Intervention insect - intervention chow	0.277	0.003

Figure 2. Beta diversity of the gut microbiota composition at baseline and after the three intervention diets. The gut microbiota composition was analyzed by 16S rRNA sequencing. PCoA plot of Bray-Curtis dissimilarity metrics of the gut microbiota composition comparing baseline (gray), pork sausage diet (n = 12, green), insect-substituted pork diet (n = 12, blue) or chow diet (n = 6, red) group (**a**). Statistical values of Bray-Curtis dissimilarity metrics for all comparisons (**b**).

The GM of the individual rats before and after intervention was examined (Figure 3). Baseline samples were characterized by high relative abundance of *Lactobacillus* and *Bifidobacterium*. Additionally, 12 of the 29 rats were also colonized by *Clostridium cluster XI* at baseline. For the chow diet group, the microbial composition changed only marginally during the intervention. The GM remained dominated by *Lactobacillus*, while *Clostridium cluster XI*, *Bifidobacterium* and *Flavonifacter* also were present in some animals. The GM of rats in the pork diet intervention group was dominated by *Lactobacillus*, *Clostridium cluster XI* and *Akkermansia*. The GM of rats in the insect diet intervention group were similar to the pork diet group dominated by *Lactobacillus*, *Clostridium cluster XI* and *Akkermansia*. One rat (Number 8) showed a different GM pattern than all other rats. Changes of abundance at genus level after intervention diet are shown in detail in Figure S2.

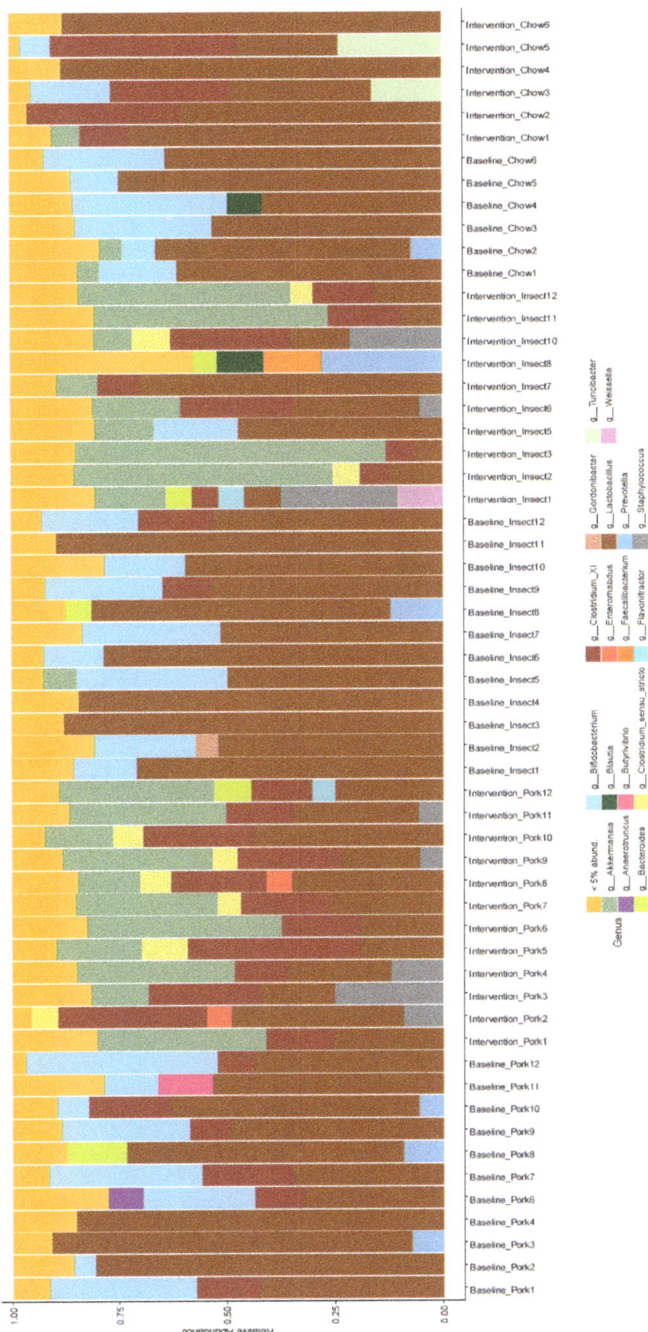

Figure 3. Gut microbiota composition at genus level for the individual rats before and after intervention diet. The gut microbiota composition were analyzed by 16S rRNA sequencing. Gut microbiota composition are shown for the individual rat at baseline and after intervention with chow diet (Chow), pork sausage diet (Pork) or insect-substituted pork diet shown as the relative abundance at genus level.

3.3. Dietary Modulations of Metabolomes

In general, PCA scores plot of the metabolomics data obtained from the various sample types collected through the gastrointestinal tract (stomach content, small intestine content, colon content) revealed a clear grouping of the three intervention diets (Figure S3). OPLS-DA models were constructed for the pairwise comparison of pork sausage diet and insect-substituted pork diets to further investigate the metabolites responsible for the diet differentiation.

Pork sausage diet was associated with a higher concentration of lactate in the stomach content, while the insect diet resulted in a higher concentration of several metabolites including different amino acids such as glutamate, alanine, the branched-chained amino acids isoleucine, leucine, and valine and the two aromatic amino acids tyrosine and phenylalanine (Figure 4A). Increased levels of creatine, creatinine, phosphorylcholine were also accompanied with the insect diet (Figure 4A). The corresponding OPLS-DA score plot is shown in Figure S4A.

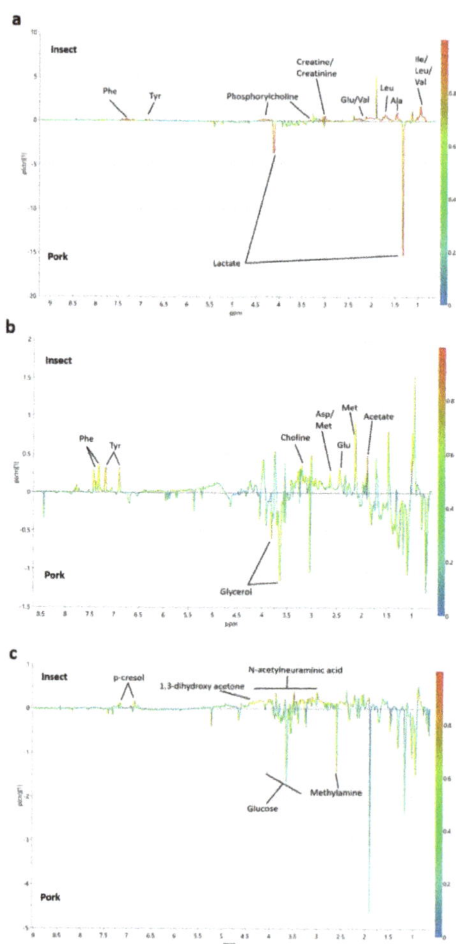

Figure 4. S-line plots from orthogonal partial least squares discriminant analysis of ^{1}H NMR spectroscopic data on insect-substituted pork sausage diet (Insect) versus pork sausage diet (Pork) analysis. Stomach content, $Q^2 = 0.96$ (**a**), small intestinal content, $Q^2 = 0.67$ (**b**) and colon content, $Q^2 = 0.59$ (**c**).

For OPLS-DA on metabolomics data from the small intestinal content, the S-line plot revealed that insect diet was associated with a higher concentration of choline and acetate as well as the amino acids aspartate, methionine, glutamate and the aromatic amino acids tyrosine and phenylalanine in small intestinal content compared to pork diet (Figure 4B). The OPLS-DA score plot is shown in Figure S4B.

For colon content, a S-line plot from an OPLS-DA model examining the metabolites discriminating the insect-substituted pork diet from the pork diet indicated that the insect diet was related to a higher level of 4-hydroxyphenylacetate, 1,3-dihydroxyacetone and N-acetylneuraminic acid while pork diet was characterized by slightly higher glucose and methylamine levels (Figure 4C). The score plot from the OPLS-DA model is shown in Figure S4C.

A robust OPLS-DA model discriminating the metabolite profile of feces after pork or insect-substituted pork diet, respectively, could not be obtained ($Q_2 = 0.34$, data not shown).

The quantified amounts of the branched-chain amino acids (BCAAs) in plasma and through the gastrointestinal tract were examined (Figure 5). The relative concentrations of valine, leucine, and isoleucine in plasma were significantly higher for the insect diet compared to both the chow and the pork sausage diet groups. In addition, the amount of BCAA in stomach and small intestinal content was significantly higher after intake of the insect-substituted pork sausages compared to the pork sausage diet (Figure 5). In the colon content, the same trend was observed, however, differences between the insect and pork diets were not significant.

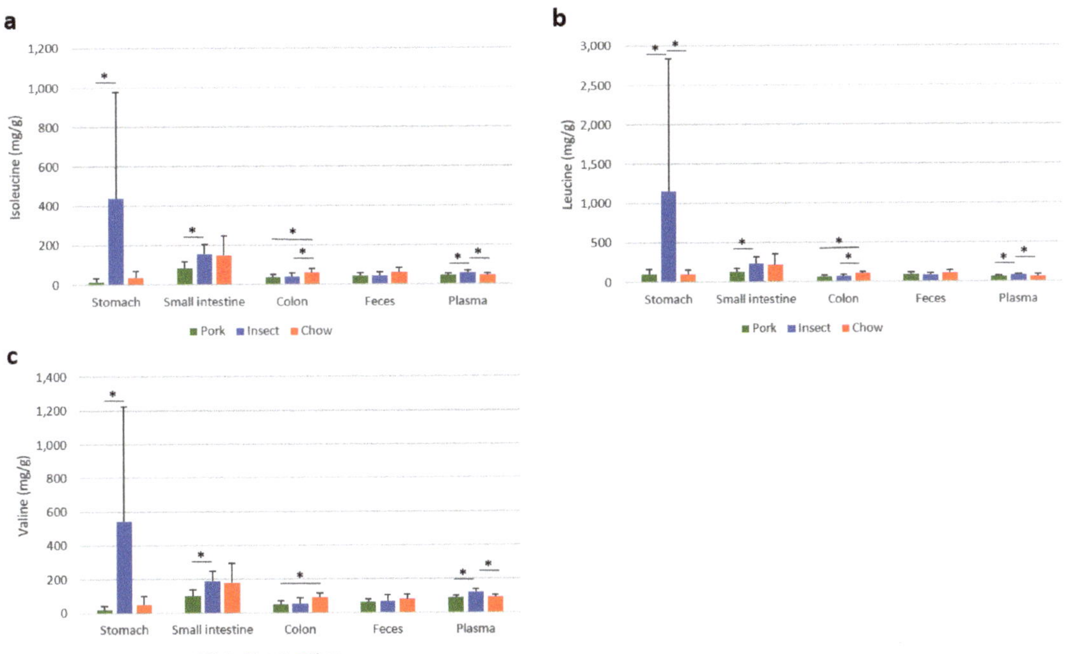

Figure 5. Concentration of branched chained amino acids (BCAA); Isoleucine (**a**), leucine (**b**) and valine (**c**) measured by ^{1}H nuclear magnetic resonance spectroscopy. The concentrations found to significantly different by ANOVA (*p*-value > 0.05) are marked with a star (*). Values reported for gastrointestinal tract are in mg/g while plasma concentrations are expressed in relative values. Chow diet (Chow, *n* = 6, red), insect-substituted pork sausage diet (Insect, *n* = 12, blue) or pork sausage diet (Pork, *n* = 12, green).

3.4. Correlation between GM and Metabolites

Correlations between the GM and metabolites were investigated using Pearson correlation which revealed a separation of bacterial genera into two groups, with the relative abundance of each group being positively correlated to a range of metabolites, as shown by Figure 6. High relative abundance of *Parasporobacterium*, *Barnesiella*, *Gordonibacteria* and *Turicibacter* were positively correlated with glucose, butyrate, uracil, hypoxanthine, and glutamine. Methanol was correlated to the abundance of *Gordonibacter*, *Clostridium cluster XVIII* and especially *Parasporobacterium*, tyrosine was correlated to *Barnesiella* and *Gordonibacteria*, while high relative abundance of *Bacteriodes* and *Parabacteriodes* correlated to high concentrations of arginine and isoleucine. Finally, *Enterorhabdus* and *Desulfovibrio* relative abundance correlated to methylamine and the amino acids leucine, isoleucine, valine, and phenylalanine.

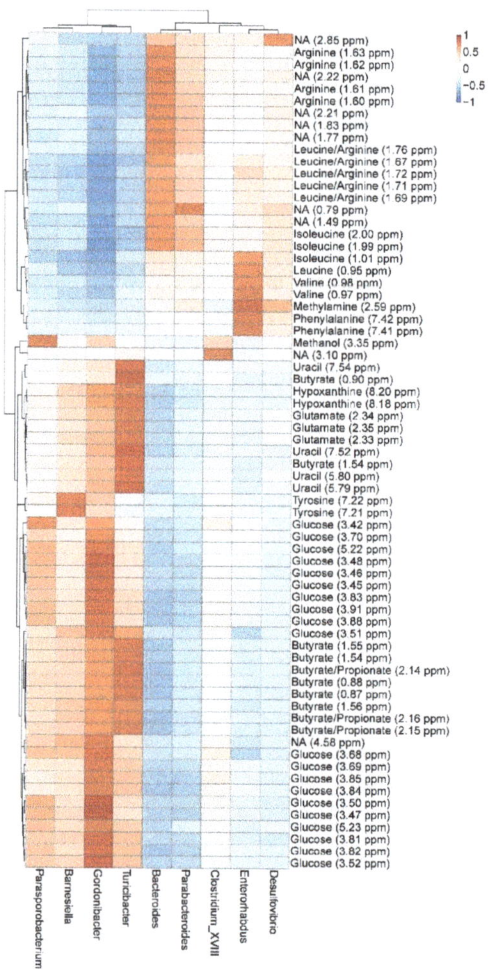

Figure 6. Heatmap of Pearson correlations between colon metabolite concentration and relative abundance of bacterial genera. The gut microbiota composition was analyzed by 16S rRNA sequencing while the metabolic pattern was analyzed by ^{1}H NMR spectroscopy. Only metabolites and bacteria genera with at a correlation coefficient >0.7 are shown. The color intensity from blue to red corresponds to the correlation coefficient from −1 to 1.

4. Discussion

In the present study, the impact of a partial substitution of meat protein with insect protein on the GM composition and metabolic processes in the gastrointestinal tract of healthy rats was investigated. Interventions with a traditional pork sausage and an insect-substituted sausage were compared with a conventional chow diet. The four week duration of the intervention was chosen to show the effects of prolonged ingestion of insect protein, mimicking an introduction of insect protein in a carnivore diet. If the changes shown in this paper will persist or increase with prolonged diet remains to be explored. Over four weeks, same energy intake and body weight gain was found irrespective of whether the rats were fed exclusively with pork protein or partly insect protein.

Analysis of GM composition revealed pronounced changes in the gut microbiome composition following interventions with pork sausage and insect-substituted sausage when compared with rats remaining on a chow diet (Figures 2 and 3). This may be expected considering that interventions with the carnivore diets (sausage products) were associated with changes in macronutrient composition. Interestingly, when comparing the GM after the interventions, Bray-Curtis dissimilarity metrics also revealed differences in GM composition between the pork diet and insect-substituted diet, respectively. Consequently, inclusion of insect as a protein-source in a carnivore diet impacted the GM to an extent where, after four weeks of intervention, it was clearly different from the rats fed the meat-based diet. A human study investigated intake of 25 g cricket powder per day for a two-week period where the cricket powder replaced cocoa powder and purple corn meal in a breakfast meal [18]. This replacement resulted in an increased fat and protein content and a lower carbohydrate content in the insect meal, but only resulted in smaller differences in the GM composition between individuals assigned to one of the two diets. Stull et al. found that insect ingestion was associated with an increase in abundance of *Bifidobacterium animalis*. In the current study, no increase of *Bifidobacterium* upon insect inclusion was observed (Figure S2). This might be related to differences in the background diet and food components that insect replaces or in the ability of *Bifidobacterium* to colonize in rats and humans [28].

Concomitantly with analysis of the GM composition, metabolic activity in the gastrointestinal tract was assessed through the application of nuclear magnetic resonance (NMR)-based metabolomics, and uniting the metabolomes from blood, stomach content, small intestinal content, colon content end feces pointed at a distinct metabolization of insect protein compared with meat protein in the gastrointestinal tract. The two carnivore diets: pork and insect, had metabolite profiles markedly distinct from the standard chow diet, which probably is a result of the differences in the macronutrient composition of the diets. The chow diet had a considerably higher content of carbohydrates and a lower amount of fat compared to the two carnivore diets. This also corresponds to earlier findings that high-carbohydrate-low-fat and low-carbohydrate-high-fat diets can be discriminated based on the metabolic profile of the plasma [29].

In the small intestine, ingestion of insect protein was associated with a larger amount of the aromatic amino acids tyrosine and phenylalanine compared with ingestion of meat protein. These aromatic amino acids are known to exert a high capability for stimulating fermentation in the gut [30]. In colon content, 4-hydroxyphenylactate was increased after ingestion of insect diet compared to pork diet. Further, 4-hydrophenylactate is involved in tyrosine metabolism and can most likely be linked to the higher concentration of tyrosine in the small intestine. Correlation analysis between GM and metabolites revealed that tyrosine levels in colon content correlated with the relative abundance of *Barnesiella* and *Gordonibacteria*. Thus, differences in tyrosine metabolism in the gut upon insect ingestion can possibly be ascribed to changes favoring the abundance of these specific bacteria.

In blood plasma, higher levels of all branched-chain amino acids (BCAAs) leucine, isoleucine and valine were found in rats ingesting the insect-substituted feed compared with rats ingesting the conventional meat product (Figure 5). These results corroborate a former study in pigs, which found lower ileal digestibilities of amino acids for diets including insect meal compared with a control diet [16]. Even though it is not possible to confirm that these BCAAs in plasma are of microbial origin, this appears plausible as the two carnivore diets had a similar content of BCAAs. Other studies have also suggested that it is the overall dietary patterns rather than the dietary intake of BCAAs that contributes to BCAA plasma concentrations [31,32]. Levels of circulating BCAAs have been linked to an increased future risk of type 2 diabetes [33]. Consequently, the present findings might have implications for human health, but further studies are warranted to establish a potential association between insect intake and type 2 diabetes risk.

Recently, there is an intense focus on identifying new dietary protein sources as part of the great food transformation introduced in an EAT Lancet report [34]. The intention is that new foods sources should nurture human health and support environmental sustainability. The present study reveals the necessity of establishing actions to thoroughly examine how new dietary sources, commenced for sustainability reasons, impact endogenous metabolism and GM to capture their true value in terms of nurturing human health. Collectively, our data suggest that insect protein ingestion results in a different pattern of protein residues and metabolism in the large intestine and apparently an increased host absorption of generated BCAAs. Thus, while insect protein instantly appears as an attractive replacer of meat from a sustainable point of view, our study shows that it is important to approach new protein sources with a holistic approach to capture effects with implications for both human health and sustainability.

5. Conclusions

The present study shows that partial substitution of meat with insect protein in a traditional pork sausage over four-weeks, with same energy intake and body weight gain influence the gut microbiota composition as well as the plasma and gastrointestinal metabolite profile. Our data indicate that protein residues comes into the colon after inclusion of insect protein in diet compared to a diet exclusively containing pork meat protein. The increased amount of protein residues in the colon was reflected in a different metabolization of aromatic amino acid residues. Insect ingestion was also accompanied by a host absorption of generated BCAAs as reflected in higher plasma concentrations of these amino acids. Our study thus proposes that introduction of insects in a common Western omnivore diet alters the gut microbiome with consequences for endogenous metabolism.

Supplementary Materials: The following are available online at https://www.mdpi.com/article/10.3390/foods10081814/s1, Table S1: Formulation of sausages diets. Ingredients added to common sausage emulsion of pork meat, pork fat and water, Figure S1: Alfa diversity of the gut microbiota composition at baseline and after the tree intervention diets. The Shannon index at baseline and after diet intervention (4 weeks) with either pork sausage diet, insect-substituted pork diet or chow diet (A). Statistical values of Shannon index for all comparisons (B), Figure S2: Change in abundance of different bacterial genera in the gut microbiota after diet intervention. The abundance at genus level are shown before (baseline) and after the different diet interventions (Intervention) with (A) pork sausage (pork), (B) insect-supplemented pork sausage (insect) or (C) chow diet (chow), Figure S3: Score plot of principal component analysis of stomach content. The first two principal components of rats on chow diet , pork sausage diet (blue) or insect-substituted pork sausage diet (green). The first principal component explaining 51% of the total variance separates the groups and by also taking the second component, explaining 27.5% of the variance, into account a clear separation is observed between the three groups, Figure S4: Score plot of OPLS-DA of stomach content of rats on insect-substituted pork sausage diet (green) versus pork sausage diet (blue). Q2 = 0.96, Figure S5: Score plot of principal component analysis of small intestinal content. The first two of three principal components of rats on chow diet , pork sausage diet (green) or insect-substituted pork sausage diet (blue). The first principal component (55.2%) separates the chow group from the others while the pork and insect diet groups are separated along both PC1 and PC2 (18.6%), Figure S6: Score plot of orthogonal partial least squares discriminant analysis of small intestinal content of rats on pork diet (green) versus insect-substituted pork sausage diet (blue). Q2 = 0.67, Figure S7: Score plot of principal component analysis of colon content. Score plot of principal component analysis with 7 principal components displaying the two first components of colon content of rats on chow diet , pork sausage diet (green) or insect-substituted pork sausage diet (blue). The first principal component (39.4%) separates the chow samples from the other groups while PC2 (14.8%) separates the pork from the insect group, Figure S8: Score plot of OPLS-DA of colon content of rats on pork diet (green) versus insect-substituted pork sausage diet (blue). Q2 = 0.59.

Author Contributions: Conceptualization, S.K.L. and H.C.B.; methodology, S.K.L., Y.Z., A.K.H., W.K., D.S.N. and H.C.B.; formal analysis, S.K.L., Y.Z., W.K., K.R.C.; writing—original draft preparation, S.K.L., H.C.B., D.S.N.; writing—review and editing, Y.Z., K.R.C., A.K.H., W.K.; supervision, H.C.B.; project administration, H.C.B.; funding acquisition, H.C.B. All authors have read and agreed to the published version of the manuscript.

Funding: The present study was part of Sofie Kaas Lanng's PhD project, which was funded by iFOOD, Centre for Innovative Food Research, Aarhus University, Denmark and HCB's Eliteforsk grant (# 6161-00016B). Yichang Zhang is funded by the China Scholarship Council.

Institutional Review Board Statement: The rat intervention study was carried out in accordance with Directive 2010/63/EU of the European Parliament and of the Council of 22 September 2010 on the protection of animals used for scientific purposes, as well as the Danish Animal Experimentation Act (LBK 474 15/05/2014). Specific approval was granted by the Animal Experiments Inspectorate under the Ministry of Environment and Food in Denmark (License No 2017-15-0201-01262).

Informed Consent Statement: Not applicable.

Data Availability Statement: Authors will make data available upon request.

Acknowledgments: The authors thank Nina Eggers for technical assistance with NMR analyses and Helene Farlov and Mette Nelander for taking care of the rats.

Conflicts of Interest: The authors declare no conflict of interest.

References

1. Gerland, P.; Raftery, A.E.; Sevcikova, H.; Li, N.; Gu, D.; Spoorenberg, T.; Alkema, L.; Fosdick, B.K.; Chunn, J.; Lalic, N.; et al. World population stabilization unlikely this century. *Science* **2014**, *346*, 234–237. [CrossRef] [PubMed]
2. Godfray, H.C.; Beddington, J.R.; Crute, I.R.; Haddad, L.; Lawrence, D.; Muir, J.F.; Pretty, J.; Robinson, S.; Thomas, S.M.; Toulmin, C. Food security: The challenge of feeding 9 billion people. *Science* **2010**, *327*, 812–818. [CrossRef] [PubMed]
3. Godfray, H.C.J.; Aveyard, P.; Garnett, T.; Hall, J.W.; Key, T.J.; Lorimer, J.; Pierrehumbert, R.T.; Scarborough, P.; Springmann, M.; Jebb, S.A. Meat consumption, health, and the environment. *Science* **2018**, *361*, 6399. [CrossRef] [PubMed]
4. Scholz-Ahrens, K.E.; Ahrens, F.; Barth, C.A. Nutritional and health attributes of milk and milk imitations. *Eur. J. Nutr.* **2020**, *59*, 19–34. [CrossRef] [PubMed]
5. Feng, Y.; Chen, X.M.; Zhao, M.; He, Z.; Sun, L.; Wang, C.Y.; Ding, W.F. Edible insects in China: Utilization and prospects. *Insect Sci.* **2018**, *25*, 184–198. [CrossRef]
6. Elhassan, M.; Wendin, K.; Olsson, V.; Langton, M. Quality Aspects of Insects as Food-Nutritional, Sensory, and Related Concepts. *Foods* **2019**, *8*, 95. [CrossRef]
7. Oonincx, D.G.; van Itterbeeck, J.; Heetkamp, M.J.; van den Brand, H.; van Loon, J.J.; van Huis, A. An exploration on greenhouse gas and ammonia production by insect species suitable for animal or human consumption. *PLoS ONE* **2010**, *5*, e14445. [CrossRef]
8. Huis, A.V. Potential of Insects as Food and Feed in Assuring Food Security. *Annu. Rev. Entomol.* **2013**, *58*, 563–583. [CrossRef]
9. Churchward-Venne, T.A.; Pinckaers, P.J.M.; van Loon, J.J.A.; van Loon, L.J.C. Consideration of insects as a source of dietary protein for human consumption. *Nutr. Rev.* **2017**, *75*, 1035–1045. [CrossRef]
10. Rumpold, B.A.; Schluter, O.K. Nutritional composition and safety aspects of edible insects. *Mol. Nutr. Food Res.* **2013**, *57*, 802–823. [CrossRef]
11. Fuso, A.; Barbi, S.; Macavei, L.I.; Luparelli, A.V.; Maistrello, L.; Montorsi, M.; Sforza, S.; Caligiani, A. Effect of the Rearing Substrate on Total Protein and Amino Acid Composition in Black Soldier Fly. *Foods* **2021**, *10*, 1773. [CrossRef]
12. Nyangena, D.N.; Mutungi, C.; Imathiu, S.; Kinyuru, J.; Affognon, H.; Ekesi, S.; Nakimbugwe, D.; Fiaboe, K.K.M. Effects of Traditional Processing Techniques on the Nutritional and Microbiological Quality of Four Edible Insect Species Used for Food and Feed in East Africa. *Foods* **2020**, *9*, 574. [CrossRef]
13. Hermans, W.J.H.; Senden, J.M.; Churchward-Venne, T.A.; Paulussen, K.J.M.; Fuchs, C.J.; Smeets, J.S.J.; van Loon, J.J.A.; Verdijk, L.B.; van Loon, L.J.C. Insects are a viable protein source for human consumption: From insect protein digestion to postprandial muscle protein synthesis in vivo in humans: A double-blind randomized trial. *Am. J. Clin. Nutr.* **2021**. [CrossRef]
14. Pinckaers, P.J.M.; Weijzen, M.E.G.; Houben, L.H.P.; Zorenc, A.H.; Kouw, I.W.K.; de Groot, L.C.; Verdijk, L.B.; Snijders, T.; van Loon, L.J.C. The Muscle Protein Synthetic Response Following Ingestion of Corn Protein, Milk Protein and Their Protein Blend in Young Males. *Curr. Dev. Nutr.* **2020**, *4*, 651. [CrossRef]

15. Vangsoe, M.T.; Joergensen, M.S.; Heckmann, L.L.; Hansen, M. Effects of Insect Protein Supplementation during Resistance Training on Changes in Muscle Mass and Strength in Young Men. *Nutrients* **2018**, *10*, 335. [CrossRef] [PubMed]

16. Meyer, S.; Gessner, D.K.; Braune, M.S.; Friedhoff, T.; Most, E.; Horing, M.; Liebisch, G.; Zorn, H.; Eder, K.; Ringseis, R. Comprehensive evaluation of the metabolic effects of insect meal from Tenebrio molitor L. in growing pigs by transcriptomics, metabolomics and lipidomics. *J. Anim. Sci. Biotechnol.* **2020**, *11*, 20. [CrossRef] [PubMed]

17. Vangsoe, M.T.; Thogersen, R.; Bertram, H.C.; Heckmann, L.L.; Hansen, M. Ingestion of Insect Protein Isolate Enhances Blood Amino Acid Concentrations Similar to Soy Protein in A Human Trial. *Nutrients* **2018**, *10*, 1357. [CrossRef]

18. Stull, V.J.; Finer, E.; Bergmans, R.S.; Febvre, H.P.; Longhurst, C.; Manter, D.K.; Patz, J.A.; Weir, T.L. Impact of Edible Cricket Consumption on Gut Microbiota in Healthy Adults, a Double-blind, Randomized Crossover Trial. *Sci. Rep.* **2018**, *8*, 10762. [CrossRef]

19. Jensen, L.D.; Miklos, R.; Dalsgaard, T.K.; Heckmann, L.H.; Nørgaard, J.V. Nutritional evaluation of common (Tenebrio molitor) and lesser (Alphitobius diaperinus) mealworms in rats and processing effect on the lesser mealworm. *J. Insects Food Feed.* **2019**, *5*, 257–266. [CrossRef]

20. Mähler Convenor, M.; Berard, M.; Feinstein, R.; Gallagher, A.; Illgen-Wilcke, B.; Pritchett-Corning, K.; Raspa, M. FELASA recommendations for the health monitoring of mouse, rat, hamster, guinea pig and rabbit colonies in breeding and experimental units. *Lab. Anim.* **2014**, *48*, 178–192. [CrossRef]

21. Savorani, F.; Tomasi, G.; Engelsen, S.B. Icoshift: A versatile tool for the rapid alignment of 1D NMR spectra. *J. Magn. Reson.* **2010**, *202*, 190–202. [CrossRef]

22. Rasmussen, T.S.; de Vries, L.; Kot, W.; Hansen, L.H.; Castro-Mejía, J.L.; Vogensen, F.K.; Hansen, A.K.; Nielsen, D.S. Mouse Vendor Influence on the Bacterial and Viral Gut Composition Exceeds the Effect of Diet. *Viruses* **2019**, *11*, 435. [CrossRef] [PubMed]

23. Edgar, R.C. UNOISE2: Improved error-correction for Illumina 16S and ITS amplicon sequencing. *BioRxiv* **2016**, 081257. [CrossRef]

24. Edgar, R.C. SINTAX: A simple non-Bayesian taxonomy classifier for 16S and ITS sequences. *BioRxiv* **2016**, 074161. [CrossRef]

25. McMurdie, P.J.; Holmes, S. phyloseq: An R Package for Reproducible Interactive Analysis and Graphics of Microbiome Census Data. *PLoS ONE* **2013**, *8*, e61217. [CrossRef]

26. Oksanen, J.; Friendly, M.; Kindt, R.; Legendre, P.; McGlinn, D.; Peter, R.; Minchin, R.; O'Hara, B.; Gavin, L.; Simpson, P.; et al. Vegan: Community Ecology Package. Available online: http://CRAN.R-project.org/package=vegan (accessed on 1 January 2019).

27. Mandal, S.; Treuren, W.; White, R.; Eggesbø, M.; Knight, R.; Peddada, S. Analysis of composition of microbiomes: A novel method for studying microbial composition. *Microb. Ecol. Health Dis.* **2015**, *26*, 27663. [CrossRef] [PubMed]

28. Nagpal, R.; Wang, S.; Solberg Woods, L.C.; Seshie, O.; Chung, S.T.; Shively, C.A.; Register, T.C.; Craft, S.; McClain, D.A.; Yadav, H. Comparative Microbiome Signatures and Short-Chain Fatty Acids in Mouse, Rat, Non-human Primate, and Human Feces. *Front. Microbiol.* **2018**, *9*, 2897. [CrossRef]

29. Esko, T.; Hirschhorn, J.N.; Feldman, H.A.; Hsu, Y.-H.H.; Deik, A.A.; Clish, C.B.; Ebbeling, C.B.; Ludwig, D.S. Metabolomic profiles as reliable biomarkers of dietary composition. *Am. J. Clin. Nutr.* **2017**, *105*, 547–554. [CrossRef] [PubMed]

30. Smith, E.A.; Macfarlane, G.T. Enumeration of amino acid fermenting bacteria in the human large intestine: Effects of pH and starch on peptide metabolism and dissimilation of amino acids. *FEMS Microbiol. Ecol.* **1998**, *25*, 355–368. [CrossRef]

31. Merz, B.; Frommherz, L.; Rist, M.J.; Kulling, S.E.; Bub, A.; Watzl, B. Dietary Pattern and Plasma BCAA-Variations in Healthy Men and Women-Results from the KarMeN Study. *Nutrients* **2018**, *10*, 623. [CrossRef]

32. Wang, F.; Wan, Y.; Yin, K.; Wei, Y.; Wang, B.; Yu, X.; Ni, Y.; Zheng, J.; Huang, T.; Song, M.; et al. Lower Circulating Branched-Chain Amino Acid Concentrations Among Vegetarians are Associated with Changes in Gut Microbial Composition and Function. *Mol. Nutr. Food Res.* **2019**, *63*, 1900612. [CrossRef] [PubMed]

33. Wang, T.J.; Larson, M.G.; Vasan, R.S.; Cheng, S.; Rhee, E.P.; McCabe, E.; Lewis, G.D.; Fox, C.S.; Jacques, P.F.; Fernandez, C.; et al. Metabolite profiles and the risk of developing diabetes. *Nat. Med.* **2011**, *17*, 448–453. [CrossRef] [PubMed]

34. Willett, W.; Rockström, J.; Loken, B.; Springmann, M.; Lang, T.; Vermeulen, S.; Garnett, T.; Tilman, D.; DeClerck, F.; Wood, A.; et al. Food in the Anthropocene: The EAT–Lancet Commission on healthy diets from sustainable food systems. *Lancet* **2019**, *393*, 447–492. [CrossRef]

MDPI

St. Alban-Anlage 66

4052 Basel

Switzerland

www.mdpi.com

Foods Editorial Office

E-mail: foods@mdpi.com

www.mdpi.com/journal/foods